Developmental Mathematics

Fifth Edition

and

TASP Version

Student's Solutions Manual

Judith A. Penna

Indiana University—Purdue University at Indianapolis

Marvin L. Bittinger

Indiana University—Purdue University at Indianapolis

Judith A. Beecher

Indiana University—Purdue University at Indianapolis

An imprint of Addison Wesley Longman, Inc.

Reading, Massachusetts • Menlo Park, California • New York • Harlow, England
Don Mills, Ontario • Sydney • Mexico City • Madrid • Amsterdam

Reproduced by Addison Wesley Longman from camera-ready copy supplied by the author.

ISBN 0-201-63681-6

1 2 3 4 5 6 7 8 9 10 PHTH 04 03 02 01 00

Table of Contents

Operations on the Whole Numbers

Exercise Set 1.1

1. 5742 = 5 thousands + 7 hundreds + 4 tens + 2 ones

3. 27,342 = 2 ten thousands + 7 thousands + 3 hundreds + 4 tens + 2 ones

5. 5609 = 5 thousands + 6 hundreds + 0 tens + 9 ones, or 5 thousands + 6 hundreds + 9 ones

7. 2300 = 2 thousands + 3 hundreds + 0 tens + 0 ones, or 2 thousands + 3 hundreds

9. 2 thousands + 4 hundreds + 7 tens + 5 ones = 2475

11. 6 ten thousands + 8 thousands + 9 hundreds + 3 tens + 9 ones = 68,939

13. 7 thousands + 3 hundreds + 0 tens + 4 ones = 7304

15. 1 thousand + 9 ones = 1009

17. A word name for 85 is eighty-five.

19. 88,000

Eighty-eight thousand

21. 123,765

One hundred twenty-three thousand,
seven hundred sixty-five

23. 7, 754, 211, 577

Seven billion,
seven hundred fifty-four million,
two hundred eleven thousand,
five hundred seventy-seven

25. 1, 867,000

One million,
eight hundred sixty-seven thousand

27. 1, 583, 141,000

One billion,
five hundred eighty-three million,
one hundred forty-one thousand

29. Two million,
two hundred thirty-three thousand,
eight hundred twelve

Standard notation is 2, 233, 812.

31. Eight billion

Standard notation is 8,000,000,000.

33. Nine trillion,
four hundred sixty billion,

Standard notation is 9,460,000,000,000.

35. Two million,
nine hundred seventy-four thousand,
six hundred

Standard notation is 2, 974, 600.

37. 2 3 [5] , 8 8 8

The digit 5 means 5 thousands.

39. 4 8 8, [5] 2 6

The digit 5 means 5 hundreds.

41. 8 9, [3] 0 2

The digit 3 tells the number of hundreds.

43. 8 9, 3 [0] 2

The digit 0 tells the number of tens.

45. 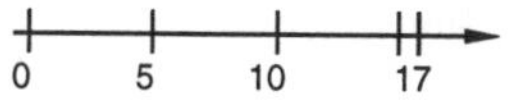

Since 0 is to the left of 17, 0 < 17.

47. 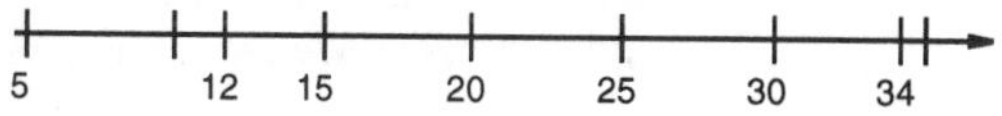

Since 34 is to the right of 12, 34 > 12.

49. 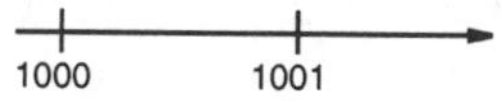

Since 1000 is to the left of 1001, 1000 < 1001.

51. 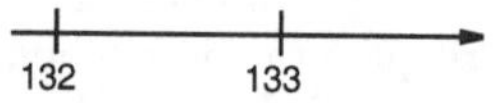

Since 133 is to the right of 132, 133 > 132.

53. 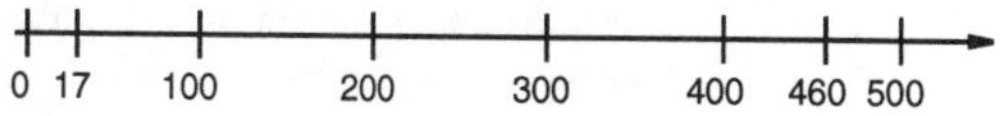

Since 460 is to the right of 17, 460 > 17.

55. 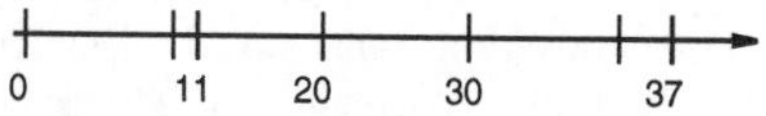

Since 37 is to the right of 11, 37 > 11.

57. 

59. All digits are 9's. Answers may vary. For an 8-digit readout, for example, it would be 99,999,999. This number has three periods.

Exercise Set 1.2

1.

7 e-mail messages Tuesday		8 e-mail messages Wednesday		15 e-mail messages altogether
8	+	7	=	15

3. 12 in. + 6 in. + 8 in. + 14 in. = 40 in.

5. 450 yd + 318 yd + 318 yd = 1086 yd

7.
$$\begin{array}{r} 364 \\ +\ \ 23 \\ \hline 387 \end{array}$$
Add ones, add tens, then add hundreds.

9.
$$\begin{array}{r} {\scriptstyle 1}\ \ \ \ \ \\ 1716 \\ +3482 \\ \hline 5198 \end{array}$$
Add ones: We get 8. Add tens: We get 9 tens. Add hundreds: We get 11 hundreds, or 1 thousand + 1 hundred. Write 1 in the hundreds column and 1 above the thousands. Add thousands: We get 5 thousands.

11.
$$\begin{array}{r} {\scriptstyle 1}\ \ \ \\ 909 \\ +101 \\ \hline 1010 \end{array}$$
Add ones: We get 10 ones, or 1 ten + 0 ones. Write 0 in the ones column and 1 above the tens. Add tens: We get 1 ten. Add hundreds: We get 10 hundreds.

13.
$$\begin{array}{r} {\scriptstyle 1}\ \ \ \ \ \\ 356 \\ +4910 \\ \hline 5266 \end{array}$$
Add ones: We get 6. Add tens: We get 6. Add hundreds: We get 12 hundreds, or 1 thousand + 2 hundreds. Write 2 in the hundreds column and 1 above the thousands. Add thousands: We get 5.

15.
$$\begin{array}{r} {\scriptstyle 1\,1}\ \ \ \\ 5093 \\ +3217 \\ \hline 8310 \end{array}$$
Add ones: We get 10 ones, or 1 ten + 0 ones. Write 0 in the ones column and 1 above the tens. Add tens: We get 11. Write 1 in the tens column and 1 above the hundreds. Add hundreds: We get 3 hundreds. Add thousands: We get 8 thousands.

17.
$$\begin{array}{r} {\scriptstyle 1\,1\,1}\ \ \ \ \\ 23,443 \\ +10,989 \\ \hline 34,432 \end{array}$$
Add ones: We get 12 ones, or 1 ten + 2 ones. Write 2 in the ones column and 1 above the tens. Add tens: We get 13 tens. Write 3 in the tens column and 1 above the hundreds. Add hundreds: We get 14 hundreds. Write 4 in the hundreds column and 1 above the thousands. Add thousands: We get 4 thousands. Add ten thousands: We get 3 ten thousands.

19.
$$\begin{array}{r} {\scriptstyle 2\,4}\ \ \\ 327 \\ 428 \\ 569 \\ 787 \\ +209 \\ \hline 2320 \end{array}$$
Add ones: We get 40. Write 0 in the ones column and 4 above the tens. Add tens: We get 22 tens. Write 2 in the tens column and 2 above the hundreds. Add hundreds: We get 23 hundreds.

21.
$$\begin{array}{r} {\scriptstyle 1\ \ 1}\ \ \ \\ 3420 \\ 8719 \\ 4312 \\ +6203 \\ \hline 22,654 \end{array}$$
Add ones: We get 14. Write 4 in the ones column and 1 above the tens. Add tens: We get 5 tens. Add hundreds: We get 16 hundreds. Write 6 in the hundreds column and 1 above the thousands. Add thousands: We get 22 thousands.

23. $188 + \square = 564$; $564 - 188 = \square$

25. $7 - 4 = 3$

↑ This number gets added (after 3).

↓ $7 = 3 + 4$

(By the commutative law of addition, $7 = 4 + 3$ is also correct.)

27. $13 - 8 = 5$

↑ This number gets added (after 5).

↓ $13 = 5 + 8$

(By the commutative law of addition, $13 = 8 + 5$ is also correct.)

29. $6 + 9 = 15$ — ↑ This addend gets subtracted from the sum. ↓ $6 = 15 - 9$

$6 + 9 = 15$ — ↑ This addend gets subtracted from the sum. ↓ $9 = 15 - 6$

31. $8 + 7 = 15$ — ↑ This addend gets subtracted from the sum. ↓ $8 = 15 - 7$

$8 + 7 = 15$ — ↑ This addend gets subtracted from the sum. ↓ $7 = 15 - 8$

33. We first write an addition sentence. Keep in mind that all numbers are in millions.

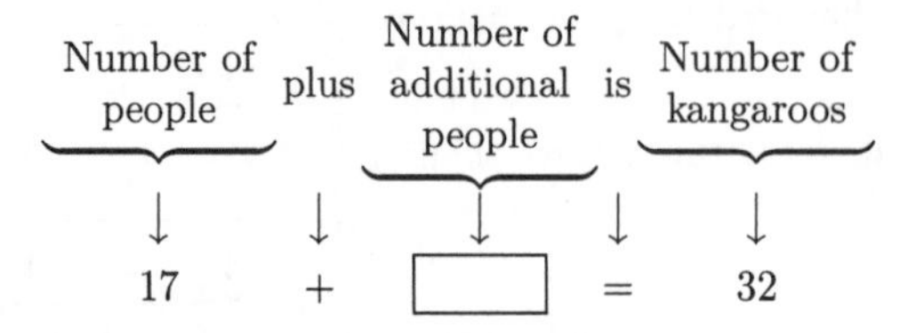

Now we write a related subtraction sentence.

$17 + \square = 32$

$\square = 32 - 17$ The addend 17 gets subtracted.

35.
$$\begin{array}{r} {\scriptstyle 7\ 16} \\ \not 8\ \not 6 \\ -\ 4\ 7 \\ \hline 3\ 9 \end{array}$$
We cannot subtract 7 ones from 6 ones. Borrow 1 ten to get 16 ones. Subtract ones, then subtract tens.

37.
$$\begin{array}{r} {\scriptstyle 11} \\ {\scriptstyle 5\ \not 1\ 15} \\ \not 6\ \not 2\ \not 5 \\ -\ 3\ 2\ 7 \\ \hline 2\ 9\ 8 \end{array}$$
We cannot subtract 7 ones from 5 ones. Borrow 1 ten to get 15 ones. Subtract ones. We cannot subtract 2 tens from 1 ten. Borrow 1 hundred to get 11 tens. Subtract tens, then subtract hundreds.

39.
$$\begin{array}{r} 8\ 6\ 6 \\ -\ 3\ 3\ 3 \\ \hline 5\ 3\ 3 \end{array}$$
Subtract ones, subtract tens, then subtract hundreds.

41.
$$\begin{array}{r} {\scriptstyle 17} \\ {\scriptstyle 8\ \not 7\ 12} \\ 3\ \not 9\ \not 8\ \not 2 \\ -\ 2\ 4\ 8\ 9 \\ \hline 1\ 4\ 9\ 3 \end{array}$$
We cannot subtract 9 ones from 2 ones. Borrow 1 ten to get 12 ones. Subtract ones. We cannot subtract 8 tens from 7 tens. Borrow 1 hundred to get 17 tens. Subtract tens, subtract hundreds, then subtract thousands.

43.
$$\begin{array}{r} {\scriptstyle 11\ 15\ 13} \\ {\scriptstyle \not 1\ \not 5\ \not 3\ 17} \\ \not 1\ \not 2,\ \not 6\ \not 4\ \not 7 \\ -\ \ 4\ 8\ 9\ 9 \\ \hline 7\ 7\ 4\ 8 \end{array}$$

45.
$$\begin{array}{r} {\scriptstyle 13} \\ {\scriptstyle \not 3\ 10} \\ \not 1\ \not 4\ \not 0 \\ -\ \ 5\ 6 \\ \hline 8\ 4 \end{array}$$

47.
$$\begin{array}{r} {\scriptstyle 6\ 9\ 9\ 10} \\ \not 7\ \not 0\ \not 0\ \not 0 \\ -\ 2\ 7\ 9\ 4 \\ \hline 4\ 2\ 0\ 6 \end{array}$$
We have 7 thousands or 700 tens. We borrow 1 ten to get 10 ones. We then have 699 tens. Subtract ones, then tens, then hundreds, then thousands.

49.
$$\begin{array}{r} {\scriptstyle 7\ 9\ 9\ 10} \\ 4\ \not 8,\ \not 0\ \not 0\ \not 0 \\ -3\ 7,\ 6\ 9\ 5 \\ \hline 1\ 0,\ 3\ 0\ 5 \end{array}$$
We have 8 thousands or 800 tens. We borrow 1 ten to get 10 ones. We then have 799 tens. Subtract ones, then tens, then hundreds, then thousands, then ten thousands.

51. 7 thousands + 9 hundreds + 9 tens + 2 ones = 7992

53. 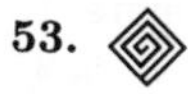

Exercise Set 1.3

1. Think of a rectangular array consisting of 21 rows with 21 objects in each row.

$21 \cdot 21 = 441$

3. If we think of filling the rectangle with square feet, we have a rectangular array.

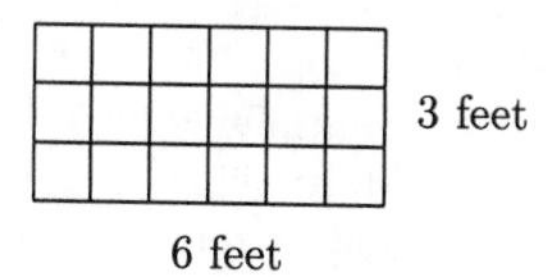

$$A = l \times w = 6 \times 3 = 18 \text{ square feet}$$

5. $A = l \times w = 11 \text{ yd} \times 11 \text{ yd} = 121$ square yards

7.
$$\begin{array}{r} 1\,0\,0 \\ \times\ \ 9\,6 \\ \hline 6\,0\,0 \\ 9\,0\,0\,0 \\ \hline 9\,6\,0\,0 \end{array}$$

9.
$$\begin{array}{r} {\scriptstyle 2} \\ 9\,4 \\ \times\ \ 6 \\ \hline 5\,6\,4 \end{array}$$
Multiplying by 6

11.
$$\begin{array}{r} {\scriptstyle 2} \\ 5\,0\,9 \\ \times\ \ \ 3 \\ \hline 1\,5\,2\,7 \end{array}$$
Multiplying by 3

13.
$$\begin{array}{r} {\scriptstyle 1\ 2\ 6} \\ 9\,2\,2\,9 \\ \times\ \ \ \ 7 \\ \hline 6\,4,6\,0\,3 \end{array}$$
Multiplying by 7

15.
$$\begin{array}{r} {\scriptstyle 2} \\ 5\,3 \\ \times\ \ 9\,0 \\ \hline 4\,7\,7\,0 \end{array}$$
Multiplying by 9 tens (We write 0 and then multiply 53 by 9.)

17.
$$\begin{array}{r} {\scriptstyle 2} \\ {\scriptstyle 3} \\ 8\,5 \\ \times\,4\,7 \\ \hline 5\,9\,5 \\ 3\,4\,0\,0 \\ \hline 3\,9\,9\,5 \end{array}$$
Multiplying by 7
Multiplying by 40
Adding

19.
$$\begin{array}{r} {\scriptstyle 2} \\ 6\,4\,0 \\ \times\ \ 7\,2 \\ \hline 1\,2\,8\,0 \\ 4\,4\,8\,0\,0 \\ \hline 4\,6,0\,8\,0 \end{array}$$
Multiplying by 2
Multiplying by 70
Adding

21.
$$\begin{array}{rl}
{\scriptstyle 1\,1} & \\
{\scriptstyle 1\,1} & \\
444 & \\
\times\ 33 & \\
\hline
1332 & \text{Multiplying by 3} \\
13320 & \text{Multiplying by 30} \\
\hline
14,652 & \text{Adding}
\end{array}$$

23.
$$\begin{array}{rl}
{\scriptstyle 3} & \\
{\scriptstyle 7} & \\
509 & \\
\times\ 408 & \\
\hline
4072 & \text{Multiplying by 8} \\
203600 & \text{Multiplying by 4 hundreds (We write 00} \\
\hline
207,672 & \text{and then multiply 509 by 4.)}
\end{array}$$

25.
$$\begin{array}{rl}
{\scriptstyle 4\,2} & \\
{\scriptstyle 1} & \\
{\scriptstyle 3\,1} & \\
853 & \\
\times\ 936 & \\
\hline
5118 & \text{Multiplying by 6} \\
25590 & \text{Multiplying by 30} \\
767700 & \text{Multiplying by 900} \\
\hline
798,408 & \text{Adding}
\end{array}$$

27.
$$\begin{array}{rl}
{\scriptstyle 1\ \ 2} & \\
{\scriptstyle 1} & \\
{\scriptstyle 1} & \\
{\scriptstyle 1\,1\,3} & \\
6428 & \\
\times\ 3224 & \\
\hline
25712 & \text{Multiplying by 4} \\
128560 & \text{Multiplying by 20} \\
1285600 & \text{Multiplying by 200} \\
19284000 & \text{Multiplying by 3000} \\
\hline
20,723,872 & \text{Adding}
\end{array}$$

29.
$$\begin{array}{rl}
{\scriptstyle 1\,3} & \\
3482 & \\
\times\ 104 & \\
\hline
13928 & \text{Multiplying by 4} \\
348200 & \text{Multiplying by 1 hundred (We write 00} \\
\hline
362,128 & \text{and then multiply 3482 by 1.)}
\end{array}$$

31.
$$\begin{array}{rl}
{\scriptstyle 2} & \\
{\scriptstyle 4} & \\
5006 & \\
\times\ 4008 & \\
\hline
40048 & \text{Multiplying by 8} \\
20024000 & \text{Multiplying by 4 thousands (We} \\
\hline
20,064,048 & \text{write 000 and then multiply 5006 by} \\
 & \text{4.)}
\end{array}$$

33.
$$\begin{array}{rl}
{\scriptstyle 2\ \ 3} & \\
{\scriptstyle 3\ \ 4} & \\
5608 & \\
\times\ 4500 & \text{Multiplying by 5 hundreds (We write} \\
\hline
2804000 & \text{00 and then multiply 5608 by 5.)} \\
22432000 & \text{Multiplying by 4000} \\
\hline
25,236,000 & \text{Adding}
\end{array}$$

35. Think of an array with 4 rows. The number of pounds in each row will go to a mule.

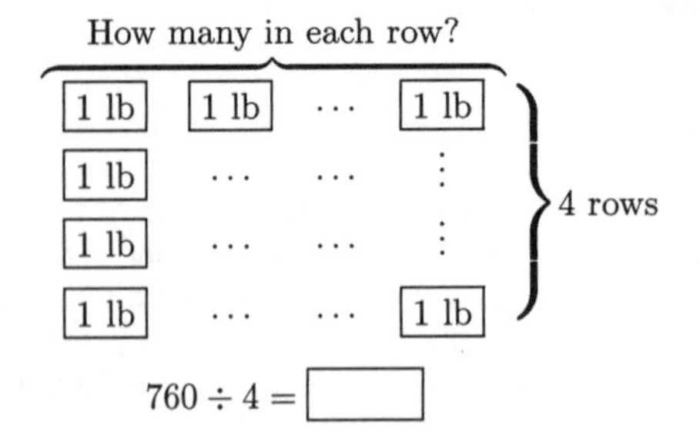

37. $18 \div 3 = 6$ The 3 moves to the right. A related multiplication sentence is $18 = 6 \cdot 3$. (By the commutative law of multiplication, there is also another multiplication sentence: $18 = 3 \cdot 6$.)

39. $22 \div 22 = 1$ The 22 on the right of the $\div$ symbol moves to the right. A related multiplication sentence is $22 = 1 \cdot 22$. (By the commutative law of multiplication, there is also another multiplication sentence: $22 = 22 \cdot 1$.)

41. $9 \times 5 = 45$

Move a factor to the other side and then write a division.

$9 \times 5 = 45$ $\qquad$ $9 \times 5 = 45$

$9 = 45 \div 5$ $\qquad$ $5 = 45 \div 9$

43. Two related division sentences for $37 \cdot 1 = 37$ are:

$37 = 37 \div 1$ $\qquad$ $(37 \cdot 1 = 37)$

and

$1 = 37 \div 37$ $\qquad$ $(37 \cdot 1 = 37)$

45.
$$\begin{array}{r}
55 \\
5\,\overline{)\,277} \\
250 \\
\hline
27 \\
25 \\
\hline
2
\end{array}$$

Think: 2 hundreds $\div$ 5. There are no hundreds in the quotient.
Think: 27 tens $\div$ 5. Estimate 5 tens.
Think: 27 ones $\div$ 5. Estimate 5 ones.

The answer is 55 R 2.

47.
$$\begin{array}{r}
108 \\
8\,\overline{)\,864} \\
800 \\
\hline
64 \\
64 \\
\hline
0
\end{array}$$

Think: 8 hundreds $\div$ 8. Estimate 1 hundred.
Think: 6 tens $\div$ 8. There are no tens in the quotient (other than the tens in 100). Write a 0 to show this.
Think: 64 ones $\div$ 8. Estimate 8 ones.

The answer is 108.

49.
$$\begin{array}{r}
307 \\
4\,\overline{)\,1228} \\
1200 \\
\hline
28 \\
28 \\
\hline
0
\end{array}$$

Think: 12 hundreds $\div$ 4. Estimate 3 hundreds.
Think: 2 tens $\div$ 4. There are no tens in the quotient (other than the tens in 300). Write a 0 to show this.
Think: 28 ones $\div$ 4. Estimate 7 ones.

The answer is 307.

51.

```
   9 2
8)7 3 8
  7 2 0
    1 8
    1 6
      2
```

Think: 73 tens ÷ 8. Estimate 9 tens.

Think: 18 ones ÷ 8. Estimate 2 ones.

The answer is 92 R 2.

53.

```
   1 7 0 3
5)8 5 1 5
  5 0 0 0
  3 5 1 5
  3 5 0 0
      1 5
      1 5
        0
```

Think: 8 thousands ÷ 5. Estimate 1 thousand.

Think: 35 hundreds ÷ 5. Estimate 7 hundreds.

Think: 1 ten ÷ 5. There are no tens in the quotient (other than the tens in 1700). Write a 0 to show this.

Think: 15 ones ÷ 5. Estimate 3 ones.

The answer is 1703.

55.

```
      2 9
3 0)8 7 5
    6 0 0
    2 7 5
    2 7 0
        5
```

Think: 87 tens ÷ 30. Estimate 2 tens.

Think: 275 ones ÷ 30. Estimate 9 ones.

The answer is 29 R 5.

57.

```
          8
8 5)7 6 7 2
    6 8 0 0
    [8 7] 2
```

Round 85 to 90.

Think: 767 tens ÷ 90. Estimate 8 tens.

Since 87 is larger than the divisor, the estimate is too low.

```
        9 0
8 5)7 6 7 2
    7 6 5 0
        2 2
```

Think: 767 tens ÷ 90. Estimate 9 tens.

Think: 22 ones ÷ 90. There are no ones in the quotient (other than the ones in 90). Write a 0 to show this.

The answer is 90 R 22.

59.

```
              3
1 1 1)3 2 1 9
      3 3 3 0
```

Round 111 to 100.

Think: 321 tens ÷ 100. Estimate 3 tens.

Since we cannot subtract 3330 from 3219, the estimate is too high.

```
            2 9
1 1 1)3 2 1 9
      2 2 2 0
        9 9 9
        9 9 9
            0
```

Think: 321 tens ÷ 100. Estimate 2 tens.

Think: 999 ones ÷ 100. Estimate 9 ones.

The answer is 29.

61.

```
        4
2 4)8 8 8 0
    9 6 0 0
```

Round 24 to 20.

Think: 88 hundreds ÷ 20. Estimate 4 hundreds.

Since we cannot subtract 9600 from 8880, the estimate is too high.

```
        3 8
2 4)8 8 8 0
    7 2 0 0
    1 6 8 0
    1 9 2 0
```

Think: 88 hundreds ÷ 20. Estimate 3 hundreds.

Think: 168 tens ÷ 20. Estimate 8 tens.

Since we cannot subtract 1920 from 1680, the estimate is too high.

```
      3 7 0
2 4)8 8 8 0
    7 2 0 0
    1 6 8 0
    1 6 8 0
          0
```

Think: 168 tens ÷ 20. Estimate 7 tens.

Think: 0 ones ÷ 20. There are no ones in the quotient (other than the ones in 370). Write a 0 to show this.

The answer is 370.

63.

```
             5
2 8)1 7, 0 6 7
    1 4  0 0 0
     [3 0] 6 7
```

Round 28 to 30.

Think: 170 hundreds ÷ 30. Estimate 5 hundreds.

Since 30 is larger than the divisor, 28, the estimate is too low.

```
         6 0 8
2 8)1 7, 0 6 7
    1 6  8 0 0
         2 6 7
         2 2 4
         [4 3]
```

Think: 170 hundreds ÷ 30. Estimate 6 hundreds.

Think: 26 tens ÷ 30. There are no tens in the quotient (other than the tens in 600.) Write a zero to show this.

Think: 267 ones ÷ 30. Estimate 8 ones.

Since 43 is larger than the divisor, 28, the estimate is too low.

```
         6 0 9
2 8)1 7, 0 6 7
    1 6  8 0 0
         2 6 7
         2 5 2
           1 5
```

Think: 267 ones ÷ 30. Estimate 9 ones.

The answer is 609 R 15.

65. Round 48 to the nearest ten.

```
4 [8]
↑
```

The digit 4 is in the tens place. Consider the next digit to the right. Since the digit, 8, is 5 or higher, round 4 tens up to 5 tens. Then change the digit to the right of the tens digit to zero.

The answer is 50.

67. Round 67 to the nearest ten.

6 [7]
↑

The digit 6 is in the tens place. Consider the next digit to the right. Since the digit, 7, is 5 or higher, round 6 tens up to 7 tens. Then change the digit to the right of the tens digit to zero.

The answer is 70.

69. Round 731 to the nearest ten.

7 3 [1]
↑

The digit 3 is in the tens place. Consider the next digit to the right. Since the digit, 1, is 4 or lower, round down, meaning that 3 tens stays as 3 tens. Then change the digit to the right of the tens digit to zero.

The answer is 730.

71. Round 895 to the nearest ten.

8 9 [5]
↑

The digit 9 is in the tens place. Consider the next digit to the right. Since the digit, 5, is 5 or higher, we round up. The 89 tens become 90 tens. Then change the digit to the right of the tens digit to zero.

The answer is 900.

73. Round 146 to the nearest hundred.

1 [4] 6
↑

The digit 1 is in the hundreds place. Consider the next digit to the right. Since the digit, 4, is 4 or lower, round down, meaning that 1 hundred stays as 1 hundred. Then change all digits to the right of the hundreds digit to zeros.

The answer is 100.

75. Round 957 to the nearest hundred.

9 [5] 7
↑

The digit 9 is in the hundreds place. Consider the next digit to the right. Since the digit, 5, is 5 or higher, round up. The 9 hundreds become 10 hundreds. Then change all digits to the right of the hundreds digit to zeros.

The answer is 1000.

77. Round 9079 to the nearest hundred.

9 0 [7] 9
↑

The digit 0 is in the hundreds place. Consider the next digit to the right. Since the digit, 7, is 5 or higher, round 0 hundreds up to 1 hundred. Then change all digits to the right of the hundreds digit to zeros.

The answer is 9100.

79. Round 32,850 to the nearest hundred.

3 2, 8 [5] 0
↑

The digit 8 is in the hundreds place. Consider the next digit to the right. Since the digit, 5, is 5 or higher, round 8 hundreds up to 9 hundreds. Then change all digits to the right of the hundreds digit to zero.

The answer is 32,900.

81. Round 5876 to the nearest thousand.

5 [8] 7 6
↑

The digit 5 is in the thousands place. Consider the next digit to the right. Since the digit, 8, is 5 or higher, round 5 thousands up to 6 thousands. Then change all digits to the right of the thousands digit to zeros.

The answer is 6000.

83. Round 7500 to the nearest thousand.

7 [5] 0 0
↑

The digit 7 is in the thousands place. Consider the next digit to the right. Since the digit, 5, is 5 or higher, round 7 thousands up to 8 thousands. Then change all the digits to the right of the thousands digit to zeros.

The answer is 8000.

85. Round 45,340 to the nearest thousand.

4 5, [3] 4 0
↑

The digit 5 is in the thousands place. Consider the next digit to the right. Since the digit, 3, is 4 or lower, round down, meaning that 5 thousands stays as 5 thousands. Then change all the digits to the right of the thousands digit to zeros.

The answer is 45,000.

87. Round 373,405 to the nearest thousand.

3 7 3, [4] 0 5
↑

The digit 3 is in the thousands place. Consider the next digit to the right. Since the digit, 4, is 4 or lower, round down, meaning that 3 thousands stays as 3 thousands. Then change all the digits to the right of the thousands digit to zeros.

The answer is 373,000.

89.

	Rounded to the nearest ten	
78	80	
+ 97	+ 100	
	180	← Estimated answer

91.

	Rounded to the nearest ten	
8074	8070	
− 2347	− 2350	
	5720	← Estimated answer

93. Rounded to the nearest hundred

$$\begin{array}{r} 7348 \\ +9247 \\ \hline \end{array} \qquad \begin{array}{r} 7300 \\ +9200 \\ \hline 16,500 \end{array} \leftarrow \text{Estimated answer}$$

95. Rounded to the nearest hundred

$$\begin{array}{r} 6852 \\ -1748 \\ \hline \end{array} \qquad \begin{array}{r} 6900 \\ -1700 \\ \hline 5200 \end{array} \leftarrow \text{Estimated answer}$$

97. Rounded to the nearest thousand

$$\begin{array}{r} 9643 \\ 4821 \\ 8943 \\ +7004 \\ \hline \end{array} \qquad \begin{array}{r} 10,000 \\ 5000 \\ 9000 \\ +7000 \\ \hline 31,000 \end{array} \leftarrow \text{Estimated answer}$$

99. Rounded to the nearest thousand

$$\begin{array}{r} 92,149 \\ -22,555 \\ \hline \end{array} \qquad \begin{array}{r} 92,000 \\ -23,000 \\ \hline 69,000 \end{array} \leftarrow \text{Estimated answer}$$

101. Rounded to the nearest ten

$$\begin{array}{r} 45 \\ \times 67 \\ \hline \end{array} \qquad \begin{array}{r} 50 \\ \times 70 \\ \hline 3500 \end{array} \leftarrow \text{Estimated answer}$$

103. Rounded to the nearest ten

$$\begin{array}{r} 34 \\ \times 29 \\ \hline \end{array} \qquad \begin{array}{r} 30 \\ \times 30 \\ \hline 900 \end{array} \leftarrow \text{Estimated answer}$$

105. Rounded to the nearest hundred

$$\begin{array}{r} 876 \\ \times 345 \\ \hline \end{array} \qquad \begin{array}{r} 900 \\ \times 300 \\ \hline 270,000 \end{array} \leftarrow \text{Estimated answer}$$

107. Rounded to the nearest hundred

$$\begin{array}{r} 432 \\ \times 199 \\ \hline \end{array} \qquad \begin{array}{r} 400 \\ \times 200 \\ \hline 80,000 \end{array} \leftarrow \text{Estimated answer}$$

109. Rounded to the nearest thousand

$$\begin{array}{r} 5608 \\ \times 4576 \\ \hline \end{array} \qquad \begin{array}{r} 6000 \\ \times 5000 \\ \hline 30,000,000 \end{array} \leftarrow \text{Estimated answer}$$

111. Rounded to the nearest thousand

$$\begin{array}{r} 7888 \\ \times 6224 \\ \hline \end{array} \qquad \begin{array}{r} 8000 \\ \times 6000 \\ \hline 48,000,000 \end{array} \leftarrow \text{Estimated answer}$$

113. 7882 = 7 thousands + 8 hundreds + 8 tens + 2 ones

115. $21 - 16 = 5$

↑ This number gets added (after 5).

↓

$21 = 5 + 16$

(By the commutative law of addition, $21 = 16 + 5$ is also correct.)

117. $47 + 9 = 56$

↑ This addend gets subtracted from the sum. ↓

$47 = 56 - 9$

$47 + 9 = 56$

↑ This addend gets subtracted from the sum. ↓

$9 = 56 - 47$

119. ◈

121. We divide 1231 by 42:

$$\begin{array}{r} 29 \\ 42\overline{)1231} \\ 840 \\ \hline 391 \\ 378 \\ \hline 13 \end{array}$$

The answer is 29 R 13. Since 13 students will be left after 29 buses are filled, then 30 buses are needed.

123. Use a calculator to perform the computations in this exercise.

First find the total area of each floor:

$A = l \times w = 172 \times 84 = 14,448$ square feet

Find the area lost to the elevator and the stairwell:

$A = l \times w = 35 \times 20 = 700$ square feet

Subtract to find the area available as office space on each floor:

$14,448 - 700 = 13,748$ square feet

Finally, multiply by the number of floors, 18, to find the total area available as office space:

$18 \times 13,748 = 247,464$ square feet

Exercise Set 1.4

1. $x + 0 = 14$

We replace x by different numbers until we get a true equation. If we replace x by 14, we get a true equation: $14 + 0 = 14$. No other replacement makes the equation true, so the solution is 14.

3. $y \cdot 17 = 0$

We replace y by different numbers until we get a true equation. If we replace y by 0, we get a true equation: $0 \cdot 17 = 0$. No other replacement makes the equation true, so the solution is 0.

5.
$$\begin{aligned} 13 + x &= 42 \\ 13 + x - 13 &= 42 - 13 \quad \text{Subtracting 13 on both sides} \\ 0 + x &= 29 \quad \text{13 plus } x \text{ minus 13 is } 0 + x. \\ x &= 29 \end{aligned}$$

7.
$$\begin{aligned} 12 &= 12 + m \\ 12 - 12 &= 12 + m - 12 \quad \text{Subtracting 12 on both sides} \\ 0 &= 0 + m \quad \text{12 plus } m \text{ minus 12 is } 0 + m. \\ 0 &= m \end{aligned}$$

9.
$$\begin{aligned} 3 \cdot x &= 24 \\ \frac{3 \cdot x}{3} &= \frac{24}{3} \quad \text{Dividing by 3 on both sides} \\ x &= 8 \quad \text{3 times } x \text{ divided by 3 is } x. \end{aligned}$$

11.
$$\begin{aligned} 112 &= n \cdot 8 \\ \frac{112}{8} &= \frac{n \cdot 8}{8} \quad \text{Dividing by 8 on both sides} \\ 14 &= n \end{aligned}$$

13. $45 \times 23 = x$

To solve the equation we carry out the calculation.

$$\begin{array}{r} 4\,5 \\ \times\ 2\,3 \\ \hline 1\,3\,5 \\ 9\,0\,0 \\ \hline 1\,0\,3\,5 \end{array}$$

The solution is 1035.

15. $t = 125 \div 5$

To solve the equation we carry out the calculation.

$$\begin{array}{r} 2\,5 \\ 5\,\overline{)\,1\,2\,5} \\ 1\,0\,0 \\ \hline 2\,5 \\ 2\,5 \\ \hline 0 \end{array}$$

The solution is 25.

17. $p = 908 - 458$

To solve the equation we carry out the calculation.

$$\begin{array}{r} 9\,0\,8 \\ -\ 4\,5\,8 \\ \hline 4\,5\,0 \end{array}$$

The solution is 450.

19. $x = 12,345 + 78,555$

To solve the equation we carry out the calculation.

$$\begin{array}{r} 1\,2,\,3\,4\,5 \\ +\ 7\,8,\,5\,5\,5 \\ \hline 9\,0,\,9\,0\,0 \end{array}$$

The solution is 90,900.

21.
$$\begin{aligned} 3 \cdot m &= 96 \\ \frac{3 \cdot m}{3} &= \frac{96}{3} \quad \text{Dividing by 3 on both sides} \\ m &= 32 \end{aligned}$$

23.
$$\begin{aligned} 715 &= 5 \cdot z \\ \frac{715}{5} &= \frac{5 \cdot z}{5} \quad \text{Dividing by 5 on both sides} \\ 143 &= z \end{aligned}$$

25.
$$\begin{aligned} 10 + x &= 89 \\ 10 + x - 10 &= 89 - 10 \\ x &= 79 \end{aligned}$$

27.
$$\begin{aligned} 61 &= 16 + y \\ 61 - 16 &= 16 + y - 16 \\ 45 &= y \end{aligned}$$

29.
$$\begin{aligned} 6 \cdot p &= 1944 \\ \frac{6 \cdot p}{6} &= \frac{1944}{6} \\ p &= 324 \end{aligned}$$

31.
$$\begin{aligned} 5 \cdot x &= 3715 \\ \frac{5 \cdot x}{5} &= \frac{3715}{5} \\ x &= 743 \end{aligned}$$

33.
$$\begin{aligned} 47 + n &= 84 \\ 47 + n - 47 &= 84 - 47 \\ n &= 37 \end{aligned}$$

35.
$$\begin{aligned} x + 78 &= 144 \\ x + 78 - 78 &= 144 - 78 \\ x &= 66 \end{aligned}$$

37.
$$\begin{aligned} 165 &= 11 \cdot n \\ \frac{165}{11} &= \frac{11 \cdot n}{11} \\ 15 &= n \end{aligned}$$

39.
$$\begin{aligned} 624 &= t \cdot 13 \\ \frac{624}{13} &= \frac{t \cdot 13}{13} \\ 48 &= t \end{aligned}$$

41.
$$\begin{aligned} x + 214 &= 389 \\ x + 214 - 214 &= 389 - 214 \\ x &= 175 \end{aligned}$$

43.
$$\begin{aligned} 567 + x &= 902 \\ 567 + x - 567 &= 902 - 567 \\ x &= 335 \end{aligned}$$

45. $18 \cdot x = 1872$

$$\frac{18 \cdot x}{18} = \frac{1872}{18}$$

$$x = 104$$

47. $40 \cdot x = 1800$

$$\frac{40 \cdot x}{40} = \frac{1800}{40}$$

$$x = 45$$

49. $2344 + y = 6400$

$$2344 + y - 2344 = 6400 - 2344$$

$$y = 4056$$

51. $8322 + 9281 = x$

$17,603 = x$ Doing the addition

53. $234 \cdot 78 = y$

$18,252 = y$ Doing the multiplication

55. $58 \cdot m = 11,890$

$$\frac{58 \cdot m}{58} = \frac{11,890}{58}$$

$$m = 205$$

57. $7 + 8 = 15$ — This number (7) gets subtracted from the sum: $7 = 15 - 8$

$7 + 8 = 15$ — This number (8) gets subtracted from the sum: $8 = 15 - 7$

59. Since 123 is to the left of 789 on a number line, $123 < 789$.

61.

```
      1 4 2
  9 ) 1 2 8 3
        9 0 0     Think: 12 hundreds ÷ 9. Estimate 1 hundred.
        3 8 3     Think: 38 tens ÷ 9. Estimate 4 tens.
        3 6 0
          2 3     Think: 23 ones ÷ 9. Estimate 2 ones.
          1 8
            5
```

The answer is 142 R 5.

63. ◈

65. $23,465 \cdot x = 8,142,355$

$$\frac{23,465 \cdot x}{23,465} = \frac{8,142,355}{23,465}$$

$x = 347$ Using a calculator to divide

Exercise Set 1.5

1. ***Familiarize.*** We visualize the situation. Let s = the total sales during the given period. We are combining amounts, so we use addition.

\$3572	+	\$2718	+	\$2809	+	\$3177	=	s
in January		in February		in March		in April		Total sales

Translate. We translate to an equation.

$$3572 + 2718 + 2809 + 3177 = s$$

Solve. We carry out the addition.

```
   2 1 2
   3 5 7 2
   2 7 1 8
   2 8 0 9
 + 3 1 7 7
 ---------
 1 2, 2 7 6
```

Thus, $12,276 = s$, or $s = 12,276$.

Check. We can repeat the calculation. We can also estimate by rounding, say to the nearest thousand.

$3572 + 2718 + 2809 + 3177 \approx 4000 + 3000 + 3000 + 3000 = 13,000 \approx 12,276$.

Since the estimated answer is close to the calculated answer, our result is probably correct.

State. The total sales from January through April were \$12,276.

3. ***Familiarize.*** We visualize the situation. Let t = the total sales for 1993 and 1994, in millions of dollars. We are combining amounts, so we use addition.

\$3534	+	\$3470	=	t
in 1993		in 1994		Total sales

Translate. We translate to an equation.

$$3534 + 3470 = t$$

Solve. We carry out the addition.

```
   1 1
   3 5 3 4
 + 3 4 7 0
 ---------
   7 0 0 4
```

Thus, $7004 = t$, or $t = 7004$.

Check. We can repeat the calculation. We can also estimate by rounding, say to the nearest hundred.

$3534 + 3470 \approx 3500 + 3500 = 7000 \approx 7004$

Since the estimated answer is close to the calculated answer, our result is probably correct.

State. Total sales for 1993 and 1994 were \$7004 million, or \$7,004,000,000.

5. ***Familiarize.*** We visualize the situation. Let s = the amount by which sales in 1994 exceeded sales in 1993, in millions of dollars.

1994 sales $3470	Excess s
1993 sales $3534	

Translate. This is a "how much more" situation. We translate to an equation.

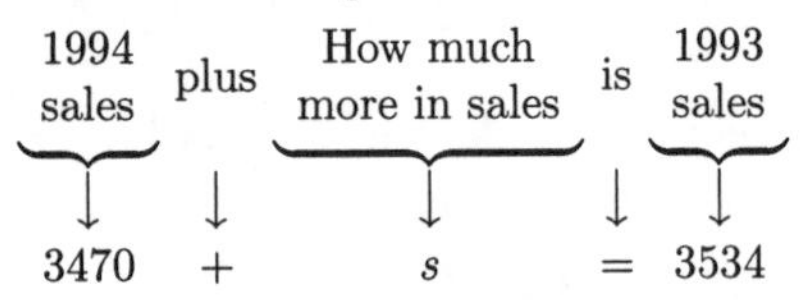

Solve. We solve the equation

$$3470 + s = 3534$$

$$3470 + s - 3470 = 3534 - 3470 \quad \text{Subtracting 3470 on both sides}$$

$$s = 64$$

Check. We can check by adding the difference, 64, to the subtrahend, 3470: $3470 + 64 = 3534$. We get the original minuend, 3534, so our answer checks.

State. Sales in 1993 were \$64 million or \$64,000,000, more than sales in 1994.

7. ***Familiarize.*** We visualize the situation. Let l = the length of the Amazon River, in miles.

Length of Amazon l	Excess length of Nile 138 miles
Length of Nile 4145 miles	

Translate. This is a "how much more" situation. We translate to an equation.

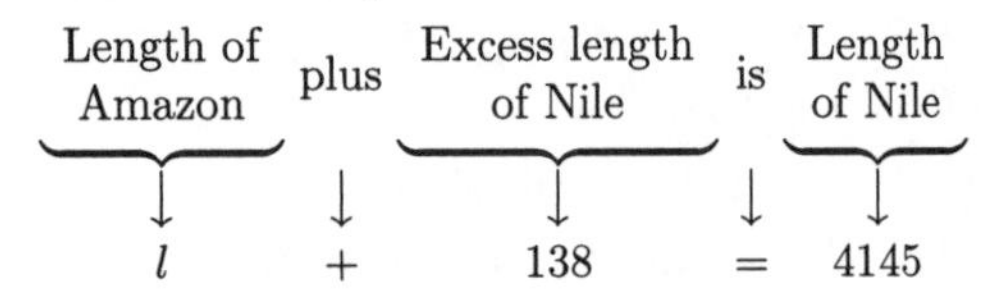

Solve. We solve the equation

$$l + 138 = 4145$$

$$l + 138 - 138 = 4145 - 138 \quad \text{Subtracting 138 on both sides}$$

$$l = 4007$$

Check. We can check by adding the difference, 4007, to the subtrahend, 138: $138 + 4007 = 4145$. Our answer checks.

We could also have estimated: $138 + 4000 \approx 140 + 4000 = 4140 \approx 4145$. Since the estimate is close to the calculated answer, our result is probably correct.

State. The length of the Amazon is 4007 miles.

9. ***Familiarize.*** We first make a drawing.

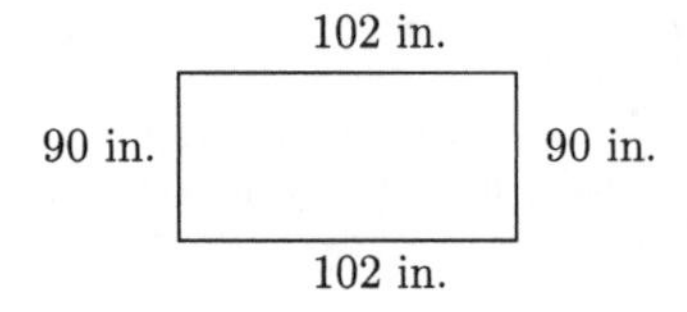

Let P = the perimeter of the sheet, in inches.

Translate. The perimeter is the sum of the lengths of the sides of the sheet. We translate to an equation.

$$P = 102 + 90 + 102 + 90$$

Solve. We carry out the addition.

$$P = 102 + 90 + 102 + 90 = 384$$

Check. We can repeat the calculation. We can also estimate by rounding, say to the nearest hundred.

$$102 + 90 + 102 + 90 \approx 100 + 100 + 100 + 100 = 400 \approx 384$$

Our answer checks.

State. The perimeter of the sheet is 384 in.

11. ***Familiarize.*** The drawing in the text visualizes the situation. Let p = the number of sheets in 9 reams of paper. Repeated addition works well here.

Translate. We translate to an equation.

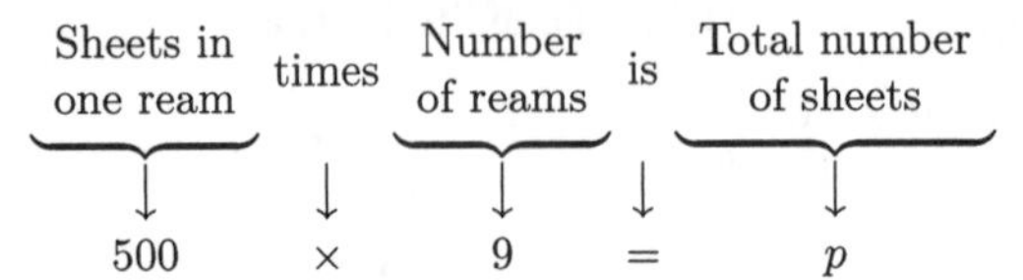

Solve. We carry out the multiplication.

$$\begin{array}{r} 500 \\ \times \quad 9 \\ \hline 4500 \end{array}$$

Thus, $4500 = p$, or $p = 4500$.

Check. We can repeat our calculation. The answer checks.

State. There are 4500 sheets in 9 reams of paper.

13. ***Familiarize.*** We visualize the situation. Let n = the number by which the number of Elvis impersonators in 1995 exceeded the number in 1977.

Number in 1977 48	n
Number in 1995 7328	

Translate. This is a "how much more" situation. We translate to an equation.

Number in 1977 plus Excess number in 1995 is Number in 1995

$$48 + n = 7328$$

Solve. We solve the equation

$$48 + n = 7328$$

$$48 + n - 48 = 7328 - 48 \quad \text{Subtracting 48 on both sides}$$

$$n = 7280$$

Check. We can check by adding the difference, 7280, to the subtrahend, 48: $7280 + 48 = 7328$. Our answer checks.

We could also have estimated: $7328 - 48 \approx 7300 - 50 = 7250 \approx 7280$. Since the estimate is close to the calculated answer, our result is probably correct.

State. In 1995 there were 7280 more Elvis impersonators than in 1977.

15. ***Familiarize.*** We first draw a picture. We let x = the amount of each payment.

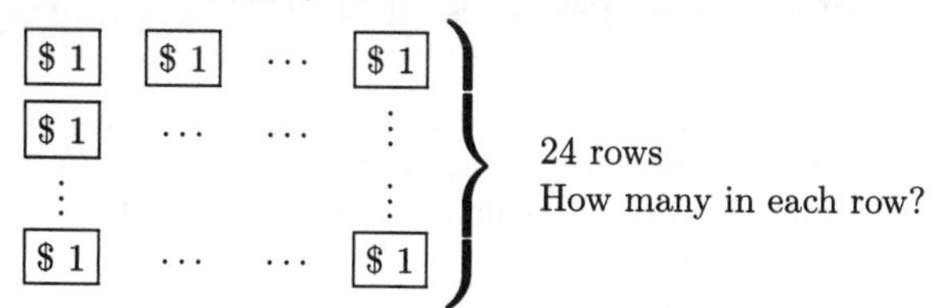

Translate. We translate to an equation.

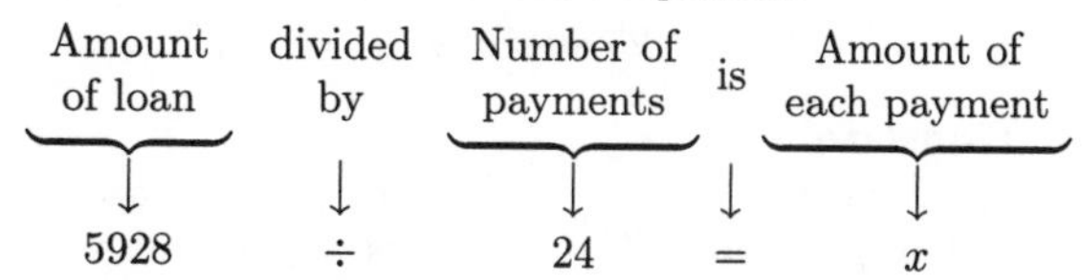

Solve. We carry out the division.

$$\begin{array}{r} 247 \\ 24\overline{)5928} \\ \underline{4800} \\ 1128 \\ \underline{960} \\ 168 \\ \underline{168} \\ 0 \end{array}$$

Thus, $247 = x$, or $x = 247$.

Check. We can check by multiplying 247 by 24: $24 \cdot 247 = 5928$. The answer checks.

State. Each payment is \$247.

17. ***Familiarize.*** We first draw a picture. Let w = the number of full weeks the episodes can run.

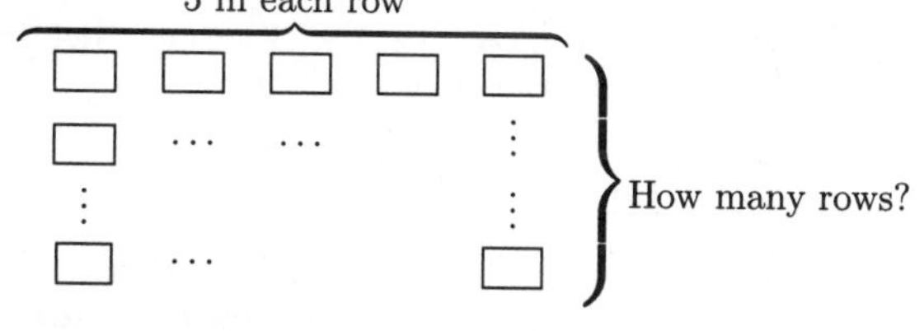

Translate. We translate to an equation.

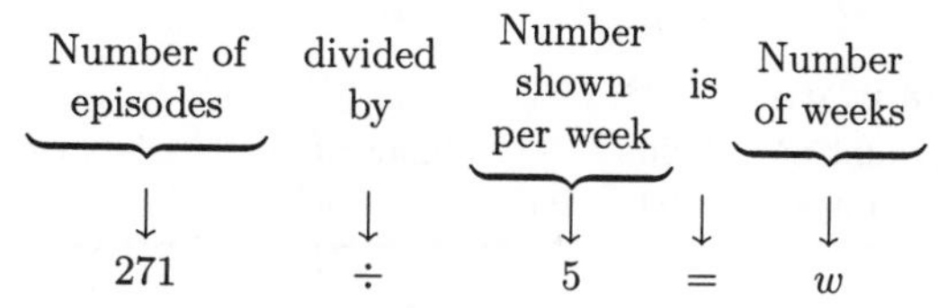

Solve. We carry out the division.

$$\begin{array}{r} 54 \\ 5\overline{)271} \\ \underline{250} \\ 21 \\ \underline{20} \\ 1 \end{array}$$

Check. We can check by multiplying the number of weeks by 5 and adding the remainder, 1:

$$5 \cdot 54 = 270, \qquad 270 + 1 = 271$$

State. 54 full weeks will pass before the station must start over. There will be 1 episode left over.

19. ***Familiarize.*** We first draw a picture. Let h = the number of hours in a week. Repeated addition works well here.

24 hours + 24 hours + ⋯ + 24 hours

7 addends

Translate. We translate to an equation.

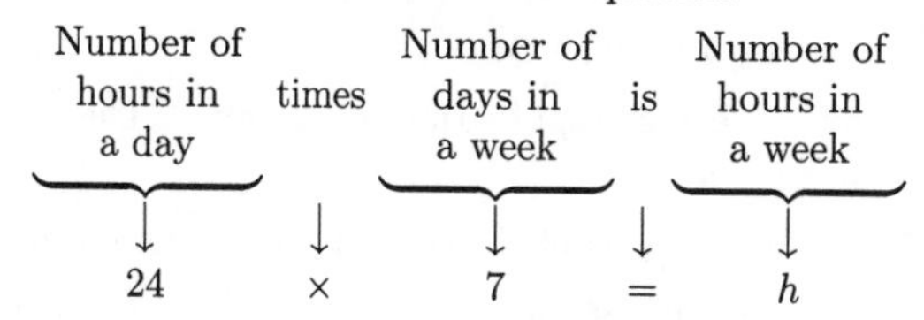

Solve. We carry out the multiplication.

$$\begin{array}{r} 24 \\ \times\ \ 7 \\ \hline 168 \end{array}$$

Thus, $168 = h$, or $h = 168$.

Check. We can repeat the calculation. We an also estimate:

$$24 \times 7 \approx 20 \times 10 = 200 \approx 168$$

Our answer checks.

State. There are 168 hours in a week.

21. ***Familiarize.*** Let a = the amount left in the account. We will start with \$568, subtract the amount of each check, and add the amount of the deposit.

Translate. We translate to an equation.

$$a = 568 - 46 - 87 - 129 + 94$$

Solve. We carry out the calculations.

$$a = 568 - 46 - 87 - 129 + 94 = 400$$

Check. We can repeat the calculations. We can also estimate:

$$568 - 46 - 87 - 129 + 94 \approx 570 - 50 - 90 - 130 + 100 = 400$$

Our answer checks.

State. \$400 is left in the account.

23. ***Familiarize.*** We first draw a picture. Let A = the area and P = the perimeter of the court, in feet.

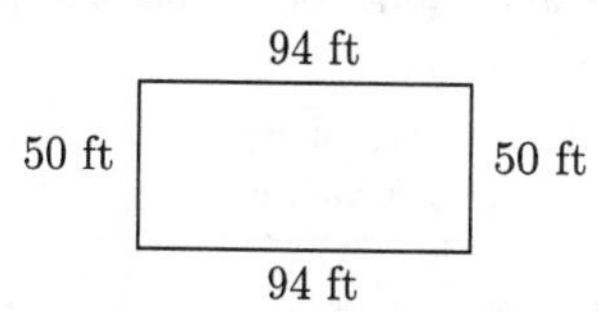

Translate. We write one equation to find the area and another to find the perimeter.

a) Using the formula for the area of a rectangle, we have

$$A = l \cdot w = 94 \cdot 50$$

b) Recall that the perimeter is the distance around the court.

$$P = 94 + 50 + 94 + 50$$

Solve. We carry out the calculations.

a)
```
    5 0
  × 9 4
  -----
  2 0 0
4 5 0 0
-------
4 7 0 0
```
Thus, $A = 4700$.

b) $P = 94 + 50 + 94 + 50 = 288$

Check. We can repeat the calculation. The answers check.

State. a) The area of the court is 4700 square feet.

b) The perimeter of the court is 288 ft.

25. ***Familiarize.*** We first draw a picture. We let $c =$ the number of cartons needed.

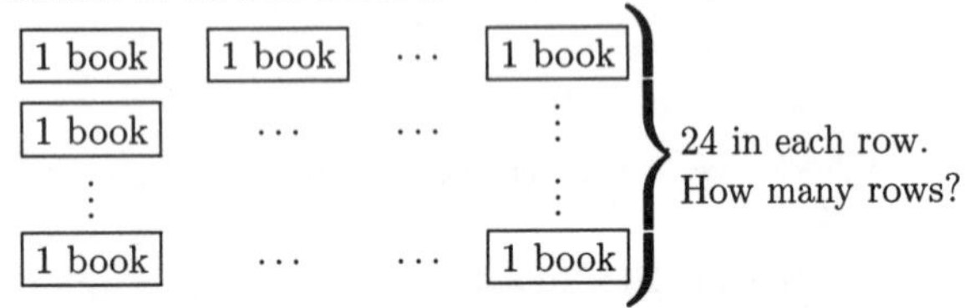

Translate.

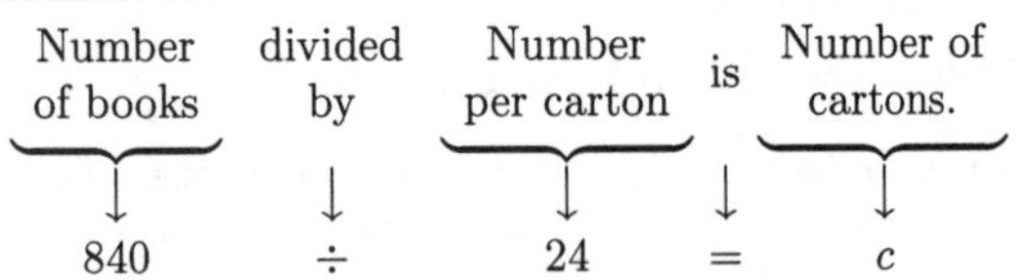

Solve. We carry out the division.

```
      3 5
 2 4)8 4 0
     7 2 0
     -----
     1 2 0
     1 2 0
     -----
         0
```

Thus, $35 = c$, or $c = 35$.

Check. We can check by multiplying: $24 \cdot 35 = 840$. The answer checks.

State. It will take 35 cartons to ship 840 books.

27. ***Familiarize.*** We visualize the situation as we did in Exercise 25. Let $c =$ the number of cartons that can be filled.

Translate.

Number of books	divided by	Number per carton	is	Number of full cartons.
↓	↓	↓	↓	↓
1355	÷	24	=	c

Solve. We carry out the division.

```
        5 6
 2 4)1 3 5 5
     1 2 0 0
     -------
       1 5 5
       1 4 4
       -----
         1 1
```

Check. We can check by multiplying the number of cartons by 24 and adding the remainder, 11:

$24 \cdot 56 = 1344, \qquad 1344 + 11 = 1355$

Our answer checks.

State. 56 cartons can be filled. There will be 11 books left over.

29. ***Familiarize.*** First we find the distance in reality between two cities that are 25 in. apart on the map. We make a drawing. Let $d =$ the distance between the cities, in miles. Repeated addition works well here.

64 miles + 64 miles + ⋯ + 64 miles

25 addends

Translate.

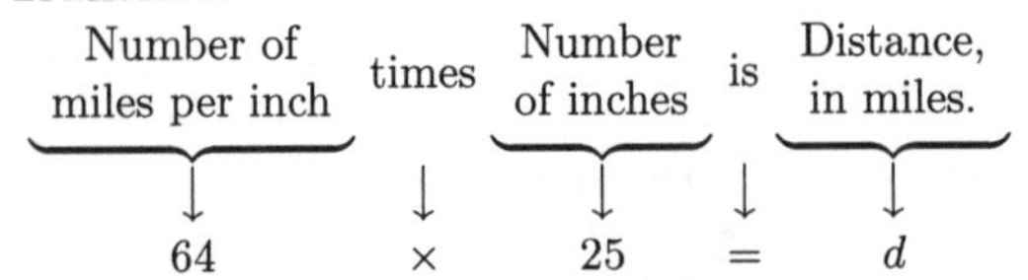

Solve. We carry out the multiplication.

```
      2 5
    × 6 4
    -----
    1 0 0
  1 5 0 0
  -------
  1 6 0 0
```

Thus, $1600 = d$, or $d = 1600$.

Check. We can repeat the calculation or estimate the product. Our answer checks.

State. Two cities that are 25 in. apart on the map are 1600 miles apart in reality.

Next we find distance on the map between two cities that, in reality, are 1728 mi apart.

Familiarize. We visualize the situation. Let $m =$ the distance between the cities on the map.

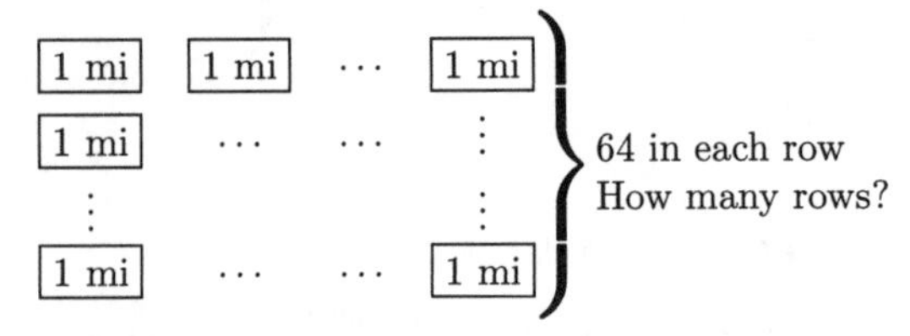

Translate.

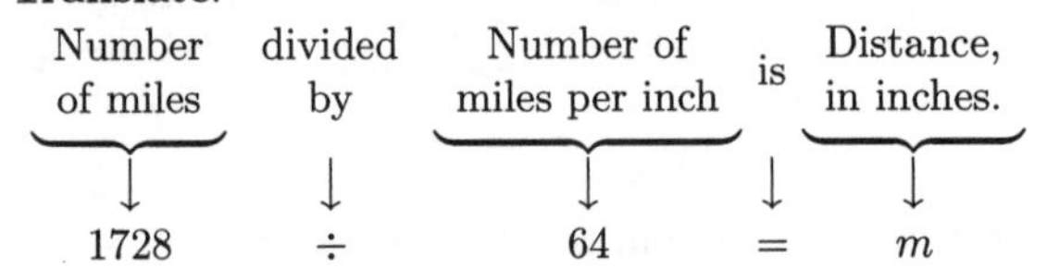

Solve. We carry out the division.

```
          2 7
 6 4)1 7 2 8
     1 2 8 0
     -------
       4 4 8
       4 4 8
       -----
           0
```

Thus, $27 = m$, or $m = 27$.

Check. We can check by multiplying: $64 \cdot 27 = 1728$.

Our answer checks.

State. The cities are 27 in. apart on the map.

31. ***Familiarize***. We first make a drawing. Let r = the number of rows.

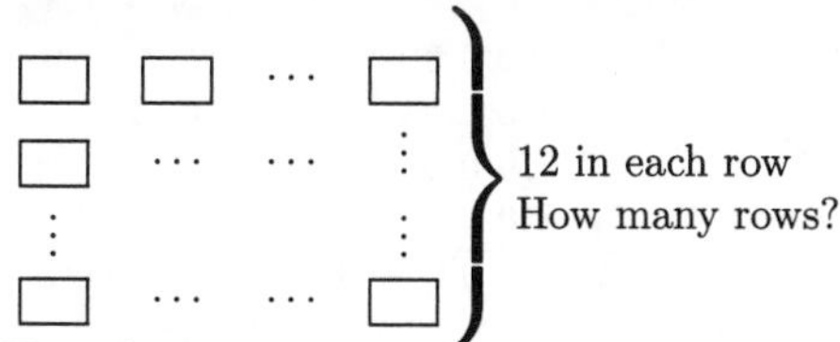

Translate.

$$\underbrace{\text{Number of holes}}_{\downarrow\atop 216}\ \ \underbrace{\text{divided by}}_{\downarrow\atop \div}\ \ \underbrace{\text{Number per row}}_{\downarrow\atop 12}\ \ \underbrace{\text{is}}_{\downarrow\atop =}\ \ \underbrace{\text{Number of rows.}}_{\downarrow\atop r}$$

Solve. We carry out the division.

$$\begin{array}{r} 18 \\ 12\overline{)216} \\ \underline{120} \\ 96 \\ \underline{96} \\ 0 \end{array}$$

Thus, $18 = r$, or $r = 18$.

Check. We can check by multiplying: $12 \cdot 18 = 216$. Our answer checks.

State. There are 18 rows.

33. ***Familiarize***. This is a multistep problem.

We must find the total price of the 5 video games. Then we must find how many 10's there are in the total price. Let p = the total price of the games.

To find the total price of the 5 video games we can use repeated addition.

$$\underbrace{\boxed{\$44} + \boxed{\$44} + \boxed{\$44} + \boxed{\$44} + \boxed{\$44}}_{\text{5 addends}}$$

Translate.

$$\underbrace{\text{Price per game}}_{\downarrow\atop 44}\ \ \underbrace{\text{times}}_{\downarrow\atop \cdot}\ \ \underbrace{\text{Number of games}}_{\downarrow\atop 5}\ \ \underbrace{\text{is}}_{\downarrow\atop =}\ \ \underbrace{\text{Total price of games}}_{\downarrow\atop p}$$

Solve. First we carry out the multiplication.

$$44 \cdot 5 = p$$
$$220 = p$$

The total price of the 5 video games is \$220. Repeated addition can be used again to find how many 10's there are in \$220. We let x = the number of \$10 bills required.

\$220			
\$10	\$10	$\cdots$	\$10

Translate to an equation and solve.

$$10 \cdot x = 220$$
$$\frac{10 \cdot x}{10} = \frac{220}{10}$$
$$x = 22$$

Check. We repeat the calculations. The answer checks.

State. It took 22 ten dollar bills.

35. ***Familiarize***. This is a multistep problem.

We must find the total cost of the 4 shirts and the total cost of the 6 pairs of pants. The total cost of the clothing is the sum of these two totals.

Repeated addition works well in finding the total cost of the 4 shirts and the total cost of the 6 pairs of pants. We let x = the total cost of the shirts and y = the total cost of the pants.

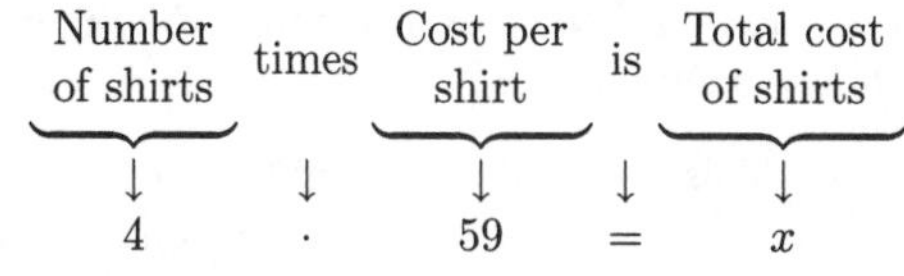

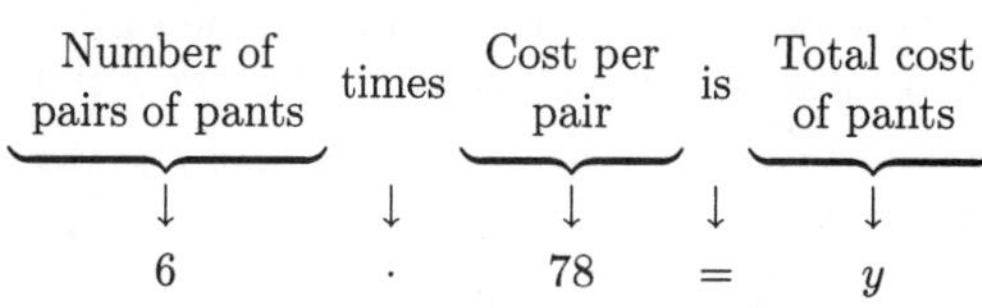

Translate. We translate to two equations.

$$\underbrace{\text{Number of shirts}}_{\downarrow\atop 4}\ \ \underbrace{\text{times}}_{\downarrow\atop \cdot}\ \ \underbrace{\text{Cost per shirt}}_{\downarrow\atop 59}\ \ \underbrace{\text{is}}_{\downarrow\atop =}\ \ \underbrace{\text{Total cost of shirts}}_{\downarrow\atop x}$$

$$\underbrace{\text{Number of pairs of pants}}_{\downarrow\atop 6}\ \ \underbrace{\text{times}}_{\downarrow\atop \cdot}\ \ \underbrace{\text{Cost per pair}}_{\downarrow\atop 78}\ \ \underbrace{\text{is}}_{\downarrow\atop =}\ \ \underbrace{\text{Total cost of pants}}_{\downarrow\atop y}$$

Solve. To solve these equations, we carry out the multiplications.

$$\begin{array}{r} 59 \\ \times\ \ 4 \\ \hline 236 \end{array} \qquad \text{Thus, } x = \$236.$$

$$\begin{array}{r} 78 \\ \times\ \ 6 \\ \hline 468 \end{array} \qquad \text{Thus } y = \$468.$$

We let a = the total amount spent.

$$\underbrace{\text{Total cost of shirts}}_{\downarrow\atop 236}\ \ \underbrace{\text{plus}}_{\downarrow\atop +}\ \ \underbrace{\text{Total cost of pants}}_{\downarrow\atop 468}\ \ \underbrace{\text{is}}_{\downarrow\atop =}\ \ \underbrace{\text{Amount spent}}_{\downarrow\atop a}$$

To solve the equation, carry out the addition.

$$\begin{array}{r} 236 \\ +\,468 \\ \hline 704 \end{array}$$

Check. We repeat the calculations. The answer checks.

State. The total cost of the clothing is \$704.

37. ***Familiarize***. This is a multistep problem.

We must find how many 100's there are in 3500. Then we must find that number times 15.

First we draw a picture

One pound			
3500 calories			
100 cal	100 cal	···	100 cal
15 min	15 min	···	15 min

In Example 9 it was determined that there are 35 100's in 3500. We let t = the time you have to do aerobic exercises to lose a pound.

Translate. We know that to do aerobic exercises for 15 min will burn 100 calories, so we need to do this 35 times to burn off one pound. We translate to an equation.

$$35 \times 15 = t$$

Solve. Carry out the multiplication.

$$35 \times 15 = t$$
$$525 = t$$

Check. Suppose you do aerobic exercises for 525 minutes. If we divide 525 by 15, we get 35, and 35 times 100 is 3500, the number of calories that must be burned off to lose one pound. The answer checks.

State. You must do aerobic exercises for 525 min, or 8 hr 45 min, to lose one pound.

39. ***Familiarize.*** This is a multistep problem. First we find the area of one side of one card. Then we double this to find the area of the front and back of one card. Finally, we find the area of the front and back sides of 100 cards.

Translate. We begin by using the formula for the area of a rectangle.

$$A = l \cdot w = 5 \cdot 3$$

Solve. First we carry out the multiplication.

$$A = 5 \cdot 3 = 15$$

This is the area of one side of one card. We multiply by 2 to find the area, x, of the front and back of one card:

$$x = 2 \cdot 15 = 30$$

Now let t = the total writing area of the front and back sides of 100 cards.

$$t = 100 \cdot 30 = 3000$$

Check. We repeat the calculations. The answer checks.

State. The total writing area on the front and back sides of a package of 100 index cards is 3000 square inches.

41. Round 234,562 to the nearest hundred.

2 3 4, 5 [6] 2
↑

The digit 5 is in the hundreds place. Consider the next digit to the right. Since the digit, 6, is 5 or higher, round 5 hundreds up to 6 hundreds. Then change all digits to the right of the hundreds place to zeros.

The answer is 234,600.

43.

	Rounded to the nearest thousand
2 7 8 3	3 0 0 0
4 6 0 2	5 0 0 0
5 7 9 7	6 0 0 0
+ 8 1 1 1	+ 8 0 0 0
	2 2, 0 0 0 ← Estimated answer

45.

	Rounded to the nearest hundred
7 8 7	8 0 0
× 3 6 3	× 4 0 0
	3 2 0, 0 0 0 ← Estimated answer

47.

49. ***Familiarize.*** This is a multistep problem. First we will find the differences in the distances traveled in 1 second. Then we will find the differences for 18 seconds. Let d = the difference in the number of miles light would travel per second in a vacuum and in ice. Let g = the difference in the number of miles light would travel per second in a vacuum and in glass.

Translate. Each is a "how much more" situation.

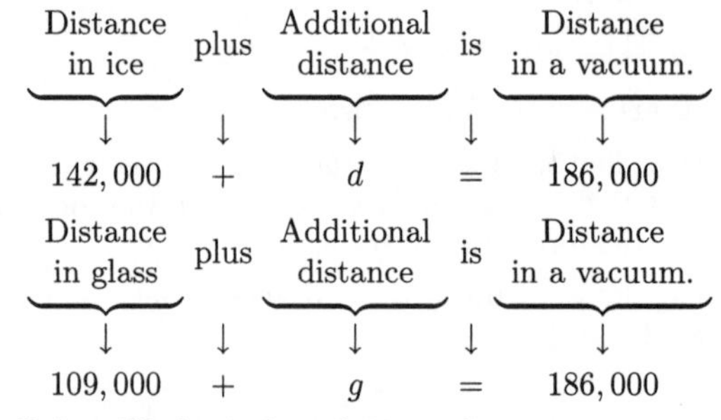

Solve. We begin by solving each equation.

$$142,000 + d = 186,000$$
$$142,000 + d - 142,000 = 186,000 - 142,000$$
$$d = 44,000$$

$$109,000 + g = 186,000$$
$$109,000 + g - 109,000 = 186,000 - 109,000$$
$$g = 77,000$$

Now to find the differences in the distances in 18 seconds, we multiply each solution by 18.

For ice: $18 \cdot 44,000 = 792,000$

For glass: $18 \cdot 77,000 = 1,386,000$

Check. We repeat the calculations. Our answers check.

State. In 18 seconds light travels 792,000 miles farther in ice and 1,386,000 miles farther in glass than in a vacuum.

Exercise Set 1.6

1. Exponential notation for $3 \cdot 3 \cdot 3 \cdot 3$ is 3^4.

3. Exponential notation for $5 \cdot 5$ is 5^2.

5. Exponential notation for $7 \cdot 7 \cdot 7 \cdot 7 \cdot 7$ is 7^5.

7. Exponential notation for $10 \cdot 10 \cdot 10$ is 10^3.

9. $7^2 = 7 \cdot 7 = 49$

11. $9^3 = 9 \cdot 9 \cdot 9 = 729$

13. $12^4 = 12 \cdot 12 \cdot 12 \cdot 12 = 20,736$

15. $11^2 = 11 \cdot 11 = 121$

17. $12 + (6+4) = 12 + 10$ Doing the calculation inside the parentheses
$= 22$ Adding

19. $52 - (40 - 8) = 52 - 32$ Doing the calculation inside the parentheses
$= 20$ Subtracting

21. $1000 \div (100 \div 10)$
$= 1000 \div 10$ Doing the calculation inside the parentheses
$= 100$ Dividing

23. $(256 \div 64) \div 4 = 4 \div 4$ Doing the calculation inside the parentheses
$= 1$ Dividing

25. $(2+5)^2 = 7^2$ Doing the calculation inside the parentheses
$= 49$ Evaluating the exponential expression

27. $(11-8)^2 - (18-16)^2$
$= 3^2 - 2^2$ Doing the calculations inside the parentheses
$= 9 - 4$ Evaluating the exponential expressions
$= 5$ Subtracting

29. $16 \cdot 24 + 50 = 384 + 50$ Doing all multiplications and divisions in order from left to right
$= 434$ Doing all additions and subtractions in order from left to right

31. $83 - 7 \cdot 6 = 83 - 42$ Doing all multiplications and divisions in order from left to right
$= 41$ Doing all additions and subtractions in order from left to right

33. $10 \cdot 10 - 3 \times 4$
$= 100 - 12$ Doing all multiplications and divisions in order from left to right
$= 88$ Doing all additions and subtractions in order from left to right

35. $4^3 \div 8 - 4$
$= 64 \div 8 - 4$ Evaluating the exponential expression
$= 8 - 4$ Doing all multiplications and divisions in order from left to right
$= 4$ Doing all additions and subtractions in order from left to right

37. $17 \cdot 20 - (17 + 20)$
$= 17 \cdot 20 - 37$ Carrying out the operation inside parentheses
$= 340 - 37$ Doing all multiplications and divisions in order from left to right
$= 303$ Doing all additions and subtractions in order from left to right

39. $6 \cdot 10 - 4 \cdot 10$
$= 60 - 40$ Doing all multiplications and divisions in order from left to right
$= 20$ Doing all additions and subtractions in order from left to right

41. $300 \div 5 + 10$
$= 60 + 10$ Doing all multiplications and divisions in order from left to right
$= 70$ Doing all additions and subtractions in order from left to right

43. $3 \cdot (2+8)^2 - 5 \cdot (4-3)^2$
$= 3 \cdot 10^2 - 5 \cdot 1^2$ Carrying out operations inside parentheses
$= 3 \cdot 100 - 5 \cdot 1$ Evaluating the exponential expressions
$= 300 - 5$ Doing all multiplications and divisions in order from left to right
$= 295$ Doing all additions and subtractions in order from left to right

45. $4^2 + 8^2 \div 2^2 = 16 + 64 \div 4$
$= 16 + 16$
$= 32$

47. $10^3 - 10 \cdot 6 - (4 + 5 \cdot 6) = 10^3 - 10 \cdot 6 - (4 + 30)$
$= 10^3 - 10 \cdot 6 - 34$
$= 1000 - 10 \cdot 6 - 34$
$= 1000 - 60 - 34$
$= 940 - 34$
$= 906$

49. $6 \times 11 - (7+3) \div 5 - (6-4) = 6 \times 11 - 10 \div 5 - 2$
$= 66 - 2 - 2$
$= 64 - 2$
$= 62$

51. $120 - 3^3 \cdot 4 \div (5 \cdot 6 - 6 \cdot 4)$
$= 120 - 3^3 \cdot 4 \div (30 - 24)$
$= 120 - 3^3 \cdot 4 \div 6$
$= 120 - 27 \cdot 4 \div 6$
$= 120 - 108 \div 6$
$= 120 - 18$
$= 102$

53. We add the numbers and then divide by the number of addends.

$$(\$64 + \$97 + \$121) \div 3 = \$282 \div 3 \\ = \$94$$

55. $8 \times 13 + \{42 \div [18 - (6+5)]\}$
$= 8 \times 13 + \{42 \div [18 - 11]\}$
$= 8 \times 13 + \{42 \div 7\}$
$= 8 \times 13 + 6$
$= 104 + 6$
$= 110$

57. $[14 - (3+5) \div 2] - [18 \div (8-2)]$
$= [14 - 8 \div 2] - [18 \div 6]$
$= [14 - 4] - 3$
$= 10 - 3$
$= 7$

59. $(82 - 14) \times [(10 + 45 \div 5) - (6 \cdot 6 - 5 \cdot 5)]$
$= (82 - 14) \times [(10 + 9) - (36 - 25)]$
$= (82 - 14) \times [19 - 11]$
$= 68 \times 8$
$= 544$

61. $4 \times \{(200 - 50 \div 5) - [(35 \div 7) \cdot (35 \div 7) - 4 \times 3]\}$
$= 4 \times \{(200 - 10) - [5 \cdot 5 - 4 \times 3]\}$
$= 4 \times \{190 - [25 - 12]\}$
$= 4 \times \{190 - 13\}$
$= 4 \times 177$
$= 708$

63.
$$x + 341 = 793$$
$$x + 341 - 341 = 793 - 341$$
$$x = 452$$

The solution is 452.

65. ***Familiarize.*** We first make a drawing.

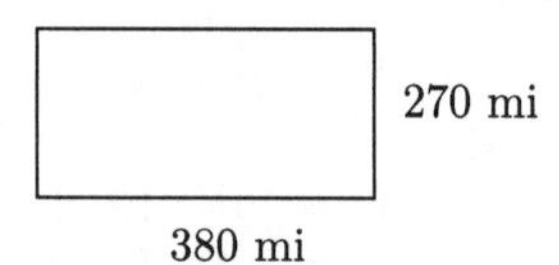

Translate. We use the formula for the area of a rectangle.

$$A = l \cdot w = 380 \cdot 270$$

Solve. We carry out the multiplication.

$$A = 380 \cdot 270 = 102,600$$

Check. We repeat the calculation. The answer checks.

State. The area is 102,600 square miles.

67. ◈

69. $15(23 - 4 \cdot 2)^3 \div (3 \cdot 25)$
$= 15(23 - 8)^3 \div 75$ Multiplying inside parentheses
$= 15 \cdot 15^3 \div 75$ Subtracting inside parentheses
$= 15 \cdot 3375 \div 75$ Evaluating the exponential expression
$= 50,625 \div 75$ Doing all multiplication and divisions in order from left to right
$= 675$

71. $1 + 5 \cdot 4 + 3 = 1 + 20 + 3$
$= 24$ Correct answer

To make the incorrect answer correct we add parentheses:

$1 + 5 \cdot (4 + 3) = 36$

73. $12 \div 4 + 2 \cdot 3 - 2 = 3 + 6 - 2$
$= 7$ Correct answer

To make the incorrect answer correct we add parentheses:

$12 \div (4 + 2) \cdot 3 - 2 = 4$

Exercise Set 1.7

1. We first find some factorizations:

$18 = 1 \cdot 18$ $18 = 3 \cdot 6$
$18 = 2 \cdot 9$

Factors: 1, 2, 3, 6, 9, 18

3. We first find some factorizations:

$54 = 1 \cdot 54$ $54 = 3 \cdot 18$
$54 = 2 \cdot 27$ $54 = 6 \cdot 9$

Factors: 1, 2, 3, 6, 9, 18, 27, 54

5. We first find some factorizations:

$4 = 1 \cdot 4$ $4 = 2 \cdot 2$

Factors: 1, 2, 4

7. The only factorization is $7 = 1 \cdot 7$.

Factors: 1, 7

9. The only factorization is $1 = 1 \cdot 1$.

Factor: 1

11. We first find some factorizations:

$98 = 1 \cdot 98$ $98 = 7 \cdot 14$
$98 = 2 \cdot 49$

Factors: 1, 2, 7, 14, 49, 98

13.
$1 \cdot 4 = 4$ $6 \cdot 4 = 24$
$2 \cdot 4 = 8$ $7 \cdot 4 = 28$
$3 \cdot 4 = 12$ $8 \cdot 4 = 32$
$4 \cdot 4 = 16$ $9 \cdot 4 = 36$
$5 \cdot 4 = 20$ $10 \cdot 4 = 40$

15.
$1 \cdot 20 = 20$ $6 \cdot 20 = 120$
$2 \cdot 20 = 40$ $7 \cdot 20 = 140$
$3 \cdot 20 = 60$ $8 \cdot 20 = 160$
$4 \cdot 20 = 80$ $9 \cdot 20 = 180$
$5 \cdot 20 = 100$ $10 \cdot 20 = 200$

17.
$1 \cdot 3 = 3$ $6 \cdot 3 = 18$
$2 \cdot 3 = 6$ $7 \cdot 3 = 21$
$3 \cdot 3 = 9$ $8 \cdot 3 = 24$
$4 \cdot 3 = 12$ $9 \cdot 3 = 27$
$5 \cdot 3 = 15$ $10 \cdot 3 = 30$

19.
$1 \cdot 12 = 12$ $6 \cdot 12 = 72$
$2 \cdot 12 = 24$ $7 \cdot 12 = 84$
$3 \cdot 12 = 36$ $8 \cdot 12 = 96$
$4 \cdot 12 = 48$ $9 \cdot 12 = 108$
$5 \cdot 12 = 60$ $10 \cdot 12 = 120$

21.

$1 \cdot 10 = 10$	$6 \cdot 10 = 60$
$2 \cdot 10 = 20$	$7 \cdot 10 = 70$
$3 \cdot 10 = 30$	$8 \cdot 10 = 80$
$4 \cdot 10 = 40$	$9 \cdot 10 = 90$
$5 \cdot 10 = 50$	$10 \cdot 10 = 100$

23.

$1 \cdot 9 = 9$	$6 \cdot 9 = 54$
$2 \cdot 9 = 18$	$7 \cdot 9 = 63$
$3 \cdot 9 = 27$	$8 \cdot 9 = 72$
$4 \cdot 9 = 36$	$9 \cdot 9 = 81$
$5 \cdot 9 = 45$	$10 \cdot 9 = 90$

25. We divide 26 by 6.

$$\begin{array}{r} 4 \\ 6\,\overline{)\,2\,6} \\ \underline{2\,4} \\ 2 \end{array}$$

Since the remainder is not 0, 26 is not divisible by 6.

27. We divide 1880 by 8.

$$\begin{array}{r} 2\,3\,5 \\ 8\,\overline{)\,1\,8\,8\,0} \\ \underline{1\,6\,0\,0} \\ 2\,8\,0 \\ \underline{2\,4\,0} \\ 4\,0 \\ \underline{4\,0} \\ 0 \end{array}$$

Since the remainder is 0, 1880 is divisible by 8.

29. We divide 256 by 16.

$$\begin{array}{r} 1\,6 \\ 1\,6\,\overline{)\,2\,5\,6} \\ \underline{1\,6\,0} \\ 9\,6 \\ \underline{9\,6} \\ 0 \end{array}$$

Since the remainder is 0, 256 is divisible by 16.

31. We divide 4227 by 9.

$$\begin{array}{r} 4\,6\,9 \\ 9\,\overline{)\,4\,2\,2\,7} \\ \underline{3\,6\,0\,0} \\ 6\,2\,7 \\ \underline{5\,4\,0} \\ 8\,7 \\ \underline{8\,1} \\ 6 \end{array}$$

Since the remainder is not 0, 4227 is not divisible by 9.

33. We divide 8650 by 16.

$$\begin{array}{r} 5\,4\,0 \\ 1\,6\,\overline{)\,8\,6\,5\,0} \\ \underline{8\,0\,0\,0} \\ 6\,5\,0 \\ \underline{6\,4\,0} \\ 1\,0 \end{array}$$

Since the remainder is not 0, 8650 is not divisible by 16.

35. 1 is neither prime nor composite.

37. The number 9 has factors 1, 3, and 9.

Since 9 is not 1 and not prime, it is composite.

39. The number 11 is prime. It has only the factors 1 and 11.

41. The number 29 is prime. It has only the factors 1 and 29.

43.

$$\begin{array}{r} 2 \\ 2\,\overline{)\,4} \\ 2\,\overline{)\,8} \end{array} \quad \leftarrow \text{ 2 is prime.}$$

$8 = 2 \cdot 2 \cdot 2$

45.

$$\begin{array}{r} 7 \\ 2\,\overline{)\,1\,4} \end{array} \quad \leftarrow \text{ 7 is prime.}$$

$14 = 2 \cdot 7$

47.

$$\begin{array}{r} 7 \\ 3\,\overline{)\,2\,1} \\ 2\,\overline{)\,4\,2} \end{array} \quad \leftarrow \text{ 7 is prime.}$$

$42 = 2 \cdot 3 \cdot 7$

49.

$$\begin{array}{r} 5 \\ 5\,\overline{)\,2\,5} \end{array} \quad \leftarrow \text{ 5 is prime.}$$

(25 is not divisible by 2 or 3. We move to 5.)

$25 = 5 \cdot 5$

51.

$$\begin{array}{r} 5 \\ 5\,\overline{)\,2\,5} \\ 2\,\overline{)\,5\,0} \end{array} \quad \leftarrow \text{ 5 is prime.}$$

(25 is not divisible by 2 or 3. We move to 5.)

$50 = 2 \cdot 5 \cdot 5$

53.

$$\begin{array}{r} 1\,3 \\ 1\,3\,\overline{)\,1\,6\,9} \end{array} \quad \leftarrow \text{ 13 is prime.}$$

(169 is not divisible by 2, 3, 5, 7 or 11. We move to 13.)

$169 = 13 \cdot 13$

55.

$$\begin{array}{r} 5 \\ 5\,\overline{)\,2\,5} \\ 2\,\overline{)\,5\,0} \\ 2\,\overline{)\,1\,0\,0} \end{array} \quad \leftarrow \text{ 5 is prime.}$$

(25 is not divisible by 2 or 3. We move to 5.)

$100 = 2 \cdot 2 \cdot 5 \cdot 5$

We can also use a factor tree.

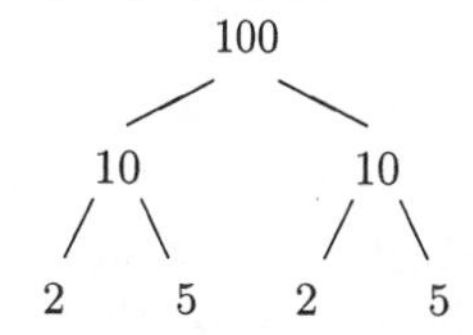

57.

$$\begin{array}{r} 7 \\ 5\,\overline{)\,3\,5} \end{array} \quad \leftarrow \text{ 7 is prime.}$$

(35 is not divisible by 2 or 3. We move to 5.)

$35 = 5 \cdot 7$

59.

$$\begin{array}{r} 3 \\ 3\,\overline{)\,9} \\ 2\,\overline{)\,1\,8} \\ 2\,\overline{)\,3\,6} \\ 2\,\overline{)\,7\,2} \end{array} \quad \leftarrow \text{ 3 is prime.}$$

(9 is not divisible by 2. We move to 3.)

$72 = 2 \cdot 2 \cdot 2 \cdot 3 \cdot 3$

We can also use a factor tree, as shown in Example 10 in the text.

61.

```
      1 1   ← 11 is prime.
   7 ⌈ 7 7   (77 is not divisible by 2, 3, or 5.
              We move to 7.)
```

$77 = 7 \cdot 11$

63.

```
       1 0 3  ← 103 is prime.
    7 ⌈ 7 2 1
  2 ⌈ 1 4 4 2
  2 ⌈ 2 8 8 4
```

$2884 = 2 \cdot 2 \cdot 7 \cdot 103$

We can also use a factor tree.

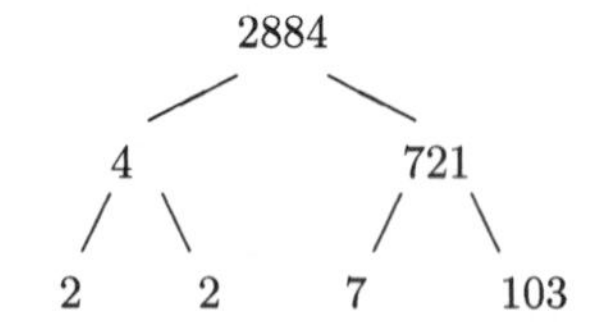

65.

```
      1 7   ← 17 is prime.
   3 ⌈ 5 1   (51 is not divisible by 2. We move
              to 3.)
```

$51 = 3 \cdot 17$

67.

```
    1 3
  ×   2
  -----
    2 6
```

69.

```
     3
    2 5
  × 1 7
  -----
   1 7 5    Multiplying by 7
   2 5 0    Multiplying by 10
  ------
   4 2 5    Adding
```

71.

```
        0
  2 2 ⌈ 0
        0
        -
        0
```

The answer is 0.

73.

```
         1
  2 2 ⌈ 2 2
        2 2
        ---
          0
```

The answer is 1.

75. ***Familiarize***. This is a multistep problem. Find the total cost of the shirts and the total cost of the pants and then find the sum of the two.

We let p = the total cost of the shirts and p = the total cost of the pants.

Translate. We write two equations.

Number of shirts	times	Cost of one shirt	is	Total cost of shirts
↓	↓	↓	↓	↓
7	$\cdot$	48	=	s

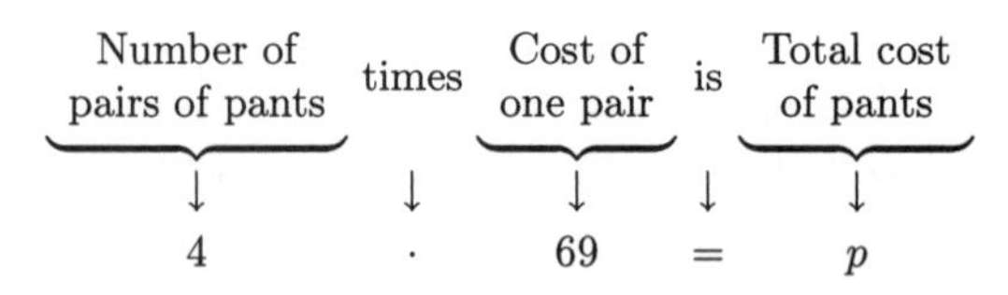

Solve. We carry out the multiplication.

$7 \cdot 48 = s$

$336 = s$ Doing the multiplication

The total cost of the 7 shirts is \$336.

$4 \cdot 69 = p$

$276 = p$ Doing the multiplication

The total cost of the 4 pairs of pants is \$276.

Now we find the total amount spent. We let t = this amount.

Total cost of shirts	plus	Total cost of pants	is	Total amount spent
↓	↓	↓	↓	↓
336	+	276	=	t

To solve the equation, carry out the addition.

```
    3 3 6
  + 2 7 6
  -------
    6 1 2
```

Check. We can repeat the calculations. The answer checks.

State. The total cost is \$612.

77. ◈

79. Row 1: 48, 90, 432, 63; row 2: 7, 2, 2, 10, 8, 6, 21, 10; row 3: 9, 18, 36, 14, 12, 11, 21; row 4: 29, 19, 42

Exercise Set 1.8

1. A number is divisible by 2 if its <u>ones digit</u> is even.

4$\underline{6}$ is divisible by 2 because $\underline{6}$ is even.
22$\underline{4}$ is divisible by 2 because $\underline{4}$ is even.
1$\underline{9}$ is not divisible by 2 because $\underline{9}$ is not even.
55$\underline{5}$ is not divisible by 2 because $\underline{5}$ is not even.
30$\underline{0}$ is divisible by 2 because $\underline{0}$ is even.
3$\underline{6}$ is divisible by 2 because $\underline{6}$ is even.
45,27$\underline{0}$ is divisible by 2 because $\underline{0}$ is even.
444$\underline{4}$ is divisible by 2 because $\underline{4}$ is even.
8$\underline{5}$ is not divisible by 2 because $\underline{5}$ is not even.
71$\underline{1}$ is not divisible by 2 because $\underline{1}$ is not even.
13,25$\underline{1}$ is not divisible by 2 because $\underline{1}$ is not even.
254,76$\underline{5}$ is not divisible by 2 because $\underline{5}$ is not even.
25$\underline{6}$ is divisible by 2 because $\underline{6}$ is even.
806$\underline{4}$ is divisible by 2 because $\underline{4}$ is even.
186$\underline{7}$ is not divisible by 2 because $\underline{7}$ is not even.
21,56$\underline{8}$ is divisible by 2 because $\underline{8}$ is even.

3. A number is divisible by 4 if the number named by the last two digits is divisible by 4.

46 is not divisible by 4 because 46 is not divisible by 4.
224 is divisible by 4 because 24 is divisible by 4.
19 is not divisible by 4 because 19 is not divisible by 4.
555 is not divisible by 4 because 55 is not divisible by 4.
300 is divisible by 4 because 00 is divisible by 4.
36 is divisible by 4 because 36 is divisible by 4.
45,270 is not divisible by 4 because 70 is not divisible by 4.
4444 is divisible by 4 because 44 is divisible by 4.
85 is not divisible by 4 because 85 is not divisible by 4.
711 is not divisible by 4 because 11 is not divisible by 4.
13,251 is not divisible by 4 because 51 is not divisible by 4.
254,765 is not divisible by 4 because 65 is not divisible by 4.
256 is divisible by 4 because 56 is divisible by 4.
8064 is divisible by 4 because 64 is divisible by 4.
1867 is not divisible by 4 because 67 is not divisible by 4.
21,568 is divisible by 4 because 68 is divisible by 4.

5. For a number to be divisible by 6, the sum of the digits must be divisible by 3 and the ones digit must be 0, 2, 4, 6 or 8 (even). It is most efficient to determine if the ones digit is even first and then, if so, to determine if the sum of the digits is divisible by 3.

46 is not divisible by 6 because 46 is not divisible by 3.

$4+6=10$
↑
Not divisible by 3

224 is not divisible by 6 because 224 is not divisible by 3.

$2+2+4=8$
↑
Not divisible by 3

19 is not divisible by 6 because 19 is not even.

19
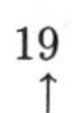
Not even

555 is not divisible by 6 because 555 is not even.

555
↑
Not even

300 is divisible by 6.

300 ↑ Even — $3+0+0=3$ ↑ Divisible by 3

36 is divisible by 6.

36 ↑ Even — $3+6=9$ ↑ Divisible by 3

45,270 is divisible by 6.

45,270 ↑ Even — $4+5+2+7+0=18$ ↑ Divisible by 3

4444 is not divisible by 6 because 4444 is not divisible by 3.

$4+4+4+4=16$
↑
Not divisible by 3

85 is not divisible by 6 because 85 is not even.

85
↑
Not even

711 is not divisible by 6 because 711 is not even.

711
↑
Not even

13,251 is not divisible by 6 because 13,251 is not even.

13,251
↑
Not even

254,765 is not divisible by 6 because 254,765 is not even.

254,765
↑
Not even

256 is not divisible by 6 because 256 is not divisible by 3.

$2+5+6=13$
↑
Not divisible by 3

8064 is divisible by 6.

8064 ↑ Even — $8+0+6+4=18$ ↑ Divisible by 3

1867 is not divisible by 6 because 1867 is not even.

1867
↑
Not even

21,568 is not divisible by 6 because 21,568 is not divisible by 3.

$2+1+5+6+8=22$
↑
Not divisible by 3

7. A number is divisible by 9 if the sum of the digits is divisible by 9.

46 is not divisible by 9 because $4+6=10$ and 10 is not divisible by 9.

224 is not divisible by 9 because $2+2+4=8$ and 8 is not divisible by 9.

19 is not divisible by 9 because $1+9=10$ and 10 is not divisible by 9.

555 is not divisible by 9 because $5+5+5=15$ and 15 is not divisible by 9.

300 is not divisible by 9 because $3+0+0=3$ and 3 is not divisible by 9.

36 is divisible by 9 because $3 + 6 = 9$ and 9 is divisible by 9.

45,270 is divisible by 9 because $4 + 5 + 2 + 7 + 0 = 18$ and 18 is divisible by 9.

4444 is not divisible by 9 because $4 + 4 + 4 + 4 = 16$ and 16 is not divisible by 9.

85 is not divisible by 9 because $8 + 5 = 13$ and 13 is not divisible by 9.

711 is divisible by 9 because $7 + 1 + 1 = 9$ and 9 is divisible by 9.

13,251 is not divisible by 9 because $1 + 3 + 2 + 5 + 1 = 12$ and 12 is not divisible by 9.

254,765 is not divisible by 9 because $2+5+4+7+6+5 = 29$ and 29 is not divisible by 9.

256 is not divisible by 9 because $2 + 5 + 6 = 13$ and 13 is not divisible by 9.

8064 is divisible by 9 because $8 + 0 + 6 + 4 = 18$ and 18 is divisible by 9.

1867 is not divisible by 9 because $1 + 8 + 6 + 7 = 22$ and 22 is not divisible by 9.

21,568 is not divisible by 9 because $2 + 1 + 5 + 6 + 8 = 22$ and 22 is not divisible by 9.

9. A number is divisible by 3 if the sum of the digits is divisible by 3.

56 is not divisible by 3 because $5 + 6 = 11$ and 11 is not divisible by 3.

324 is divisible by 3 because $3 + 2 + 4 = 9$ and 9 is divisible by 3.

784 is not divisible by 3 because $7 + 8 + 4 = 19$ and 19 is not divisible by 3.

55,555 is not divisible by 3 because $5 + 5 + 5 + 5 + 5 = 25$ and 25 is not divisible by 3.

200 is not divisible by 3 because $2 + 0 + 0 = 2$ and 2 is not divisible by 3.

42 is divisible by 3 because $4 + 2 = 6$ and 6 is divisible by 3.

501 is divisible by 3 because $5 + 0 + 1 = 6$ and 6 is divisible by 3.

3009 is divisible by 3 because $3 + 0 + 0 + 9 = 12$ and 12 is divisible by 3.

75 is divisible by 3 because $7 + 5 = 12$ and 12 is divisible by 3.

812 is not divisible by 3 because $8 + 1 + 2 = 11$ and 11 is not divisible by 3.

2345 is not divisible by 3 because $2 + 3 + 4 + 5 = 14$ and 14 is not divisible by 3.

2001 is divisible by 3 because $2 + 0 + 0 + 1 = 3$ and 3 is divisible by 3.

35 is not divisible by 3 because $3 + 5 = 8$ and 8 is not divisible by 3.

402 is divisible by 3 because $4 + 0 + 2 = 6$ and 6 is divisible by 3.

111,111 is divisible by 3 because $1 + 1 + 1 + 1 + 1 + 1 = 6$ and 6 is divisible by 3.

1005 is divisible by 3 because $1 + 0 + 0 + 5 = 6$ and 6 is divisible by 3.

11. A number is divisible by 5 if the ones digit is 0 or 5.

5$\underline{6}$ is not divisible by 5 because the ones digit (6) is not 0 or 5.

32$\underline{4}$ is not divisible by 5 because the ones digit (4) is not 0 or 5.

78$\underline{4}$ is not divisible by 5 because the ones digit (4) is not 0 or 5.

55,55$\underline{5}$ is divisible by 5 because the ones digit is 5.

20$\underline{0}$ is divisible by 5 because the ones digit is 0.

4$\underline{2}$ is not divisible by 5 because the ones digit (2) is not 0 or 5.

50$\underline{1}$ is not divisible by 5 because the ones digit (1) is not 0 or 5.

300$\underline{9}$ is not divisible by 5 because the ones digit (9) is not 0 or 5.

7$\underline{5}$ is divisible by 5 because the ones digit is 5.

81$\underline{2}$ is not divisible by 5 because the ones digit (2) is not 0 or 5.

234$\underline{5}$ is divisible by 5 because the ones digit is 5.

200$\underline{1}$ is not divisible by 5 because the ones digit (1) is not 0 or 5.

3$\underline{5}$ is divisible by 5 because the ones digit is 5.

40$\underline{2}$ is not divisible by 5 because the ones digit (2) is not 0 or 5.

111,11$\underline{1}$ is not divisible by 5 because the ones digit (1) is not 0 or 5.

100$\underline{5}$ is divisible by 5 because the ones digit is 5.

13. A number is divisible by 9 if the sum of the digits is divisible by 9.

56 is not divisible by 9 because $5 + 6 = 11$ and 11 is not divisible by 9.

324 is divisible by 9 because $3 + 2 + 4 = 9$ and 9 is divisible by 9.

784 is not divisible by 9 because $7 + 8 + 4 = 19$ and 19 is not divisible by 9.

55,555 is not divisible by 9 because $5 + 5 + 5 + 5 + 5 = 25$ and 25 is not divisible by 9.

200 is not divisible by 9 because $2 + 0 + 0 = 2$ and 2 is not divisible by 9.

42 is not divisible by 9 because $4 + 2 = 6$ and 6 is not divisible by 9.

501 is not divisible by 9 because $5 + 0 + 1 = 6$ and 6 is not divisible by 9.

3009 is not divisible by 9 because $3 + 0 + 0 + 9 = 12$ and 12 is not divisible by 9.

75 is not divisible by 9 because $7 + 5 = 12$ and 12 is not divisible by 9.

812 is not divisible by 9 because $8+1+2=11$ and 11 is not divisible by 9.

2345 is not divisible by 9 because $2+3+4+5=14$ and 14 is not divisible by 9.

2001 is not divisible by 9 because $2+0+0+1=3$ and 3 is not divisible by 9.

35 is not divisible by 9 because $3+5=8$ and 8 is not divisible by 9.

402 is not divisible by 9 because $4+0+2=6$ and 6 is not divisible by 9.

111,111 is not divisible by 9 because $1+1+1+1+1+1=6$ and 6 is not divisible by 9.

1005 is not divisible by 9 because $1+0+0+5=6$ and 6 is not divisible by 9.

15. A number is divisible by 10 if the ones digit is 0.

Of the numbers under consideration, the only one whose ones digit is 0 is 200. Therefore, 200 is divisible by 10. None of the other numbers is divisible by 10.

17.
$$\begin{aligned} 56+x &= 194 \\ 56+x-56 &= 194-56 \quad \text{Subtracting 56 on both sides} \\ x &= 138 \end{aligned}$$

The solution is 138.

19.
$$\begin{aligned} 18\cdot t &= 1008 \\ \frac{18\cdot t}{18} &= \frac{1008}{18} \quad \text{Dividing by 18 on both sides} \\ t &= 56 \end{aligned}$$

The solution is 56.

21.
$$\begin{array}{r} 2\,3\,4 \\ 9\,\overline{)\,2\,1\,0\,6} \\ \underline{1\,8\,0\,0} \\ 3\,0\,6 \\ \underline{2\,7\,0} \\ 3\,6 \\ \underline{3\,6} \\ 0 \end{array}$$

The answer is 234.

23. ***Familiarize.*** We visualize the situation. Let g = the number of gallons of gasoline the automobile will use to travel 1485 mi.

1 mi	1 mi	...	1 mi
1 mi	...	...	⋮
⋮			⋮
1 mi	...	...	1 mi

33 in each row
How many rows?

Translate. We translate to an equation.

Number of miles	divided by	Miles per gallon	is	Number of gallons
↓	↓	↓	↓	↓
1485	÷	33	=	g

Solve. We carry out the division.

$$\begin{array}{r} 4\,5 \\ 3\,3\,\overline{)\,1\,4\,8\,5} \\ \underline{1\,3\,2\,0} \\ 1\,6\,5 \\ \underline{1\,6\,5} \\ 0 \end{array}$$

Thus $45=g$, or $g=45$.

Check. We can repeat the calculation. The answer checks.

State. The automobile will use 45 gallons of gasoline to travel 1485 mi.

25.

27. $780\underline{0}$ is divisible by 2 because the ones digit (0) is even.

$7800 \div 2 = 3900$ so $7800 = 2\cdot 3900$.

$390\underline{0}$ is divisible by 2 because the ones digit (0) is even.

$3900\div 2 = 1950$ so $3900 = 2\cdot 1950$ and $7800 = 2\cdot 2\cdot 1950$.

$195\underline{0}$ is divisible by 2 because the ones digit (0) is even.

$1950\div 2 = 975$ so $1950 = 2\cdot 975$ and $7800 = 2\cdot 2\cdot 2\cdot 975$.

$97\underline{5}$ is not divisible by 2 because the ones digit (5) is not even. Move on to 3.

975 is divisible by 3 because the sum of the digits $(9+7+5=21)$ is divisible by 3.

$975\div 3 = 325$ so $975 = 3\cdot 325$ and $7800 = 2\cdot 2\cdot 2\cdot 3\cdot 325$.

Since 975 is not divisible by 2, none of its factors is divisible by 2. Therefore, we no longer need to check for divisibility by 2.

325 is not divisible by 3 because the sum of the digits $(3+2+5=10)$ is not divisible by 3. Move on to 5.

$32\underline{5}$ is divisible by 5 because the ones digit is 5.

$325\div 5 = 65$ so $325 = 5\cdot 65$ and $7800 = 2\cdot 2\cdot 2\cdot 3\cdot 5\cdot 65$.

Since 325 is not divisible by 3, none of its factors is divisible by 3. Therefore we no longer need to check for divisibility by 3.

$6\underline{5}$ is divisible by 5 because the ones digit is 5.

$65\div 5 = 13$ so $65 = 5\cdot 13$ and $7800 = 2\cdot 2\cdot 2\cdot 3\cdot 5\cdot 5\cdot 13$.

13 is prime so the prime factorization of 7800 is $2\cdot 2\cdot 2\cdot 3\cdot 5\cdot 5\cdot 13$.

29. $277\underline{2}$ is divisible by 2 because the ones digit (2) is even.

$2772\div 2 = 1386$ so $2772 = 2\cdot 1386$.

$138\underline{6}$ is divisible by 2 because the ones digit (6) is even.

$1386\div 2 = 693$ so $1386 = 2\cdot 693$ and $2772 = 2\cdot 2\cdot 693$.

$69\underline{3}$ is not divisible by 2 because the ones digit (3) is not even. We move to 3.

693 is divisible by 3 because the sum of the digits $(6+9+3=18)$ is divisible by 3.

$693\div 3 = 231$ so $693 = 3\cdot 231$ and $2772 = 2\cdot 2\cdot 3\cdot 231$.

Since 693 is not divisible by 2, none of its factors is divisible by 2. Therefore, we no longer need to check divisibility by 2.

231 is divisible by 3 because the sum of the digits $(2+3+1=6)$ is divisible by 3.

$231 \div 3 = 77$ so $231 = 3 \cdot 77$ and $2772 = 2 \cdot 2 \cdot 3 \cdot 3 \cdot 77$.

77 is not divisible by 3 since the sum of the digits $(7+7=14)$ is not divisible by 3. We move to 5.

77 is not divisible by 5 because the ones digit (7) is not 0 or 5. We move to 7.

We have not stated a test for divisibility by 7 so we will just try dividing by 7.

$$7\overline{)77} \quad 11 \leftarrow 11 \text{ is prime}$$

$77 \div 7 = 11$ so $77 = 7 \cdot 11$ and the prime factorization of 2772 is $2 \cdot 2 \cdot 3 \cdot 3 \cdot 7 \cdot 11$.

31. We begin with 7 since it is the largest number. We can place the largest number remaining, 6, to the right of 7 since $7+6=13$ and 13 is a prime number.

7 6 _ _ _ _ _

We can place the largest number remaining, 5, to the right of 6 because $6+5=11$ and 11 is a prime number.

7 6 5 _ _ _ _

The largest number remaining, 4, cannot be placed to the right of 5 because $5+4=9$ and 9 is not a prime number. Similarly, we cannot place 3 to the right of 5 because $5+3=8$ and 8 is not a prime number. We can place 2 to the right of 5 because $5+2=7$ and 7 is a prime number.

7 6 5 2 _ _ _

The largest number remaining, 4, cannot be placed to the right of 2 because $2+4=6$ and 6 is not a prime number. We can place 3 to the right of 2 because $2+3=5$ and 5 is a prime number.

7 6 5 2 3 _ _

We can place the largest remaining number, 4, to the right of 3 because $3+4=7$ and 7 is a prime number. Then we place the only number remaining, 1.

7 6 5 2 3 4 1

The number is 7,652,341.

Exercise Set 1.9

In this section we will find the LCM using the multiples method in Exercises 1 - 19 and the factorization method in Exercises 21 - 43.

1. a) 4 is a multiple of 2, so it is the LCM.

c) The LCM = 4.

3. a) 25 is not a multiple of 10.

b) Check multiples:

$2 \cdot 25 = 50$ A multiple of 10

c) The LCM = 50.

5. a) 40 is a multiple of 20, so it is the LCM.

c) The LCM = 40.

7. a) 27 is not a multiple of 18.

b) Check multiples:

$2 \cdot 27 = 54$ A multiple of 18

c) The LCM = 54.

9. a) 50 is not a multiple of 30.

b) Check multiples:

$2 \cdot 50 = 100$ Not a multiple of 30
$3 \cdot 50 = 150$ A multiple of 30

c) The LCM = 150.

11. a) 40 is not a multiple of 30.

b) Check multiples:

$2 \cdot 40 = 80$ Not a multiple of 30
$3 \cdot 40 = 120$ A multiple of 30

c) The LCM = 120.

13. a) 24 is not a multiple of 18.

b) Check multiples:

$2 \cdot 24 = 48$ Not a multiple of 18
$3 \cdot 24 = 72$ A multiple of 18

c) The LCM = 72.

15. a) 70 is not a multiple of 60.

b) Check multiples:

$2 \cdot 70 = 140$ Not a multiple of 60
$3 \cdot 70 = 210$ Not a multiple of 60
$4 \cdot 70 = 280$ Not a multiple of 60
$5 \cdot 70 = 350$ Not a multiple of 60
$6 \cdot 70 = 420$ A multiple of 60

c) The LCM = 420.

17. a) 36 is not a multiple of 16.

b) Check multiples:

$2 \cdot 36 = 72$ Not a multiple of 16
$3 \cdot 36 = 108$ Not a multiple of 16
$4 \cdot 36 = 144$ A multiple of 16

c) The LCM = 144.

19. a) 36 is not a multiple of 32.

b) Check multiples:

$2 \cdot 36 = 72$ Not a multiple of 32
$3 \cdot 36 = 108$ Not a multiple of 32
$4 \cdot 36 = 144$ Not a multiple of 32
$5 \cdot 36 = 180$ Not a multiple of 32
$6 \cdot 36 = 216$ Not a multiple of 32
$7 \cdot 36 = 252$ Not a multiple of 32
$8 \cdot 36 = 288$ A multiple of 32

c) The LCM = 288.

21. Note that each of the numbers 2, 3, and 5 is prime. They have no common prime factor. When this happens, the LCM is just the product of the numbers.

The LCM is $2 \cdot 3 \cdot 5$, or 30.

23. Note that each of the numbers 3, 5, and 7 is prime. They have no common prime factor. When this happens, the LCM is just the product of the numbers.

The LCM is $3 \cdot 5 \cdot 7$, or 105.

25. a) Find the prime factorization of each number.

$24 = 2 \cdot 2 \cdot 2 \cdot 3$
$36 = 2 \cdot 2 \cdot 3 \cdot 3$
$12 = 2 \cdot 2 \cdot 3$

b) Create a product by writing factors, using each the greatest number of times it occurs in any one factorization.

Consider the factor 2. The greatest number of times 2 occurs in any one factorization is three. We write 2 as a factor three times.

$2 \cdot 2 \cdot 2 \cdot\ ?$

Consider the factor 3. The greatest number of times 3 occurs in any one factorization is two. We write 3 as a factor two times.

$2 \cdot 2 \cdot 2 \cdot 3 \cdot 3 \cdot\ ?$

Since there are no other prime factors in any of the factorizations, the LCM is $2 \cdot 2 \cdot 2 \cdot 3 \cdot 3$, or 72.

27. a) Find the prime factorization of each number.

$5 = 5$ (5 is prime.)
$12 = 2 \cdot 2 \cdot 3$
$15 = 3 \cdot 5$

b) Create a product by writing each factor the greatest number of times it occurs in any one factorization.

The greatest number of times 2 occurs in any one factorization is two times.

The greatest number of times 3 occurs in any one factorization is one time.

The greatest number of times 5 occurs in any one factorization is one time.

Since there are no other prime factors in any of the factorizations, the LCM is $2 \cdot 2 \cdot 3 \cdot 5$, or 60.

29. a) Find the prime factorization of each number.

$9 = 3 \cdot 3$
$12 = 2 \cdot 2 \cdot 3$
$6 = 2 \cdot 3$

b) Create a product by writing each factor the greatest number of times it occurs in any one factorization.

The greatest number of times 2 occurs in any one factorization is two times.

The greatest number of times 3 occurs in any one factorization is two times.

Since there are no other prime factors in any of the factorizations, the LCM is $2 \cdot 2 \cdot 3 \cdot 3$, or 36.

31. a) Find the prime factorization of each number.

$180 = 2 \cdot 2 \cdot 3 \cdot 3 \cdot 5$
$100 = 2 \cdot 2 \cdot 5 \cdot 5$
$450 = 2 \cdot 3 \cdot 3 \cdot 5 \cdot 5$

b) Create a product by writing each factor the greatest number of times it occurs in any one factorization.

The greatest number of times 2 occurs in any one factorization is two times.

The greatest number of times 3 occurs in any one factorization is two times.

The greatest number of times 5 occurs in any one factorization is two times.

Since there are no other prime factors in any of the factorizations, the LCM is $2 \cdot 2 \cdot 3 \cdot 3 \cdot 5 \cdot 5$, or 900.

We can also find the LCM using exponents.

$180 = 2^2 \cdot 3^2 \cdot 5^1$
$100 = 2^2 \cdot 5^2$
$450 = 2^1 \cdot 3^2 \cdot 5^2$

The largest exponents of 2, 3, 5 in any of the factorizations are each 2. Thus, the LCM $= 2^2 \cdot 3^2 \cdot 5^2$, or 900.

33. Note that 8 is a factor of 48. If one number is a factor of another, the LCM is the greater number.

The LCM is 48.

The factorization method will also work here if you do not recognize at the outset that 8 is a factor of 48.

35. Note that 5 is a factor of 50. If one number is a factor of another, the LCM is the greater number.

The LCM is 50.

37. Note that 11 and 13 are prime. They have no common prime factor. When this happens, the LCM is just the product of the numbers.

The LCM is $11 \cdot 13$, or 143.

39. a) Find the prime factorization of each number.

$12 = 2 \cdot 2 \cdot 3$
$35 = 5 \cdot 7$

b) Note that the two numbers have no common prime factor. When this happens, the LCM is just the product of the numbers.

The LCM is $12 \cdot 35$, or 420.

41. a) Find the prime factorization of each number.

$54 = 3 \cdot 3 \cdot 3 \cdot 2$
$63 = 3 \cdot 3 \cdot 7$

b) Create a product by writing each factor the greatest number of times it occurs in any one factorization.

The greatest number of times 2 occurs in any one factorization is one time.

The greatest number of times 3 occurs in any one factorization is three times.

The greatest number of times 7 occurs in any one factorization is one time.
Since there are no other prime factors in any of the factorizations, the LCM is $2 \cdot 3 \cdot 3 \cdot 3 \cdot 7$, or 378.

43. a) Find the prime factorization of each number.

$$81 = 3 \cdot 3 \cdot 3 \cdot 3$$
$$90 = 2 \cdot 3 \cdot 3 \cdot 5$$

b) Create a product by writing each factor the greatest number of times it occurs in any one factorization.

The greatest number of times 2 occurs in any one factorization is one time.

The greatest number of times 3 occurs in any one factorization is four times.

The greatest number of times 5 occurs in any one factorization is one time.

Since there are no other prime factors in any of the factorizations, the LCM is $2 \cdot 3 \cdot 3 \cdot 3 \cdot 3 \cdot 5$, or 810.

45. We find the LCM of the number of years it takes Jupiter and Saturn to make a complete revolution around the sun.

Jupiter: $12 = 2 \cdot 2 \cdot 3$

Saturn: $30 = 2 \cdot 3 \cdot 5$

The LCM $= 2 \cdot 2 \cdot 3 \cdot 5$, or 60. Thus, Jupiter and Saturn will appear in the same direction in the night sky once every 60 years.

47. 9001 is to the left of 10,001 on the number line, so $9001 < 10,001$.

49.

$$\begin{array}{r} {\scriptstyle 1} \\ {\scriptstyle 1\,1} \\ 3\,4\,5 \\ \times \quad 2\,3 \\ \hline 1\,0\,3\,5 \\ 6\,9\,0\,0 \\ \hline 7\,9\,3\,5 \end{array}$$

51. 2 ten thousands + 4 thousands + 6 hundreds + 0 tens + 5 ones, or 2 ten thousands + 4 thousands + 6 hundreds + 5 ones

53. ◈

55. a) 324 is not a multiple of 288.

b) Check multiples, using a calculator.

$2 \cdot 324 = 648$	Not a multiple of 288
$3 \cdot 324 = 972$	Not a multiple of 288
$4 \cdot 324 = 1296$	Not a multiple of 288
$5 \cdot 324 = 1620$	Not a multiple of 288
$6 \cdot 324 = 1944$	Not a multiple of 288
$7 \cdot 324 = 2268$	Not a multiple of 288
$8 \cdot 324 = 2592$	A multiple of 288

c) The LCM = 2592.

57. From Example 8 we know that the LCM of 27, 90, and 84 is $2 \cdot 3 \cdot 3 \cdot 5 \cdot 3 \cdot 2 \cdot 7$, so the LCM of 27, 90, 84, 210, 108, and 50 must contain at least these factors. We write the prime factorizations of 210, 108, and 50:

$$210 = 2 \cdot 3 \cdot 5 \cdot 7$$
$$108 = 2 \cdot 2 \cdot 3 \cdot 3 \cdot 3$$
$$50 = 2 \cdot 5 \cdot 5$$

Neither of the four factorizations above contains the other three.

Begin with the LCM of 27, 90, and 84, $2 \cdot 3 \cdot 3 \cdot 5 \cdot 3 \cdot 2 \cdot 7$. Neither 210 nor 108 contains any factors that are missing in this factorization. Next we look for factors of 50 that are missing. Since 50 contains a second factor of 5, we multiply by 5:

$$2 \cdot 3 \cdot 3 \cdot 5 \cdot 3 \cdot 2 \cdot 7 \cdot 5$$

The LCM is $2 \cdot 3 \cdot 3 \cdot 5 \cdot 3 \cdot 2 \cdot 7 \cdot 5$, or 18,900.

59. The width of the carton will be the common width, 5 in. The length of the carton must be a multiple of both 6 and 8. The shortest length carton will be the least common multiple of 6 and 8.

$$6 = 2 \cdot 3$$
$$8 = 2 \cdot 2 \cdot 2$$

LCM is $2 \cdot 2 \cdot 2 \cdot 3$, or 24.

The shortest carton is 24 in. long.

Chapter 2

Fractional Notation

Exercise Set 2.1

1. The dollar is divided into 4 parts of the same size, and 2 of them are shaded. This is $2 \cdot \frac{1}{4}$ or $\frac{2}{4}$. Thus, $\frac{2}{4}$ (two-fourths) of the dollar is shaded.

3. The yard is divided into 8 parts of the same size, and 1 of them is shaded. Thus, $\frac{1}{8}$ (one-eighth) of the yard is shaded.

5. We have 2 quarts, each divided into 3 parts. We take 4 of those parts. This is $4 \cdot \frac{1}{3}$ or $\frac{4}{3}$. Thus, $\frac{4}{3}$ of a quart is shaded.

7. The triangle is divided into 4 triangles of the same size, and 3 of them are shaded. This is $3 \cdot \frac{1}{4}$ or $\frac{3}{4}$. Thus, $\frac{3}{4}$ (three-fourths) of the triangle is shaded.

9. The pie is divided into 8 parts of the same size, and 4 of them are shaded. This is $4 \cdot \frac{1}{8}$, or $\frac{4}{8}$. Thus, $\frac{4}{8}$ (four-eighths) of the pie is shaded.

11. The acre is divided into 12 parts of the same size, and 6 of them are shaded. This is $6 \cdot \frac{1}{12}$, or $\frac{6}{12}$. Thus, $\frac{6}{12}$ (six-twelfths) of the acre is shaded.

13. Remember: $\frac{n}{1} = n$

$\frac{18}{1} = 18$

15. Remember: $\frac{0}{n} = 0$, for n that is not 0.

$\frac{0}{8} = 0$

Think of dividing an object into 8 parts and taking none of them. We get 0.

17. Remember: $\frac{n}{n} = 1$, for n that is not 0.

$\frac{20}{20} = 1$

If we divide an object into 20 parts and take 20 of them, we get all of the object (1 whole object).

19. $\frac{5}{6-6} = \frac{5}{0}$

Remember: $\frac{n}{0}$ is not defined for any whole number n. Thus, $\frac{5}{6-6}$ is not defined.

21. Remember: $\frac{n}{0}$ is not defined for any whole number n.

$\frac{729}{0}$ is not defined.

23. Remember: $\frac{n}{n} = 1$, for n that is not 0.

$\frac{87}{87} = 1$

If we divide an object into 87 parts and take 87 of them, we get all of the object (1 whole object).

25. $\frac{1}{2} \cdot \frac{1}{3} = \frac{1 \cdot 1}{2 \cdot 3} = \frac{1}{6}$

27. $5 \times \frac{1}{8} = \frac{5 \times 1}{8} = \frac{5}{8}$

29. $\frac{2}{3} \times \frac{1}{5} = \frac{2 \times 1}{3 \times 5} = \frac{2}{15}$

31. $\frac{2}{5} \cdot \frac{2}{3} = \frac{2 \cdot 2}{5 \cdot 3} = \frac{4}{15}$

33. $\frac{3}{4} \cdot \frac{3}{4} = \frac{3 \cdot 3}{4 \cdot 4} = \frac{9}{16}$

35. $\frac{2}{3} \cdot \frac{7}{13} = \frac{2 \cdot 7}{3 \cdot 13} = \frac{14}{39}$

37. $7 \cdot \frac{3}{4} = \frac{7 \cdot 3}{4} = \frac{21}{4}$

39. $\frac{7}{8} \cdot \frac{7}{8} = \frac{7 \cdot 7}{8 \cdot 8} = \frac{49}{64}$

41. Since $2 \cdot 5 = 10$, we multiply by $\frac{5}{5}$.

$$\frac{1}{2} = \frac{1}{2} \cdot \frac{5}{5} = \frac{1 \cdot 5}{2 \cdot 5} = \frac{5}{10}$$

43. Since $8 \cdot 4 = 32$, we multiply by $\frac{4}{4}$.

$$\frac{5}{8} = \frac{5}{8} \cdot \frac{4}{4} = \frac{5 \cdot 4}{8 \cdot 4} = \frac{20}{32}$$

45. Since $3 \cdot 15 = 45$, we multiply by $\frac{15}{15}$.

$$\frac{5}{3} = \frac{5}{3} \cdot \frac{15}{15} = \frac{5 \cdot 15}{3 \cdot 15} = \frac{75}{45}$$

47. Since $22 \cdot 6 = 132$, we multiply by $\frac{6}{6}$.

$$\frac{7}{22} = \frac{7}{22} \cdot \frac{6}{6} = \frac{7 \cdot 6}{22 \cdot 6} = \frac{42}{132}$$

49. $\frac{6}{8} = \frac{3\cdot 2}{4\cdot 2}$ ⟵ Factor the numerator / ⟵ Factor the denominator

$= \frac{3}{4}\cdot\frac{2}{2}$ ⟵ Factor the fraction

$= \frac{3}{4}\cdot 1$ ⟵ $\frac{2}{2}=1$

$= \frac{3}{4}$ ⟵ Removing a factor of 1

51. $\frac{2}{15} = \frac{1\cdot 3}{5\cdot 3}$ ⟵ Factor the numerator / ⟵ Factor the denominator

$= \frac{1}{5}\cdot\frac{3}{3}$ ⟵ Factor the fraction

$= \frac{1}{5}\cdot 1$ ⟵ $\frac{3}{3}=1$

$= \frac{1}{5}$ ⟵ Removing a factor of 1

53. $\frac{24}{8}=\frac{3\cdot 8}{1\cdot 8}=\frac{3}{1}\cdot\frac{8}{8}=\frac{3}{1}\cdot 1=\frac{3}{1}=3$

55. $\frac{18}{24}=\frac{3\cdot 6}{4\cdot 6}=\frac{3}{4}\cdot\frac{6}{6}=\frac{3}{4}\cdot 1=\frac{3}{4}$

57. $\frac{14}{16}=\frac{7\cdot 2}{8\cdot 2}=\frac{7}{8}\cdot\frac{2}{2}=\frac{7}{8}\cdot 1=\frac{7}{8}$

59. $\frac{17}{51}=\frac{1\cdot 17}{3\cdot 17}=\frac{1}{3}\cdot\frac{17}{17}=\frac{1}{3}\cdot 1=\frac{1}{3}$

61. $\frac{150}{25}=\frac{6\cdot 25}{1\cdot 25}=\frac{6}{1}\cdot\frac{25}{25}=\frac{6}{1}\cdot 1=\frac{6}{1}=6$

We could also simplify $\frac{150}{25}$ by doing the division $150\div 25$. That is, $\frac{150}{25}=150\div 25=6$.

63. ◈

Exercise Set 2.2

1. $\frac{2}{3}\cdot\frac{1}{2}=\frac{2\cdot 1}{3\cdot 2}=\frac{2}{2}\cdot\frac{1}{3}=1\cdot\frac{1}{3}=\frac{1}{3}$

3. $\frac{1}{4}\cdot\frac{2}{3}=\frac{1\cdot 2}{4\cdot 3}=\frac{1\cdot 2}{2\cdot 2\cdot 3}=\frac{2}{2}\cdot\frac{1}{2\cdot 3}=\frac{1}{2\cdot 3}=\frac{1}{6}$

5. $\frac{12}{5}\cdot\frac{9}{8}=\frac{12\cdot 9}{5\cdot 8}=\frac{4\cdot 3\cdot 9}{5\cdot 2\cdot 4}=\frac{4}{4}\cdot\frac{3\cdot 9}{5\cdot 2}=\frac{3\cdot 9}{5\cdot 2}=\frac{27}{10}$

7. $\frac{10}{9}\cdot\frac{7}{5}=\frac{10\cdot 7}{9\cdot 5}=\frac{5\cdot 2\cdot 7}{9\cdot 5}=\frac{5}{5}\cdot\frac{2\cdot 7}{9}=\frac{2\cdot 7}{9}=\frac{14}{9}$

9. $9\cdot\frac{1}{9}=\frac{9\cdot 1}{9}=\frac{9\cdot 1}{9\cdot 1}=1$

11. $\frac{7}{5}\cdot\frac{5}{7}=\frac{7\cdot 5}{5\cdot 7}=\frac{7\cdot 5}{7\cdot 5}=1$

13. $24\cdot\frac{1}{6}=\frac{24\cdot 1}{6}=\frac{24}{6}=\frac{4\cdot 6}{1\cdot 6}=\frac{4}{1}\cdot\frac{6}{6}=\frac{4}{1}=4$

15. $12\cdot\frac{3}{4}=\frac{12\cdot 3}{4}=\frac{4\cdot 3\cdot 3}{4\cdot 1}=\frac{4}{4}\cdot\frac{3\cdot 3}{1}=\frac{3\cdot 3}{1}=9$

17. $\frac{7}{10}\cdot 28=\frac{7\cdot 28}{10}=\frac{7\cdot 2\cdot 14}{2\cdot 5}=\frac{2}{2}\cdot\frac{7\cdot 14}{5}=\frac{7\cdot 14}{5}=\frac{98}{5}$

19. $240\cdot\frac{1}{8}=\frac{240\cdot 1}{8}=\frac{240}{8}=\frac{8\cdot 30}{8\cdot 1}=\frac{8}{8}\cdot\frac{30}{1}=\frac{30}{1}=30$

21. $\frac{4}{10}\cdot\frac{5}{10}=\frac{4\cdot 5}{10\cdot 10}=\frac{2\cdot 2\cdot 5\cdot 1}{2\cdot 5\cdot 2\cdot 5}=\frac{2\cdot 2\cdot 5}{2\cdot 2\cdot 5}\cdot\frac{1}{5}=\frac{1}{5}$

23. $\frac{8}{10}\cdot\frac{45}{100}=\frac{8\cdot 45}{10\cdot 100}=\frac{2\cdot 2\cdot 2\cdot 5\cdot 9}{2\cdot 5\cdot 2\cdot 5\cdot 2\cdot 5}$

$=\frac{2\cdot 2\cdot 2\cdot 5}{2\cdot 2\cdot 2\cdot 5}\cdot\frac{9}{5\cdot 5}=\frac{9}{5\cdot 5}=\frac{9}{25}$

25. $\frac{11}{24}\cdot\frac{3}{5}=\frac{11\cdot 3}{24\cdot 5}=\frac{11\cdot 3}{3\cdot 8\cdot 5}=\frac{3}{3}\cdot\frac{11}{8\cdot 5}=\frac{11}{8\cdot 5}=\frac{11}{40}$

27. $\frac{10}{21}\cdot\frac{3}{4}=\frac{10\cdot 3}{21\cdot 4}=\frac{2\cdot 5\cdot 3}{3\cdot 7\cdot 2\cdot 2}$

$=\frac{2\cdot 3}{2\cdot 3}\cdot\frac{5}{7\cdot 2}=\frac{5}{7\cdot 2}=\frac{5}{14}$

29. $\frac{5}{6}$ Interchange the numerator and denominator.

The reciprocal of $\frac{5}{6}$ is $\frac{6}{5}$. $\left(\frac{5}{6}\cdot\frac{6}{5}=\frac{30}{30}=1\right)$

31. Think of 6 as $\frac{6}{1}$.

$\frac{6}{1}$ Interchange the numerator and denominator.

The reciprocal of $\frac{6}{1}$ is $\frac{1}{6}$. $\left(\frac{6}{1}\cdot\frac{1}{6}=\frac{6}{6}=1\right)$

33. $\frac{1}{6}$ Interchange the numerator and denominator.

The reciprocal of $\frac{1}{6}$ is 6. $\left(\frac{6}{1}=6;\ \frac{1}{6}\cdot\frac{6}{1}=\frac{6}{6}=1\right)$

(Note that we also found that 6 and $\frac{1}{6}$ are reciprocals in Exercise 31.)

35. $\frac{10}{3}$ Interchange the numerator and denominator.

The reciprocal of $\frac{10}{3}$ is $\frac{3}{10}$. $\left(\frac{10}{3}\cdot\frac{3}{10}=\frac{30}{30}=1\right)$

37. $\frac{3}{5}\div\frac{3}{4}=\frac{3}{5}\cdot\frac{4}{3}$ Multiplying the dividend $\left(\frac{3}{5}\right)$ by the reciprocal of the divisor $\left(\text{The reciprocal of } \frac{3}{4} \text{ is } \frac{4}{3}.\right)$

$=\frac{3\cdot 4}{5\cdot 3}$ Multiplying numerators and denominators

$=\frac{3}{3}\cdot\frac{4}{5}=\frac{4}{5}$ Simplifying

39. $\frac{3}{5} \div \frac{9}{4} = \frac{3}{5} \cdot \frac{4}{9}$ Multiplying the dividend $\left(\frac{3}{5}\right)$ by the reciprocal of the divisor $\left(\text{The reciprocal of } \frac{9}{4} \text{ is } \frac{4}{9}.\right)$

$= \frac{3 \cdot 4}{5 \cdot 9}$ Multiplying numerators and denominators

$= \frac{3 \cdot 4}{5 \cdot 3 \cdot 3}$

$= \frac{3}{3} \cdot \frac{4}{5 \cdot 3}$ Simplifying

$= \frac{4}{5 \cdot 3} = \frac{4}{15}$

41. $\frac{4}{3} \div \frac{1}{3} = \frac{4}{3} \cdot 3 = \frac{4 \cdot 3}{3} = \frac{3}{3} \cdot 4 = 4$

43. $\frac{1}{3} \div \frac{1}{6} = \frac{1}{3} \cdot 6 = \frac{1 \cdot 6}{3} = \frac{1 \cdot 2 \cdot 3}{1 \cdot 3} = \frac{1 \cdot 3}{1 \cdot 3} \cdot 2 = 2$

45. $\frac{3}{8} \div 3 = \frac{3}{8} \cdot \frac{1}{3} = \frac{3 \cdot 1}{8 \cdot 3} = \frac{3}{3} \cdot \frac{1}{8} = \frac{1}{8}$

47. $\frac{12}{7} \div 4 = \frac{12}{7} \cdot \frac{1}{4} = \frac{12 \cdot 1}{7 \cdot 4} = \frac{4 \cdot 3 \cdot 1}{7 \cdot 4} = \frac{4}{4} \cdot \frac{3 \cdot 1}{7} = \frac{3 \cdot 1}{7} = \frac{3}{7}$

49. $12 \div \frac{3}{2} = 12 \cdot \frac{2}{3} = \frac{12 \cdot 2}{3} = \frac{3 \cdot 4 \cdot 2}{3 \cdot 1} = \frac{3}{3} \cdot \frac{4 \cdot 2}{1}$

$= \frac{4 \cdot 2}{1} = \frac{8}{1} = 8$

51. $28 \div \frac{4}{5} = 28 \cdot \frac{5}{4} = \frac{28 \cdot 5}{4} = \frac{4 \cdot 7 \cdot 5}{4 \cdot 1} = \frac{4}{4} \cdot \frac{7 \cdot 5}{1}$

$= \frac{7 \cdot 5}{1} = 35$

53. $\frac{5}{8} \div \frac{5}{8} = \frac{5}{8} \cdot \frac{8}{5} = \frac{5 \cdot 8}{8 \cdot 5} = \frac{5 \cdot 8}{5 \cdot 8} = 1$

55. $\frac{8}{15} \div \frac{4}{5} = \frac{8}{15} \cdot \frac{5}{4} = \frac{8 \cdot 5}{15 \cdot 4} = \frac{2 \cdot 4 \cdot 5}{3 \cdot 5 \cdot 4} = \frac{4 \cdot 5}{4 \cdot 5} \cdot \frac{2}{3} = \frac{2}{3}$

57. $\frac{9}{5} \div \frac{4}{5} = \frac{9}{5} \cdot \frac{5}{4} = \frac{9 \cdot 5}{5 \cdot 4} = \frac{5}{5} \cdot \frac{9}{4} = \frac{9}{4}$

59. $120 \div \frac{5}{6} = 120 \cdot \frac{6}{5} = \frac{120 \cdot 6}{5} = \frac{5 \cdot 24 \cdot 6}{5 \cdot 1} = \frac{5}{5} \cdot \frac{24 \cdot 6}{1}$

$= \frac{24 \cdot 6}{1} = 144$

61. $\frac{4}{5} \cdot x = 60$

$x = 60 \div \frac{4}{5}$ Dividing on both sides by $\frac{4}{5}$

$x = 60 \cdot \frac{5}{4}$ Multiplying by the reciprocal

$= \frac{60 \cdot 5}{4} = \frac{4 \cdot 15 \cdot 5}{4 \cdot 1} = \frac{4}{4} \cdot \frac{15 \cdot 5}{1} = \frac{15 \cdot 5}{1} = 75$

63. $\frac{5}{3} \cdot y = \frac{10}{3}$

$y = \frac{10}{3} \div \frac{5}{3}$ Dividing on both sides by $\frac{5}{3}$

$y = \frac{10}{3} \cdot \frac{3}{5}$ Multiplying by the reciprocal

$= \frac{10 \cdot 3}{3 \cdot 5} = \frac{2 \cdot 5 \cdot 3}{3 \cdot 5 \cdot 1} = \frac{5 \cdot 3}{5 \cdot 3} \cdot \frac{2}{1} = \frac{2}{1} = 2$

65. $x \cdot \frac{25}{36} = \frac{5}{12}$

$x = \frac{5}{12} \div \frac{25}{36} = \frac{5}{12} \cdot \frac{36}{25} = \frac{5 \cdot 36}{12 \cdot 25} = \frac{5 \cdot 3 \cdot 12}{12 \cdot 5 \cdot 5}$

$= \frac{5 \cdot 12}{5 \cdot 12} \cdot \frac{3}{5} = \frac{3}{5}$

67. $n \cdot \frac{8}{7} = 360$

$n = 360 \div \frac{8}{7} = 360 \cdot \frac{7}{8} = \frac{360 \cdot 7}{8} = \frac{8 \cdot 45 \cdot 7}{8 \cdot 1}$

$= \frac{8}{8} \cdot \frac{45 \cdot 7}{1} = \frac{45 \cdot 7}{1} = 315$

69.

$$\begin{array}{r} 67 \\ 4 \overline{)\,268} \\ \underline{240} \\ 28 \\ \underline{28} \\ 0 \end{array}$$

The answer is 67.

71.

$$\begin{array}{r} 285 \\ 24 \overline{)\,6842} \\ \underline{4800} \\ 2042 \\ \underline{1920} \\ 122 \\ \underline{120} \\ 2 \end{array}$$

The answer is 285 R 2.

73. $4 \cdot x = 268$

$\frac{4 \cdot x}{4} = \frac{268}{4}$ Dividing by 4 on both sides

$x = 67$

The solution is 67.

75. $y + 502 = 9001$

$y + 502 - 502 = 9001 - 502$ Subtracting 502 on both sides

$y = 8499$

The solution is 8499.

77.

79. Let n = the number.

$\frac{1}{3} \cdot n = \frac{1}{4}$

$n = \frac{1}{4} \div \frac{1}{3} = \frac{1}{4} \cdot \frac{3}{1} = \frac{1 \cdot 3}{4 \cdot 1} = \frac{3}{4}$

The number is $\frac{3}{4}$. Now we find $\frac{1}{2}$ of $\frac{3}{4}$.

$$\frac{1}{2}\cdot\frac{3}{4}=\frac{1\cdot 3}{2\cdot 4}=\frac{3}{8}$$

One-half of the number is $\frac{3}{8}$.

Exercise Set 2.3

1. $\frac{7}{8}+\frac{1}{8}=\frac{7+1}{8}=\frac{8}{8}=1$

3. $\frac{1}{8}+\frac{5}{8}=\frac{1+5}{8}=\frac{6}{8}=\frac{3\cdot 2}{4\cdot 2}=\frac{3}{4}\cdot\frac{2}{2}=\frac{3}{4}\cdot 1=\frac{3}{4}$

5. $\frac{2}{3}+\frac{5}{6}$ 3 is a factor of 6, so the LCD is 6.

$=\frac{2}{3}\cdot\frac{2}{2}+\frac{5}{6}$ ⟵ This fraction already has the LCD as denominator.

Think: $3\times\square=6$. The answer is 2, so we multiply by 1, using $\frac{2}{2}$.

$=\frac{4}{6}+\frac{5}{6}=\frac{9}{6}$

$=\frac{3}{2}$ Simplifying

7. $\frac{1}{8}+\frac{1}{6}$ $8=2\cdot 2\cdot 2$ and $6=2\cdot 3$, so the LCD is $2\cdot 2\cdot 2\cdot 3$, or 24

$=\frac{1}{8}\cdot\frac{3}{3}+\frac{1}{6}\cdot\frac{4}{4}$

Think: $6\times\square=24$. The answer is 4, so we multiply by 1, using $\frac{4}{4}$.

Think: $8\times\square=24$. The answer is 3, so we multiply by 1, using $\frac{3}{3}$.

$=\frac{3}{24}+\frac{4}{24}$

$=\frac{7}{24}$

9. $\frac{4}{5}+\frac{7}{10}$ 5 is a factor of 10, so the LCD is 10.

$=\frac{4}{5}\cdot\frac{2}{2}+\frac{7}{10}$ ⟵ This fraction already has the LCD as denominator.

Think: $5\times\square=10$. The answer is 2, so we multiply by 1, using $\frac{2}{2}$.

$=\frac{8}{10}+\frac{7}{10}=\frac{15}{10}$

$=\frac{3}{2}$ Simplifying

11. $\frac{5}{12}+\frac{3}{8}$ $12=2\cdot 2\cdot 3$ and $8=2\cdot 2\cdot 2$, so the LCD is $2\cdot 2\cdot 2\cdot 3$, or 24.

$=\frac{5}{12}\cdot\frac{2}{2}+\frac{3}{8}\cdot\frac{3}{3}$

Think: $8\times\square=24$. The answer is 3, so we multiply by 1, using $\frac{3}{3}$.

Think: $12\times\square=24$. The answer is 2, so we multiply by 1, using $\frac{2}{2}$.

$=\frac{10}{24}+\frac{9}{24}=\frac{19}{24}$

13. $\frac{3}{20}+\frac{3}{4}$ 4 is a factor of 20, so the LCD is 20.

$=\frac{3}{20}+\frac{3}{4}\cdot\frac{5}{5}$ Multiplying by 1

$=\frac{3}{20}+\frac{15}{20}=\frac{18}{20}=\frac{9}{10}$

15. $\frac{5}{6}+\frac{7}{9}$ $6=2\cdot 3$ and $9=3\cdot 3$, so the LCD is $2\cdot 3\cdot 3$, or 18.

$=\frac{5}{6}\cdot\frac{3}{3}+\frac{7}{9}\cdot\frac{2}{2}$ Multiplying by 1

$=\frac{15}{18}+\frac{14}{18}=\frac{29}{18}$

17. $\frac{7}{8}+\frac{0}{1}$ 1 is a factor of 8, so the LCD is 8.

$=\frac{7}{8}+\frac{0}{1}\cdot\frac{8}{8}$

$=\frac{7}{8}+\frac{0}{8}=\frac{7}{8}$

Note that if we had observed at the outset that $\frac{0}{1}=0$, the computation becomes $\frac{7}{8}+0=\frac{7}{8}$.

19. $\frac{3}{8}+\frac{1}{6}$ $8=2\cdot 2\cdot 2$ and $6=2\cdot 3$, so the LCD is $2\cdot 2\cdot 2\cdot 3$, or 24.

$=\frac{3}{8}\cdot\frac{3}{3}+\frac{1}{6}\cdot\frac{4}{4}$

$=\frac{9}{24}+\frac{4}{24}=\frac{13}{24}$

21. $\frac{5}{8}+\frac{5}{6}$ $8=2\cdot 2\cdot 2$ and $6=2\cdot 3$, so the LCD is $2\cdot 2\cdot 2\cdot 3$, or 24.

$=\frac{5}{8}\cdot\frac{3}{3}+\frac{5}{6}\cdot\frac{4}{4}$

$=\frac{15}{24}+\frac{20}{24}=\frac{35}{24}$

23. $\frac{9}{10}+\frac{3}{100}$ 10 is a factor of 100, so the LCD is 100.

$=\frac{9}{10}\cdot\frac{10}{10}+\frac{3}{100}$

$=\frac{90}{100}+\frac{3}{100}=\frac{93}{100}$

25. $\frac{5}{12}+\frac{7}{24}$ 12 is a factor of 24, so the LCD is 24.

$=\frac{5}{12}\cdot\frac{2}{2}+\frac{7}{24}$

$=\frac{10}{24}+\frac{7}{24}=\frac{17}{24}$

27. $\frac{8}{10}+\frac{7}{100}+\frac{4}{1000}$ 10 and 100 are factors of 1000, so the LCD is 1000.

$=\frac{8}{10}\cdot\frac{100}{100}+\frac{7}{100}\cdot\frac{10}{10}+\frac{4}{1000}$

$=\frac{800}{1000}+\frac{70}{1000}+\frac{4}{1000}=\frac{874}{1000}$

$=\frac{437}{500}$

29. $\frac{3}{8}+\frac{5}{12}+\frac{8}{15}$

$=\frac{3}{2\cdot 2\cdot 2}+\frac{5}{2\cdot 2\cdot 3}+\frac{8}{3\cdot 5}$ Factoring the denominators

The LCM is $2\cdot 2\cdot 2\cdot 3\cdot 5$, or 120.

$=\frac{3}{2\cdot 2\cdot 2}\cdot\frac{3\cdot 5}{3\cdot 5}+\frac{5}{2\cdot 2\cdot 3}\cdot\frac{2\cdot 5}{2\cdot 5}+\frac{8}{3\cdot 5}\cdot\frac{2\cdot 2\cdot 2}{2\cdot 2\cdot 2}$

In each case we multiply by 1 to obtain the LCD in the denominator.

$=\frac{3\cdot 3\cdot 5}{2\cdot 2\cdot 2\cdot 3\cdot 5}+\frac{5\cdot 2\cdot 5}{2\cdot 2\cdot 3\cdot 2\cdot 5}+\frac{8\cdot 2\cdot 2\cdot 2}{3\cdot 5\cdot 2\cdot 2\cdot 2}$

$=\frac{45}{120}+\frac{50}{120}+\frac{64}{120}$

$=\frac{159}{120}=\frac{53}{40}$

31. $\frac{15}{24}+\frac{7}{36}+\frac{91}{48}$

$=\frac{15}{2\cdot 2\cdot 2\cdot 3}+\frac{7}{2\cdot 2\cdot 3\cdot 3}+\frac{91}{2\cdot 2\cdot 2\cdot 2\cdot 3}$

Factoring the denominators.

The LCM is $2\cdot 2\cdot 2\cdot 2\cdot 3\cdot 3$, or 144.

$=\frac{15}{2\cdot 2\cdot 2\cdot 3}\cdot\frac{2\cdot 3}{2\cdot 3}+\frac{7}{2\cdot 2\cdot 3\cdot 3}\cdot\frac{2\cdot 2}{2\cdot 2}+$

$\frac{91}{2\cdot 2\cdot 2\cdot 2\cdot 3}\cdot\frac{3}{3}$

In each case we multiply by 1 to obtain the LCD in the denominator.

$=\frac{15\cdot 2\cdot 3}{2\cdot 2\cdot 2\cdot 3\cdot 2\cdot 3}+\frac{7\cdot 2\cdot 2}{2\cdot 2\cdot 3\cdot 3\cdot 2\cdot 2}+\frac{91\cdot 3}{2\cdot 2\cdot 2\cdot 2\cdot 3\cdot 3}$

$=\frac{90}{144}+\frac{28}{144}+\frac{273}{144}=\frac{391}{144}$

33. When denominators are the same, subtract the numerators and keep the denominator.

$\frac{5}{6}-\frac{1}{6}=\frac{5-1}{6}=\frac{4}{6}=\frac{2\cdot 2}{2\cdot 3}=\frac{2}{2}\cdot\frac{2}{3}=\frac{2}{3}$

35. When denominators are the same, subtract the numerators and keep the denominator.

$\frac{11}{12}-\frac{2}{12}=\frac{11-2}{12}=\frac{9}{12}=\frac{3\cdot 3}{3\cdot 4}=\frac{3}{3}\cdot\frac{3}{4}=\frac{3}{4}$

37. The LCM of 4 and 8 is 8.

$\frac{3}{4}-\frac{1}{8}=\frac{3}{4}\cdot\frac{2}{2}-\frac{1}{8}$ ←This fraction already has the LCM as the denominator.

Think: $4\times\square=8$. The answer is 2, so we multiply by 1, using $\frac{2}{2}$.

$=\frac{6}{8}-\frac{1}{8}=\frac{5}{8}$

39. The LCM of 8 and 12 is 24.

$\frac{1}{8}-\frac{1}{12}=\frac{1}{8}\cdot\frac{3}{3}-\frac{1}{12}\cdot\frac{2}{2}$

Think: $12\times\square=24$. The answer is 2, so we multiply by 1, using $\frac{2}{2}$.

Think: $8\times\square=24$. The answer is 3, so we multiply by 1, using $\frac{3}{3}$.

$=\frac{3}{24}-\frac{2}{24}=\frac{1}{24}$

41. The LCM of 3 and 6 is 6.

$$\frac{4}{3}-\frac{5}{6}=\frac{4}{3}\cdot\frac{2}{2}-\frac{5}{6}$$
$$=\frac{8}{6}-\frac{5}{6}=\frac{3}{6}$$
$$=\frac{1\cdot 3}{2\cdot 3}=\frac{1}{2}\cdot\frac{3}{3}$$
$$=\frac{1}{2}$$

43. The LCM of 12 and 15 is 60.

$$\frac{5}{12}-\frac{2}{15}=\frac{5}{12}\cdot\frac{5}{5}-\frac{2}{15}\cdot\frac{4}{4}$$
$$=\frac{25}{60}-\frac{8}{60}=\frac{17}{60}$$

45. The LCM of 10 and 100 is 100.

$$\frac{6}{10}-\frac{7}{100}=\frac{6}{10}\cdot\frac{10}{10}-\frac{7}{100}$$
$$=\frac{60}{100}-\frac{7}{100}=\frac{53}{100}$$

47. The LCM of 15 and 25 is 75.

$$\frac{7}{15}-\frac{3}{25}=\frac{7}{15}\cdot\frac{5}{5}-\frac{3}{25}\cdot\frac{3}{3}$$
$$=\frac{35}{75}-\frac{9}{75}=\frac{26}{75}$$

49. The LCM of 12 and 8 is 24.

$$\frac{5}{12}-\frac{3}{8}=\frac{5}{12}\cdot\frac{2}{2}-\frac{3}{8}\cdot\frac{3}{3}$$
$$=\frac{10}{24}-\frac{9}{24}$$
$$=\frac{1}{24}$$

51. The LCM of 8 and 16 is 16.

$$\frac{7}{8}-\frac{1}{16}=\frac{7}{8}\cdot\frac{2}{2}-\frac{1}{16}$$
$$=\frac{14}{16}-\frac{1}{16}$$
$$=\frac{13}{16}$$

53. The LCM of 25 and 15 is 75.

$$\frac{17}{25}-\frac{4}{15}=\frac{17}{25}\cdot\frac{3}{3}-\frac{4}{15}\cdot\frac{5}{5}$$
$$=\frac{51}{75}-\frac{20}{75}$$
$$=\frac{31}{75}$$

55. The LCM of 25 and 150 is 150.

$$\frac{23}{25}-\frac{112}{150}=\frac{23}{25}\cdot\frac{6}{6}-\frac{112}{150}$$
$$=\frac{138}{150}-\frac{112}{150}=\frac{26}{150}$$
$$=\frac{2\cdot 13}{2\cdot 75}=\frac{2}{2}\cdot\frac{13}{75}$$
$$=\frac{13}{75}$$

57. Since there is a common denominator, compare the numerators.

$$5<6, \text{ so } \frac{5}{8}<\frac{6}{8}.$$

59. The LCD is 12.

$\frac{1}{3}\cdot\frac{4}{4}=\frac{4}{12}$ We multiply by 1 to get the LCD.

$\frac{1}{4}\cdot\frac{3}{3}=\frac{3}{12}$ We multiply by 1 to get the LCD.

Since $4>3$, it follows that $\frac{4}{12}>\frac{3}{12}$, so $\frac{1}{3}>\frac{1}{4}$.

61. The LCD is 21.

$\frac{2}{3}\cdot\frac{7}{7}=\frac{14}{21}$ We multiply by 1 to get the LCD.

$\frac{5}{7}\cdot\frac{3}{3}=\frac{15}{21}$ We multiply by 1 to get the LCD.

Since $14<15$, it follows that $\frac{14}{21}<\frac{15}{21}$, so $\frac{2}{3}<\frac{5}{7}$.

63. The LCD is 30.

$$\frac{4}{5}\cdot\frac{6}{6}=\frac{24}{30}$$
$$\frac{5}{6}\cdot\frac{5}{5}=\frac{25}{30}$$

Since $24<25$, it follows that $\frac{24}{30}<\frac{25}{30}$, so $\frac{4}{5}<\frac{5}{6}$.

65. The LCD is 20.

The denominator of $\frac{19}{20}$ is the LCD.

$$\frac{4}{5}\cdot\frac{4}{4}=\frac{16}{20}$$

Since $19>16$, it follows that $\frac{19}{20}>\frac{16}{20}$, so $\frac{19}{20}>\frac{4}{5}$.

67. The LCD is 20.

The denominator of $\frac{19}{20}$ is the LCD.

$$\frac{9}{10}\cdot\frac{2}{2}=\frac{18}{20}$$

Since $19>18$, it follows that $\frac{19}{20}>\frac{18}{20}$, so $\frac{19}{20}>\frac{9}{10}$.

69. The LCD is $21 \cdot 13$, or 273.

$$\frac{31}{21}\cdot\frac{13}{13}=\frac{403}{273}$$
$$\frac{41}{13}\cdot\frac{21}{21}=\frac{861}{273}$$

Since $403 < 861$, it follows that $\frac{403}{273} < \frac{861}{273}$, so $\frac{31}{21} < \frac{41}{13}$.

71.
$$x+\frac{1}{30}=\frac{1}{10}$$
$$x+\frac{1}{30}-\frac{1}{30}=\frac{1}{10}-\frac{1}{30}\quad \text{Subtracting } \frac{1}{30} \text{ on both sides}$$
$$x+0=\frac{1}{10}\cdot\frac{3}{3}-\frac{1}{30}\quad \text{The LCD is 30. We multiply by 1 to get the LCD.}$$
$$x=\frac{3}{30}-\frac{1}{30}=\frac{2}{30}$$
$$x=\frac{1\cdot 2}{2\cdot 15}=\frac{1}{15}\cdot\frac{2}{2}$$
$$x=\frac{1}{15}$$

The solution is $\frac{1}{15}$.

73.
$$\frac{2}{3}+t=\frac{4}{5}$$
$$\frac{2}{3}+t-\frac{2}{3}=\frac{4}{5}-\frac{2}{3}\quad \text{Subtracting } \frac{2}{3} \text{ on both sides.}$$
$$t+0=\frac{4}{5}\cdot\frac{3}{3}-\frac{2}{3}\cdot\frac{5}{5}\quad \text{The LCD is 15. We multiply by 1 to get the LCD.}$$
$$t=\frac{12}{15}-\frac{10}{15}=\frac{2}{15}$$

The solution is $\frac{2}{15}$.

75.
$$m+\frac{5}{6}=\frac{9}{10}$$
$$m+\frac{5}{6}-\frac{5}{6}=\frac{9}{10}-\frac{5}{6}$$
$$m+0=\frac{9}{10}\cdot\frac{3}{3}-\frac{5}{6}\cdot\frac{5}{5}$$
$$m=\frac{27}{30}-\frac{25}{30}=\frac{2}{30}$$
$$m=\frac{1\cdot 2}{2\cdot 15}=\frac{1}{15}\cdot\frac{2}{2}$$
$$m=\frac{1}{15}$$

The solution is $\frac{1}{15}$.

77.
```
      15
   8 9 5 10
   9 0 6 0
 - 4 3 8 7
 ---------
   4 6 7 3
```

79.
```
        2 0 4
 3 5 ) 7 1 4 0
       7 0 0 0
       -------
         1 4 0
         1 4 0
         -----
             0
```

The answer is 204.

81. ◈

83. ***Familiarize***. We visualize the situation. We let h = the elevation at which the climber finished.

Translate.

First climb minus First descent plus Second climb minus Second descent is Final elevation

$$\frac{3}{5} - \frac{1}{4} + \frac{1}{3} - \frac{1}{7} = h$$

Solve. We carry out the calculation. The LCD is $5\cdot 4\cdot 3\cdot 7$, or 420.

$$\frac{3}{5}-\frac{1}{4}+\frac{1}{3}-\frac{1}{7}=h$$
$$\frac{3}{5}\cdot\frac{4\cdot 3\cdot 7}{4\cdot 3\cdot 7}-\frac{1}{4}\cdot\frac{5\cdot 3\cdot 7}{5\cdot 3\cdot 7}+\frac{1}{3}\cdot\frac{5\cdot 4\cdot 7}{5\cdot 4\cdot 7}-\frac{1}{7}\cdot\frac{5\cdot 4\cdot 3}{5\cdot 4\cdot 3}=h$$
$$\frac{252}{5\cdot 4\cdot 3\cdot 7}-\frac{105}{5\cdot 4\cdot 3\cdot 7}+\frac{140}{5\cdot 4\cdot 3\cdot 7}-\frac{60}{5\cdot 4\cdot 3\cdot 7}=h$$
$$\frac{252-105+140-60}{5\cdot 4\cdot 3\cdot 7}=h$$
$$\frac{227}{5\cdot 4\cdot 3\cdot 7}=h$$
$$\frac{227}{420}=h$$

Check. We repeat the calculation.

State. The climber's final elevation is $\frac{227}{420}$ km.

Exercise Set 2.4

1. $5\frac{2}{3} = \frac{17}{3}$

[a] Multiply: $5 \cdot 3 = 15$.

[b] Add: $15 + 2 = 17$.

[c] Keep the denominator.

3. $9\frac{5}{6} = \frac{59}{6}$ $\quad (9 \cdot 6 = 54,\ 54 + 5 = 59)$

5. $12\frac{3}{4} = \frac{51}{4}$ $\quad (12 \cdot 4 = 48,\ 48 + 3 = 51)$

7. To convert $\frac{18}{5}$ to a mixed numeral, we divide.

$$\begin{array}{r} 3 \\ 5\overline{)18} \\ \underline{15} \\ 3 \end{array} \qquad \frac{18}{5} = 3\frac{3}{5}$$

9.
$$\begin{array}{r} 5 \\ 10\overline{)57} \\ \underline{50} \\ 7 \end{array} \qquad \frac{57}{10} = 5\frac{7}{10}$$

11.
$$\begin{array}{r} 43 \\ 8\overline{)345} \\ \underline{320} \\ 25 \\ \underline{24} \\ 1 \end{array} \qquad \frac{345}{8} = 43\frac{1}{8}$$

13.
$$\begin{array}{r} 2\frac{7}{8} \\ +3\frac{5}{8} \\ \hline 5\frac{12}{8} \end{array} = 5 + \frac{12}{8} = 5 + 1\frac{1}{2} = 6\frac{1}{2}$$

To find a mixed numeral for $\frac{12}{8}$ we divide:

$$\begin{array}{r} 1 \\ 8\overline{)12} \\ \underline{8} \\ 4 \end{array} \qquad \frac{12}{8} = 1\frac{4}{8} = 1\frac{1}{2}$$

15. The LCD is 12.

$$\begin{array}{rcr} 1\boxed{\frac{1}{4} \cdot \frac{3}{3}} & = & 1\frac{3}{12} \\ +1\boxed{\frac{2}{3} \cdot \frac{4}{4}} & = & +1\frac{8}{12} \\ \hline & & 2\frac{11}{12} \end{array}$$

17. The LCD is 12.

$$\begin{array}{rcr} 8\boxed{\frac{3}{4} \cdot \frac{3}{3}} & = & 8\frac{9}{12} \\ +5\boxed{\frac{5}{6} \cdot \frac{2}{2}} & = & +5\frac{10}{12} \\ \hline & & 13\frac{19}{12} \end{array} = 13 + \frac{19}{12} = 13 + 1\frac{7}{12} = 14\frac{7}{12}$$

19. The LCD is 10.

$$\begin{array}{rcr} 12\boxed{\frac{4}{5} \cdot \frac{2}{2}} & = & 12\frac{8}{10} \\ +8\frac{7}{10} & = & +8\frac{7}{10} \\ \hline & & 20\frac{15}{10} \end{array} = 20 + \frac{15}{10} = 20 + 1\frac{5}{10} = 21\frac{5}{10} = 21\frac{1}{2}$$

21. The LCD is 8.

$$\begin{array}{rcr} 14\frac{5}{8} & = & 14\frac{5}{8} \\ +13\boxed{\frac{1}{4} \cdot \frac{2}{2}} & = & +13\frac{2}{8} \\ \hline & & 27\frac{7}{8} \end{array}$$

23. The LCD is 24.

$$\begin{array}{rcr} 7\boxed{\frac{1}{8} \cdot \frac{3}{3}} & = & 7\frac{3}{24} \\ 9\boxed{\frac{2}{3} \cdot \frac{8}{8}} & = & 9\frac{16}{24} \\ +10\boxed{\frac{3}{4} \cdot \frac{6}{6}} & = & +10\frac{18}{24} \\ \hline & & 26\frac{37}{24} \end{array} = 26 + \frac{37}{24} = 26 + 1\frac{13}{24} = 27\frac{13}{24}$$

25.
$$\begin{array}{rcr} 4\frac{1}{5} & = & 3\frac{6}{5} \\ -2\frac{3}{5} & = & -2\frac{3}{5} \\ \hline & & 1\frac{3}{5} \end{array}$$

Since $\frac{1}{5}$ is smaller than $\frac{3}{5}$, we cannot subtract until we borrow:

$$4\frac{1}{5} = 3+\frac{5}{5}+\frac{1}{5} = 3+\frac{6}{5} = 3\frac{6}{5}$$

27. The LCD is 10.

$$\begin{array}{rcr} 6\boxed{\frac{3}{5}\cdot\frac{2}{2}} & = & 6\,\frac{6}{10} \\ -2\boxed{\frac{1}{2}\cdot\frac{5}{5}} & = & -2\,\frac{5}{10} \\ \hline & & 4\,\frac{1}{10} \end{array}$$

29.
$$\begin{array}{rcr} 34 & = & 33\frac{8}{8} \\ -\,18\frac{5}{8} & = & -\,18\frac{5}{8} \\ \hline & & 15\frac{3}{8} \end{array}$$
$\left(34 = 33+1 = 33+\frac{8}{8} = 33\frac{8}{8}\right)$

31. The LCD is 12.

$$\begin{array}{rcrcr} 21\boxed{\frac{1}{6}\cdot\frac{2}{2}} & = & 21\,\frac{2}{12} & = & 20\,\frac{14}{12} \\ -13\boxed{\frac{3}{4}\cdot\frac{3}{3}} & = & -13\,\frac{9}{12} & = & -13\,\frac{9}{12} \\ \hline & & & & 7\,\frac{5}{12} \end{array}$$

(Since $\frac{2}{12}$ is smaller than $\frac{9}{12}$, we cannot subtract until we borrow: $21\frac{2}{12} = 20+\frac{12}{12}+\frac{2}{12} = 20+\frac{14}{12} = 20\frac{14}{12}$.)

33. The LCD is 8.

$$\begin{array}{rcrcr} 14\,\frac{1}{8} & = & 14\,\frac{1}{8} & = & 13\,\frac{9}{8} \\ -\,\boxed{\frac{3}{4}\cdot\frac{2}{2}} & = & -\,\frac{6}{8} & = & -\,\frac{6}{8} \\ \hline & & & & 13\,\frac{3}{8} \end{array}$$

(Since $\frac{1}{8}$ is smaller than $\frac{6}{8}$, we cannot subtract until we borrow: $14\frac{1}{8} = 13+\frac{8}{8}+\frac{1}{8} = 13+\frac{9}{8} = 13\frac{9}{8}$.)

35. The LCD is 18.

$$\begin{array}{rcrcr} 25\boxed{\frac{1}{9}\cdot\frac{2}{2}} & = & 25\,\frac{2}{18} & = & 24\,\frac{20}{18} \\ -13\boxed{\frac{5}{6}\cdot\frac{3}{3}} & = & -13\,\frac{15}{18} & = & -13\,\frac{15}{18} \\ \hline & & & & 11\,\frac{5}{18} \end{array}$$

(Since $\frac{2}{18}$ is smaller than $\frac{15}{18}$, we cannot subtract until we borrow: $25\frac{2}{18} = 24+\frac{18}{18}+\frac{2}{18} = 24+\frac{20}{18} = 24\frac{20}{18}$.)

37. $8 \cdot 2\frac{5}{6}$

$= \frac{8}{1}\cdot\frac{17}{6}$ Writing fractional notation

$= \frac{8\cdot 17}{1\cdot 6} = \frac{2\cdot 4\cdot 17}{1\cdot 2\cdot 3} = \frac{2}{2}\cdot\frac{4\cdot 17}{1\cdot 3} = \frac{68}{3} = 22\frac{2}{3}$

39. $3\frac{5}{8} \cdot \frac{2}{3}$

$= \frac{29}{8}\cdot\frac{2}{3}$ Writing fractional notation

$= \frac{29\cdot 2}{8\cdot 3} = \frac{29\cdot 2}{2\cdot 4\cdot 3} = \frac{2}{2}\cdot\frac{29}{4\cdot 3} = \frac{29}{12} = 2\frac{5}{12}$

41. $3\frac{1}{2}\cdot 2\frac{1}{3} = \frac{7}{2}\cdot\frac{7}{3} = \frac{49}{6} = 8\frac{1}{6}$

43. $3\frac{2}{5}\cdot 2\frac{7}{8} = \frac{17}{5}\cdot\frac{23}{8} = \frac{391}{40} = 9\frac{31}{40}$

45. $4\frac{7}{10}\cdot 5\frac{3}{10} = \frac{47}{10}\cdot\frac{53}{10} = \frac{2491}{100} = 24\frac{91}{100}$

47. $20\frac{1}{2}\cdot 10\frac{1}{5}\cdot 4\frac{2}{3} = \frac{41}{2}\cdot\frac{51}{5}\cdot\frac{14}{3} = \frac{41\cdot 51\cdot 14}{2\cdot 5\cdot 3} =$

$\frac{41\cdot 3\cdot 17\cdot 2\cdot 7}{2\cdot 5\cdot 3} = \frac{2\cdot 3}{2\cdot 3}\cdot\frac{41\cdot 17\cdot 7}{5} = \frac{4879}{5} = 975\frac{4}{5}$

49. $20 \div 3\frac{1}{5}$

$= 20\div\frac{16}{5}$ Writing fractional notation

$= 20\cdot\frac{5}{16}$ Multiplying by the reciprocal

$= \frac{20\cdot 5}{16} = \frac{4\cdot 5\cdot 5}{4\cdot 4} = \frac{4}{4}\cdot\frac{5\cdot 5}{4} = \frac{25}{4} = 6\frac{1}{4}$

51. $8\frac{2}{5}\div 7$

$= \frac{42}{5}\div 7$ Writing fractional notation

$= \frac{42}{5}\cdot\frac{1}{7}$ Multiplying by the reciprocal

$= \frac{42\cdot 1}{5\cdot 7} = \frac{6\cdot 7}{5\cdot 7} = \frac{7}{7}\cdot\frac{6}{5} = \frac{6}{5} = 1\frac{1}{5}$

53. $4\frac{3}{4}\div 1\frac{1}{3} = \frac{19}{4}\div\frac{4}{3} = \frac{19}{4}\cdot\frac{3}{4} = \frac{19\cdot 3}{4\cdot 4} = \frac{57}{16} = 3\frac{9}{16}$

55. $1\frac{7}{8}\div 1\frac{2}{3} = \frac{15}{8}\div\frac{5}{3} = \frac{15}{8}\cdot\frac{3}{5} = \frac{15\cdot 3}{8\cdot 5} = \frac{5\cdot 3\cdot 3}{8\cdot 5}$

$= \frac{5}{5}\cdot\frac{3\cdot 3}{8} = \frac{3\cdot 3}{8} = \frac{9}{8} = 1\frac{1}{8}$

57. $5\frac{1}{10}\div 4\frac{3}{10} = \frac{51}{10}\div\frac{43}{10} = \frac{51}{10}\cdot\frac{10}{43} = \frac{51\cdot 10}{10\cdot 43}$

$= \frac{10}{10}\cdot\frac{51}{43} = \frac{51}{43} = 1\frac{8}{43}$

59. $20\frac{1}{4} \div 90 = \frac{81}{4} \div 90 = \frac{81}{4} \cdot \frac{1}{90} = \frac{81 \cdot 1}{4 \cdot 90} = \frac{9 \cdot 9 \cdot 1}{4 \cdot 9 \cdot 10}$

$= \frac{9}{9} \cdot \frac{9 \cdot 1}{4 \cdot 10} = \frac{9}{40}$

61. ***Familiarize.*** Let c = the number of 16-oz cartons that were filled.

Translate.

Total number of ounces	divided by	Ounces in each carton	is	Number of cartons
↓	↓	↓	↓	↓
4578	$\div$	16	$=$	c

Solve. We carry out the division.

```
       2 8 6
  1 6 )4 5 7 8
       3 2 0 0
       1 3 7 8
       1 2 8 0
           9 8
           9 6
             2
```

Check. Multiply the number of cartons by 16 and then add the remainder.

$16 \cdot 286 = 4576, \qquad 4576 + 2 = 4578$

We get the total number of ounces of milk, so the answer checks.

State. 286 16-oz cartons were filled; 2 oz of milk was left over.

63. ◈

Exercise Set 2.5

1. ***Familiarize.*** Recall that area is length times width. We draw a picture. We will let A = the area of the table top.

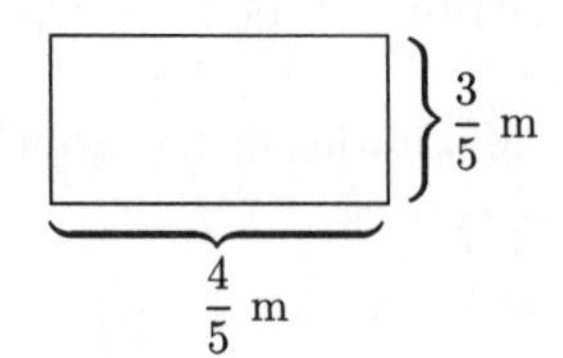

Translate. Then we translate.

Area	is	length	times	width
↓	↓	↓	↓	↓
A	$=$	$\frac{4}{5}$	$\times$	$\frac{3}{5}$

Solve. The sentence tells us what to do. We multiply.

$$\frac{4}{5} \times \frac{3}{5} = \frac{4 \times 3}{5 \times 5} = \frac{12}{25}$$

Check. We repeat the calculation. The answer checks.

State. The area is $\frac{12}{25}$ m^2.

3. ***Familiarize.*** We visualize the situation. We let a = the amount received for working $\frac{3}{4}$ of a day.

1 day $36	
3/4 day a	

Translate. We write an equation.

Pay for 3/4 of a day	is	$\frac{3}{4}$	of	$36
↓	↓	↓	↓	↓
a	$=$	$\frac{3}{4}$	$\cdot$	36

Solve. We carry out the multiplication.

$a = \frac{3}{4} \cdot 36 = \frac{3 \cdot 36}{4}$

$= \frac{3 \cdot 9 \cdot 4}{1 \cdot 4} = \frac{3 \cdot 9}{1} \cdot \frac{4}{4}$

$= 27$

Check. We can repeat the calculation. We can also determine that the answer seems reasonable since we multiplied 36 by a number less than 1 and the result is less than 36. The answer checks.

State. $27 is received for working $\frac{3}{4}$ of a day.

5. ***Familiarize.*** We draw a picture.

$\frac{2}{3}$ in.

1 in.
240 miles

We let n = the number of miles represented by $\frac{2}{3}$ in.

Translate. The multiplication sentence

$$n = \frac{2}{3} \cdot 240$$

corresponds to the situation.

Solve. We multiply and simplify:

$n = \frac{2}{3} \cdot 240 = \frac{2 \cdot 240}{3} = \frac{2 \cdot 3 \cdot 80}{1 \cdot 3}$

$= \frac{3}{3} \cdot \frac{2 \cdot 80}{1} = \frac{2 \cdot 80}{1}$

$= 160$

Check. We can repeat the calculation. We can also determine that the answer seems reasonable since we multiplied 240 by a number less than 1 and the result is less than 240.

State. $\frac{2}{3}$ in. on the map represents 160 miles.

7. ***Familiarize.*** We draw a picture. We let n = the number of pairs of basketball shorts that can be made.

$\frac{3}{4}$ yd	$\frac{3}{4}$ yd	$\cdots$	$\frac{3}{4}$ yd

n pairs of shorts

Translate. The multiplication that corresponds to the situation is

$$\frac{3}{4} \cdot n = 24.$$

Solve. We solve the equation by dividing on both sides by $\frac{3}{4}$ and carrying out the division:

$$n = 24 \div \frac{3}{4} = 24 \cdot \frac{4}{3} = \frac{24 \cdot 4}{3} = \frac{3 \cdot 8 \cdot 4}{3 \cdot 1} = \frac{3}{3} \cdot \frac{8 \cdot 4}{1}$$
$$= \frac{8 \cdot 4}{1} = 32$$

Check. We repeat the calculation. The answer checks.

State. 32 pairs of basketball shorts can be made from 24 yd of fabric.

9. ***Familiarize.*** This is a multistep problem. First we find the length of the total trip. Then we find how many kilometers were left to drive. We draw a picture. We let $n =$ the length of the total trip.

$\frac{5}{8}$ of the trip

180 km

n km

Translate. We translate to an equation.

Fraction of trip completed	times	Total length of trip	is	Amount already traveled
↓	↓	↓	↓	↓
$\frac{5}{8}$	$\cdot$	n	$=$	180

Solve. We solve the equation as follows:

$$\frac{5}{8} \cdot n = 180$$
$$n = 180 \div \frac{5}{8} = 180 \cdot \frac{8}{5} = \frac{5 \cdot 36 \cdot 8}{5 \cdot 1}$$
$$= \frac{5}{5} \cdot \frac{36 \cdot 8}{1} = \frac{36 \cdot 8}{1} = 288$$

The total trip was 288 km.

Now we find how many kilometers were left to travel. Let $t =$ this number.

Length of total trip	minus	Distance traveled	is	Distance left to travel
↓	↓	↓	↓	↓
288	$-$	180	$=$	t

We carry out the subtraction:

$$288 - 180 = t$$
$$108 = t$$

Check. We repeat the calculation. The results check.

State. The total trip was 288 km. There were 108 km left to travel.

11. ***Familiarize.*** We draw a picture. We let $D =$ the total distance Russ walked.

$\frac{7}{6}$ mi $\quad$ $\frac{3}{4}$ mi

D

Translate. An addition sentence corresponds to this situation.

Distance to friend's house	plus	Distance to class	is	Total distance
↓	↓	↓	↓	↓
$\frac{7}{6}$	$+$	$\frac{3}{4}$	$=$	D

Solve. To solve the equation, carry out the addition. Since $6 = 2 \cdot 3$ and $4 = 2 \cdot 2$, the LCM of the denominators is $2 \cdot 2 \cdot 3$, or 12.

$$\frac{7}{6} \cdot \frac{2}{2} + \frac{3}{4} \cdot \frac{3}{3} = D$$
$$\frac{14}{12} + \frac{9}{12} = D$$
$$\frac{23}{12} = D$$

Check. We repeat the calculation. We also note that the sum is larger than either of the original distances, so the answer seems reasonable.

State. Russ walked $\frac{23}{12}$ mi.

13. ***Familiarize.*** This is a multistep problem. First we find the total weight of the cubic meter of concrete mix. We visualize the situation, letting $w =$ the total weight.

420 kg	150 kg	120 kg
w		

Translate. An addition sentence corresponds to this situation.

Weight of cement	plus	Weight of stone	plus	Weight of sand	is	Total weight
↓	↓	↓	↓	↓	↓	↓
420	$+$	150	$+$	120	$=$	w

Solve. We carry out the addition.

$$420 + 150 + 120 = w$$
$$690 = w$$

Since the mix contains 420 kg of cement, the part that is cement is $\frac{420}{690} = \frac{14 \cdot 30}{23 \cdot 30} = \frac{14}{23} \cdot \frac{30}{30} = \frac{14}{23}$.

Since the mix contains 150 kg of stone, the part that is stone is $\frac{150}{690} = \frac{5 \cdot 30}{23 \cdot 30} = \frac{5}{23} \cdot \frac{30}{30} = \frac{5}{23}$.

Since the mix contains 120 kg of sand, the part that is sand is $\frac{120}{690} = \frac{4 \cdot 30}{23 \cdot 30} = \frac{4}{23} \cdot \frac{30}{30} = \frac{4}{23}$.

We add these amounts: $\frac{14}{23} + \frac{5}{23} + \frac{4}{23} =$

$\frac{14+5+4}{23} = \frac{23}{23} = 1$.

Check. We repeat the calculations. We also note that since the total of the fractional parts is 1, the answer is probably correct.

State. The total weight of the cubic meter of concrete mix is 690 kg. Of this, $\frac{14}{23}$ is cement, $\frac{5}{23}$ is stone, and $\frac{4}{23}$ is sand. The result when we add these amounts is 1.

15. ***Familiarize.*** We visualize the situation. Let t = the number of hours Monica listened to Brahms.

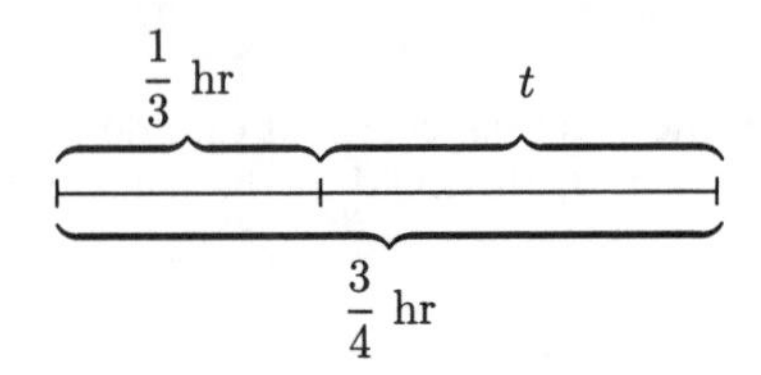

Translate. This is a "how much more" situation that can be translated as follows:

Time spent listening to Beethoven	plus	Time spent listening to Brahms	is	Total listening time
↓	↓	↓	↓	↓
$\frac{1}{3}$	$+$	t	$=$	$\frac{3}{4}$

Solve. We subtract $\frac{1}{3}$ on both sides of the equation.

$\frac{1}{3} + t - \frac{1}{3} = \frac{3}{4} - \frac{1}{3}$

$t + 0 = \frac{3}{4} \cdot \frac{3}{3} - \frac{1}{3} \cdot \frac{4}{4}$ The LCD is 12. We multiply by 1 to get the LCD.

$t = \frac{9}{12} - \frac{4}{12} = \frac{5}{12}$

Check. We return to the original problem and add.

$\frac{1}{3} + \frac{5}{12} = \frac{1}{3} \cdot \frac{4}{4} + \frac{5}{12} = \frac{4}{12} + \frac{5}{12} = \frac{9}{12} = \frac{3}{3} \cdot \frac{3}{4} = \frac{3}{4}$

The answer checks.

State. Monica spent $\frac{5}{12}$ hr listening to Brahms.

17. ***Familiarize.*** We visualize the situation. Let d = the number of inches by which the new tread depth exceeds the more typical depth.

$\frac{11}{32}$ in. d

$\frac{3}{8}$ in.

Translate. This is a "how much more" situation.

Typical tread depth	plus	Excess depth	is	New tread depth
↓	↓	↓	↓	↓
$\frac{11}{32}$	$+$	d	$=$	$\frac{3}{8}$

Solve. We subtract $\frac{11}{32}$ on both sides of the equation.

$\frac{11}{32} + d - \frac{11}{32} = \frac{3}{8} - \frac{11}{32}$

$d + 0 = \frac{3}{8} \cdot \frac{4}{4} - \frac{11}{32}$ The LCD is 32. We multiply by 1 to get the LCD.

$d = \frac{12}{32} - \frac{11}{32}$

$d = \frac{1}{32}$

Check. We return to the original problem and add.

$\frac{11}{32} + \frac{1}{32} = \frac{12}{32} = \frac{4 \cdot 3}{4 \cdot 8} = \frac{4}{4} \cdot \frac{3}{8} = \frac{3}{8}$

The answer checks.

State. The new tread depth is $\frac{1}{32}$ in. deeper than the more typical depth of $\frac{11}{32}$ in.

19. ***Familiarize.*** We let h = the woman's excess height.

Translate. We have a "how much more" situation.

Height of son	plus	How much more height	is	Height of woman
↓	↓	↓	↓	↓
$59\frac{7}{12}$	$+$	h	$=$	66

Solve. We solve the equation as follows:

$h = 66 - 59\frac{7}{12}$

$$\begin{array}{rcr} 66 & = & 65\frac{12}{12} \\ -\ 59\frac{7}{12} & = & -\ 59\frac{7}{12} \\ \hline & & 6\frac{5}{12} \end{array}$$

Check. We add the woman's excess height to her son's height:

$6\frac{5}{12} + 59\frac{7}{12} = 65\frac{12}{12} = 66$

The answer checks.

State. The woman is $6\frac{5}{12}$ in. taller.

21. ***Familiarize.*** Let d = the difference between the high and low prices of the stock.

Translate. We write an equation.

$$\underbrace{\text{High price}}_{\downarrow\; 37\frac{5}{8}} \;\underset{\downarrow\;-}{\text{minus}}\; \underbrace{\text{Low price}}_{\downarrow\; 20\frac{1}{2}} \;\underset{\downarrow\;=}{\text{is}}\; \underbrace{\text{Difference in prices}}_{\downarrow\; d}$$

Solve. To solve we carry out the subtraction. The LCD is 8.

$$\begin{array}{rcr} 37\,\frac{5}{8} & = & 37\,\frac{5}{8} \\ -\,20\boxed{\frac{1}{2}\cdot\frac{4}{4}} & = & -\,20\,\frac{4}{8} \\ \hline & & 17\,\frac{1}{8} \end{array}$$

Check. We add the difference in price to the low price:

$$17\frac{1}{8}+20\frac{1}{2}=17\frac{1}{8}+20\frac{4}{8}=37\frac{5}{8}$$

This checks.

State. The difference between the high and low prices of the stock was $\$17\frac{1}{8}$.

23. ***Familiarize.*** We let $w =$ the total weight of the meat.

Translate. We write an equation.

$$\underbrace{\text{Weight of one package}}_{\downarrow\; 1\frac{2}{3}} \;\underset{\downarrow\;+}{\text{plus}}\; \underbrace{\text{Weight of second package}}_{\downarrow\; 5\frac{3}{4}} \;\underset{\downarrow\;=}{\text{is}}\; \underbrace{\text{Total weight}}_{\downarrow\; w}$$

Solve. We carry out the addition. The LCD is 12.

$$\begin{array}{rcl} 1\boxed{\frac{2}{3}\cdot\frac{4}{4}} & = & 1\,\frac{8}{12} \\ +5\boxed{\frac{3}{4}\cdot\frac{3}{3}} & = & +5\,\frac{9}{12} \\ \hline & & 6\,\frac{17}{12} = 6+\frac{17}{12} \\ & & \quad = 6+1\frac{5}{12} \\ & & \quad = 7\frac{5}{12} \end{array}$$

Check. We repeat the calculation. We also note that the answer is larger than either of the individual weights, so the answer seems reasonable.

State. The total weight of the meat was $7\frac{5}{12}$ lb.

25. The length of each of the five sides is $5\frac{3}{4}$ yd. We add to find the distance around the figure.

$$5\frac{3}{4}+5\frac{3}{4}+5\frac{3}{4}+5\frac{3}{4}+5\frac{3}{4}=25\frac{15}{4}=25+3\frac{3}{4}=28\frac{3}{4}$$

The distance is $28\frac{3}{4}$ yd.

27. ***Familiarize.*** We make a drawing. We let $t =$ the number of hours the designer worked on the third day.

$$\vdash 2\frac{1}{2}\text{ hr} \;+\; 4\frac{1}{5}\text{ hr} \;+\; t \dashv$$

$$\vdash \quad 10\frac{1}{2}\text{ hr} \quad \dashv$$

Translate. We write an addition sentence.

$$2\frac{1}{2}+4\frac{1}{5}+t=10\frac{1}{2}$$

Solve. This is a two-step problem.

First we add $2\frac{1}{2}+4\frac{1}{5}$ to find the time worked on the first two days. The LCD is 10.

$$\begin{array}{rcr} 2\boxed{\frac{1}{2}\cdot\frac{5}{5}} & = & 2\,\frac{5}{10} \\ +\,4\boxed{\frac{1}{5}\cdot\frac{2}{2}} & = & +\,4\,\frac{2}{10} \\ \hline & & 6\,\frac{7}{10} \end{array}$$

Then we subtract $6\frac{7}{10}$ from $10\frac{1}{2}$ to find the time worked on the third day. The LCD is 10.

$$6\frac{7}{10}+t=10\frac{1}{2}$$

$$t=10\frac{1}{2}-6\frac{7}{10}$$

$$\begin{array}{rcrcr} 10\boxed{\frac{1}{2}\cdot\frac{5}{5}} & = & 10\,\frac{5}{10} & = & 9\,\frac{15}{10} \\ -\;6\,\frac{7}{10} & = & -\;6\,\frac{7}{10} & = & -\,6\,\frac{7}{10} \\ \hline & & & & 3\,\frac{8}{10} = 3\frac{4}{5} \end{array}$$

Check. We repeat the calculations.

State. The designer worked $3\frac{4}{5}$ hr the third day.

29. We see that d and the two smallest distances combined are the same as the largest distance. We translate and solve.

$$2\frac{3}{4}+d+2\frac{3}{4}=12\frac{7}{8}$$

$$d=12\frac{7}{8}-2\frac{3}{4}-2\frac{3}{4}$$

$$=10\frac{1}{8}-2\frac{3}{4}\quad \text{Subtracting } 2\frac{3}{4} \text{ from } 12\frac{7}{8}$$

$$=7\frac{3}{8}\quad \text{Subtracting } 2\frac{3}{4} \text{ from } 10\frac{1}{8}$$

The length of d is $7\frac{3}{8}$ ft.

31. ***Familiarize.*** We let $t =$ the number of inches of tape used in 60 sec of recording.

Translate. We write an equation.

$$\underbrace{\text{Inches per second}}_{1\frac{3}{8}} \cdot \underbrace{\text{Number of seconds}}_{60} = \underbrace{\text{Tape used}}_{t}$$

Solve. We carry out the multiplication.

$$t = 1\frac{3}{8}\cdot 60 = \frac{11}{8}\cdot 60$$
$$= \frac{11\cdot 4\cdot 15}{2\cdot 4} = \frac{11\cdot 15}{2}\cdot\frac{4}{4}$$
$$= \frac{165}{2} = 82\frac{1}{2}$$

Check. We repeat the calculation.

State. $82\frac{1}{2}$ in. of tape are used in 60 sec of recording in short-play mode.

33. ***Familiarize***. We let n = the number of cubic feet occupied by 250 lb of water.

Translate. We write an equation.

$$\underbrace{\text{Total weight}}_{250} \div \underbrace{\text{Weight per cubic foot}}_{62\frac{1}{2}} = \underbrace{\text{Number of cubic feet}}_{n}$$

Solve. To solve the equation we carry out the division.

$$n = 250 \div 62\frac{1}{2} = 250 \div \frac{125}{2}$$
$$= 250\cdot\frac{2}{125} = \frac{2\cdot 125\cdot 2}{125\cdot 1}$$
$$= \frac{125}{125}\cdot\frac{2\cdot 2}{1} = 4$$

Check. We repeat the calculation.

State. 4 cubic feet would be occupied.

35. ***Familiarize***. Visualize the situation as a rectangular array containing 24 hours with $1\frac{1}{2}$ hours in each row. We must determine how many rows the array has. (The last row may be incomplete.) We let n = the number of orbits made every 24 hours.

Translate. The division that corresponds to this situation is

$$24 \div 1\frac{1}{2} = n.$$

Solve. We carry out the division.

$$n = 24 \div 1\frac{1}{2} = 24 \div \frac{3}{2} = 24\cdot\frac{2}{3} = \frac{24\cdot 2}{3} = \frac{3\cdot 8\cdot 2}{3\cdot 1} =$$
$$\frac{3}{3}\cdot\frac{8\cdot 2}{1} = 16$$

Check. We check by multiplying the number of orbits by the time it takes to make one orbit.

$$16\cdot 1\frac{1}{2} = 16\cdot\frac{3}{2} = \frac{16\cdot 3}{2} = \frac{2\cdot 8\cdot 3}{2\cdot 1} = \frac{2}{2}\cdot\frac{8\cdot 3}{1} = 24$$

The answer checks.

State. The shuttle makes 16 orbits every 24 hr.

37. ***Familiarize***. We let m = the number of miles per gallon the car got.

Translate. We write an equation.

$$\underbrace{\text{Total number of miles traveled}}_{213} \div \underbrace{\text{Number of gallons of gas used}}_{14\frac{2}{10}} = \underbrace{\text{Miles per gallon}}_{m}$$

Solve. To solve the equation we carry out the division.

$$m = 213 \div 14\frac{2}{10} = 213 \div \frac{142}{10}$$
$$= 213\cdot\frac{10}{142} = \frac{3\cdot 71\cdot 2\cdot 5}{2\cdot 71\cdot 1}$$
$$= \frac{2\cdot 71}{2\cdot 71}\cdot\frac{3\cdot 5}{1} = 15$$

Check. We repeat the calculation.

State. The car got 15 miles per gallon of gas.

39. ***Familiarize***. Let c = the number of cassettes that can be placed on each shelf.

Translate. A division corresponds to this situation.

$$c = 27 \div 1\frac{1}{8}$$

Solve. We carry out the division.

$$c = 27 \div 1\frac{1}{8} = 27 \div \frac{9}{8} = 27\cdot\frac{8}{9} = \frac{27\cdot 8}{9} =$$
$$\frac{3\cdot 3\cdot 3\cdot 8}{3\cdot 3\cdot 1} = \frac{3\cdot 3}{3\cdot 3}\cdot\frac{3\cdot 8}{1} = \frac{24}{1} = 24$$

Check.We can check by multiplying the number of cassettes by the width of a cassette.

$$24\cdot 1\frac{1}{8} = 24\cdot\frac{9}{8} = \frac{24\cdot 9}{8} = \frac{3\cdot 8\cdot 3\cdot 3}{8\cdot 1} =$$
$$\frac{8}{8}\cdot\frac{3\cdot 3\cdot 3}{1} = 27$$

The answer checks.

State. Irene can place 24 cassettes on the shelf.

41. ***Familiarize***. The figure is composed of two rectangles. One has dimensions s by $\frac{1}{2}\cdot s$, or $6\frac{7}{8}$ in. by $\frac{1}{2}\cdot 6\frac{7}{8}$ in. The other has dimensions $\frac{1}{2}\cdot s$ by $\frac{1}{2}\cdot s$, or $\frac{1}{2}\cdot 6\frac{7}{8}$ in. by $\frac{1}{2}\cdot 6\frac{7}{8}$ in. The total area is the sum of the areas of these two rectangles. We let A = the total area.

Translate. We write an equation.

$$A = \left(6\frac{7}{8}\right)\cdot\left(\frac{1}{2}\cdot 6\frac{7}{8}\right) + \left(\frac{1}{2}\cdot 6\frac{7}{8}\right)\cdot\left(\frac{1}{2}\cdot 6\frac{7}{8}\right)$$

Solve. We carry out each multiplication and then add.

$$A = \left(6\frac{7}{8}\right)\cdot\left(\frac{1}{2}\cdot 6\frac{7}{8}\right)+\left(\frac{1}{2}\cdot 6\frac{7}{8}\right)\cdot\left(\frac{1}{2}\cdot 6\frac{7}{8}\right)$$
$$= \frac{55}{8}\cdot\left(\frac{1}{2}\cdot\frac{55}{8}\right)+\left(\frac{1}{2}\cdot\frac{55}{8}\right)\cdot\left(\frac{1}{2}\cdot\frac{55}{8}\right)$$
$$= \frac{55}{8}\cdot\frac{55}{16}+\frac{55}{16}\cdot\frac{55}{16}$$
$$= \frac{3025}{128}+\frac{3025}{256}=\frac{3025}{128}\cdot\frac{2}{2}+\frac{3025}{256}$$
$$= \frac{6050}{256}+\frac{3025}{256}=\frac{9075}{256}$$
$$= 35\frac{115}{256}$$

Check. We repeat the calculation.

State. The area is $35\frac{115}{256}$ sq in.

43. ***Familiarize***. We make a drawing.

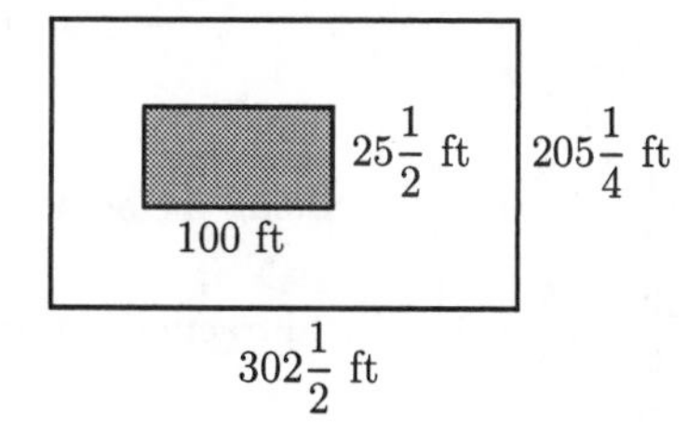

Translate. We let $A =$ the area of the lot not covered by the building.

Area left over	is	Area of lot	minus	Area of building
↓	↓	↓	↓	↓
A	$=$	$\left(302\frac{1}{2}\right)\cdot\left(205\frac{1}{4}\right)$	$-$	$(100)\cdot\left(25\frac{1}{2}\right)$

Solve. We do each multiplication and then find the difference.

$$A = \left(302\frac{1}{2}\right)\cdot\left(205\frac{1}{4}\right)-(100)\cdot\left(25\frac{1}{2}\right)$$
$$= \frac{605}{2}\cdot\frac{821}{4}-\frac{100}{1}\cdot\frac{51}{2}$$
$$= \frac{605\cdot 821}{2\cdot 4}-\frac{100\cdot 51}{1\cdot 2}$$
$$= \frac{605\cdot 821}{2\cdot 4}-\frac{2\cdot 50\cdot 51}{1\cdot 2}=\frac{605\cdot 821}{2\cdot 4}-\frac{2}{2}\cdot\frac{50\cdot 51}{1}$$
$$= \frac{496,705}{8}-2550=62,088\frac{1}{8}-2550$$
$$= 59,538\frac{1}{8}$$

Check. We repeat the calculation.

State. The area left over is $59,538\frac{1}{8}$ sq ft.

45. ***Familiarize***. We make a drawing. We let $A =$ the area.

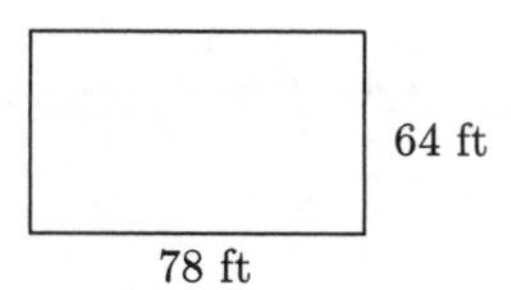

Translate. Using the formula for area, we have

$A = l\cdot w = 78\cdot 64.$

Solve. We carry out the multiplication.

$$\begin{array}{r} 78 \\ \times\ 64 \\ \hline 312 \\ 4680 \\ \hline 4992 \end{array}$$

Thus, $A = 4992$.

Check. We repeat the calculation. The answer checks.

State. The area is 4992 ft^2.

47.
$$\begin{array}{r} 34 \\ -\ 23 \\ \hline 11 \end{array}$$

49.
$$\begin{array}{r} \scriptstyle 7\ 9\ 13 \\ \not{8}\not{0}\not{3} \\ -\ 617 \\ \hline 186 \end{array}$$

51. $30\cdot x = 150$

$\frac{30\cdot x}{30}=\frac{150}{30}$ Dividing by 30 on both sides

$x = 5$

The solution is 5.

53. First, we solve an equation to find the portion of the tape used at the 4-hr speed.

Tape speed	times	Portion used	is	Time used
↓	↓	↓	↓	↓
4	$\cdot$	x	$=$	$\frac{1}{2}$

$$x = \frac{1}{2}\div 4$$
$$x = \frac{1}{2}\cdot\frac{1}{4}$$
$$x = \frac{1}{8}$$

Thus, $\frac{1}{8}$ of the tape was used at the 4-hr speed.

Now we solve an equation to find the portion of the tape used at the 2-hr speed.

Tape speed	times	Portion used	is	Time used
↓	↓	↓	↓	↓
2	$\cdot$	y	$=$	$\frac{3}{4}$

$$y = \frac{3}{4} \div 2$$
$$y = \frac{3}{4} \cdot \frac{1}{2}$$
$$y = \frac{3}{8}$$

Thus, $\frac{3}{8}$ of the tape is used at the 2-hr speed.

Next, we solve an equation to find the portion of the tape that is unused.

Portion used at 4-hr speed	+	Portion used at 2-hr speed	+	Unused portion	=	Entire tape
↓	↓	↓	↓	↓	↓	↓
$\frac{1}{8}$	+	$\frac{3}{8}$	+	p	=	1

$$p = 1 - \frac{1}{8} - \frac{3}{8}$$
$$p = \frac{4}{8} = \frac{1}{2}$$

One-half of the tape is unused.

To find how much time is left on the tape at the 6-hr speed, we solve an equation.

Unused portion	of	Tape speed	is	Time left
↓	↓	↓	↓	↓
$\frac{1}{2}$	$\cdot$	6	=	t

$$\frac{6}{2} = t$$
$$3 = t$$

There are three hours left at the 6-hr speed.

Exercise Set 2.6

1. $\frac{1}{2} \cdot \frac{1}{3} \cdot \frac{1}{4}$

$= \frac{1}{6} \cdot \frac{1}{4}$ Doing the multiplications in

$= \frac{1}{24}$ order from left to right

3. $\frac{5}{8} \div \frac{1}{4} - \frac{2}{3} \cdot \frac{4}{5}$

$= \frac{5}{8} \cdot \frac{4}{1} - \frac{2}{3} \cdot \frac{4}{5}$ Dividing

$= \frac{5 \cdot \not{4}}{2 \cdot \not{4} \cdot 1} - \frac{2}{3} \cdot \frac{4}{5}$

$= \frac{5}{2} - \frac{2}{3} \cdot \frac{4}{5}$ Removing a factor of 1

$= \frac{5}{2} - \frac{2 \cdot 4}{3 \cdot 5}$ Multiplying

$= \frac{5}{2} - \frac{8}{15}$

$= \frac{5}{2} \cdot \frac{15}{15} - \frac{8}{15} \cdot \frac{2}{2}$ Multiplying by 1 to obtain the LCD

$= \frac{75}{30} - \frac{16}{30}$

$= \frac{59}{30}$, or $1\frac{29}{30}$ Subtracting

5. $28\frac{1}{8} - 5\frac{1}{4} + 3\frac{1}{2}$

$= 28\frac{1}{8} - 5\frac{2}{8} + 3\frac{1}{2}$ Doing the additions and

$= 27\frac{9}{8} - 5\frac{2}{8} + 3\frac{1}{2}$ subtractions in order

$= 22\frac{7}{8} + 3\frac{1}{2}$ from left to right

$= 22\frac{7}{8} + 3\frac{4}{8}$

$= 25\frac{11}{8}$

$= 26\frac{3}{8}$, or $\frac{211}{8}$

7. $\frac{7}{8} \div \frac{1}{2} \cdot \frac{1}{4}$

$= \frac{7}{8} \cdot \frac{2}{1} \cdot \frac{1}{4}$ Dividing

$= \frac{7 \cdot \not{2}}{\not{2} \cdot 4 \cdot 1} \cdot \frac{1}{4}$

$= \frac{7}{4} \cdot \frac{1}{4}$ Removing a factor of 1

$= \frac{7}{16}$ Multiplying

9. $\left(\frac{2}{3}\right)^2 - \frac{1}{3} \cdot 1\frac{1}{4}$

$= \frac{4}{9} - \frac{1}{3} \cdot 1\frac{1}{4}$ Evaluating the exponental expression

$= \frac{4}{9} - \frac{1}{3} \cdot \frac{5}{4}$

$= \frac{4}{9} - \frac{5}{12}$ Multiplying

$= \frac{4}{9} \cdot \frac{4}{4} - \frac{5}{12} \cdot \frac{3}{3}$

$= \frac{16}{36} - \frac{15}{36}$

$= \frac{1}{36}$ Subtracting

11. $\frac{1}{2} - \left(\frac{1}{2}\right)^2 + \left(\frac{1}{2}\right)^3$

$= \frac{1}{2} - \frac{1}{4} + \frac{1}{8}$ Evaluating the exponental expressions

$= \frac{2}{4} - \frac{1}{4} + \frac{1}{8}$ Doing the additions and

$= \frac{1}{4} + \frac{1}{8}$ subtractions in order

$= \frac{2}{8} + \frac{1}{8}$ from left to right

$= \frac{3}{8}$

13. Add the numbers and divide by the number of addends.

$\left(\frac{1}{6} + \frac{1}{8} + \frac{3}{4}\right) \div 3$

$= \left(\frac{4}{24} + \frac{3}{24} + \frac{18}{24}\right) \div 3$

$= \frac{25}{24} \div 3$

$= \frac{25}{24} \cdot \frac{1}{3}$

$= \frac{25}{72}$

15. Add the numbers and divide by the number of addends.

$\left(3\frac{1}{2} + 9\frac{3}{8}\right) \div 2$

$= \left(3\frac{4}{8} + 9\frac{3}{8}\right) \div 2$

$= 12\frac{7}{8} \div 2$

$= \frac{103}{8} \div 2$

$= \frac{103}{8} \cdot \frac{1}{2}$

$= \frac{103}{16}$, or $6\frac{7}{16}$

17. $\left(\frac{1}{2} + \frac{1}{3}\right)^2 \cdot 144 - \frac{5}{8} \div 10\frac{1}{2}$

$= \left(\frac{3}{6} + \frac{2}{6}\right)^2 \cdot 144 - \frac{5}{8} \div 10\frac{1}{2}$

$= \left(\frac{5}{6}\right)^2 \cdot 144 - \frac{5}{8} \div 10\frac{1}{2}$

$= \frac{25}{36} \cdot 144 - \frac{5}{8} \div 10\frac{1}{2}$

$= \frac{25 \cdot \cancel{36} \cdot 4}{\cancel{36} \cdot 1} - \frac{5}{8} \div 10\frac{1}{2}$

$= 100 - \frac{5}{8} \div 10\frac{1}{2}$

$= 100 - \frac{5}{8} \div \frac{21}{2}$

$= 100 - \frac{5}{8} \cdot \frac{2}{21}$

$= 100 - \frac{5 \cdot \cancel{2}}{\cancel{2} \cdot 4 \cdot 21}$

$= 100 - \frac{5}{84}$

$= 99\frac{79}{84}$, or $\frac{8395}{84}$

19. $\frac{2}{47}$

Because 2 is very small compared to 47, $\frac{2}{47} \approx 0$.

21. $\frac{7}{100}$

Because 7 is very small compared to 100, $\frac{7}{100} \approx 0$.

23. $\frac{6}{11}$

Because $2 \cdot 6 = 12$ and 12 is close to 11, the denominator is about twice the numerator. Thus, $\frac{6}{11} \approx \frac{1}{2}$.

25. $\frac{\square}{11}$

A fraction is close to $\frac{1}{2}$ when the denominator is about twice the numerator. Since $2 \cdot 5 = 10$ and $2 \cdot 6 = 12$ and both 10 and 12 are close to 11, we know that $\frac{5}{11}$ and $\frac{6}{11}$ are both close to $\frac{1}{2}$. We also want a fraction that is greater than $\frac{1}{2}$, so we choose 6 for the numerator and obtain $\frac{6}{11}$. Answers may vary.

27. $\frac{\square}{23}$

A fraction is close to $\frac{1}{2}$ when the denominator is about twice the numerator. Since $2 \cdot 11 = 22$ and $2 \cdot 12 = 24$, and both 22 and 24 are close to 23, we know that $\frac{11}{23}$ and $\frac{12}{23}$ are both close to $\frac{1}{2}$. We also want a fraction that is greater than $\frac{1}{2}$, so we choose 12 for the numerator and obtain $\frac{12}{23}$. Answers may vary.

29. $\frac{8}{\square}$

A fraction is close to $\frac{1}{2}$ when the denominator is about twice the numerator. Since $2 \cdot 8 = 16$, we know a number close to 16 will yield a fraction close to $\frac{1}{2}$. We also want a fraction that is greater than $\frac{1}{2}$, we can choose 15 for the denominator and obtain $\frac{8}{15}$. Answers may vary.

31. $\frac{7}{\square}$

If the denominator were 7, the fraction would be equivalent to 1. Then a denominator of 6 will make the fraction close to but greater than 1. Answers may vary.

33. $\frac{13}{\square}$

If the denominator were 13, the fraction would be equivalent to 1. Then a denominator of 12 will make the fraction close to but greater than 1. Answers may vary.

35. $\frac{\square}{18}$

If the numerator were 18, the fraction would be equivalent to 1. Then a numerator of 19 will make the fraction close to but greater than 1. Answers may vary.

37. $2\frac{7}{8}$

Since $\frac{7}{8} \approx 1$, we have $2\frac{7}{8} = 2 + \frac{7}{8} \approx 2 + 1$, or 3.

39. $\frac{2}{3} + \frac{7}{13} + \frac{5}{9} \approx 1 + \frac{1}{2} + \frac{1}{2} = 2$

41. $24 \div 7\frac{8}{9} \approx 24 \div 8 = 3$

43. $76\frac{3}{14} + 23\frac{19}{20} \approx 76 + 24 = 100$

45. $\frac{43}{100} + \frac{1}{10} - \frac{11}{1000} \approx \frac{1}{2} + 0 - 0 = \frac{1}{2}$

47. $7\frac{29}{80} + 10\frac{12}{13} \cdot 24\frac{2}{17} \approx 7\frac{1}{2} + 11 \cdot 24 =$

$7\frac{1}{2} + 264 = 271\frac{1}{2}$

49. $16\frac{1}{5} \div 2\frac{1}{11} + 25\frac{9}{10} - 4\frac{11}{23} \approx$

$16 \div 2 + 26 - 4\frac{1}{2} = 8 + 26 - 4\frac{1}{2} =$

$34 - 4\frac{1}{2} = 29\frac{1}{2}$

51. ***Familiarize.*** This is a multistep problem. First we find the total amount of the checks. Then we find the amount left in the account after the checks are written. Let $t =$ the total amount of the two checks.

Translate.

$$\underbrace{\text{Amount of first check}}\ \text{plus}\ \underbrace{\text{Amount of second check}}\ \text{is}\ \underbrace{\text{Total amount of checks}}$$

$$\downarrow \qquad \downarrow \qquad \downarrow \qquad \downarrow \qquad \downarrow$$

$$329 \quad + \quad 52 \quad = \quad t$$

Solve. We carry out the addition.

$$\begin{array}{r} \scriptstyle 1 \\ 3\,2\,9 \\ +\ \ 5\,2 \\ \hline 3\,8\,1 \end{array}$$

Thus, $381 = t$, or $t = 381$.

Now let $a =$ the amount left in the account.

$$\underbrace{\text{Original amount}}\ \text{less}\ \underbrace{\text{Check total}}\ \text{is}\ \underbrace{\text{Amount left}}$$

$$\downarrow \qquad \downarrow \qquad \downarrow \qquad \downarrow \qquad \downarrow$$

$$3458 \quad - \quad 381 \quad = \quad a$$

We carry out the subtraction.

$$\begin{array}{r} \scriptstyle 3\ 15\ \ \\ 3\,\not{4}\,\not{5}\,8 \\ -3\,8\,1 \\ \hline 3\,0\,7\,7 \end{array}$$

We have $3077 = a$, or $a = 3077$.

Check. We can repeat the calculations. The answer checks.

State. There is \$3077 left in the account.

53.
$$\begin{array}{r} \scriptstyle 12\ \ \ \\ \scriptstyle 1\ 9\ \not{2}\ 17 \\ \not{2}\,\not{0}\,\not{3}\,\not{7} \\ -\,1\,1\,8\,9 \\ \hline 8\,4\,8 \end{array}$$

55.
$$\begin{array}{r} \scriptstyle 16\ 10\ 10\ \ \ \\ \scriptstyle 5\ \not{6}\ \ \not{0}\ \ \not{0}\ 13 \\ \not{6}\,\not{7},\,\not{1}\,\not{1}\,\not{3} \\ -\,2\,9,\,8\,7\,4 \\ \hline 3\,7,\,2\,3\,9 \end{array}$$

57. ◈

59. Use a calculator to find decimal notation for each fraction.

$\frac{3}{4} = 0.75$; $\frac{17}{21} \approx 0.8095$; $\frac{13}{15} \approx 0.8667$; $\frac{7}{9} \approx 0.7778$;

$\frac{15}{17} \approx 0.8824$; $\frac{13}{12} \approx 1.0833$; $\frac{19}{22} \approx 0.8636$

Now arrange the fractions in order from smallest to largest.

$\frac{3}{4}, \frac{7}{9}, \frac{17}{21}, \frac{19}{22}, \frac{13}{15}, \frac{15}{17}, \frac{13}{12}$

Chapter 3

Decimal Notation

Exercise Set 3.1

1. 499.06

a) Write a word name for the whole number. — Four hundred forty-nine

b) Write "and" for the decimal point. — Four hundred forty-nine and

c) Write a word name for the number to the right of the decimal point, followed by the place value of the last digit. — Four hundred forty-nine and six hundredths

A word name for 449.06 is four hundred forty-nine and six hundredths.

3. $1.5599

a) Write a word name for the whole number. — One

b) Write "and" for the decimal point. — One and

c) Write a word name for the number to the right of the decimal point, followed by the place value of the last digit. — One and five thousand five hundred ninety-nine ten thousandths

A word name for 1.5599 is one and five thousand five hundred ninety-nine ten thousandths.

5. Thirty-four — and — eight hundred ninety-one thousandths

34 . 891

7. Write "and 48 cents" as "and $\frac{48}{100}$ dollars." A word name for $326.48 is three hundred twenty-six and $\frac{48}{100}$ dollars.

9. Write "and 72 cents" as "and $\frac{72}{100}$ dollars." A word name for $36.72 is thirty-six and $\frac{72}{100}$ dollars.

11. 8.3 (1 place) → 8.3. (Move 1 place.) → $\frac{83}{10}$ (1 zero)

$8.3 = \frac{83}{10}$

13. 3.56 (2 places) → 3.56. (Move 2 places.) → $\frac{356}{100}$ (2 zeros)

$3.56 = \frac{356}{100}$

15. 46.03 (2 places) → 46.03. (Move 2 places.) → $\frac{4603}{100}$ (2 zeros)

$46.03 = \frac{4603}{100}$

17. 0.00013 (5 places) → 0.00013. (Move 5 places.) → $\frac{13}{100,000}$ (5 zeros)

$0.00013 = \frac{13}{100,000}$

19. 1.0008 (4 places) → 1.0008. (Move 4 places.) → $\frac{10,008}{10,000}$ (4 zeros)

$1.0008 = \frac{10,008}{10,000}$

21. 20.003 (3 places) → 20.003. (Move 3 places.) → $\frac{20,003}{1000}$ (3 zeros)

$20.003 = \frac{20,003}{1000}$

23. $\frac{8}{10}$ (1 zero) → 0.8. (Move 1 place.)

$\frac{8}{10} = 0.8$

25. $\frac{889}{100}$ (2 zeros) → 8.89. (Move 2 places.)

$\frac{889}{100} = 8.89$

27. $\frac{3798}{1000}$ 3.798.

3 zeros   Move 3 places.

$\frac{3798}{1000} = 3.798$

29. $\frac{78}{10,000}$ 0.0078.

4 zeros   Move 4 places.

$\frac{78}{10,000} = 0.0078$

31. $\frac{19}{100,000}$ 0.00019.

5 zeros   Move 5 places.

$\frac{19}{100,000} = 0.00019$

33. $\frac{376,193}{1,000,000}$ 0.376193.

6 zeros   Move 6 places.

$\frac{376,193}{1,000,000} = 0.376193$

35. $99\frac{44}{100} = 99 + \frac{44}{100} = 99 \text{ and } \frac{44}{100} = 99.44$

37. $3\frac{798}{1000} = 3 + \frac{798}{1000} = 3 \text{ and } \frac{798}{1000} = 3.798$

39. $2\frac{1739}{10,000} = 2 + \frac{1739}{10,000} = 2 \text{ and } \frac{1739}{10,000} = 2.1739$

41. $8\frac{953,073}{1,000,000} = 8 + \frac{953,073}{1,000,000} =$

$8 \text{ and } \frac{953,073}{1,000,000} = 8.953073$

43. To compare two numbers in decimal notation, start at the left and compare corresponding digits moving from left to right. When two digits differ, the number with the larger digit is the larger of the two numbers.

0.06

Different; 5 is larger than 0.

0.58

Thus, 0.58 is larger.

45. 0.905

Starting at the left, these digits are the first to differ; 1 is larger than 0.

0.91

Thus, 0.91 is larger.

47. 0.0009

Starting at the left, these digits are the first to differ, and 1 is larger than 0.

0.001

Thus, 0.001 is larger.

49. 234.07

Starting at the left, these digits are the first to differ, and 5 is larger than 4.

235.07

Thus, 235.07 is larger.

51. $\frac{4}{100} = 0.04$ so we compare 0.004 and 0.04.

0.004

Starting at the left, these digits are the first to differ, and 4 is larger than 0.

0.04

Thus, 0.04 or $\frac{4}{100}$ is larger.

53. 0.4320

Starting at the left, these digits are the first to differ, and 5 is larger than 0.

0.4325

Thus, 0.4325 is larger.

55.

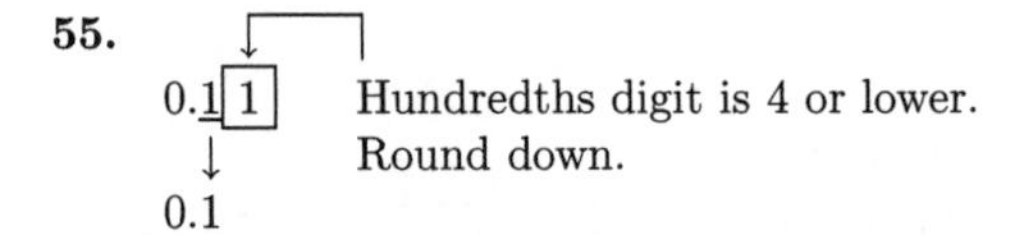

0.1 1   Hundredths digit is 4 or lower.
Round down.

0.1

57. 0.4 9   Hundredths digit is 5 or higher.
Round up.

0.5

59. 2.7 4 49   Hundredths digit is 4 or lower.
Round down.

2.7

61. 123.6 5   Hundredths digit is 5 or higher.
Round up.

123.7

63. 0.89 3   Thousandths digit is 4 or lower.
Round down.

0.89

65. 0.66 6 6   Thousandths digit is 5 or higher.
Round up.

0.67

67. 0.99 5   Thousandths digit is 5 or higher.
Round up.

1.00

(When we make the hundredths digit a 10, we carry 1 to the tenths place. This then requires us to carry 1 to the ones place.)

69. 0.09|4| Thousandths digit is 4 or lower. Round down.
↓
0.09

71. 0.324|6| Ten-thousandths digit is 5 or higher. Round up.
↓
0.325

73. 17.001|5| Ten-thousandths digit is 5 or higher. Round up.
↓
17.002

75. 10.101|1| Ten-thousandths digit is 4 or lower. Round down.
↓
10.101

77. 9.998|9| Ten-thousandths digit is 5 or higher. Round up.
↓
9.999

79. 8|0|9.4732 Tens digit is 4 or lower. Round down.
↓
800

81. 809.473|2| Ten-thousandths digit is 4 or lower. Round down.
↓
809.473

83. 809.|4|732 Tenths digit is 4 or lower. Round down.
↓
809

85. 34.5438|9| Hundred-thousandths digit is 5 or higher. Round up.
↓
34.5439

87. 34.54|3|89 Thousandths digit is 4 or lower. Round down.
↓
34.54

89. 34.|5|4389 Tenths digit is 5 or higher. Round up.
↓
35

91. Round 617|2| to the nearest ten.
↑

The digit 7 is in the tens place. Since the next digit to the right, 2, is 4 or lower, round down, meaning that 7 tens stays as 7 tens. Then change the digit to the right of the tens digit to zero.

The answer is 6170.

93. Round 6|1|72 to the nearest thousand.
↑

The digit 6 is in the thousands place. Since the next digit to the right, 1, is 4 or lower, round down, meaning that 6 thousands stays as 6 thousands. Then change all digits to the right of the thousands digit to zeros.

The answer is 6000.

95. $\frac{681}{1000}+\frac{149}{1000}=\frac{830}{1000}=\frac{83\cdot 10}{100\cdot 10}=\frac{83}{100}\cdot\frac{10}{10}=\frac{83}{100}$

97. $\frac{267}{100}-\frac{85}{100}=\frac{182}{100}=\frac{91\cdot 2}{50\cdot 2}=\frac{91}{50}\cdot\frac{2}{2}=\frac{91}{50}$

99. ◈

101. 6.78346|1902| ←Drop all decimal places past the fifth place.
↓
6.78346

103. 0.03030|3030303| ←Drop all decimal places past the fifth place.
↓
0.03030

Exercise Set 3.2

1.

$$\begin{array}{r} {}^{1} \\ 316.25 \\ +\ \ 18.12 \\ \hline 334.37 \end{array}$$

Add hundredths.
Add tenths.
Write a decimal point in the answer.
Add ones.
Add tens.
Add hundreds.

3.

$$\begin{array}{r} 659.403 \\ +\ 916.812 \\ \hline 1576.215 \end{array}$$

Add thousandths.
Add hundredths.
Add tenths.
Write a decimal point in the answer.
Add ones.
Add tens.
Add hundreds.

5.

$$\begin{array}{r} 9.104 \\ +\ 123.456 \\ \hline 132.560 \end{array}$$

7.

$$\begin{array}{r} 81.008 \\ +\ \ 3.409 \\ \hline 84.417 \end{array}$$

9.

$$\begin{array}{r} 20.0124 \\ +\ 30.0124 \\ \hline 50.0248 \end{array}$$

11. Line up the decimal points.

$$\begin{array}{r} 39.000 \\ +\ \ 1.007 \\ \hline 40.007 \end{array}$$

Writing 2 extra zeros

13. Line up the decimal points.

```
         1
      0.3 4 0      Writing an extra zero
      3.5 0 0      Writing 2 extra zeros
      0.1 2 7
+ 7 6 8.0 0 0      Writing in the decimal point
-------------      and 3 extra zeros
  7 7 1.9 6 7      Adding
```

15.

```
  1   1 1
  1 7.0 0 0 0      Writing in the decimal point.
    3.2 4 0 0      You may find it helpful to
    0.2 5 6 0      write extra zeros.
+   0.3 6 8 9
-------------
  2 0.8 6 4 9
```

17.

```
  1 2 1     1
        2.7 0 3 0
      7 8.3 3 0 0
      2 8.0 0 0 9
+ 1 1 8.4 3 4 1
---------------
  2 2 7.4 6 8 0
```

19.

```
  1 2 1     1
          9 9.6 0 0 1
    7 2 8 5.1 8 0 0
      5 0 0.0 4 2 0
+     8 7 0.0 0 0 0
-------------------
  8 7 5 4.8 2 2 1
```

21.

```
  4 12
  5.2      Borrow ones to subtract tenths.
- 3.9      Subtract tenths.
-----
  1.3      Write a decimal point in the answer.
           Subtract ones.
```

23.

```
  4 11 2 11    Borrow tenths to subtract hundredths.
  5 1 .3 1     Subtract hundredths.
-   2 .2 9     Subtract tenths.
----------
  4 9 .0 2     Write a decimal point in the answer.
               Borrow tens to subtract ones.
               Subtract ones.
               Subtract tens.
```

25.

```
  4 8.7 6
-   3.1 5
---------
  4 5.6 1
```

27.

```
      11
  8   1  13
  9 2 .3 4 1
-   6 .4 2
------------
  8 5 .9 2 1
```

29.

```
     4 9 9 10
  2. 5 0 0 0     Writing 3 extra zeros
- 0. 0 0 2 5
------------
  2. 4 9 7 5
```

31.

```
     3 9 10
  3. 4 0 0     Writing 2 extra zeros
- 0. 0 0 3
----------
  3. 3 9 7
```

33. Line up the decimal points. Write an extra zero if desired.

```
    17 11
  1 7  1 10
  2 8 .2 0
- 1 9 .3 5
----------
    8 .8 5
```

35.

```
   3 10
  3 4.0 7
- 3 0.7
--------
    3.3 7
```

37.

```
      4 10
  8.4 5 0
- 7.4 0 5
---------
  1.0 4 5
```

39.

```
  5 10
  6.0 0 3
- 2.3
--------
  3.7 0 3
```

41.

```
     9  9  9 10
  1. 0  0  0 0     Writing in the decimal point
- 0. 0  0  9 8     and 4 extra zeros
--------------
  0. 9  9  0 2     Subtracting
```

43.

```
     9 9   9 10
  1  0 0.  0 0
-        0. 3 4
--------------
       9 9. 6 6
```

45.

```
  6 14
  7. 4 8
- 2. 6
--------
  4. 8 8
```

47.

```
    2 9 9 10
    3. 0 0 0
  - 2. 0 0 6
  ----------
    0. 9 9 4
```

49.

```
     8  9 9 10
    1 9. 0 0 0
  -   1. 1 9 8
  ------------
    1 7. 8 0 2
```

51.

```
      4 9 10
    6 5. 0 0
  - 1 3. 8 7
  ----------
    5 1. 1 3
```

53.

```
    8 10
  3. 9 0 7
- 1. 4 1 6
----------
  2. 4 9 1
```

55.

```
            8 17
  3 2. 7 9  7 8
-   0. 0 5  9 2
---------------
  3 2. 7 3  8 6
```

57.

```
  2  9 10 6 14
  3. 0  0 7 4
- 1. 3  4 0 8
-------------
  1. 6  6 6 6
```

59.

```
            18
          4  8  9 17
  2 3 4 5. 9 0  7 8 6
-          0. 9 9 9
---------------------
  2 3 4 4. 9 0  8 8 6
```

61.
$$x + 17.5 = 29.15$$
$$x + 17.5 - 17.5 = 29.15 - 17.5 \qquad \text{Subtracting 17.5 on both sides}$$
$$x = 11.65$$
$$\begin{array}{r} 2\,\overset{8}{\not 9}.\,\overset{11}{\not 1}\,5 \\ -\,1\,7.\,5 \\ \hline 1\,1.\,6\,5 \end{array}$$

63.
$$3.205 + m = 22.456$$
$$3.205 + m - 3.205 = 22.456 - 3.205 \qquad \text{Subtracting 3.205 on both sides}$$
$$m = 19.251$$
$$\begin{array}{r} \overset{1}{\not 2}\,\overset{12}{\not 2}.\,4\,5\,6 \\ -\quad 3.\,2\,0\,5 \\ \hline 1\,9.\,2\,5\,1 \end{array}$$

65.
$$17.95 + p = 402.63$$
$$17.95 + p - 17.95 = 402.63 - 17.95 \qquad \text{Subtracting 17.95 on both sides}$$
$$p = 384.68$$
$$\begin{array}{r} \overset{3}{\not 4}\,\overset{9}{\not 0}\,\overset{\overset{11}{\not 1}}{\not 2}.\,\overset{\overset{15}{\not 5}}{\not 6}\,\overset{13}{\not 3} \\ -\quad 1\,7.\,9\,5 \\ \hline 3\,8\,4.\,6\,8 \end{array}$$

67.
$$13,083.3 = x + 12,500.33$$
$$13,083.3 - 12,500.33 = x + 12,500.33 - 12,500.33 \qquad \text{Subtracting 12,500.33 on both sides}$$
$$582.97 = x$$
$$\begin{array}{r} 1\,\overset{2}{\not 3},\,\overset{10}{\not 0}\,8\,\overset{2}{\not 3}.\,\overset{\overset{12}{\not 2}}{\not 3}\,\overset{10}{\not 0} \\ -\,1\,2,\,5\,0\,0.\,3\,3 \\ \hline 5\,8\,2.\,9\,7 \end{array}$$

69.
$$x + 2349 = 17,684.3$$
$$x + 2349 - 2349 = 17,684.3 - 2349 \qquad \text{Subtracting 2349 on both sides}$$
$$x = 15,335.3$$
$$\begin{array}{r} 1\,7,\,6\,\overset{7}{\not 8}\,\overset{14}{\not 4}.\,3 \\ -\quad 2\,3\,4\,9.\,0 \\ \hline 1\,5,\,3\,3\,5.\,3 \end{array}$$

71. 34,[5]67 — Hundreds digit is 5 or higher. Round up.
↓
35,000

73.
$$\frac{13}{24} - \frac{3}{8} = \frac{13}{24} - \frac{3}{8}\cdot\frac{3}{3}$$
$$= \frac{13}{24} - \frac{9}{24}$$
$$= \frac{13-9}{24} = \frac{4}{24}$$
$$= \frac{4\cdot 1}{4\cdot 6} = \frac{4}{4}\cdot\frac{1}{6}$$
$$= \frac{1}{6}$$

75.
$$\begin{array}{r} 8\,\overset{7}{\not 8}\,\overset{9}{\not 0}\,\overset{15}{\not 5} \\ -\,2\,6\,3\,9 \\ \hline 6\,1\,6\,6 \end{array}$$

77. ***Familiarize***. We draw a picture.

$\frac{1}{3}$ lb	$\frac{1}{3}$ lb	....	$\frac{1}{3}$ lb

←——— $5\frac{1}{2}$ lb ———→

We let s = the number of servings that can be prepared from $5\frac{1}{2}$ lb of flounder fillet.

Translate. The situation corresponds to a division sentence.
$$s = 5\frac{1}{2} \div \frac{1}{3}$$

Solve. We carry out the division.
$$s = 5\frac{1}{2} \div \frac{1}{3} = \frac{11}{2} \div \frac{1}{3}$$
$$= \frac{11}{2}\cdot\frac{3}{1} = \frac{33}{2}$$
$$= 16\frac{1}{2}$$

Check. We check by multiplying. If $16\frac{1}{2}$ servings are prepared, then
$$16\frac{1}{2}\cdot\frac{1}{3} = \frac{33}{2}\cdot\frac{1}{3} = \frac{3\cdot 11\cdot 1}{2\cdot 3} = \frac{3}{3}\cdot\frac{11\cdot 1}{2} = \frac{11}{2} = 5\frac{1}{2}\text{ lb}$$
of flounder is used. Our answer checks.

State. $16\frac{1}{2}$ servings can be prepared from $5\frac{1}{2}$ lb of flounder fillet.

79.

81. First, "undo" the incorrect addition by subtracting 235.7 from the incorrect answer:
$$\begin{array}{r} 8\,1\,7.\,2 \\ -\,2\,3\,5.\,7 \\ \hline 5\,8\,1.\,5 \end{array}$$

The original minuend was 581.5. Now subtract 235.7 from this as the student originally intended:
$$\begin{array}{r} 5\,8\,1.\,5 \\ -\,2\,3\,5.\,7 \\ \hline 3\,4\,5.\,8 \end{array}$$

The correct answer is 345.8.

Exercise Set 3.3

1.
$$\begin{array}{rl} 8.\,6 & \text{(1 decimal place)} \\ \times\quad 7 & \text{(0 decimal places)} \\ \hline 6\,0.\,2 & \text{(1 decimal place)} \end{array}$$

3.
$$\begin{array}{rl} 0.\,8\,4 & \text{(2 decimal places)} \\ \times\quad 8 & \text{(0 decimal places)} \\ \hline 6.\,7\,2 & \text{(2 decimal places)} \end{array}$$

5.
$$\begin{array}{rl} 6.\,3 & \text{(1 decimal place)} \\ \times\,0.\,0\,4 & \text{(2 decimal places)} \\ \hline 0.\,2\,5\,2 & \text{(3 decimal places)} \end{array}$$

7.
$$\begin{array}{rl} 8\,7 & \text{(0 decimal places)} \\ \times\,0.\,0\,0\,6 & \text{(3 decimal places)} \\ \hline 0.\,5\,2\,2 & \text{(3 decimal places)} \end{array}$$

9. $1\underline{0} \times 23.76$ 23.7.6

1 zero Move 1 place to the right.

$10 \times 23.76 = 237.6$

11. $1\underline{000} \times 583.686852$ 583.686.852

3 zeros Move 3 places to the right.

$1000 \times 583.686852 = 583,686.852$

13. $7.8 \times 1\underline{00}$ 7.80.

2 zeros Move 2 places to the right.

$7.8 \times 100 = 780$

15. $0.\underline{1} \times 89.23$ 8.9.23

1 decimal place Move 1 place to the left.

$0.1 \times 89.23 = 8.923$

17. $0.\underline{001} \times 97.68$ 0.097.68

3 decimal places Move 3 places to the left.

$0.001 \times 97.68 = 0.09768$

19. $78.2 \times 0.\underline{01}$ 0.78.2

2 decimal places Move 2 places to the left.

$78.2 \times 0.01 = 0.782$

21.
$$\begin{array}{rl} 32.6 & \text{(1 decimal place)} \\ \times\ 16 & \text{(0 decimal places)} \\ \hline 1956 & \\ 3260 & \\ \hline 521.6 & \text{(1 decimal place)} \end{array}$$

23.
$$\begin{array}{rl} 0.984 & \text{(3 decimal places)} \\ \times\ 3.3 & \text{(1 decimal place)} \\ \hline 2952 & \\ 29520 & \\ \hline 3.2472 & \text{(4 decimal places)} \end{array}$$

25.
$$\begin{array}{rl} 374 & \text{(0 decimal places)} \\ \times\ 2.4 & \text{(1 decimal place)} \\ \hline 1496 & \\ 7480 & \\ \hline 897.6 & \text{(1 decimal place)} \end{array}$$

27.
$$\begin{array}{rl} 749 & \text{(0 decimal places)} \\ \times\ 0.43 & \text{(2 decimal places)} \\ \hline 2247 & \\ 29960 & \\ \hline 322.07 & \text{(2 decimal places)} \end{array}$$

29.
$$\begin{array}{rl} 0.87 & \text{(2 decimal places)} \\ \times\ 64 & \text{(0 decimal places)} \\ \hline 348 & \\ 5220 & \\ \hline 55.68 & \text{(2 decimal places)} \end{array}$$

31.
$$\begin{array}{rl} 46.50 & \text{(2 decimal places)} \\ \times\ 75 & \text{(0 decimal places)} \\ \hline 23250 & \\ 325500 & \\ \hline 3487.50 & \text{(2 decimal places)} \end{array}$$

Since the last decimal place is 0, we could also write this answer as 3487.5.

33.
$$\begin{array}{rl} 81.7 & \text{(1 decimal place)} \\ \times\ 0.612 & \text{(3 decimal places)} \\ \hline 1634 & \\ 8170 & \\ 490200 & \\ \hline 50.0004 & \text{(4 decimal places)} \end{array}$$

35.
$$\begin{array}{rl} 10.105 & \text{(3 decimal places)} \\ \times\ 11.324 & \text{(3 decimal places)} \\ \hline 40420 & \\ 202100 & \\ 3031500 & \\ 10105000 & \\ 101050000 & \\ \hline 114.429020 & \text{(6 decimal places)} \end{array}$$

or 114.42902

37.
$$\begin{array}{rl} 12.3 & \text{(1 decimal place)} \\ \times\ 1.08 & \text{(2 decimal places)} \\ \hline 984 & \\ 12300 & \\ \hline 13.284 & \text{(3 decimal places)} \end{array}$$

39.
$$\begin{array}{rl} 32.4 & \text{(1 decimal place)} \\ \times\ 2.8 & \text{(1 decimal place)} \\ \hline 2592 & \\ 6480 & \\ \hline 90.72 & \text{(2 decimal places)} \end{array}$$

41.
$$\begin{array}{rl} 0.00342 & \text{(5 decimal places)} \\ \times\ 0.84 & \text{(2 decimal places)} \\ \hline 1368 & \\ 27360 & \\ \hline 0.0028728 & \text{(7 decimal places)} \end{array}$$

43.
$$\begin{array}{rl} 0.347 & \text{(3 decimal places)} \\ \times\ 2.09 & \text{(2 decimal places)} \\ \hline 3123 & \\ 69400 & \\ \hline 0.72523 & \text{(5 decimal places)} \end{array}$$

45.
$$\begin{array}{rl} 3.005 & \text{(3 decimal places)} \\ \times\ 0.623 & \text{(3 decimal places)} \\ \hline 9015 & \\ 60100 & \\ 1803000 & \\ \hline 1.872115 & \text{(6 decimal places)} \end{array}$$

47. $1\underline{000} \times 45.678$ 45.678.

3 zeros Move 3 places to the right.

$1000 \times 45.678 = 45,678$

49. Move 2 places to the right.

$28.88.¢

Change from $ sign in front to ¢ sign at end.

$\$28.88 = 2888¢$

51. Move 2 places to the right.

$0.66.¢

Change from $ sign in front to ¢ sign at end.

$\$0.66 = 66¢$

53. Move 2 places to the left.

$0.34.¢

Change from ¢ sign at end to $ sign in front.

$34¢ = \$0.34$

55. Move 2 places to the left.

$34.45.¢

Change from ¢ sign at end to $ sign in front.

$3345¢ = \$34.45$

57. $\$3.6 \text{ billion} = \$3.6 \times 1,\underbrace{000,000,000}_{9 \text{ zeros}}$

$3.600000000.

Move 9 places to the right.

$\$3.6 \text{ billion} = \$3,600,000,000$

59. $\$196.8 \text{ million} = \$196.8 \times 1,\underbrace{000,000}_{6 \text{ zeros}}$

$196.800000.

Move 6 places to the right.

$\$196.8 \text{ million} = \$196,800,000$

61.
$$2\frac{1}{3} \cdot 4\frac{4}{5} = \frac{7}{3} \cdot \frac{24}{5} = \frac{7 \cdot 3 \cdot 8}{3 \cdot 5}$$
$$= \frac{3}{3} \cdot \frac{7 \cdot 8}{5} = \frac{56}{5}$$
$$= 11\frac{1}{5}$$

63.
$$\begin{array}{r} 342 \\ 24\overline{)8208} \\ \underline{7200} \\ 1008 \\ \underline{960} \\ 48 \\ \underline{48} \\ 0 \end{array}$$

The answer is 342.

65.
$$\begin{array}{r} 4566 \\ 7\overline{)31,962} \\ \underline{28\,000} \\ 3962 \\ \underline{3500} \\ 462 \\ \underline{420} \\ 42 \\ \underline{42} \\ 0 \end{array}$$

The answer is 4566.

67. ◈

69. (1 trillion) · (1 billion)

$$= 1,\underbrace{000,000,000,000}_{12 \text{ zeros}} \times 1,\underbrace{000,000,000}_{9 \text{ zeros}}$$
$$= 1,\underbrace{000,000,000,000,000,000,000}_{21 \text{ zeros}}$$
$$= 10^{21}$$

Exercise Set 3.4

1.
$$\begin{array}{r} 2.99 \\ 2\overline{)5.98} \\ \underline{4\,00} \\ 1\,98 \\ \underline{1\,80} \\ 18 \\ \underline{18} \\ 0 \end{array}$$

Divide as though dividing whole numbers. Place the decimal point directly above the decimal point in the dividend.

3.
$$\begin{array}{r} 23.78 \\ 4\overline{)95.12} \\ \underline{80\,00} \\ 15\,12 \\ \underline{12\,00} \\ 3\,12 \\ \underline{2\,80} \\ 32 \\ \underline{32} \\ 0 \end{array}$$

Divide as though dividing whole numbers. Place the decimal point directly above the decimal point in the dividend.

5.
$$\begin{array}{r} 7.48 \\ 12\overline{)89.76} \\ \underline{84\,00} \\ 5\,76 \\ \underline{4\,80} \\ 96 \\ \underline{96} \\ 0 \end{array}$$

7.
$$\begin{array}{r} 7.2 \\ 33\overline{)237.6} \\ \underline{2310} \\ 66 \\ \underline{66} \\ 0 \end{array}$$

9.
$$\begin{array}{r} 1.143 \\ 8\overline{)9.144} \\ \underline{8000} \\ 1144 \\ \underline{800} \\ 344 \\ \underline{320} \\ 24 \\ \underline{24} \\ 0 \end{array}$$

11.
$$\begin{array}{r} 4.041 \\ 3\overline{)12.123} \\ \underline{12000} \\ 123 \\ \underline{120} \\ 3 \\ \underline{3} \\ 0 \end{array}$$

13.
$$\begin{array}{r} 0.07 \\ 5\overline{)0.35} \\ \underline{35} \\ 0 \end{array}$$

15.
$$\begin{array}{r} 70. \\ 0.12_{\wedge}\overline{)8.40_{\wedge}} \\ \underline{840} \\ 0 \end{array}$$

Multiply the divisor by 100 (move the decimal point 2 places). Multiply the same way in the dividend (move 2 places). Then divide.

17.
$$\begin{array}{r} 20. \\ 3.4_{\wedge}\overline{)68.0_{\wedge}} \\ \underline{680} \\ 0 \end{array}$$

Put a decimal point at the end of the whole number. Multiply the divisor by 10 (move the decimal point 1 place). Multiply the same way in the dividend (move 1 place), adding an extra 0. Then divide.

19.
$$\begin{array}{r} 0.4 \\ 15\overline{)6.0} \\ \underline{60} \\ 0 \end{array}$$

Put a decimal point at the end of the whole number. Write an extra 0 to the right of the decimal point. Then divide.

21.
$$\begin{array}{r} 0.41 \\ 36\overline{)14.76} \\ \underline{1440} \\ 36 \\ \underline{36} \\ 0 \end{array}$$

23.
$$\begin{array}{rl} 8.5 & \\ 3.2_{\wedge}\overline{)27.2_{\wedge}0} & \\ \underline{256} & \\ 160 & \text{Write an extra 0.} \\ \underline{160} & \\ 0 & \end{array}$$

25.
$$\begin{array}{r} 9.3 \\ 4.2_{\wedge}\overline{)39.0_{\wedge}6} \\ \underline{3780} \\ 126 \\ \underline{126} \\ 0 \end{array}$$

27.
$$\begin{array}{rl} 0.625 & \\ 8\overline{)5.000} & \\ \underline{48} & \\ 20 & \text{Write an extra 0.} \\ \underline{16} & \\ 40 & \text{Write an extra 0.} \\ \underline{40} & \\ 0 & \end{array}$$

29.
$$\begin{array}{r} 0.26 \\ 0.47_{\wedge}\overline{)0.12_{\wedge}22} \\ \underline{940} \\ 282 \\ \underline{282} \\ 0 \end{array}$$

31.
$$\begin{array}{r} 15.625 \\ 4.8_{\wedge}\overline{)75.0_{\wedge}000} \\ \underline{480} \\ 270 \\ \underline{240} \\ 300 \\ \underline{288} \\ 120 \\ \underline{96} \\ 240 \\ \underline{240} \\ 0 \end{array}$$

33.
$$\begin{array}{r} 2.34 \\ 0.032_{\wedge}\overline{)0.074_{\wedge}88} \\ \underline{6400} \\ 1088 \\ \underline{960} \\ 128 \\ \underline{128} \\ 0 \end{array}$$

35.
$$\begin{array}{r} 0.47 \\ 82\overline{)38.54} \\ \underline{3280} \\ 574 \\ \underline{574} \\ 0 \end{array}$$

37. $\dfrac{213.4567}{1000}$ 0.213.4567

3 zeros Move 3 places to the left.

$$\frac{213.4567}{1000} = 0.2134567$$

39. $\dfrac{213.4567}{10}$ 21.3.4567

1 zero Move 1 place to the left.

$$\frac{213.4567}{10} = 21.34567$$

41. $\frac{1.0237}{0.001}$ 1.023.7 (Move 3 places to the right.)

3 decimal places

$\frac{1.0237}{0.001} = 1023.7$

43. $4.2 \cdot x = 39.06$

$\frac{4.2 \cdot x}{4.2} = \frac{39.06}{4.2}$ Dividing on both sides by 4.2

$x = 9.3$

$$\begin{array}{r} 09.3 \\ 4.2_\wedge \overline{)\,39.0_\wedge 6} \\ 378\,0 \\ \hline 12\,6 \\ 12\,6 \\ \hline 0 \end{array}$$

The solution is 9.3.

45. $1000 \cdot y = 9.0678$

$\frac{1000 \cdot y}{1000} = \frac{9.0678}{1000}$ Dividing on both sides by 1000

$y = 0.0090678$ Moving the decimal point 3 places to the left

The solution is 0.0090678.

47. $1048.8 = 23 \cdot t$

$\frac{1048.8}{23} = \frac{23 \cdot t}{23}$ Dividing on both sides by 23

$45.6 = t$

$$\begin{array}{r} 45.6 \\ 23 \overline{)\,1048.8} \\ 920\,0 \\ \hline 128\,8 \\ 115\,0 \\ \hline 13\,8 \\ 13\,8 \\ \hline 0 \end{array}$$

The solution is 45.6.

49. $14 \times (82.6 + 67.9) = 14 \times (150.5)$ Doing the calculation inside the parentheses

$= 2107$ Multiplying

51. $0.003 + 3.03 \div 0.01 = 0.003 + 303$ Dividing first

$= 303.003$ Adding

53. $42 \times (10.6 + 0.024)$

$= 42 \times 10.624$ Doing the calculation inside the parentheses

$= 446.208$ Multiplying

55. $4.2 \times 5.7 + 0.7 \div 3.5$

$= 23.94 + 0.2$ Doing the multiplications and divisions in order from left to right

$= 24.14$ Adding

57. $9.0072 + 0.04 \div 0.1^2$

$= 9.0072 + 0.04 \div 0.01$ Evaluating the exponential expression

$= 9.0072 + 4$ Dividing

$= 13.0072$ Adding

59. $(8 - 0.04)^2 \div 4 + 8.7 \times 0.4$

$= (7.96)^2 \div 4 + 8.7 \times 0.4$ Doing the calculation inside the parentheses

$= 63.3616 \div 4 + 8.7 \times 0.4$ Evaluating the exponential expression

$= 15.8404 + 3.48$ Doing the multiplications and divisions in order from left to right

$= 19.3204$ Adding

61. $86.7 + 4.22 \times (9.6 - 0.03)^2$

$= 86.7 + 4.22 \times (9.57)^2$ Doing the calculation inside the parentheses

$= 86.7 + 4.22 \times 91.5849$ Evaluating the exponential expression

$= 86.7 + 386.488278$ Multiplying

$= 473.188278$ Adding

63. $4 \div 0.4 + 0.1 \times 5 - 0.1^2$

$= 4 \div 0.4 + 0.1 \times 5 - 0.01$ Evaluating the exponential expression

$= 10 + 0.5 - 0.01$ Doing the multiplications and divisions in order from left to right

$= 10.49$ Adding and subtracting in order from left to right

65. $5.5^2 \times [(6 - 4.2) \div 0.06 + 0.12]$

$= 5.5^2 \times [1.8 \div 0.06 + 0.12]$ Doing the calculation in the innermost parentheses first

$= 5.5^2 \times [30 + 0.12]$ Doing the calculation inside the parentheses

$= 5.5^2 \times 30.12$

$= 30.25 \times 30.12$ Evaluating the exponential expression

$= 911.13$ Multiplying

67. $200 \times \{[(4 - 0.25) \div 2.5] - (4.5 - 4.025)\}$

$= 200 \times \{[3.75 \div 2.5] - 0.475\}$ Doing the calculations in the innermost parentheses first

$= 200 \times \{1.5 - 0.475\}$ Again, doing the calculations in the innermost parentheses

$= 200 \times 1.025$ Subtracting inside the parentheses

$= 205$ Multiplying

69. We add the numbers and then divide by the number of addends.

($1276.59 + $1350.49 + $1123.78 + $1402.56) ÷ 4

= $5153.42 ÷ 4

= $1288.355

≈ $1288.36

71. We add the temperature for the years 1992 through 1996 and then divide by the number of addends, 5:

$(59.23 + 59.36 + 59.56 + 59.72 + 59.58) \div 5 = 297.45 \div 5 = 59.49$

The average temperature for the years 1992 through 1996 was 59.49°F.

73. $10\frac{1}{2} + 4\frac{5}{8} = 10\frac{4}{8} + 4\frac{5}{8}$

$= 14\frac{9}{8} = 15\frac{1}{8}$

75. $\frac{36}{42} = \frac{6 \cdot 6}{6 \cdot 7} = \frac{6}{6} \cdot \frac{6}{7} = \frac{6}{7}$

77.

$$\begin{array}{r} 19 \\ 3\overline{)\,57} \\ 3\overline{)\,171} \\ 2\overline{)\,342} \\ 2\overline{)\,684} \end{array}$$

$684 = 2 \cdot 2 \cdot 3 \cdot 3 \cdot 19$

79.

81. Use a calculator.

$9.0534 - 2.041^2 \times 0.731 \div 1.043^2$

$= 9.0534 - 4.165681 \times 0.731 \div 1.087849$ Evaluating the exponential expressions

$= 9.0534 - 3.045112811 \div 1.087849$

$= 9.0534 - 2.799205415$ Multiplying and dividing in order from left to right

$= 6.254194585$

83. $439.57 \times 0.01 \div 1000 \times \underline{\quad} = 4.3957$

$4.3957 \div 1000 \times \underline{\quad} = 4.3957$

$0.0043957 \times \underline{\quad} = 4.3957$

We need to multiply 0.0043957 by a number that moves the decimal point 3 places to the right. Thus, we need to multiply by 1000. This is the missing value.

85. $0.0329 \div 0.001 \times 10^4 \div \underline{\quad} = 3290$

$0.0329 \div 0.001 \times 10{,}000 \div \underline{\quad} = 3290$

$32.9 \times 10{,}000 \div \underline{\quad} = 3290$

$329{,}000 \div \underline{\quad} = 3290$

We need to divide 329,000 by a number that moves the decimal point 2 places to the left. Thus, we need to divide by 100. This is the missing value.

Exercise Set 3.5

1. $\frac{3}{5} = \frac{3}{5} \cdot \frac{2}{2}$ We use $\frac{2}{2}$ for 1 to get a denominator of 10.

$= \frac{6}{10} = 0.6$

3. $\frac{13}{40} = \frac{13}{40} \cdot \frac{25}{25}$ We use $\frac{25}{25}$ for 1 to get a denominator of 1000.

$= \frac{325}{1000} = 0.325$

5. $\frac{1}{5} = \frac{1}{5} \cdot \frac{2}{2} = \frac{2}{10} = 0.2$

7. $\frac{17}{20} = \frac{17}{20} \cdot \frac{5}{5} = \frac{85}{100} = 0.85$

9. $\frac{19}{40} = \frac{19}{40} \cdot \frac{25}{25} = \frac{475}{1000} = 0.475$

11. $\frac{39}{40} = \frac{39}{40} \cdot \frac{25}{25} = \frac{975}{1000} = 0.975$

13. $\frac{13}{25} = \frac{13}{25} \cdot \frac{4}{4} = \frac{52}{100} = 0.52$

15. $\frac{2502}{125} = \frac{2502}{125} \cdot \frac{8}{8} = \frac{20{,}016}{1000} = 20.016$

17. $\frac{1}{4} = \frac{1}{4} \cdot \frac{25}{25} = \frac{25}{100} = 0.25$

19. $\frac{23}{40} = \frac{23}{40} \cdot \frac{25}{25} = \frac{575}{1000} = 0.575$

21. $\frac{18}{25} = \frac{18}{25} \cdot \frac{4}{4} = \frac{72}{100} = 0.72$

23. $\frac{19}{16} = \frac{19}{16} \cdot \frac{625}{625} = \frac{11{,}875}{10{,}000} = 1.1875$

25. $\frac{4}{15} = 4 \div 15$

$$\begin{array}{r} 0.266 \\ 15\overline{)\,4.000} \\ \underline{3\,0} \\ 1\,00 \\ \underline{90} \\ 100 \\ \underline{90} \\ 10 \end{array}$$

Since 10 keeps reappearing as a remainder, the digits repeat and

$\frac{4}{15} = 0.2666\ldots$ or $0.2\overline{6}$.

27. $\frac{1}{3} = 1 \div 3$

$$\begin{array}{r} 0.333 \\ 3\overline{)\,1.000} \\ \underline{9} \\ 10 \\ \underline{9} \\ 10 \\ \underline{9} \\ 1 \end{array}$$

Since 1 keeps reappearing as a remainder, the digits repeat and

$\frac{1}{3} = 0.333\ldots$ or $0.\overline{3}$.

29. $\frac{4}{3} = 4 \div 3$

$$\begin{array}{r} 1.33 \\ 3\,\overline{)\,4.00} \\ \underline{3} \\ 1\,0 \\ \underline{9} \\ 1\,0 \\ \underline{9} \\ 1 \end{array}$$

Since 1 keeps reappearing as a remainder, the digits repeat and

$\frac{4}{3} = 1.333\ldots$ or $1.\overline{3}$.

31. $\frac{7}{6} = 7 \div 6$

$$\begin{array}{r} 1.166 \\ 6\,\overline{)\,7.000} \\ \underline{6} \\ 1\,0 \\ \underline{6} \\ 4\,0 \\ \underline{3\,6} \\ 4\,0 \\ \underline{3\,6} \\ 4 \end{array}$$

Since 4 keeps reappearing as a remainder, the digits repeat and

$\frac{7}{6} = 1.166\ldots$ or $1.1\overline{6}$.

33. $\frac{4}{7} = 4 \div 7$

$$\begin{array}{r} 0.5714285 \\ 7\,\overline{)\,4.0000000} \\ \underline{3\,5} \\ 5\,0 \\ \underline{4\,9} \\ 1\,0 \\ \underline{7} \\ 3\,0 \\ \underline{2\,8} \\ 2\,0 \\ \underline{1\,4} \\ 6\,0 \\ \underline{5\,6} \\ 4\,0 \\ \underline{3\,5} \\ 5 \end{array}$$

Since 5 reappears as a remainder, the sequence repeats and

$\frac{4}{7} = 0.571428571428\ldots$ or $0.\overline{571428}$.

35. $\frac{11}{12} = 11 \div 12$

$$\begin{array}{r} 0.9166 \\ 12\,\overline{)\,11.0000} \\ \underline{10\,8} \\ 2\,0 \\ \underline{1\,2} \\ 8\,0 \\ \underline{7\,2} \\ 8\,0 \\ \underline{7\,2} \\ 8 \end{array}$$

Since 8 keeps reappearing as a remainder, the digits repeat and $\frac{11}{12} = 0.91666\ldots$ or $0.91\overline{6}$.

37. Round 0.2 [6] 6 6 ... to the nearest tenth.
Hundredths digit is 5 or higher.
0.3 Round up.

Round 0.2 6 [6] 6 ... to the nearest hundredth.
Thousandths digit is 5 or higher.
0.27 Round up.

Round 0.2 6 6 [6] ... to the nearest thousandth.
Ten-thousandths digit is 5 or higher.
0.267 Round up.

39. Round 0.3 [3] 3 3 ... to the nearest tenth.
Hundredths digit is 4 or lower.
0.3 Round down.

Round 0.3 3 [3] 3 ... to the nearest hundredth.
Thousandths digit is 4 or lower.
0.33 Round down.

Round 0.3 3 3 [3] ... to the nearest thousandth.
Ten-thousandths digit is 4 or lower.
0.333 Round down.

41. Round 1.3 [3] 3 3 ... to the nearest tenth.
Hundredths digit is 4 or lower.
1.3 Round down.

Round 1.3 3 [3] 3 ... to the nearest hundredth.
Thousandths digit is 4 or lower.
1.33 Round down.

Round 1.3 3 3 [3] ... to the nearest thousandth.
Ten-thousandths digit is 4 or lower.
1.333 Round down.

43. Round 1.1 [6] 6 6 ... to the nearest tenth.
Hundredths digit is 5 or higher.
1.2 Round up.

Round 1. 1 6 [6] 6 ... to the nearest hundredth.
Thousandths digit is 5 or higher.
1. 1 7 Round up.

Round 1. 1 6 6 [6] ... to the nearest thousandth.
Ten-thousandths digit is 5 or higher.
1. 1 6 7 Round up.

45. $0.\overline{571428}$

Round to the nearest tenth.

0.5[7]1428571428...
Hundredths digit is 5 or higher.
0.6 Round up.

Round to the nearest hundredth.

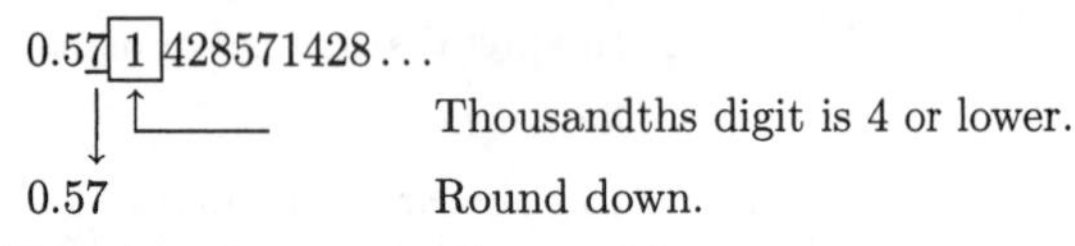

Round to the nearest thousandth.

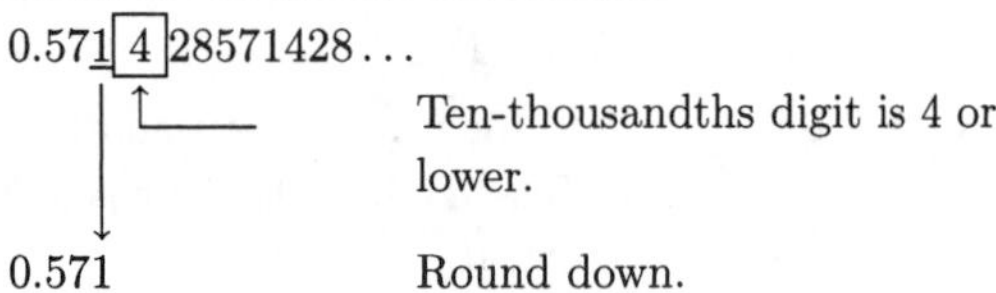

47. Round 0. 9 [1] 6 6 ... to the nearest tenth.
Hundredths digit is 4 or lower.
0. 9 Round down.

Round 0. 9 1 [6] 6 ... to the nearest hundredth.
Thousandths digit is 5 or higher.
0. 9 2 Round up.

Round 0. 9 1 6 [6] ... to the nearest thousandth.
Ten-thousandths digit is 5 or higher.
0. 9 1 7 Round up.

49. Round 0. 1 [8] 1 8 ... to the nearest tenth.
Hundredths digit is 5 or higher.
0. 2 Round up.

Round 0. 1 8 [1] 8 ... to the nearest hundredth.
Thousandths digit is 4 or lower.
0. 1 8 Round down.

Round 0. 1 8 1 [8] ... to the nearest thousandth.
Ten-thousandths digit is 5 or higher.
0. 1 8 2 Round up.

51. Round 0. 2 [7] 7 7 ... to the nearest tenth.
Hundredths digit is 5 or higher.
0. 3 Round up.

Round 0. 2 7 [7] 7 ... to the nearest hundredth.
Thousandths digit is 5 or higher.
0. 2 8 Round up.

Round 0. 2 7 7 [7] ... to the nearest thousandth.
Ten-thousandths digit is 5 or higher.
0. 2 7 8 Round up.

53. We will use the first method discussed in the text.

$$\begin{aligned}\frac{7}{8}\times 12.64 &= \frac{7}{8}\times\frac{1264}{100} = \frac{7\cdot 1264}{8\cdot 100}\\ &= \frac{7\cdot 2\cdot 2\cdot 2\cdot 2\cdot 79}{2\cdot 2\cdot 2\cdot 2\cdot 2\cdot 5\cdot 5}\\ &= \frac{2\cdot 2\cdot 2\cdot 2}{2\cdot 2\cdot 2\cdot 2}\cdot\frac{7\cdot 79}{2\cdot 5\cdot 5}\\ &= 1\cdot\frac{7\cdot 79}{2\cdot 5\cdot 5}\\ &= \frac{7\cdot 79}{2\cdot 5\cdot 5} = \frac{553}{50},\ \text{or } 11.06\end{aligned}$$

55. $2\frac{3}{4} + 5.65 = 2.75 + 5.65$ Writing $2\frac{3}{4}$ using decimal notation

$= 8.4$ Adding

57. We will use the second method discussed in the text.

$$\begin{aligned}\frac{47}{9}\times 79.95 &= 5.\overline{2}\times 79.95\\ &\approx 5.222\times 79.95 = 417.4989\end{aligned}$$

Note that this answer is not as accurate as those found using either of the other methods, due to rounding.

59. $\frac{1}{2} - 0.5 = 0.5 - 0.5$ Writing $\frac{1}{2}$ using decimal notation

$= 0$

61. $4.875 - 2\frac{1}{16} = 4.875 - 2.0625$ Writing $2\frac{1}{16}$ using decimal notation

$= 2.8125$

63. We will use the third method discussed in the text.

$$\begin{aligned}\frac{5}{6}\times 0.0765 + \frac{5}{4}\times 0.1124 &= \frac{5}{6}\times\frac{0.0765}{1} + \frac{5}{4}\times\frac{0.1124}{1}\\ &= \frac{5\times 0.0765}{6\times 1} + \frac{5\times 0.1124}{4\times 1}\\ &= \frac{0.3825}{6} + \frac{0.562}{4}\\ &= 0.06375 + 0.1405\\ &= 0.20425\end{aligned}$$

65. We use the rules for order of operations, doing the multiplication first and then the division. Then we add.

$$\begin{aligned}\frac{4}{5}\times 384.8 + 24.8\div\frac{8}{3} &= 307.84 + 24.8\cdot\frac{3}{8}\\ &= 307.84 + 9.3\\ &= 317.14\end{aligned}$$

67. We do the multiplications in order from left to right. Then we subtract.

$$\begin{aligned}\frac{7}{8}\times 0.86 - 0.76\times\frac{3}{4} &= 0.7525 - 0.76\times\frac{3}{4}\\ &= 0.7525 - 0.57\\ &= 0.1825\end{aligned}$$

69. $3.375\times 5\frac{1}{3} = 3.375\times\frac{16}{3}$ Writing $5\frac{1}{3}$ using fractional notation

$= 18$ Multiplying

71. $6.84\div 2\frac{1}{2} = 6.84\div 2.5$ Writing $2\frac{1}{2}$ using decimal notation

$= 2.736$ Dividing

73. $9\cdot 2\frac{1}{3} = \frac{9}{1}\cdot\frac{7}{3} = \frac{9\cdot 7}{1\cdot 3} = \frac{3\cdot 3\cdot 7}{1\cdot 3} = \frac{3}{3}\cdot\frac{3\cdot 7}{1} = 21$

75.

$$\begin{array}{rcl} 20 & = & 19\frac{5}{5}\\ -16\frac{3}{5} & = & -16\frac{3}{5}\\ \hline & & 3\frac{2}{5}\end{array}$$

77. ***Familiarize***. We draw a picture and let c = the total number of cups of liquid ingredients.

$\frac{2}{3}$ cup	$\frac{1}{4}$ cup	$\frac{1}{8}$ cup
c		

Translate. The problem can be translated to an equation as follows:

$$\underbrace{\text{Amount of water}}_{\frac{2}{3}}\ \underset{+}{\text{plus}}\ \underbrace{\text{Amount of milk}}_{\frac{1}{4}}\ \underset{+}{\text{plus}}\ \underbrace{\text{Amount of oil}}_{\frac{1}{8}}\ \underset{=}{\text{is}}\ \underbrace{\text{Amount of liquid}}_{c}$$

Solve. We carry out the addition. Since $3 = 3$, $4 = 2\cdot 2$, and $8 = 2\cdot 2\cdot 2$, the LCM of the denominators is $3\cdot 2\cdot 2\cdot 2$, or 24.

$$\begin{aligned}\frac{2}{3}+\frac{1}{4}+\frac{1}{8} &= c\\ \frac{2}{3}\cdot\frac{8}{8}+\frac{1}{4}\cdot\frac{6}{6}+\frac{1}{8}\cdot\frac{3}{3} &= c\\ \frac{16}{24}+\frac{6}{24}+\frac{3}{24} &= c\\ \frac{25}{24} &= c\end{aligned}$$

Check. We repeat the calculation. We also note that the sum is larger than any of the individual amounts, as expected.

State. The recipe calls for $\frac{25}{24}$ cups, or $1\frac{1}{24}$ cups, of liquid ingredients.

79. ◈

81. Using a calculator we find that

$\frac{1}{7} = 1\div 7 = 0.\overline{142857}.$

83. Using a calculator we find that

$\frac{3}{7} = 3\div 7 = 0.\overline{428571}.$

85. Using a calculator we find that

$\frac{5}{7} = 5\div 7 = 0.\overline{714285}.$

87. Using a calculator we find that

$\frac{1}{9} = 1\div 9 = 0.\overline{1}.$

89. Using a calculator we find that

$\frac{1}{999} = 0.\overline{001}.$

Exercise Set 3.6

1. We are estimating the sum

$$\$109.95 + \$249.95.$$

We round both numbers to the nearest ten. The estimate is

$$\$110 + \$250 = \$360.$$

Answer (d) is correct.

3. We are estimating the difference

$$\$299 - \$249.95.$$

We round both numbers to the nearest ten. The estimate is

$$\$300 - \$250 = \$50.$$

Answer (c) is correct.

5. We are estimating the product

$$9\times \$299.$$

We round \$299 to the nearest ten. The estimate is

$$9\times \$300 = \$2700.$$

Answer (a) is correct.

7. We are estimating the quotient

$$\$1700\div \$299.$$

Rounding \$299, we get \$300. Since \$1700 is close to \$1800, which is a multiple of \$300, we estimate

$$\$1800\div \$300,$$

so the answer is about 6.

Answer (c) is correct.

9. This is about $0.0 + 1.3 + 0.3$, so the answer is about 1.6.

11. This is about $6 + 0 + 0$, so the answer is about 6.

13. This is about $52 + 1 + 7$, so the answer is about 60.

15. This is about $2.7 - 0.4$, so the answer is about 2.3.

17. This is about $200 - 20$, so the answer is about 180.

19. This is about 50×8, rounding 49 to the nearest ten and 7.89 to the nearest one, so the answer is about 400. Answer (a) is correct.

21. This is about 100×0.08, rounding 98.4 to the nearest ten and 0.083 to the nearest hundredth, so the answer is about 8. Answer (c) is correct.

23. This is about $4 \div 4$, so the answer is about 1. Answer (b) is correct.

25. This is about $75 \div 25$, so the answer is about 3. Answer (b) is correct.

27. We estimate the quotient by rounding the total revenue to the nearest million and the average revenue to the nearest thousand.

$53.6 \text{ million} \div 6716 = 53,600,000 \div 6716 \approx$

$54,000,000 \div 7000 \approx 7700$

About 7700 screens were showing the movie.

29.

```
        3
    3 ) 9
   3 ) 2 7
  2 ) 5 4
 2 ) 1 0 8
```

$108 = 2 \cdot 2 \cdot 3 \cdot 3 \cdot 3$

31.

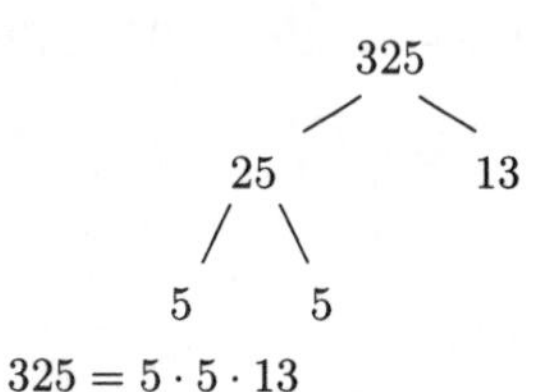

$325 = 5 \cdot 5 \cdot 13$

33. $\dfrac{125}{400} = \dfrac{25 \cdot 5}{25 \cdot 16} = \dfrac{25}{25} \cdot \dfrac{5}{16} = \dfrac{5}{16}$

35. $\dfrac{72}{81} = \dfrac{9 \cdot 8}{9 \cdot 9} = \dfrac{9}{9} \cdot \dfrac{8}{9} = \dfrac{8}{9}$

37.

39. We round each factor to the nearest ten. The estimate is $180 \times 60 = 10,800$. The estimate is close to the result given, so the decimal point was placed correctly.

41. We round each number on the left to the nearest one. The estimate is $19 - 1 \times 4 = 19 - 4 = 15$. The estimate is not close to the result given, so the decimal point was not placed correctly.

Exercise Set 3.7

1. ***Familiarize.*** Repeated addition fits this situation. We let C = the cost of 8 pairs of socks.

$$\underbrace{\boxed{\$4.95} + \boxed{\$4.95} + \cdots + \boxed{\$4.95}}_{\text{8 addends}}$$

Translate.

$$\underbrace{\text{Price per pair}}_{4.95} \quad \underbrace{\text{times}}_{\times} \quad \underbrace{\text{Number of pairs}}_{8} \quad \underbrace{\text{is}}_{=} \quad \underbrace{\text{Total cost}}_{C}$$

Solve. We carry out the multiplication.

```
   4. 9 5
 ×      8
 ---------
  3 9. 6 0
```

Thus, $C = 39.60$.

Check. We obtain a partial check by rounding and estimating:

$$4.95 \times 8 \approx 5 \times 8 = 40 \approx 39.60.$$

State. Eight pairs of socks cost \$39.60.

3. ***Familiarize.*** Repeated addition fits this situation. We let c = the cost of 17.7 gal of gasoline.

Translate.

$$\underbrace{\text{Cost per gallon}}_{1.199} \quad \underbrace{\text{times}}_{\cdot} \quad \underbrace{\text{Number of gallons}}_{17.7} \quad \underbrace{\text{is}}_{=} \quad \underbrace{\text{Total cost}}_{c}$$

Solve. We carry out the multiplication.

```
      1. 1 9 9
  ×     1 7. 7
  ------------
        8 3 9 3
      8 3 9 3 0
    1 1 9 9 0 0
  -------------
    2 1. 2 2 2 3
```

Thus, $c = 21.2223$.

Check. We obtain a partial check by rounding and estimating:

$$1.199 \times 17.7 \approx 1 \times 20 = 20 \approx 21.2223.$$

State. We round \$21.2223 to the nearest cent and find that the cost of the gasoline is \$21.22.

5. ***Familiarize.*** We visualize the situation. We let c = the amount of change.

<table>
<tr><td colspan="2">$20</td></tr>
<tr><td>$16.99</td><td>c</td></tr>
</table>

Translate. This is a "take-away" situation.

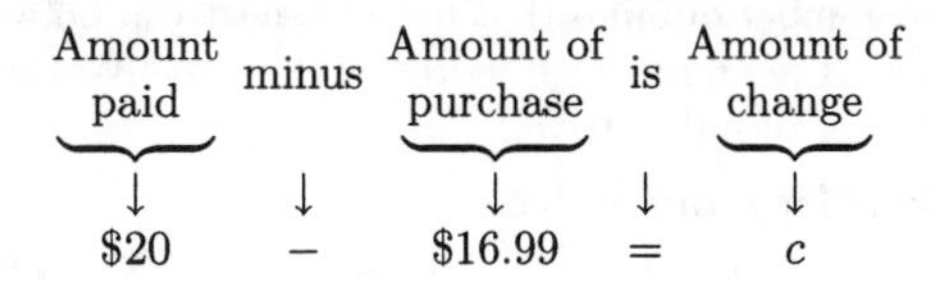

Solve. To solve the equation we carry out the subtraction.

```
   1 9 9 10
   2 0. 0 0
 − 1 6. 9 9
 ----------
     3. 0 1
```

Thus, $c = \$3.01$.

Check. We check by adding 3.01 to 16.99 to get 20. This checks.

State. The change was $3.01.

7. ***Familiarize.*** We visualize the situation. We let $n =$ the new temperature.

98.6°	4.2°
n	

Translate. We are combining amounts.

Normal body temperature	plus	Degrees temperature rises	is	New temperature
↓	↓	↓	↓	↓
98.6	+	4.2	=	n

Solve. To solve the equation we carry out the addition.

```
   1
   9 8.6
 +   4.2
 -------
 1 0 2.8
```

Thus, $n = 102.8$.

Check. We can check by repeating the addition. We can also check by rounding:

$$98.6 + 4.2 \approx 99 + 4 = 103 \approx 102.8$$

State. The new temperature was 102.8°F.

9. ***Familiarize.*** We visualize the situation. Let $w =$ each winner's share.

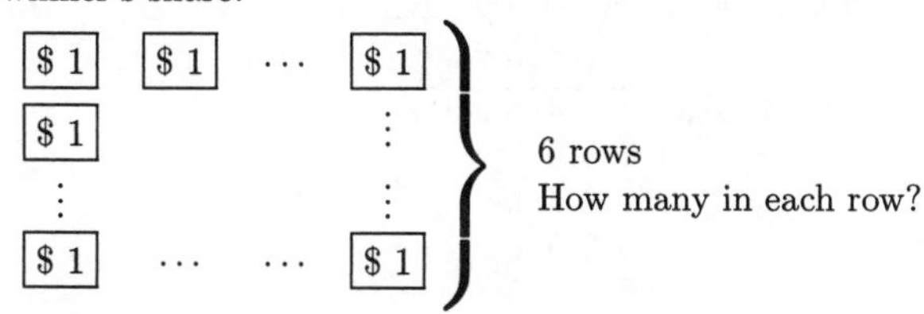

Translate.

Total prize	÷	Number of winners	=	Each winner's share
↓	↓	↓	↓	↓
127,315	÷	6	=	w

Solve. We carry out the division.

```
      2 1, 2 1 9. 1 6 6
   6 | 1 2 7, 3 1 5. 0 0 0
       1 2 0 0 0 0
       -----------
           7 3 1 5
           6 0 0 0
           -------
           1 3 1 5
           1 2 0 0
           -------
             1 1 5
               6 0
             -----
               5 5
               5 4
               ---
                 1 0
                   6
                 ---
                   4 0
                   3 6
                   ---
                     4 0
                     3 6
                     ---
                       4
```

Rounding to the nearest cent, or hundredth, we get $w = 21,219.17$.

Check. We can repeat the calculation. The answer checks.

State. Each winner's share is $21,219.17.

11. ***Familiarize.*** We draw a picture, letting $A =$ the area.

A	312.6 ft
800.4 ft	

Translate. We use the formula $A = l \cdot w$.

$A = 800.4 \times 312.6$

Solve. We carry out the multiplication.

```
            3 1 2. 6
          × 8 0 0. 4
          ----------
          1 2 5 0 4
  2 5 0 0 8 0 0 0
  -------------------
  2 5 0, 2 0 5. 0 4
```

Thus, $A = 250,205.04$.

Check. We obtain a partial check by rounding and estimating:

$$800.4 \times 312.6 \approx 800 \times 300 = 240,000 \approx 250,205.04$$

State. The area is 250,205.04 sq ft.

13. ***Familiarize.*** We visualize the situation. We let $m =$ the odometer reading at the end of the trip.

22,456.8 mi	234.7 mi
m	

Translate. We are combining amounts.

Reading before trip	plus	Miles driven	is	Reading at end of trip
↓	↓	↓	↓	↓
22,456.8	+	234.7	=	m

Solve. To solve the equation we carry out the addition.

$$\begin{array}{r} \scriptstyle 1\ 1 \quad \\ 2\,2{,}4\,5\,6.8 \\ +\quad 2\,3\,4.7 \\ \hline 2\,2{,}6\,9\,1.5 \end{array}$$

Thus, $m = 22,691.5$.

Check. We can check by repeating the addition. We can also check by rounding:

$$22,456.8 + 234.7 \approx 22,460 + 230 = 22,690 \approx 22,691.5$$

State. The odometer reading at the end of the trip was 22,691.5.

15. ***Familiarize***. We visualize the situation. We let n = the number by which hamburgers exceed hot dogs, in billions.

24.8 billion	
15.9 billion	n

Translate. This is a "how-much-more" situation.

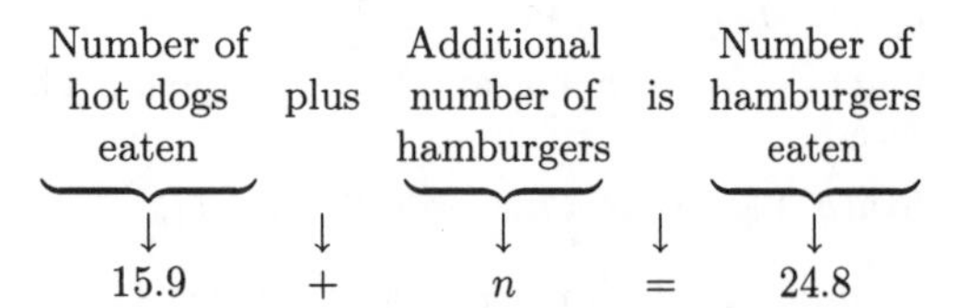

Solve. We subtract 15.9 on both sides.

$n = 24.8 - 15.9$

$n = 8.9$

$$\begin{array}{r} \scriptstyle 1\ \ 13\ 18 \\ \not{2}\,\not{4}.\not{8} \\ -\ 1\,5.9 \\ \hline 8.9 \end{array}$$

Check. We check by adding 8.9 to 15.9 to get 24.8. This checks.

State. Americans eat 8.9 billion more hamburgers than hot dogs.

17. ***Familiarize***. We visualize the situation. We let d = the difference in the speeds.

0.85	d
1.15	

Translate. This is a "how much more situation."

$$\underbrace{\text{Lower speed}}_{\downarrow \atop 0.85} \quad \underset{\downarrow \atop +}{\text{plus}} \quad \underbrace{\text{Additional speed}}_{\downarrow \atop d} \quad \underset{\downarrow \atop =}{\text{is}} \quad \underbrace{\text{Higher speed}}_{\downarrow \atop 1.15}$$

Solve. We subtract 0.85 on both sides.

$d = 1.15 - 0.85$

$d = 0.3$

$$\begin{array}{r} \scriptstyle 0\ 11 \quad \\ \not{1}.\not{1}5 \\ -\ 0.85 \\ \hline 0.30 \end{array}$$

Check. We check by adding 0.3 to 0.85 to get 1.15. The answer checks.

State. The difference in speeds was mach 0.3.

19. ***Familiarize.*** This is a two-step problem. First, we find the number of miles that have been driven between fillups. This is a "how-much-more" situation. We let n = the number of miles driven.

Translate and Solve.

$$\underbrace{\text{First odometer reading}}_{\downarrow \atop 26,342.8} \quad \underset{\downarrow \atop +}{\text{plus}} \quad \underbrace{\text{Number of miles driven}}_{\downarrow \atop n} \quad \underset{\downarrow \atop =}{\text{is}} \quad \underbrace{\text{Second odometer reading}}_{\downarrow \atop 26,736.7}$$

To solve the equation we subtract 26,342.8 on both sides.

$n = 26,736.7 - 26,342.8$

$n = 393.9$

$$\begin{array}{r} 2\,6,\,7\,3\,6.7 \\ -\ 2\,6,\,3\,4\,2.8 \\ \hline 3\,9\,3.9 \end{array}$$

Second, we divide the total number of miles driven by the number of gallons. This gives us m = the number of miles per gallon.

$$393.9 \div 19.5 = m$$

To find the number m, we divide.

```
            2 0.2
1 9.5^)3 9 3. 9^0
       3 9 0 0
           3 9 0
           3 9 0
               0
```

Thus, $m = 20.2$.

Check. To check, we first multiply the number of miles per gallon times the number of gallons:

$$19.5 \times 20.2 = 393.9$$

Then we add 393.9 to 26,342.8:

$$26,342.8 + 393.9 = 26,736.7$$

The number 20.2 checks.

State. The driver gets 20.2 miles per gallon.

21. ***Familiarize.*** We visualize a rectangular array consisting of 748.45 objects with 62.5 objects in each row. We want to find n, the number of rows.

Translate. We think (Total number of pounds) $\div$ (Pounds per cubic foot) = (Number of cubic feet).

$$748.45 \div 62.5 = n$$

Solve. We carry out the division.

```
             1 1.9 7 5 2
6 2.5^)7 4 8. 4^5 0 0 0
       6 2 5 0 0
       1 2 3 4 5
         6 2 5 0
         6 0 9 5
         5 6 2 5
           4 7 0 0
           4 3 7 5
             3 2 5 0
             3 1 2 5
               1 2 5 0
               1 2 5 0
                     0
```

Thus, $n = 11.9752$.

Check. We obtain a partial check by rounding and estimating:

$748.45 \div 62.5 \approx 700 \div 70 = 10 \approx 11.9752$

State. The tank holds 11.9752 cubic feet of water.

23. ***Familiarize.*** This is a two-step problem. First, we find the number of games that can be played in one hour. Think of an array containing 60 minutes (1 hour = 60 minutes) with 1.5 minutes in each row. We want to find how many rows there are. We let g represent this number.

Translate and Solve. We think (Number of minutes) ÷ (Number of minutes per game) = (Number of games).

$60 \div 1.5 = g$

To solve the equation we carry out the division.

$$\begin{array}{r} 4\,0. \\ 1.5_\wedge\overline{)\,6\,0.\,0_\wedge} \\ \underline{6\,0\,0} \\ 0 \\ \underline{0} \\ 0 \end{array}$$

Thus, $g = 40$.

Second, we find the cost t of playing 40 video games. Repeated addition fits this situation. (We express 25¢ as \$0.25.)

$$\underbrace{\text{Cost of one game}}_{\downarrow \atop 0.25} \quad \underset{\times}{\overset{\text{times}}{\downarrow}} \quad \underbrace{\text{Number of games played}}_{\downarrow \atop 40} \quad \underset{=}{\overset{\text{is}}{\downarrow}} \quad \underbrace{\text{Total cost}}_{\downarrow \atop t}$$

To solve the equation we carry out the multiplication.

$$\begin{array}{r} 0.\,2\,5 \\ \times \quad 4\,0 \\ \hline 1\,0.\,0\,0 \end{array}$$

Thus, $t = 10$.

Check. To check, we first divide the total cost by the cost per game to find the number of games played:

$10 \div 0.25 = 40$

Then we multiply 40 by 1.5 to find the total time:

$1.5 \times 40 = 60$

The number 10 checks.

State. It costs \$10 to play video games for one hour.

25. ***Familiarize.*** We let d = the distance around the figure.

Translate. We are combining lengths.

$$\underbrace{\text{The sum of the lengths of the 5 sides}}_{\downarrow \atop 8.9+23.8+4.7+22.1+18.6} \quad \underset{=}{\overset{\text{is}}{\downarrow}} \quad \underbrace{\text{the distance around the figure.}}_{\downarrow \atop d}$$

Solve. To solve we carry out the addition.

$$\begin{array}{r} {\scriptstyle 2\;3} \\ 8.9 \\ 2\,3.8 \\ 4.7 \\ 2\,2.1 \\ +\,1\,8.6 \\ \hline 7\,8.1 \end{array}$$

Thus, $d = 78.1$.

Check. To check we can repeat the addition. We can also check by rounding:

$8.9+23.8+4.7+22.1+18.6 \approx 9+24+5+22+19 = 79 \approx 78.1$

State. The distance around the figure is 78.1 cm.

27. ***Familiarize.*** This is a multistep problem. First we find the sum s of the two 0.8 cm segments. Then we use this length to find d.

Translate and Solve.

$$\underbrace{\text{Length of one small segment}}_{\downarrow \atop 0.8} \quad \underset{+}{\overset{\text{plus}}{\downarrow}} \quad \underbrace{\text{Length of other small segment}}_{\downarrow \atop 0.8} \quad \underset{=}{\overset{\text{is}}{\downarrow}} \quad \underbrace{\text{Total length.}}_{\downarrow \atop s}$$

To solve we carry out the addition.

$$\begin{array}{r} {\scriptstyle 1} \\ 0.\,8 \\ +\,0.\,8 \\ \hline 1.\,6 \end{array}$$

Thus, $s = 1.6$.

Now we find d.

$$\underbrace{\text{Total length of smaller segments}}_{\downarrow \atop 1.6} \quad \underset{+}{\overset{\text{plus}}{\downarrow}} \quad \underbrace{\text{length of } d}_{\downarrow \atop d} \quad \underset{=}{\overset{\text{is}}{\downarrow}} \quad \underbrace{3.91\text{ cm}}_{\downarrow \atop 3.91}$$

To solve we subtract 1.6 on both sides of the equation.

$d = 3.91 - 1.6$

$d = 2.31$

$$\begin{array}{r} 3.9\,1 \\ -\,1.6\,0 \\ \hline 2.3\,1 \end{array}$$

Check. We repeat the calculations.

State. The length d is 2.31 cm.

29. ***Familiarize.*** This is a two-step problem. First, we find how many minutes there are in 2 hr. We let m represent this number. Repeated addition fits this situation (Remember that 1 hr = 60 min.)

Translate and Solve.

$$\underbrace{\text{Number of minutes in 1 hour}}_{\downarrow \atop 60} \quad \underset{\cdot}{\overset{\text{times}}{\downarrow}} \quad \underbrace{\text{Number of hours}}_{\downarrow \atop 2} \quad \underset{=}{\overset{\text{is}}{\downarrow}} \quad \underbrace{\text{Total number of minutes}}_{\downarrow \atop m}$$

To solve the equation we carry out the multiplication.

$$\begin{array}{r} 6\,0 \\ \times \quad 2 \\ \hline 1\,2\,0 \end{array}$$

Thus, $m = 120$.

Next, we find how many calories are burned in 120 minutes. We let t represent this number. Repeated addition fits this situation also.

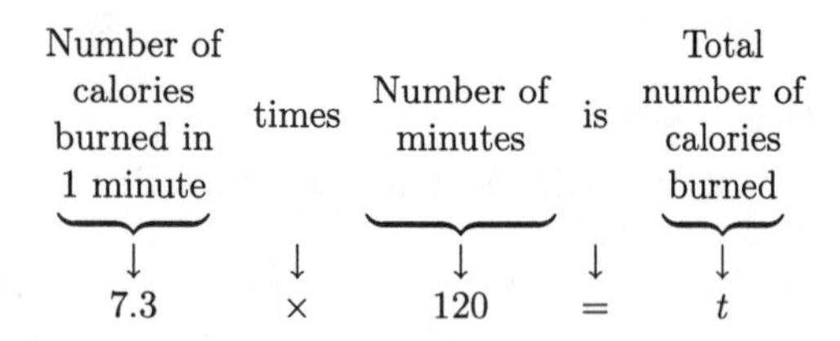

To solve the equation we carry out the multiplication.

$$\begin{array}{r} 1\,2\,0 \\ \times\ \ 7.\,3 \\ \hline 3\,6\,0 \\ 8\,4\,0\,0 \\ \hline 8\,7\,6.\,0 \end{array}$$

Thus, $t = 876$.

Check. To check, we first divide the total number of calories by the number of calories burned in one minute to find the total number of minutes the person mowed:

$876 \div 7.3 = 120$

Then we divide 120 by 60 to find the number of hours:

$120 \div 60 = 2$

The number 876 checks.

State. In 2 hr of mowing, 876 calories would be burned.

31. ***Familiarize.*** This is a multistep problem. We will first find the total amount of the checks. Then we will find how much is left in the account after the checks are written. Finally, we will use this amount and the amount of the deposit to find the balance in the account after all the changes. We will let $c =$ the total amount of the checks.

Translate and Solve. We are combining amounts.

First check	plus	Second check	plus	Third check	is	Total amount of checks
↓	↓	↓	↓	↓	↓	↓
23.82	+	507.88	+	98.32	=	c

To solve the equation we carry out the addition.

$$\begin{array}{r} {\scriptstyle 1\ 2\ \ 2\,1\ \ } \\ 2\,3.8\,2 \\ 5\,0\,7.8\,8 \\ +\ \ \ 9\,8.3\,2 \\ \hline 6\,3\,0.0\,2 \end{array}$$

Thus, $c = 630.02$.

Now we let $a =$ the amount in the account after the checks are written.

Original amount	less	Check amount	is	New amount
↓	↓	↓	↓	↓
1123.56	−	630.02	=	a

To solve the equation we carry out the subtraction.

$$\begin{array}{r} {\scriptstyle 10\ \ \ \ \ \ \ \ \ \ \ } \\ {\scriptstyle \not{0}\ 12\ \ \ \ \ \ \ } \\ \not{1}\ \not{1}\ \not{2}\ 3.5\,6 \\ -\ \ \ 6\ \ 3\ \ 0.0\,2 \\ \hline 4\ \ 9\ \ 3.5\,4 \end{array}$$

Thus, $a = 493.54$.

Finally, we let $f =$ the amount in the account after the paycheck is deposited.

Amount after checks	plus	Amount of deposit	is	Final amount
↓	↓	↓	↓	↓
493.54	+	678.20	=	f

We carry out the addition.

$$\begin{array}{r} {\scriptstyle 1\ 1\ \ \ \ \ \ \ \ \ } \\ 4\,9\,3.5\,4 \\ +\,6\,7\,8.2\,0 \\ \hline 1\,1\,7\,1.7\,4 \end{array}$$

Thus, $f = 1171.74$.

Check. We repeat the calculations.

State. There is \$1171.74 in the account after the changes.

33. ***Familiarize.*** We make and label a drawing. The question deals with a rectangle and a square, so we also list the relevant area formulas. We let $g =$ the area covered by grass.

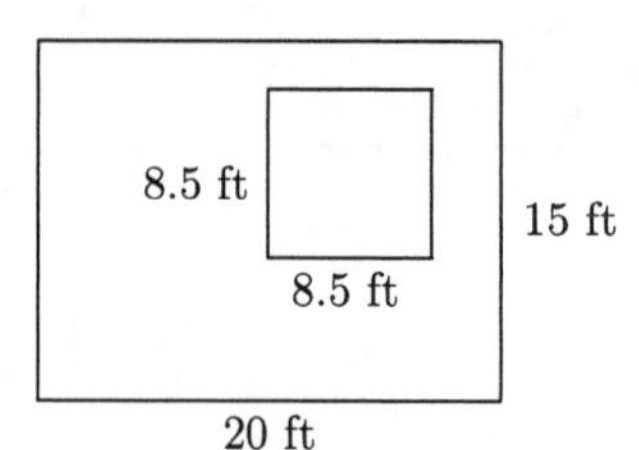

Area of a rectangle with length l and width w: $A = l \times w$

Area of a square with side s: $A = s^2$

Translate. We subtract the area of the square from the area of the rectangle.

Area of rectangle	minus	Area of square	is	Area covered by grass
↓	↓	↓	↓	↓
20×15	−	$(8.5)^2$	=	g

Solve. We carry out the computations.

$$\begin{aligned} 20 \times 15 - (8.5)^2 &= g \\ 20 \times 15 - 72.25 &= g \\ 300 - 72.25 &= g \\ 227.75 &= g \end{aligned}$$

Check. We can repeat the calculations. Also note that 227.75 is less than the area of the yard but more than the area of the flower garden. This agrees with the impression given by our drawing.

State. Grass covers 227.75 ft^2 of the yard.

35. ***Familiarize.*** The batting average is a fraction whose numerator is the number of hits and whose denominator is the number of at bats. We let $a =$ the batting average.

Translate. We think (Number of hits) $\div$ (Number of at bats) = (Batting average).

$168 \div 551 = a$

Solve. We carry out the division.

```
          0. 3 0 4 9
551 ) 1 6 8. 0 0 0 0
      1 6 5 3
          2 7 0 0
          2 2 0 4
            4 9 6 0
            4 9 5 9
                  1
```

We stop dividing at this point, because we will round to the nearest thousandth. Thus, $a \approx 0.305$.

Check. We can obtain a partial check by rounding and estimating:

$168 \div 551 \approx 200 \div 600 = 0.\overline{3} \approx 0.305$.

State. 0.305 of the at bats were hits.

37. ***Familiarize.*** This is a two-step problem. First we find the mileage cost of driving 120 miles at 27¢ per mile. We let c represent this number.

Translate and Solve.

Cost per mile	times	Number of miles	is	Mileage cost
↓	↓	↓	↓	↓
0.27	$\cdot$	120	=	c

To solve the equation we carry out the multiplication.

```
    1 2 0
  × 0. 2 7
    8 4 0
  2 4 0 0
  3 2. 4 0
```

Thus $c = 32.40$.

Next we add the rental cost for one day to the mileage cost to find the total cost. We let y represent this number.

Cost for one day	plus	Mileage cost	is	Total cost for one day
↓	↓	↓	↓	↓
24.95	+	32.40	=	y

To solve the equation we carry out the addition.

```
   2 4. 9 5
 + 3 2. 4 0
   5 7. 3 5
```

Thus $y = 57.35$.

Check. To check, we first subtract the daily charge from the total cost to get the total mileage cost:

$57.35 - 24.95 = 32.40$

Then we divide the total mileage cost by the cost per mile

$32.4 \div 0.27 = 120$

The number 57.35 checks.

State. The total cost of driving 120 miles in 1 day is $57.35.

39. ***Familiarize.*** Repeated addition fits this situation. We let t = the total savings in 1 year.

Translate.

Number of weeks	times	Savings per week	is	Total savings
↓	↓	↓	↓	↓
52	$\cdot$	6.72	=	t

Solve. To solve the equation we carry out the multiplication.

```
      6. 7 2
  ×     5 2
    1 3 4 4
  3 3 6 0 0
  3 4 9. 4 4
```

Check. We can obtain a partial check by rounding and estimating:

$52 \cdot 6.72 \approx 50 \cdot 7 = 350 \approx 349.44$.

State. The family would save $349.44 in 1 year.

41. ***Familiarize.*** This is a two-step problem. First we find the number of eggs in 20 dozen (1 dozen = 12). We let n represent this number.

Translate and Solve. We think (Number of dozens) $\cdot$ (Number in a dozen) = (Number of eggs).

$20 \cdot 12 = n$

$240 = n$

Second, we find the cost c of one egg. We think (Total cost) $\div$ (Number of eggs) = (Cost of one egg).

$\$13.80 \div 240 = c$

We carry out the division.

```
          0.0 5 7 5
240 ) 1 3.8 0 0 0
      1 2 0 0
        1 8 0 0
        1 6 8 0
          1 2 0 0
          1 2 0 0
                0
```

Thus, $c = 0.0575 \approx 0.058$ (rounded to the nearest tenth of a cent).

Check. We repeat the calculations.

State. Each egg cost about $0.058, or 5.8¢.

43. ***Familiarize.*** This is a three-step problem. We will find the area S of a standard soccer field and the area F of a standard football field using the formula Area = $l \cdot w$. Then we will find E, the amount by which the area of a soccer field exceeds the area of a football field.

Translate and Solve.

$S = l \cdot w = 114.9 \times 74.4 = 8548.56$

$F = l \cdot w = 120 \times 53.3 = 6396$

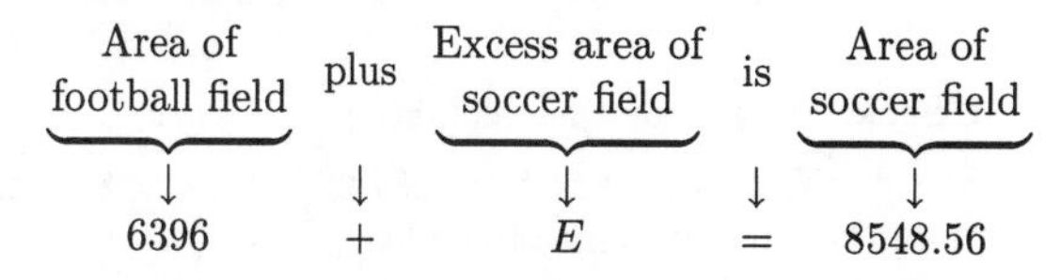

To solve the equation we subtract 6396 on both sides.

$E = 8548.56 - 6396$
$E = 2152.56$

$$\begin{array}{r} \overset{4\ 14}{8\,\not{5}\,\not{4}\,8.56} \\ -\ 6\,3\,9\,6.00 \\ \hline 2\,1\,5\,2.56 \end{array}$$

Check. We can obtain a partial check by rounding and estimating:

$114.9 \times 74.4 \approx 110 \times 75 = 8250 \approx 8548.56$

$120 \times 53.3 \approx 120 \times 50 = 6000 \approx 6396$

$8250 - 6000 = 2250 \approx 2152.56$

State. The area of a soccer field is 2152.56 sq yd greater than the area of a football field.

45. ***Familiarize.*** This is a three-step problem. First we find m = the number of months in 30 years.

Translate and Solve.

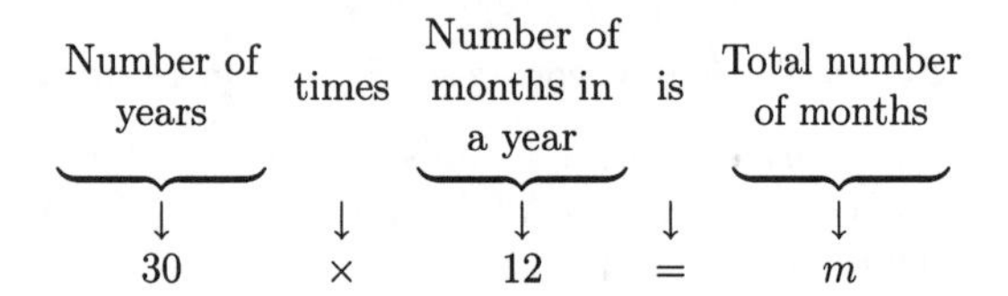

We carry out the multiplication.

$$\begin{array}{r} 1\,2 \\ \times\ 3\,0 \\ \hline 3\,6\,0 \end{array}$$

Thus, $m = 360$.

Next we find a = the amount paid back.

Monthly payment times Number of months is Amount paid back

$880.52 \times 360 = a$

We carry out the multiplication.

$$\begin{array}{r} 8\,8\,0.5\,2 \\ \times\ \ \ 3\,6\,0 \\ \hline 5\,2\,8\,3\,1\,2\,0 \\ 2\,6\,4\,1\,5\,6\,0\,0 \\ \hline 3\,1\,6,9\,8\,7.2\,0 \end{array}$$

Thus, $a = 316,987.20$.

Finally, we find p = the amount by which the amount paid back exceeds the amount of the loan. This is a "how-much-more" situation.

Amount of loan plus Excess amount is Amount paid back

$120,000 + p = 316,987.20$

We subtract 120,000 on both sides.

$p = 316,987.20 - 120,000$
$p = 196,987.20$

$$\begin{array}{r} 3\,1\,6,9\,8\,7.2\,0 \\ -\ 1\,2\,0,0\,0\,0.0\,0 \\ \hline 1\,9\,6,9\,8\,7.2\,0 \end{array}$$

Check. We repeat the calculations.

State. You pay back $316,987.20. This is $196,987.20 more than the amount of the loan.

47. ***Familiarize.*** We visualize the situation. We let p = the number by which the O'Hare passengers exceed the San Francisco passengers in millions.

67.3 million	
36.3 million	p

Translate. We have a "how much more " situation.

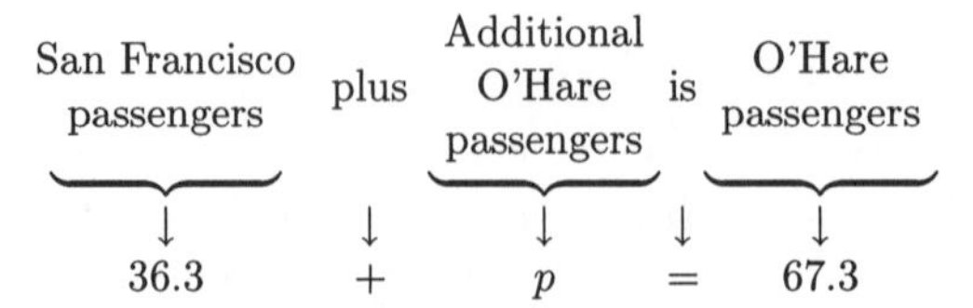

Solve. We subtract 36.3 on both sides of the equation:

$p = 67.3 - 36.3$
$p = 31$

$$\begin{array}{r} 6\,7.3 \\ -\ 3\,6.3 \\ \hline 3\,1.0 \end{array}$$

Check. We add 31 to 36.3 to get 67.3. This checks.

State. O'Hare handles 31 million more passengers per year than San Francisco.

49. ***Familiarize.*** We visualize the situation. We let t = the number of passengers handled by Dallas/Ft. Worth and San Francisco together in millions.

54.3 million	36.3 million
t	

Translate. We are combining amounts.

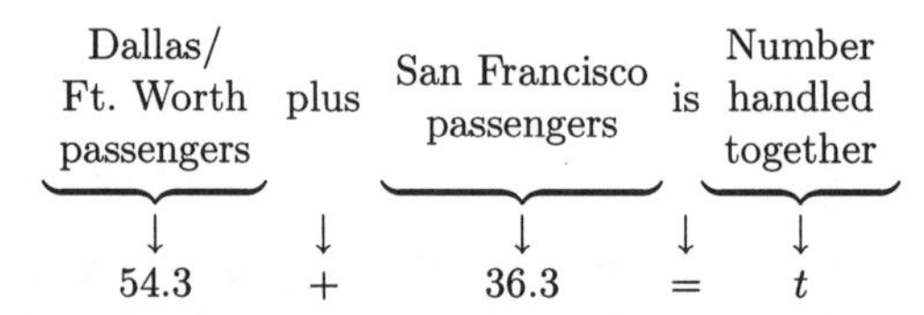

Solve. We carry out the addition.

$$\begin{array}{r} 5\,4.3 \\ +\ 3\,6.3 \\ \hline 9\,0.6 \end{array}$$

Thus $t = 90.6$.

Check. We can repeat the addition. We can also check by rounding: $54.3 + 36.3 \approx 50 + 40 = 90 \approx 90.6$.

State. Dallas/Ft. Worth and San Francisco handle 90.6 million passengers per year together.

51. ***Familiarize.*** We visualize the situation. We let t = the number of degrees by which the temperature of the bath water exceeds normal body temperature.

98.6°F	t
100°F	

Translate. We have a "how-much-more" situation.

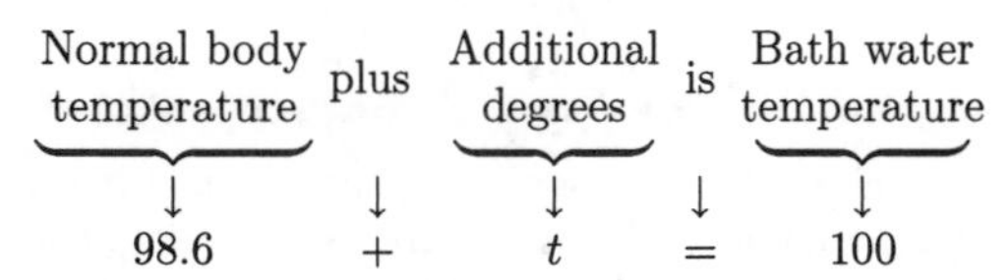

Solve. To solve we subtract 98.6 on both sides of the equation.

$t = 100 - 98.6$
$t = 1.4$

$$\begin{array}{r} \overset{9\ 9\ 10}{\not{1}\not{0}\not{0}.\not{0}} \\ -\ 9\ 8.\ 6 \\ \hline 1.\ 4 \end{array}$$

Check. To check we add 1.4 to 98.6 to get 100. This checks.

State. The temperature of the bath water is 1.4°F above normal body temperature.

53. ***Familiarize.*** This is a multistep problem. We will find the correct change and the amount of the actual change and compare them to determine if the change was correct. We let $c =$ the correct change and $a =$ the actual change.

Translate. We write two equations.

$$\underbrace{\$20}\ \text{less}\ \underbrace{\text{Cost of book}}\ \text{is}\ \underbrace{\text{Correct change}}$$
$$20 - 10.75 = c$$

$$\underbrace{\text{Amount in \$5 bills}}\ \text{plus}\ \underbrace{\text{Amount in \$1 bills}}\ \text{plus}\ \underbrace{\text{Amount in dimes}}\ \text{plus}$$
$$5 + (1+1+1) + 0.10 +$$
$$\underbrace{\text{Amount in nickels}}\ \text{is}\ \underbrace{\text{Actual change}}$$
$$(0.05 + 0.05) = a$$

Solve. To solve the first equation we carry out the subtraction.

$$\begin{array}{r} \overset{1\ 9\ 9\ 10}{\not{2}\not{0}.\not{0}\not{0}} \\ -1\ 0.\ 7\ 5 \\ \hline 9.\ 2\ 5 \end{array}$$

Thus, $c = 9.25$.

To solve the second equation we carry out the addition.

$$5 + 1 + 1 + 1 + 0.10 + 0.05 + 0.05 = 8.20$$

Thus, $a = 8.20$.

The correct change is different from the actual change.

Check. We repeat the calculations.

State. The change was not correct.

55. ***Familiarize.*** We let $C =$ the cost of the home in San Francisco. Using the table in Example 8, find the indexes of Fresno and San Francisco.

Translate. Using the formula given in Example 8, we translate to an equation.

$$C = \$125,000 \div 82 \times 286$$

Solve. We carry out the computation.

$C = \$125,000 \div 82 \times 286$
$\approx \$1524.390 \times 286$ Dividing
$\approx \$435,976$ Multiplying and rounding to the nearest one

Check. We can repeat the computations. We can also estimate:

$C = \$125,000 \div 82 \times 286$
$\approx \$125,000 \div 100 \times 300$
$\approx \$375,000$

The answer checks.

State. A home selling for $125,000 in Fresno would cost about $435,976 in San Francisco.

57. ***Familiarize.*** We let $C =$ the cost of the home in Tampa. Using the table in Example 8, find the indexes of Indianapolis and Tampa.

Translate. Using the formula given in Example 8, we translate to an equation.

$$C = \$96,000 \div 79 \times 72$$

Solve. We carry out the computation.

$C = \$96,000 \div 79 \times 72$
$\approx \$1215.190 \times 72$ Dividing
$\approx \$87,494$ Multiplying and rounding to the nearest one

Check. We can repeat the computations. We can also estimate:

$C = \$96,000 \div 79 \times 72$
$\approx \$96,000 \div 80 \times 70$
$\approx \$84,000$

The answer checks.

State. A home selling for $96,000 in Indianapolis would cost about $87,494 in Tampa.

59. ***Familiarize.*** We let $C =$ the cost of the home in Atlanta. Using the table in Example 8, find the indexes of San Francisco and Atlanta.

Translate. Using the formula given in Example 8, we translate to an equation.

$$C = \$240,000 \div 286 \times 81$$

Solve. We carry out the computation.

$C = \$240,000 \div 286 \times 81$
$\approx \$839.161 \times 81$ Dividing
$\approx \$67,972$ Multiplying and rounding to the nearest one

Check. We can repeat the computations. We can also estimate:

$C = \$240,000 \div 286 \times 81$
$\approx \$240,000 \div 300 \times 100$
$\approx \$80,000$

The answer checks.

State. A home selling for \$240,000 in San Francisco would cost about \$67,972 in Atlanta.

61.
$$\begin{array}{r} {\scriptstyle 1\,1\,1} \\ 4\,5\,6\,9 \\ +\,1\,7\,6\,6 \\ \hline 6\,3\,3\,5 \end{array}$$

63.
$$\begin{array}{rcr} 4\boxed{\frac{1}{3}\cdot\frac{2}{2}} & = & 4\frac{2}{6} \\ +\,2\boxed{\frac{1}{2}\cdot\frac{3}{3}} & = & +\,2\frac{3}{6} \\ \hline & & 6\frac{5}{6} \end{array}$$

65.
$$\begin{array}{r} {\scriptstyle 1\,1} \\ 8\,0\,9\,9 \\ +\,5\,6\,6\,7 \\ \hline 1\,3,7\,6\,6 \end{array}$$

67.
$$\begin{aligned} \frac{2}{3}-\frac{5}{8} &= \frac{2}{3}\cdot\frac{8}{8}-\frac{5}{8}\cdot\frac{3}{3} \\ &= \frac{16}{24}-\frac{15}{24}=\frac{16-15}{24} \\ &= \frac{1}{24} \end{aligned}$$

69.
$$\begin{aligned} \frac{5}{6}-\frac{7}{10} &= \frac{5}{6}\cdot\frac{5}{5}-\frac{7}{10}\cdot\frac{3}{3} \\ &= \frac{25}{30}-\frac{21}{30} \\ &= \frac{25-21}{30}=\frac{4}{30} \\ &= \frac{2\cdot 2}{2\cdot 15}=\frac{2}{2}\cdot\frac{2}{15} \\ &= \frac{2}{15} \end{aligned}$$

71. ***Familiarize***. Visualize the situation as a rectangular array containing 469 revolutions with $16\frac{3}{4}$ revolutions in each row. We must determine how many rows the array has. (The last row may be incomplete.) We let t = the time the wheel rotates.

Translate. The division that corresponds to the situation is

$$469 \div 16\frac{3}{4} = t.$$

Solve. We carry out the division.

$$t = 469 \div 16\frac{3}{4} = 469 \div \frac{67}{4} = 469 \cdot \frac{4}{67} =$$

$$\frac{67\cdot 7\cdot 4}{67\cdot 1} = \frac{67}{67}\cdot\frac{7\cdot 4}{1} = 28$$

Check. We check by multiplying the time by the number of revolutions per minute.

$$16\frac{3}{4}\cdot 28 = \frac{67}{4}\cdot 28 = \frac{67\cdot 7\cdot 4}{4\cdot 1} = \frac{4}{4}\cdot\frac{67\cdot 7}{1} = 469$$

The answer checks.

State. The water wheel rotated for 28 min.

73. 

75. ***Familiarize***. This is a multistep problem. First we will find the total number of cards purchased. Then we will find the number of half-dozens in this number. Finally, we will find the purchase price. Let t = the total number of cards, h = the number of half-dozens, and p = the purchase price. Recall that 1 dozen = 12, so $\frac{1}{2}$ dozen $= \frac{1}{2}\cdot 12 = 6$.

Translate. We write an equation to find the total number of cards purchased.

$$\underbrace{\text{Number of packs}}_{\downarrow\; 6} \;\times\; \underbrace{\text{Number per pack}}_{\downarrow\; 12} \;=\; \underbrace{\text{Total number of cards}}_{\downarrow\; t}$$

Solve. We carry out the multiplication.

$$t = 6 \times 12 = 72$$

Next we divide by 6 to find the number of half-dozens in 72:

$$h = 72 \div 6 = 12$$

Finally, we multiply the number of half-dozens by the price per half dozen to find the purchase price:

$$p = 12 \times \$0.12 = \$1.44$$

Check. We can repeat the calculations. The answer checks.

State. The purchase price of the cards is \$1.44.

Chapter 4

Percent Notation

Exercise Set 4.1

1. The ratio of 4 to 5 is $\frac{4}{5}$.

3. The ratio of 56.78 to 98.35 is $\frac{56.78}{98.35}$.

5. If four of every five fatal accidents involving a Corvette do not involve another vehicle, then $5 - 4$, or 1, involves a Corvette and at least one other vehicle. Thus, the ratio of fatal accidents involving just a Corvette to those involving a Corvette and at least one other vehicle is $\frac{4}{1}$.

7. The ratio of 18 to 24 is $\frac{18}{24} = \frac{3 \cdot 6}{4 \cdot 6} = \frac{3}{4} \cdot \frac{6}{6} = \frac{3}{4}$.

9. The ratio is $\frac{0.48}{0.64} = \frac{0.48}{0.64} \cdot \frac{100}{100} = \frac{48}{64} = \frac{3 \cdot 16}{4 \cdot 16} = \frac{3}{4} \cdot \frac{16}{16} = \frac{3}{4}$.

11. The ratio is $\frac{6.4}{20.2} = \frac{6.4}{20.2} \cdot \frac{10}{10} = \frac{64}{202} = \frac{2 \cdot 32}{2 \cdot 101} = \frac{2}{2} \cdot \frac{32}{101} = \frac{32}{101}$.

13. $\frac{120 \text{ km}}{3 \text{ hr}}$, or $40 \frac{\text{km}}{\text{hr}}$

15. $\frac{440 \text{ m}}{40 \text{ sec}}$, or $11 \frac{\text{m}}{\text{sec}}$

17. $\frac{342 \text{ yd}}{2.25 \text{ days}}$, or $152 \frac{\text{yd}}{\text{day}}$

$$
\begin{array}{r}
152. \\
2.25_{\wedge}\overline{)\,342.00_{\wedge}} \\
\underline{225\,00} \\
117\,00 \\
\underline{112\,50} \\
4\,50 \\
\underline{4\,50} \\
0
\end{array}
$$

19. $\frac{\$5.75}{10 \text{ min}} = \frac{575¢}{10 \text{ min}} = 57.5 \frac{¢}{\text{min}}$

21. $\frac{623 \text{ gal}}{1000 \text{ sq ft}} = 0.623 \text{ gal/ft}^2$

23. $\frac{310 \text{ km}}{2.5 \text{ hr}} = 124 \frac{\text{km}}{\text{hr}}$

25. $\frac{2660 \text{ mi}}{4.75 \text{ hr}} = 560 \frac{\text{mi}}{\text{hr}}$

27. We can use cross-products:

$5 \cdot 9 = 45$ $\quad$ $\frac{5}{6} \quad \frac{7}{9}$ $\quad$ $6 \cdot 7 = 42$

Since the cross-products are not the same, $45 \neq 42$, we know that the numbers are not proportional.

29. We can use cross-products:

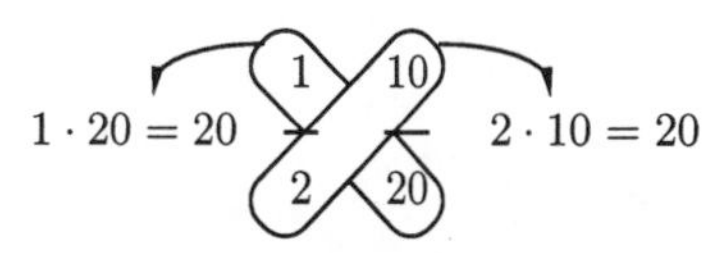

Since the cross-products are the same, $20 = 20$, we know that $\frac{1}{2} = \frac{10}{20}$, so the numbers are proportional.

31. We can use cross-products:

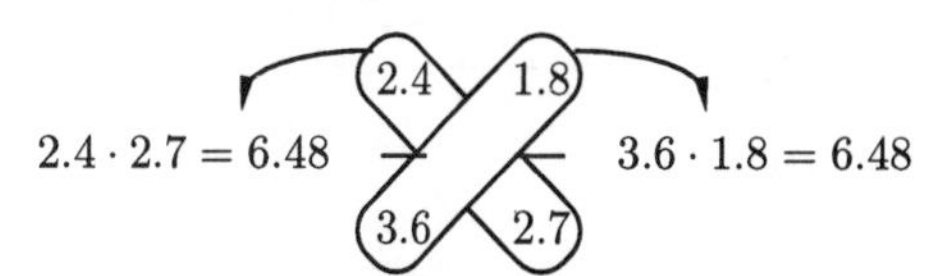

Since the cross-products are the same, $6.48 = 6.48$, we know that $\frac{2.4}{3.6} = \frac{1.8}{2.7}$, so the numbers are proportional.

33. We can use cross-products:

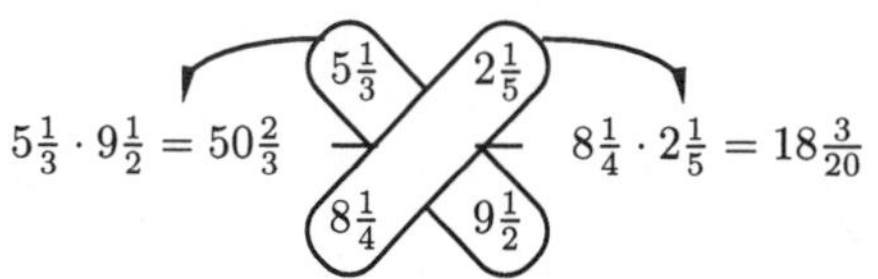

Since the cross-products are not the same, $50\frac{2}{3} \neq 18\frac{3}{20}$, we know that the numbers are not proportional.

35.

$$\frac{18}{4} = \frac{x}{10}$$

$18 \cdot 10 = 4 \cdot x$ $\quad$ Equating cross-products

$\frac{18 \cdot 10}{4} = x$ $\quad$ Dividing by 4

$\frac{180}{4} = x$ $\quad$ Multiplying

$45 = x$ $\quad$ Dividing

37.

$$\frac{t}{12} = \frac{5}{6}$$
$$6 \cdot t = 12 \cdot 5$$
$$t = \frac{12 \cdot 5}{6}$$
$$t = \frac{60}{6}$$
$$t = 10$$

39.

$$\frac{2}{5} = \frac{8}{n}$$
$$2 \cdot n = 5 \cdot 8$$
$$n = \frac{5 \cdot 8}{2}$$
$$n = \frac{40}{2}$$
$$n = 20$$

41. $\frac{16}{12} = \frac{24}{x}$

$16 \cdot x = 12 \cdot 24$

$x = \frac{12 \cdot 24}{16}$

$x = \frac{288}{16}$

$x = 18$

43. $\frac{t}{0.16} = \frac{0.15}{0.40}$

$0.40 \times t = 0.16 \times 0.15$

$t = \frac{0.16 \times 0.15}{0.40}$

$t = \frac{0.024}{0.40}$

$t = 0.06$

45. $\frac{100}{25} = \frac{20}{n}$

$100 \cdot n = 25 \cdot 20$

$n = \frac{25 \cdot 20}{100}$

$n = \frac{500}{100}$

$n = 5$

47. $\frac{\frac{1}{4}}{\frac{1}{2}} = \frac{\frac{1}{2}}{x}$

$\frac{1}{4} \cdot x = \frac{1}{2} \cdot \frac{1}{2}$

$x = \frac{\frac{1}{2} \cdot \frac{1}{2}}{\frac{1}{4}}$

$x = \frac{\frac{1}{4}}{\frac{1}{4}}$

$x = 1$

49. $\frac{1.28}{3.76} = \frac{4.28}{y}$

$1.28 \times y = 3.76 \times 4.28$

$y = \frac{3.76 \times 4.28}{1.28}$

$y = \frac{16.0928}{1.28}$

$y = 12.5725$

51. ***Familiarize***. Let $d =$ the distance traveled in 42 days, in miles.

Translate. We translate to a proportion, keeping the distance in the numerators.

$$\begin{array}{r} \text{Distance} \rightarrow \\ \text{Time} \rightarrow \end{array} \frac{234}{14} = \frac{d}{42} \begin{array}{l} \leftarrow \text{Distance} \\ \leftarrow \text{Time} \end{array}$$

Solve.

$234 \cdot 42 = 14 \cdot d$ Equating cross-products

$\frac{234 \cdot 42}{14} = d$ Dividing by 14

$\frac{234 \cdot 3 \cdot 14}{14} = d$ Factoring

$702 = d$ Multiplying and dividing

Check. We substitute into the proportion and check cross-products.

$\frac{234}{14} = \frac{702}{42}$; $234 \cdot 42 = 9828$; $14 \cdot 702 = 9828$

Since the cross-products are the same, the answer checks.

State. Monica would travel 702 mi in 42 days.

53. ***Familiarize***. Let $t =$ the number of trees required to produce 375 pounds of coffee.

Translate. We translate to a proportion.

$$\begin{array}{r} \text{Trees} \rightarrow \\ \text{Pounds} \rightarrow \end{array} \frac{14}{17} = \frac{t}{375} \begin{array}{l} \leftarrow \text{Trees} \\ \leftarrow \text{Pounds} \end{array}$$

Solve.

$14 \cdot 375 = 17 \cdot t$

$\frac{14 \cdot 375}{17} = t$

$\frac{5250}{17} = t$, or

$308\frac{14}{17} = t$

Because it doesn't make sense to talk about a fractional part of a tree, we round the answer up to 309.

Check. We substitute in the proportion and check cross-products.

$\frac{14}{17} = \frac{5250/17}{375}$, $14 \cdot 375 = 5250$; $17 \cdot \frac{5250}{17} = 5250$

Since the cross-products are the same, the answer checks.

State. 309 trees are required to produce 3.75 pounds of coffee.

55. ***Familiarize***. Let $d =$ the number of defective bulbs in a lot of 22,000.

Translate. We translate to a proportion.

$$\begin{array}{r} \text{Defective bulbs} \rightarrow \\ \text{Bulbs in lot} \rightarrow \end{array} \frac{18}{200} = \frac{d}{22,000} \begin{array}{l} \leftarrow \text{Defective bulbs} \\ \leftarrow \text{Bulbs in lot} \end{array}$$

Solve.

$18 \cdot 22,000 = 200 \cdot d$

$\frac{18 \cdot 22,000}{200} = d$

$1980 = d$

Check. We substitute in the proportion and check cross-products.

$\frac{18}{200} = \frac{1980}{22,000}$; $18 \cdot 22,000 = 396,000$;

$200 \cdot 1980 = 396,000$

Since the cross-products are the same, the answer checks.

State. There would be 1980 defective bulbs in a lot of 22,000.

57. ***Familiarize***. We let $g =$ the number of gallons of sealant Bonnie should buy.

Translate. We translate to a proportion.

$$\begin{array}{r} \text{Area} \rightarrow \\ \text{Sealant} \rightarrow \end{array} \frac{450}{2} = \frac{1200}{g} \begin{array}{l} \leftarrow \text{Area} \\ \leftarrow \text{Sealant} \end{array}$$

Solve.

$$450 \cdot g = 2 \cdot 1200$$
$$g = \frac{2 \cdot 1200}{450}$$
$$g = \frac{16}{3}, \text{ or}$$
$$g = 5\frac{1}{3}$$

Check. We use a different approach. We find the area that can be waterproofed with 1 gal of sealant and then divide 1200 by that number:

$$450 \div 2 = 225 \text{ and } 1200 \div 225 = 5.\overline{3}, \text{ or } 5\frac{1}{3}$$

The answer checks.

State. Bonnie needs 5 entire gallons of sealant and $\frac{1}{3}$ of a sixth gallon, so she should buy 6 gal of sealant.

59. ***Familiarize.*** Using the label on the drawing in the text, we let x = the height of the largest painting, in feet.

Translate. We translate to a proportion.

$$\begin{matrix}\text{Length} \rightarrow \\ \text{Height} \rightarrow\end{matrix} \frac{7}{4} = \frac{14}{x} \begin{matrix}\leftarrow \text{Length} \\ \leftarrow \text{Height}\end{matrix}$$

Solve.

$$7 \cdot x = 4 \cdot 14 \quad \text{Equating cross-products}$$
$$x = \frac{4 \cdot 14}{7}$$
$$x = 8 \quad \text{Multiplying and dividing}$$

Check. We substitute in the proportion and check cross-products.

$$\frac{7}{4} = \frac{14}{8};\ 7 \cdot 8 = 56;\ 4 \cdot 14 = 56$$

Since the cross-products are the same, the answer checks.

State. The larger painting is 8 ft high.

61. ***Familiarize.*** Let w = the number of inches of water to which $5\frac{1}{2}$ ft of snow will melt.

Translate.

$$\begin{matrix}\text{Snow} \rightarrow \\ \text{Water} \rightarrow\end{matrix} \frac{1\frac{1}{2}}{2} = \frac{5\frac{1}{2}}{w} \begin{matrix}\leftarrow \text{Snow} \\ \leftarrow \text{Water}\end{matrix}$$

Solve.

$$1\frac{1}{2} \cdot w = 2 \cdot 5\frac{1}{2}$$
$$\frac{3}{2} \cdot w = \frac{2}{1} \cdot \frac{11}{2}$$
$$w = \frac{2}{1} \cdot \frac{11}{2} \cdot \frac{2}{3} \quad \text{Dividing by } \frac{3}{2}$$
$$w = \frac{2 \cdot 11 \cdot 2}{1 \cdot 2 \cdot 3}$$
$$w = \frac{22}{3}, \text{ or } 7\frac{1}{3}$$

Check. We substitute in the proportion and check cross-products.

$$\frac{1\frac{1}{2}}{2} = \frac{5\frac{1}{2}}{7\frac{1}{3}};\ 1\frac{1}{2} \cdot 7\frac{1}{3} = \frac{3}{2} \cdot \frac{22}{3} = 11;$$
$$2 \cdot 5\frac{1}{2} = 2 \cdot \frac{11}{2} = 11$$

Since the cross-products are the same, the answer checks.

State. Thus, $5\frac{1}{2}$ ft of snow will melt to $7\frac{1}{3}$ in. of water.

63. ***Familiarize.*** Let ERA = the earned run average.

Translate. We translate to a proportion.

$$\begin{matrix}\text{Earned runs} \rightarrow \\ \text{Innings} \rightarrow\end{matrix} \frac{ERA}{9} = \frac{74}{245} \begin{matrix}\leftarrow \text{Earned runs} \\ \leftarrow \text{Innings}\end{matrix}$$

Solve.

$$245 \cdot ERA = 9 \cdot 74$$
$$ERA = \frac{9 \cdot 74}{245}$$
$$ERA \approx 2.72 \quad \text{Rounding to the nearest hundredth}$$

Check. We substitute in the proportion and check cross-products.

$$\frac{2.72}{9} = \frac{74}{245};\ 2.72(245) = 666.4;\ 9 \cdot 74 = 666$$

Since $666.4 \approx 666$, the answer checks.

State. Greg Maddux's earned run average was about 2.72.

65. ***Familiarize.*** Let ERA = the earned run average.

Translate. We translate to a proportion.

$$\begin{matrix}\text{Earned runs} \rightarrow \\ \text{Innings} \rightarrow\end{matrix} \frac{ERA}{9} = \frac{125}{213} \begin{matrix}\leftarrow \text{Earned runs} \\ \leftarrow \text{Innings}\end{matrix}$$

Solve.

$$213 \cdot ERA = 9 \cdot 125$$
$$ERA = \frac{9 \cdot 125}{213}$$
$$ERA \approx 5.28 \quad \text{Rounding to the nearest hundredth}$$

Check. We substitute in the proportion and check cross-products.

$$\frac{5.28}{9} = \frac{125}{213};\ 5.28(213) = 1124.64;\ 9 \cdot 125 = 1125$$

Since $1124.64 \approx 1125$, the answer checks.

State. Kevin Ritz's earned run average was about 5.28.

67. ***Familiarize.*** We visualize the situation. We let p = the amount of precipitation in inches and c = the amount of precipitation in centimeters.

31.1 in., or 78.994 cm	2.6 in., or 6.604 cm
p in., or c cm	

Translate and Solve. We are combining amounts.

$$\underbrace{\text{Amount of rain}} \quad \text{plus} \quad \underbrace{\text{Amount of snow}} \quad \text{is} \quad \underbrace{\text{Total precipitation}}$$

a) Taking the amounts in inches, we translate to an equation:

$$31.1 + 2.6 = p$$

To solve we carry out the addition.

$$\begin{array}{r} 3\,1.1 \\ +\quad 2.6 \\ \hline 3\,3.7 \end{array}$$

Thus, $p = 33.7$ in.

b) Taking the amounts in centimeters, we translate to an equation:

$$78.994 + 6.604 = c$$

To solve we carry out the addition.

$$\begin{array}{r} {\scriptstyle 1\ 1}\quad\ \ \\ 7\,8.9\,9\,4 \\ +\quad 6.6\,0\,4 \\ \hline 8\,5.5\,9\,8 \end{array}$$

Thus, $c = 85.598$ cm.

Check. We can repeat the addition. We can also check by rounding:

$$31.1 + 2.6 \approx 31 + 3 = 34 \approx 33.7;$$

$$78.994 + 6.604 \approx 79 + 7 = 86 \approx 85.598.$$

State. a) The total average annual precipitation in Dallas, Texas, is 33.7 in.

b) The total average annual precipitation in Dallas, Texas, is 85.598 cm.

69. 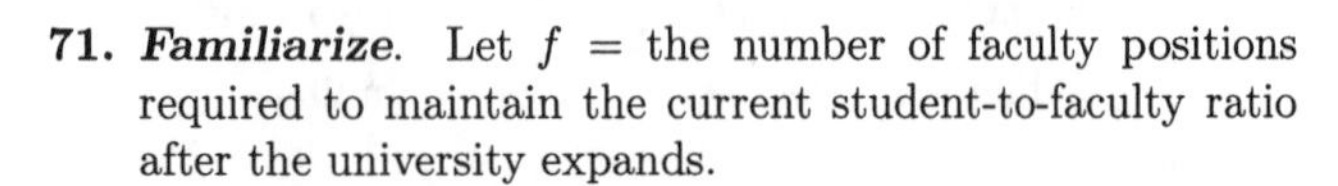

71. ***Familiarize***. Let f = the number of faculty positions required to maintain the current student-to-faculty ratio after the university expands.

Translate. We translate to a proportion.

$$\begin{array}{rcl} \text{Students} \rightarrow & \dfrac{2700}{217} = \dfrac{2900}{f} & \leftarrow \text{Students} \\ \text{Faculty} \rightarrow & & \leftarrow \text{Faculty} \end{array}$$

Solve.

$$2700 \cdot f = 217 \cdot 2900$$

$$f = \frac{217 \cdot 2900}{2700}$$

$$f = \frac{6293}{27}, \text{ or } 233\frac{2}{27}$$

Since it is impossible to create a fractional part of a position, we round up to the nearest whole position. Thus, 234 positions will be required after the university expands. We subtract to find how many new positions should be created:

$$234 - 217 = 17$$

Check. We substitute in the proportion and check cross-products.

$$\frac{2700}{217} = \frac{2900}{6293/27}; \ 2700 \cdot \frac{6293}{27} = 629,300;$$

$$217 \cdot 2900 = 629,300$$

State. 17 new faculty positions should be created.

73. ***Familiarize***. This is a multistep problem. First we will find the area A of the wall. Then we will find p, the number of gallons of paint Sue should buy.

Translate. We substitute 100 for the length and 30 for the width in the formula for the area of a rectangle to find the area of the wall.

$$A = l \times w$$

$$A = 100 \times 30$$

Solve. We carry out the multiplication.

$$A = 100 \times 30 = 3000$$

Now we translate to a proportion and solve it.

$$\begin{array}{rcl} \text{Area} \rightarrow & \dfrac{950}{2} = \dfrac{3000}{p} & \leftarrow \text{Area} \\ \text{Paint} \rightarrow & & \leftarrow \text{Paint} \end{array}$$

$$950 \cdot p = 2 \cdot 3000$$

$$p = \frac{2 \cdot 3000}{950}$$

$$p = \frac{2 \cdot 50 \cdot 60}{19 \cdot 50}$$

$$p = \frac{2 \cdot 60}{19}$$

$$p = \frac{120}{19}, \text{ or } 6\frac{6}{19}$$

Assuming that Sue is buying paint in one gallon cans, she will have to buy 7 gal of paint.

Check. Repeat the calculations.

State. Sue should buy 7 gal of paint.

Exercise Set 4.2

1.
$90\% = \dfrac{90}{100}$ — A ratio of 90 to 100

$90\% = 90 \times \dfrac{1}{100}$ — Replacing % with $\times\dfrac{1}{100}$

$90\% = 90 \times 0.01$ — Replacing % with $\times 0.01$

3.
$12.5\% = \dfrac{12.5}{100}$ — A ratio of 12.5 to 100

$12.5\% = 12.5 \times \dfrac{1}{100}$ — Replacing % with $\times\dfrac{1}{100}$

$12.5\% = 12.5 \times 0.01$ — Replacing % with $\times 0.01$

5. 67%

a) Replace the percent symbol with $\times 0.01$.

67×0.01

b) Move the decimal point two places to the left.

0.67.

Thus, $67\% = 0.67$.

7. 45.6%

a) Replace the percent symbol with $\times 0.01$.

45.6×0.01

b) Move the decimal point two places to the left.

0.45.6
↑__|

Thus, 45.6% = 0.456.

9. 59.01%

a) Replace the percent symbol with $\times 0.01$.

59.01 × 0.01

b) Move the decimal point two places to the left.

0.59.01
↑__|

Thus, 59.01% = 0.5901.

11. 10%

a) Replace the percent symbol with $\times 0.01$.

10 × 0.01

b) Move the decimal point two places to the left.

0.10.
↑__|

Thus, 10% = 0.1.

13. 1%

a) Replace the percent symbol with $\times 0.01$.

1 × 0.01

b) Move the decimal point two places to the left.

0.01.
↑__|

Thus, 1% = 0.01.

15. 200%

a) Replace the percent symbol with $\times 0.01$.

200 × 0.01

b) Move the decimal point two places to the left.

2.00.
↑__|

Thus, 200% = 2.

17. 0.1%

a) Replace the percent symbol with $\times 0.01$.

0.1 × 0.01

b) Move the decimal point two places to the left.

0.00.1
↑__|

Thus, 0.1% = 0.001.

19. 0.09%

a) Replace the percent symbol with $\times 0.01$.

0.09 × 0.01

b) Move the decimal point two places to the left.

0.00.09
↑__|

Thus, 0.09% = 0.0009.

21. 0.18%

a) Replace the percent symbol with $\times 0.01$.

0.18 × 0.01

b) Move the decimal point two places to the left.

0.00.18
↑__|

Thus, 0.18% = 0.0018.

23. 23.19%

a) Replace the percent symbol with $\times 0.01$.

23.19 × 0.01

b) Move the decimal point two places to the left.

0.23.19
↑__|

Thus, 23.19% = 0.2319.

25. 40%

a) Replace the percent symbol with $\times 0.01$.

40 × 0.01

b) Move the decimal point two places to the left.

0.40.
↑__|

Thus, 40% = 0.4.

27. 2.5%

a) Replace the percent symbol with $\times 0.01$.

2.5 × 0.01

b) Move the decimal point two places to the left.

0.02.5
↑__|

Thus, 2.5% = 0.025.

29. 62.2%

a) Replace the percent symbol with $\times 0.01$.

62.2 × 0.01

b) Move the decimal point two places to the left.

0.62.2
↑__|

Thus, 62.2% = 0.622.

31. 0.47

a) Move the decimal point two places to the right.

0.47.
|__↑

b) Write a percent symbol: 47%

Thus, 0.47 = 47%.

33. 0.03

a) Move the decimal point two places to the right.

0.03.
|__↑

b) Write a percent symbol: 3%

Thus, 0.03 = 3%.

35. 8.7

a) Move the decimal point two places to the right.

8.70.

b) Write a percent symbol: 870%

Thus, 8.7 = 870%.

37. 0.334

a) Move the decimal point two places to the right.

0.33.4

b) Write a percent symbol: 33.4%

Thus, 0.334 = 33.4%.

39. 0.75

a) Move the decimal point two places to the right.

0.75.

b) Write a percent symbol: 75%

Thus, 0.75 = 75%.

41. 0.4

a) Move the decimal point two places to the right.

0.40.

b) Write a percent symbol: 40%

Thus, 0.4 = 40%.

43. 0.006

a) Move the decimal point two places to the right.

0.00.6

b) Write a percent symbol: 0.6%

Thus, 0.006 = 0.6%.

45. 0.017

a) Move the decimal point two places to the right.

0.01.7

b) Write a percent symbol: 1.7%

Thus, 0.017 = 1.7%.

47. 0.2718

a) Move the decimal point two places to the right.

0.27.18

b) Write a percent symbol: 27.18%

Thus, 0.2718 = 27.18%.

49. 0.0239

a) Move the decimal point two places to the right.

0.02.39

b) Write a percent symbol: 2.39%

Thus, 0.0239 = 2.39%.

51. 0.000104

a) Move the decimal point two places to the right.

0.00.0104

b) Write a percent symbol: 0.0104%

Thus, 0.000104 = 0.0104%.

53. 0.24

a) Move the decimal point two places to the right.

0.24.

b) Write a percent symbol: 24%

Thus, 0.24 = 24%.

55. 0.581

a) Move the decimal point two places to the right.

0.58.1

b) Write a percent symbol: 58.1%

Thus, 0.581 = 58.1%.

57. To convert $\frac{100}{3}$ to a mixed numeral, we divide.

$$\begin{array}{r} 33 \\ 3\overline{)100} \\ \underline{90} \\ 10 \\ \underline{9} \\ 1 \end{array} \qquad \frac{100}{3} = 33\frac{1}{3}$$

59. To convert $\frac{75}{8}$ to a mixed numeral, we divide.

$$\begin{array}{r} 9 \\ 8\overline{)75} \\ \underline{72} \\ 3 \end{array} \qquad \frac{75}{8} = 9\,\frac{3}{8}$$

61. To convert $\frac{2}{3}$ to decimal notation, we divide.

$$\begin{array}{r} 0.66 \\ 3\overline{)2.00} \\ \underline{1\,8} \\ 20 \\ \underline{18} \\ 2 \end{array}$$

Since 2 keeps reappearing as a remainder, the digits repeat and

$$\frac{2}{3} = 0.66\ldots \quad \text{or} \quad 0.\overline{6}.$$

63. To convert $\frac{5}{6}$ to decimal notation, we divide.

$$\begin{array}{r} 0.8\,3 \\ 6\,\overline{)\,5.0\,0} \\ 4\,8 \\ \hline 2\,0 \\ 1\,8 \\ \hline 2 \end{array}$$

Since 2 keeps reappearing as a remainder, the digits repeat and

$$\frac{5}{6} = 0.833\ldots \quad \text{or} \quad 0.8\overline{3}.$$

65. 

Exercise Set 4.3

1. We use the definition of percent as a ratio.

$$\frac{41}{100} = 41\%$$

3. We use the definition of percent as a ratio.

$$\frac{5}{100} = 5\%$$

5. We multiply by 1 to get 100 in the denominator.

$$\frac{2}{10} = \frac{2}{10} \cdot \frac{10}{10} = \frac{20}{100} = 20\%$$

7. We multiply by 1 to get 100 in the denominator.

$$\frac{3}{10} = \frac{3}{10} \cdot \frac{10}{10} = \frac{30}{100} = 30\%$$

9. $\frac{1}{2} = \frac{1}{2} \cdot \frac{50}{50} = \frac{50}{100} = 50\%$

11. Find decimal notation by division.

$$\begin{array}{r} 0.6\,2\,5 \\ 8\,\overline{)\,5.0\,0\,0} \\ 4\,8 \\ \hline 2\,0 \\ 1\,6 \\ \hline 4\,0 \\ 4\,0 \\ \hline 0 \end{array}$$

$$\frac{5}{8} = 0.625$$

Convert to percent notation.

0.62.5

$$\frac{5}{8} = 62.5\%, \text{or } 62\tfrac{1}{2}\%$$

13. $\frac{4}{5} = \frac{4}{5} \cdot \frac{20}{20} = \frac{80}{100} = 80\%$

15. Find decimal notation by division.

$$\begin{array}{r} 0.6\,6\,6 \\ 3\,\overline{)\,2.0\,0\,0} \\ 1\,8 \\ \hline 2\,0 \\ 1\,8 \\ \hline 2\,0 \\ 1\,8 \\ \hline 2 \end{array}$$

We get a repeating decimal: $\frac{2}{3} = 0.66\overline{6}$

Convert to percent notation.

$0.66.\overline{6}$

$$\frac{2}{3} = 66.\overline{6}\%, \text{or } 66\tfrac{2}{3}\%$$

17.

$$\begin{array}{r} 0.1\,6\,6 \\ 6\,\overline{)\,1.0\,0\,0} \\ 6 \\ \hline 4\,0 \\ 3\,6 \\ \hline 4\,0 \\ 3\,6 \\ \hline 4 \end{array}$$

We get a repeating decimal: $\frac{1}{6} = 0.16\overline{6}$

Convert to percent notation.

$0.16.\overline{6}$

$$\frac{1}{6} = 16.\overline{6}\%, \text{or } 16\tfrac{2}{3}\%$$

19. $\frac{4}{25} = \frac{4}{25} \cdot \frac{4}{4} = \frac{16}{100} = 16\%$

21. $\frac{1}{20} = \frac{1}{20} \cdot \frac{5}{5} = \frac{5}{100} = 5\%$

23. $\frac{17}{50} = \frac{17}{50} \cdot \frac{2}{2} = \frac{34}{100} = 34\%$

25. $\frac{9}{25} = \frac{9}{25} \cdot \frac{4}{4} = \frac{36}{100} = 36\%$

27. $\frac{21}{100} = 21\%$ Using the definition of percent as a ratio

29. $\frac{6}{25} = \frac{6}{25} \cdot \frac{4}{4} = \frac{24}{100} = 24\%$

31. $85\% = \frac{85}{100}$ Definition of percent

$$\left.\begin{aligned} &= \frac{5 \cdot 17}{5 \cdot 20} \\ &= \frac{5}{5} \cdot \frac{17}{20} \\ &= \frac{17}{20} \end{aligned}\right\} \text{Simplifying}$$

33. $62.5\% = \frac{62.5}{100}$ Definition of percent

$= \frac{62.5}{100} \cdot \frac{10}{10}$ Multiplying by 1 to eliminate the decimal point in the numerator

$= \frac{625}{1000}$

$= \frac{5 \cdot 125}{8 \cdot 125}$

$= \frac{5}{8} \cdot \frac{125}{125}$ Simplifying

$= \frac{5}{8}$

35. $33\frac{1}{3}\% = \frac{100}{3}\%$ Converting from mixed numeral to fractional notation

$= \frac{100}{3} \times \frac{1}{100}$ Definition of percent

$= \frac{100 \cdot 1}{3 \cdot 100}$ Multiplying

$= \frac{1}{3} \cdot \frac{100}{100}$ Simplifying

$= \frac{1}{3}$

37. $16.\overline{6}\% = 16\frac{2}{3}\%$ $(16.\overline{6} = 16\frac{2}{3})$

$= \frac{50}{3}\%$ Converting from mixed numeral to fractional notation

$= \frac{50}{3} \times \frac{1}{100}$ Definition of percent

$= \frac{50 \cdot 1}{3 \cdot 50 \cdot 2}$ Multiplying

$= \frac{1}{2 \cdot 3} \cdot \frac{50}{50}$ Simplifying

$= \frac{1}{6}$

39. $7.25\% = \frac{7.25}{100} = \frac{7.25}{100} \cdot \frac{100}{100}$

$= \frac{725}{10,000} = \frac{29 \cdot 25}{400 \cdot 25} = \frac{29}{400} \cdot \frac{25}{25}$

$= \frac{29}{400}$

41. $0.8\% = \frac{0.8}{100} = \frac{0.8}{100} \cdot \frac{10}{10}$

$= \frac{8}{1000} = \frac{1 \cdot 8}{125 \cdot 8} = \frac{1}{125} \cdot \frac{8}{8}$

$= \frac{1}{125}$

43. $25\frac{3}{8}\% = \frac{203}{8}\%$

$= \frac{203}{8} \times \frac{1}{100}$ Definition of percent

$= \frac{203}{800}$

45. $78\frac{2}{9}\% = \frac{704}{9}\%$

$= \frac{704}{9} \times \frac{1}{100}$ Definition of percent

$= \frac{4 \cdot 176 \cdot 1}{9 \cdot 4 \cdot 25}$

$= \frac{4}{4} \cdot \frac{176 \cdot 1}{9 \cdot 25}$

$= \frac{176}{225}$

47. $64\frac{7}{11}\% = \frac{711}{11}\%$

$= \frac{711}{11} \times \frac{1}{100}$

$= \frac{711}{1100}$

49. $150\% = \frac{150}{100} = \frac{3 \cdot 50}{2 \cdot 50} = \frac{3}{2} \cdot \frac{50}{50} = \frac{3}{2}$

51. $0.0325\% = \frac{0.0325}{100} = \frac{0.0325}{100} \cdot \frac{10,000}{10,000} = \frac{325}{1,000,000} = \frac{25 \cdot 13}{25 \cdot 40,000} = \frac{25}{25} \cdot \frac{13}{40,000} = \frac{13}{40,000}$

53. Note that $33.\overline{3}\% = 33\frac{1}{3}\%$ and proceed as in Exercise 35; $33.\overline{3}\% = \frac{1}{3}$.

55. $55\% = \frac{55}{100} = \frac{5 \cdot 11}{5 \cdot 20} = \frac{5}{5} \cdot \frac{11}{20} = \frac{11}{20}$

57. $38\% = \frac{38}{100} = \frac{2 \cdot 19}{2 \cdot 50} = \frac{2}{2} \cdot \frac{19}{50} = \frac{19}{50}$

59. $11\% = \frac{11}{100}$

61. $25\% = \frac{25}{100} = \frac{25 \cdot 1}{25 \cdot 4} = \frac{25}{25} \cdot \frac{1}{4} = \frac{1}{4}$

63. $5.69\% = \frac{5.69}{100} = \frac{5.69}{100} \cdot \frac{100}{100} = \frac{569}{10,000}$

65. $\frac{1}{8} = 1 \div 8$

```
     0.1 2 5
  8 ⟌1.0 0 0
       8
       2 0
       1 6
         4 0
         4 0
           0
```

$\mathbf{\frac{1}{8} = 0.125 = 12\frac{1}{2}\%}$**, or 12.5%**

$$\frac{1}{6} = 1 \div 6$$

$$\begin{array}{r} 0.1\,6\,6 \\ 6\,\overline{)\,1.0\,0\,0} \\ \underline{6} \\ 4\,0 \\ \underline{3\,6} \\ 4\,0 \\ \underline{3\,6} \\ 4 \end{array}$$

We get a repeating decimal: $0.1\overline{6}$

$0.16.\overline{6}$ $\qquad 0.1\overline{6} = 16.\overline{6}\%$

$$\mathbf{\frac{1}{6} = 0.1\overline{6} = 16.\overline{6}\%, \text{ or } 16\frac{2}{3}\%}$$

$$20\% = \frac{20}{100} = \frac{1}{5} \cdot \frac{20}{20} = \frac{1}{5}$$

$0.20.$ $\qquad 20\% = 0.2$

$$\mathbf{\frac{1}{5} = 0.2 = 20\%}$$

$0.25.$ $\qquad 0.25 = 25\%$

$$25\% = \frac{25}{100} = \frac{1}{4} \cdot \frac{25}{25} = \frac{1}{4}$$

$$\mathbf{\frac{1}{4} = 0.25 = 25\%}$$

$$33\frac{1}{3}\% = \frac{100}{3}\% = \frac{100}{3} \times \frac{1}{100} = \frac{100}{300} = \frac{1}{3} \cdot \frac{100}{100} = \frac{1}{3}$$

$0.33.\overline{3}$ $\qquad 33.\overline{3}\% = 0.33\overline{3}, \text{ or } 0.\overline{3}$

$$\mathbf{\frac{1}{3} = 0.\overline{3} = 33\frac{1}{3}\%, \text{ or } 33.\overline{3}\%}$$

$$37.5\% = \frac{37.5}{100} = \frac{37.5}{100} \cdot \frac{10}{10} = \frac{375}{1000} = \frac{3}{8} \cdot \frac{125}{125} = \frac{3}{8}$$

$0.37.5$ $\qquad 37.5\% = 0.375$

$$\mathbf{\frac{3}{8} = 0.375 = 37\frac{1}{2}\%, \text{ or } 37.5\%}$$

$$40\% = \frac{40}{100} = \frac{2}{5} \cdot \frac{20}{20} = \frac{2}{5}$$

$0.40.$ $\qquad 40\% = 0.4$

$$\mathbf{\frac{2}{5} = 0.4 = 40\%}$$

67. $0.50.$ $\qquad 0.5 = 50\%$

$$50\% = \frac{50}{100} = \frac{1}{2} \cdot \frac{50}{50} = \frac{1}{2}$$

$$\mathbf{\frac{1}{2} = 0.5 = 50\%}$$

$$\frac{1}{3} = 1 \div 3$$

$$\begin{array}{r} 0.3 \\ 3\,\overline{)\,1.0} \\ \underline{9} \\ 1 \end{array}$$

We get a repeating decimal: $0.\overline{3}$

$0.33.\overline{3}$ $\qquad 0.\overline{3} = 33.\overline{3}\%$

$$\mathbf{\frac{1}{3} = 0.\overline{3} = 33.\overline{3}\%, \text{ or } 33\frac{1}{3}\%}$$

$$25\% = \frac{25}{100} = \frac{25}{25} \cdot \frac{1}{4} = \frac{1}{4}$$

$0.25.$ $\qquad 25\% = 0.25$

$$\mathbf{\frac{1}{4} = 0.25 = 25\%}$$

$$16\frac{2}{3}\% = \frac{50}{3}\% = \frac{50}{3} \times \frac{1}{100} = \frac{50 \cdot 1}{3 \cdot 2 \cdot 50} = \frac{50}{50} \cdot \frac{1}{6} = \frac{1}{6}$$

$$\frac{1}{6} = 1 \div 6$$

$$\begin{array}{r} 0.1\,6 \\ 6\,\overline{)\,1.0\,0} \\ \underline{6} \\ 4\,0 \\ \underline{3\,6} \\ 4 \end{array}$$

We get a repeating decimal: $0.1\overline{6}$

$$\mathbf{\frac{1}{6} = 0.1\overline{6} = 16\frac{2}{3}\%, \text{ or } 16.\overline{6}\%}$$

$0.12.5$ $\qquad 0.125 = 12.5\%$

$$12.5\% = \frac{12.5}{100} = \frac{12.5}{100} \cdot \frac{10}{10} = \frac{125}{1000} = \frac{125}{125} \cdot \frac{1}{8} = \frac{1}{8}$$

$$\mathbf{\frac{1}{8} = 0.125 = 12.5\%, \text{ or } 12\frac{1}{2}\%}$$

$$\frac{3}{4} = \frac{3}{4} \cdot \frac{25}{25} = \frac{75}{100} = 75\%$$

$0.75.$ $\qquad 75\% = 0.75$

$$\mathbf{\frac{3}{4} = 0.75 = 75\%}$$

$0.8\overline{3} = 0.83.\overline{3}$ $\quad$ $0.8\overline{3} = 83.\overline{3}\%$

$83.\overline{3}\% = 83\frac{1}{3}\% = \frac{250}{3}\% = \frac{250}{3} \times \frac{1}{100} = \frac{5 \cdot 50}{3 \cdot 2 \cdot 50} =$
$\frac{5}{6} \cdot \frac{50}{50} = \frac{5}{6}$

$\mathbf{\frac{5}{6} = 0.8\overline{3} = 83.\overline{3}\%,\ or\ 83\frac{1}{3}\%}$

$\frac{3}{8} = 3 \div 8$

$$\begin{array}{r} 0.375 \\ 8\overline{)3.000} \\ 24 \\ \hline 60 \\ 56 \\ \hline 40 \\ 40 \\ \hline 0 \end{array}$$

$\frac{3}{8} = 0.375$

$0.37.5$ $\quad$ $0.375 = 37.5\%$

$\mathbf{\frac{3}{8} = 0.375 = 37.5\%,\ or\ 37\frac{1}{2}\%}$

69. $13 \cdot x = 910$

$\frac{13 \cdot x}{13} = \frac{910}{13}$

$x = 70$

71. $0.05 \times b = 20$

$\frac{0.05 \times b}{0.05} = \frac{20}{0.05}$

$b = 400$

73. $\frac{1}{2} \cdot x = 2$

$2 \cdot \frac{1}{2} \cdot x = 2 \cdot 2$ Multiplying by 2 on both sides

$x = 4$

The solution is 4.

75.
$$\begin{array}{r} 3 \\ 13\overline{)40} \\ 39 \\ \hline 1 \end{array}$$

$\frac{40}{13} = 3\frac{1}{13}$

77.
$$\begin{array}{r} 83 \\ 3\overline{)250} \\ 240 \\ \hline 10 \\ 9 \\ \hline 1 \end{array}$$

$\frac{250}{3} = 83\frac{1}{3}$

79.
$$\begin{array}{r} 43 \\ 8\overline{)345} \\ 320 \\ \hline 25 \\ 24 \\ \hline 1 \end{array}$$

$\frac{345}{8} = 43\frac{1}{8}$

81.
$$\begin{array}{r} 18 \\ 4\overline{)75} \\ 40 \\ \hline 35 \\ 32 \\ \hline 3 \end{array}$$

$\frac{75}{4} = 18\frac{3}{4}$

83. ◈

85. Use a calculator.

$\frac{41}{369} = 0.11.\overline{1} = 11.\overline{1}\%$

87. $\frac{14}{9}\% = \frac{14}{9} \times \frac{1}{100} = \frac{2 \cdot 7 \cdot 1}{9 \cdot 2 \cdot 50} = \frac{2}{2} \cdot \frac{7}{450} = \frac{7}{450}$

To find decimal notation for $\frac{7}{450}$ we divide.

$$\begin{array}{r} 0.0155 \\ 450\overline{)7.0000} \\ 450 \\ \hline 2500 \\ 2250 \\ \hline 2500 \\ 2250 \\ \hline 250 \end{array}$$

We get a repeating decimal: $\frac{14}{9}\% = 0.01\overline{5}$

Exercise Set 4.4

1. What is 32% of 78?

$a = 32\% \times 78$

3. 89 is what percent of 99?

$89 = n \times 99$

5. 13 is 25% of what?

$13 = 25\% \times b$

7. What is 85% of 276?

Translate: $a = 85\% \cdot 276$

Solve: The letter is by itself. To solve the equation we convert 85% to decimal notation and multiply.

$$\begin{array}{r} 2\,7\,6 \\ \times\, 0.\,8\,5 \\ \hline 1\,3\,8\,0 \\ 2\,2\,0\,8\,0 \\ \hline a = 2\,3\,4.\,6\,0 \end{array} \qquad (85\% = 0.85)$$

234.6 is 85% of 276. The answer is 234.6.

9. 150% of 30 is what?

Translate: $150\% \times 30 = a$

Solve: Convert 150% to decimal notation and multiply.

$$\begin{array}{r} 3\,0 \\ \times\, 1.\,5 \\ \hline 1\,5\,0 \\ 3\,0\,0 \\ \hline a = 4\,5.\,0 \end{array} \qquad (150\% = 1.5)$$

150% of 30 is 45. The answer is 45.

11. What is 6% of \$300?

Translate: $a = 6\% \cdot \$300$

Solve: Convert 6% to decimal notation and multiply.

$$\begin{array}{r} \$\ 3\,0\,0 \\ \times\, 0.\,0\,6 \\ \hline a = \$\ 1\,8.\,0\,0 \end{array} \qquad (6\% = 0.06)$$

\$18 is 6% of \$300. The answer is \$18.

13. 3.8% of 50 is what?

Translate: $3.8\% \cdot 50 = a$

Solve: Convert 3.8% to decimal notation and multiply.

$$\begin{array}{r} 5\,0 \\ \times\ 0.\,0\,3\,8 \\ \hline 4\,0\,0 \\ 1\,5\,0\,0 \\ \hline a = 1.\,9\,0\,0 \end{array} \qquad (3.8\% = 0.038)$$

3.8% of 50 is 1.9. The answer is 1.9.

15. \$39 is what percent of \$50?

Translate: $39 = n \times 50$

Solve: To solve the equation we divide on both sides by 50 and convert the answer to percent notation.

$$n \cdot 50 = 39$$

$$\frac{n \cdot 50}{50} = \frac{39}{50}$$

$$n = 0.78 = 78\%$$

\$39 is 78% of \$50. The answer is 78%.

17. 20 is what percent of 10?

Translate: $20 = n \times 10$

Solve: To solve the equation we divide on both sides by 10 and convert the answer to percent notation.

$$n \cdot 10 = 20$$

$$\frac{n \cdot 10}{10} = \frac{20}{10}$$

$$n = 2 = 200\%$$

20 is 200% of 10. The answer is 200%.

19. What percent of \$300 is \$150?

Translate: $n \times 300 = 150$

Solve: $n \cdot 300 = 150$

$$\frac{n \cdot 300}{300} = \frac{150}{300}$$

$$n = 0.5 = 50\%$$

50% of \$300 is \$150. The answer is 50%.

21. What percent of 80 is 100?

Translate: $n \times 80 = 100$

Solve: $n \cdot 80 = 100$

$$\frac{n \cdot 80}{80} = \frac{100}{80}$$

$$n = 1.25 = 125\%$$

125% of 80 is 100. The answer is 125%.

23. 20 is 50% of what?

Translate: $20 = 50\% \times b$

Solve: To solve the equation we divide on both sides by 50%:

$$\frac{20}{50\%} = \frac{50\% \times b}{50\%}$$

$$\frac{20}{0.5} = b \qquad (50\% = 0.5)$$

$$40 = b$$

$$\begin{array}{r} 4\,0\,. \\ 0.\,5_{\wedge}\overline{)\,2\,0.\,0_{\wedge}} \\ 2\,0\,0 \\ \hline 0 \\ 0 \\ \hline 0 \end{array}$$

20 is 50% of 40. The answer is 40.

25. 40% of what is \$16?

Translate: $40\% \times b = 16$

Solve: To solve the equation we divide on both sides by 40%:

$$\frac{40\% \times b}{40\%} = \frac{16}{40\%}$$

$$b = \frac{16}{0.4} \qquad (40\% = 0.4)$$

$$b = 40$$

$$\begin{array}{r} 4\,0\,. \\ 0.\,4_{\wedge}\overline{)\,1\,6.\,0_{\wedge}} \\ 1\,6\,0 \\ \hline 0 \\ 0 \\ \hline 0 \end{array}$$

40% of \$40 is \$16. The answer is \$40.

27. 56.32 is 64% of what?

Translate: $56.32 = 64\% \times b$

Solve: $\frac{56.32}{64\%} = \frac{64\% \times b}{64\%}$

$\frac{56.32}{0.64} = b$

$88 = b$

$$0.64_{\wedge}\overline{)56.32_{\wedge}} = 88.$$

$$\begin{array}{r} 5120 \\ \hline 512 \\ 512 \\ \hline 0 \end{array}$$

56.32 is 64% of 88. The answer is 88.

29. 70% of what is 14?

Translate: $70\% \times b = 14$

Solve: $\frac{70\% \times b}{70\%} = \frac{14}{70\%}$

$b = \frac{14}{0.7}$

$b = 20$

$$0.7_{\wedge}\overline{)14.0_{\wedge}} = 20.$$

$$\begin{array}{r} 140 \\ \hline 0 \\ 0 \\ \hline 0 \end{array}$$

70% of 20 is 14. The answer is 20.

31. What is $62\frac{1}{2}\%$ of 10?

Translate: $a = 62\frac{1}{2}\% \times 10$

Solve: $a = 0.625 \times 10$ $\quad (62\frac{1}{2}\% = 0.625)$

$a = 6.25$ $\quad$ Multiplying

6.25 is $62\frac{1}{2}\%$ of 10. The answer is 6.25.

33. What is 8.3% of $10,200?

Translate: $a = 8.3\% \times 10,200$

Solve: $a = 8.3\% \times 10,200$

$a = 0.083 \times 10,200$ $\quad (8.3\% = 0.083)$

$a = 846.6$ $\quad$ Multiplying

$846.60 is 8.3% of $10,200. The answer is $846.60.

35. $0.\underline{09} = \frac{9}{1\underline{00}}$

2 decimal places $\quad$ 2 zeros

37. $0.\underline{875} = \frac{875}{1\underline{000}}$

3 decimal places $\quad$ 3 zeros

$\frac{875}{1000} = \frac{7 \cdot 125}{8 \cdot 125} = \frac{7}{8} \cdot \frac{125}{125} = \frac{7}{8}$

Thus, $0.875 = \frac{875}{1000}$, or $\frac{7}{8}$.

39. $\frac{89}{1\underline{00}}$ $\quad$ 0.89.

2 zeros $\quad$ Move 2 places

$\frac{89}{100} = 0.89$

41. $\frac{3}{1\underline{0}}$ $\quad$ 0.3.

1 zero $\quad$ Move 1 place

$\frac{3}{10} = 0.3$

43. ◈

45. Estimate: Round 7.75% to 8% and $10,880 to $11,000. Then translate:

What is 8% of $11,000?

↓ ↓ ↓ ↓ ↓

$a = 8\% \times 11,000$

We convert 8% to decimal notation and multiply.

$$\begin{array}{r} 11,000 \\ \times \quad 0.08 \\ \hline 880.00 \end{array} \quad (8\% = 0.08)$$

$880 is about 7.75% of $10,880. (Answers may vary.)

Calculate: First we translate.

What is 7.75% of $10,880?

↓ ↓ ↓ ↓ ↓

$a = 7.75\% \times 10,880$

Use a calculator to multiply:

$0.0775 \times 10,880 = 843.2$

$843.20 is 7.75% of $10,880.

47. We reword the problem as two questions, translate each to an equation, and solve the equations.

What is 40% of 270?

↓ ↓ ↓ ↓ ↓

$a = 40\% \times 270$

$a = 0.4 \times 270$

$a = 108$

What is 50% of 270?

↓ ↓ ↓ ↓ ↓

$a = 50\% \times 270$

$a = 0.5 \times 270$

$a = 135$

108 tons to 135 tons of the trash is recyclable.

Exercise Set 4.5

1. What is $\quad$ 37% $\quad$ of 74?

↓ ↓ ↓

amount $\quad$ number of hundredths $\quad$ base

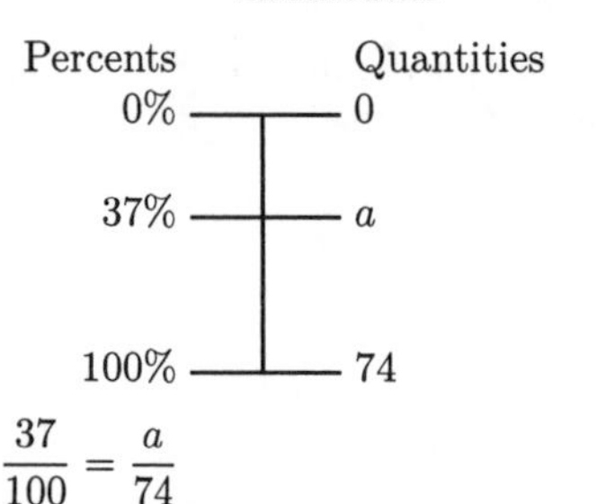

$\frac{37}{100} = \frac{a}{74}$

3. 4.3 is what percent of 5.9?

4.3	is	what	percent of 5.9?
↓		↓	↓
amount		number of hundredths	base

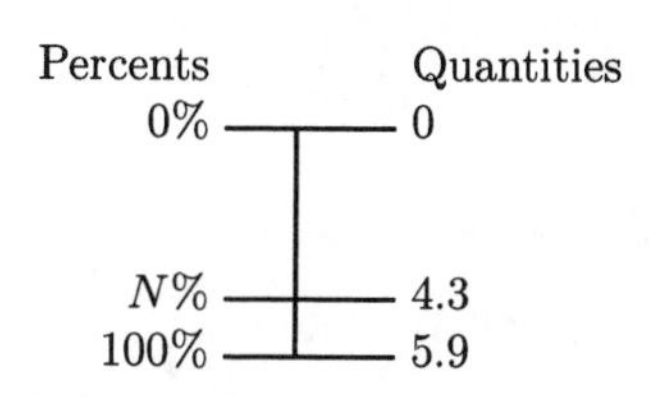

$$\frac{N}{100} = \frac{4.3}{5.9}$$

5. 14 is 25% of what?

14	is	25%	of what?
↓		↓	↓
amount		number of hundredths	base

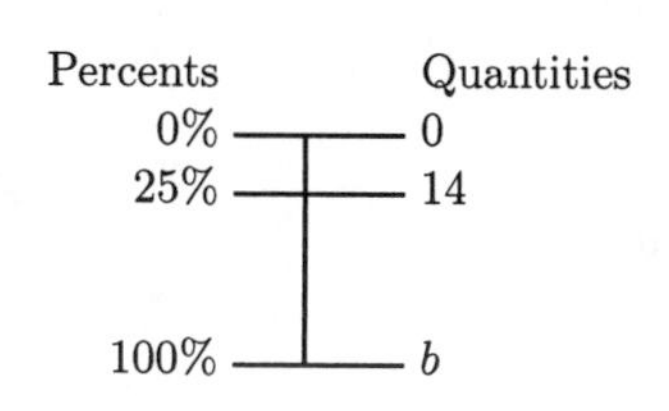

$$\frac{25}{100} = \frac{14}{b}$$

7. What is 76% of 90?

What	is	76%	of 90?
↓		↓	↓
amount		number of hundredths	base

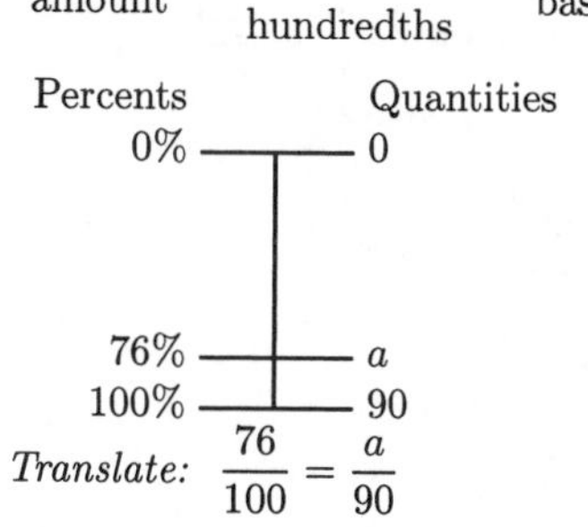

Translate: $\dfrac{76}{100} = \dfrac{a}{90}$

Solve: $76 \cdot 90 = 100 \cdot a$ Equating cross-products

$$\frac{76 \cdot 90}{100} = \frac{100 \cdot a}{100}$$ Dividing by 100

$$\frac{6840}{100} = a$$

$$68.4 = a$$ Simplifying

68.4 is 76% of 90. The answer is 68.4.

9. 70% of 660 is what?

70%	of 660	is	what?
↓	↓		↓
number of hundredths	base		amount

Percents Quantities
0% — 0
70% — a
100% — 660

Translate: $\dfrac{70}{100} = \dfrac{a}{660}$

Solve: $70 \cdot 660 = 100 \cdot a$ Equating cross-products

$$\frac{70 \cdot 660}{100} = \frac{100 \cdot a}{100}$$ Dividing by 100

$$\frac{46,200}{100} = a$$

$$462 = a$$ Simplifying

70% of 660 is 462. The answer is 462.

11. What is 4% of 1000?

What	is	4%	of 1000?
↓		↓	↓
amount		number of hundredths	base

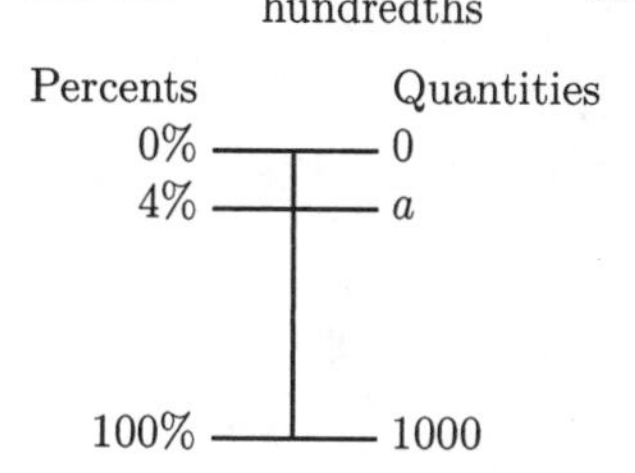

Translate: $\dfrac{4}{100} = \dfrac{a}{1000}$

Solve: $4 \cdot 1000 = 100 \cdot a$

$$\frac{4 \cdot 1000}{100} = \frac{100 \cdot a}{100}$$

$$\frac{4000}{100} = a$$

$$40 = a$$

40 is 4% of 1000. The answer is 40.

13. 4.8% of 60 is what?

4.8%	of 60	is	what?
↓	↓		↓
number of hundredths	base		amount

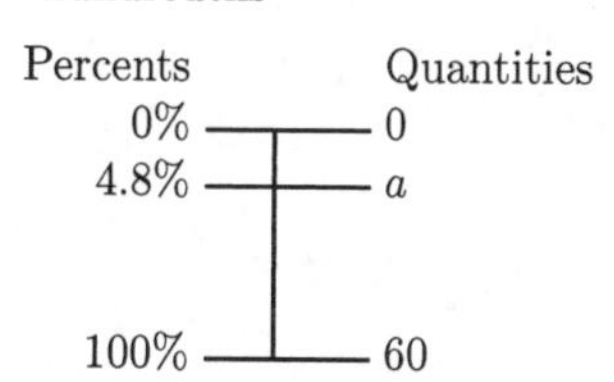

Translate: $\dfrac{4.8}{100} = \dfrac{a}{60}$

Solve: $4.8 \cdot 60 = 100 \cdot a$

$$\frac{4.8 \cdot 60}{100} = \frac{100 \cdot a}{100}$$

$$\frac{288}{100} = a$$

$$2.88 = a$$

4.8% of 60 is 2.88. The answer is 2.88.

15. 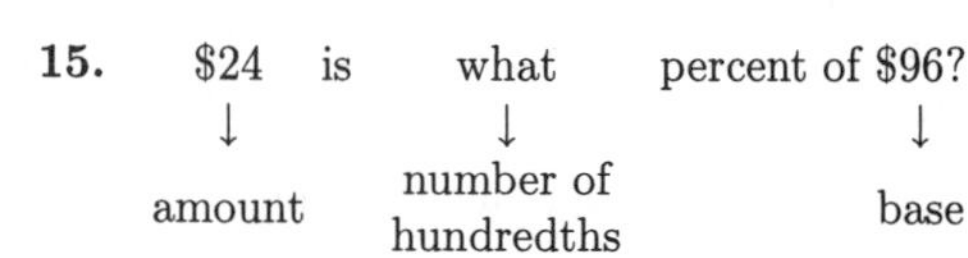

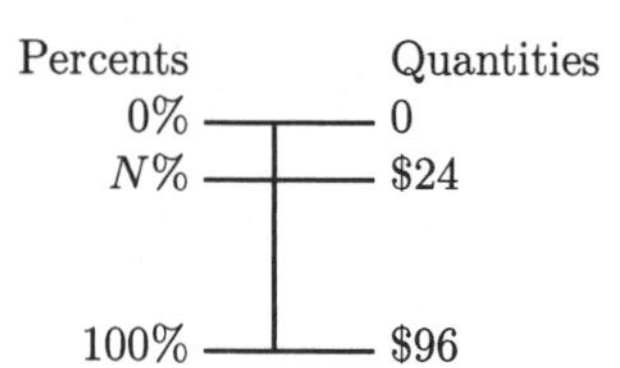

Translate: $\dfrac{N}{100} = \dfrac{24}{96}$

Solve: $96 \cdot N = 100 \cdot 24$

$$\frac{96N}{N} = \frac{100 \cdot 24}{96}$$

$$N = \frac{100 \cdot 24}{96}$$

$$N = 25$$

\$24 is 25% of \$96. The answer is 25%.

17. 102 is what percent of 100?

102 ↓ amount; what ↓ number of hundredths; 100 ↓ base

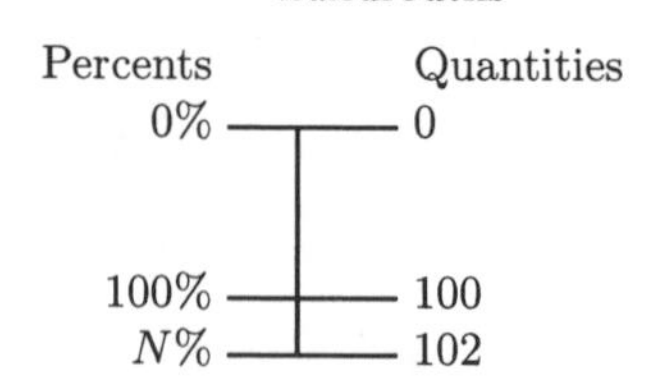

Translate: $\dfrac{N}{100} = \dfrac{102}{100}$

Solve: $100 \cdot N = 100 \cdot 102$

$$\frac{100 \cdot N}{100} = \frac{100 \cdot 102}{100}$$

$$N = \frac{100 \cdot 102}{100}$$

$$N = 102$$

102 is 102% of 100. The answer is 102%.

19. What percent of \$480 is \$120?

What ↓ number of hundredths; \$480 ↓ base; \$120 ↓ amount

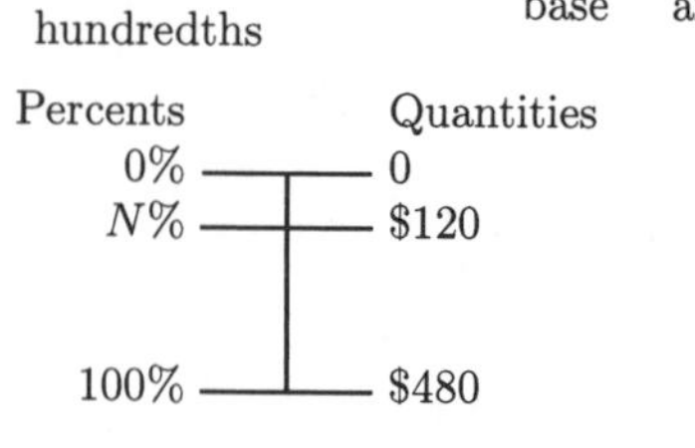

Translate: $\dfrac{N}{100} = \dfrac{120}{480}$

Solve: $480 \cdot N = 100 \cdot 120$

$$\frac{480 \cdot N}{480} = \frac{100 \cdot 120}{480}$$

$$N = \frac{100 \cdot 120}{480}$$

$$N = 25$$

25% of \$480 is \$120. The answer is 25%.

21. What percent of 160 is 150?

What ↓ number of hundredths; 160 ↓ base; 150 ↓ amount

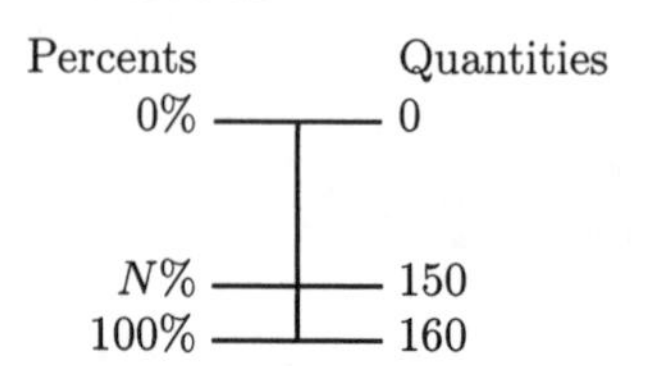

Translate: $\dfrac{N}{100} = \dfrac{150}{160}$

Solve: $160 \cdot N = 100 \cdot 150$

$$\frac{160 \cdot N}{160} = \frac{100 \cdot 150}{160}$$

$$N = \frac{100 \cdot 150}{160}$$

$$N = 93.75$$

93.75% of 160 is 150. The answer is 93.75%.

23. \$18 is 25% of what?

\$18 ↓ amount; 25% ↓ number of hundredths; what ↓ base

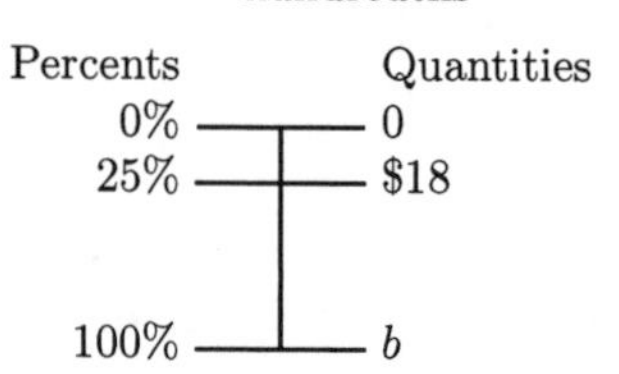

Translate: $\dfrac{25}{100} = \dfrac{18}{b}$

Solve: $25 \cdot b = 100 \cdot 18$

$$\frac{25 \cdot b}{b} = \frac{100 \cdot 18}{25}$$

$$b = \frac{100 \cdot 18}{25}$$

$$b = 72$$

\$18 is 25% of \$72. The answer is \$72.

25. 60% of what is \$54?

60% ↓ number of hundredths; what ↓ base; \$54 ↓ amount

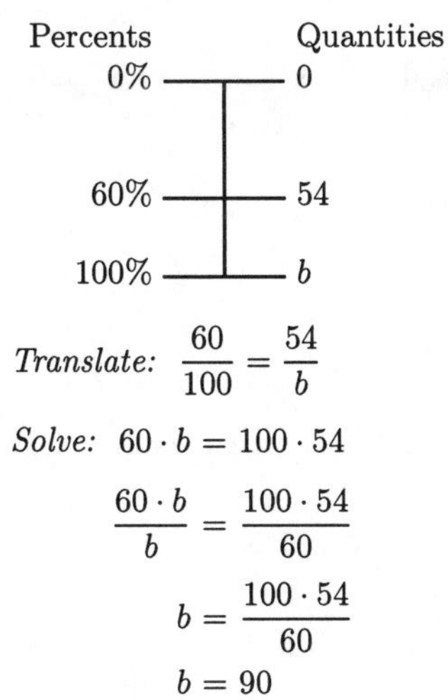

Translate: $\frac{60}{100} = \frac{54}{b}$

Solve: $60 \cdot b = 100 \cdot 54$

$$\frac{60 \cdot b}{b} = \frac{100 \cdot 54}{60}$$

$$b = \frac{100 \cdot 54}{60}$$

$$b = 90$$

60% of 90 is 54. The answer is 90.

27. 65.12 is 74% of what?
65.12 → amount; 74% → number of hundredths; what → base

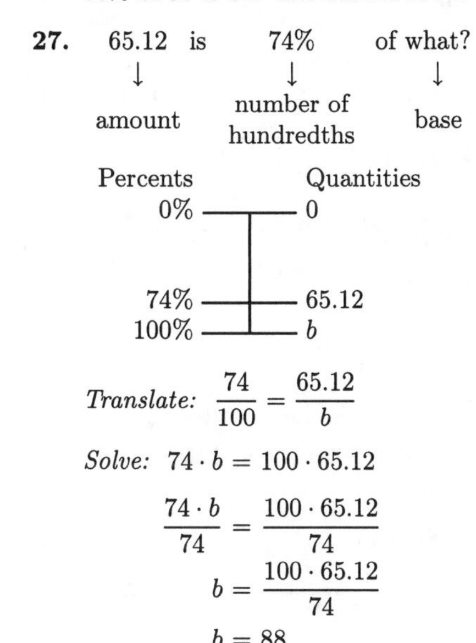

Translate: $\frac{74}{100} = \frac{65.12}{b}$

Solve: $74 \cdot b = 100 \cdot 65.12$

$$\frac{74 \cdot b}{74} = \frac{100 \cdot 65.12}{74}$$

$$b = \frac{100 \cdot 65.12}{74}$$

$$b = 88$$

65.12 is 74% of 88. The answer is 88.

29. 80% of what is 16?
80% → number of hundredths; what → base; 16 → amount

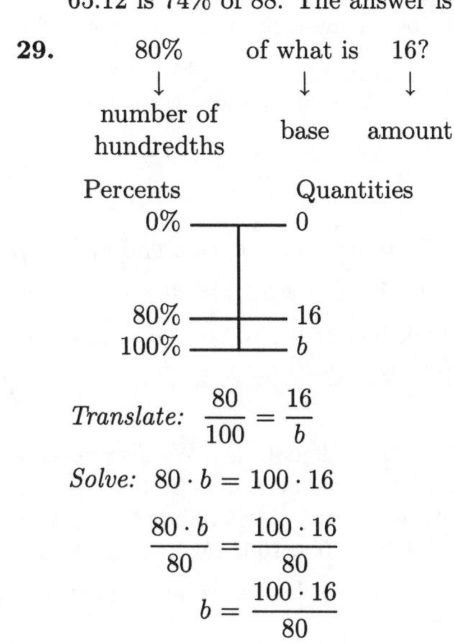

Translate: $\frac{80}{100} = \frac{16}{b}$

Solve: $80 \cdot b = 100 \cdot 16$

$$\frac{80 \cdot b}{80} = \frac{100 \cdot 16}{80}$$

$$b = \frac{100 \cdot 16}{80}$$

$$b = 20$$

80% of 20 is 16. The answer is 20.

31. What is $62\frac{1}{2}\%$ of 40?
What → amount; $62\frac{1}{2}\%$ → number of hundredths; 40 → base

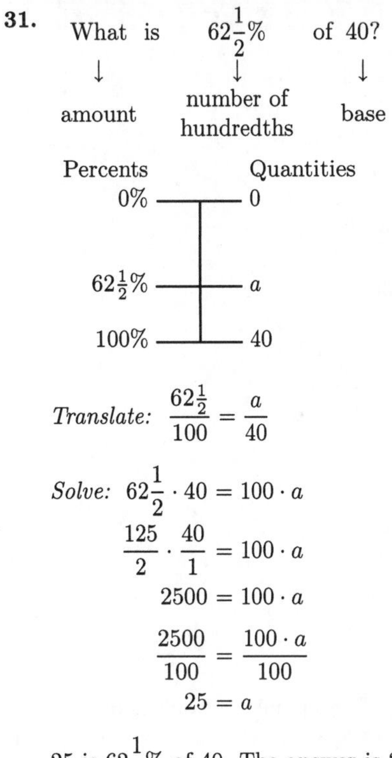

Translate: $\frac{62\frac{1}{2}}{100} = \frac{a}{40}$

Solve: $62\frac{1}{2} \cdot 40 = 100 \cdot a$

$$\frac{125}{2} \cdot \frac{40}{1} = 100 \cdot a$$

$$2500 = 100 \cdot a$$

$$\frac{2500}{100} = \frac{100 \cdot a}{100}$$

$$25 = a$$

25 is $62\frac{1}{2}\%$ of 40. The answer is 25.

33. What is 9.4% of $8300?
What → amount; 9.4% → number of hundredths; $8300 → base

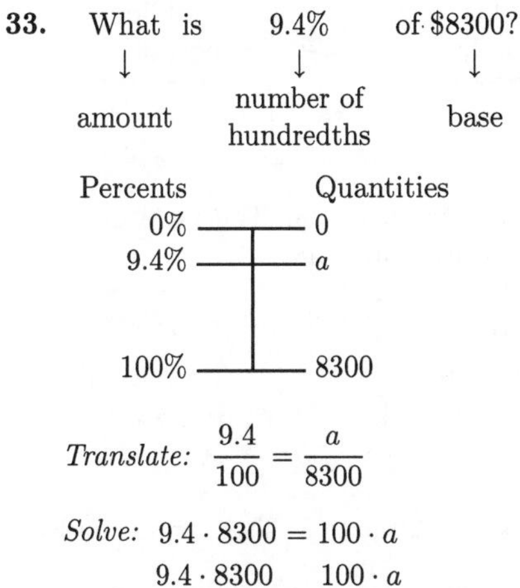

Translate: $\frac{9.4}{100} = \frac{a}{8300}$

Solve: $9.4 \cdot 8300 = 100 \cdot a$

$$\frac{9.4 \cdot 8300}{100} = \frac{100 \cdot a}{100}$$

$$\frac{78,020}{100} = a$$

$$780.2 = a$$

$780.20 is 9.4% of $8300. The answer is $780.20.

35. $\frac{x}{188} = \frac{2}{47}$

$$47 \cdot x = 188 \cdot 2$$

$$x = \frac{188 \cdot 2}{47}$$

$$x = \frac{4 \cdot 47 \cdot 2}{47}$$

$$x = 8$$

37.
$$\frac{4}{7} = \frac{x}{14}$$
$$4 \cdot 14 = 7 \cdot x$$
$$\frac{4 \cdot 14}{7} = x$$
$$\frac{4 \cdot 2 \cdot 7}{7} = x$$
$$8 = x$$

39.
$$\frac{5000}{t} = \frac{3000}{60}$$
$$5000 \cdot 60 = 3000 \cdot t$$
$$\frac{5000 \cdot 60}{3000} = t$$
$$\frac{5 \cdot 1000 \cdot 3 \cdot 20}{3 \cdot 1000} = t$$
$$100 = t$$

41.
$$\frac{x}{1.2} = \frac{36.2}{5.4}$$
$$5.4 \cdot x = 1.2(36.2)$$
$$x = \frac{1.2(36.2)}{5.4}$$
$$x = 8.0\overline{4}$$

43. ***Familiarize.*** Let $q =$ the number of quarts of liquid ingredients the recipe calls for.

Translate.

Buttermilk	plus	Skim milk	plus	Oil	is	Total liquid ingredients
↓	↓	↓	↓	↓	↓	↓
$\frac{1}{2}$	$+$	$\frac{1}{3}$	$+$	$\frac{1}{16}$	$=$	q

Solve. We carry out the addition. The LCM of the denominators is 48, so the LCD is 48.

$$\frac{1}{2} \cdot \frac{24}{24} + \frac{1}{3} \cdot \frac{16}{16} + \frac{1}{16} \cdot \frac{3}{3} = q$$
$$\frac{24}{48} + \frac{16}{48} + \frac{3}{48} = q$$
$$\frac{43}{48} = q$$

Check. We repeat the calculation. The answer checks.

State. The recipe calls for $\frac{43}{48}$ qt of liquid ingredients.

45. ◈

47. Estimate: Round 8.85% to 9%, and $12,640 to $12,600.

What is	9%	of $12,600?
↓	↓	↓
amount	number of hundredths	base

Percents	Quantities
0%	0
9%	a
100%	$12,600

Translate: $\frac{9}{100} = \frac{a}{12,600}$

Solve:
$$9 \cdot 12,600 = 100 \cdot a$$
$$\frac{9 \cdot 12,600}{100} = \frac{100 \cdot a}{100}$$
$$\frac{113,400}{100} = a$$
$$1134 = a$$

$1134 is about 8.85% of $12,640. (Answers may vary.)

Calculate:

What is	8.85%	of $12,640?
↓	↓	↓
amount	number of hundredths	base

Percents	Quantities
0%	0
8.85%	a
100%	$12,640

Translate: $\frac{8.85}{100} = \frac{a}{12,640}$

Solve:
$$8.85 \cdot 12,640 = 100 \cdot a$$
$$\frac{8.85 \cdot 12,640}{100} = \frac{100 \cdot a}{100}$$
$$\frac{111,864}{100} = a \quad \text{Use a calculator to multiply and divide.}$$
$$1118.64 = a$$

$1118.64 is 8.85% of $12,640.

Exercise Set 4.6

1. ***Familiarize.*** First we find the number of bowlers who would be expected to be left-handed. Let a represent this number.

Translate. We rephrase the question and translate.

What is 17% of 120?
↓ ↓ ↓ ↓ ↓
$a = 17\% \times 120$

Solve: We convert 17% to decimal notation and multiply.

$$a = 17\% \times 120 = 0.17 \times 120 = 20.4 \approx 20$$

We can subtract to find the number of bowlers who would not be expected to be left-handed:

$$120 - 20 = 100$$

Check. We can repeat the calculations. We also observe that, since 17% of bowlers are expected to be left-handed, 83% would not be expected to be left-handed. Because 83% of $120 = 0.83 \times 120 = 99.6 \approx 100$, our answer checks.

State. You would expect 20 bowlers to be left-handed and 100 not to be left-handed.

3. ***Familiarize***. First we find the number of moviegoers in the 12-29 age group. Let a represent this number.

Translate. We rephrase the question and translate.

$$\begin{array}{ccccc} \text{What} & \text{is} & 67\% & \text{of} & 800? \\ \downarrow & \downarrow & \downarrow & \downarrow & \downarrow \\ a & = & 67\% & \times & 800 \end{array}$$

Solve. We convert 67% to decimal notation and multiply.

$a = 67\% \times 800 = 0.67 \times 800 = 536$

We can subtract to find the number who are not in this age group.

$800 - 536 = 264$

Check. We can repeat the calculations. Also, observe that, since 67% of moviegoers are in the 12-29 age group, 33% are not. Because 33% of $800 = 0.33 \times 800 = 264$, our answer checks.

State. 536 moviegoers were in the 12-29 age group, and 264 were not in this age group.

5. ***Familiarize***. The question asks for percents. We know that 10% of 40 is 4. Since $13 \approx 3 \times 4$, we would expect the percent of at bats that are hits to be close to 30%. Then we would also expect the percent of at bats that are not hits to be 70%. First we find n, the percent that are hits.

Translate. We rephrase the question and translate.

$$\begin{array}{ccccc} 13 & \text{is} & \underbrace{\text{what percent}} & \text{of} & 40? \\ \downarrow & \downarrow & \downarrow & \downarrow & \downarrow \\ 13 & = & n & \times & 40 \end{array}$$

Solve. We divide on both sides by 40 and convert the result to percent notation.

$$13 = n \times 40$$
$$\frac{13}{40} = \frac{n \times 40}{40}$$
$$0.325 = n$$
$$32.5\% = n \qquad \text{Finding percent notation}$$

Now we find the percent of at bats that are not hits. We let $p =$ this percent.

$$\begin{array}{ccccc} \underbrace{\begin{array}{c}100\% \\ \text{of hits}\end{array}} & \text{minus} & \underbrace{\begin{array}{c}\text{percent} \\ \text{of hits}\end{array}} & \text{is} & \underbrace{\begin{array}{c}\text{percent that} \\ \text{are not hits}\end{array}} \\ \downarrow & \downarrow & \downarrow & \downarrow & \downarrow \\ 100\% & - & 32.5\% & = & p \end{array}$$

To solve the equation we carry out the subtraction.

$p = 100\% - 32.5\% = 67.5\%$

Check. Note that the answer, 32.5%, is close to 30% as estimated in the Familiarize step. We also note that the percent of at bats that are not hits, 67.5%, is close to 70% as estimated

State. 32.5% of at bats are hits, and 67.5% are not hits.

7. ***Familiarize***. First we find the amount of the solution that is acid. We let $a =$ this amount.

Translate. We rephrase the question and translate.

$$\begin{array}{ccccc} \text{What} & \text{is} & 3\% & \text{of} & 680? \\ \downarrow & \downarrow & \downarrow & \downarrow & \downarrow \\ a & = & 3\% & \text{of} & 680 \end{array}$$

Solve. We convert 3% to decimal notation and multiply.

$a = 3\% \times 680 = 0.03 \times 680 = 20.4$

Now we find the amount that is water. We let $w =$ this amount.

$$\begin{array}{ccccc} \underbrace{\begin{array}{c}\text{Total} \\ \text{amount}\end{array}} & \text{minus} & \underbrace{\begin{array}{c}\text{Amount} \\ \text{of acid}\end{array}} & \text{is} & \underbrace{\begin{array}{c}\text{Amount} \\ \text{of water}\end{array}} \\ \downarrow & \downarrow & \downarrow & \downarrow & \downarrow \\ 680 & - & 20.4 & = & w \end{array}$$

To solve the equation we carry out the subtraction.

$w = 680 - 20.4 = 659.6$

Check. We can repeat the calculations. Also, observe that, since 3% of the solution is acid, 97% is water. Because 97% of $680 = 0.97 \times 680 = 659.6$, our answer checks.

State. The solution contains 20.4 mL of acid and 659.6 mL of water.

9. ***Familiarize***. We let $n =$ the percent of time that television sets are on.

Translate. We rephrase the question and translate.

$$\begin{array}{ccccc} 2190 & \text{is} & \underbrace{\text{what percent}} & \text{of} & 8760? \\ \downarrow & \downarrow & \downarrow & \downarrow & \downarrow \\ 2190 & = & n & \times & 8760 \end{array}$$

Solve. We divide on both sides by 8760 and convert the result to percent notation.

$$2190 = n \times 8760$$
$$\frac{2190}{8760} = \frac{n \times 8760}{8760}$$
$$0.25 = n$$
$$25\% = n$$

Check. To check we find 25% of 8760:

$25\% \times 8760 = 0.25 \times 8760 = 2190$. The answer checks.

State. Television sets are on for 25% of the year.

11. First we find the maximum heart rate for a 25 year old person.

Familiarize. Note that $220 - 25 = 195$. We let $x =$ the maximum heart rate for a 25 year old person.

Translate. We rephrase the question and translate.

$$\begin{array}{ccccc} \text{What} & \text{is} & 85\% & \text{of} & 195? \\ \downarrow & \downarrow & \downarrow & \downarrow & \downarrow \\ x & = & 85\% & \times & 195 \end{array}$$

Solve. We convert 85% to a decimal and simplify.

$x = 0.85 \times 195 = 165.75 \approx 166$

Check. We can repeat the calculations. Also, 85% of $195 \approx 0.85 \times 200 = 170 \approx 166$. The answer checks.

State. The maximum heart rate for a 25 year old person is 166 beats per minute.

Next we find the maximum heart rate for a 36 year old person.

Familiarize. Note that $220 - 36 = 184$. We let $x =$ the maximum heart rate for a 36 year old person.

Translate. We rephrase the question and translate.

What is 85% of 184?
$$x = 85\% \times 184$$

Solve. We convert 85% to a decimal and simplify.

$$x = 0.85 \times 184 = 156.4 \approx 156$$

Check. We can repeat the calculations. Also, 85% of $184 \approx 0.9 \times 180 = 162 \approx 156$. The answer checks.

State. The maximum heart rate for a 36 year old person is 156 beats per minute.

Next we find the maximum heart rate for a 48 year old person.

Familiarize. Note that $220 - 48 = 172$. We let x = the maximum heart rate for a 48 year old person.

Translate. We rephrase the question and translate.

What is 85% of 172?
$$x = 85\% \times 172$$

Solve. We convert 85% to a decimal and simplify.

$$x = 0.85 \times 172 = 146.2 \approx 146$$

Check. We can repeat the calculations. Also, 85% of $172 \approx 0.9 \times 170 = 153 \approx 146$. The answer checks.

State. The maximum heart rate for a 48 year old person is 146 beats per minute.

We find the maximum heart rate for a 55 year old person.

Familiarize. Note that $220 - 55 = 165$. We let x = the maximum heart rate for a 55 year old person.

Translate. We rephrase the question and translate.

What is 85% of 165?
$$x = 85\% \times 165$$

Solve. We convert 85% to a decimal and simplify.

$$x = 0.85 \times 165 = 140.25 \approx 140$$

Check. We can repeat the calculations. Also, 85% of $165 \approx 0.9 \times 160 = 144 \approx 140$. The answer checks.

State. The maximum heart rate for a 55 year old person is 140 beats per minute.

Finally we find the maximum heart rate for a 76 year old person.

Familiarize. Note that $220 - 76 = 144$. We let x = the maximum heart rate for a 76 year old person.

Translate. We rephrase the question and translate.

What is 85% of 144?
$$x = 85\% \times 144$$

Solve. We convert 85% to a decimal and simplify.

$$x = 0.85 \times 144 = 122.4 \approx 122$$

Check. We can repeat the calculations. Also, 85% of $144 \approx 0.9 \times 140 = 126 \approx 122$. The answer checks.

State. The maximum heart rate for a 76 year old person is 122 beats per minute.

13. ***Familiarize***. Use the drawing in the text to visualize the situation. Note that the increase in the amount was \$16.

Let n = the percent of increase.

Translate. We rephrase the question and translate.

\$16 is what percent of \$200?
$$16 = n \times 200$$

Solve. We divide by 200 on both sides and convert the result to percent notation.

$$16 = n \times 200$$
$$\frac{16}{200} = \frac{n \times 200}{200}$$
$$0.08 = n$$
$$8\% = n$$

Check. Find 8% of 200: $8\% \times 200 = 0.08 \times 200 = 16$. Since this is the amount of the increase, the answer checks.

State. The percent of increase was 8%.

15. ***Familiarize***. We use the drawing in the text to visualize the situation. Note that the reduction is \$18.

We let n = the percent of decrease.

Translate. We rephrase the question and translate.

\$18 is what percent of \$90?
$$18 = n \times 90$$

Solve. To solve the equation, we divide on both sides by 90 and convert the result to percent notation.

$$\frac{18}{90} = \frac{n \times 90}{90}$$
$$0.2 = n$$
$$20\% = n$$

Check. We find 20% of 90: $20\% \times 90 = 0.2 \times 90 = 18$. Since this is the price decrease, the answer checks.

State. The percent of decrease was 20%.

17. ***Familiarize***. We note that the amount of the raise can be found and then added to the old salary. A drawing helps us visualize the situation.

\$28,600	\$?
100%	5%

We let x = the new salary.

Translate. We rephrase the question and translate.

What is the old salary plus 5% of the old salary?
$$x = 28,600 + 5\% \times 28,600$$

Solve. We convert 5% to a decimal and simplify.

$$x = 28,600 + 0.05 \times 28,600$$
$$= 28,600 + 1430 \qquad \text{The raise is \$1430.}$$
$$= 30,030$$

Check. To check, we note that the new salary is 100% of the old salary plus 5% of the old salary, or 105% of the old salary. Since $1.05 \times 28,600 = 30,030$, our answer checks.

State. The new salary is \$30,030.

19. ***Familiarize.*** We visualize the situation.

\$18,000	
	\$?
100%	
70%	30%

This is a two-step problem. First we find the amount of the decrease. Let a represent this amount.

Translate. We rephrase the question and translate.

What is 30% of \$18,000?
↓ ↓ ↓ ↓ ↓
$a = 30\% \times 18,000$

Solve. We convert to decimal notation and multiply:

$$a = 0.3 \times 18,000 = 5400$$

To find the value after one year we subtract:

$$\$18,000 - \$5400 = \$12,600$$

Check. Note that with a 30% decrease, the reduced value should be 70% of the original value. Since 70% of $18,000 = 0.7 \times 18,000 = 12,600$, the answer checks.

State. The value of the car is \$12,600 after one year.

21. ***Familiarize.*** This is a multi-step problem. To find the population in 2000, we first find the increase over the population in 1999. Let a represent this increase in billions.

Translate. We rephrase the question and translate.

What is 1.6% of 6.0 billion?
↓ ↓ ↓ ↓ ↓
$a = 1.6\% \times 6.0$

We convert to decimal notation and multiply:

$$a = 0.016 \times 6.0 = 0.096$$

Now we add to find the population in 2000:

$$6.0 + 0.096 = 6.096$$

To find the population in 2001, we first find the increase a over the population in 2000.

What is 1.6% of 6.096 billion?
↓ ↓ ↓ ↓ ↓
$a = 1.6\% \times 6.096$

We convert to decimal notation and multiply:

$$a = 0.016 \times 6.096 = 0.097536$$

Then we add to find the population in 2001:

$$6.096 + 0.097536 = 6.193536 \approx 6.194$$

To find the population in 2002, we first find the increase a over the population in 2001.

What is 1.6% of 6.194 billion?
↓ ↓ ↓ ↓ ↓
$a = 1.6\% \times 6.194$

Again, we convert to decimal notation and multiply:

$$a = 0.016 \times 6.194 = 0.099104$$

Then we add to find the population in 2002:

$$6.194 + 0.099104 = 6.293104 \approx 6.293$$

Check. Note that the population each year is 101.6% of the population the previous year. Since

101.6% of $6.0 = 1.016 \times 6.0 = 6.096$,

101.6% of $6.096 = 1.016 \times 6.096 \approx 6.194$, and

101.6% of $6.096 = 1.016 \times 6.194 \approx 6.293$,

the results check.

State. In 2000 the world population will be 6.096 billion, in 2001 it will be about 6.194 billion, and in 2002 it will be about 6.293 billion.

23. ***Familiarize.*** Since the car depreciates 30% in the first year, its value after the first year is 100% − 30%, or 70%, of the original value. To find the decrease in value, we ask:

\$11,480 is 70% of what?

Let b = the original cost.

Translate. We rephrase the question and translate.

\$25,480 is 70% of what?
↓ ↓ ↓ ↓ ↓
$\$25,480 = 70\% \times b$

Solve.

$$25,480 = 70\% \times b$$

$$\frac{25,480}{70\%} = \frac{70\% \times b}{70\%}$$

$$\frac{25,480}{0.7} = b$$

$$36,400 = b$$

Check. We find 30% of 36,400 and then subtract this amount from 36,400:

$0.3 \times 36,400 = 10,920$ and

$36,400 - 10,920 = 25,480$

The answer checks.

State. The original cost was \$36,400.

25. ***Familiarize.*** This is a multistep problem. First we find the amount of each tip. Then we add that amount to the corresponding cost of the meal. Let x, y, and z represent the amounts of the tips on the \$15, \$34, and \$49 meals, respectively.

Translate. We rephrase the questions and translate.

What is 15% of \$15?
↓ ↓ ↓ ↓ ↓
$x = 15\% \times 15$

What is 15% of \$34?
↓ ↓ ↓ ↓ ↓
$y = 15\% \times 34$

What is 15% of \$49?
↓ ↓ ↓ ↓ ↓
$z = 15\% \times 49$

Solve. We convert to decimal notation and multiply.

$x = 0.15 \times 15 = 2.25$
$y = 0.15 \times 34 = 5.10$
$z = 0.15 \times 49 = 7.35$

Now we add to find the amounts charged.

For the \$15 meal: \$15 + \$2.25 = \$17.25

For the \$34 meal: \$34 + \$5.10 = \$39.10

For the \$49 meal: \$49 + \$7.35 = \$56.35

Check. Note that the amount charged in each case is 115% of the cost of the meal. Since

115% of \$15 = 1.15 × \$15 = \$17.25,

115% of \$34 = 1.15 × \$34 = \$39.10, and

115% of \$49 = 1.15 × \$49 = \$56.35,

the answer checks.

State. The total amounts charged are \$17.25 for the \$15 meal, \$39.10 for the \$34 meal, and \$56.35 for the \$49 meal.

27. a) ***Familiarize.*** Note that the increase in deaths from 1994 to 1995 is $17,274 - 16,589$, or 685. Let n = the percent of increase.

Translate. We rephrase the question and translate.

685 is what percent of 16,589?

$$685 = n \times 16,589$$

Solve. We divide on both sides by 16,589 and convert to percent notation.

$$\frac{685}{16,589} = \frac{n \times 16,589}{16,589}$$
$$0.041 \approx n$$
$$4.1\% \approx n$$

Check. Note that the number of deaths in 1995 will be 104.1% of the number in 1994. Since $1.041 \times 16,589 = 17,269.149 \approx 17,274$, the answer checks.

State. The percent of increase in alcohol-related deaths from 1994 to 1995 was about 4.1%.

b) ***Familiarize.*** Note that the decrease in deaths from 1986 to 1994 is $24,045 - 16,589$, or 7456. Let n = the percent of decrease.

Translate. We rephrase the question and translate.

7456 is what percent of 24,045?

$$7456 = n \times 24,045$$

Solve. We divide on both sides by 7456 and convert to percent notation.

$$\frac{7456}{24,045} = \frac{n \times 24,045}{24,045}$$
$$0.310 \approx n$$
$$31.0\% \approx n$$

Check. We find 31.0% of 24,045 and subtract this number from 24,045.

$$0.310 \times 24,045 \approx 7454$$

$$24,045 - 7454 = 16,591 \approx 16,589$$

The answer checks.

State. The percent of decrease in alcohol-related deaths from 1986 to 1994 was about 31.0%.

29. ***Familiarize.*** First we use the formula $A = l \times w$ to find the area of the strike zone:

$$A = 40 \times 17 = 680 \text{ in}^2$$

When a 2-in. border is added to the outside of the strike zone, the dimensions of the larger zone are 19 in. by 44 in. The area of this zone is

$$A = 44 \times 21 = 924 \text{ in}^2$$

We subtract to find the increase in area:

$$924 \text{ in}^2 - 680 \text{ in}^2 = 244 \text{ in}^2$$

We let N = the percent of increase in the area.

Translate. We rephrase the question and translate.

244 is what percent of 680?

$$244 = N \times 680$$

Solve. We divide by 680 on both sides and convert to percent notation.

$$\frac{244}{680} = \frac{N \times 680}{680}$$
$$0.359 = N$$
$$35.9\% = N$$

Check. We repeat the calculations.

State. The area of the strike zone is increased by 35.9%.

31. $\dfrac{25}{11} = 25 \div 11$

$$\begin{array}{r} 2.27 \\ 11 \overline{)25.00} \\ \underline{22} \\ 30 \\ \underline{22} \\ 80 \\ \underline{77} \\ 3 \end{array}$$

Since the remainders begin to repeat, we have a repeating decimal.

$$\frac{25}{11} = 2.\overline{27}$$

33. $\dfrac{27}{8} = 27 \div 8$

$$\begin{array}{r} 3.375 \\ 8 \overline{)27.000} \\ \underline{24} \\ 30 \\ \underline{24} \\ 60 \\ \underline{56} \\ 40 \\ \underline{40} \\ 0 \end{array}$$

$$\frac{27}{8} = 3.375$$

We could also do this conversion as follows:

$$\frac{27}{8} = \frac{27}{8} \cdot \frac{125}{125} = \frac{3375}{1000} = 3.375$$

35. $\frac{23}{25} = \frac{23}{25} \cdot \frac{4}{4} = \frac{92}{100} = 0.92$

37. $\frac{14}{32} = 14 \div 32$

```
       0. 4 3 7 5
  3 2 )1 4. 0 0 0 0
       1 2 8
         1 2 0
           9 6
           2 4 0
           2 2 4
             1 6 0
             1 6 0
                 0
```

$$\frac{14}{32} = 0.4375$$

(Note that we could have simplified the fraction first, getting $\frac{7}{16}$ and then found the quotient $7 \div 16$.)

39. Since 10,000 has 4 zeros, we move the decimal point in the number in the numerator 4 places to the left.

$$\frac{34,809}{10,000} = 3.4809$$

41.

43. Let S = the original salary. After a 3% raise, the salary becomes $103\% \cdot S$, or $1.03S$. After a 6% raise, the new salary is $1.06\% \cdot 1.03S$, or $1.06(1.03S)$. Finally, after a 9% raise, the salary is $109\% \cdot 1.06(1.03S)$, or $1.09(1.06)(1.03S)$. Multiplying, we get $1.09(1.06)(1.03S) = 1.190062S$. This is equivalent to $119.0062\% \cdot S$, so the original salary has increased by 19.0062%, or about 19%.

45. ***Familiarize.*** We will express 4 ft, 8 in. as 56 in. (4 ft + 8 in. $= 4 \cdot 12$ in. + 8 in. = 48 in. + 8 in. = 56 in.) We let h = Cynthia's final adult height.

Translate. We rephrase the question and translate.

$$\underbrace{56\text{ in.}} \text{ is } 84.4\% \text{ of what?}$$
$$\downarrow \quad \downarrow \quad \downarrow \quad \downarrow \quad \downarrow$$
$$56 \quad = 84.4\% \times \quad h$$

Solve. First we convert 84.4% to a decimal.

$$56 = 0.844 \times h$$
$$\frac{56}{0.844} = \frac{0.844 \times h}{0.844}$$
$$66 \approx h$$

Check. We find 84.4% of 66: $0.844 \times 66 \approx 56$. The answer checks.

State. Cynthia's final adult height will be about 66 in., or 5 ft, 6 in.

47. ***Familiarize.*** If p is 120% of q, then $p = 1.2q$. Let n = the percent of p that q represents.

Translate. We rephrase the question and translate. We use $1.2q$ for p.

$$q \text{ is } \underbrace{\text{what percent}} \text{ of } p?$$
$$\downarrow \ \downarrow \qquad \downarrow \qquad \downarrow \ \downarrow$$
$$q = \qquad n \qquad \times\ 1.2q$$

Solve.

$$q = n \times 1.2q$$
$$\frac{q}{1.2q} = \frac{n \times 1.2q}{1.2q}$$
$$\frac{1}{1.2} = n$$
$$0.8\overline{3} = n$$
$$83.\overline{3}\%, \text{ or } 83\frac{1}{3}\% = n$$

Check. We find $83\frac{1}{3}\%$ of $1.2q$:

$$0.8\overline{3} \times 1.2q = q$$

The answer checks.

State. q is $83.\overline{3}\%$, or $83\frac{1}{3}\%$, of p.

Exercise Set 4.7

1. a) We first find the cost of the telephones. It is

$$5 \times \$53 = \$265.$$

b) The sales tax on items costing \$265 is

$$\underbrace{\text{Sales tax rate}} \times \underbrace{\text{Purchase price}}$$
$$\downarrow \qquad \downarrow \qquad \downarrow$$
$$6.25\% \qquad \times \qquad \$265,$$

or 0.0625×265, or 16.5625. Thus the tax is \$16.56.

c) The total price is given by the purchase price plus the sales tax:

$$\$265 + \$16.56, \text{ or } \$281.56.$$

To check, note that the total price is the purchase price plus 6.25% of the purchase price. Thus the total price is 106.25% of the purchase price. Since $1.0625 \times \$265 =$ \$281.56 (rounded to the nearest cent), we have a check. The total price is \$281.56.

3. *Rephrase:* $\underbrace{\text{Sales tax}}$ is $\underbrace{\text{what percent}}$ of $\underbrace{\text{purchase price?}}$

Translate: $48 = r \times 960$

To solve the equation, we divide on both sides by 960.

$$\frac{48}{960} = \frac{r \times 960}{960}$$
$$0.05 = r$$
$$5\% = r$$

The sales tax rate is 5%.

5. *Rephrase:* $\underbrace{\text{Sales tax}}$ is $\underbrace{\text{what percent}}$ of $\underbrace{\text{purchase price?}}$

Translate: $35.80 = r \times 895$

To solve the equation, we divide on both sides by 895.

$$\frac{35.80}{895} = \frac{r \times 895}{895}$$
$$0.04 = r$$
$$4\% = r$$

The sales tax rate is 4%.

7. *Rephrase:* $\underbrace{\text{Sales tax}}$ is 5% of what?

Translate: $100 = 5\% \times b$, or $100 = 0.05 \times b$

To solve the equation, we divide on both sides by 0.05.

$$\frac{100}{0.05} = \frac{0.05 \times b}{0.05}$$
$$2000 = b$$

$$\begin{array}{r} 2000. \\ 0.05_\wedge \overline{)\,100.00_\wedge} \\ \underline{10000} \\ 0 \end{array}$$

The purchase price is $2000.

9. a) We first find the cost of the shower units. It is

$2 \times \$332.50 = \$665.$

b) The total tax rate is the city tax rate plus the state tax rate, or $1\% + 6\% = 7\%$. The sales tax paid on items costing $665 is

$\underbrace{\text{Sales tax rate}} \times \underbrace{\text{Purchase price}}$

$7\% \times \$665,$

or 0.07×665, or 46.55. Thus the tax is $46.55.

c) The total price is given by the purchase price plus the sales tax:

$\$665 + \$46.55 = \$711.55.$

To check, note that the total price is the purchase price plus 7% of the purchase price. Thus the total price is 107% of the purchase price. Since $1.07 \times 665 = 711.55$, we have a check. The total amount paid for the 2 shower units is $711.55.

11. *Rephrase:* $\underbrace{\text{Sales tax}}$ is $\underbrace{\text{what percent}}$ of $\underbrace{\text{purchase price?}}$

Translate: $1030.40 = r \times 18,400$

To solve the equation, we divide on both sides by 18,400.

$$\frac{1030.40}{18,400} = \frac{r \times 18,400}{18,400}$$
$$0.056 = r$$
$$5.6\% = r$$

The sales tax rate is 5.6%.

13. Commission = Commission rate × Sales

$C = 6\% \times 45,000$

This tells us what to do. We multiply.

$$\begin{array}{r} 45,000 \\ \times \quad 0.06 \\ \hline 2700.00 \end{array} \qquad (6\% = 0.06)$$

The commission is $2700.

15. Commission = Commission rate × Sales

$120 = r \times 2400$

To solve this equation we divide on both sides by 2400:

$$\frac{120}{2400} = \frac{r \times 2400}{2400}$$

We can divide, but this time we simplify by removing a factor of 1:

$$r = \frac{120}{2400} = \frac{1}{20} \cdot \frac{120}{120} = \frac{1}{20} = 0.05 = 5\%$$

The commission rate is 5%.

17. Commission = Commission rate × Sales

$392 = 40\% \times S$

To solve this equation we divide on both sides by 0.4:

$$\frac{392}{0.4} = \frac{0.4 \times S}{0.4}$$
$$980 = S$$

$$\begin{array}{r} 980. \\ 0.4_\wedge \overline{)\,392.0_\wedge} \\ \underline{3600} \\ 320 \\ \underline{320} \\ 0 \\ \underline{0} \\ 0 \end{array}$$

$980 worth of artwork was sold.

19. Commission = Commission rate × Sales

$C = 6\% \times 98,000$

This tells us what to do. We multiply.

$$\begin{array}{r} 98,000 \\ \times \quad 0.06 \\ \hline 5880.00 \end{array} \qquad (6\% = 0.06)$$

The commission is $5880.

21. Commission = Commission rate × Sales

$280.80 = r \times 2340$

To solve this equation we divide on both sides by 2340.

$$\frac{280.80}{2340} = \frac{r \times 2340}{2340}$$
$$0.12 = r$$
$$12\% = r$$

$$\begin{array}{r} 0.12 \\ 2340 \overline{)\,280.80} \\ \underline{2340} \\ 4680 \\ \underline{4680} \\ 0 \end{array}$$

The commission is 12%.

23. First we find the commission on the first \$2000 of sales.

Commission = Commission rate × Sales
$C = 5\% \times 2000$

This tells us what to do. We multiply.

$$\begin{array}{r} 2000 \\ \times\ 0.05 \\ \hline 100.00 \end{array}$$

The commission on the first \$2000 of sales is \$100.

Next we subtract to find the amount of sales over \$2000.

\$6000 − \$2000 = \$4000

Miguel had \$4000 in sales over \$2000.

Then we find the commission on the sales over \$2000.

Commission = Commission rate × Sales
$C = 8\% \times 4000$

This tells us what to do. We multiply.

$$\begin{array}{r} 4000 \\ \times\ 0.08 \\ \hline 320.00 \end{array}$$

The commission on the sales over \$2000 is \$320.

Finally we add to find the total commission.

\$100 + \$320 = \$420

The total commission is \$420.

25. Discount = Marked price − Sale price
$83 = M - 377$

We add 377 on both sides of the equation:

$$83 + 377 = M$$
$$460 = M$$

The marked price is \$460.

Discount = Rate of discount × Marked price
$83 = R \times 460$

To solve the equation we divide on both sides by 460.

$$\frac{83}{460} = \frac{R \times 460}{460}$$
$$0.18043 \approx R$$
$$18.043\% \approx R$$

To check note that a discount rate of 18.043% means that 81.957% of the marked price is paid: $0.81957 \times 460 = 377.0022 \approx 377$. Since this is the sale price, the answer checks.

The rate of discount is 18.043%.

27. Discount = Rate of discount × Marked price
$D = 10\% \times \$300$

Convert 10% to decimal notation and multiply.

$$\begin{array}{r} 300 \\ \times\ 0.1 \\ \hline 30.0 \end{array} \qquad (10\% = 0.10 = 0.1)$$

The discount is \$30.

Sale price = Marked price − Discount
$S = 300 - 30$

We subtract:

$$\begin{array}{r} 300 \\ -\ 30 \\ \hline 270 \end{array}$$

To check, note that the sale price is 90% of the marked price: $0.9 \times 300 = 270$.

The sale price is \$270.

29. Discount = Rate of discount × Marked price
$D = 15\% \times \$17$

Convert 15% to decimal notation and multiply.

$$\begin{array}{r} 17 \\ \times\ 0.15 \\ \hline 85 \\ 170 \\ \hline 2.55 \end{array} \qquad (15\% = 0.15)$$

The discount is \$2.55.

Sale price = Marked price − Discount
$S = 17 - 2.55$

We subtract:
$$\begin{array}{r} 17.00 \\ -\ 2.55 \\ \hline 14.45 \end{array}$$

To check, note that the sale price is 85% of the marked price: $0.85 \times 17 = 14.45$.

The sale price is \$14.45.

31. Discount = Rate of discount × Marked price
$12.50 = 10\% \times M$

To solve the equation we divide on both sides by 0.1.

$$\frac{12.50}{0.1} = \frac{0.1 \times M}{0.1}$$
$$125 = M$$

The marked price is \$125.

Sale price = Marked price − Discount
$S = 125.00 - 12.50$

We subtract:
$$\begin{array}{r} 125.00 \\ -\ 12.50 \\ \hline 112.50 \end{array}$$

To check, note that the sale price is 90% of the marked price: $0.9 \times 125 = 112.50$.

The sale price is \$112.50.

33. Discount = Rate of discount × Marked price
$240 = r \times 600$

To solve the equation we divide on both sides by 600.

$$\frac{240}{600} = \frac{r \times 600}{600}$$

We can simplify by removing a factor of 1:

$$r = \frac{240}{600} = \frac{2}{5} \cdot \frac{120}{120} = \frac{2}{5} = 0.4 = 40\%$$

The rate of discount is 40%.

Sale price = Marked price − Discount
$S = 600 - 240$

We subtract:
$$\begin{array}{r} 600 \\ -\ 240 \\ \hline 360 \end{array}$$

To check, note that a 40% discount rate means that 60% of the marked price is paid. Since $\frac{360}{600} = 0.6$, or 60%, we have a check.

The sale price is \$360.

35. $I = P \cdot r \cdot t$
$= \$200 \times 13\% \times 1$
$= \$200 \times 0.13$
$= \$26$

$$\begin{array}{r} 200 \\ \times\ 0.13 \\ \hline 600 \\ 2000 \\ \hline 26.00 \end{array}$$

The interest is \$26.

37. $I = P \cdot r \cdot t$
$= \$2000 \times 12.4\% \times \frac{1}{2}$
$= \frac{\$2000 \times 0.124}{2}$
$= \$124$

The interest is \$124.

(We could have instead found $\frac{1}{2}$ of 12.4% and then multiplied by 2000.)

39. $I = P \cdot r \cdot t$
$= \$4300 \times 14\% \times \frac{1}{4}$
$= \frac{\$4300 \times 0.14}{4}$
$= \$150.50$

The interest is \$150.50.

(We could have instead found $\frac{1}{4}$ of 14% and then multiplied by 4300.)

41. a) We express 90 days as a fractional part of a year and find the interest.

$I = P \cdot r \cdot t$
$= \$6500 \times 8\% \times \frac{90}{365}$
$= \$6500 \times 0.08 \times \frac{90}{365}$
$\approx \$128.22$ Using a calculator

The interest due for 90 days is \$128.22.

b) The total amount that must be paid after 90 days is the principal plus the interest.

$6500 + 128.22 = 6628.22$

The total amount due is \$6628.22.

43. a) After 1 year, the account will contain 110% of \$400.

$1.1 \times \$400 = \$440$

$$\begin{array}{r} 400 \\ \times\ 1.1 \\ \hline 400 \\ 4000 \\ \hline 440.0 \end{array}$$

b) At the end of the second year, the account will contain 110% of \$440.

$1.1 \times \$440 = \$484$

$$\begin{array}{r} 440 \\ \times\ 1.1 \\ \hline 440 \\ 4400 \\ \hline 484.0 \end{array}$$

The amount in the account after 2 years is \$484.

(Note that we could have used the formula $A = P \cdot \left(1 + \frac{r}{n}\right)^{n \cdot t}$, substituting \$400 for P, 10% for r, 1 for n, and 2 for t.)

45. a) After 1 year, the account will contain 108.8% of \$200.

$1.088 \times \$200 = \$217.60$

$$\begin{array}{r} 1.088 \\ \times\ \ 200 \\ \hline 217.600 \end{array}$$

b) At the end of the second year, the account will contain 108.8% of \$217.60.

$1.088 \times \$217.60 = \$236.7488$

$$\begin{array}{r} 217.6 \\ \times\ 1.088 \\ \hline 17408 \\ 174080 \\ 2176000 \\ \hline 236.7488 \end{array}$$

$\approx \$236.75$ Rounding to the nearest cent

The amount in the account after 2 years is \$236.75.

(Note that we could have used the formula $A = P \cdot \left(1 + \frac{r}{n}\right)^{n \cdot t}$, substituting \$200 for P, 15% for r, 1 for n, and 2 for t.)

47. We use the compound interest formula, substituting \$4000 for P, 7% for r, 2 for n, and 1 for t.

$$A = P \cdot \left(1 + \frac{r}{n}\right)^{n \cdot t}$$
$$A = \$4000 \cdot \left(1 + \frac{0.07}{2}\right)^{2 \cdot 1}$$
$$A = \$4000 \cdot (1 + 0.035)^2$$
$$A = \$4000 \cdot (1.035)^2$$
$$A = \$4284.90$$

The amount in the account after 1 year is \$4284.90.

49. We use the compound interest formula, substituting \$2000 for P, 9% for r, 2 for n, and 3 for t.

$$A = P \cdot \left(1 + \frac{r}{n}\right)^{n \cdot t}$$
$$A = \$2000 \cdot \left(1 + \frac{0.09}{2}\right)^{2 \cdot 3}$$
$$A = \$2000 \cdot (1 + 0.045)^6$$
$$A = \$2000 \cdot (1.045)^6$$
$$A = \$2604.52$$

The amount in the account after 3 years is \$2604.52.

51. We use the compound interest formula, substituting \$4000 for P, 6% for r, 12 for n, and $\frac{5}{12}$ for t.

$$A = P \cdot \left(1 + \frac{r}{n}\right)^{n \cdot t}$$

$$A = \$4000 \cdot \left(1 + \frac{0.06}{12}\right)^{12 \cdot \frac{5}{15}}$$

$$A = \$4000 \cdot (1 + 0.005)^5$$

$$A = \$4000 \cdot (1.005)^5$$

$$A \approx \$4101.01$$

The amount in the account after 5 months is \$4101.01.

53. $0.\underline{93}$ $0.93.$ $\frac{93}{1\underline{00}}$

2 places   Move 2 places.   2 zeros

$$0.93 = \frac{93}{100}$$

55. $\frac{13}{11} = 13 \div 11$

$$\begin{array}{r} 1.1818 \\ 11\overline{)13.0000} \\ \underline{11} \\ 2\,0 \\ \underline{1\,1} \\ 9\,0 \\ \underline{8\,8} \\ 2\,0 \\ \underline{1\,1} \\ 9\,0 \\ \underline{8\,8} \\ 2 \end{array}$$

We get a repeating decimal.

$$\frac{13}{11} = 1.\overline{18}$$

57. ◈

Chapter 5

Data Analysis, Graphs, and Statistics

Exercise Set 5.1

1. To find the average, add the numbers. Then divide by the number of addends.

$$\frac{16+18+29+14+29+19+15}{7} = \frac{140}{7} = 20$$

The average is 20.

To find the median, first list the numbers in order from smallest to largest. Then locate the middle number.

$$\begin{array}{c} 14, 15, 16, 18, 19, 29, 29 \\ \uparrow \\ \text{Middle number} \end{array}$$

The median is 18.

Find the mode:

The number that occurs most often is 29. The mode is 29.

3. To find the average, add the numbers. Then divide by the number of addends.

$$\frac{5+30+20+20+35+5+25}{7} = \frac{140}{7} = 20$$

The average is 20.

To find the median, first list the numbers in order from smallest to largest. Then locate the middle number.

$$\begin{array}{c} 5, 5, 20, 20, 25, 30, 35 \\ \uparrow \\ \text{Middle number} \end{array}$$

The median is 20.

Find the mode:

There are two numbers that occur most often, 5 and 20. Thus the modes are 5 and 20.

5. Find the average:

$$\frac{1.2+4.3+5.7+7.4+7.4}{5} = \frac{26}{5} = 5.2$$

The average is 5.2.

Find the median:

$$\begin{array}{c} 1.2, 4.3, 5.7, 7.4, 7.4 \\ \uparrow \\ \text{Middle number} \end{array}$$

The median is 5.7.

Find the mode:

The number that occurs most often is 7.4. The mode is 7.4.

7. Find the average:

$$\frac{234+228+234+229+234+278}{6} = \frac{1437}{6} = 239.5$$

The average is 239.5.

Find the median:

$$\begin{array}{c} 228, 229, 234, 234, 234, 278 \\ \uparrow \\ \text{Middle number} \end{array}$$

The median is halfway between 234 and 234. Although it seems clear that this is 234, we can compute it as follows:

$$\frac{234+234}{2} = \frac{468}{2} = 234$$

The median is 234.

Find the mode:

The number that occurs most often is 234. The mode is 234.

9. Find the average:

$$\frac{43^\circ+40^\circ+23^\circ+38^\circ+54^\circ+35^\circ+47^\circ}{7} = \frac{280^\circ}{7} = 40^\circ$$

The average temperature was 40°.

Find the median:

$$\begin{array}{c} 23^\circ, 35^\circ, 38^\circ, 40^\circ, 43^\circ, 47^\circ, 54^\circ \\ \uparrow \\ \text{Middle number} \end{array}$$

The median is 40°.

Find the mode:

No number repeats, so no mode exists.

11. We divide the total number of miles, 297, by the number of gallons, 9.

$$\frac{297}{9} = 33$$

The average was 33 miles per gallon.

13. To find the GPA we first add the grade point values for each hour taken. This is done by first multiplying the grade point value by the number of hours in the course and then adding as follows:

$$\begin{array}{ll} \text{B} & 3.00 \cdot 4 = 12 \\ \text{B} & 3.00 \cdot 5 = 15 \\ \text{B} & 3.00 \cdot 3 = \ \ 9 \\ \text{C} & 2.00 \cdot 4 = \underline{\ \ 8} \\ & \qquad\qquad 44 \text{ (Total)} \end{array}$$

The total number of hours taken is

$4+5+3+4$, or 16.

We divide 44 by 16 and round to the nearest tenth.

$\frac{44}{16} = 2.75 \approx 2.8$

The student's grade point average is 2.8.

15. Find the average price per pound:

$$\frac{\$7.99 + \$9.49 + \$9.99 + \$7.99 + \$10.49}{5} = \frac{\$45.95}{5} = \$9.19$$

The average price per pound of Atlantic salmon was \$9.19.

Find the median price per pound:

List the prices in order:

$$\$7.99, \$7.99, \$9.49, \$9.99, \$10.49$$

↑ Middle number

The median is \$9.49.

Find the mode:

The number that occurs most often is \$7.99. The mode is \$7.99.

17. We can find the total of the five scores needed as follows:

$$80 + 80 + 80 + 80 + 80 = 400.$$

The total of the scores on the first four tests is

$$80 + 74 + 81 + 75 = 310.$$

Thus Rich needs to get at least

$$400 - 310, \text{ or } 90$$

to get a B. We can check this as follows:

$$\frac{80 + 74 + 81 + 75 + 90}{5} = \frac{400}{5} = 80.$$

19. We can find the total number of days needed as follows:

$$266 + 266 + 266 + 266 = 1064.$$

The total number of days for Marta's first three pregnancies is

$$270 + 259 + 272 = 801.$$

Thus, Marta's fourth pregnancy must last

$$1064 - 801 = 263 \text{ days}$$

in order to equal the worldwide average.

We can check this as follows:

$$\frac{270 + 259 + 272 + 263}{4} = \frac{1064}{4} = 266.$$

21.

$$\begin{array}{r} 1\,4 \\ \times\,1\,4 \\ \hline 5\,6 \\ 1\,4\,0 \\ \hline 1\,9\,6 \end{array}$$

23.

$$\begin{array}{rl} 1.\,4 & \text{(1 decimal place)} \\ \times\,1.\,4 & \text{(1 decimal place)} \\ \hline 5\,6 & \\ 1\,4\,0 & \\ \hline 1.\,9\,6 & \text{(2 decimal places)} \end{array}$$

25. ***Familiarize.*** We let $c =$ the cost of 19 CDs.

Translate. We translate to a proportion with the number of CDs in the numerators.

$$\begin{array}{r} \text{CDs} \rightarrow \\ \text{Cost} \rightarrow \end{array} \frac{4}{239.80} = \frac{19}{c} \begin{array}{l} \leftarrow \text{CDs} \\ \leftarrow \text{Cost} \end{array}$$

Solve.

$4 \cdot c = 239.80(19)$ Equating cross-products

$c = \frac{239.80(19)}{4}$

$c = 1139.05$

Check. We substitute into the proportion and check cross-products.

$\frac{4}{239.80} = \frac{19}{1139.05};$

$4(1139.05) = 4556.2;\ 239.80(19) = 4556.2$

Since the cross-products are the same, the answer checks.

State. Nineteen comparable CDs would cost \$1139.05.

27. ◈

29. Divide the total by the number of games. Use a calculator.

$$\frac{547}{3} \approx 182.33$$

Drop the amount to the right of the decimal point.

$\underline{182}\,.\,\boxed{33}$

This is the average. ↑ ↑ Drop this amount.

The bowler's average is 182.

31. We can find the total number of home runs needed over Aaron's 22-yr career as follows:

$$22 \cdot 34\frac{7}{22} = 22 \cdot \frac{755}{22} = \frac{22 \cdot 755}{22} = \frac{22}{22} \cdot \frac{755}{1} = 755.$$

The total number of home runs during the first 21 years of Aaron's career was

$$21 \cdot 35\frac{10}{21} = 21 \cdot \frac{745}{21} = \frac{21 \cdot 745}{21} = \frac{21}{21} \cdot \frac{745}{1} = 745.$$

Then Aaron hit

$$755 - 745 = 10 \text{ home runs}$$

in his final year.

Exercise Set 5.2

1. Go down the Planet column to Jupiter. Then go across to the column headed Average Distance from Sun (in miles) and read the entry, 483,612,200. The average distance from the sun to Jupiter is 483,612,200 miles.

3. Go down the column headed Time of Revolution in Earth Time (in years) to 164.78. Then go across the Planet column. The entry there is Neptune, so Neptune has a time of revolution of 164.78 days.

5. All of the entries in the column headed Average Distance from Sun (in miles) are greater than 1,000,000. Thus, all of the planets have an average distance from the sun that is greater than 1,000,000 mi.

7. Go down the Planet column to earth and then across to the Diameter (in miles) column to find that the diameter of earth is 7926 mi. Similarly, find that the diameter of Jupiter is 88,846 mi. Then divide:

$$\frac{88,846}{7926} \approx 11$$

It would take about 11 earth diameters to equal one Jupiter diameter.

9. Find the average of all the numbers in the column headed Diameter (in miles):

$(3031 + 7520 + 7926 + 4221 + 88,846 + 74,898 + 31,763 + 31,329 + 1423)/9 = 27,884.\overline{1}$

The average of the diameters of the planets is $27,884.\overline{1}$ mi.

To find the median of the diameters of the planets we first list the diameters in order from smallest to largest:

1423, 3031, 4221, 7520, 7926, 31,329, 31,763, 74,898, 88,846.

The middle number is 7926, so the median of the diameters is 7926 mi.

Since no number appears more than once in the Diameter (in miles) column, there is no mode.

11. Go down the column headed Actual Temperature (°F) to 80°. Then go across to the Relative Humidity column headed 60%. The entry is 92, so the apparent temperature is 92°F.

13. Go down the column headed Actual Temperature (°F) to 85°. Then go across the Relative Humidity column headed 90%. The entry is 108, so the apparent temperature is 108°F.

15. The number 100 appears in the columns headed Apparent Temperature (°F) 3 times, so there are 3 temperature-humidity combinations that given an apparent temperature of 100°.

17. Go down the Relative Humidity column headed 50% and find all the entries greater than 100. The last 4 entries are greater than 100. Then go across to the column headed Actual Temperature (°F) and read the temperatures that correspond to these entries. At 50% humidity, the actual temperatures 90° and higher give an apparent temperature above 100°.

19. Go down the column headed Actual Temperature (°F) to 95°. Then read across to locate the entries greater than 100. All of the entries except the first two are greater than 100. Go up from each entry to find the corresponding relative humidity. At an actual temperature of 95°, relative humidities of 30% and higher give an apparent temperature above 100°.

21. Go down the column headed Actual Temperature (°F) to 85°, then across to 97, and up to find that the corresponding relative humidity is 50%. Similarly, go down to 85°, across to 111, and up to 100%. At an actual temperature of 85°, the humidity would have to increase by

$$100\% - 50\%, \text{ or } 50\%$$

to raise the apparent temperature from 97° to 111°.

23. To find the average global temperature in 1986, go down the column headed Year to 1986 and then across to find the entry 59.29°. Similarly, find that the average global temperature in 1987 was 59.58°.

To find the percent of increase from 1986 to 1987, we first subtract to find the amount of increase:

$$59.58 - 59.29 = 0.29$$

Then divide this number by the average global temperature in 1986:

$$\frac{0.29}{59.29} \approx 0.005, \text{ or } 0.5\%$$

The percent of increase in the temperature from 1986 to 1987 was about 0.5%.

25. Average for 1986 to 1988:

$$\frac{59.29° + 59.58° + 56.63°}{3} = \frac{178.50°}{3} = 59.50°$$

Average for 1994 to 1996:

$$\frac{59.56° + 59.72° + 59.58°}{3} = \frac{178.86°}{3} = 59.62°$$

We subtract to find by how many degrees the latter average exceeds the former:

$$59.62° - 59.50° = 0.12°$$

27. The world population in 1850 is represented by 1 symbol, so the population was 1 billion.

29. The 1999 (projected) population is represented by the most symbols, so the population was largest in 1999.

31. The smallest increase in the number of symbols is represented by $\frac{1}{2}$ symbol from 1650 to 1850 (as opposed to 1 or more symbols for each of the other pairs). Then the growth was the least between these two years.

33. The world population in 1975 is represented by 4 symbols so it was 4×1 billion, or 4 billion people. The population in 1999 is represented by 6 symbols so it was 6×1 billion, or 6 billion people. We subtract to find the difference:

$$6 \text{ billion} - 4 \text{ billion} = 2 \text{ billion}$$

The world population in 1999 was 2 billion more than in 1975.

To find the percent of increase, we divide the amount of increase, 2 billion, by the population in 1975, 4 billion:

$$\frac{2 \text{ billion}}{4 \text{ billion}} = \frac{2}{4} = 0.5, \text{ or } 50\%$$

The percent of increase in the world population from 1975 to 1999 was 50%.

35. There are more bike symbols beside 1998 than any other year. Therefore, the year in which the greatest number of bikes was sold is 1998.

37. There was positive growth between every pair of consecutive years except 1996 and 1997. The pair of years for which the amount of positive growth was the least was 1994 and 1995 (represented by an increase of 1 bike symbol as opposed to 2 or 3 symbols for each of the other pairs).

39. Sales for 1996 are represented by 7 bike symbols. Since each bike symbol stands for 1000 bikes sold, we multiply to find 7×1000, or 7000, bikes were sold in 1996.

41. We look for a row of the chart containing fewer symbols than the one immediately below it. The only such row is the one showing sales in 1997. Therefore, in 1997 there was a decline in the number of bikes sold.

43. For 1992: Note that $\$168,000,000 = 1.68 \times \$100,000,000$. Thus, we need 1 whole symbol and 0.68, or about $\frac{2}{3}$, of another symbol.

For 1993: Note that $\$312,000,000 = 3.12 \times \$100,000,000$. Thus we need 3 whole symbols and 0.12, or about $\frac{1}{10}$, of another symbol.

For 1994: Note that $\$577,000,000 = 5.77 \times \$100,000,000$. Thus, we need 5 whole symbols and 0.77, or about $\frac{3}{4}$, of another symbol.

For 1995: Note that $\$889,000,000 = 8.89 \times \$100,000,000$. Thus, we need 8 whole symbols and 0.89, or about $\frac{9}{10}$, of another symbol.

For 1996: Note that $\$1,100,000,000 = 11 \times \$100,000,000$. Thus, we need 11 whole symbols.

Now we draw the pictograph.

Lettuce Sales

1992

1993

1994

1995

1996

= $100,000,000

45. ***Familiarize***. Let a = the number of square miles of Maine that are forest.

Translate. We rephrase the question and translate.

$$\begin{array}{ccccc} \text{What} & \text{is} & 90\% & \text{of} & 30,955 \text{ mi}^2? \\ \downarrow & \downarrow & \downarrow & \downarrow & \downarrow \\ a & = & 90\% & \times & 30,955 \end{array}$$

Solve. Convert 90% to decimal notation and multiply.

$$a = 90\% \times 30,955 = 0.9 \times 30,955 = 27,859.5$$

Check. We can repeat the calculation. The answer checks.

State. 27,859.5 mi^2 of Maine is forest.

47. $4.8\% = \frac{4.8}{100} = \frac{4.8}{100} \cdot \frac{10}{10} = \frac{48}{1000} = \frac{8 \cdot 6}{8 \cdot 125} = \frac{8}{8} \cdot \frac{6}{125} = \frac{6}{125}$

49. 

Exercise Set 5.3

1. Move to the right along the bar representing 1 cup of hot cocoa with skim milk. We read that there are about 190 calories in the cup of cocoa.

3. The longest bar is for 1 slice of chocolate cake with fudge frosting. Thus, it has the highest caloric content.

5. We locate 460 calories at the bottom of the graph and then go up until we reach a bar that ends at approximately 460 calories. Now go across to the left and read the dessert, 1 cup of premium chocolate ice cream.

7. From the graph we see that 1 cup of hot cocoa made with whole milk has about 310 calories and 1 cup of hot cocoa made with skim milk has about 190 calories. We subtract to find the difference:

$$310 - 190 = 120$$

The cocoa made with whole milk has about 120 more calories than the cocoa made with skim milk.

9. From Exercise 5 we know that 1 cup of premium ice cream has about 460 calories. We multiply to find the caloric content of 2 cups:

$$2 \times 460 = 920$$

Kristin consumes about 920 calories.

11. From the graph we see that a 2-oz chocolate bar with peanuts contains about 270 calories. We multiply to find the number of extra calories Paul adds to his diet in 1 year:

$$365 \times 270 \text{ calories} = 98,550 \text{ calories}$$

Then we divide to determine the number of pounds he will gain:

$$\frac{98,550}{3500} \approx 28$$

Paul will gain about 28 pounds.

13. In the group of bars representing 1980 find the bar representing Latin America. Go to the top of that bar and then across to the left to read 920 on the vertical scale. Units on this scale are in thousands of hectares, so the forest area of Latin America in 1980 was about 920,000 hectares.

15. The heights of the pair of bars representing Latin America decrease more from 1980 to 1990 than the heights of either of the other pairs of bars. Thus, Latin America experienced the greatest loss of forest area from 1980 to 1990.

17. We go up the vertical scale to 600. Then we move to the right until we come to a bar in the group representing 1990 that ends at about 600. Moving down that bar we see that it represents Africa, so Africa had about 600 thousand hectares of forest area in 1990.

19. From Exercise 13 we know that the forest area of Latin America was about 920,000 hectares in 1980. From the graph we find that it was about 840,000 hectares in 1990. We find the average of these two numbers:

$$\frac{920,000 + 840,000}{2} = \frac{1,760,000}{2} = 880,000$$

Thus, the average forest area in Latin America for the years 1980 and 1990 was about 880,000 hectares.

21. On the horizontal scale in six equally spaced intervals indicate the names of the cities. Label this scale "City." Then label the vertical scale "Commuting Time (in minutes)." Note that the smallest time is 21.9 minutes and the largest is 30.6 minutes. We could start the vertical scale at 0 or we could start it at 20, using a jagged line to indicate the missing numbers. We choose the second option. Label the marks on the vertical scale by 5's. Finally, draw vertical bars above the cities to show the commuting times.

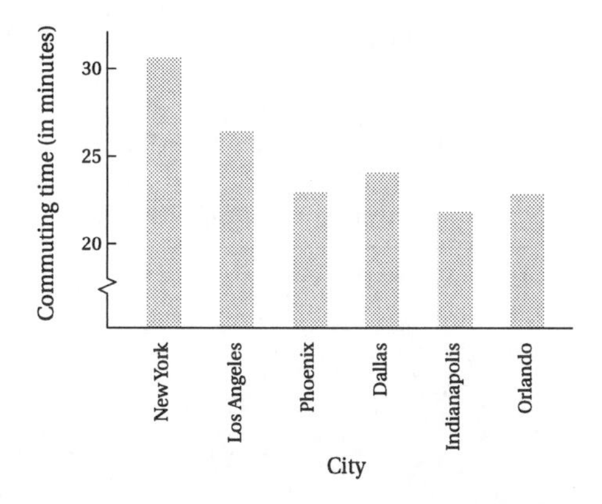

23. The shortest bar represents Indianapolis, so it has the least commuting time.

25. First list the commuting times in order from smallest to largest:

21.9, 22.9, 23.0, 24.1, 26.4, 30.6

We find the average of the two middle numbers to determine the median:

$$\frac{23.0 + 24.1}{2} = \frac{47.1}{2} = 23.55$$

The median commuting time is 23.55 min.

27. The greatest increase in the length of bars going from one year to the next occurs from 1991 to 1992. (Using the data in the table to find the increase from one year to the next confirms this.) Thus, the greatest increase in driving incidents causing death occurred between 1991 and 1992.

29. $\frac{1129 + 1297 + 1478 + 1555 + 1669 + 1708}{6} = \frac{8836}{6} = 1472.\overline{6}$

The average number of death-causing incidents was $1472.\overline{6}$.

31. The highest point on the graph lies above 1997 on the horizontal axis. Thus, the average salary was highest in 1997.

33. From the graph we see that the average salary was the lowest in 1991. It was about \$0.85 million. From Exercise 31 we know that the average salary was the highest in 1997. It was about \$1.35 million. We subtract to find the difference:

$$\$1.35 \text{ million} - \$0.85 \text{ million} = \$0.5 \text{ million}$$

The difference between the highest and lowest average salaries was about \$0.5 million.

35. The segment connecting 1994 and 1995 drops, so the average salary decreased between 1994 and 1995.

37. First indicate the years on the horizontal scale and label it "Year." The smallest ozone level is 2981 parts per billion and the largest is 3148 parts per billion. We could start the vertical scale at 0, but the graph will be more compact and easier to read if we start at a higher number, say at 2980. We do this, using a jagged line to indicate the missing numbers. Mark the vertical scale appropriately, say by 20's, and label it "Ozone level (in parts per billion)." Next, at the appropriate level above each year, mark the corresponding ozone level. Finally, draw line segments connecting the points.

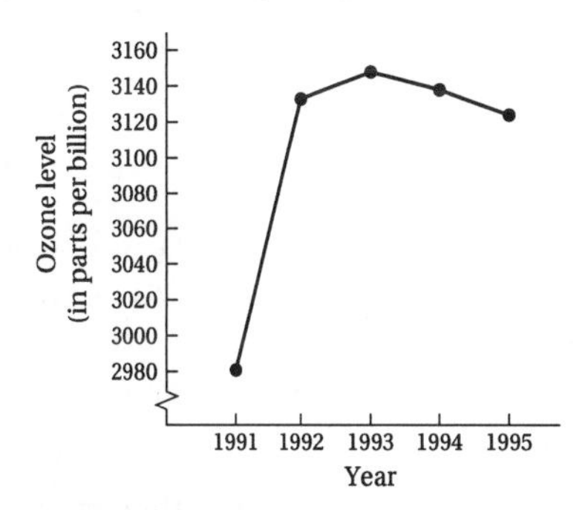

39. The graph falls most sharply from 1994 to 1995. (The data in the table confirms that the greatest decrease occurred between 1994 and 1995.) Thus, the decrease in the ozone level was the greatest between 1994 and 1995.

41. List the ozone levels in the table in order from smallest to largest:

2981, 3124, 3133, 3138, 3148

The middle number is 3133, so the median ozone level was 3133 parts per billion.

43. The segment connecting 1993 and 1994 rises most steeply, so the increase was the greatest between 1993 and 1994. (The data in the table confirms this.)

45. $\frac{38.2 + 42.4 + 44.0 + 50.4 + 54.1 + 61.0}{6} = \frac{290.1}{6} = 48.35$

The average motion-picture expense was \$48.35 million.

47. $\dfrac{38.2+42.2+44.0}{3} = \dfrac{124.6}{3} = 41.5\overline{3}$

The average motion-picture expense from 1991 through 1993 was about $41.5 million.

49. ***Familiarize.*** Let t = the number of minutes the clock will lose in 72 hr.

Translate. We translate to a proportion with the number of minutes lost in the numerators.

$$\begin{array}{lcl}\text{Minutes lost} \rightarrow & \dfrac{3}{12} = \dfrac{t}{72} & \leftarrow \text{Minutes lost} \\ \text{Hours} \rightarrow & & \leftarrow \text{Hours}\end{array}$$

Solve. We equate cross-products.

$$3 \cdot 72 = 12 \cdot t$$
$$\frac{3 \cdot 72}{12} = t$$
$$18 = t$$

Checks. We substitute into the proportion and check cross-products.

$$\frac{3}{12} = \frac{18}{72};\ 3 \cdot 72 = 216;\ 12 \cdot 18 = 216$$

Since the cross-products are the same, the answer checks.

State. The clock will lose 18 min in 72 hr.

51. ***Translate.*** $110\% \times 75 = a$

Solve. We carry out the calculation.

$$\begin{array}{r} 7\ 5 \\ \times\ 1.\ 1 \\ \hline 7\ 5 \\ 7\ 5\ 0 \\ \hline 8\ 2.\ 5 \end{array}$$

Thus, 110% of 75 is 82.5.

53. 

55. The average increase from one year to the next during the 6-yr period is $4.36 million. Add this to the expense for 1996 to get an estimate of the average expense for 1997:

$61.0 million + $4.36 million = $65.36 million ≈ $65.4 million

Answers will vary depending on the method used.

You could tell for sure by getting the actual figure from the Motion Picture Association of America.

Exercise Set 5.4

1. We see from the graph that 3.7% of all records sold are jazz.

3. We see from the graph that 9% of all records sold are country. Find 9% of 3000: $0.09 \times 3000 = 270$. Then 270 of the records are country.

5. We see from the graph that 6.8% of all records sold are classical.

7. The section of the graph representing food is the largest, so food accounts for the greatest expense.

9. We add percents:

12% (medical care) + 2% (personal care) = 14%

11. Since the circle is divided into 100 sections, we can think of it as a pie cut into 100 equally sized pieces. We shade a wedge equal in size to 4 of these pieces to represent 4%. Then we shade wedges equal in size to 30, 34, 30, and 2 of these pieces to represent 30%, 34%, 30%, and 2%, respectively. Finally, give the graph an appropriate title.

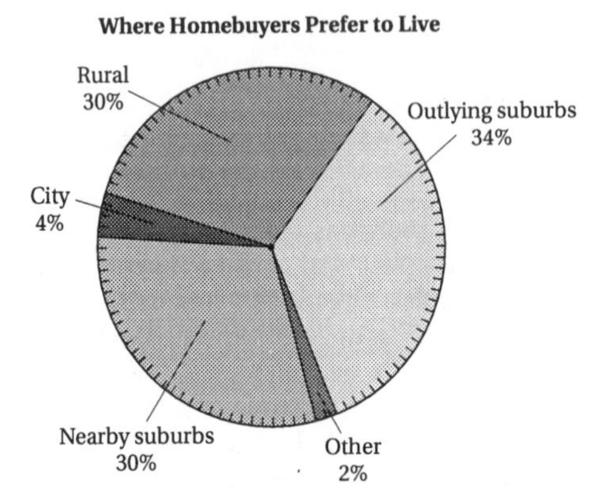

13. Since the circle is divided into 100 sections, we can think of it as a pie cut into 100 equally sized pieces. We shade a wedge equal in size to 6 of these pieces to represent 6%. Then we shade wedges equal in size to 32, 36, and 26 of these pieces to represent 32%, 36%, and 26%, respectively. Finally, we give the graph an appropriate title.

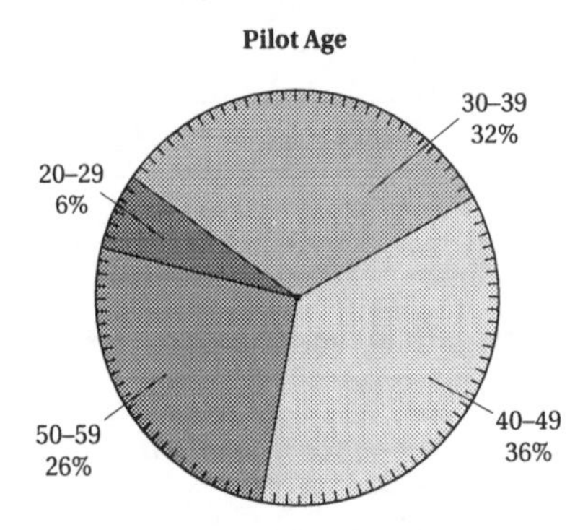

15. Since the circle is divided into 100 sections, we can think of it as a pie cut into 100 equally sized pieces. We shade a wedge equal in size to 14 of these pieces to represent 11%. Then we shade wedges equal in size to 13, 32, 24, 13, and 4 of these pieces to represent 13%, 32%, 24%, 13%, and 4%, respectively. Finally, we give the graph an appropriate title.

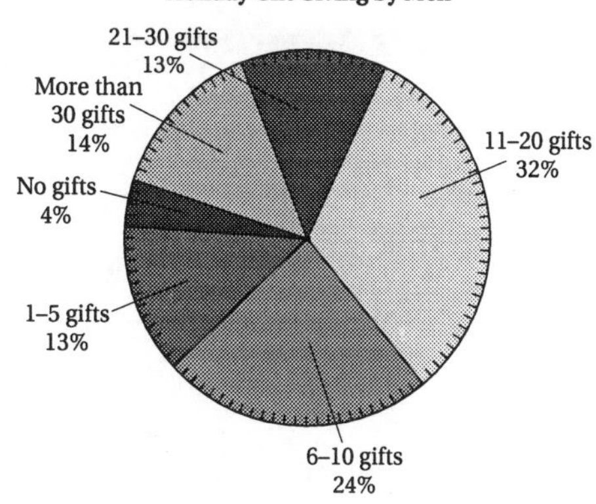

17. ***Translate.*** $a = 45\% \times 668$

Solve. We carry out the calculation.

$$\begin{array}{r} 6\ 6\ 8 \\ \times\ 0.\ 4\ 5 \\ \hline 3\ 3\ 4\ 0 \\ 2\ 6\ 7\ 2\ 0 \\ \hline 3\ 0\ 0.\ 6\ 0 \end{array}$$

Thus, 45% of 668 is 300.6.

19. ***Translate***: $23 = 20\% \times b$

Solve: We divide on both sides by 20%.

$$\frac{23}{20\%} = \frac{20\% \times b}{20\%}$$

$$\frac{23}{0.2} = b$$

$$115 = b$$

Thus, 23 is 20% of 115.

Exercise Set 5.5

1. Compare the averages of the two sets of data.

Bulb A: Average $= (983 + 964 + 1214 + 1417 + 1211 + 1521 + 1084 + 1075 + 892 + 1423 + 949 + 1322)/12 = 1171.25$

Bulb B: Average $= (979 + 1083 + 1344 + 984 + 1445 + 975 + 1492 + 1325 + 1283 + 1325 + 1352 + 1432)/12 \approx 1251.58$

Since the average life of Bulb A is 1171.25 hr and of Bulb B is about 1251.58 hr, Bulb B is better.

3. We interpolate by finding the average of the data values for 17 hours of study and 19 hours of study.

$$\frac{80 + 86}{2} = \frac{166}{2} = 83$$

The missing data value is 83.

We could have also used a graph to find this value, as in Example 2.

5. Use the line graph in Exercise Set 7.3, Exercise 37, to extrapolate. Drawing a "representative" line through the data and beyond gives an estimate of about 3112 parts per billion for the missing data value. Answers will vary according to the placement of the representative line.

7. Graph the data and use the graph to extrapolate. Drawing a "representative" line through the data and beyond gives an estimate of about \$148.8 billion for the missing data value. Answers will vary according to the accuracy of the graph and the placement of the representative line.

9. ***Familiarize.*** Let $c =$ the building costs on a 2400-ft^2 house.

Translate. We translate to a proportion with the sizes of the houses in the numerators.

$$\begin{array}{l} \text{Size} \rightarrow \\ \text{Cost} \rightarrow \end{array} \frac{2200}{118,000} = \frac{2400}{c} \begin{array}{l} \leftarrow \text{Size} \\ \leftarrow \text{Cost} \end{array}$$

Solve. We equate cross-products.

$$\frac{2200}{118,000} = \frac{2400}{c}$$

$$2200 \cdot c = 118,000 \cdot 2400$$

$$c = \frac{118,000 \cdot 2400}{2200}$$

$$c \approx 128,727$$

Check. Substitute into the proportion and check cross-products.

$\frac{2200}{118,000} = \frac{2400}{128,727}$; $2200 \cdot 128,727 = 283,199,400$;

$118,000 \cdot 2400 = 283,200,000$

The cross-products are approximately the same (we rounded the solution of the proportion in the Solve step), so the answer checks.

State. The building costs on a 2400-ft^2 house are about \$128,727.

11. ***Familiarize.*** We let $c =$ the cost of 23 CDs.

Translate. We translate to a proportion with the number of CDs in the numerators.

$$\begin{array}{l} \text{CDs} \rightarrow \\ \text{Cost} \rightarrow \end{array} \frac{4}{239.80} = \frac{23}{c} \begin{array}{l} \leftarrow \text{CDs} \\ \leftarrow \text{Cost} \end{array}$$

Solve.

$4 \cdot c = 239.80(23)$ Equating cross-products

$$c = \frac{239.80(23)}{4}$$

$$c = 1378.85$$

Check. We substitute into the proportion and check cross-products.

$\frac{4}{239.80} = \frac{23}{1378.85}$;

$4(1378.85) = 5515.40$; $239.80(23) = 5515.40$

Since the cross-products are the same, the answer checks.

State. Twenty-three comparable CDs would cost \$1378.85.

13. $\frac{5}{6} \div \frac{7}{18} = \frac{5}{6} \cdot \frac{18}{7} = \frac{5 \cdot 18}{6 \cdot 7} = \frac{5 \cdot 3 \cdot \not{6}}{\not{6} \cdot 7} = \frac{15}{7}$

15. $\frac{17}{25} \div 1000 = \frac{17}{25} \cdot \frac{1}{1000} = \frac{17 \cdot 1}{25 \cdot 1000} = \frac{17}{25,000}$

17. ◈

Chapter 6

Geometry

Exercise Set 6.1

1. The segment consists of the endpoints G and H and all points between them.

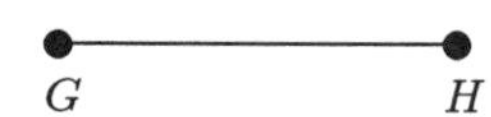

It can be named $\overline{GH}$ or $\overline{HG}$.

3. The ray with endpoint Q extends forever in the direction of point D.

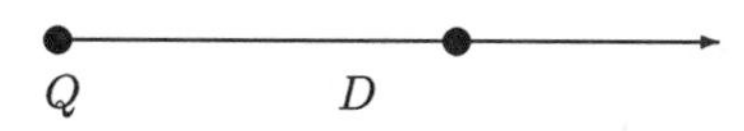

In naming a ray, the endpoint is always given first. This ray is named $\overrightarrow{QD}$.

5. l — D — E — F

The line can be named with the small letter l, or it can be named by any two points on it. This line can be named

l, $\overleftrightarrow{DE}$, $\overleftrightarrow{ED}$, $\overleftrightarrow{DF}$, $\overleftrightarrow{FD}$, $\overleftrightarrow{EF}$, or $\overleftrightarrow{FE}$.

7. The angle can be named in five different ways:
angle GHI, angle IHG, $\angle\, GHI$, $\angle\, IHG$, or $\angle\, H$.

9. Place the $\triangle$ of the protractor at the vertex of the angle, and line up one of the sides at 0°. We choose the horizontal side. Since 0° is on the inside scale, we check where the other side of the angle crosses the inside scale. It crosses at 10°. Thus, the measure of the angle is 10°.

11. Place the $\triangle$ of the protractor at the vertex of the angle, point B. Line up one of the sides at 0°. We choose the side that contains point A. Since 0° is on the outside scale, we check where the other side crosses the outside scale. It crosses at 180°. Thus, the measure of the angle is 180°.

13. Place the $\triangle$ of the protractor at the vertex of the angle, and line up one of the sides at 0°. We choose the horizontal side. Since 0° is on the inside scale, we check where the other side crosses the inside scale. It crosses at 130°. Thus, the measure of the angle is 130°.

15. The measure of the angle in Exercise 9 is 10°. Since its measure is greater than 0°and less than 90°, it is an acute angle.

17. The measure of the angle in Exercise 11 is 180°. It is a straight angle.

19. The measure of the angle in Exercise 13 is 130°. Since its measure is greater than 90°and less than 180°, it is an obtuse angle.

21. Using a protractor, we find that the lines do not intersect to form a right angle. They are not perpendicular.

23. Using a protractor, we find that the lines intersect to form a right angle. They are perpendicular.

25. All the sides are of different lengths. The triangle is a scalene triangle.

One angle is an obtuse angle. The triangle is an obtuse triangle.

27. All the sides are of different lengths. The triangle is a scalene triangle.

One angle is a right angle. The triangle is a right triangle.

29. All the sides are the same length. The triangle is an equilateral triangle.

All three angles are acute. The triangle is an acute triangle.

31. All the sides are of different lengths. The triangle is a scalene triangle.

One angle is an obtuse angle. The triangle is an obtuse triangle.

33. The polygon has 4 sides. It is a quadrilateral.

35. The polygon has 5 sides. It is a pentagon.

37. The polygon has 3 sides. It is a triangle.

39. The polygon has 5 sides. It is a pentagon.

41. The polygon has 6 sides. It is a hexagon.

43. If a polygon has n sides, the sum of its angle measures is $(n-2)\cdot 180°$. A decagon has 10 sides. Substituting 10 for n in the formula, we get

$$\begin{aligned}(n-2)\cdot 180° &= (10-2)\cdot 180° \\ &= 8\cdot 180° \\ &= 1440°.\end{aligned}$$

45. If a polygon has n sides, the sum of its angle measures is $(n-2)\cdot 180°$. A heptagon has 7 sides. Substituting 7 for n in the formula, we get

$$\begin{aligned}(n-2)\cdot 180° &= (7-2)\cdot 180° \\ &= 5\cdot 180° \\ &= 900°.\end{aligned}$$

47. If a polygon has n sides, the sum of its angle measures is $(n-2)\cdot 180°$. To find the sum of the angle measures for a 14-sided polygon, substitute 14 for n in the formula.

$$\begin{aligned}(n-2)\cdot 180° &= (14-2)\cdot 180° \\ &= 12\cdot 180° \\ &= 2160°\end{aligned}$$

49. If a polygon has n sides, the sum of its angle measures is $(n-2)\cdot 180°$. To find the sum of the angle measures for a 20-sided polygon, substitute 20 for n in the formula.

$$\begin{aligned}(n-2)\cdot 180° &= (20-2)\cdot 180°\\ &= 18\cdot 180°\\ &= 3240°\end{aligned}$$

51.
$$\begin{aligned}m(\angle A)+m(\angle B)+m(\angle C) &= 180°\\ 42°+92°+x &= 180°\\ 134°+x &= 180°\\ x &= 180°-134°\\ x &= 46°\end{aligned}$$

53.
$$\begin{aligned}m(\angle R)+m(\angle S)+m(\angle T) &= 180°\\ x+58°+79° &= 180°\\ x+137° &= 180°\\ x &= 180°-137°\\ x &= 43°\end{aligned}$$

55.
$$\begin{array}{r} 1.7\,5\\ 12\,\overline{)\,2\,1.0\,0}\\ 1\,2\,0\,0\\ \hline 9\,0\,0\\ 8\,4\,0\\ \hline 6\,0\\ 6\,0\\ \hline 0\end{array}$$

The answer is 1.75.

57. To divide by 100, move the decimal point 2 places to the left.

23.4   .23.4

$23.4 \div 100 = 0.234$

59.
$$\begin{array}{rl} 3.\,1\,4 & \text{(2 decimal places)}\\ \times\ 4.\,4\,1 & \text{(2 decimal places)}\\ \hline 3\,1\,4 & \\ 1\,2\,5\,6\,0 & \\ 1\,2\,5\,6\,0\,0 & \\ \hline 1\,3.\,8\,4\,7\,4 & \text{(4 decimal places)}\end{array}$$

Round

13. 8 4 [7] 4   to the nearest hundredth.

Thousandths digit is 5 or higher.

13. 8 5   Round up.

61. $48\times\frac{1}{12}=\frac{48\times 1}{12}=\frac{48}{12}=4$

63. ◈

Exercise Set 6.2

1.
$$\begin{aligned}\text{Perimeter} &= 4\text{ mm}+6\text{ mm}+7\text{ mm}\\ &= (4+6+7)\text{ mm}\\ &= 17\text{ mm}\end{aligned}$$

3.
$$\begin{aligned}\text{Perimeter} &= 3.5\text{ in.}+3.5\text{ in.}+4.25\text{ in.}+\\ &\quad 0.5\text{ in.}+3.5\text{ in.}\\ &= (3.5+3.5+4.25+0.5+3.5)\text{ in.}\\ &= 15.25\text{ in.}\end{aligned}$$

5. $P=4\cdot s$ Perimeter of a square
$P=4\cdot 3.25\text{ m}$
$P=13\text{ m}$

7. $P=2\cdot(l+w)$ Perimeter of a rectangle
$P=2\cdot(5\text{ ft}+10\text{ ft})$
$P=2\cdot(15\text{ ft})$
$P=30\text{ ft}$

9. $P=2\cdot(l+w)$ Perimeter of a rectangle
$P=2\cdot(34.67\text{ cm}+4.9\text{ cm})$
$P=2\cdot(39.57\text{ cm})$
$P=79.14\text{ cm}$

11. $P=4\cdot s$ Perimeter of a square
$P=4\cdot 22\text{ ft}$
$P=88\text{ ft}$

13. $P=4\cdot s$ Perimeter of a square
$P=4\cdot 45.5\text{ mm}$
$P=182\text{ mm}$

15. ***Familiarize.*** First we find the perimeter of the field. Then we multiply to find the cost of the fence wire. We make a drawing.

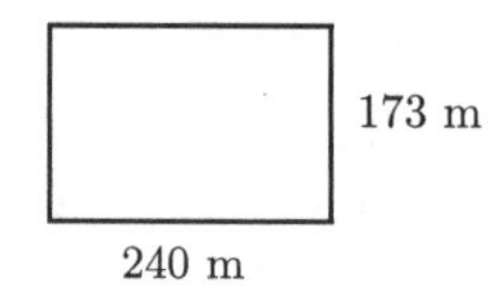

Translate. The perimeter of the field is given by

$$P=2\cdot(l+w)=2\cdot(240\text{ m}+173\text{ m}).$$

Solve. We calculate the perimeter.

$$P=2\cdot(240\text{ m}+173\text{ m})=2\cdot(413\text{ m})=826\text{ m}$$

Then we multiply to find the cost of the fence wire.

$$\begin{aligned}\text{Cost} &= \$1.45/\text{m}\times\text{Perimeter}\\ &= \$1.45/\text{m}\times 826\text{ m}\\ &= \$1197.70\end{aligned}$$

Check. Repeat the calculations.

State. The perimeter of the field is 826 m. The fencing will cost \$1197.70.

17. ***Familiarize.*** We make a drawing and let P = the perimeter.

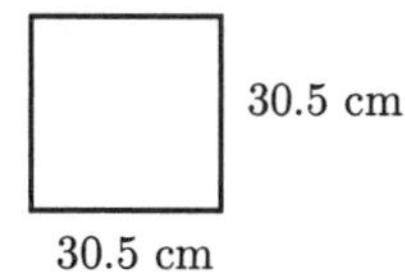

Translate. The perimeter of the square is given by

$$P = 4 \cdot s = 4 \cdot (30.5 \text{ cm}).$$

Solve. We do the calculation.

$$P = 4 \cdot (30.5 \text{ cm}) = 122 \text{ cm}.$$

Check. Repeat the calculation.

State. The perimeter of the tile is 122 cm.

19. ***Familiarize.*** We label the missing lengths on the drawing and let P = the perimeter.

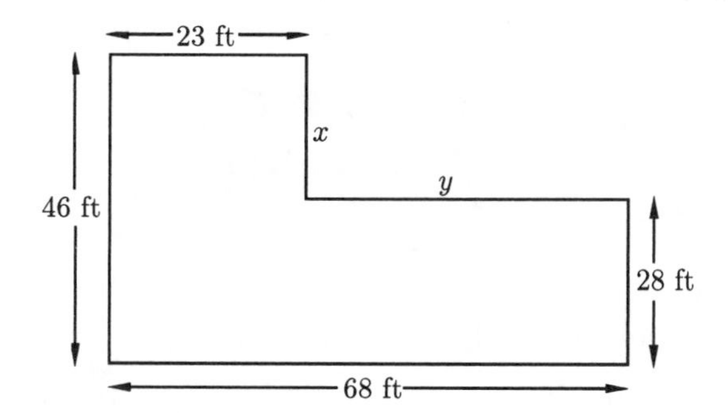

Translate. First we find the missing lengths x and y.

$$\underbrace{\text{28 ft}}_{\downarrow \atop 28} \;\; \underset{\downarrow \atop +}{\text{plus}} \;\; \underbrace{\text{how many more ft}}_{\downarrow \atop x} \;\; \underset{\downarrow \atop =}{\text{is}} \;\; \underbrace{\text{46 ft}}_{\downarrow \atop 46}$$

$$\underbrace{\text{23 ft}}_{\downarrow \atop 23} \;\; \underset{\downarrow \atop +}{\text{plus}} \;\; \underbrace{\text{how many more ft}}_{\downarrow \atop y} \;\; \underset{\downarrow \atop =}{\text{is}} \;\; \underbrace{\text{68 ft}}_{\downarrow \atop 68}$$

Solve. We solve for x and y.

$$\begin{aligned} 28 + x &= 46 \\ x &= 46 - 28 \\ x &= 18 \end{aligned} \qquad \begin{aligned} 23 + y &= 68 \\ y &= 68 - 23 \\ y &= 45 \end{aligned}$$

a) To find the perimeter we add the lengths of the sides of the house.

$$\begin{aligned} P &= 23 \text{ ft} + 18 \text{ ft} + 45 \text{ ft} + 28 \text{ ft} + 68 \text{ ft} + 46 \text{ ft} \\ &= (23 + 18 + 45 + 28 + 68 + 46) \text{ ft} \\ &= 228 \text{ ft} \end{aligned}$$

b) Next we find t, the total cost of the gutter.

$$\underbrace{\text{Cost per foot}}_{\downarrow \atop 4.59} \;\; \underset{\downarrow \atop \times}{\text{times}} \;\; \underbrace{\text{Number of feet}}_{\downarrow \atop 228} \;\; \underset{\downarrow \atop =}{\text{is}} \;\; \underbrace{\text{Total cost}}_{\downarrow \atop t}$$

We carry out the multiplication.

$$\begin{array}{r} 2\,2\,8 \\ \times\; 4\,.5\,9 \\ \hline 2\,0\,5\,2 \\ 1\,1\,4\,0\,0 \\ 9\,1\,2\,0\,0 \\ \hline 1\,0\,4\,6\,.5\,2 \end{array}$$

Thus, $t = 1046.52$.

Check. We can repeat the calculations.

State. (a) The perimeter of the house is 228 ft. (b) The total cost of the gutter is \$1046.52.

21. 56.1%

a) Replace the percent symbol with $\times$ 0.01.

56.1×0.01

b) Move the decimal point two places to the left.

0.56.1 (decimal point moved two places to the left)

Thus, 56.1% = 0.561.

23. a) First find decimal notation by division.

$$\begin{array}{r} 1.\,1\,2\,5 \\ 8\;\overline{)\;9.\,0\,0\,0} \\ 8 \\ \hline 1\;0 \\ 8 \\ \hline 2\,0 \\ 1\,6 \\ \hline 4\,0 \\ 4\,0 \\ \hline 0 \end{array}$$

$$\frac{9}{8} = 1.125$$

b) Convert the decimal notation to percent notation. Move the decimal point two places to the right and write a % symbol.

1.12.5 (decimal point moved two places to the right)

$$\frac{9}{8} = 112.5\%, \text{ or } 112\frac{1}{2}\%$$

25. $10^2 = 10 \cdot 10 = 100$

27. $4.7 \text{ million} = 4.7 \times 1,\underbrace{000,000}_{6 \text{ zeros}}$

4.700000. (decimal point moved to the right)

Move 6 places to the right.

$4.7 \text{ million} = 4,700,000$

29.

31. $18 \text{ in.} = 18 \cancel{\text{in.}} \times \dfrac{1 \text{ ft}}{12 \cancel{\text{in.}}} = \dfrac{18}{12} \times 1 \text{ ft} = \dfrac{3}{2} \text{ ft}$

$P = 2 \cdot (l + w)$

$P = 2 \cdot \left(3 \text{ ft} + \dfrac{3}{2} \text{ ft}\right)$

$P = 2 \cdot \left(\dfrac{9}{2} \text{ ft}\right)$

$P = 9 \text{ ft}$

Exercise Set 6.3

1. $A = l \cdot w$ Area of a rectangular region
$A = (5 \text{ km}) \cdot (3 \text{ km})$
$A = 5 \cdot 3 \cdot \text{ km} \cdot \text{ km}$
$A = 15 \text{ km}^2$

3. $A = l \cdot w$ Area of a rectangular region
$A = (2 \text{ in.}) \cdot (0.7 \text{ in.})$
$A = 2 \cdot 0.7 \cdot \text{ in.} \cdot \text{ in.}$
$A = 1.4 \text{ in}^2$

5. $A = s \cdot s$ Area of a square
$A = \left(2\frac{1}{2} \text{ yd}\right) \cdot \left(2\frac{1}{2} \text{ yd}\right)$
$A = \left(\frac{5}{2} \text{ yd}\right) \cdot \left(\frac{5}{2} \text{ yd}\right)$
$A = \frac{5}{2} \cdot \frac{5}{2} \cdot \text{ yd} \cdot \text{ yd}$
$A = \frac{25}{4} \text{ yd}^2, \text{ or } 6\frac{1}{4} \text{ yd}^2$

7. $A = s \cdot s$ Area of a square
$A = (90 \text{ ft}) \cdot (90 \text{ ft})$
$A = 90 \cdot 90 \cdot \text{ ft} \cdot \text{ ft}$
$A = 8100 \text{ ft}^2$

9. $A = l \cdot w$ Area of a rectangular region
$A = (10 \text{ ft}) \cdot (5 \text{ ft})$
$A = 10 \cdot 5 \cdot \text{ ft} \cdot \text{ ft}$
$A = 50 \text{ ft}^2$

11. $A = l \cdot w$ Area of a rectangular region
$A = (34.67 \text{ cm}) \cdot (4.9 \text{ cm})$
$A = 34.67 \cdot 4.9 \cdot \text{ cm} \cdot \text{ cm}$
$A = 169.883 \text{ cm}^2$

13. $A = l \cdot w$ Area of a rectangular region
$A = \left(4\frac{2}{3} \text{ in.}\right) \cdot \left(8\frac{5}{6} \text{ in.}\right)$
$A = \left(\frac{14}{3} \text{ in.}\right) \cdot \left(\frac{53}{6} \text{ in.}\right)$
$A = \frac{14}{3} \cdot \frac{53}{6} \cdot \text{ in.} \cdot \text{ in.}$
$A = \frac{2 \cdot 7 \cdot 53}{3 \cdot 2 \cdot 3} \text{ in}^2$
$A = \frac{2}{2} \cdot \frac{7 \cdot 53}{3 \cdot 3} \text{ in}^2$
$A = \frac{371}{9} \text{ in}^2, \text{ or } 41\frac{2}{9} \text{ in}^2$

15. $A = s \cdot s$ Area of a square
$A = (22 \text{ ft}) \cdot (22 \text{ ft})$
$A = 22 \cdot 22 \cdot \text{ ft} \cdot \text{ ft}$
$A = 484 \text{ ft}^2$

17. $A = s \cdot s$ Area of a square
$A = (56.9 \text{ km}) \cdot (56.9 \text{ km})$
$A = 56.9 \cdot 56.9 \cdot \text{ km} \cdot \text{ km}$
$A = 3237.61 \text{ km}^2$

19. $A = s \cdot s$ Area of a square
$A = \left(5\frac{3}{8} \text{ yd}\right) \cdot \left(5\frac{3}{8} \text{ yd}\right)$
$A = \left(\frac{43}{8} \text{ yd}\right) \cdot \left(\frac{43}{8} \text{ yd}\right)$
$A = \frac{43}{8} \cdot \frac{43}{8} \cdot \text{ yd} \cdot \text{ yd}$
$A = \frac{1849}{64} \text{ yd}^2, \text{ or } 28\frac{57}{64} \text{ yd}^2$

21. ***Familiarize.*** We draw a picture

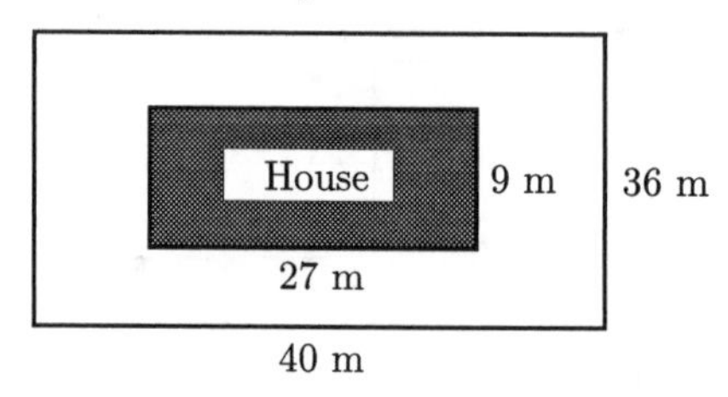

Translate. We let A = the area left over.

Area left over	is	Area of lot	minus	Area of house
A	$=$	$(40 \text{ m}) \cdot (36 \text{ m})$	$-$	$(27 \text{ m}) \cdot (9 \text{ m})$

Solve. The area of the lot is

$$(40 \text{ m}) \cdot (36 \text{ m}) = 40 \cdot 36 \cdot \text{ m} \cdot \text{ m} = 1440 \text{ m}^2.$$

The area of the house is

$$(27 \text{ m}) \cdot (9 \text{ m}) = 27 \cdot 9 \cdot \text{ m} \cdot \text{ m} = 243 \text{ m}^2.$$

The area left over is

$$A = 1440 \text{ m}^2 - 243 \text{ m}^2 = 1197 \text{ m}^2.$$

Check. Repeat the calculations.

State. The area left over for the lawn is 1197 m^2.

23. ***Familiarize.*** We use the drawing in the text.

Translate. We let A = the area of the sidewalk.

Area of sidewalk	is	Total area	minus	Area of building
A	$=$	$(113.4 \text{ m}) \times (75.4 \text{ m})$	$-$	$(110 \text{ m}) \times (72 \text{ m})$

Solve. The total area is

$$(113.4 \text{ m}) \times (75.4 \text{ m}) = 113.4 \times 75.4 \times \text{ m} \times \text{ m} = 8550.36 \text{ m}^2.$$

The area of the building is

$$(110 \text{ m}) \times (72 \text{ m}) = 110 \times 72 \times \text{ m} \times \text{ m} = 7920 \text{ m}^2.$$

The area of the sidewalk is

$$A = 8550.36 \text{ m}^2 - 7920 \text{ m}^2 = 630.36 \text{ m}^2.$$

Check. Repeat the calculations.

State. The area of the sidewalk is 630.36 m^2.

25. ***Familiarize.*** The dimensions are as follows:

Two walls are 15 ft by 8 ft.

Two walls are 20 ft by 8 ft.

The ceiling is 15 ft by 20 ft.

The total area of the walls and ceiling is the total area of the rectangles described above less the area of the windows and the door.

Translate. a) We let A = the total area of the walls and ceiling. The total area of the two 15 ft by 8 ft walls is

$$2 \cdot (15 \text{ ft}) \cdot (8 \text{ ft}) = 2 \cdot 15 \cdot 8 \cdot \text{ ft} \cdot \text{ ft} = 240 \text{ ft}^2$$

The total area of the two 20 ft by 8 ft walls is

$$2 \cdot (20 \text{ ft}) \cdot (8 \text{ ft}) = 2 \cdot 20 \cdot 8 \cdot \text{ ft} \cdot \text{ ft} = 320 \text{ ft}^2$$

The area of the ceiling is

$$(15 \text{ ft}) \cdot (20 \text{ ft}) = 15 \cdot 20 \cdot \text{ ft} \cdot \text{ ft} = 300 \text{ ft}^2$$

The area of the two windows is

$$2 \cdot (3 \text{ ft}) \cdot (4 \text{ ft}) = 2 \cdot 3 \cdot 4 \cdot \text{ ft} \cdot \text{ ft} = 24 \text{ ft}^2$$

The area of the door is

$$\begin{aligned}\left(2\frac{1}{2} \text{ ft}\right) \cdot \left(6\frac{1}{2} \text{ ft}\right) &= \left(\frac{5}{2} \text{ ft}\right) \cdot \left(\frac{13}{2} \text{ ft}\right) \\ &= \frac{5}{2} \cdot \frac{13}{2} \cdot \text{ ft} \cdot \text{ ft} \\ &= \frac{65}{4} \text{ ft}^2, \text{ or } 16\frac{1}{4} \text{ ft}^2\end{aligned}$$

Thus

$$\begin{aligned}A &= 240 \text{ ft}^2 + 320 \text{ ft}^2 + 300 \text{ ft}^2 - 24 \text{ ft}^2 - 16\frac{1}{4} \text{ ft}^2 \\ &= 819\frac{3}{4} \text{ ft}^2, \text{ or } 819.75 \text{ ft}^2\end{aligned}$$

b) We divide to find how many gallons of paint are needed.

$$819.75 \div 86.625 \approx 9.46$$

It will be necessary to buy 10 gallons of paint in order to have the required 9.46 gallons.

c) We multiply to find the cost of the paint.

$$10 \times \$17.95 = \$179.50$$

Check. We repeat the calculations.

State. (a) The total area of the walls and ceiling is 819.75 ft^2. (b) 10 gallons of paint are needed. (c) It will cost \$179.50 to paint the room.

27.

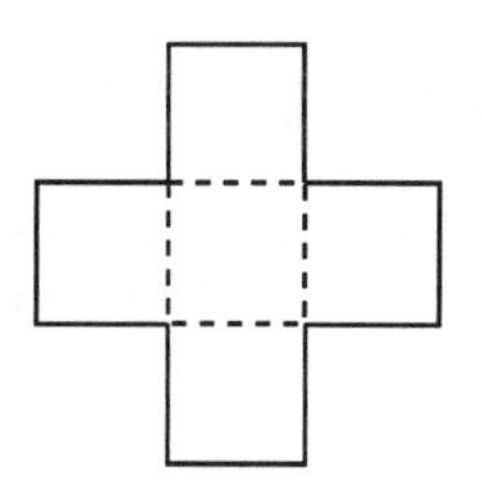

Each side is 4 cm.

The region is composed of 5 squares, each with sides of length 4 cm. The area is

$$A = 5 \cdot (s \cdot s) = 5 \cdot (4 \text{ cm} \cdot 4 \text{ cm}) = 5 \cdot 4 \cdot 4 \text{ cm} \cdot \text{ cm} = 80 \text{ cm}^2$$

29. 0.452 0.45.2 Move the decimal point 2 places to the right.

Write a % symbol: 45.2%

0.452 = 45.2%

31. We multiply by 1 to get 100 in the denominator.

$$\frac{11}{20} = \frac{11}{20} \cdot \frac{5}{5} = \frac{55}{100} = 55\%$$

33. The ratio of the amount spent in Florida to the total amount spent is $\frac{2.7}{13.1}$. This can also be expressed as follows:

$$\frac{2.7}{13.1} = \frac{2.7}{13.1} \cdot \frac{10}{10} = \frac{27}{131}$$

The ratio of the total amount spent to the amount spent in Florida is $\frac{13.1}{2.7}$, or $\frac{131}{27}$.

35. ◈

37.

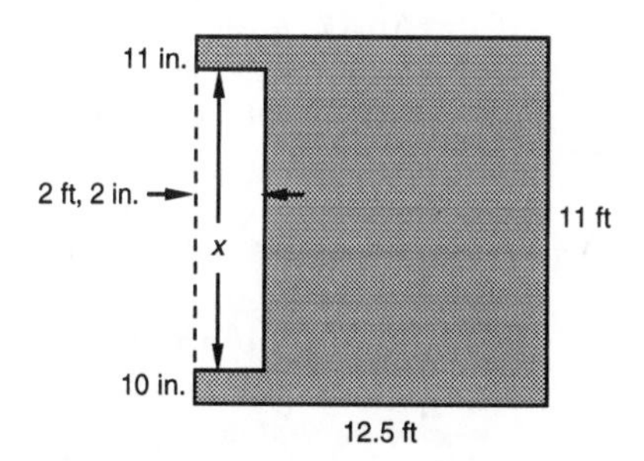

2 ft = 2 × 1 ft = 2 × 12 in. = 24 in., so 2 ft, 2 in. = 2 ft + 2 in. = 24 in. + 2 in. = 26 in.

11 ft = 11 × 1 ft = 11 × 12 in. = 132 in.

12.5 ft = 12.5 × 1 ft = 12.5 × 12 in. = 150 in.

We solve an equation to find x, in inches:

$$\begin{aligned}11 + x + 10 &= 132 \\ 21 + x &= 132 \\ 21 + x - 21 &= 132 - 21 \\ x &= 111\end{aligned}$$

Then the area of the shaded region is the area of a 150 in. by 132 in. rectangle less the area of a 111 in. by 26 in. rectangle.

$$\begin{aligned}A &= (150 \text{ in.}) \cdot (132 \text{ in.}) - (111 \text{ in.}) \cdot (26 \text{ in.}) \\ A &= 19,800 \text{ in}^2 - 2886 \text{ in}^2 \\ A &= 16,914 \text{ in}^2\end{aligned}$$

Exercise Set 6.4

1. $A = b \cdot h$ Area of a parallelogram
$A = 8 \text{ cm} \cdot 4 \text{ cm}$ Substituting 8 cm for b and 4 cm for h
$A = 32 \text{ cm}^2$

3. $A = \frac{1}{2} \cdot b \cdot h$ Area of a triangle
$A = \frac{1}{2} \cdot 15 \text{ in.} \cdot 8 \text{ in.}$ Substituting 15 in. for b and 8 in. for h
$A = 60 \text{ in}^2$

5. $A = \frac{1}{2} \cdot h \cdot (a+b)$ Area of a trapezoid
$A = \frac{1}{2} \cdot 8 \text{ ft} \cdot (6+20) \text{ ft}$ Substituting 8 ft for h, 6 ft for a, and 20 ft for b
$A = \frac{8 \cdot 26}{2} \text{ ft}^2$
$A = 104 \text{ ft}^2$

7. $A = \frac{1}{2} \cdot h \cdot (a+b)$ Area of a trapezoid
$A = \frac{1}{2} \cdot 7 \text{ in.} \cdot (4.5+8.5) \text{ in.}$ Substituting 7 in. for h, 4.5 in. for a, and 8.5 in. for b
$A = \frac{7 \cdot 13}{2} \text{ in}^2$
$A = \frac{91}{2} \text{ in}^2$
$A = 45.5 \text{ in}^2$

9. $A = b \cdot h$ Area of a parallelogram
$A = 2.3 \text{ cm} \cdot 3.5 \text{ cm}$ Substituting 2.3 cm for b and 3.5 cm for h
$A = 8.05 \text{ cm}^2$

11. $A = \frac{1}{2} \cdot h \cdot (a+b)$ Area of a trapezoid
$A = \frac{1}{2} \cdot 18 \text{ cm} \cdot (9+24) \text{ cm}$ Substituting 18 cm for h, 9 cm for a, and 24 cm for b
$A = \frac{18 \cdot 33}{2} \text{ cm}^2$
$A = 297 \text{ cm}^2$

13. $A = \frac{1}{2} \cdot b \cdot h$ Area of a triangle
$A = \frac{1}{2} \cdot 4 \text{ m} \cdot 3.5 \text{ m}$ Substituting 4 m for b and 3.5 m for h
$A = \frac{4 \cdot 3.5}{2} \text{ m}^2$
$A = 7 \text{ m}^2$

15. ***Familiarize.*** We look for the kinds of figures whose areas we can calculate using area formulas that we already know.

Translate. The shaded region consists of a square region with a triangular region removed from it. The sides of the square are 30 cm, and the triangle has base 30 cm and height 15 cm. We find the area of the square using the formula $A = s \cdot s$, and the area of the triangle using $A = \frac{1}{2} \cdot b \cdot h$. Then we subtract.

Solve. Area of the square: $A = 30 \text{ cm} \cdot 30 \text{ cm} = 900 \text{ cm}^2$.

Area of the triangle: $A = \frac{1}{2} \cdot 30 \text{ cm} \cdot 15 \text{ cm} = 225 \text{ cm}^2$.

Area of the shaded region: $A = 900 \text{ cm}^2 - 225 \text{ cm}^2 = 675 \text{ cm}^2$.

Check. We repeat the calculations.

State. The area of the shaded region is 675 cm^2.

17. ***Familiarize.*** We look for the kinds of figures whose areas we can calculate using area formulas that we already know.

Translate. The shaded region consists of 8 triangles, each with base 43 in. and height 52 in. We will find the area of one triangle using the formula $A = \frac{1}{2} \cdot b \cdot h$. Then we will multiply by 8.

Solve. $A = \frac{1}{2} \cdot 43 \text{ in.} \cdot 52 \text{ in.} = 1118 \text{ in}^2$

Then we multiply by 8: $8 \cdot 1118 \text{ in}^2 = 8944 \text{ in}^2$

Check. We repeat the calculations.

State. The area of the shaded region is 8944 in^2.

19. ***Familiarize.*** We make a drawing, shading the area left over after the triangular piece is cut from the sailcloth.

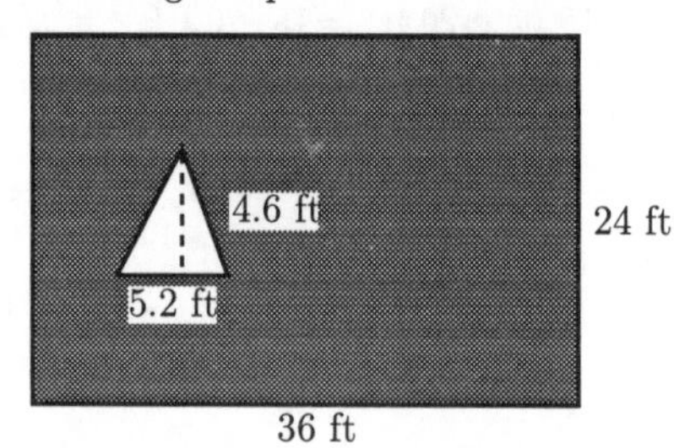

Translate. The shaded region consists of a rectangular region with a triangular region removed from it. The rectangular region has dimensions 36 ft by 24 ft, and the triangular region has base 5.2 ft and height 4.6 ft. We will find the area of the rectangular region using the formula $A = b \cdot h$, and the area of the triangle using $A = \frac{1}{2} \cdot b \cdot h$. Then we will subtract to find the area of the shaded region.

Solve. Area of the rectangle: $A = 36 \text{ ft} \cdot 24 \text{ ft} = 864 \text{ ft}^2$.

Area of the triangle: $A = \frac{1}{2} \cdot 5.2 \text{ ft} \cdot 4.6 \text{ ft} = 11.96 \text{ ft}^2$.

Area of the shaded region: $A = 864 \text{ ft}^2 - 11.96 \text{ ft}^2 = 852.04 \text{ ft}^2$.

Check. We repeat the calculation.

State. The area left over is 852.04 ft^2.

21. $35\% = \frac{35}{100} = \frac{5 \cdot 7}{5 \cdot 20} = \frac{5}{5} \cdot \frac{7}{20} = \frac{7}{20}$

23. $37\frac{1}{2}\% = \frac{75}{2}\% = \frac{75}{2} \times \frac{1}{100} = \frac{75}{2 \cdot 100} = \frac{25 \cdot 3}{2 \cdot 25 \cdot 4} = \frac{25}{25} \cdot \frac{3}{2 \cdot 4} = \frac{3}{8}$

25. $83.\overline{3}\% = 83\frac{1}{3}\% = \frac{250}{3}\% = \frac{250}{3} \times \frac{1}{100} =$

$\frac{250 \cdot 1}{3 \cdot 100} = \frac{5 \cdot 50 \cdot 1}{3 \cdot 2 \cdot 50} = \frac{50}{50} \cdot \frac{5 \cdot 1}{3 \cdot 2} = \frac{5}{6}$

27. ***Familiarize.*** Let s = the number of sheets in 15 reams of paper. Repeated addition works well here.

$$\underbrace{\boxed{500} + \boxed{500} + \cdots + \boxed{500}}_{\text{15 addends}}$$

Translate.

$$\underbrace{\text{Sheets in one ream}}_{500} \;\; \underset{\times}{\text{times}} \;\; \underbrace{\text{Number of reams}}_{15} \;\; \underset{=}{\text{is}} \;\; \underbrace{\text{Total number of sheets}}_{s}$$

Solve. We multiply.

$500 \times 15 = 7500$, so $7500 = s$, or $s = 7500$.

Check. We can repeat the calculation. The answer checks.

State. There are 7500 sheets in 15 reams of paper.

29. 

Exercise Set 6.5

1. $d = 2 \cdot r$

$d = 2 \cdot 7 \text{ cm} = 14 \text{ cm}$

$C = 2 \cdot \pi \cdot r$

$C \approx 2 \cdot \frac{22}{7} \cdot 7 \text{ cm} = \frac{2 \cdot 22 \cdot 7}{7} \text{ cm} = 44 \text{ cm}$

$A = \pi \cdot r \cdot r$

$A \approx \frac{22}{7} \cdot 7 \text{ cm} \cdot 7 \text{ cm} = \frac{22}{7} \cdot 49 \text{ cm}^2 = 154 \text{ cm}^2$

3. $d = 2 \cdot r$

$d = 2 \cdot \frac{3}{4} \text{ in.} = \frac{6}{4} \text{ in.} = \frac{3}{2} \text{ in., or } 1\frac{1}{2} \text{ in.}$

$C = 2 \cdot \pi \cdot r$

$C \approx 2 \cdot \frac{22}{7} \cdot \frac{3}{4} \text{ in.} = \frac{2 \cdot 22 \cdot 3}{7 \cdot 4} \text{ in.} = \frac{132}{28} \text{ in.} = \frac{33}{7} \text{ in.,}$

or $4\frac{5}{7}$ in.

$A = \pi \cdot r \cdot r$

$A \approx \frac{22}{7} \cdot \frac{3}{4} \text{ in.} \cdot \frac{3}{4} \text{ in.} = \frac{22 \cdot 3 \cdot 3}{7 \cdot 4 \cdot 4} \text{ in}^2 = \frac{99}{56} \text{ in}^2, \text{ or } 1\frac{43}{56} \text{ in}^2$

5. $r = \frac{d}{2}$

$r = \frac{32 \text{ ft}}{2} = 16 \text{ ft}$

$C = \pi \cdot d$

$C \approx 3.14 \cdot 32 \text{ ft} = 100.48 \text{ ft}$

$A = \pi \cdot r \cdot r$

$A \approx 3.14 \cdot 16 \text{ ft} \cdot 16 \text{ ft} \quad (r = \frac{d}{2}; r = \frac{32 \text{ ft}}{2} = 16 \text{ ft})$

$A = 3.14 \cdot 256 \text{ ft}^2$

$A = 803.84 \text{ ft}^2$

7. $r = \frac{d}{2}$

$r = \frac{1.4 \text{ cm}}{2} = 0.7 \text{ cm}$

$C = \pi \cdot d$

$C \approx 3.14 \cdot 1.4 \text{ cm} = 4.396 \text{ cm}$

$A = \pi \cdot r \cdot r$

$A \approx 3.14 \cdot 0.7 \text{ cm} \cdot 0.7 \text{ cm}$

$(r = \frac{d}{2}; r = \frac{1.4 \text{ cm}}{2} = 0.7 \text{ cm})$

$A = 3.14 \cdot 0.49 \text{ cm}^2 = 1.5386 \text{ cm}^2$

9. $r = \frac{d}{2}$

$r = \frac{6 \text{ cm}}{2} = 3 \text{ cm}$

The radius is 3 cm.

$C = \pi \cdot d$

$C \approx 3.14 \cdot 6 \text{ cm} = 18.84 \text{ cm}$

The circumference is about 18.84 cm.

$A = \pi \cdot r \cdot r$

$A \approx 3.14 \cdot 3 \text{ cm} \cdot 3 \text{ cm} = 28.26 \text{ cm}^2$

The area is about 28.26 cm^2.

11. $A = \pi \cdot r \cdot r$

$A \approx 3.14 \cdot 220 \text{ mi} \cdot 220 \text{ mi} = 151,976 \text{ mi}^2$

The broadcast area is about 151,976 mi^2.

13.

$C = \pi \cdot d$	
$7.85 \text{ cm} \approx 3.14 \cdot d$	Substituting 7.85 cm for C and 3.14 for π
$\frac{7.85 \text{ cm}}{3.14} = d$	Dividing on both sides by 3.14
$2.5 \text{ cm} = d$	

The diameter is about 2.5 cm.

$r = \frac{d}{2}$

$r = \frac{2.5 \text{ cm}}{2} = 1.25 \text{ cm}$

The radius is about 1.25 cm.

$A = \pi \cdot r \cdot r$

$A \approx 3.14 \cdot 1.25 \text{ cm} \cdot 1.25 \text{ cm} = 4.90625 \text{ cm}^2$

The area is about 4.90625 cm^2.

15. $C = \pi \cdot d$

$C \approx 3.14 \cdot 1.1 \text{ ft} = 3.454 \text{ ft}$

The circumference of the elm tree is about 3.454 ft.

17. Find the area of the larger circle (pool plus walk). Its diameter is 1 yd + 20 yd + 1 yd, or 22 yd. Thus its radius is $\frac{22}{2}$ yd, or 11 yd.

$A = \pi \cdot r \cdot r$

$A \approx 3.14 \cdot 11 \text{ yd} \cdot 11 \text{ yd} = 379.94 \text{ yd}^2$

Find the area of the pool. Its diameter is 20 yd. Thus its radius is $\frac{20}{2}$ yd, or 10 yd.

$$A = \pi \cdot r \cdot r$$
$$A \approx 3.14 \cdot 10 \text{ yd} \cdot 10 \text{ yd} = 314 \text{ yd}^2$$

We subtract to find the area of the walk:

$$A = 379.94 \text{ yd}^2 - 314 \text{ yd}^2$$
$$A = 65.94 \text{ yd}^2$$

The area of the walk is 65.94 yd^2.

19. The perimeter consists of the circumferences of three semicircles, each with diameter 8 ft, and one side of a square of length 8 ft. We first find the circumference of one semicircle. This is one-half the circumference of a circle with diameter 8 ft:

$$\frac{1}{2} \cdot \pi \cdot d \approx \frac{1}{2} \cdot 3.14 \cdot 8 \text{ ft} = 12.56 \text{ ft}$$

Then we multiply by 3:

$$3 \cdot (12.56 \text{ ft}) = 37.68 \text{ ft}$$

Finally we add the circumferences of the semicircles and the length of the side of the square:

$$37.68 \text{ ft} + 8 \text{ ft} = 45.68 \text{ ft}$$

The perimeter is 45.68 ft.

21. The perimeter consists of three-fourths of the circumference of a circle with radius 4 yd and two sides of a square with sides of length 4 yd. We first find three-fourths of the circumference of the circle:

$$\frac{3}{4} \cdot 2 \cdot \pi \cdot r \approx 0.75 \cdot 2 \cdot 3.14 \cdot 4 \text{ yd} = 18.84 \text{ yd}$$

Then we add this length to the lengths of two sides of the square:

$$18.84 \text{ yd} + 4 \text{ yd} + 4 \text{ yd} = 26.84 \text{ yd}$$

The perimeter is 26.84 yd.

23. The perimeter consists of three-fourths of the perimeter of a square with side of length 10 yd and the circumference of a semicircle with diameter 10 yd. First we find three-fourths of the perimeter of the square:

$$\frac{3}{4} \cdot 4 \cdot s = \frac{3}{4} \cdot 4 \cdot 10 \text{ yd} = 30 \text{ yd}$$

Then we find one-half of the circumference of a circle with diameter 10 yd:

$$\frac{1}{2} \cdot \pi \cdot d \approx \frac{1}{2} \cdot 3.14 \cdot 10 \text{ yd} = 15.7 \text{ yd}$$

Then we add:

$$30 \text{ yd} + 15.7 \text{ yd} = 45.7 \text{ yd}$$

The perimeter is 45.7 yd.

25. The shaded region consists of a circle of radius 8 m, with two circles each of diameter 8 m, removed. First we find the area of the large circle:

$$A = \pi \cdot r \cdot r \approx 3.14 \cdot 8 \text{ m} \cdot 8 \text{ m} = 200.96 \text{ m}^2$$

Then we find the area of one of the small circles:
The radius is $\frac{8 \text{ m}}{2} = 4$ m.

$$A = \pi \cdot r \cdot r \approx 3.14 \cdot 4 \text{ m} \cdot 4 \text{ m} = 50.24 \text{ m}^2$$

We multiply this area by 2 to find the area of the two small circles:

$$2 \cdot 50.24 \text{ m}^2 = 100.48 \text{ m}^2$$

Finally we subtract to find the area of the shaded region:

$$200.96 \text{ m}^2 - 100.48 \text{ m}^2 = 100.48 \text{ m}^2$$

The area of the shaded region is 100.48 m^2.

27. The shaded region consists of one-half of a circle with diameter 2.8 cm and a triangle with base 2.8 cm and height 2.8 cm. First we find the area of the semicircle. The radius is $\frac{2.8 \text{ cm}}{2} = 1.4$ cm.

$$A = \frac{1}{2} \cdot \pi \cdot r \cdot r \approx \frac{1}{2} \cdot 3.14 \cdot 1.4 \text{ cm} \cdot 1.4 \text{ cm} = 3.0772 \text{ cm}^2$$

Then we find the area of the triangle:

$$A = \frac{1}{2} \cdot b \cdot h = \frac{1}{2} \cdot 2.8 \text{ cm} \cdot 2.8 \text{ cm} = 3.92 \text{ cm}^2$$

Finally we add to find the area of the shaded region:

$$3.0772 \text{ cm}^2 + 3.92 \text{ cm}^2 = 6.9972 \text{ cm}^2$$

The area of the shaded region is 6.9972 cm^2.

29. The shaded area consists of a rectangle of dimensions 11.4 in. by 14.6 in., with the area of two semicircles, each of diameter 11.4 in., removed. This is equivalent to removing one circle with diameter 11.4 in. from the rectangle. First we find the area of the rectangle:

$$l \cdot w = (11.4 \text{ in.}) \cdot (14.6 \text{ in.}) = 166.44 \text{ in}^2$$

Then we find the area of the circle. The radius is $\frac{11.4 \text{ in.}}{2} = 5.7$ in.

$$\pi \cdot r \cdot r \approx 3.14 \cdot 5.7 \text{ in.} \cdot 5.7 \text{ in.} = 102.0186 \text{ in}^2$$

Finally we subtract to find the area of the shaded region:

$$166.44 \text{ in}^2 - 102.0186 \text{ in}^2 = 64.4214 \text{ in}^2$$

31. 0.875

a) Move the decimal point 2 places to the right. 0.87.5

b) Add a percent symbol 87.5%

0.875 = 87.5%

33. $0.\overline{6}$

a) Move the decimal point 2 places to the right. $0.66.\overline{6}$ (with an arrow under the digits showing the point's move)

b) Add a percent symbol $66.\overline{6}\%$

$0.\overline{6} = 66.\overline{6}\%$

35. a) Find decimal notation using long division.

$$\begin{array}{r} 0.375 \\ 8\,\overline{)\,3.000} \\ \underline{2\,4} \\ 60 \\ \underline{56} \\ 40 \\ \underline{40} \\ 0 \end{array}$$

$\frac{3}{8} = 0.375$

b) Convert the decimal notation to percent notation. Move the decimal point two places to the right, and write a % symbol.

0.37.5 (with an arrow under the digits showing the point's move)

$\frac{3}{8} = 37.5\%$

37. a) Find decimal notation using long division.

$$\begin{array}{r} 0.66 \\ 3\,\overline{)\,2.00} \\ \underline{1\,8} \\ 20 \\ \underline{18} \\ 2 \end{array}$$

$\frac{2}{3} = 0.\overline{6}$

b) Convert the decimal notation to percent notation. Move the decimal point two places to the right, and write a % symbol.

$0.66.\overline{6}$ (with an arrow under the digits showing the point's move)

$\frac{2}{3} = 66.\overline{6}\%$

39. $3\frac{7}{8} = 3 + \frac{7}{8} \approx 3 + 1 = 4$

41. $13\frac{1}{6} = 13 + \frac{1}{6} \approx 13 + 0 = 13$

43. $\frac{4}{5} + 3\frac{7}{8} = \frac{4}{5} + 3 + \frac{7}{8} \approx 1 + 3 + 1 = 5$

45. $\frac{2}{3} + \frac{7}{15} + \frac{8}{9} \approx \frac{1}{2} + \frac{1}{2} + 1 = 2$

47. $\frac{57}{100} - \frac{1}{10} + \frac{9}{1000} \approx \frac{1}{2} - 0 + 0 = \frac{1}{2}$

49. $11\frac{29}{80} + 10\frac{14}{15} \cdot 24\frac{2}{17} \approx 11\frac{1}{2} + 11 \cdot 24 =$

$11\frac{1}{2} + 264 = 275\frac{1}{2}$

51. ◈

53. Find $3927 \div 1250$ using a calculator.

$\frac{3927}{1250} = 3.1416 \approx 3.142$ Rounding

55. The height of the stack of tennis balls is three times the diameter of one ball, or $3 \cdot d$.

The circumference of one ball is given by $\pi \cdot d$.

The circumference of one ball is greater than the height of the stack of balls, because $\pi > 3$.

Exercise Set 6.6

1. $V = l \cdot w \cdot h$

$V = 12\text{ cm} \cdot 8\text{ cm} \cdot 8\text{ cm}$

$V = 12 \cdot 64\text{ cm}^3$

$V = 768\text{ cm}^3$

3. $V = l \cdot w \cdot h$

$V = 7.5\text{ in.} \cdot 2\text{ in.} \cdot 3\text{ in.}$

$V = 7.5 \cdot 6\text{ in}^3$

$V = 45\text{ in}^3$

5. $V = l \cdot w \cdot h$

$V = 10\text{ m} \cdot 5\text{ m} \cdot 1.5\text{ m}$

$V = 10 \cdot 7.5\text{ m}^3$

$V = 75\text{ m}^3$

7. $V = l \cdot w \cdot h$

$V = 6\frac{1}{2}\text{ yd} \cdot 5\frac{1}{2}\text{ yd} \cdot 10\text{ yd}$

$V = \frac{13}{2} \cdot \frac{11}{2} \cdot 10\text{ yd}^3$

$V = \frac{715}{2}\text{ yd}^3$

$V = 357\frac{1}{2}\text{ yd}^3$

9. $V = Bh = \pi \cdot r^2 \cdot h$

$\approx 3.14 \times 8\text{ in.} \times 8\text{ in.} \times 4\text{ in.}$

$\approx 803.84\text{ in}^3$

11. $V = Bh = \pi \cdot r^2 \cdot h$

$\approx 3.14 \times 5\text{ cm} \times 5\text{ cm} \times 4.5\text{ cm}$

$\approx 353.25\text{ cm}^3$

13. $V = Bh = \pi \cdot r^2 \cdot h$

$\approx \frac{22}{7} \times 210\text{ yd} \times 210\text{ yd} \times 300\text{ yd}$

$\approx 41,580,000\text{ yd}^3$

15. $V = \frac{4}{3} \cdot \pi \cdot r^3$

$\approx \frac{4}{3} \times 3.14 \times (100\text{ in.})^3$

$\approx \frac{4 \times 3.14 \times 1,000,000\text{ in}^3}{3}$

$\approx 4,186,666\frac{2}{3}\text{ in}^3$

17. $$\begin{aligned} V &= \frac{4}{3} \cdot \pi \cdot r^3 \\ &\approx \frac{4}{3} \times 3.14 \times (3.1 \text{ m})^3 \\ &\approx \frac{4 \times 3.14 \times 29.791 \text{ m}^3}{3} \\ &\approx 124.72 \text{ m}^3 \end{aligned}$$

19. $$\begin{aligned} V &= \frac{4}{3} \cdot \pi \cdot r^3 \\ &\approx \frac{4}{3} \times \frac{22}{7} \times (7 \text{ km})^3 \\ &\approx \frac{4 \times 22 \times 343 \text{ km}^3}{3 \times 7} \\ &\approx 1437\frac{1}{3} \text{ km}^3 \end{aligned}$$

21. $$\begin{aligned} V &= \frac{1}{3} \cdot \pi \cdot r^2 \cdot h \\ &\approx \frac{1}{3} \times 3.14 \times 33 \text{ ft} \times 33 \text{ ft} \times 100 \text{ ft} \\ &\approx 113,982 \text{ ft}^3 \end{aligned}$$

23. $$\begin{aligned} V &= \frac{1}{3} \cdot \pi \cdot r^2 \cdot h \\ &\approx \frac{1}{3} \times \frac{22}{7} \times 1.4 \text{ cm} \times 1.4 \text{ cm} \times 12 \text{ cm} \\ &\approx 24.64 \text{ cm}^3 \end{aligned}$$

25. We must find the radius of the base in order to use the formula for the volume of a circular cylinder.

$$\begin{aligned} r &= \frac{d}{2} = \frac{14 \text{ yd}}{2} = 7 \text{ yd} \\ V &= Bh = \pi \cdot r^2 \cdot h \\ &\approx \frac{22}{7} \times 7 \text{ yd} \times 7 \text{ yd} \times 220 \text{ yd} \\ &\approx 33,880 \text{ yd}^3 \end{aligned}$$

27. We must find the radius of the silo in order to use the formula for the volume of a circular cylinder.

$$\begin{aligned} r &= \frac{d}{2} = \frac{6 \text{ m}}{2} = 3 \text{ m} \\ V &= Bh = \pi \cdot r^2 \cdot h \\ &\approx 3.14 \times 3 \text{ m} \times 3 \text{ m} \times 13 \text{ m} \\ &\approx 367.38 \text{ m}^3 \end{aligned}$$

29. First we find the radius of the ball:

$$r = \frac{d}{2} = \frac{6.5 \text{ cm}}{2} = 3.25 \text{ cm}$$

Then we find the volume, using the formula for the volume of a sphere.

$$\begin{aligned} V &= \frac{4}{3} \cdot \pi \cdot r^3 \\ &\approx \frac{4}{3} \cdot 3.14 \cdot (3.25 \text{ cm})^3 \\ &\approx 143.72 \text{ cm}^3 \end{aligned}$$

31. First we find the radius of the earth:

$$\frac{3980 \text{ mi}}{2} = 1990 \text{ mi}$$

Then we find the volume, using the formula for the volume of a sphere.

$$\begin{aligned} V &= \frac{4}{3} \cdot \pi \cdot r^3 \\ &\approx \frac{4}{3} \cdot 3.14 \cdot (1990 \text{ mi})^3 \\ &\approx 32,993,441,150 \text{ mi}^3 \end{aligned}$$

33. $$\begin{aligned} \text{Interest} &= P \cdot r \cdot t \\ &= \$600 \times 6.4\% \times \frac{1}{2} \\ &= \frac{\$600 \times 0.064}{2} \\ &= \$19.20 \end{aligned}$$

The interest is \$19.20.

35. $10^3 = 10 \cdot 10 \cdot 10 = 1000$

37. $7^2 = 7 \cdot 7 = 49$

39. *Rephrase:* Sales tax is what percent of purchase price?

Translate: $878 = r \times 17,560$

To solve the equation we divide on both sides by 17,560.

$$\begin{aligned} \frac{878}{17,560} &= \frac{r \times 17,560}{17,560} \\ 0.05 &= r \\ 5\% &= r \end{aligned}$$

The sales tax rate is 5%.

41. ◈

43. First find the volume of one one-dollar bill in cubic inches:

$$\begin{aligned} V &= l \cdot w \cdot h \\ V &= 6.0625 \text{ in.} \times 2.3125 \text{ in.} \times 0.0041 \text{ in.} \\ V &= 0.05748 \text{ in}^3 \quad \text{Rounding} \end{aligned}$$

Then multiply to find the volume of one million one-dollar bills in cubic inches:

$$1,000,000 \times 0.05748 \text{ in}^3 = 57,480 \text{ in}^3$$

Thus the volume of one million one-dollar bills is about $57,480 \text{ in}^3$.

45. First find the radius of the tank in inches:

$$r = \frac{d}{2} = \frac{16 \text{ in.}}{2} = 8 \text{ in.}$$

Then convert 8 in. to feet:

$$8 \text{ in.} = 8 \text{ in.} \times \frac{1 \text{ ft}}{12 \text{ in.}} = \frac{8}{12} \times \frac{\text{in.}}{\text{in.}} \times \text{ft} = \frac{2}{3} \text{ ft}$$

Find the volume of the tank:

$$V = Bh = \pi \cdot r^2 \cdot h$$
$$\approx 3.14 \times \frac{2}{3}\text{ ft} \times \frac{2}{3}\text{ ft} \times 5\text{ ft} \approx 6.9\overline{7}\text{ ft}^3$$

The volume of the tank is $6.9\overline{7}$ ft^3.

We multiply to find the number of gallons the tank will hold:

$$6.9\overline{7} \times 7.5\text{ gal} = 52.\overline{3}\text{ gal}$$

The tank will hold $52.\overline{3}$ gallons of water.

Exercise Set 6.7

1. Two angles are complementary if the sum of their measures is 90°.

$$90° - 11° = 79°.$$

The measure of a complement is 79°.

3. Two angles are complementary if the sum of their measures is 90°.

$$90° - 67° = 23°.$$

The measure of a complement is 23°.

5. Two angles are supplementary if the sum of their measures is 180°.

$$180° - 3° = 177°.$$

The measure of a supplement is 177°.

7. Two angles are supplementary if the sum of their measures is 180°.

$$180° - 139° = 41°.$$

The measure of a supplement is 41°.

9. The segments have different lengths. They are not congruent.

11. $m\angle G = m\angle R$, so $\angle G \cong \angle R$.

13. Since $\angle 2$ and $\angle 5$ are vertical angles, $m\angle 2 = 67°$. Likewise, $\angle 1$ and $\angle 4$ are vertical angles, so $m\angle 4 = 80°$.

$$m\angle 1 + m\angle 2 + m\angle 3 = 180°$$
$$80° + 67° + m\angle 3 = 180° \quad \text{Substituting}$$
$$147° + m\angle 3 = 180°$$
$$m\angle 3 = 180° - 147°$$
$$m\angle 3 = 33°$$

Since $\angle 3$ and $\angle 6$ are vertical angles, $m\angle 6 = 33°$.

15. a) The pairs of corresponding angles are

$\angle 1$ and $\angle 3$,

$\angle 2$ and $\angle 4$,

$\angle 8$ and $\angle 6$,

$\angle 7$ and $\angle 5$.

b) The interior angles are $\angle 2$, $\angle 3$, $\angle 6$, and $\angle 7$.

c) The pairs of alternate interior angles are

$\angle 2$ and $\angle 6$,

$\angle 3$ and $\angle 7$.

17. $\angle 4$ and $\angle 6$ are vertical angles, so $m\angle 6 = 125°$.

$\angle 4$ and $\angle 2$ are corresponding angles. By Property 1, $m\angle 2 = 125°$.

$\angle 6$ and $\angle 8$ are corresponding angles. By Property 1, $m\angle 8 = 125°$.

$\angle 2$ and $\angle 3$ are interior angles on the same side of the transversal. Using Property 4 and $m\angle 2 = 125°$, $m\angle 3 = 55°$.

$\angle 6$ and $\angle 7$ are interior angles on the same side of the transversal. Using Property 4 and $m\angle 6 = 125°$, $m\angle 7 = 55°$.

$\angle 3$ and $\angle 5$ are vertical angles, so $m\angle 5 = 55°$.

$\angle 7$ and $\angle 1$ are vertical angles, so $m\angle 1 = 55°$.

19. Considering the transversal $\overleftrightarrow{BC}$, $\angle ABE$ and $\angle DCE$ are alternate interior angles. By Property 2, $\angle ABE \cong \angle DCE$. Then $m\angle ABE = m\angle DCE = 95°$.

Considering the transversal $\overleftrightarrow{AD}$, $\angle BAE$ and $\angle CDE$ are alternate interior angles. By Property 2, $\angle BAE \cong \angle CDE$. We cannot determine the measure of these angles.

$\angle AEB$ and $\angle DEC$ are vertical angles, so $\angle AEB \cong \angle DEC$. We cannot determine the measure of these angles.

$\angle BED$ and $\angle AEC$ are also vertical angles, so $\angle BED \cong \angle AEC$. We cannot determine their measures.

21. Considering the transversal $\overleftrightarrow{CE}$, $\angle AEC$ and $\angle DCE$ are alternate interior angles. By Property 2, $\angle AEC \cong \angle DCE$. Then $m\angle AEC = m\angle DCE = 50°$.

Considering the transversal $\overleftrightarrow{DE}$, $\angle BED$ and $\angle EDC$ are alternate interior angles. By Property 2, $\angle BED \cong \angle EDC$. Then $m\angle BED = m\angle EDC = 41°$.

Exercise Set 6.8

1. The notation tells us the way in which the vertices of the two triangles are matched.

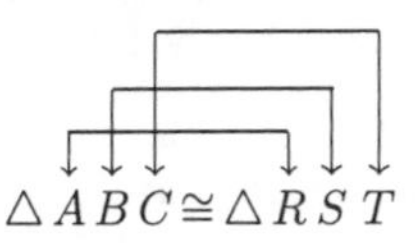

$\triangle ABC \cong \triangle RST$ means

$\angle A \cong \angle R$ and $\overline{AB} \cong \overline{RS}$
$\angle B \cong \angle S$ $\quad \overline{AC} \cong \overline{RT}$
$\angle C \cong \angle T$ $\quad \overline{BC} \cong \overline{ST}$

3. The notation tells us the way in which the vertices of the two triangles are matched.

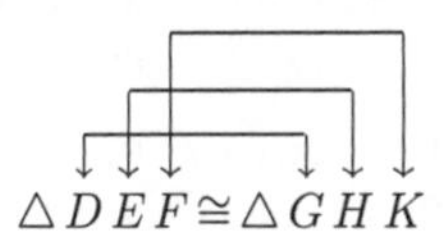

$\triangle DEF \cong \triangle GHK$ means

$\angle D \cong \angle G$ and $\overline{DE} \cong \overline{GH}$
$\angle E \cong \angle H$ $\quad \overline{DF} \cong \overline{GK}$
$\angle F \cong \angle K$ $\quad \overline{EF} \cong \overline{HK}$

5. The notation tells us the way in which the vertices of the two triangles are matched.

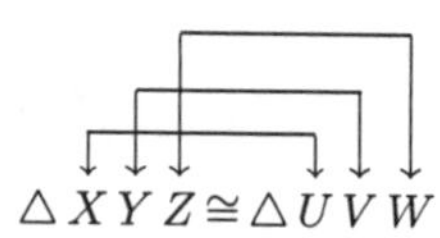

$\triangle XYZ \cong \triangle UVW$ means

$\angle X \cong \angle U$ and $\overline{XY} \cong \overline{UV}$
$\angle Y \cong \angle V$ $\quad \overline{XZ} \cong \overline{UW}$
$\angle Z \cong \angle W$ $\quad \overline{YZ} \cong \overline{VW}$

7. The notation tells us the way in which the vertices of the two triangles are matched.

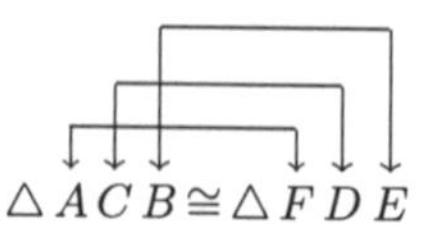

$\triangle ACB \cong \triangle FDE$ means

$\angle A \cong \angle F$ and $\overline{AC} \cong \overline{FD}$
$\angle C \cong \angle D$ $\quad \overline{AB} \cong \overline{FE}$
$\angle B \cong \angle E$ $\quad \overline{CB} \cong \overline{DE}$

9. The notation tells us the way in which the vertices of the two triangles are matched.

$\triangle MNO \cong \triangle QPS$

$\triangle MNO \cong \triangle QPS$ means

$\angle M \cong \angle Q$ and $\overline{MN} \cong \overline{QP}$
$\angle N \cong \angle P$ $\quad \overline{MO} \cong \overline{QS}$
$\angle O \cong \angle S$ $\quad \overline{NO} \cong \overline{PS}$

11. We cannot determine from the information given that two sides of one triangle and the included angle are congruent to two sides and the included angle of the other triangle. Therefore, we cannot use the SAS Property.

13. Two sides of one triangle and the included angle are congruent to two sides and the included angle of the other triangle. They are congruent by the SAS Property.

15. Two sides of one triangle and the included angle are congruent to two sides and the included angle of the other triangle. They are congruent by the SAS Property.

17. We cannot determine from the information given that three sides of one triangle are congruent to three sides of the other triangle. Therefore, we cannot use the SSS Property.

19. Three sides of one triangle are congruent to three sides of the other triangle. They are congruent by the SSS Property.

21. Three sides of one triangle are congruent to three sides of the other triangle. They are congruent by the SSS Property.

23. Two angles and the included side are of one triangle are congruent to two angles and the included side of the other triangle. They are congruent by the ASA Property.

25. Two angles and the included side are of one triangle are congruent to two angles and the included side of the other triangle. They are congruent by the ASA Property.

27. The vertical angles are congruent so two angles and the included side are of one triangle are congruent to two angles and the included side of the other triangle. They are congruent by the ASA Property.

29. Two angles and the included side are of one triangle are congruent to two angles and the included side of the other triangle. They are congruent by the ASA Property.

31. Two sides of one triangle and the included angle are congruent to two sides and the included angle of the other triangle. They are congruent by the SAS Property.

33. Three sides of one triangle are congruent to three sides of the other triangle. In addition, two sides of one triangle and the included angle are congruent to two sides and the included angle of the other triangle. Therefore, we can use either the SSS Property or the SAS Property to show that they are congruent.

35. Since R is the midpoint of $\overline{PT}$, $\overline{PR} \cong \overline{TR}$.

Since R is the midpoint of $\overline{QS}$, $\overline{RQ} \cong \overline{RS}$.

$\angle PRQ$ and $\angle TRS$ are vertical angles, so $\angle PRQ \cong \angle TRS$.

Two sides and the included angle of $\triangle PRQ$ are congruent to two sides and the included angle of $\triangle TRS$, so $\triangle PRQ \cong \triangle TRS$ by the SAS Property.

37. Since $GL \perp KM$, $m\angle GLK = m\angle GLM = 90°$. Then $\angle GLK \cong \angle GLM$.

Since L is the midpoint of $\overline{KM}$, $\overline{KL} \cong \overline{LM}$.

$\overline{GL} \cong \overline{GL}$.

Two sides and the included angle of $\triangle KLG$ are congruent to two sides and the included angle of $\triangle MLG$, so $\triangle KLG \cong \triangle MLG$ by the SAS Property.

39. The information given tells us that $\overline{AE} \cong \overline{CB}$ and $\overline{AB} \cong \overline{CD}$.

Since B is the midpoint of $\overline{ED}$, $\overline{EB} \cong \overline{BD}$.

Three sides of $\triangle AEB$ are congruent to three sides of $\triangle CDB$, so $\triangle AEB \cong \triangle CDB$ by the SSS Property.

41. The information given tells us that $\overline{HK} \cong \overline{KJ}$ and $\overline{GK} \cong \overline{LK}$.

Since $\overline{GK} \perp \overline{LJ}$, $m\angle HKL = m\angle GKJ = 90°$.

Then $\angle HKL \cong \angle GKJ$.

Two sides and the included angle of $\triangle LKH$ are congruent to two sides and the included angle of $\triangle GKJ$, so $\triangle LKH \cong \triangle GKJ$ by the SAS Property. This means that the remaining corresponding parts of the two triangles are congruent. That is, $\angle HLK \cong \angle JGK$, $\angle LHK \cong \angle GJK$, and $\overline{LH} \cong \overline{GJ}$.

43. Two angles and the included side of $\triangle PED$ are congruent to two angles and the included side of $\triangle PFG$, so $\triangle PED \cong \triangle PFG$ by the ASA Property. Then corresponding parts of the two triangles are congruent, so $\overline{EP} \cong \overline{FP}$. Therefore, P is the midpoint of $\overline{EF}$.

45. $\angle A$ and $\angle C$ are opposite angles, so $m\angle A = 70°$ by Property 2.

$\angle C$ and $\angle B$ are consecutive angles, so the are supplementary by Property 4. Then

$$m\angle B = 180° - m\angle C$$
$$m\angle B = 180° - 70°$$
$$m\angle B = 110°.$$

$\angle B$ and $\angle D$ are opposite angles, so $m\angle D = 110°$ by Property 2.

47. $\angle M$ and $\angle K$ are opposite angles, so $m\angle M = 71°$ by Property 2.

$\angle K$ and $\angle L$ are consecutive angles, so the are supplementary by Property 4. Then

$$m\angle L = 180° - m\angle K$$
$$m\angle L = 180° - 71°$$
$$m\angle L = 109°.$$

$\angle J$ and $\angle L$ are opposite angles, so $m\angle J = 109°$ by Property 2.

49. $\overline{ON}$ and $\overline{TU}$ are opposite sides of the parallelogram. So are $\overline{OT}$ and $\overline{NU}$. The opposite sides of a parallelogram are congruent (Property 3), so $TU = 9$ and $NU = 15$.

51. $\overline{JM}$ and $\overline{KL}$ are opposite sides of the parallelogram. So are $\overline{JK}$ and $\overline{ML}$. The opposite sides of a parallelogram are congruent (Property 3). Then $KL = 3\frac{1}{2}$ and $JK + LM = 22 - 3\frac{1}{2} - 3\frac{1}{2} = 15$. Thus, $JK = LM = \frac{1}{2} \cdot 15 = 7\frac{1}{2}$.

53. The diagonals of a parallelogram bisect each other (Property 5). Then

$$AC = 2 \cdot AB = 2 \cdot 14 = 28$$
$$ED = 2 \cdot BD = 2 \cdot 19 = 38$$

Exercise Set 6.9

1. Vertex R is matched with vertex A, vertex S is matched with vertex B, and vertex T is matched with vertex C. Then

$$\begin{array}{lll} \overline{RS} \longleftrightarrow \overline{AB} & \text{and} & \angle R \longleftrightarrow \angle A \\ \overline{ST} \longleftrightarrow \overline{BC} & & \angle S \longleftrightarrow \angle B \\ \overline{TR} \longleftrightarrow \overline{CA} & & \angle T \longleftrightarrow \angle C \end{array}$$

3. Vertex C is matched with vertex W, vertex B is matched with vertex J, and vertex S is matched with vertex Z. Then

$$\begin{array}{lll} \overline{CB} \longleftrightarrow \overline{WJ} & \text{and} & \angle C \longleftrightarrow \angle W \\ \overline{BS} \longleftrightarrow \overline{JZ} & & \angle B \longleftrightarrow \angle J \\ \overline{SC} \longleftrightarrow \overline{ZW} & & \angle S \longleftrightarrow \angle Z \end{array}$$

5. The notation tells us the way in which the vertices are matched.

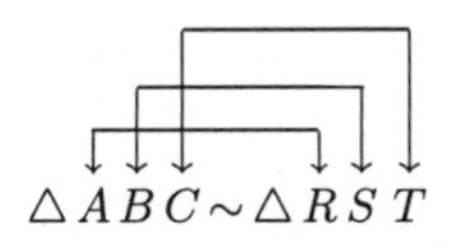

$\triangle ABC \sim \triangle RST$ means

$$\begin{array}{l} \angle A \cong \angle R \\ \angle B \cong \angle S \\ \angle C \cong \angle T \end{array} \quad \text{and} \quad \frac{AB}{RS} = \frac{AC}{RT} = \frac{BC}{ST}.$$

7. The notation tells us the way in which the vertices are matched.

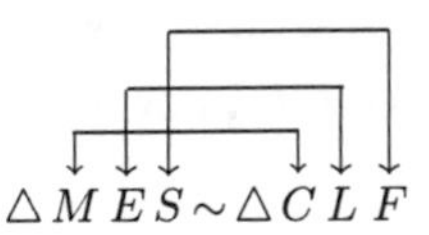

$\triangle MES \sim \triangle CLF$ means

$$\begin{array}{l} \angle M \cong \angle C \\ \angle E \cong \angle L \\ \angle S \cong \angle F \end{array} \quad \text{and} \quad \frac{ME}{CL} = \frac{MS}{CF} = \frac{ES}{LF}.$$

9. If we match P with N, S with D, and Q with M, the corresponding angles will be congruent. That is, $\triangle PSQ \sim \triangle NDM$. Then

$$\frac{PS}{ND} = \frac{PQ}{NM} = \frac{SQ}{DM}.$$

11. If we match T with G, A with F, and W with C, the corresponding angles will be congruent. That is, $\triangle TAW \sim \triangle GFC$. Then

$$\frac{TA}{GF} = \frac{TW}{GC} = \frac{AW}{FC}.$$

13. Since $\triangle ABC \sim \triangle PQR$, the corresponding sides are proportional. Then

$$\begin{array}{rcl} \frac{3}{6} = \frac{4}{PR} & \text{and} & \frac{3}{6} = \frac{5}{QR} \\ 3(PR) = 6 \cdot 4 & & 3(QR) = 6 \cdot 5 \\ 3(PR) = 24 & & 3(QR) = 30 \\ PR = 8 & & QR = 10 \end{array}$$

15. Recall that if a transversal intersects two parallel lines, then the alternate interior angles are congruent. Thus,

$$\angle A \cong \angle B \text{ and } \angle D \cong \angle C.$$

Since $\angle AED$ and $\angle CEB$ are vertical angles, they are congruent. Thus,

$$\angle AED \cong \angle CEB.$$

Then $\triangle AED \sim \triangle CEB$, and the lengths of the corresponding sides are proportional.

$$\frac{AD}{CB} = \frac{ED}{EC}$$
$$\frac{7}{21} = \frac{6}{EC}$$
$$7 \cdot EC = 126$$
$$EC = 18$$

17. If we use the sun's rays to represent the third side of a triangle in a drawing of the situation, we see that we have similar triangles. We let $h =$ the height of the tree.

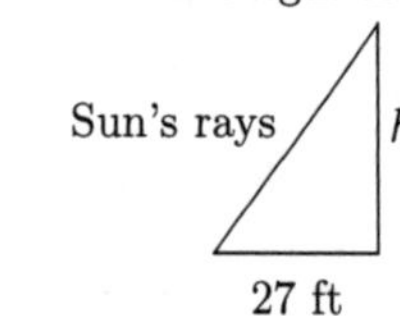

The ratio of h to 4 is the same as the ratio of 27 to 3. We have the proportion

$$\frac{h}{4} = \frac{27}{3}.$$

Solve: $3 \cdot h = 4 \cdot 27$

$$h = \frac{4 \cdot 27}{3}$$

$$h = 36$$

The tree is 36 ft tall.

19. Since the ratio of d to 25 ft is the same as the ratio of 40 ft to 10 ft, we have the proportion

$$\frac{d}{25} = \frac{40}{10}.$$

Solve: $10 \cdot d = 25 \cdot 40$

$$d = \frac{25 \cdot 40}{10}$$

$$d = 100$$

The distance across the river is 100 ft.

21. $2\frac{4}{5} \times 10\frac{1}{2} = \frac{14}{5} \times \frac{21}{2} = \frac{14 \times 21}{5 \times 2} =$

$\frac{2 \times 7 \times 21}{5 \times 2} = \frac{\cancel{2} \times 7 \times 21}{5 \times \cancel{2}} = \frac{147}{5} = 29\frac{2}{5}$

23. $8 \times 9\frac{3}{4} = \frac{8}{1} \times \frac{39}{4} = \frac{8 \times 39}{1 \times 4} = \frac{2 \times 4 \times 39}{1 \times 4} =$

$\frac{2 \times \cancel{4} \times 39}{1 \times \cancel{4}} = \frac{78}{1} = 78$

25. ◈

Chapter 7

Introduction to Real Numbers and Algebraic Expressions

Exercise Set 7.1

1. Substitute 34 for n: $600(34) = 20,400$, so \$20,400 is collected if 34 students enroll.

Substitute 78 for n: $600(78) = 46,800$, so \$46,800 is collected if 78 students enroll.

Substitute 250 for n: $600(250) = 150,000$, so \$150,000 is collected if 250 students enroll.

3. Substitute 45 m for b and 86 m for h, and carry out the multiplication:

$$\begin{aligned} A = \frac{1}{2}bh &= \frac{1}{2}(45 \text{ m})(86 \text{ m}) \\ &= \frac{1}{2}(45)(86)(\text{m})(\text{m}) \\ &= 1935 \text{ m}^2 \end{aligned}$$

5. Substitute 65 for r and 4 for t, and carry out the multiplication:

$$d = rt = 65 \cdot 4 = 260 \text{ mi}$$

7. $8x = 8 \cdot 7 = 56$

9. $\dfrac{a}{b} = \dfrac{24}{3} = 8$

11. $\dfrac{3p}{q} = \dfrac{3 \cdot 2}{6} = \dfrac{6}{6} = 1$

13. $\dfrac{x+y}{5} = \dfrac{10+20}{5} = \dfrac{30}{5} = 6$

15. $\dfrac{x-y}{8} = \dfrac{20-4}{8} = \dfrac{16}{8} = 2$

17. $b + 7$, or $7 + b$

19. $c - 12$

21. $4 + q$, or $q + 4$

23. $a + b$, or $b + a$

25. $y - x$

27. $x + w$, or $w + x$

29. $n - m$

31. $s + r$, or $r + s$

33. $2z$

35. $3m$

37. Let x represent the number. Then we have $89\%x$, or $0.89x$.

39. The distance traveled is the product of the speed and the time. Thus the driver traveled $55t$ miles.

41. We use a factor tree.

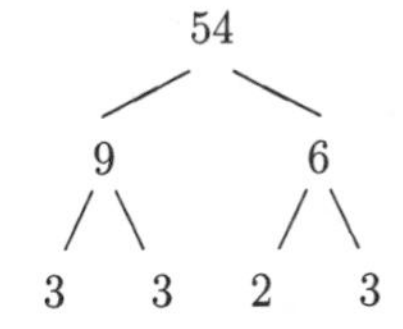

The prime factorization is $2 \cdot 3 \cdot 3 \cdot 3$.

43. We use the list of primes. The first prime that is a factor of 108 is 2.

$$108 = 2 \cdot 54$$

We keep dividing by 2 until it is no longer possible to do so.

$$108 = 2 \cdot 2 \cdot 27$$

Now we do the same thing for the next prime, 3.

$$108 = 2 \cdot 2 \cdot 3 \cdot 3 \cdot 3$$

This is the prime factorization of 108.

45. $6 = 2 \cdot 3$

$18 = 2 \cdot 3 \cdot 3$

The LCM is $2 \cdot 3 \cdot 3$, or 18.

47. $10 = 2 \cdot 5$

$20 = 2 \cdot 2 \cdot 5$

$30 = 2 \cdot 3 \cdot 5$

The LCM is $2 \cdot 2 \cdot 3 \cdot 5$, or 60.

49. ◈

51. $x + 3y$

53. $2x - 3$

Exercise Set 7.2

1. The integer -1286 corresponds to 1286 ft below sea level; the integer 13,804 corresponds to 13,804 ft above sea level.

3. The integer 24 corresponds to 24° above zero; the integer -2 corresponds to 2° below zero.

5. The integer $-5,200,000,000,000$ corresponds to the total public debt of \$5.2 trillion.

7. The number $\frac{10}{3}$ can be named $3\frac{1}{3}$, or $3.3\overline{3}$. The graph is $\frac{1}{3}$ of the way from 3 to 4.

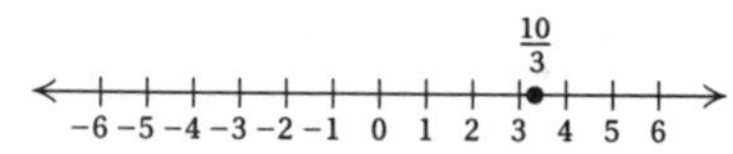

9. The graph of -5.2 is $\frac{2}{10}$ of the way from -5 to -6.

−5.2

−6 −5 −4 −3 −2 −1 0 1 2 3 4 5 6

11. We first find decimal notation for $\frac{7}{8}$. Since $\frac{7}{8}$ means $7 \div 8$, we divide.

$$\begin{array}{r} 0.875 \\ 8\overline{)7.000} \\ \underline{6\,4} \\ 60 \\ \underline{56} \\ 40 \\ \underline{40} \\ 0 \end{array}$$

Thus $\frac{7}{8} = 0.875$, so $-\frac{7}{8} = -0.875$.

13. $\frac{5}{6}$ means $5 \div 6$, so we divide.

$$\begin{array}{r} 0.833\ldots \\ 6\overline{)5.000} \\ \underline{4\,8} \\ 20 \\ \underline{18} \\ 20 \\ \underline{18} \\ 0 \end{array}$$

We have $\frac{5}{6} = 0.8\overline{3}$.

15. $\frac{7}{6}$ means $7 \div 6$, so we divide.

$$\begin{array}{r} 1.166. \\ 6\overline{)7.000} \\ \underline{6} \\ 1\,0 \\ \underline{6} \\ 40 \\ \underline{36} \\ 40 \\ \underline{36} \\ 4 \end{array}$$

We have $\frac{7}{6} = 1.1\overline{6}$.

17. $\frac{2}{3}$ means $2 \div 3$, so we divide.

$$\begin{array}{r} 0.666 \\ 3\overline{)2.000} \\ \underline{1\,8} \\ 20 \\ \underline{18} \\ 20 \\ \underline{18} \\ 2 \end{array}$$

We have $\frac{2}{3} = 0.\overline{6}$.

19. We first find decimal notation for $\frac{1}{2}$. Since $\frac{1}{2}$ means $1 \div 2$, we divide.

$$\begin{array}{r} 0.5 \\ 2\overline{)1.0} \\ \underline{1\,0} \\ 0 \end{array}$$

Thus $\frac{1}{2} = 0.5$, so $-\frac{1}{2} = -0.5$

21. $\frac{1}{10}$ means $1 \div 10$, so we divide.

$$\begin{array}{r} 0.1 \\ 10\overline{)1.0} \\ \underline{1\,0} \\ 0 \end{array}$$

We have $\frac{1}{10} = 0.1$

23. Since 8 is to the right of 0, we have $8 > 0$.

25. Since -8 is to the left of 3, we have $-8 < 3$.

27. Since -8 is to the left of 8, we have $-8 < 8$.

29. Since -8 is to the left of -5, we have $-8 < -5$.

31. Since -5 is to the right of -11, we have $-5 > -11$.

33. Since -6 is to the left of -5, we have $-6 < -5$.

35. Since 2.14 is to the right of 1.24, we have $2.14 > 1.24$.

37. Since -14.5 is to the left of 0.011, we have $-14.5 < 0.011$.

39. Since -12.88 is to the left of -6.45, we have $-12.88 < -6.45$.

41. Convert to decimal notation $\frac{5}{12} = 0.4166\ldots$ and $\frac{11}{25} = 0.44$. Since $0.4166\ldots$ is to the left of 0.44, $\frac{5}{12} < \frac{11}{25}$.

43. $-3 \geq -11$ is true since $-3 > -11$ is true.

45. $0 \geq 8$ is false since neither $0 > 8$ nor $0 = 8$ is true.

47. $x < -6$ has the same meaning as $-6 > x$.

49. $y \geq -10$ has the same meaning as $-10 \leq y$.

51. The distance of -3 from 0 is 3, so $|-3| = 3$.

53. The distance of 10 from 0 is 10, so $|10| = 10$.

55. The distance of 0 from 0 is 0, so $|0| = 0$.

57. The distance of -24 from 0 is 24, so $|-24| = 24$.

59. The distance of $-\frac{2}{3}$ from 0 is $\frac{2}{3}$, so $\left|-\frac{2}{3}\right| = \frac{2}{3}$.

61. The distance of $\frac{0}{4}$ from 0 is $\frac{0}{4}$, or 0, so $\left|\frac{0}{4}\right| = 0$.

63. 63% 0.63.

Move the decimal point 2 places to the left.

$63\% = 0.63$

65. 110% 1.10.

Move the decimal point 2 places to the left.

$110\% = 1.1$

67. $\frac{3}{4} = 0.75 = 75\%$

69. From Exercise 13 we know that $\frac{5}{6} = 0.8\overline{3}$, or $0.83\overline{3}$, so $\frac{5}{6} = 83.\overline{3}\%$, or $83\frac{1}{3}\%$.

71.

73. $-\frac{2}{3}, \frac{1}{2}, -\frac{3}{4}, -\frac{5}{6}, \frac{3}{8}, \frac{1}{6}$ can be written in decimal notation as $-0.\overline{6}$, 0.5, -0.75, $-0.8\overline{3}$, 0.375, $0.1\overline{6}$, respectively. Listing from least to greatest, we have

$$-\frac{5}{6}, -\frac{3}{4}, -\frac{2}{3}, \frac{1}{6}, \frac{3}{8}, \frac{1}{2}.$$

75. $0.1\overline{1} = \frac{0.3\overline{3}}{3} = \frac{\frac{1}{3}}{3} = \frac{1}{3} \cdot \frac{1}{3} = \frac{1}{9}$

77. First consider $0.5\overline{5}$.

$0.5\overline{5} = 0.3\overline{3} \cdot \frac{5}{3} = \frac{1}{3} \cdot \frac{5}{3} = \frac{5}{9}$

Then, $5.5\overline{5} = 5 + 0.5\overline{5} = 5 + \frac{5}{9} = 5\frac{5}{9}$.

Exercise Set 7.3

1. $2 + (-9)$ The absolute values are 2 and 9. The difference is $9 - 2$, or 7. The negative number has the larger absolute value, so the answer is negative. $2 + (-9) = -7$

3. $-11 + 5$ The absolute values are 11 and 5. The difference is $11 - 5$, or 6. The negative number has the larger absolute value, so the answer is negative. $-11 + 5 = -6$

5. $-8 + 8$ A negative and a positive number. The numbers have the same absolute value. The sum is 0. $-8 + 8 = 0$

7. $-3 + (-5)$ Two negatives. Add the absolute values, getting 8. Make the answer negative. $-3 + (-5) = -8$

9. $-7 + 0$ One number is 0. The answer is the other number. $-7 + 0 = -7$

11. $0 + (-27)$ One number is 0. The answer is the other number. $0 + (-27) = -27$

13. $17 + (-17)$ A negative and a positive number. The numbers have the same absolute value. The sum is 0. $17 + (-17) = 0$

15. $-17 + (-25)$ Two negatives. Add the absolute values, getting 42. Make the answer negative. $-17 + (-25) = -42$

17. $18 + (-18)$ A positive and a negative number. The numbers have the same absolute value. The sum is 0. $18 + (-18) = 0$

19. $-28 + 28$ A negative and a positive number. The numbers have the same absolute value. The sum is 0. $-28 + 28 = 0$

21. $8 + (-5)$ The absolute values are 8 and 5. The difference is $8 - 5$, or 3. The positive number has the larger absolute value, so the answer is positive. $8 + (-5) = 3$

23. $-4 + (-5)$ Two negatives. Add the absolute values, getting 9. Make the answer negative. $-4 + (-5) = -9$

25. $13 + (-6)$ The absolute values are 13 and 6. The difference is $13 - 6$, or 7. The positive number has the larger absolute value, so the answer is positive. $13 + (-6) = 7$

27. $-25 + 25$ A negative and a positive number. The numbers have the same absolute value. The sum is 0. $-25 + 25 = 0$

29. $53 + (-18)$ The absolute values are 53 and 18. The difference is $53 - 18$, or 35. The positive number has the larger absolute value, so the answer is positive. $53 + (-18) = 35$

31. $-8.5 + 4.7$ The absolute values are 8.5 and 4.7. The difference is $8.5 - 4.7$, or 3.8. The negative number has the larger absolute value, so the answer is negative. $-8.5 + 4.7 = -3.8$

33. $-2.8 + (-5.3)$ Two negatives. Add the absolute values, getting 8.1. Make the answer negative. $-2.8 + (-5.3) = -8.1$

35. $-\frac{3}{5} + \frac{2}{5}$ The absolute values are $\frac{3}{5}$ and $\frac{2}{5}$. The difference is $\frac{3}{5} - \frac{2}{5}$, or $\frac{1}{5}$. The negative number has the larger absolute value, so the answer is negative. $-\frac{3}{5} + \frac{2}{5} = -\frac{1}{5}$

37. $-\frac{2}{9} + \left(-\frac{5}{9}\right)$ Two negatives. Add the absolute values, getting $\frac{7}{9}$. Make the answer negative. $-\frac{2}{9} + \left(-\frac{5}{9}\right) = -\frac{7}{9}$

39. $-\frac{5}{8} + \frac{1}{4}$ The absolute values are $\frac{5}{8}$ and $\frac{1}{4}$. The difference is $\frac{5}{8} - \frac{2}{8}$, or $\frac{3}{8}$. The negative number has the larger absolute value, so the answer is negative. $-\frac{5}{8} + \frac{1}{4} = -\frac{3}{8}$

41. $-\frac{5}{8} + \left(-\frac{1}{6}\right)$ Two negatives. Add the absolute values, getting $\frac{15}{24} + \frac{4}{24}$, or $\frac{19}{24}$. Make the answer negative.

$-\frac{5}{8} + \left(-\frac{1}{6}\right) = -\frac{19}{24}$

43. $-\frac{3}{8}+\frac{5}{12}$ The absolute values are $\frac{3}{8}$ and $\frac{5}{12}$. The difference is $\frac{10}{24}-\frac{9}{24}$, or $\frac{1}{24}$. The positive number has the larger absolute value, so the answer is positive. $-\frac{3}{8}+\frac{5}{12}=\frac{1}{24}$

45. $76+(-15)+(-18)+(-6)$

a) Add the negative numbers: $-15+(-18)+(-6)=-39$

b) Add the results: $76+(-39)=37$

47. $-44+\left(-\frac{3}{8}\right)+95+\left(-\frac{5}{8}\right)$

a) Add the negative numbers: $-44+\left(-\frac{3}{8}\right)+\left(-\frac{5}{8}\right)=-45$

b) Add the results: $-45+95=50$

49. We add from left to right.

$$\begin{aligned} & 98+(-54)+113+(-998)+44+(-612)\\ =\ & 44+113+(-998)+44+(-612)\\ =\ & 157+(-998)+44+(-612)\\ =\ & -841+44+(-612)\\ =\ & -797+(-612)\\ =\ & -1409 \end{aligned}$$

51. The additive inverse of 24 is -24 because $24+(-24)=0$.

53. The additive inverse of -26.9 is 26.9 because $-26.9+26.9=0$.

55. If $x=8$, then $-x=-8$. (The opposite of 8 is -8.)

57. If $x=-\frac{13}{8}$ then $-x=-\left(-\frac{13}{8}\right)=\frac{13}{8}$. (The opposite of $-\frac{13}{8}$ is $\frac{13}{8}$.)

59. If $x=-43$ then $-(-x)=-(-(-43))=-43$. (The opposite of the opposite of -43 is -43.)

61. If $x=\frac{4}{3}$ then $-(-x)=-\left(-\frac{4}{3}\right)=\frac{4}{3}$. (The opposite of the opposite of $\frac{4}{3}$ is $\frac{4}{3}$.)

63. $-(-24)=24$ (The opposite of -24 is 24.)

65. $-\left(-\frac{3}{8}\right)=\frac{3}{8}$ (The opposite of $-\frac{3}{8}$ is $\frac{3}{8}$.)

67. 57% 0.57.

Move the decimal point two places to the left.

57% = 0.57.

69. 52.9% 0.52.9

Move the decimal point two places to the left.

52.9% = 0.529.

71. $\frac{5}{4}=1.25=125\%$

73. $\frac{13}{25}=0.52=52\%$

75. ◈

77. When x is positive, the opposite of x, $-x$, is negative, so $-x$ is negative for all positive numbers x.

79. Use a calculator.

$-3496+(-2987)=-6483$

81. If a is positive, $-a$ is negative. Thus $-a+b$, the sum of two negatives, is negative.

Exercise Set 7.4

1. $2-9=2+(-9)=-7$

3. $0-4=0+(-4)=-4$

5. $-8-(-2)=-8+2=-6$

7. $-11-(-11)=-11+11=0$

9. $12-16=12+(-16)=-4$

11. $20-27=20+(-27)=-7$

13. $-9-(-3)=-9+3=-6$

15. $-40-(-40)=-40+40=0$

17. $7-7=7+(-7)=0$

19. $7-(-7)=7+7=14$

21. $8-(-3)=8+3=11$

23. $-6-8=-6+(-8)=-14$

25. $-4-(-9)=-4+9=5$

27. $1-8=1+(-8)=-7$

29. $-6-(-5)=-6+5=-1$

31. $8-(-10)=8+10=18$

33. $0-10=0+(-10)=-10$

35. $-5-(-2)=-5+2=-3$

37. $-7-14=-7+(-14)=-21$

39. $0-(-5)=0+5=5$

41. $-8-0=-8+0=-8$

43. $7-(-5)=7+5=12$

45. $2-25=2+(-25)=-23$

47. $-42-26=-42+(-26)=-68$

49. $-71-2=-71+(-2)=-73$

51. $24-(-92)=24+92=116$

53. $-50-(-50)=-50+50=0$

55. $-\frac{3}{8}-\frac{5}{8}=-\frac{3}{8}+\left(-\frac{5}{8}\right)=-\frac{8}{8}=-1$

57. $\frac{3}{4}-\frac{2}{3}=\frac{3}{4}+\left(-\frac{2}{3}\right)=\frac{9}{12}+\left(-\frac{8}{12}\right)=\frac{1}{12}$

59. $-\frac{3}{4}-\frac{2}{3}=-\frac{3}{4}+\left(-\frac{2}{3}\right)=-\frac{9}{12}+\left(-\frac{8}{12}\right)=-\frac{17}{12}$

61. $-\frac{5}{8}-\left(-\frac{3}{4}\right)=-\frac{5}{8}+\frac{3}{4}=-\frac{5}{8}+\frac{6}{8}=\frac{1}{8}$

63. $6.1-(-13.8)=6.1+13.8=19.9$

65. $-2.7-5.9=-2.7+(-5.9)=-8.6$

67. $0.99-1=0.99+(-1)=-0.01$

69. $-79-114=-79+(-114)=-193$

71. $0-(-500)=0+500=500$

73. $-2.8-0=-2.8+0=-2.8$

75. $7-10.53=7+(-10.53)=-3.53$

77. $\frac{1}{6}-\frac{2}{3}=\frac{1}{6}+\left(-\frac{2}{3}\right)=\frac{1}{6}+\left(-\frac{4}{6}\right)=-\frac{3}{6}$, or $-\frac{1}{2}$

79. $-\frac{4}{7}-\left(-\frac{10}{7}\right)=-\frac{4}{7}+\frac{10}{7}=\frac{6}{7}$

81. $-\frac{7}{10}-\frac{10}{15}=-\frac{7}{10}+\left(-\frac{10}{15}\right)=-\frac{21}{30}+\left(-\frac{20}{30}\right)=-\frac{41}{30}$

83. $\frac{1}{5}-\frac{1}{3}=\frac{1}{5}+\left(-\frac{1}{3}\right)=\frac{3}{15}+\left(-\frac{5}{15}\right)=-\frac{2}{15}$

85. $18-(-15)-3-(-5)+2=18+15+(-3)+5+2=37$

87. $-31+(-28)-(-14)-17=(-31)+(-28)+14+(-17)=-62$

89. $-34-28+(-33)-44=(-34)+(-28)+(-33)+(-44)=-139$

91. $-93-(-84)-41-(-56)=(-93)+84+(-41)+56=6$

93. $-5-(-30)+30+40-(-12)=(-5)+30+30+40+12=107$

95. $132-(-21)+45-(-21)=132+21+45+21=219$

97. We subtract the smaller number from the larger:

$$-8648-(-11,033)=-8648+11,033=2385$$

The difference in elevation is 2385 m.

99. The number -476.89 represents the original debt. The number 128.95 represents the credit received when the sweater is returned. We add:

$$-476.89+128.95=-347.94$$

You owe \$347.94.

101. We subtract the smaller number from the larger:

$$44-(-56)=44+56=100$$

The temperature dropped 100°F.

103. $5^3=5\times5\times5=125$

105.
$$\begin{aligned}256\div64\div2^3+100&=256\div64\div8+100\\&=4\div8+100\\&=\frac{1}{2}+100\\&=100\frac{1}{2}\text{, or }100.5\end{aligned}$$

107. 58.3% $\quad$ 0.58.3 (decimal point moved two places)

Move the decimal point two places to the left.

$58.3\%=0.583$

109. ◈

111. Use a calculator.

$123,907-433,789=-309,882$

113. False. $3-0=3, 0-3=-3, 3-0\neq0-3$

115. True

117. True by definition of opposites.

119. The changes during weeks 1 to 5 are represented by the integers -13, -16, 36, -11, and 19, respectively. We add to find the total rise or fall:

$$-13+(-16)+36+(-11)+19=15$$

The market rose 15 points during the 5 week period.

Exercise Set 7.5

1. -8

3. -48

5. -24

7. -72

9. 16

11. 42

13. -120

15. -238

17. 1200

19. 98

21. -72

23. -12.4

25. 30

27. 21.7

29. $\frac{2}{3}\cdot\left(-\frac{3}{5}\right)=-\left(\frac{2\cdot3}{3\cdot5}\right)=-\left(\frac{2}{5}\cdot\frac{3}{3}\right)=-\frac{2}{5}$

31. $-\frac{3}{8}\cdot\left(-\frac{2}{9}\right)=\frac{3\cdot2}{8\cdot9}=\frac{3\cdot2\cdot1}{4\cdot2\cdot3\cdot3}=\frac{3\cdot2}{3\cdot2}\cdot\frac{1}{4\cdot3}=$

33. -17.01

35. $-\frac{5}{9}\cdot\frac{3}{4}=-\left(\frac{5\cdot3}{9\cdot4}\right)=-\frac{5\cdot3}{3\cdot3\cdot4}=-\frac{5}{3\cdot4}\cdot\frac{3}{3}=\ldots2$

37. $7\cdot(-4)\cdot(-3)\cdot5=7\cdot12\cdot5=7\cdot60=420$

39. $-\frac{2}{3}\cdot\frac{1}{2}\cdot\left(-\frac{6}{7}\right)=-\frac{2}{6}\cdot\left(-\frac{6}{7}\right)=\frac{2\cdot6}{7\cdot6}=\frac{2}{7}\cdot\frac{6}{6}=\frac{2}{7}$

41. $-3 \cdot (-4) \cdot (-5) = 12 \cdot (-5) = -60$

43. $-2 \cdot (-5) \cdot (-3) \cdot (-5) = 10 \cdot 15 = 150$

45. $-\dfrac{2}{45}$

47. $-7 \cdot (-21) \cdot 13 = 147 \cdot 13 = 1911$

49. $-4 \cdot (-1.8) \cdot 7 = (7.2) \cdot 7 = 50.4$

51. $-\dfrac{1}{9} \cdot \left(-\dfrac{2}{3}\right) \cdot \left(\dfrac{5}{7}\right) = \dfrac{2}{27} \cdot \dfrac{5}{7} = \dfrac{10}{189}$

53. $4 \cdot (-4) \cdot (-5) \cdot (-12) = -16 \cdot (60) = -960$

55. $0.07 \cdot (-7) \cdot 6 \cdot (-6) = 0.07 \cdot 6 \cdot (-7) \cdot (-6) = 0.42 \cdot (42) = 17.64$

57. $\left(-\dfrac{5}{6}\right)\left(\dfrac{1}{8}\right)\left(-\dfrac{3}{7}\right)\left(-\dfrac{1}{7}\right) = \left(-\dfrac{5}{48}\right)\left(\dfrac{3}{49}\right) = -\dfrac{5 \cdot 3}{16 \cdot 3 \cdot 49} = -\dfrac{5}{16 \cdot 49} \cdot \dfrac{3}{3} = -\dfrac{5}{784}$

59. 0, The product of 0 and any real number is 0.

61. $(-8)(-9)(-10) = 72(-10) = -720$

63. $(-6)(-7)(-8)(-9)(-10) = 42 \cdot 72 \cdot (-10) = 3024 \cdot (-10) = -30,240$

65.

$(-3x)^2 = (-3 \cdot 7)^2$	Substituting
$= (-21)^2$	Multiplying inside the parentheses
$= (-21)(-21)$	Evaluating the power
$= 441$	

$-3x^2 = -3(7)^2$	Substituting
$= -3 \cdot 49$	Evaluating the power
$= -147$	

67. When $x = 2$:

$5x^2 = 5(2)^2$	Substituting
$= 5 \cdot 4$	Evaluating the power
$= 20$	

When $x = -2$:

$5x^2 = 5(-2)^2$	Substituting
$= 5 \cdot 4$	Evaluating the power
$= 20$	

69. $36 = 2 \cdot 2 \cdot 3 \cdot 3$

$60 = 2 \cdot 2 \cdot 3 \cdot 5$

$\text{LCM} = 2 \cdot 2 \cdot 3 \cdot 3 \cdot 5$, or 180

$\dfrac{26}{39} = \dfrac{2 \cdot \cancel{13}}{3 \cdot \cancel{13}} = \dfrac{2}{3}$

$\dfrac{264}{84} = \dfrac{\cancel{2} \cdot \cancel{2} \cdot 2 \cdot 3 \cdot \cancel{11}}{\cancel{2} \cdot \cancel{2} \cdot \cancel{11} \cdot 11} = \dfrac{6}{11}$

75

77. diver rises $7 \cdot 9$, or 63 m.

T ew elevation is -95 m + 63 m, or -32 m, or 32 m be the surface.

79. a) a d b have different signs;

b) eit a or b or both must be zero;

c) a an have the same sign

Exercise Set 7.6

1. $48 \div (-6) = -8$ Check: $-8(-6) = 48$

3. $\dfrac{28}{-2} = -14$ Check: $-14(-2) = 28$

5. $\dfrac{-24}{8} = -3$ Check: $-3 \cdot 8 = -24$

7. $\dfrac{-36}{-12} = 3$ Check: $3(-12) = -36$

9. $\dfrac{-72}{9} = -8$ Check: $-8 \cdot 9 = -72$

11. $-100 \div (-50) = 2$ Check: $2(-50) = -100$

13. $-108 \div 9 = -12$ Check: $9(-12) = -108$

15. $\dfrac{200}{-25} = -8$ Check: $-8(-25) = 200$

17. Undefined

19. $\dfrac{-23}{-2} = \dfrac{23}{2}$ Check: $\dfrac{23}{2}(-2) = -23$

21. The reciprocal of $\dfrac{15}{7}$ is $\dfrac{7}{15}$ because $\dfrac{15}{7} \cdot \dfrac{7}{15} = 1$.

23. The reciprocal of $-\dfrac{47}{13}$ is $-\dfrac{13}{47}$ because $\left(-\dfrac{47}{13}\right) \cdot \left(-\dfrac{13}{47}\right) = 1$.

25. The reciprocal of 13 is $\dfrac{1}{13}$ because $13 \cdot \dfrac{1}{13} = 1$.

27. The reciprocal of 4.3 is $\dfrac{1}{4.3}$ because $4.3 \cdot \dfrac{1}{4.3} = 1$.

29. The reciprocal of $-\dfrac{1}{7.1}$ is -7.1 because $\left(-\dfrac{1}{7.1}\right)(-7.1) = 1$.

31. The reciprocal of $\dfrac{p}{q}$ is $\dfrac{q}{p}$ because $\dfrac{p}{q} \cdot \dfrac{q}{p} = 1$.

33. The reciprocal of $\dfrac{1}{4y}$ is $4y$ because $\dfrac{1}{4y} \cdot 4y = 1$.

35. The reciprocal of $\dfrac{2a}{3b}$ is $\dfrac{3b}{2a}$ because $\dfrac{2a}{3b} \cdot \dfrac{3b}{2a} = 1$.

37. $4 \cdot \dfrac{1}{17}$

39. $8 \cdot \left(-\dfrac{1}{13}\right)$

41. $13.9 \cdot \left(-\dfrac{1}{1.5}\right)$

43. $x \cdot y$

45. $(3x + 4)\left(\dfrac{1}{5}\right)$

47. $(5a - b)\left(\dfrac{1}{5a + b}\right)$

49. $\dfrac{3}{4} \div \left(-\dfrac{2}{3}\right) = \dfrac{3}{4} \cdot \left(-\dfrac{3}{2}\right) = -\dfrac{9}{8}$

51. $-\frac{5}{4} \div \left(-\frac{3}{4}\right) = -\frac{5}{4} \cdot \left(-\frac{4}{3}\right) = \frac{20}{12} = \frac{5 \cdot 4}{3 \cdot 4} = \frac{5}{3}$

53. $-\frac{2}{7} \div \left(-\frac{4}{9}\right) = -\frac{2}{7} \cdot \left(-\frac{9}{4}\right) = \frac{18}{28} = \frac{9 \cdot 2}{14 \cdot 2} = \frac{9}{14}$

55. $-\frac{3}{8} \div \left(-\frac{8}{3}\right) = -\frac{3}{8} \cdot \left(-\frac{3}{8}\right) = \frac{9}{64}$

57. $-6.6 \div 3.3 = -2$ Do the long division. Make the answer negative.

59. $\frac{-11}{-13} = \frac{11}{13}$ The opposite of a number divided by the opposite of another number is the quotient of the two numbers.

61. $\frac{48.6}{-3} = -16.2$ Do the long division. Make the answer negative.

63. $\frac{-9}{17-17} = \frac{-9}{0}$ Division by 0 is undefined.

65. $\frac{264}{468} = \frac{4 \cdot 66}{4 \cdot 117} = \frac{\cancel{4} \cdot \cancel{3} \cdot 22}{\cancel{4} \cdot \cancel{3} \cdot 39} = \frac{22}{39}$

67. $2^3 - 5 \cdot 3 + 8 \cdot 10 \div 2$

$= 8 - 5 \cdot 3 + 8 \cdot 10 \div 2$ Evaluating the power

$= 8 - 15 + 80 \div 2$ Multiplying and dividing

$= 8 - 15 + 40$ in order from left to right

$= -7 + 40$ Adding and subtracting

$= 33$ in order from left to right

69. $\frac{7}{8} = 0.875 = 87.5\%$

71. $\frac{12}{25} \div \frac{32}{75} = \frac{12}{25} \cdot \frac{75}{32}$

$= \frac{12 \cdot 75}{25 \cdot 32}$

$= \frac{3 \cdot \cancel{4} \cdot 3 \cdot \cancel{25}}{\cancel{25} \cdot \cancel{4} \cdot 8}$

$= \frac{9}{8}$

73. ◈

75. ◈

77. $-b$ is positive and a is negative, so $\frac{-b}{a}$ is the quotient of a positive and a negative number and, thus, is negative.

79. $\frac{-b}{a}$ is negative (see Exercise 77), so $-\left(\frac{-b}{a}\right)$ is the opposite of a negative number and, thus, is positive.

81. $-b$ and $-a$ are both positive, so $\frac{-b}{-a}$ is the quotient of two positive numbers and, thus, is positive. Then, $-\left(\frac{-b}{-a}\right)$ is the opposite of a positive number and, thus, is negative.

Exercise Set 7.7

1. Note that $5y = 5 \cdot y$. We multiply by 1, using y/y as an equivalent expression for 1:

$$\frac{3}{5} = \frac{3}{5} \cdot 1 = \frac{3}{5} \cdot \frac{y}{y} = \frac{3y}{5y}$$

3. Note that $15x = 3 \cdot 5x$. We multiply by 1, using $5x/5x$ as an equivalent expression for 1:

$$\frac{2}{3} = \frac{2}{3} \cdot 1 = \frac{2}{3} \cdot \frac{5x}{5x} = \frac{10x}{15x}$$

5. $-\frac{24a}{16a} = -\frac{3 \cdot 8a}{2 \cdot 8a}$

$= -\frac{3}{2} \cdot \frac{8a}{8a}$

$= -\frac{3}{2} \cdot 1$ $\left(\frac{8a}{8a} = 1\right)$

$= -\frac{3}{2}$ Identity property of 1

7. $-\frac{42ab}{36ab} = -\frac{7 \cdot 6ab}{6 \cdot 6ab}$

$= -\frac{7}{6} \cdot \frac{6ab}{6ab}$

$= -\frac{7}{6} \cdot 1$ $\left(\frac{6ab}{6ab} = 1\right)$

$= -\frac{7}{6}$ Identity property of 1

9. $8 + y$, commutative law of addition

11. nm, commutative law of multiplication

13. $xy + 9$, commutative law of addition

$9 + yx$, commutative law of multiplication

15. $c + ab$, commutative law of addition

$ba + c$, commutative law of multiplication

17. $(a + b) + 2$, associative law of addition

19. $8(xy)$, associative law of multiplication

21. $a + (b + 3)$, associative law of addition

23. $(3a)b$, associative law of multiplication

25. a) $(a + b) + 2 = a + (b + 2)$, associative law of addition

b) $(a + b) + 2 = (b + a) + 2$, commutative law of additi

c) $(a + b) + 2 = (b + a) + 2$ Using the commutative law first,

$= b + (a + 2)$ then the associative law

There are other correct answers.

27. a) $5 + (v + w) = (5 + v) + w$, associative law of a tion

b) $5 + (v + w) = 5 + (w + v)$, commutative law o ddition

c) $5 + (v + w) = 5 + (w + v)$ Using the comm ative law first,

$= (5 + w) + v$ then the associ ve law

There are other correct answers.

29. a) $(xy)3 = x(y3)$, associative law of multiplication

b) $(xy)3 = (yx)3$, commutative law of multiplication

c) $(xy)3 = (yx)3$ Using the commutative law first,

$= y(x3)$ then the associative law

There are other correct answers.

31. a) $7(ab) = (7a)b$

b) $7(ab) = (7a)b = b(7a)$

c) $7(ab) = 7(ba) = (7b)a$

There are other correct answers.

33. $2(b+5) = 2 \cdot b + 2 \cdot 5 = 2b + 10$

35. $7(1+t) = 7 \cdot 1 + 7 \cdot t = 7 + 7t$

37. $6(5x+2) = 6 \cdot 5x + 6 \cdot 2 = 30x + 12$

39. $7(x+4+6y) = 7 \cdot x + 7 \cdot 4 + 7 \cdot 6y = 7x + 28 + 42y$

41. $7(x-3) = 7 \cdot x - 7 \cdot 3 = 7x - 21$

43. $-3(x-7) = -3 \cdot x - (-3) \cdot 7 = -3x - (-21) = -3x + 21$

45. $\frac{2}{3}(b-6) = \frac{2}{3} \cdot b - \frac{2}{3} \cdot 6 = \frac{2}{3}b - 4$

47. $7.3(x-2) = 7.3 \cdot x - 7.3 \cdot 2 = 7.3x - 14.6$

49. $-\frac{3}{5}(x - y + 10) = -\frac{3}{5} \cdot x - \left(-\frac{3}{5}\right) \cdot y + \left(-\frac{3}{5}\right) \cdot 10 =$
$-\frac{3}{5}x - \left(-\frac{3}{5}y\right) + (-6) = -\frac{3}{5}x + \frac{3}{5}y - 6$

51. $-9(-5x - 6y + 8) = -9(-5x) - (-9)6y + (-9)8$
$= 45x - (-54y) + (-72) = 45x + 54y - 72$

53. $-4(x - 3y - 2z) = -4 \cdot x - (-4)3y - (-4)2z$
$= -4x - (-12y) - (-8z) = -4x + 12y + 8z$

55. $3.1(-1.2x + 3.2y - 1.1) = 3.1(-1.2x) + (3.1)3.2y - 3.1(1.1)$
$= -3.72x + 9.92y - 3.41$

57. $4x + 3z$ Parts are separated by plus signs. The terms are $4x$ and $3z$.

. $7x + 8y - 9z = 7x + 8y + (-9z)$ Separating parts with plus signs

The terms are $7x$, $8y$, and $-9z$.

$x + 4 = 2 \cdot x + 2 \cdot 2 = 2(x+2)$

6 $+ 5y = 5 \cdot 6 + 5 \cdot y = 5(6+y)$

65. $+ 21y = 7 \cdot 2x + 7 \cdot 3y = 7(2x + 3y)$

67. 5 $10 + 15y = 5 \cdot x + 5 \cdot 2 + 5 \cdot 3y = 5(x + 2 + 3y)$

69. $8x$ $4 = 8 \cdot x - 8 \cdot 3 = 8(x-3)$

71. $32 -$ $= 4 \cdot 8 - 4 \cdot y = 4(8-y)$

73. $8x + 1$ $- 22 = 2 \cdot 4x + 2 \cdot 5y - 2 \cdot 11 = 2(4x + 5y - 11)$

75. $ax - a$ $a \cdot x - a \cdot 1 = a(x-1)$

77. $ax - ay - az = a \cdot x - a \cdot y - a \cdot z = a(x - y - z)$

79. $18x - 12y + 6 = 6 \cdot 3x - 6 \cdot 2y + 6 \cdot 1 = 6(3x - 2y + 1)$

81. $\frac{2}{3}x - \frac{5}{3}y + \frac{1}{3} = \frac{1}{3} \cdot 2x - \frac{1}{3} \cdot 5y + \frac{1}{3} \cdot 1 =$
$\frac{1}{3}(2x - 5y + 1)$

83. $9a + 10a = (9+10)a = 19a$

85. $10a - a = 10a - 1 \cdot a = (10-1)a = 9a$

87. $2x + 9z + 6x = 2x + 6x + 9z = (2+6)x + 9z = 8x + 9z$

89. $7x + 6y^2 + 9y^2 = 7x + (6+9)y^2 = 7x + 15y^2$

91.
$$\begin{aligned} 41a + 90 - 60a - 2 &= 41a - 60a + 90 - 2 \\ &= (41 - 60)a + (90 - 2) \\ &= -19a + 88 \end{aligned}$$

93.
$$\begin{aligned} &23 + 5t + 7y - t - y - 27 \\ &= 23 - 27 + 5t - 1 \cdot t + 7y - 1 \cdot y \\ &= (23 - 27) + (5-1)t + (7-1)y \\ &= -4 + 4t + 6y, \text{ or } 4t + 6y - 4 \end{aligned}$$

95. $\frac{1}{2}b + \frac{1}{2}b = \left(\frac{1}{2} + \frac{1}{2}\right)b = 1b = b$

97. $2y + \frac{1}{4}y + y = 2y + \frac{1}{4}y + 1 \cdot y = \left(2 + \frac{1}{4} + 1\right)y = 3\frac{1}{4}y$, or $\frac{13}{4}y$

99. $11x - 3x = (11-3)x = 8x$

101. $6n - n = (6-1)n = 5n$

103. $y - 17y = (1-17)y = -16y$

105.
$$\begin{aligned} &-8 + 11a - 5b + 6a - 7b + 7 \\ &= 11a + 6a - 5b - 7b - 8 + 7 \\ &= (11+6)a + (-5-7)b + (-8+7) \\ &= 17a - 12b - 1 \end{aligned}$$

107. $9x + 2y - 5x = (9-5)x + 2y = 4x + 2y$

109. $11x + 2y - 4x - y = (11-4)x + (2-1)y = 7x + y$

111. $2.7x + 2.3y - 1.9x - 1.8y = (2.7 - 1.9)x + (2.3 - 1.8)y =$
$0.8x + 0.5y$

113.
$$\begin{aligned} &\frac{13}{2}a + \frac{9}{5}b - \frac{2}{3}a - \frac{3}{10}b - 42 \\ &= \left(\frac{13}{2} - \frac{2}{3}\right)a + \left(\frac{9}{5} - \frac{3}{10}\right)b - 42 \\ &= \left(\frac{39}{6} - \frac{4}{6}\right)a + \left(\frac{18}{10} - \frac{3}{10}\right)b - 42 \\ &= \frac{35}{6}a + \frac{15}{10}b - 42 \\ &= \frac{35}{6}a + \frac{3}{2}b - 42 \end{aligned}$$

115.
$$\begin{aligned} \frac{11}{12} + \frac{15}{16} &= \frac{11}{12} \cdot \frac{4}{4} + \frac{15}{16} \cdot \frac{3}{3} \quad \text{LCD is 48} \\ &= \frac{44}{48} + \frac{45}{48} \\ &= \frac{89}{48} \end{aligned}$$

117. $16 = 2 \cdot 2 \cdot 2 \cdot 2$

$18 = 2 \cdot 3 \cdot 3$

$24 = 2 \cdot 2 \cdot 2 \cdot 3$

The LCM is $2 \cdot 2 \cdot 2 \cdot 2 \cdot 3 \cdot 3$, or 144.

119. $\frac{1}{8} - \frac{1}{3} = \frac{1}{8} + \left(-\frac{1}{3}\right) = \frac{3}{24} + \left(-\frac{8}{24}\right) = -\frac{5}{24}$

121. ◈

123. No; for any replacement other than 5 the two expressions do not have the same value. For example, let $t = 2$. Then $3 \cdot 2 + 5 = 6 + 5 = 11$, but $3 \cdot 5 + 2 = 15 + 2 = 17$.

125. Yes; commutative law of addition

127. $q + qr + qrs + qrst$ There are no like terms.
$= q \cdot 1 + q \cdot r + q \cdot rs + q \cdot rst$
$= q(1 + r + rs + rst)$ Factoring

Exercise Set 7.8

1. $-(2x + 7) = -2x - 7$ Changing the sign of each term

3. $-(5x - 8) = -5x + 8$ Changing the sign of each term

5. $-4a + 3b - 7c$

7. $-6x + 8y - 5$

9. $-3x + 5y + 6$

11. $8x + 6y + 43$

13. $9x - (4x + 3) = 9x - 4x - 3$ Removing parentheses by changing the sign of every term
$= 5x - 3$ Collecting like terms

15. $2a - (5a - 9) = 2a - 5a + 9 = -3a + 9$

17. $2x + 7x - (4x + 6) = 2x + 7x - 4x - 6 = 5x - 6$

19. $2x - 4y - 3(7x - 2y) = 2x - 4y - 21x + 6y = -19x + 2y$

21. $15x - y - 5(3x - 2y + 5z)$
$= 15x - y - 15x + 10y - 25z$ Multiplying each term in parentheses by -5
$= 9y - 25z$

23. $(3x + 2y) - 2(5x - 4y) = 3x + 2y - 10x + 8y = -7x + 10y$

25. $(12a - 3b + 5c) - 5(-5a + 4b - 6c)$
$= 12a - 3b + 5c + 25a - 20b + 30c$
$= 37a - 23b + 35c$

27. $[9 - 2(5 - 4)] = [9 - 2 \cdot 1]$ Computing $5 - 4$
$= [9 - 2]$ Computing $2 \cdot 1$
$= 7$

29. $8[7 - 6(4 - 2)] = 8[7 - 6(2)] = 8[7 - 12] = 8[-5] = -40$

31. $[4(9 - 6) + 11] - [14 - (6 + 4)]$
$= [4(3) + 11] - [14 - 10]$
$= [12 + 11] - [14 - 10]$
$= 23 - 4$
$= 19$

33. $[10(x + 3) - 4] + [2(x - 1) + 6]$
$= [10x + 30 - 4] + [2x - 2 + 6]$
$= [10x + 26] + [2x + 4]$
$= 10x + 26 + 2x + 4$
$= 12x + 30$

35. $[7(x + 5) - 19] - [4(x - 6) + 10]$
$= [7x + 35 - 19] - [4x - 24 + 10]$
$= [7x + 16] - [4x - 14]$
$= 7x + 16 - 4x + 14$
$= 3x + 30$

37. $3\{[7(x - 2) + 4] - [2(2x - 5) + 6]\}$
$= 3\{[7x - 14 + 4] - [4x - 10 + 6]\}$
$= 3\{[7x - 10] - [4x - 4]\}$
$= 3\{7x - 10 - 4x + 4\}$
$= 3\{3x - 6\}$
$= 9x - 18$

39. $4\{[5(x - 3) + 2] - 3[2(x + 5) - 9]\}$
$= 4\{[5x - 15 + 2] - 3[2x + 10 - 9]\}$
$= 4\{[5x - 13] - 3[2x + 1]\}$
$= 4\{5x - 13 - 6x - 3\}$
$= 4\{-x - 16\}$
$= -4x - 64$

41. $8 - 2 \cdot 3 - 9 = 8 - 6 - 9$ Multiplying
$= 2 - 9$ Doing all additions and subtractions in order from left to right
$= -7$

43. $(8 - 2 \cdot 3) - 9 = (8 - 6) - 9$ Multiplying inside the parentheses
$= 2 - 9$ Subtracting inside the parentheses
$= -7$

45. $[(-24) \div (-3)] \div \left(-\frac{1}{2}\right) = 8 \div \left(-\frac{1}{2}\right) = 8 \cdot (-2) = -16$

47. $16 \cdot (-24) + 50 = -384 + 50 = -334$

49. $2^4 + 2^3 - 10 = 16 + 8 - 10 = 24 - 10 = 14$

51. $5^3 + 26 \cdot 71 - (16 + 25 \cdot 3) = 5^3 + 26 \cdot 71 - (16 + 75) =$
$5^3 + 26 \cdot 71 - 91 = 125 + 26 \cdot 71 - 91 = 125 + 1846 - 91 =$
$1971 - 91 = 1880$

53. $4 \cdot 5 - 2 \cdot 6 + 4 = 20 - 12 + 4 = 8 + 4 = 12$

55. $4^3/8 = 64/8 = 8$

57. $8(-7) + 6(-5) = -56 - 30 = -86$

59. $19 - 5(-3) + 3 = 19 + 15 + 3 = 34 + 3 = 37$

61. $9 \div (-3) + 16 \div 8 = -3 + 2 = -1$

63. $6 - 4^2 = 6 - 16 = -10$

65. $(3 - 8)^2 = (-5)^2 = 25$

67. $12 - 20^3 = 12 - 8000 = -7988$

69. $2 \cdot 10^3 - 5000 = 2 \cdot 1000 - 5000 = 2000 - 5000 = -3000$

71. $6[9 - (3 - 4)] = 6[9 - (-1)] = 6[9 + 1] = 6[10] = 60$

73. $-1000 \div (-100) \div 10 = 10 \div 10 = 1$

75. $8 - (7 - 9) = 8 - (-2) = 8 + 2 = 10$

77. $\dfrac{10 - 6^2}{9^2 + 3^2} = \dfrac{10 - 36}{81 + 9} = \dfrac{-26}{90} = -\dfrac{13}{45}$

79. $\dfrac{3(6-7) - 5 \cdot 4}{6 \cdot 7 - 8(4-1)} = \dfrac{3(-1) - 5 \cdot 4}{42 - 8 \cdot 3} = \dfrac{-3 - 20}{42 - 24} = -\dfrac{23}{18}$

81. $\dfrac{2^3 - 3^2 + 12 \cdot 5}{-32 \div (-16) \div (-4)} = \dfrac{8 - 9 + 12 \cdot 5}{-32 \div (-16) \div (-4)} =$

$\dfrac{8 - 9 + 60}{2 \div (-4)} = \dfrac{8 - 9 + 60}{-\frac{1}{2}} = \dfrac{-1 + 60}{-\frac{1}{2}} = \dfrac{59}{-\frac{1}{2}} =$

$59(-2) = -118$

83. We divide by the first prime number, 2, until it is no longer possible to do so.

$$236 = 2 \cdot 118$$
$$236 = 2 \cdot 2 \cdot 59$$

Each factor in $2 \cdot 2 \cdot 59$ is a prime number, so this is the prime factorization.

85. $\dfrac{2}{3} \div \dfrac{5}{12} = \dfrac{2}{3} \cdot \dfrac{12}{5} = \dfrac{2 \cdot 12}{3 \cdot 5} = \dfrac{2 \cdot \cancel{3} \cdot 4}{\cancel{3} \cdot 5} = \dfrac{8}{5}$

87. $3^4 = 3 \cdot 3 \cdot 3 \cdot 3 = 81$

89. $10^2 = 10 \cdot 10 = 100$

91.

93. $6y + 2x - 3a + c = 6y - (-2x) - 3a - (-c) = 6y - (-2x + 3a - c)$

95. $6m + 3n - 5m + 4b = 6m - (-3n) - 5m - (-4b) =$

$6m - (-3n + 5m - 4b)$

97.
$$\begin{aligned} &\{x - [f - (f - x)] + [x - f]\} - 3x \\ &= \{x - [f - f + x] + [x - f]\} - 3x \\ &= \{x - [x] + [x - f]\} - 3x \\ &= \{x - x + x - f\} - 3x = x - f - 3x = -2x - f \end{aligned}$$

99. a) $x^2 + 3 = 7^2 + 3 = 49 + 3 = 52;$
$x^2 + 3 = (-7)^2 + 3 = 49 + 3 = 52;$
$x^2 + 3 = (-5.013)^2 + 3 = 25.130169 + 3 = 28.130169$

b) $1 - x^2 = 1 - 5^2 = 1 - 25 = -24;$
$1 - x^2 = 1 - (-5)^2 = 1 - 25 = -24;$
$1 - x^2 = 1 - (-10.455)^2 = 1 - 109.307025 =$
-108.307025

Chapter 8

Solving Equations and Inequalities

Exercise Set 8.1

1. $x + 17 = 32$ Writing the equation

$15 + 17 \;?\; 32$ Substituting 15 for x

32 TRUE

Since the left-hand and right-hand sides are the same, 15 is a solution of the equation.

3. $x - 7 = 12$ Writing the equation

$21 - 7 \;?\; 12$ Substituting 21 for x

14 FALSE

Since the left-hand and right-hand sides are not the same, 21 is not a solution of the equation.

5. $6x = 54$ Writing the equation

$6(-7) \;?\; 54$ Substituting

-42 FALSE

-7 is not a solution of the equation.

7. $\frac{x}{6} = 5$ Writing the equation

$\frac{30}{6} \;?\; 5$ Substituting

5 TRUE

5 is a solution of the equation.

9. $5x + 7 = 107$

$5 \cdot 19 + 7 \;?\; 107$ Substituting

$95 + 7$

102 FALSE

19 is not a solution of the equation.

11. $7(y - 1) = 63$

$7(-11 - 1) \;?\; 63$ Substituting

$7(-12)$

-84 FALSE

-11 is not a solution of the equation.

13.
$$x + 2 = 6$$
$$x + 2 - 2 = 6 - 2 \quad \text{Subtracting 2 on both sides}$$
$$x = 4 \quad \text{Simplifying}$$

Check: $x + 2 = 6$

$4 + 2 \;?\; 6$

6 TRUE

The solution is 4.

15.
$$x + 15 = -5$$
$$x + 15 - 15 = -5 - 15 \quad \text{Subtracting 15 on both sides}$$
$$x = -20$$

Check: $x + 15 = -5$

$-20 + 15 \;?\; -5$

-5 TRUE

The solution is -20.

17.
$$x + 6 = -8$$
$$x + 6 - 6 = -8 - 6$$
$$x = -14$$

Check: $x + 6 = -8$

$-14 + 6 \;?\; -8$

-8 TRUE

The solution is -14.

19.
$$x + 16 = -2$$
$$x + 16 - 16 = -2 - 16$$
$$x = -18$$

Check: $x + 16 = -2$

$-18 + 16 \;?\; -2$

-2 TRUE

The solution is -18.

21.
$$x - 9 = 6$$
$$x - 9 + 9 = 6 + 9$$
$$x = 15$$

Check: $x - 9 = 6$

$15 - 9 \;?\; 6$

6 TRUE

The solution is 15.

23.
$$x - 7 = -21$$
$$x - 7 + 7 = -21 + 7$$
$$x = -14$$

Check: $x - 7 = -21$

$-14 - 7 \;?\; -21$

-21 TRUE

The solution is -14.

25.
$$5 + t = 7$$
$$-5 + 5 + t = -5 + 7$$
$$t = 2$$

Check: $5 + t = 7$

$5 + 2 \;?\; 7$

7 TRUE

The solution is 2.

27.
$$-7+y=13$$
$$7+(-7)+y=7+13$$
$$y=20$$

Check: $-7+y=13$

$-7+20 \ ? \ 13$

$13 \mid$ TRUE

The solution is 20.

29.
$$-3+t=-9$$
$$3+(-3)+t=3+(-9)$$
$$t=-6$$

Check: $-3+t=-9$

$-3+(-6) \ ? \ -9$

$-9 \mid$ TRUE

The solution is -6.

31.
$$x+\frac{1}{2}=7$$
$$x+\frac{1}{2}-\frac{1}{2}=7-\frac{1}{2}$$
$$x=6\frac{1}{2}$$

Check: $x+\frac{1}{2}=7$

$6\frac{1}{2}+\frac{1}{2} \ ? \ 7$

$7 \mid$ TRUE

The solution is $6\frac{1}{2}$.

33.
$$12=a-7.9$$
$$12+7.9=a-7.9+7.9$$
$$19.9=a$$

Check: $12=a-7.9$

$12 \ ? \ 19.9-7.9$

$\mid 12$ TRUE

The solution is 19.9.

35.
$$r+\frac{1}{3}=\frac{8}{3}$$
$$r+\frac{1}{3}-\frac{1}{3}=\frac{8}{3}-\frac{1}{3}$$
$$r=\frac{7}{3}$$

Check: $r+\frac{1}{3}=\frac{8}{3}$

$\frac{7}{3}+\frac{1}{3} \ ? \ \frac{8}{3}$

$\frac{8}{3} \mid$ TRUE

The solution is $\frac{7}{3}$.

37.
$$m+\frac{5}{6}=-\frac{11}{12}$$
$$m+\frac{5}{6}-\frac{5}{6}=-\frac{11}{12}-\frac{5}{6}$$
$$m=-\frac{11}{12}-\frac{5}{6}\cdot\frac{2}{2}$$
$$m=-\frac{11}{12}-\frac{10}{12}$$
$$m=-\frac{21}{12}=-\frac{\not{3}\cdot 7}{\not{3}\cdot 4}$$
$$m=-\frac{7}{4}$$

Check: $m+\frac{5}{6}=-\frac{11}{12}$

$-\frac{7}{4}+\frac{5}{6} \ ? \ -\frac{11}{12}$

$-\frac{21}{12}+\frac{10}{12}$

$-\frac{11}{12} \mid$ TRUE

The solution is $-\frac{7}{4}$.

39.
$$x-\frac{5}{6}=\frac{7}{8}$$
$$x-\frac{5}{6}+\frac{5}{6}=\frac{7}{8}+\frac{5}{6}$$
$$x=\frac{7}{8}\cdot\frac{3}{3}+\frac{5}{6}\cdot\frac{4}{4}$$
$$x=\frac{21}{24}+\frac{20}{24}$$
$$x=\frac{41}{24}$$

Check: $x-\frac{5}{6}=\frac{7}{8}$

$\frac{41}{24}-\frac{5}{6} \ ? \ \frac{7}{8}$

$\frac{41}{24}-\frac{20}{24} \mid \frac{21}{24}$

$\frac{21}{24} \mid$ TRUE

The solution is $\frac{41}{24}$.

41.
$$-\frac{1}{5}+z=-\frac{1}{4}$$
$$\frac{1}{5}-\frac{1}{5}+z=\frac{1}{5}-\frac{1}{4}$$
$$z=\frac{1}{5}\cdot\frac{4}{4}-\frac{1}{4}\cdot\frac{5}{5}$$
$$z=\frac{4}{20}-\frac{5}{20}$$
$$z=-\frac{1}{20}$$

Check:
$$\begin{array}{c|c} \multicolumn{2}{c}{-\frac{1}{5}+z=-\frac{1}{4}} \\ \hline -\frac{1}{5}+\left(-\frac{1}{20}\right) \;?\; & -\frac{1}{4} \\ -\frac{4}{20}+\left(-\frac{1}{20}\right) & -\frac{5}{20} \\ -\frac{5}{20} & \end{array} \quad \text{TRUE}$$

The solution is $-\frac{1}{20}$.

43.
$$\begin{aligned} x+2.3 &= 7.4 \\ x+2.3-2.3 &= 7.4-2.3 \\ x &= 5.1 \end{aligned}$$

Check:
$$\begin{array}{c|c} \multicolumn{2}{c}{x+2.3=7.4} \\ \hline 5.1+2.3 \;?\; & 7.4 \\ 7.4 & \end{array} \quad \text{TRUE}$$

The solution is 5.1.

45.
$$\begin{aligned} 7.6 &= x-4.8 \\ 7.6+4.8 &= x-4.8+4.8 \\ 12.4 &= x \end{aligned}$$

Check:
$$\begin{array}{c|c} \multicolumn{2}{c}{7.6=x-4.8} \\ \hline 7.6 \;?\; & 12.4-4.8 \\ & 7.6 \end{array} \quad \text{TRUE}$$

The solution is 12.4.

47.
$$\begin{aligned} -9.7 &= -4.7+y \\ 4.7+(-9.7) &= 4.7+(-4.7)+y \\ -5 &= y \end{aligned}$$

Check:
$$\begin{array}{c|c} \multicolumn{2}{c}{-9.7=-4.7+y} \\ \hline -9.7 \;?\; & -4.7+(-5) \\ & -9.7 \end{array} \quad \text{TRUE}$$

The solution is -5.

49.
$$\begin{aligned} 5\tfrac{1}{6}+x &= 7 \\ -5\tfrac{1}{6}+5\tfrac{1}{6}+x &= -5\tfrac{1}{6}+7 \\ x &= -\frac{31}{6}+\frac{42}{6} \\ x &= \frac{11}{6}, \text{ or } 1\tfrac{5}{6} \end{aligned}$$

Check:
$$\begin{array}{c|c} \multicolumn{2}{c}{5\tfrac{1}{6}+x=7} \\ \hline 5\tfrac{1}{6}+1\tfrac{5}{6} \;?\; & 7 \\ 7 & \end{array} \quad \text{TRUE}$$

The solution is $\frac{11}{6}$, or $1\frac{5}{6}$.

51.
$$\begin{aligned} q+\frac{1}{3} &= -\frac{1}{7} \\ q+\frac{1}{3}-\frac{1}{3} &= -\frac{1}{7}-\frac{1}{3} \\ q &= -\frac{1}{7}\cdot\frac{3}{3}-\frac{1}{3}\cdot\frac{7}{7} \\ q &= -\frac{3}{21}-\frac{7}{21} \\ q &= -\frac{10}{21} \end{aligned}$$

Check:
$$\begin{array}{c|c} \multicolumn{2}{c}{q+\frac{1}{3}=-\frac{1}{7}} \\ \hline -\frac{10}{21}+\frac{1}{3} \;?\; & -\frac{1}{7} \\ -\frac{10}{21}+\frac{7}{21} & -\frac{3}{21} \\ -\frac{3}{21} & \end{array} \quad \text{TRUE}$$

The solution is $-\frac{10}{21}$.

53. $-3+(-8)$ Two negative numbers. We add the absolute values, getting 11, and make the answer negative.

$-3+(-8)=-11$

55. $-\frac{2}{3}\cdot\frac{5}{8}=-\frac{2\cdot 5}{3\cdot 8}=-\frac{\cancel{2}\cdot 5}{3\cdot\cancel{2}\cdot 4}=-\frac{5}{12}$

57. $\frac{2}{3}\div\left(-\frac{4}{9}\right)=\frac{2}{3}\cdot\left(-\frac{9}{4}\right)=-\frac{2\cdot 9}{3\cdot 4}=-\frac{\cancel{2}\cdot\cancel{3}\cdot 3}{\cancel{3}\cdot\cancel{2}\cdot 2}=-\frac{3}{2}$

59. The translation is $50-x$.

61. ◈

63.
$$\begin{aligned} -356.788 &= -699.034+t \\ 699.034+(-356.788) &= 699.034+(-699.034)+t \\ 342.246 &= t \end{aligned}$$

The solution is 342.246.

65.
$$\begin{aligned} x+\frac{4}{5} &= -\frac{2}{3}-\frac{4}{15} \\ x+\frac{4}{5} &= -\frac{2}{3}\cdot\frac{5}{5}-\frac{4}{15} \quad \text{Adding on the right side} \\ x+\frac{4}{5} &= -\frac{10}{15}-\frac{4}{15} \\ x+\frac{4}{5} &= -\frac{14}{15} \\ x+\frac{4}{5}-\frac{4}{5} &= -\frac{14}{15}-\frac{4}{5} \\ x &= -\frac{14}{15}-\frac{4}{5}\cdot\frac{3}{3} \\ x &= -\frac{14}{15}-\frac{12}{15} \\ x &= -\frac{26}{15} \end{aligned}$$

The solution is $-\frac{26}{15}$.

67. $16 + x - 22 = -16$

$x - 6 = -16$ Adding on the left side

$x - 6 + 6 = -16 + 6$

$x = -10$

The solution is -10.

69. $x + 3 = 3 + x$

$x + 3 - 3 = 3 + x - 3$

$x = x$

$x = x$ is true for all real numbers. Thus the solution is all real numbers.

71. $-\frac{3}{2} + x = -\frac{5}{17} - \frac{3}{2}$

$\frac{3}{2} - \frac{3}{2} + x = \frac{3}{2} - \frac{5}{17} - \frac{3}{2}$

$x = \left(\frac{3}{2} - \frac{3}{2}\right) - \frac{5}{17}$

$x = -\frac{5}{17}$

The solution is $-\frac{5}{17}$.

73. $|x| + 6 = 19$

$|x| + 6 - 6 = 19 - 6$

$|x| = 13$

x represents a number whose distance from 0 is 13. Thus $x = -13$ or $x = 13$.

The solutions are -13 and 13.

Exercise Set 8.2

1. $6x = 36$

$\frac{6x}{6} = \frac{36}{6}$ Dividing by 6 on both sides

$1 \cdot x = 6$ Simplifying

$x = 6$ Identity property of 1

Check: $6x = 36$

$6 \cdot 6 \;?\; 36$

36 TRUE

The solution is 6.

3. $5x = 45$

$\frac{5x}{5} = \frac{45}{5}$ Dividing by 5 on both sides

$1 \cdot x = 9$ Simplifying

$x = 9$ Identity property of 1

Check: $5x = 45$

$5 \cdot 9 \;?\; 45$

45 TRUE

The solution is 9.

5. $84 = 7x$

$\frac{84}{7} = \frac{7x}{7}$ Dividing by 7 on both sides

$12 = 1 \cdot x$

$12 = x$

Check: $84 = 7x$

$84 \;?\; 7 \cdot 12$

84 TRUE

The solution is 12.

7. $-x = 40$

$-1 \cdot x = 40$

$\frac{-1 \cdot x}{-1} = \frac{40}{-1}$

$1 \cdot x = -40$

$x = -40$

Check: $-x = 40$

$-(-40) \;?\; 40$

40 TRUE

The solution is -40.

9. $-x = -1$

$-1 \cdot x = -1$

$\frac{-1 \cdot x}{-1} = \frac{-1}{-1}$

$1 \cdot x = 1$

$x = 1$

Check: $-x = -1$

$-(1) \;?\; -1$

-1 TRUE

The solution is 1.

11. $7x = -49$

$\frac{7x}{7} = \frac{-49}{7}$

$1 \cdot x = -7$

$x = -7$

Check: $7x = -49$

$7(-7) \;?\; -49$

-49 TRUE

The solution is -7.

13. $-12x = 72$

$\frac{-12x}{-12} = \frac{72}{-12}$

$1 \cdot x = -6$

$x = -6$

Check: $-12x = 72$

$-12(-6) \;?\; 72$

72 TRUE

The solution is -6.

15. $-21x = -126$

$\frac{-21x}{-21} = \frac{-126}{-21}$

$1 \cdot x = 6$

$x = 6$

Check: $-21x = -126$

$-21 \cdot 6 \;?\; -126$

-126 TRUE

The solution is 6.

17. $\frac{t}{7} = -9$

$7 \cdot \frac{1}{7}t = 7 \cdot (-9)$

$1 \cdot t = -63$

$t = -63$

Check: $\frac{t}{7} = -9$

$\frac{-63}{7} \;?\; -9$

-9 TRUE

The solution is -63.

19. $\frac{3}{4}x = 27$

$\frac{4}{3} \cdot \frac{3}{4}x = \frac{4}{3} \cdot 27$

$1 \cdot x = \frac{4 \cdot \cancel{3} \cdot 3 \cdot 3}{\cancel{3} \cdot 1}$

$x = 36$

Check: $\frac{3}{4}x = 27$

$\frac{3}{4} \cdot 36 \;?\; 27$

27 TRUE

The solution is 36.

21. $\frac{-t}{3} = 7$

$3 \cdot \frac{1}{3} \cdot (-t) = 3 \cdot 7$

$-t = 21$

$-1 \cdot (-1 \cdot t) = -1 \cdot 21$

$1 \cdot t = -21$

$t = -21$

Check: $\frac{-t}{3} = 7$

$\frac{-(-21)}{3} \;?\; 7$

$\frac{21}{3}$

7 TRUE

The solution is -21.

23. $-\frac{m}{3} = \frac{1}{5}$

$-\frac{1}{3} \cdot m = \frac{1}{5}$

$-3 \cdot \left(-\frac{1}{3} \cdot m\right) = -3 \cdot \frac{1}{5}$

$m = -\frac{3}{5}$

Check: $-\frac{m}{3} = \frac{1}{5}$

$-\frac{-\frac{3}{5}}{3} \;?\; \frac{1}{5}$

$-\left(-\frac{3}{5} \div 3\right)$

$-\left(-\frac{3}{5} \cdot \frac{1}{3}\right)$

$-\left(-\frac{1}{5}\right)$

$\frac{1}{5}$ TRUE

The solution is $-\frac{3}{5}$.

25. $-\frac{3}{5}r = \frac{9}{10}$

$-\frac{5}{3} \cdot \left(-\frac{3}{5}r\right) = -\frac{5}{3} \cdot \frac{9}{10}$

$1 \cdot r = -\frac{\cancel{5} \cdot \cancel{3} \cdot 3}{\cancel{3} \cdot \cancel{5} \cdot 2}$

$r = -\frac{3}{2}$

Check: $-\frac{3}{5}r = \frac{9}{10}$

$-\frac{3}{5} \cdot \left(-\frac{3}{2}\right) \;?\; \frac{9}{10}$

$\frac{9}{10}$ TRUE

The solution is $-\frac{3}{2}$.

27. $-\frac{3}{2}r = -\frac{27}{4}$

$-\frac{2}{3} \cdot \left(-\frac{3}{2}r\right) = -\frac{2}{3} \cdot \left(-\frac{27}{4}\right)$

$1 \cdot r = \frac{\cancel{2} \cdot \cancel{3} \cdot 3 \cdot 3}{\cancel{3} \cdot \cancel{2} \cdot 2}$

$r = \frac{9}{2}$

Check: $-\frac{3}{2}r = -\frac{27}{4}$

$-\frac{3}{2} \cdot \frac{9}{2} \;?\; -\frac{27}{4}$

$-\frac{27}{4}$ TRUE

The solution is $\frac{9}{2}$.

29. $6.3x = 44.1$

$\frac{6.3x}{6.3} = \frac{44.1}{6.3}$

$1 \cdot x = 7$

$x = 7$

Check:

$$\begin{array}{c|l} \multicolumn{2}{c}{6.3x = 44.1} \\ \hline 6.3 \cdot 7 \ ? & 44.1 \\ 44.1 & \qquad \text{TRUE} \end{array}$$

The solution is 7.

31.

$$-3.1y = 21.7$$
$$\frac{-3.1y}{-3.1} = \frac{21.7}{-3.1}$$
$$1 \cdot y = -7$$
$$y = -7$$

Check:

$$\begin{array}{c|l} \multicolumn{2}{c}{3.1y = 21.7} \\ \hline -3.1(-7) \ ? & 21.7 \\ 21.7 & \qquad \text{TRUE} \end{array}$$

The solution is -7.

33.

$$38.7m = 309.6$$
$$\frac{38.7m}{38.7} = \frac{309.6}{38.7}$$
$$1 \cdot m = 8$$
$$m = 8$$

Check:

$$\begin{array}{c|l} \multicolumn{2}{c}{38.7m = 309.6} \\ \hline 38.7 \cdot 8 \ ? & 309.6 \\ 309.6 & \qquad \text{TRUE} \end{array}$$

The solution is 8.

35.

$$-\frac{2}{3}y = -10.6$$
$$-\frac{3}{2} \cdot (-\frac{2}{3}y) = -\frac{3}{2} \cdot (-10.6)$$
$$1 \cdot y = \frac{31.8}{2}$$
$$y = 15.9$$

Check:

$$\begin{array}{c|l} \multicolumn{2}{c}{-\frac{2}{3}y = -10.6} \\ \hline -\frac{2}{3} \cdot (15.9) \ ? & -10.6 \\ -\frac{31.8}{3} & \\ -10.6 & \qquad \text{TRUE} \end{array}$$

The solution is 15.9.

37. $3x + 4x = (3 + 4)x = 7x$

39. $-4x + 11 - 6x + 18x = (-4 - 6 + 18)x + 11 = 8x + 11$

41. $3x - (4 + 2x) = 3x - 4 - 2x = x - 4$

43. $8y - 6(3y + 7) = 8y - 18y - 42 = -10y - 42$

45. The translation is $8r$.

47. ◈

49.

$$-0.2344m = 2028.732$$
$$\frac{-0.2344m}{-0.2344} = \frac{2028.732}{-0.2344}$$
$$1 \cdot m = -8655$$
$$m = -8655$$

The solution is -8655.

51. For all x, $0 \cdot x = 0$. There is no solution to $0 \cdot x = 9$.

53.

$$2|x| = -12$$
$$\frac{2|x|}{2} = \frac{-12}{2}$$
$$1 \cdot |x| = -6$$
$$|x| = -6$$

Absolute value cannot be negative. The equation has no solution.

55.

$$3x = \frac{b}{a}$$
$$\frac{1}{3} \cdot 3x = \frac{1}{3} \cdot \frac{b}{a}$$
$$x = \frac{b}{3a}$$

The solution is $\frac{b}{3a}$.

57.

$$\frac{a}{b}x = 4$$
$$\frac{b}{a} \cdot \frac{a}{b}x = \frac{b}{a} \cdot 4$$
$$x = \frac{4b}{a}$$

The solution is $\frac{4b}{a}$.

Exercise Set 8.3

1.

$5x + 6 = 31$	
$5x + 6 - 6 = 31 - 6$	Subtracting 6 on both sides
$5x = 25$	Simplifying
$\frac{5x}{5} = \frac{25}{5}$	Dividing by 5 on both sides
$x = 5$	Simplifying

Check:

$$\begin{array}{c|l} \multicolumn{2}{c}{5x + 6 = 31} \\ \hline 5 \cdot 5 + 6 \ ? & 31 \\ 25 + 6 & \\ 31 & \qquad \text{TRUE} \end{array}$$

The solution is 5.

3.

$8x + 4 = 68$	
$8x + 4 - 4 = 68 - 4$	Subtracting 4 on both sides
$8x = 64$	Simplifying
$\frac{8x}{8} = \frac{64}{8}$	Dividing by 8 on both sides
$x = 8$	Simplifying

Check: $\begin{array}{c|c} 8x+4 & 68 \\ \hline 8\cdot 8+4 & ?\ 68 \\ 64+4 & \\ 68 & \end{array}$ TRUE

The solution is 8.

5. $4x-6=34$

$4x-6+6=34+6$ Adding 6 on both sides

$4x=40$

$\frac{4x}{4}=\frac{40}{4}$ Dividing by 4 on both sides

$x=10$

Check: $\begin{array}{c|c} 4x-6 & 34 \\ \hline 4\cdot 10-6 & ?\ 34 \\ 40-6 & \\ 34 & \end{array}$ TRUE

The solution is 10.

7. $3x-9=33$

$3x-9+9=33+9$

$3x=42$

$\frac{3x}{3}=\frac{42}{3}$

$x=14$

Check: $\begin{array}{c|c} 3x-9 & 33 \\ \hline 3\cdot 14-9 & ?\ 33 \\ 42-9 & \\ 33 & \end{array}$ TRUE

The solution is 14.

9. $7x+2=-54$

$7x+2-2=-54-2$

$7x=-56$

$\frac{7x}{7}=\frac{-56}{7}$

$x=-8$

Check: $\begin{array}{c|c} 7x+2 & -54 \\ \hline 7(-8)+2 & ?\ -54 \\ -56+2 & \\ -54 & \end{array}$ TRUE

The solution is -8.

11. $-45=6y+3$

$-45-3=6y+3-3$

$-48=6y$

$\frac{-48}{6}=\frac{6y}{6}$

$-8=y$

Check: $\begin{array}{c|c} -45 & 6y+3 \\ \hline -45\ ? & 6(-8)+3 \\ & -48+3 \\ & -45 \end{array}$ TRUE

The solution is -8.

13. $-4x+7=35$

$-4x+7-7=35-7$

$-4x=28$

$\frac{-4x}{-4}=\frac{28}{-4}$

$x=-7$

Check: $\begin{array}{c|c} -4x+7 & 35 \\ \hline -4(-7)+7 & ?\ 35 \\ 28+7 & \\ 35 & \end{array}$ TRUE

The solution is -7.

15. $-7x-24=-129$

$-7x-24+24=-129+24$

$-7x=-105$

$\frac{-7x}{-7}=\frac{-105}{-7}$

$x=15$

Check: $\begin{array}{c|c} -7x-24 & -129 \\ \hline -7\cdot 15-24 & ?\ -129 \\ -105-24 & \\ -129 & \end{array}$ TRUE

The solution is 15.

17. $5x+7x=72$

$12x=72$ Collecting like terms

$\frac{12x}{12}=\frac{72}{12}$ Dividing by 12 on both sides

$x=6$

Check: $\begin{array}{c|c} 5x+7x & 72 \\ \hline 5\cdot 6+7\cdot 6 & ?\ 72 \\ 30+42 & \\ 72 & \end{array}$ TRUE

The solution is 6.

19. $8x+7x=60$

$15x=60$ Collecting like terms

$\frac{15x}{15}=\frac{60}{15}$ Dividing by 15 on both sides

$x=4$

Check: $\begin{array}{c|c} 8x+7x & 60 \\ \hline 8\cdot 4+7\cdot 4 & ?\ 60 \\ 32+28 & \\ 60 & \end{array}$ TRUE

The solution is 4.

21. $4x+3x=42$

$7x=42$

$\frac{7x}{7}=\frac{42}{7}$

$x=6$

Check:
$$\begin{array}{r|l} \multicolumn{2}{c}{4x + 3x = 42} \\ \hline 4 \cdot 6 + 3 \cdot 6 & ?\ 42 \\ 24 + 18 & \\ 42 & \quad \text{TRUE} \end{array}$$

The solution is 6.

23.
$$\begin{aligned} -6y - 3y &= 27 \\ -9y &= 27 \\ \frac{-9y}{-9} &= \frac{27}{-9} \\ y &= -3 \end{aligned}$$

Check:
$$\begin{array}{r|l} \multicolumn{2}{c}{-6y - 3y = 27} \\ \hline -6(-3) - 3(-3) & ?\ 27 \\ 18 + 9 & \\ 27 & \quad \text{TRUE} \end{array}$$

The solution is -3.

25.
$$\begin{aligned} -7y - 8y &= -15 \\ -15y &= -15 \\ \frac{-15y}{-15} &= \frac{-15}{-15} \\ y &= 1 \end{aligned}$$

Check:
$$\begin{array}{r|l} \multicolumn{2}{c}{-7y - 8y = -15} \\ \hline -7 \cdot 1 - 8 \cdot 1 & ?\ -15 \\ -7 - 8 & \\ -15 & \quad \text{TRUE} \end{array}$$

The solution is 1.

27.
$$\begin{aligned} x + \frac{1}{3}x &= 8 \\ \left(1 + \frac{1}{3}\right)x &= 8 \\ \frac{4}{3}x &= 8 \\ \frac{3}{4} \cdot \frac{4}{3}x &= \frac{3}{4} \cdot 8 \\ x &= 6 \end{aligned}$$

Check:
$$\begin{array}{r|l} \multicolumn{2}{c}{x + \frac{1}{3}x = 8} \\ \hline 6 + \frac{1}{3} \cdot 6 & ?\ 8 \\ 6 + 2 & \\ 8 & \quad \text{TRUE} \end{array}$$

The solution is 6.

29.
$$\begin{aligned} 10.2y - 7.3y &= -58 \\ 2.9y &= -58 \\ \frac{2.9y}{2.9} &= \frac{-58}{2.9} \\ y &= -20 \end{aligned}$$

Check:
$$\begin{array}{r|l} \multicolumn{2}{c}{10.2y - 7.3y = -58} \\ \hline 10.2(-20) - 7.3(-20) & ?\ -58 \\ -204 + 146 & \\ -58 & \quad \text{TRUE} \end{array}$$

The solution is -20.

31.
$$\begin{aligned} 8y - 35 &= 3y \\ 8y &= 3y + 35 && \text{Adding 35 and simplifying} \\ 8y - 3y &= 35 && \text{Subtracting } 3y \text{ and simplifying} \\ 5y &= 35 && \text{Collecting like terms} \\ \frac{5y}{5} &= \frac{35}{5} && \text{Dividing by 5} \\ y &= 7 \end{aligned}$$

Check:
$$\begin{array}{r|l} \multicolumn{2}{c}{8y - 35 = 3y} \\ \hline 8 \cdot 7 - 35 & ?\ 3 \cdot 7 \\ 56 - 35 & 21 \\ 21 & \quad \text{TRUE} \end{array}$$

The solution is 7.

33.
$$\begin{aligned} 8x - 1 &= 23 - 4x \\ 8x + 4x &= 23 + 1 && \text{Adding 1 and } 4x \text{ and simplifying} \\ 12x &= 24 && \text{Collecting like terms} \\ \frac{12x}{12} &= \frac{24}{12} && \text{Dividing by 12} \\ x &= 2 \end{aligned}$$

Check:
$$\begin{array}{r|l} \multicolumn{2}{c}{8x - 1 = 23 - 4x} \\ \hline 8 \cdot 2 - 1 & ?\ 23 - 4 \cdot 2 \\ 16 - 1 & 23 - 8 \\ 15 & 15 \quad \text{TRUE} \end{array}$$

The solution is 2.

35.
$$\begin{aligned} 2x - 1 &= 4 + x \\ 2x - x &= 4 + 1 && \text{Adding 1 and } -x \\ x &= 5 && \text{Collecting like terms} \end{aligned}$$

Check:
$$\begin{array}{r|l} \multicolumn{2}{c}{2x - 1 = 4 + x} \\ \hline 2 \cdot 5 - 1 & ?\ 4 + 5 \\ 10 - 1 & 9 \\ 9 & \quad \text{TRUE} \end{array}$$

The solution is 5.

37.
$$\begin{aligned} 6x + 3 &= 2x + 11 \\ 6x - 2x &= 11 - 3 \\ 4x &= 8 \\ \frac{4x}{4} &= \frac{8}{4} \\ x &= 2 \end{aligned}$$

Check:
$$\begin{array}{r|l} \multicolumn{2}{c}{6x + 3 = 2x + 11} \\ \hline 6 \cdot 2 + 3 & ?\ 2 \cdot 2 + 11 \\ 12 + 3 & 4 + 11 \\ 15 & 15 \quad \text{TRUE} \end{array}$$

The solution is 2.

39.
$$\begin{aligned} 5-2x &= 3x-7x+25 \\ 5-2x &= -4x+25 \\ 4x-2x &= 25-5 \\ 2x &= 20 \\ \frac{2x}{2} &= \frac{20}{2} \\ x &= 10 \end{aligned}$$

Check:
$$\begin{array}{r|l} \multicolumn{2}{c}{5-2x = 3x-7x+25} \\ \hline 5-2\cdot 10 \ ? & 3\cdot 10-7\cdot 10+25 \\ 5-20 & 30-70+25 \\ -15 & -40+25 \\ & -15 \quad \text{TRUE} \end{array}$$

The solution is 10.

41.
$$\begin{aligned} 4+3x-6 &= 3x+2-x \\ 3x-2 &= 2x+2 \qquad \text{Collecting like terms on each side} \\ 3x-2x &= 2+2 \\ x &= 4 \end{aligned}$$

Check:
$$\begin{array}{r|l} \multicolumn{2}{c}{4+3x-6 = 3x+2-x} \\ \hline 4+3\cdot 4-6 \ ? & 3\cdot 4+2-4 \\ 4+12-6 & 12+2-4 \\ 16-6 & 14-4 \\ 10 & 10 \quad \text{TRUE} \end{array}$$

The solution is 4.

43.
$$\begin{aligned} 4y-4+y+24 &= 6y+20-4y \\ 5y+20 &= 2y+20 \\ 5y-2y &= 20-20 \\ 3y &= 0 \\ y &= 0 \end{aligned}$$

Check:
$$\begin{array}{r|l} \multicolumn{2}{c}{4y-4+y+24 = 6y+20-4y} \\ \hline 4\cdot 0-4+0+24 \ ? & 6\cdot 0+20-4\cdot 0 \\ 0-4+0+24 & 0+20-0 \\ 20 & 20 \quad \text{TRUE} \end{array}$$

The solution is 0.

45. $\frac{7}{2}x+\frac{1}{2}x = 3x+\frac{3}{2}+\frac{5}{2}x$

The least common multiple of all the denominators is 2. We multiply by 2 on both sides.

$$\begin{aligned} 2\left(\frac{7}{2}x+\frac{1}{2}x\right) &= 2\left(3x+\frac{3}{2}+\frac{5}{2}x\right) \\ 2\cdot\frac{7}{2}x+2\cdot\frac{1}{2}x &= 2\cdot 3x+2\cdot\frac{3}{2}+2\cdot\frac{5}{2}x \\ 7x+x &= 6x+3+5x \\ 8x &= 11x+3 \\ 8x-11x &= 3 \\ -3x &= 3 \\ \frac{-3x}{-3} &= \frac{3}{-3} \\ x &= -1 \end{aligned}$$

Check:
$$\begin{array}{r|l} \multicolumn{2}{c}{\frac{7}{2}x+\frac{1}{2}x = 3x+\frac{3}{2}+\frac{5}{2}x} \\ \hline \frac{7}{2}(-1)+\frac{1}{2}(-1) \ ? & 3(-1)+\frac{3}{2}+\frac{5}{2}(-1) \\ -\frac{7}{2}-\frac{1}{2} & -3+\frac{3}{2}-\frac{5}{2} \\ -4 & -\frac{8}{2} \\ & -4 \quad \text{TRUE} \end{array}$$

The solution is -1.

47. $\frac{2}{3}+\frac{1}{4}t = \frac{1}{3}$

The least common multiple of all the denominators is 12. We multiply by 12 on both sides.

$$\begin{aligned} 12\left(\frac{2}{3}+\frac{1}{4}t\right) &= 12\cdot\frac{1}{3} \\ 12\cdot\frac{2}{3}+12\cdot\frac{1}{4}t &= 12\cdot\frac{1}{3} \\ 8+3t &= 4 \\ 3t &= 4-8 \\ 3t &= -4 \\ \frac{3t}{3} &= \frac{-4}{3} \\ t &= -\frac{4}{3} \end{aligned}$$

Check:
$$\begin{array}{r|l} \multicolumn{2}{c}{\frac{2}{3}+\frac{1}{4}t = \frac{1}{3}} \\ \hline \frac{2}{3}+\frac{1}{4}\left(-\frac{4}{3}\right) \ ? & \frac{1}{3} \\ \frac{2}{3}-\frac{1}{3} & \\ \frac{1}{3} & \quad \text{TRUE} \end{array}$$

The solution is $-\frac{4}{3}$.

49.
$$\begin{aligned} \frac{2}{3}+3y &= 5y-\frac{2}{15}, \quad \text{LCM is 15} \\ 15\left(\frac{2}{3}+3y\right) &= 15\left(5y-\frac{2}{15}\right) \\ 15\cdot\frac{2}{3}+15\cdot 3y &= 15\cdot 5y-15\cdot\frac{2}{15} \\ 10+45y &= 75y-2 \\ 10+2 &= 75y-45y \\ 12 &= 30y \\ \frac{12}{30} &= \frac{30y}{30} \\ \frac{2}{5} &= y \end{aligned}$$

Check:

$$\begin{array}{c|c}
\multicolumn{2}{c}{\frac{2}{3}+3y = 5y-\frac{2}{15}} \\ \hline
\frac{2}{3}+3\cdot\frac{2}{5} \ ? & 5\cdot\frac{2}{5}-\frac{2}{15} \\
\frac{2}{3}+\frac{6}{5} & 2-\frac{2}{15} \\
\frac{10}{15}+\frac{18}{15} & \frac{30}{15}-\frac{2}{15} \\
\frac{28}{15} & \frac{28}{15} \quad \text{TRUE}
\end{array}$$

The solution is $\frac{2}{5}$.

51.

$$\frac{5}{3}+\frac{2}{3}x = \frac{25}{12}+\frac{5}{4}x+\frac{3}{4}, \quad \text{LCM is 12}$$
$$12\left(\frac{5}{3}+\frac{2}{3}x\right) = 12\left(\frac{25}{12}+\frac{5}{4}x+\frac{3}{4}\right)$$
$$12\cdot\frac{5}{3}+12\cdot\frac{2}{3}x = 12\cdot\frac{25}{12}+12\cdot\frac{5}{4}x+12\cdot\frac{3}{4}$$
$$20+8x = 25+15x+9$$
$$20+8x = 15x+34$$
$$20-34 = 15x-8x$$
$$-14x = 7x$$
$$\frac{-14}{7} = \frac{7x}{7}$$
$$-2 = x$$

Check:

$$\begin{array}{c|c}
\multicolumn{2}{c}{\frac{5}{3}+\frac{2}{3}x = \frac{25}{12}+\frac{5}{4}x+\frac{3}{4}} \\ \hline
\frac{5}{3}+\frac{2}{3}(-2) \ ? & \frac{25}{12}+\frac{5}{4}(-2)+\frac{3}{4} \\
\frac{5}{3}-\frac{4}{3} & \frac{25}{12}-\frac{5}{2}+\frac{3}{4} \\
\frac{1}{3} & \frac{25}{12}-\frac{30}{12}+\frac{9}{12} \\
 & \frac{4}{12} \\
 & \frac{1}{3} \quad \text{TRUE}
\end{array}$$

The solution is -2.

53.

$$2.1x+45.2 = 3.2-8.4x$$

Greatest number of decimal places is 1

$$10(2.1x+45.2) = 10(3.2-8.4x)$$

Multiplying by 10 to clear decimals

$$10(2.1x)+10(45.2) = 10(3.2)-10(8.4x)$$
$$21x+452 = 32-84x$$
$$21x+84x = 32-452$$
$$105x = -420$$
$$\frac{105x}{105} = \frac{-420}{105}$$
$$x = -4$$

Check:

$$\begin{array}{c|c}
\multicolumn{2}{c}{2.1x+45.2 = 3.2-8.4x} \\ \hline
2.1(-4)+45.2 \ ? & 3.2-8.4(-4) \\
-8.4+45.2 & 3.2+33.6 \\
36.8 & 36.8 \quad \text{TRUE}
\end{array}$$

The solution is -4.

55.

$$1.03-0.62x = 0.71-0.22x$$

Greatest number of decimal places is 2

$$100(1.03-0.62x) = 100(0.71-0.22x)$$

Multiplying by 100 to clear decimals

$$100(1.03)-100(0.62x) = 100(0.71)-100(0.22x)$$
$$103-62x = 71-22x$$
$$32 = 40x$$
$$\frac{32}{40} = \frac{40x}{40}$$
$$\frac{4}{5} = x, \text{ or}$$
$$0.8 = x$$

Check:

$$\begin{array}{c|c}
\multicolumn{2}{c}{1.03-0.62x = 0.71-0.22x} \\ \hline
1.03-0.62(0.8) \ ? & 0.71-0.22(0.8) \\
1.03-0.496 & 0.71-0.176 \\
0.534 & 0.534 \quad \text{TRUE}
\end{array}$$

The solution is $\frac{4}{5}$, or 0.8.

57.

$$\frac{2}{7}x-\frac{1}{2}x = \frac{3}{4}x+1, \text{ LCM is 28}$$
$$28\left(\frac{2}{7}x-\frac{1}{2}x\right) = 28\left(\frac{3}{4}x+1\right)$$
$$28\cdot\frac{2}{7}x-28\cdot\frac{1}{2}x = 28\cdot\frac{3}{4}x+28\cdot 1$$
$$8x-14x = 21x+28$$
$$-6x = 21x+28$$
$$-6x-21x = 28$$
$$-27x = 28$$
$$x = -\frac{28}{27}$$

Check:

$$\begin{array}{c|c}
\multicolumn{2}{c}{\frac{2}{7}x-\frac{1}{2}x = \frac{3}{4}x+1} \\ \hline
\frac{2}{7}\left(-\frac{28}{27}\right)-\frac{1}{2}\left(-\frac{28}{27}\right) \ ? & \frac{3}{4}\left(-\frac{28}{27}\right)+1 \\
-\frac{8}{27}+\frac{14}{27} & -\frac{21}{27}+1 \\
\frac{6}{27} & \frac{6}{27} \quad \text{TRUE}
\end{array}$$

The solution is $-\frac{28}{27}$.

59.

$$3(2y-3) = 27$$
$$6y-9 = 27 \quad \text{Using a distributive law}$$
$$6y = 27+9 \quad \text{Adding 9}$$
$$6y = 36$$
$$y = 6 \quad \text{Dividing by 6}$$

Check: $3(2y-3)=27$

$$\begin{array}{r|l} 3(2\cdot 6-3) & ?\ 27 \\ 3(12-3) & \\ 3\cdot 9 & \\ 27 & \quad \text{TRUE} \end{array}$$

The solution is 6.

61. $40 = 5(3x+2)$

$40 = 15x + 10$ Using a distributive law

$40 - 10 = 15x$

$30 = 15x$

$2 = x$

Check: $40 = 5(3x+2)$

$$\begin{array}{r|l} 40 & ?\ 5(3\cdot 2+2) \\ & 5(6+2) \\ & 5\cdot 8 \\ & 40 \quad \text{TRUE} \end{array}$$

The solution is 2.

63. $2(3+4m)-9=45$

$6+8m-9=45$ Collecting like terms

$8m-3=45$

$8m=45+3$

$8m=48$

$m=6$

Check: $2(3+4m)-9=45$

$$\begin{array}{r|l} 2(3+4\cdot 6)-9 & ?\ 45 \\ 2(3+24)-9 & \\ 2\cdot 27-9 & \\ 54-9 & \\ 45 & \quad \text{TRUE} \end{array}$$

The solution is 6.

65. $5r-(2r+8)=16$

$5r-2r-8=16$

$3r-8=16$ Collecting like terms

$3r=16+8$

$3r=24$

$r=8$

Check: $5r-(2r+8)=16$

$$\begin{array}{r|l} 5\cdot 8-(2\cdot 8+8) & ?\ 16 \\ 40-(16+8) & \\ 40-24 & \\ 16 & \quad \text{TRUE} \end{array}$$

The solution is 8.

67. $6-2(3x-1)=2$

$6-6x+2=2$

$8-6x=2$

$8-2=6x$

$6=6x$

$1=x$

Check: $6-2(3x-1)=2$

$$\begin{array}{r|l} 6-2(3\cdot 1-1) & ?\ 2 \\ 6-2(3-1) & \\ 6-2\cdot 2 & \\ 6-4 & \\ 2 & \quad \text{TRUE} \end{array}$$

The solution is 1.

69. $5(d+4)=7(d-2)$

$5d+20=7d-14$

$20+14=7d-5d$

$34=2d$

$17=d$

Check: $5(d+4)=7(d-2)$

$$\begin{array}{r|l} 5(17+4) & ?\ 7(17-2) \\ 5\cdot 21 & 7\cdot 15 \\ 105 & 105 \quad \text{TRUE} \end{array}$$

The solution is 17.

71. $8(2t+1)=4(7t+7)$

$16t+8=28t+28$

$16t-28t=28-8$

$-12t=20$

$t=-\frac{20}{12}$

$t=-\frac{5}{3}$

Check: $8(2t+1)=4(7t+7)$

$$\begin{array}{r|l} 8\left(2\left(-\frac{5}{3}\right)+1\right) & ?\ 4\left(7\left(-\frac{5}{3}\right)+7\right) \\ 8\left(-\frac{10}{3}+1\right) & 4\left(-\frac{35}{3}+7\right) \\ 8\left(-\frac{7}{3}\right) & 4\left(-\frac{14}{3}\right) \\ -\frac{56}{3} & -\frac{56}{3} \quad \text{TRUE} \end{array}$$

The solution is $-\frac{5}{3}$.

73. $3(r-6)+2=4(r+2)-21$

$3r-18+2=4r+8-21$

$3r-16=4r-13$

$13-16=4r-3r$

$-3=r$

Check: $3(r-6)+2=4(r+2)-21$

$$\begin{array}{r|l} 3(-3-6)+2 & ?\ 4(-3+2)-21 \\ 3(-9)+2 & 4(-1)-21 \\ -27+2 & -4-21 \\ -25 & -25 \quad \text{TRUE} \end{array}$$

The solution is -3.

75.
$$\begin{aligned}19-(2x+3)&=2(x+3)+x\\19-2x-3&=2x+6+x\\16-2x&=3x+6\\16-6&=3x+2x\\10&=5x\\2&=x\end{aligned}$$

Check:
$$\begin{array}{r|l}\multicolumn{2}{c}{19-(2x+3)=2(x+3)+x}\\\hline 19-(2\cdot2+3)\ ? & 2(2+3)+2\\19-(4+3) & 2\cdot5+2\\19-7 & 10+2\\12 & 12\quad\text{TRUE}\end{array}$$

The solution is 2.

77.
$$\begin{aligned}2[4-2(3-x)]-1&=4[2(4x-3)+7]-25\\2[4-6+2x]-1&=4[8x-6+7]-25\\2[-2+2x]-1&=4[8x+1]-25\\-4+4x-1&=32x+4-25\\4x-5&=32x-21\\-5+21&=32x-4x\\16&=28x\\\frac{16}{28}&=x\\\frac{4}{7}&=x\end{aligned}$$

The check is left to the student.

The solution is $\frac{4}{7}$.

79.
$$\begin{aligned}0.7(3x+6)&=1.1-(x+2)\\2.1x+4.2&=1.1-x-2\\10(2.1x+4.2)&=10(1.1-x-2)\quad\text{Clearing decimals}\\21x+42&=11-10x-20\\21x+42&=-10x-9\\21x+10x&=-9-42\\31x&=-51\\x&=-\frac{51}{31}\end{aligned}$$

The check is left to the student.

The solution is $-\frac{51}{31}$.

81.
$$\begin{aligned}a+(a-3)&=(a+2)-(a+1)\\a+a-3&=a+2-a-1\\2a-3&=1\\2a&=1+3\\2a&=4\\a&=2\end{aligned}$$

Check:
$$\begin{array}{r|l}\multicolumn{2}{c}{a+(a-3)=(a+2)-(a+1)}\\\hline 2+(2-3)\ ? & (2+2)-(2+1)\\2-1 & 4-3\\1 & 1\quad\text{TRUE}\end{array}$$

The solution is 2.

83. Do the long division. The answer is negative.

$$\begin{array}{r}6.5\\3.4_{\wedge}\overline{)\,22.1_{\wedge}0}\\\underline{20\,4}\\17\,0\\\underline{17\,0}\\0\end{array}$$

$-22.1\div3.4=-6.5$

85. Since -15 is to the left of -13 on the number line, -15 is less than -13, so $-15<-13$.

87. $-22.1+3.4$ The absolute values are 22.1 and 3.4. The difference is 18.7. The negative number has the larger absolute value, so the answer is negative.

$-22.1+3.4=-18.7$

89. The translation is $c\div8$, or $\frac{c}{8}$.

91.

93. Since we are using a calculator we will not clear the decimals.

$$\begin{aligned}0.008+9.62x-42.8&=0.944x+0.0083-x\\9.62x-42.792&=-0.056x+0.0083\\9.62x+0.056x&=0.0083+42.792\\9.676x&=42.8003\\x&=\frac{42.8003}{9.676}\\x&\approx4.4233464\end{aligned}$$

The solution is approximately 4.4233464.

95.
$$\begin{aligned}0&=y-(-14)-(-3y)\\0&=y+14+3y\\0&=4y+14\\-14&=4y\\\frac{-14}{4}&=y\\-\frac{7}{2}&=y\end{aligned}$$

The solution is $-\frac{7}{2}$.

97.
$$\begin{aligned}\frac{5+2y}{3}&=\frac{25}{12}+\frac{5y+3}{4},\ \text{LCM is 12}\\12\Big(\frac{5+2y}{3}\Big)&=12\Big(\frac{25}{12}+\frac{5y+3}{4}\Big)\\4(5+2y)&=25+3(5y+3)\\20+8y&=25+15y+9\\-7y&=14\\y&=-2\end{aligned}$$

The solution is -2.

99.
$$\begin{aligned}-2y+5y&=6y\\3y&=6y\\0&=3y\\0&=y\end{aligned}$$

The solution is 0.

101. $\frac{1}{3}(6x+24)-20 = -\frac{1}{4}(12x-72)$

$$\begin{aligned} 2x+8-20 &= -3x+18 \quad \text{Multiplying} \\ 2x-12 &= -3x+18 \\ 5x &= 30 \\ x &= 6 \end{aligned}$$

The solution is 6.

103. $\frac{3}{4}\left(3x-\frac{1}{2}\right)-\frac{2}{3}=\frac{1}{3}$

$$\begin{aligned} \frac{3}{4}\left(3x-\frac{1}{2}\right) &= 1 \quad \text{Adding } \frac{2}{3} \text{ on both sides} \\ 3\left(3x-\frac{1}{2}\right) &= 4 \quad \text{Multiplying by 4} \\ 9x-\frac{3}{2} &= 4 \\ 9x &= \frac{11}{2} \\ x &= \frac{1}{9}\cdot\frac{11}{2} \\ x &= \frac{11}{18} \end{aligned}$$

The solution is $\frac{11}{18}$.

105. Addition principle:

$$\begin{aligned} 4x-8 &= 32 \\ 4x &= 40 \\ x &= 10 \end{aligned}$$

Multiplication principle:

$$\begin{aligned} 4x-8 &= 32 \\ \frac{1}{4}(4x-8) &= \frac{1}{4}\cdot 32 \\ x-2 &= 8 \\ x &= 10 \end{aligned}$$

Exercise Set 8.4

1. ***Familiarize.*** Let x = the number. Then "two times a number" translates to $2x$, and "two times a number added to 85" translates to $2x+85$.

Translate.

$$\underbrace{\text{Two times a number}}_{\downarrow \atop 2x}\ \underbrace{\text{added to}}_{\downarrow \atop +}\ \underset{\downarrow \atop 85}{85}\ \underset{\downarrow \atop =}{\text{is}}\ \underset{\downarrow \atop 117}{117.}$$

Solve. We solve the equation.

$$\begin{aligned} 2x+85 &= 117 \\ 2x+85-85 &= 117-85 \\ 2x &= 32 \\ \frac{2x}{2} &= \frac{32}{2} \\ x &= 16 \end{aligned}$$

Check. Two times 16 is 32. Adding 32 to 85 we get 117. The answer checks.

State. The number is 16.

3. ***Familiarize.*** Let n = the number. Then "twice a number" translates to $2n$, and "three less than twice a number" translates to $2n-3$.

Translate. We reword the problem.

$$\underbrace{\text{Twice a number}}_{\downarrow \atop 2n}\ \underset{\downarrow \atop -}{\text{less}}\ \underset{\downarrow \atop 3}{3}\ \underset{\downarrow \atop =}{\text{is}}\ \underset{\downarrow \atop -4}{-4.}$$

Solve. We solve the equation.

$$\begin{aligned} 2n-3 &= -4 \\ 2n-3+3 &= -4+3 \\ 2n &= -1 \\ \frac{2n}{2} &= \frac{-1}{2} \\ n &= -\frac{1}{2} \end{aligned}$$

Check. Twice $-\frac{1}{2}$ is -1, and 3 less than -1 is $-1-3$, or -4. The answer checks.

State. The number is $-\frac{1}{2}$.

5. ***Familiarize.*** We let x = the number. Then "four times a certain number" translates to $4x$, and "when 17 is subtracted from four times a certain number" translates to $4x-17$.

Translate. We reword the problem.

$$\underbrace{\text{Four times a number}}_{\downarrow \atop 4x}\ \underbrace{\text{less}}_{\downarrow \atop -}\ \underset{\downarrow \atop 17}{17}\ \underset{\downarrow \atop =}{\text{is}}\ \underset{\downarrow \atop 211}{211.}$$

Solve. We solve the equation.

$$\begin{aligned} 4x-17 &= 211 \\ 4x-17+17 &= 211+17 \\ 4x &= 228 \\ \frac{4x}{4} &= \frac{228}{4} \\ x &= 57 \end{aligned}$$

Check. Four times 57 is 228, and when 17 is subtracted from 228 we get 211. The answer checks.

State. The number is 57.

7. ***Familiarize.*** Using the labels on the drawing in the text, we let x = the length of the shorter piece, in inches, and $3x$ = the length of the longer piece, in inches.

Translate. We reword the problem.

$$\underbrace{\text{The length of the shorter piece}}_{\downarrow \atop x}\ \underset{\downarrow \atop +}{\text{plus}}\ \underbrace{\text{the length of the longer piece}}_{\downarrow \atop 3x}\ \underset{\downarrow \atop =}{\text{is}}\ \underset{\downarrow \atop 240}{\text{240 ft.}}$$

Solve. We solve the equation.

$$\begin{aligned} x+3x &= 240 \\ 4x &= 240 \quad \text{Collecting like terms} \\ \frac{4x}{4} &= \frac{240}{4} \\ x &= 60 \end{aligned}$$

If x is 60, then $3x = 3\cdot 60$, or 180.

Check. 180 is three times 60, and $60+180=240$. The answer checks.

State. The lengths of the pieces are 60 in. and 180 in.

9. ***Familiarize***. Let h = the height of the Statue of Liberty.

Translate.

$$\underbrace{\text{Height of Statue of Liberty}}_{h}\ \underbrace{\text{plus}}_{+}\ \underbrace{\text{Additional height}}_{669}\ \underbrace{\text{is}}_{=}\ \underbrace{\text{Height of Eiffel Tower}}_{974}$$

Solve. We solve the equation.

$$h + 669 = 974$$
$$h + 669 - 669 = 974 - 669 \quad \text{Subtracting 669}$$
$$h = 305$$

Check. If we add 669 ft to 305 ft, we get 974 ft. The answer checks.

State. The height of the Statue of Liberty is 305 ft.

11. ***Familiarize***. Let c = the cost of one 18-oz box of Wheaties. Then four boxes cost $4c$.

Translate.

$$\underbrace{\text{The cost of four boxes}}_{4c}\ \underbrace{\text{was}}_{=}\ \underbrace{\$11.56}_{11.56}$$

Solve. We solve the equation.

$$4c = 11.56$$
$$\frac{4c}{4} = \frac{11.56}{4}$$
$$c = 2.89$$

Check. If one box cost \$2.89, then four boxes cost 4(\$2.89), or \$11.56. The result checks.

State. One box cost \$2.89.

13. ***Familiarize***. Let y = the number.

Translate. We reword the problem.

$$\underbrace{\text{Two times a number}}_{2\cdot y}\ \underbrace{\text{plus}}_{+}\ \underbrace{16}_{16}\ \underbrace{\text{is}}_{=}\ \underbrace{\tfrac{2}{3}\text{ of the number}}_{\frac{2}{3}\cdot y}$$

Solve. We solve the equation.

$$2y + 16 = \frac{2}{3}y$$
$$3(2y + 16) = 3\cdot\frac{2}{3}y \quad \text{Clearing the fraction}$$
$$6y + 48 = 2y$$
$$6y + 48 - 6y = 2y - 6y$$
$$48 = -4y$$
$$\frac{48}{-4} = \frac{-4y}{-4}$$
$$-12 = y$$

Check. We double -12 and get -24. Adding 16, we get -8. Also, $\frac{2}{3}(-12) = -8$. The answer checks.

State. The number is -12.

15. ***Familiarize***. Let d = the musher's distance from Nome, in miles. Then $2d$ = the distance from Anchorage, in miles. This is the number of miles the musher has completed. The sum of the two distances is the length of the race, 1049 miles.

Translate.

$$\underbrace{\text{Distance from Nome}}_{d}\ \underbrace{\text{plus}}_{+}\ \underbrace{\text{distance from Anchorage}}_{2d}\ \underbrace{\text{is}}_{=}\ \underbrace{1049\text{ mi.}}_{1049}$$

Carry out. We solve the equation.

$$d + 2d = 1049$$
$$3d = 1049 \quad \text{Collecting like terms}$$
$$\frac{3d}{3} = \frac{1049}{3}$$
$$d = \frac{1049}{3}$$

If $d = \frac{1049}{3}$, then $2d = 2\cdot\frac{1049}{3} = \frac{2098}{3} = 699\frac{1}{3}$.

Check. $\frac{2098}{3}$ is twice $\frac{1049}{3}$, and $\frac{1049}{3} + \frac{2098}{3} = \frac{3147}{3} = 1049$. The result checks.

State. The musher has traveled $699\frac{1}{3}$ miles.

17. ***Familiarize.*** Using the labels on the drawing in the text, we let x = the smaller number and $x + 1$ = the larger number.

Translate. We reword the problem.

$$\underbrace{\text{First number}}_{x}\ +\ \underbrace{\text{second number}}_{(x+1)}\ \underbrace{\text{is}}_{=}\ \underbrace{573}_{573}$$

Solve. We solve the equation.

$$x + (x + 1) = 573$$
$$2x + 1 = 573 \quad \text{Collecting like terms}$$
$$2x + 1 - 1 = 573 - 1 \quad \text{Subtracting 1}$$
$$2x = 572$$
$$\frac{2x}{2} = \frac{572}{2} \quad \text{Dividing by 2}$$
$$x = 286$$

If x is 286, then $x + 1$ is 287.

Check. 286 and 287 are consecutive integers, and their sum is 573. The answer checks.

State. The page numbers are 286 and 287.

19. ***Familiarize.*** Let a = the first number. Then $a + 1$ = the second number, and $a + 2$ = the third number.

Translate. We reword the problem.

$$\underbrace{\text{First number}}_{a}\ +\ \underbrace{\text{second number}}_{(a+1)}\ +\ \underbrace{\text{third number}}_{(a+2)}\ \underbrace{\text{is}}_{=}\ \underbrace{126}_{114}$$

Solve. We solve the equation.

$$
\begin{aligned}
a + (a+1) + (a+2) &= 126 \\
3a + 3 &= 126 \quad \text{Collecting like terms} \\
3a + 3 - 3 &= 126 - 3 \\
3a &= 123 \\
\frac{3a}{3} &= \frac{123}{3} \\
a &= 41
\end{aligned}
$$

If a is 41, then $a+1$ is 42 and $a+2$ is 43.

Check. 41, 42, and 43 are consecutive integers, and their sum is 126. The answer checks.

State. The numbers are 41, 42, and 43.

21. ***Familiarize.*** Let x = the first odd integer. Then $x+2$ = the next odd integer and $(x+2)+2$, or $x+4$ = the third odd integer.

Translate. We reword the problem.

$$
\underbrace{\text{First odd integer}}_{\downarrow\; x} \; + \; \underbrace{\text{second odd integer}}_{\downarrow\; (x+2)} \; + \; \underbrace{\text{third odd integer}}_{\downarrow\; (x+4)} \; \underbrace{\text{is}}_{\downarrow\; =} \; 189
$$

Solve. We solve the equation.

$$
\begin{aligned}
x + (x+2) + (x+4) &= 189 \\
3x + 6 &= 189 \quad \text{Collecting like terms} \\
3x + 6 - 6 &= 189 - 6 \\
3x &= 183 \\
\frac{3x}{3} &= \frac{183}{3} \\
x &= 61
\end{aligned}
$$

If x is 61, then $x+2$ is 63 and $x+4$ is 65.

Check. 61, 63, and 65 are consecutive odd integers, and their sum is 189. The answer checks.

State. The integers are 61, 63, and 65.

23. ***Familiarize.*** Using the labels on the drawing in the text, we let w = the width and $3w+6$ = the length. The perimeter P of a rectangle is given by the formula $2l+2w = P$, where l = the length and w = the width.

Translate. Substitute $3w+6$ for l and 124 for P:

$$
\begin{aligned}
2l + 2w &= P \\
2(3w+6) + 2w &= 124
\end{aligned}
$$

Solve. We solve the equation.

$$
\begin{aligned}
2(3w+6) + 2w &= 124 \\
6w + 12 + 2w &= 124 \\
8w + 12 &= 124 \\
8w + 12 - 12 &= 124 - 12 \\
8w &= 112 \\
\frac{8w}{8} &= \frac{112}{8} \\
w &= 14
\end{aligned}
$$

The possible dimensions are $w = 14$ ft and $l = 3w + 6 = 3(14) + 6$, or 48 ft.

Check. The length, 48 ft, is 6 ft more than three times the width, 14 ft. The perimeter is $2(48 \text{ ft}) + 2(14 \text{ ft}) = 96 \text{ ft} + 28 \text{ ft} = 124$ ft. The answer checks.

State. The width is 14 ft, and the length is 48 ft.

25. ***Familiarize.*** Let n = the number of visits required for a total parking cost of \$27.00. The parking cost for each $1\frac{1}{2}$ hour visit is \$1.50 for the first hour plus \$1.00 for part of a second hour, or \$2.50. Then the total parking cost for n visits is $2.50n$ dollars.

Translate. We reword the problem.

$$
\underbrace{\text{Total parking cost}}_{\downarrow\; 2.50n} \; \underbrace{\text{is}}_{\downarrow\; =} \; \underbrace{\$27.00.}_{\downarrow\; 27.00}
$$

Solve. We solve the equation.

$$
\begin{aligned}
2.5n &= 27 \\
10(2.5n) &= 10(27) \quad \text{Clearing the decimal} \\
25n &= 270 \\
\frac{25n}{25} &= \frac{270}{25} \\
n &= 10.8
\end{aligned}
$$

If the total parking cost is \$27.00 for 10.8 visits, then the cost will be more than \$27.00 for 11 or more visits.

Check. The parking cost for 10 visits is \$2.50(10), or \$25, and the parking cost for 11 visits is \$2.50(11), or \$27.50. Since 11 is the smallest number for which the parking cost exceeds \$27.00, the answer checks.

State. The minimum number of weekly visits for which it is worthwhile to buy a parking pass is 11.

27. ***Familiarize.*** Let x = the measure of the first angle. Then $3x$ = the measure of the second angle, and $x+40$ = the measure of the third angle. Recall that the sum of measures of the angles of a triangle is 180°.

Translate.

$$
\underbrace{\text{Measure of first angle}}_{\downarrow\; x} \; + \; \underbrace{\text{measure of second angle}}_{\downarrow\; 3x} \; + \; \underbrace{\text{measure of third angle}}_{\downarrow\; (x+40)} \; \underbrace{\text{is}}_{\downarrow\; =} \; 180.
$$

Solve. We solve the equation.

$$
\begin{aligned}
x + 3x + (x+40) &= 180 \\
5x + 40 &= 180 \\
5x + 40 - 40 &= 180 - 40 \\
5x &= 140 \\
\frac{5x}{5} &= \frac{140}{5} \\
x &= 28
\end{aligned}
$$

Possible answers for the angle measures are as follows:

First angle: $x = 28°$

Second angle: $3x = 3(28) = 84°$

Third angle: $x + 40 = 28 + 40 = 68°$

Check. Consider 28°, 84°, and 68°. The second angle is three times the first, and the third is 40° more than the first. The sum, 28° + 84° + 68°, is 180°. These numbers check.

State. The measures of the angles are 28°, 84°, and 68°.

29. ***Familiarize.*** Using the labels on the drawing in the text, we let x = the measure of the first angle, $x + 5$ = the measure of the second angle, and $3x + 10$ = the measure of the third angle. Recall that the sum of measures of the angles of a triangle is 180°.

Translate.

$$\underbrace{\text{Measure of first angle}} + \underbrace{\text{measure of second angle}} + \underbrace{\text{measure of third angle}} \text{ is } 180.$$

$$x + (x+5) + (3x+10) = 180$$

Solve. We solve the equation.

$$x + (x+5) + (3x+10) = 180$$
$$5x + 15 = 180$$
$$5x + 15 - 15 = 180 - 15$$
$$5x = 165$$
$$\frac{5x}{5} = \frac{165}{5}$$
$$x = 33$$

Possible answers for the angle measures are as follows:

First angle: $x = 33°$

Second angle: $x + 5 = 33 + 5 = 38°$

Third angle: $3x + 10 = 3(33) + 10 = 109°$

Check. The second angle is 5° more than the first, and the third is 10° more than 3 times the first. The sum, 33° + 38° + 109°, is 180°. The numbers check.

State. The measures of the angles are 33°, 38°, and 109°.

31. a) The year 1999 corresponds to $x = 9$. Substitute 9 for x in the equation.

$y = 66.2x + 460.2 = 66.2(9) + 460.2 = 1056$

The revenue in 1999 was $1056 million.

The year 2000 corresponds to $x = 10$. Substitute 10 for x in the equation.

$y = 66.2x + 460.2 = 66.2(10) + 460.2 = 1122.2$

The revenue in 2000 will be $1122.2 million.

The year 2010 corresponds to $x = 20$. Substitute 20 for x in the equation.

$y = 66.2x + 460.2 = 66.2(20) + 460.2 = 1784.2$

The revenue in 2010 will be $1784.2 million.

b) To find the year in which the total revenue will be about $1254.6 million, substitute 1254.6 for y and solve for x.

$$1254.6 = 66.2x + 460.2$$
$$1254.6 - 460.2 = 66.2x + 460.2 - 460.2$$
$$794.4 = 66.2x$$
$$\frac{794.4}{66.2} = \frac{66.2x}{66.2}$$
$$12 = x$$

The solution of the equation, 12, corresponds to the year 2002. Thus, in 2002 the total revenue will be about $1254.6 million.

33. $-\frac{4}{5} - \frac{3}{8} = -\frac{4}{5} + \left(-\frac{3}{8}\right)$

$= -\frac{32}{40} + \left(-\frac{15}{40}\right)$

$= -\frac{47}{40}$

35. $-\frac{4}{5} \cdot \frac{3}{8} = -\frac{4 \cdot 3}{5 \cdot 8}$

$= -\frac{4 \cdot 3}{5 \cdot 2 \cdot 4}$

$= -\frac{\not{4} \cdot 3}{5 \cdot 2 \cdot \not{4}}$

$= -\frac{3}{10}$

37. $-25.6 \div (-16)$

First we do the long division $25.6 \div 16$.

```
      1.6
 16 ) 25.6
      16 0
       9 6
       9 6
         0
```

The two original numbers have the same sign so the answer is positive.

$-25.6 \div (-16) = 1.6$

39. $-25.6 - (-16) = -25.6 + 16 = -9.6$

41. ◈

43. ***Familiarize.*** Let a = the original number of apples. Then $\frac{1}{3}a$, $\frac{1}{4}a$, $\frac{1}{8}a$, and $\frac{1}{5}a$ are given to four people, respectively. The fifth and sixth people get 10 apples and 1 apple, respectively.

Translate. We reword the problem.

$$\underbrace{\text{The total number of apples}} \text{ is } a$$

$$\frac{1}{3}a + \frac{1}{4}a + \frac{1}{8}a + \frac{1}{5}a + 10 + 1 = a$$

Solve. We solve the equation.

$$\frac{1}{3}a + \frac{1}{4}a + \frac{1}{8}a + \frac{1}{5}a + 10 + 1 = a, \text{ LCD is 120}$$

$$120\left(\frac{1}{3}a + \frac{1}{4}a + \frac{1}{8}a + \frac{1}{5}a + 11\right) = 120 \cdot a$$

$$40a + 30a + 15a + 24a + 1320 = 120a$$

$$109a + 1320 = 120a$$

$$1320 = 11a$$

$$120 = a$$

Check. If the original number of apples was 120, then the first four people got $\frac{1}{3} \cdot 120$, $\frac{1}{4} \cdot 120$, $\frac{1}{8} \cdot 120$, and $\frac{1}{5} \cdot 120$, or 40, 30, 15, and 24 apples, respectively. Adding all the apples we get 40 + 30 + 15 + 24 + 10 + 1, or 120. The result checks.

State. There were originally 120 apples in the basket.

45. Divide the largest triangle into three triangles, each with a vertex at the center of the circle and with height x as shown.

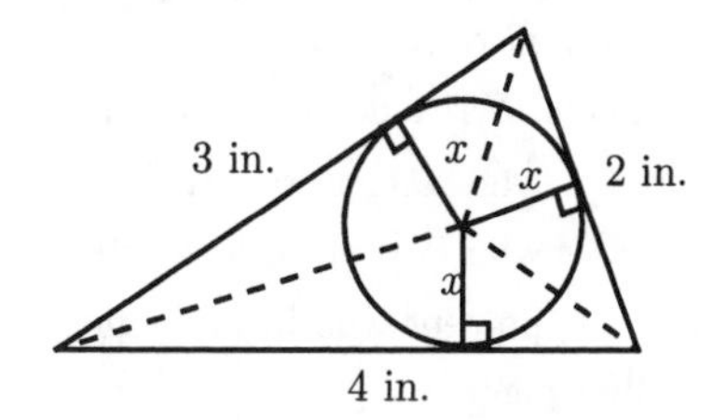

Then the sum of the areas of the three smaller triangles is the area of the original triangle. We have:

$$\frac{1}{2} \cdot 3x + \frac{1}{2} \cdot 2x + \frac{1}{2} \cdot 4x = 2.9047$$

$$2\left(\frac{1}{2} \cdot 3x + \frac{1}{2} \cdot 2x + \frac{1}{2} \cdot 4x\right) = 2(2.9047)$$

$$3x + 2x + 4x = 5.8094$$

$$9x = 5.8094$$

$$x \approx 0.65$$

Thus, x is about 0.65 in.

Exercise Set 8.5

1. ***Familiarize.*** Let x = the percent.

Translate.

$$\underbrace{\text{What percent}}_{x} \ \underset{\cdot}{\text{of}} \ \underbrace{180}_{180} \ \underset{=}{\text{is}} \ \underbrace{36?}_{36}$$

Solve. We solve the equation.

$$x \cdot 180 = 36$$

$$\frac{x \cdot 180}{180} = \frac{36}{180}$$

$$x = \frac{1}{5}$$

$$x = 20\%$$

Check. We check by finding 20% of 180:

$$20\% \cdot 180 = 0.2 \cdot 180 = 36$$

State. The answer is 20%.

3. ***Familiarize.*** Let y = the number we are taking 30% of.

Translate.

45 is 30% of what?

↓ ↓ ↓ ↓ ↓

$45 = 30\% \cdot y$

Solve. We solve the equation.

$$45 = 30\% \cdot y$$

$$45 = 0.3y \qquad \text{Converting to decimal notation}$$

$$\frac{45}{0.3} = \frac{y}{0.3}$$

$$150 = y$$

Check. We find 30% of 150:

$$30\% \cdot 150 = 0.3 \times 150 = 45$$

State. The answer is 150.

5. ***Familiarize.*** Let y = the unknown number.

Translate.

What is 65% of 840?

↓ ↓ ↓ ↓ ↓

$y = 65\% \cdot 840$

Solve. We solve the equation.

$$y = 65\% \cdot 840$$

$$y = 0.65 \times 840$$

$$y = 546$$

Check. The check is the computation we used to solve the equation:

$$65\% \cdot 840 = 0.65 \times 840 = 546$$

State. The answer is 546.

7. ***Familiarize.*** Let y = the percent.

Translate.

$$\underset{30}{30} \ \underset{=}{\text{is}} \ \underbrace{\text{what percent}}_{y} \ \underset{\cdot}{\text{of}} \ \underset{125}{125?}$$

Solve. We solve the equation.

$$30 = y \cdot 125$$

$$\frac{30}{125} = \frac{y \cdot 125}{125}$$

$$0.24 = y$$

$$24\% = y$$

Check. We find 24% of 125:

$$24\% \cdot 125 = 0.24 \cdot 125 = 30$$

State. The answer is 24%.

9. ***Familiarize.*** Let $y =$ the number we are taking 12% of.

Translate.

12% of what number is 0.3?

$$12\% \cdot y = 0.3$$

Solve. We solve the equation.

$$12\% \cdot y = 0.3$$
$$0.12y = 0.3 \quad \text{Converting to decimal notation}$$
$$\frac{y}{0.12} = \frac{0.3}{0.12}$$
$$y = 2.5$$

Check. We find 12% of 2.5:

$$12\% \cdot 2.5 = 0.12(2.5) = 0.3$$

State. The answer is 2.5.

11. ***Familiarize.*** Let $y =$ the percent.

Translate.

2 is what percent of 40?

$$2 = y\% \cdot 40$$

Solve. We solve the equation.

$$2 = y\% \cdot 40$$
$$2 = y \times 0.01 \times 40$$
$$2 = y(0.4)$$
$$\frac{2}{0.4} = \frac{y(0.4)}{0.4}$$
$$5 = y$$

Check. We find 5% of 40:

$$5\% \cdot 40 = 0.05 \times 40 = 2$$

State. The answer is 5%.

13. ***Familiarize.*** Write down the information.

Total sales: \$39 billion

Percent of McDonald's sales: 43%

Let $t =$ the total hamburger sales, in billions of dollars, by McDonald's.

Translate. We reword the problem.

What is 43% of \$39 billion?

$$t = 43\% \cdot 39$$

Solve. We solve the equation.

$$t = 43\% \cdot 39$$
$$t = 0.43 \times 39$$
$$t = 16.77$$

Check. The check is the computation we used to solve the equation.

$$43\% \cdot 39 = 0.43 \times 39 = 16.77$$

The answer checks.

State. McDonald's total hamburger sales were \$16.77 billion.

15. a) ***Familiarize.*** Write down the information.

Number of brochures mailed: 10,500

Percent opened: 78%

Let $n =$ the number of brochures that can be expected to be opened and read.

Translate. We reword the problem.

What is 78% of 10,500?

$$n = 78\% \cdot 10{,}500$$

Solve. We solve the equation.

$$n = 78\% \cdot 10{,}500$$
$$n = 0.78 \times 10{,}500$$
$$n = 8190$$

Check. The check is the computation we used to solve the equation.

$$78\% \cdot 10{,}500 = 0.78 \times 10{,}500 = 8190$$

The answer checks.

State. The business can expect 8190 brochures to be opened and read.

b) ***Familiarize.*** Write down the information.

Number who receive the brochure: 10,500

Number who buy the video: 189

Let $x =$ the percent who buy the video.

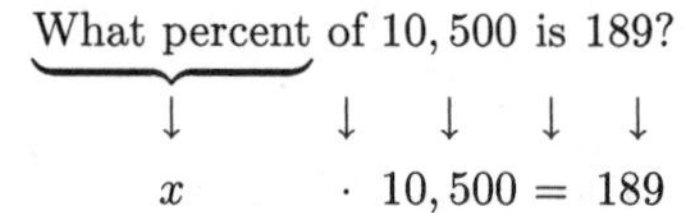

Solve. We solve the equation.

$$x \cdot 10{,}500 = 189$$
$$\frac{x \cdot 10{,}500}{10{,}500} = \frac{189}{10{,}500}$$
$$x = 0.018$$
$$x = 1.8\%$$

Check. We find 1.8% of 10,500:

$$1.8\% \times 10{,}500 = 0.018 \times 10{,}500 = 189$$

The answer checks.

State. Of the 10,500 people who receive the brochure, 1.8% buy the video.

17. a) ***Familiarize.*** Write down the information.

Amount of tip: \$4

Cost of meal: \$25

Let $x =$ the percent of the meal's cost that the tip represents.

Translate. We reword the problem.

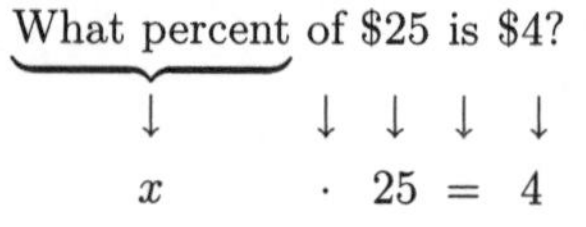

Solve. We solve the equation.

$$x \cdot 25 = 4$$
$$\frac{x \cdot 25}{25} = \frac{4}{25}$$
$$x = 0.16$$
$$x = 16\%$$

Check. We find 16% of \$25:

$$16\% \cdot \$25 = 0.16 \times \$25 = \$4$$

The answer checks.

State. The tip was 16% of the cost of the meal.

b) We add to find the total cost of the meal, including tip:

$$\$25 + \$4 = \$29$$

19. a) ***Familiarize***. Write down the information.

Percent tipped: 15%

Cost of meal: \$25

Let $t =$ the amount of the tip.

Translate. We reword the problem.

What is 15% of \$25?
$\downarrow \ \downarrow \ \downarrow \ \downarrow \ \downarrow$
$t = 15\% \cdot 25$

Solve. We solve the equation.

$$t = 15\% \cdot 25$$
$$t = 0.15 \times 25$$
$$t = 3.75$$

Check. The check is the computation we used to solve the equation.

$$15\% \cdot 25 = 0.15 \times 25 = 3.75$$

The answer checks.

State. The tip was \$3.75.

b) We add to find the total cost of the meal, including tip:

$$\$25 + \$3.75 = \$28.75$$

21. a) ***Familiarize***. Write down the information.

Percent of tip: 15%

Amount of tip: \$4.32

Let $c =$ the cost of the meal before the tip.

Translate. We reword the problem.

15% of what is \$4.32?
$\downarrow \ \downarrow \ \downarrow \ \downarrow \ \downarrow$
$15\% \cdot c = 4.32$

Solve. We solve the equation.

$$15\% \cdot c = 4.32$$
$$0.15 \cdot c = 4.32$$
$$\frac{0.15 \cdot c}{0.15} = \frac{4.32}{0.15}$$
$$c = 28.8$$

Check. We find 15% of \$28.80.

$$15\% \cdot \$28.80 = 0.15 \times \$28.80 = \$4.32$$

The answer checks.

State. The cost of the meal before the tip was \$28.80.

b) We add to find the total cost of the meal, including tip:

$$\$28.80 + \$4.32 = \$33.12$$

23. ***Familiarize***. Let $c =$ the cost of the meal before the tip. We know that the cost of the meal before the tip plus the tip, 15% of the cost, is the total cost, \$41.40.

Translate.

$\underbrace{\text{Cost of meal}}$ plus tip is \$41.40
$\downarrow \quad \downarrow \ \downarrow \ \downarrow \ \downarrow$
$c + 15\%c = 41.40$

Solve. We solve the equation.

$$c + 15\%c = 41.40$$
$$c + 0.15c = 41.40$$
$$1c + 0.15c = 41.40$$
$$1.15c = 41.40$$
$$\frac{1.15c}{1.15} = \frac{41.40}{1.15}$$
$$c = 36$$

Check. We find 15% of \$36 and add it to \$36:

$15\% \times \$36 = 0.15 \times \$36 = \$5.40$ and $\$36 + \$5.40 = \$41.40$.

The answer checks.

State. The cost of the meal before the tip was added was \$36.

25. ***Familiarize***. Write down the information.

Number of women who had babies in good or excellent health: 16

Percent of women who had babies in good or excellent health: 8%

Let $w =$ the number of women in the original study.

Translate. We reword the problem.

8% of what is 16?
$\downarrow \ \downarrow \ \downarrow \ \downarrow \ \downarrow$
$8\% \cdot w = 16$

Solve. We solve the equation.

$$8\% \cdot w = 16$$
$$0.08 \cdot w = 16$$
$$\frac{0.08 \cdot w}{0.08} = \frac{16}{0.08}$$
$$w = 200$$

Check. We find 8% of 200:

$$8\% \cdot 200 = 0.08 \cdot 200 = 16$$

The answer checks.

State. There were 200 women in the original study.

27. ***Familiarize***. Write down the information.

Nonsmoker's premium: \$166 per year

Smoker's premium: 170% of nonsmoker's premium

Let $p =$ the premium for a smoker.

Translate. We reword the problem.

What is 170% of \$166?

$\downarrow \quad \downarrow \quad \downarrow \quad \downarrow \quad \downarrow$

$p \quad = 170\% \quad \cdot \quad 166$

Solve. We solve the equation.

$$p = 170\% \cdot 166$$
$$p = 1.7 \cdot 166$$
$$p = 282.2$$

Check. The check is the computation we used to solve the equation:

$$170\% \cdot 166 = 1.7 \cdot 166 = 282.2$$

State. The premium for a smoker is \$282.20 per year.

29. ***Familiarize***. Write down the information.

Fat percentage: 19.8%

Body weight: 214 lb

Let $p =$ the part of the body weight, in pounds, that is fat.

Translate. We reword the problem.

What is 19.8% of 214?

$\downarrow \quad \downarrow \quad \downarrow \quad \downarrow \quad \downarrow$

$p \quad = 19.8\% \quad \cdot \quad 214$

Solve. We solve the equation.

$$p = 19.8\% \cdot 214$$
$$p = 0.198 \cdot 214$$
$$p = 42.372 \approx 42.4$$

Check. The check is the computation we used to solve the equation:

$$19.8\% \cdot 214 = 0.198 \cdot 214 = 42.372 \approx 42.4$$

State. About 42.4 lb of the author's body weight is fat.

31. ***Familiarize***. Write down the information.

Total number of lightning strikes: 3327

Percent killed under trees: 17%

Let $x =$ the number of deaths that occurred under trees.

Translate. We reword the problem.

What is 17% of 3327?

$\downarrow \quad \downarrow \quad \downarrow \quad \downarrow \quad \downarrow$

$x \quad = 17\% \quad \cdot \quad 3327$

Solve. We solve the equation.

$$x = 17\% \cdot 3327$$
$$x = 0.17 \cdot 3327$$
$$x = 565.59 \approx 566$$

Check. The check is the computation we used to solve the equation.

$$17\% \cdot 3327 = 0.17 \cdot 3327 = 565.59 \approx 566$$

State. About 566 deaths occurred under trees.

33. ***Familiarize***. Let $p =$ the average price of a ticket before the increase. The price before the increase plus 15% of that price is the new average price, \$17.69.

Translate.

$$\underbrace{\text{Average price before increase}}_{\downarrow \; p} \quad \underbrace{\text{plus}}_{\downarrow \; +} \quad \underbrace{\text{amount of increase}}_{\downarrow \; 15\%p} \quad \underbrace{\text{is}}_{\downarrow \; =} \quad \underbrace{\$17.69}_{\downarrow \; 17.69}$$

Solve. We solve the equation.

$$p + 15\%p = 17.69$$
$$p + 0.15p = 17.69$$
$$1p + 0.15p = 17.69$$
$$1.15p = 17.69$$
$$\frac{1.15p}{1.15} = \frac{17.69}{1.15}$$
$$p \approx 15.38$$

Check.We find 15% of \$15.38 and add it to \$15.38:

$15\% \times \$15.38 = 0.15 \times \$15.38 \approx \$2.31$ and $\$15.38 + \$2.31 = \$17.69$.

The answer checks.

State. The average price the preceding season was about \$15.38.

35. ***Familiarize***. Recall that the formula for simple interest is $I = Prt$, where $I =$ the interest, $P =$ the principal, or the amount invested, $r =$ the interest rate, and $t =$ the time the principal is invested. Let $a =$ the amount originally invested. We know that the amount in the account at the end of the year is the principal a plus the interest, 6% of the principal.

$$\underbrace{\text{Principal}}_{\downarrow \; a} \quad \underbrace{\text{plus}}_{\downarrow \; +} \quad \underbrace{\text{interest}}_{\downarrow \; 6\%a} \quad \underbrace{\text{is}}_{\downarrow \; =} \quad \underbrace{\text{total amount}}_{\downarrow \; 8268}$$

Solve. We solve the equation.

$$a + 6\%a = 8268$$
$$a + 0.06a = 8268$$
$$1a + 0.06a = 8268$$
$$1.06a = 8268$$
$$\frac{1.06a}{1.06} = \frac{8268}{1.06}$$
$$a = 7800$$

Check. We find 6% of \$7800 and add it to \$7800:

$6\% \cdot \$7800 = 0.06 \cdot \$7800 = \$468$ and $\$7800 + \$468 = \$8268$.

The answer checks.

State. \$7800 was originally invested.

37. ***Familiarize***. Let $p =$ the original price. Then $40\%p$ is the amount of the reduction.

Translate. We reword the problem.

$$\underbrace{\text{Original price}}_{p} \; - \; \underbrace{\text{reduction}}_{40\%p} \; \text{is} \; \underbrace{\text{sale price}}_{34.80}$$

$$p \quad - \quad 40\%p \quad = \quad 34.80$$

Solve. We solve the equation.

$$p - 40\%p = 34.80$$
$$p - 0.4p = 34.80$$
$$1p - 0.4p = 34.80$$
$$0.6p = 34.80$$
$$\frac{0.6p}{0.6} = \frac{34.80}{0.6}$$
$$p = 58$$

Check. We find 40% of \$58 and subtract it from \$58:

$$40\% \times \$58 = 0.4 \times \$58 = \$23.20$$

Then \$58.00 − \$23.20 = \$34.80, so \$58 checks.

State. The original price was \$58.

39.

```
           1 8 1 . 5 2
0.0 5^⟌ 9.0 7^6 0
         5
         4 0
         4 0
           7
           5
           2 6
           2 5
             1 0
             1 0
               0
```

The answer is 181.52.

41.

```
     1 1 1
     1. 0 8 9 0
   1 0. 8 9 0 0
 +   0. 1 0 8 9
   1 2. 0 8 7 9
```

43. Substitute 58 for x and 42 for y and carry out the subtraction.

$$x - y = 58 - 42 = 16$$

45. Substitute 25 for a and 15 for b and carry out the computation.

$$\frac{6a}{b} = \frac{6 \cdot 25}{15} = \frac{150}{15} = 10$$

47. ◈

49. ***Familiarize.*** Write down the information. Since 6 ft = 6×1 ft = 6×12 in. = 72 in., we can express 6 ft, 4 in. as 72 in. + 4 in., or 76 in.

Percent of adult height: 96.1%

Height at age 15: 76 in.

Let h = Jaraan's final adult height.

Translate. We reword the problem.

96.1% of what is $\underbrace{\text{76 in.}}$?

$$96.1\% \quad \cdot \quad h \quad = \quad 76$$

Solve. We solve the equation.

$$96.1\% \cdot h = 76$$
$$0.961 \cdot h = 76$$
$$\frac{0.961 \cdot h}{0.961} = \frac{76}{0.961}$$
$$h \approx 79$$

Note that 79 in. = 72 in. + 7 in. = 6 ft, 7 in.

Check. We find 96.1% of 79:

$$96.1\% \cdot 79 = 0.961 \cdot 79 \approx 76$$

The answer checks.

State. Jaraan's final adult height will be about 6 ft, 7 in.

51. ***Familiarize.*** Let p = the price of the gasoline as registered on the pump. Then the sales tax will be $9\%p$.

Translate. We reword the problem.

$$\underbrace{\text{Price on pump}}_{p} \; \text{plus} \; \underbrace{\text{sales tax}}_{9\%p} \; \text{is} \; \underbrace{\$10}_{10}$$

$$p \quad + \quad 9\%p \quad = \quad 10$$

Solve. We solve the equation.

$$p + 9\%p = 10$$
$$1p + 0.09p = 10$$
$$1.09p = 10$$
$$\frac{1.09p}{1.09} = \frac{10}{1.09}$$
$$p \approx 9.17$$

Check. We find 9% of \$9.17 and add it to \$9.17:

$$9\% \times \$9.17 = 0.09 \times \$9.17 \approx \$0.83$$

Then \$9.17 + \$0.83 = \$10, so \$9.17 checks.

State. The attendant should have filled the tank until the pump read \$9.17, not \$9.10.

Exercise Set 8.6

1.
$$A = bh$$
$$\frac{A}{b} = \frac{bh}{b} \quad \text{Dividing by } b$$
$$\frac{A}{b} = h$$

3.
$$P = 2l + 2w$$
$$P - 2l = 2l + 2w - 2l \quad \text{Subtracting } 2l$$
$$P - 2l = 2w$$
$$\frac{P - 2l}{2} = \frac{2w}{2} \quad \text{Dividing by 2}$$
$$\frac{P - 2l}{2} = w, \text{ or}$$
$$\frac{1}{2}P - l = w$$

5.
$$A = \frac{a + b}{2}$$
$$2A = a + b \quad \text{Multiplying by 2}$$
$$2A - b = a \quad \text{Subtracting } b$$

7. $F = ma$

$\frac{F}{m} = \frac{ma}{m}$ Dividing by m

$\frac{F}{m} = a$

9. $E = mc^2$

$\frac{E}{m} = \frac{mc^2}{m}$ Dividing by m

$\frac{E}{m} = c^2$

11. $Ax + By = c$

$Ax = c - By$ Subtracting By

$\frac{Ax}{A} = \frac{c - By}{A}$ Dividing by A

$x = \frac{c - By}{A}$

13. $v = \frac{3k}{t}$

$tv = t \cdot \frac{3k}{t}$ Multiplying by t

$tv = 3k$

$\frac{tv}{v} = \frac{3k}{v}$ Dividing by v

$t = \frac{3k}{v}$

15. a) We substitute 1900 for a and calculate b.

$b = 30a = 30 \cdot 1900 = 57,000$

The minimum furnace output is 57,000 Btu's.

b) $b = 30a$

$\frac{b}{30} = \frac{30a}{30}$ Dividing by 30

$\frac{b}{30} = a$

17. a) We substitute 21,345 for n and calculate F.

$F = \frac{n}{15} = \frac{21,345}{15} = 1423$

The number of full-time-equivalent students is 1423.

b) $F = \frac{n}{15}$

$15F = n$ Multiplying by 15

19. a) We substitute 120 for w, 67 for h, and 23 for a and calculate K.

$K = 917 + 6(w + h - a)$

$K = 917 + 6(120 + 67 - 23)$

$K = 917 + 6(164)$

$K = 917 + 984$

$K = 1901$ calories

b) Solve for a:

$K = 917 + 6(w + h - a)$

$K = 917 + 6w + 6h - 6a$

$K + 6a = 917 + 6w + 6h$

$6a = 917 + 6w + 6h - K$

$a = \frac{917 + 6w + 6h - K}{6}$

Solve for h:

$K = 917 + 6(w + h - a)$

$K = 917 + 6w + 6h - 6a$

$K - 917 - 6w + 6a = 6h$

$\frac{K - 917 - 6w + 6a}{6} = h$

Solve for w:

$K = 917 + 6(w + h - a)$

$K = 917 + 6w + 6h - 6a$

$K - 917 - 6h + 6a = 6w$

$\frac{K - 917 - 6h + 6a}{6} = w$

21. We divide:

$$\begin{array}{r} 0.9\,2 \\ 2\,5 \overline{)\,2\,3.0\,0} \\ 2\,2\,5 \\ \hline 5\,0 \\ 5\,0 \\ \hline 0 \end{array}$$

Decimal notation for $\frac{23}{25}$ is 0.92.

23. $-45.8 - (-32.6) = -45.8 + 32.6 = -13.2$

25. $-\frac{2}{3} + \frac{5}{6} = -\frac{2}{3} \cdot \frac{2}{2} + \frac{5}{6}$

$= -\frac{4}{6} + \frac{5}{6}$

$= \frac{1}{6}$

27. ◈

29. $A = \frac{1}{2}ah + \frac{1}{2}bh$

$2A = 2\left(\frac{1}{2}ah + \frac{1}{2}bh\right)$ Clearing the fractions

$2A = ah + bh$

$2A - ah = bh$ Subtracting ah

$\frac{2A - ah}{h} = b$ Dividing by h

$A = \frac{1}{2}ah + \frac{1}{2}bh$

$2A = ah + bh$ Clearing fractions as above

$2A = h(a + b)$ Factoring

$\frac{2A}{a + b} = h$ Dividing by $a + b$

31. $A = lw$

When l and w both double, we have

$$2l \cdot 2w = 4lw = 4A,$$

so A quadruples.

33. $A = \frac{1}{2}bh$

When b increases by 4 units we have

$$\frac{1}{2}(b+4)h = \frac{1}{2}bh + 2h = A + 2h,$$

so A increases by $2h$ units.

Exercise Set 8.7

1. $x > -4$

a) Since $4 > -4$ is true, 4 is a solution.

b) Since $0 > -4$ is true, 0 is a solution.

c) Since $-4 > -4$ is false, -4 is not a solution.

d) Since $6 > -4$ is true, 6 is a solution.

e) Since $5.6 > -4$ is true, 5.6 is a solution.

3. $x \geq 6.8$

a) Since $-6 \geq 6.8$ is false, -6 is not a solution.

b) Since $0 \geq 6.8$ is false, 0 is not a solution.

c) Since $6 \geq 6.8$ is false, 6 is not a solution.

d) Since $8 \geq 6.8$ is true, 8 is a solution.

e) Since $-3\frac{1}{2} \geq 6.8$ is false, $-3\frac{1}{2}$ is not a solution.

5. The solutions of $x > 4$ are those numbers greater than 4. They are shown on the graph by shading all points to the right of 4. The open circle at 4 indicates that 4 is not part of the graph.

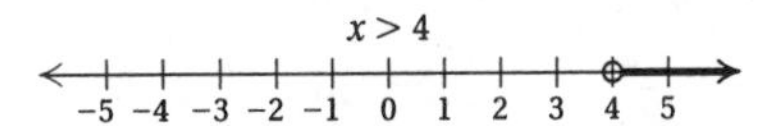

7. The solutions of $t < -3$ are those numbers less than -3. They are shown on the graph by shading all points to the left of -3. The open circle at -3 indicates that -3 is not part of the graph.

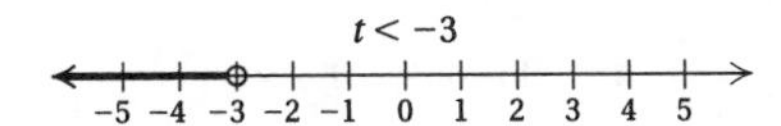

9. The solutions of $m \geq -1$ are are shown by shading the point for -1 and all points to the right of -1. The closed circle at -1 indicates that -1 is part of the graph.

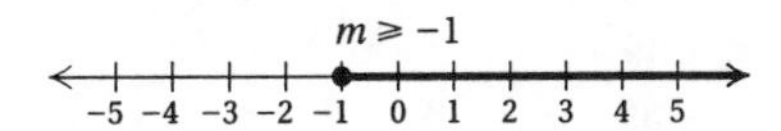

11. In order to be a solution of the inequality $-3 < x \leq 4$, a number must be a solution of both $-3 < x$ and $x \leq 4$. The solution set is graphed as follows:

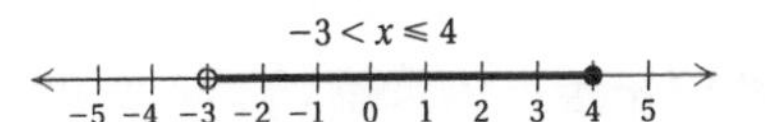

The open circle at -3 means that -3 is not part of the graph. The closed circle at 4 means that 4 is part of the graph.

13. In order to be a solution of the inequality $0 < x < 3$, a number must be a solution of both $0 < x$ and $x < 3$. The solution set is graphed as follows:

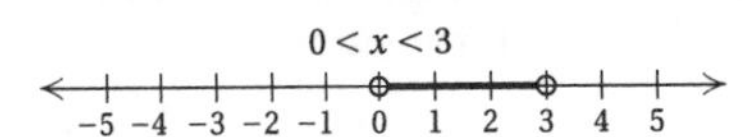

The open circles at 0 and at 3 mean that 0 and 3 are not part of the graph.

15.

$$\begin{aligned} x + 7 &> 2 \\ x + 7 - 7 &> 2 - 7 \quad \text{Subtracting 7} \\ x &> -5 \quad \text{Simplifying} \end{aligned}$$

The solution set is $\{x|x > -5\}$.

The graph is as follows:

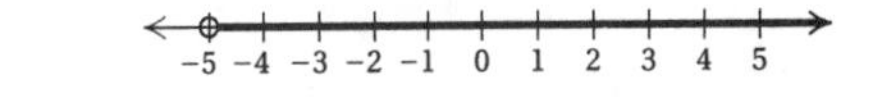

17.

$$\begin{aligned} x + 8 &\leq -10 \\ x + 8 - 8 &\leq -10 - 8 \quad \text{Subtracting 8} \\ x &\leq -18 \quad \text{Simplifying} \end{aligned}$$

The solution set is $\{x|x \leq -18\}$.

The graph is as follows:

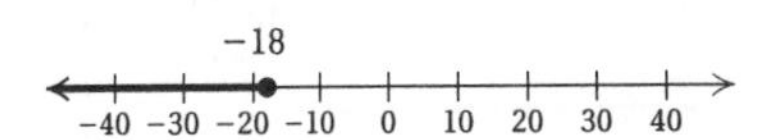

19.

$$\begin{aligned} y - 7 &> -12 \\ y - 7 + 7 &> -12 + 7 \quad \text{Adding 7} \\ y &> -5 \quad \text{Simplifying} \end{aligned}$$

The solution set is $\{y|y > -5\}$.

21.

$$\begin{aligned} 2x + 3 &> x + 5 \\ 2x + 3 - 3 &> x + 5 - 3 \quad \text{Subtracting 3} \\ 2x &> x + 2 \quad \text{Simplifying} \\ 2x - x &> x + 2 - x \quad \text{Subtracting } x \\ x &> 2 \quad \text{Simplifying} \end{aligned}$$

The solution set is $\{x|x > 2\}$.

23.

$$\begin{aligned} 3x + 9 &\leq 2x + 6 \\ 3x + 9 - 9 &\leq 2x + 6 - 9 \quad \text{Subtracting 9} \\ 3x &\leq 2x - 3 \quad \text{Simplifying} \\ 3x - 2x &\leq 2x - 3 - 2x \quad \text{Subtracting } 2x \\ x &\leq -3 \quad \text{Simplifying} \end{aligned}$$

The solution set is $\{x|x \leq -3\}$.

25.

$$\begin{aligned} 5x - 6 &< 4x - 2 \\ 5x - 6 + 6 &< 4x - 2 + 6 \\ 5x &< 4x + 4 \\ 5x - 4x &< 4x + 4 - 4x \\ x &< 4 \end{aligned}$$

The solution set is $\{x|x < 4\}$.

27.
$$\begin{aligned} -9+t &> 5 \\ -9+t+9 &> 5+9 \\ t &> 14 \end{aligned}$$

The solution set is $\{t|t > 14\}$.

29.
$$\begin{aligned} y+\frac{1}{4} &\leq \frac{1}{2} \\ y+\frac{1}{4}-\frac{1}{4} &\leq \frac{1}{2}-\frac{1}{4} \\ y &\leq \frac{2}{4}-\frac{1}{4} \quad \text{Obtaining a common denominator} \\ y &\leq \frac{1}{4} \end{aligned}$$

The solution set is $\left\{y\middle|y \leq \frac{1}{4}\right\}$.

31.
$$\begin{aligned} x-\frac{1}{3} &> \frac{1}{4} \\ x-\frac{1}{3}+\frac{1}{3} &> \frac{1}{4}+\frac{1}{3} \\ x &> \frac{3}{12}+\frac{4}{12} \quad \text{Obtaining a common denominator} \\ x &> \frac{7}{12} \end{aligned}$$

The solution set is $\left\{x\middle|x > \frac{7}{12}\right\}$.

33.
$$\begin{aligned} 5x &< 35 \\ \frac{5x}{5} &< \frac{35}{5} \quad \text{Dividing by 5} \\ x &< 7 \end{aligned}$$

The solution set is $\{x|x < 7\}$. The graph is as follows:

−8 −6 −4 −2 0 2 4 6 8

35.
$$\begin{aligned} -12x &> -36 \\ \frac{-12x}{-12} &< \frac{-36}{-12} \quad \text{Dividing by } -12 \end{aligned}$$

↑ The symbol has to be reversed.

$$x < 3 \quad \text{Simplifying}$$

The solution set is $\{x|x < 3\}$. The graph is as follows:

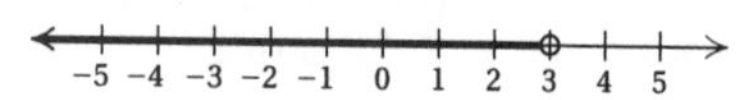

37.
$$\begin{aligned} 5y &\geq -2 \\ \frac{5y}{5} &\geq \frac{-2}{5} \quad \text{Dividing by 5} \\ y &\geq -\frac{2}{5} \end{aligned}$$

The solution set is $\left\{y\middle|y \geq -\frac{2}{5}\right\}$.

39.
$$\begin{aligned} -2x &\leq 12 \\ \frac{-2x}{-2} &\geq \frac{12}{-2} \quad \text{Dividing by } -2 \end{aligned}$$

↑ The symbol has to be reversed.

$$x \geq -6 \quad \text{Simplifying}$$

The solution set is $\{x|x \geq -6\}$.

41.
$$\begin{aligned} -4y &\geq -16 \\ \frac{-4y}{-4} &\leq \frac{-16}{-4} \quad \text{Dividing by } -4 \end{aligned}$$

↑ The symbol has to be reversed.

$$y \leq 4 \quad \text{Simplifying}$$

The solution set is $\{y|y \leq 4\}$.

43.
$$\begin{aligned} -3x &< -17 \\ \frac{-3x}{-3} &> \frac{-17}{-3} \quad \text{Dividing by } -3 \end{aligned}$$

↑ The symbol has to be reversed.

$$x > \frac{17}{3} \quad \text{Simplifying}$$

The solution set is $\left\{x\middle|x > \frac{17}{3}\right\}$.

45.
$$\begin{aligned} -2y &> \frac{1}{7} \\ -\frac{1}{2}\cdot(-2y) &< -\frac{1}{2}\cdot\frac{1}{7} \end{aligned}$$

↑ The symbol has to be reversed.

$$y < -\frac{1}{14}$$

The solution set is $\left\{y\middle|y < -\frac{1}{14}\right\}$.

47.
$$\begin{aligned} -\frac{6}{5} &\leq -4x \\ -\frac{1}{4}\cdot\left(-\frac{6}{5}\right) &\geq -\frac{1}{4}\cdot(-4x) \\ \frac{6}{20} &\geq x \\ \frac{3}{10} &\geq x, \text{ or } x \leq \frac{3}{10} \end{aligned}$$

The solution set is $\left\{x\middle|\frac{3}{10} \geq x\right\}$, or $\left\{x\middle|x \leq \frac{3}{10}\right\}$.

49.
$$\begin{aligned} 4+3x &< 28 \\ -4+4+3x &< -4+28 \quad \text{Adding } -4 \\ 3x &< 24 \quad \text{Simplifying} \\ \frac{3x}{3} &< \frac{24}{3} \quad \text{Dividing by 3} \\ x &< 8 \end{aligned}$$

The solution set is $\{x|x < 8\}$.

51.
$$\begin{aligned} 3x-5 &\leq 13 \\ 3x-5+5 &\leq 13+5 \quad \text{Adding 5} \\ 3x &\leq 18 \\ \frac{3x}{3} &\leq \frac{18}{3} \quad \text{Dividing by 3} \\ x &\leq 6 \end{aligned}$$

The solution set is $\{x|x \leq 6\}$.

53.
$$\begin{aligned} 13x-7 &< -46 \\ 13x-7+7 &< -46+7 \\ 13x &< -39 \\ \frac{13x}{13} &< \frac{-39}{13} \\ x &< -3 \end{aligned}$$

The solution set is $\{x|x < -3\}$.

55.
$$30 > 3 - 9x$$
$$30 - 3 > 3 - 9x - 3 \quad \text{Subtracting 3}$$
$$27 > -9x$$
$$\frac{27}{-9} < \frac{-9x}{-9} \quad \text{Dividing by } -9$$
The symbol has to be reversed.
$$-3 < x$$
The solution set is $\{x|-3 < x\}$, or $\{x|x > -3\}$.

57.
$$4x + 2 - 3x \leq 9$$
$$x + 2 \leq 9 \quad \text{Collecting like terms}$$
$$x + 2 - 2 \leq 9 - 2$$
$$x \leq 7$$
The solution set is $\{x|x \leq 7\}$.

59.
$$-3 < 8x + 7 - 7x$$
$$-3 < x + 7 \quad \text{Collecting like terms}$$
$$-3 - 7 < x + 7 - 7$$
$$-10 < x$$
The solution set is $\{x|-10 < x\}$, or $\{x|x > -10\}$.

61.
$$6 - 4y > 4 - 3y$$
$$6 - 4y + 4y > 4 - 3y + 4y \quad \text{Adding } 4y$$
$$6 > 4 + y$$
$$-4 + 6 > -4 + 4 + y \quad \text{Adding } -4$$
$$2 > y, \text{ or } y < 2$$
The solution set is $\{y|2 > y\}$, or $\{y|y < 2\}$.

63.
$$5 - 9y \leq 2 - 8y$$
$$5 - 9y + 9y \leq 2 - 8y + 9y$$
$$5 \leq 2 + y$$
$$-2 + 5 \leq -2 + 2 + y$$
$$3 \leq y, \text{ or } y \geq 3$$
The solution set is $\{y|3 \leq y\}$, or $\{y|y \geq 3\}$.

65.
$$19 - 7y - 3y < 39$$
$$19 - 10y < 39 \quad \text{Collecting like terms}$$
$$-19 + 19 - 10y < -19 + 39$$
$$-10y < 20$$
$$\frac{-10y}{-10} > \frac{20}{-10}$$
The symbol has to be reversed.
$$y > -2$$
The solution set is $\{y|y > -2\}$.

67.
$$2.1x + 45.2 > 3.2 - 8.4x$$
$$10(2.1x + 45.2) > 10(3.2 - 8.4x) \quad \text{Multiplying by 10 to clear decimals}$$
$$21x + 452 > 32 - 84x$$
$$21x + 84x > 32 - 452 \quad \text{Adding } 84x \text{ and subtracting 452}$$
$$105x > -420$$
$$x > -4 \quad \text{Dividing by 105}$$
The solution set is $\{x|x > -4\}$.

69.
$$\frac{x}{3} - 2 \leq 1$$
$$3\left(\frac{x}{3} - 2\right) \leq 3 \cdot 1 \quad \text{Multiplying by 3 to to clear the fraction}$$
$$x - 6 \leq 3 \quad \text{Simplifying}$$
$$x \leq 9 \quad \text{Adding 6}$$
The solution set is $\{x|x \leq 9\}$.

71.
$$\frac{y}{5} + 1 \leq \frac{2}{5}$$
$$5\left(\frac{y}{5} + 1\right) \leq 5 \cdot \frac{2}{5} \quad \text{Clearing fractions}$$
$$y + 5 \leq 2$$
$$y \leq -3 \quad \text{Subtracting 5}$$
The solution set is $\{y|y \leq -3\}$.

73.
$$3(2y - 3) < 27$$
$$6y - 9 < 27 \quad \text{Removing parentheses}$$
$$6y < 36 \quad \text{Adding 9}$$
$$y < 6 \quad \text{Dividing by 6}$$
The solution set is $\{y|y < 6\}$.

75.
$$2(3 + 4m) - 9 \geq 45$$
$$6 + 8m - 9 \geq 45 \quad \text{Removing parentheses}$$
$$8m - 3 \geq 45 \quad \text{Collecting like terms}$$
$$8m \geq 48 \quad \text{Adding 3}$$
$$m \geq 6 \quad \text{Dividing by 8}$$
The solution set is $\{m|m \geq 6\}$.

77.
$$8(2t + 1) > 4(7t + 7)$$
$$16t + 8 > 28t + 28$$
$$16t - 28t > 28 - 8$$
$$-12t > 20$$
$$t < -\frac{20}{12} \quad \text{Dividing by } -12 \text{ and reversing the symbol}$$
$$t < -\frac{5}{3}$$
The solution set is $\{t|t < -\frac{5}{3}\}$.

79.
$$3(r - 6) + 2 < 4(r + 2) - 21$$
$$3r - 18 + 2 < 4r + 8 - 21$$
$$3r - 16 < 4r - 13$$
$$-16 + 13 < 4r - 3r$$
$$-3 < r, \text{ or } r > -3$$
The solution set is $\{r|r > -3\}$.

81.
$$0.8(3x + 6) \geq 1.1 - (x + 2)$$
$$2.4x + 4.8 \geq 1.1 - x - 2$$
$$10(2.4x + 4.8) \geq 10(1.1 - x - 2) \quad \text{Clearing decimals}$$
$$24x + 48 \geq 11 - 10x - 20$$
$$24x + 48 \geq -10x - 9 \quad \text{Collecting like terms}$$
$$24x + 10x \geq -9 - 48$$
$$34x \geq -57$$
$$x \geq -\frac{57}{34}$$
The solution set is $\left\{x\middle|x \geq -\frac{57}{34}\right\}$.

83. $\frac{5}{3}+\frac{2}{3}x<\frac{25}{12}+\frac{5}{4}x+\frac{3}{4}$

The number 12 is the least common multiple of all the denominators. We multiply by 12 on both sides.

$$12\left(\frac{5}{3}+\frac{2}{3}x\right)<12\left(\frac{25}{12}+\frac{5}{4}x+\frac{3}{4}\right)$$
$$12\cdot\frac{5}{3}+12\cdot\frac{2}{3}x<12\cdot\frac{25}{12}+12\cdot\frac{5}{4}x+12\cdot\frac{3}{4}$$
$$20+8x<25+15x+9$$
$$20+8x<34+15x$$
$$20-34<15x-8x$$
$$-14<7x$$
$$-2<x,\text{ or } x>-2$$

The solution set is $\{x|x>-2\}$.

85. $-56+(-18)$ Two negative numbers. Add the absolute values and make the answer negative.

$-56+(-18)=-74$

87. $-\frac{3}{4}+\frac{1}{8}$ One negative and one positive number. Find the difference of the absolute values. Then make the answer negative, since the negative number has the larger absolute value.

$-\frac{3}{4}+\frac{1}{8}=-\frac{6}{8}+\frac{1}{8}=-\frac{5}{8}$

89. $-56-(-18)=-56+18=-38$

91. $-2.3-7.1=-2.3+(-7.1)=-9.4$

93.
$$\begin{aligned}5-3^2+(8-2)^2\cdot 4&=5-3^2+6^2\cdot 4\\&=5-9+36\cdot 4\\&=5-9+144\\&=-4+144\\&=140\end{aligned}$$

95. $5(2x-4)-3(4x+1)=10x-20-12x-3=$
$-2x-23$

97. ◈

99. $|x|<3$

a) Since $|0|=0$ and $0<3$ is true, 0 is a solution.

b) Since $|-2|=2$ and $2<3$ is true, -2 is a solution.

c) Since $|-3|=3$ and $3<3$ is false, -3 is not a solution.

d) Since $|4|=4$ and $4<3$ is false, 4 is not a solution.

e) Since $|3|=3$ and $3<3$ is false, 3 is not a solution.

f) Since $|1.7|=1.7$ and $1.7<3$ is true, 1.7 is a solution.

g) Since $|-2.8|=2.8$ and $2.8<3$ is true, -2.8 is a solution.

101.
$x+3\le 3+x$
$x-x\le 3-3$ Subtracting x and 3
$0\le 0$

We get an inequality that is true for all values of x, so the inequality is true for all real numbers.

Exercise Set 8.8

1. $x>8$

3. $y\le -4$

5. $n\ge 1300$

7. $a\le 500$

9. $3x+2<13$, or $2+3x<13$

11. ***Familiarize.*** The average of the five scores is their sum divided by the number of quizzes, 5. We let s represent the student's score on the last quiz.

Translate. The average of the five scores is given by

$$\frac{73+75+89+91+s}{5}.$$

Since this average must be at least 85, this means that it must be greater than or equal to 85. Thus, we can translate the problem to the inequality

$$\frac{73+75+89+91+s}{5}\ge 85.$$

Solve. We first multiply by 5 to clear the fraction.

$$5\left(\frac{73+75+89+91+s}{5}\right)\ge 5\cdot 85$$
$$73+75+89+91+s\ge 425$$
$$328+s\ge 425$$
$$s\ge 425-328$$
$$s\ge 97$$

Check. Suppose s is a score greater than or equal to 97. Then by successively adding 73, 75, 89, and 91 on both sides of the inequality we get

$$73+75+89+91+s\ge 425$$

so

$$\frac{73+75+89+91+s}{5}\ge\frac{425}{5},\text{ or } 85.$$

State. Any score which is at least 97 will give an average quiz grade of 85. The solution set is $\{s|s\ge 97\}$.

13. ***Familiarize.*** $R=-0.075t+3.85$

In the formula R represents the world record and t represents the years since 1930. When $t=0$ (1930), the record was $-0.075\cdot 0+3.85$, or 3.85 minutes. When $t=2$ (1932), the record was $-0.075(2)+3.85$, or 3.7 minutes. For what values of t will $-0.075t+3.85$ be less than 3.5?

Translate. The record is to be less than 3.5. We have the inequality

$$R<3.5.$$

To find the t values which satisfy this condition we substitute $-0.075t+3.85$ for R.

$$-0.075t+3.85<3.5$$

Solve.

$$-0.075t+3.85<3.5$$
$$-0.075t<3.5-3.85$$
$$-0.075t<-0.35$$
$$t>\frac{-0.35}{-0.075}$$
$$t>4\frac{2}{3}$$

Check. With inequalities it is impossible to check each solution. But we can check to see if the solution set we obtained seems reasonable.

When $t = 4\frac{1}{2}$, $R = -0.075(4.5) + 3.85$, or 3.5125.

When $t = 4\frac{2}{3}$, $R = -0.075\left(\frac{14}{3}\right) + 3.85$, or 3.5.

When $t = 4\frac{3}{4}$, $R = -0.075(4.75) + 3.85$, or 3.49375.

Since $r = 3.5$ when $t = 4\frac{2}{3}$ and R decreases as t increases, R will be less than 3.5 when t is greater than $4\frac{2}{3}$.

State. The world record will be less than 3.5 minutes more than $4\frac{2}{3}$ years after 1930. If we let $Y =$ the year, then the solution set is $\{Y|Y \geq 1935\}$.

15. *Familiarize.* As in the drawing in the text, we let $L =$ the length of the envelope. Recall that the area of a rectangle is the product of the length and the width.

Translate.

Length	times	width	is at least	$17\frac{1}{2}$ in^2
↓	↓	↓	↓	↓
L	$\cdot$	$3\frac{1}{2}$	$\geq$	$17\frac{1}{2}$

Solve.

$$L \cdot 3\frac{1}{2} \geq 17\frac{1}{2}$$
$$L \cdot \frac{7}{2} \geq \frac{35}{2}$$
$$L \cdot \frac{7}{2} \cdot \frac{2}{7} \geq \frac{35}{2} \cdot \frac{2}{7}$$
$$L \geq 5$$

The solution set is $\{L|L \geq 5\}$.

Check. We can obtain a partial check by substituting a number greater than or equal to 5 in the inequality. For example, when $L = 6$:

$$L \cdot 3\frac{1}{2} = 6 \cdot 3\frac{1}{2} = 6 \cdot \frac{7}{2} = 21 \geq 17\frac{1}{2}$$

The result appears to be correct.

State. Lengths of 5 in. or more will satisfy the constraints. The solution set is $\{L|L \geq 5 \text{ in.}\}$.

17. *Familiarize.* Let n represent the number.

Translate.

The number	plus	15	is less than	4	times	the number.
↓	↓	↓	↓	↓	↓	↓
n	$+$	15	$<$	4	$\cdot$	n

Solve.

$$n + 15 < 4n$$
$$15 < 3n$$
$$5 < n, \text{ or } n > 5$$

Check. With inequalities it is impossible to check each solution. But we can check to see if the solution set we obtained seems reasonable.

When $n = 4$, we have $4 + 15 < 4 \cdot 4$, or $19 < 16$. This is false.

When $n = 5$, we have $5 + 15 < 4 \cdot 5$, or $20 < 20$. This is false.

When $n = 6$, we have $6 + 15 < 4 \cdot 6$, or $21 < 24$. This is true.

Since the inequality is false for the numbers less than or equal to 5 that we tried and true for the number greater than 5, it would appear that $n > 5$ is correct.

State. All numbers greater than 5 are solutions. The solution set is $\{n|n > 5\}$.

19. *Familiarize.* We let $d =$ the number of days after the calf's birth. Then, for the first few weeks, the calf gains $2d$ lb in d days.

Translate.

Birth weight	plus	weight gain	is more than	125 lb
↓	↓	↓	↓	↓
75	$+$	$2d$	$>$	125

Solve.

$$75 + 2d > 125$$
$$2d > 50 \quad \text{Subtracting 75}$$
$$d > 25 \quad \text{Dividing by 2}$$

The solution set is $\{d|d > 25\}$.

Check. We can obtain a partial check by substituting a number less than or equal to 25 in the inequality. For example, when $d = 20$:

$$75 + 2d = 75 + 2 \cdot 20 = 75 + 40 = 115 < 125$$

Our result appears to be correct.

State. The calf's weight is more than 125 lb more than 25 days after its birth. The solution set is $\{d|d > 25\}$.

21. *Familiarize.* We first make a drawing. We let b represent the length of the base. Then the lengths of the other sides are $b - 2$ and $b + 3$.

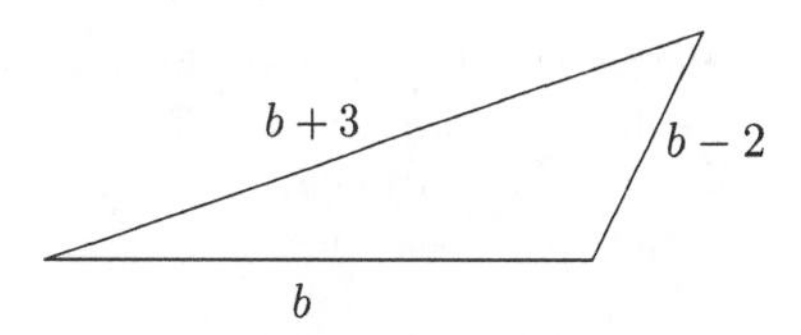

The perimeter is the sum of the lengths of the sides or $b + b - 2 + b + 3$, or $3b + 1$.

Translate.

The perimeter	is greater than	19 cm.
↓	↓	↓
$3b + 1$	$>$	19

Solve.

$$3b + 1 > 19$$
$$3b > 18$$
$$b > 6$$

Check. We check to see if the solution seems reasonable.

When $b = 5$, the perimeter is $3 \cdot 5 + 1$, or 16 cm.

When $b = 6$, the perimeter is $3 \cdot 6 + 1$, or 19 cm.

When $b = 7$, the perimeter is $3 \cdot 7 + 1$, or 22 cm.

From these calculations, it would appear that the solution is correct.

State. For lengths of the base greater than 6 cm the perimeter will be greater than 19 cm. The solution set is $\{b|b > 6 \text{ cm}\}$.

23. ***Familiarize.*** The average number of calls per week is the sum of the calls for the three weeks divided by the number of weeks, 3. We let c represent the number of calls made during the third week.

Translate. The average of the three weeks is given by

$$\frac{17 + 22 + c}{3}.$$

Since the average must be at least 20, this means that it must be greater than or equal to 20. Thus, we can translate the problem to the inequality

$$\frac{17 + 22 + c}{3} \geq 20.$$

Solve. We first multiply by 3 to clear the fraction.

$$3\left(\frac{17 + 22 + c}{3}\right) \geq 3 \cdot 20$$
$$17 + 22 + c \geq 60$$
$$39 + c \geq 60$$
$$c \geq 21$$

Check. Suppose c is a number greater than or equal to 21. Then by adding 17 and 22 on both sides of the inequality we get

$$17 + 22 + c \geq 17 + 22 + 21$$
$$17 + 22 + c \geq 60$$

so

$$\frac{17 + 22 + c}{3} \geq \frac{60}{3}, \text{ or } 20.$$

State. Any number of calls which is at least 21 will maintain an average of at least 20 for the three-week period. The solution set is $\{c|c \geq 21\}$.

25. ***Familiarize.*** We let $t =$ the length of the service call, in hours. Then the hourly charge is $60t$. The total cost of the service call is the sum of the \$70 flat fee, the hourly charge, and the \$35 cost of the freon.

Translate.

Flat fee	plus	hourly charge	plus	cost of freon	is at most	total charge
↓	↓	↓	↓	↓	↓	↓
70	+	$60t$	+	35	$\leq$	150

Solve.

$$70 + 60t + 35 \leq 150$$
$$60t + 105 \leq 150 \quad \text{Collecting like terms}$$
$$60t \leq 45 \quad \text{Subtracting 105}$$
$$t \leq 0.75 \quad \text{Dividing by 60}$$

Check. We can obtain a partial check by substituting a number greater than 0.75 in the inequality. For example, when $t = 1$:

$$70 + 60t + 35 = 70 + 60 \cdot 1 + 35 = 70 + 60 + 35 = 165 > 150.$$

Our result appears to be correct.

State. A service call that is no more than 0.75 hr in length will allow the family to stay within its budget. The solution set is $\{t|t \leq 0.75 \text{ hr}\}$.

27. ***Familiarize.*** If the reduced fat peanut butter has 25% less fat than regular peanut butter, then it has no more than 75% of the fat content of the regular peanut butter. Let $f =$ the number of grams of fat in a serving of regular peanut butter.

Translate.

Fat in reduced fat peanut butter	is no more than	75%	of	fat in regular peanut butter
↓	↓	↓	↓	↓
12	$\leq$	75%	$\cdot$	f

Solve.

$$12 \leq 75\% \cdot f$$
$$12 \leq 0.75f$$
$$\frac{12}{0.75} \leq f$$
$$16 \leq f$$

Check. We can obtain a partial check by substituting a number less than 16 in the inequality. For example, when $f = 15$:

$$75\% \cdot 15 = 0.75(15) = 11.25 < 12$$

The result appears to be correct.

State. Regular Skippy peanut butter contains at least 16 g of fat per serving. The solution set is $\{f|f \geq 16 \text{ g}\}$.

29. $-3 + 2(-5)^2(-3) - 7 = -3 + 2(25)(-3) - 7$
$= -3 + 50(-3) - 7$
$= -3 - 150 - 7$
$= -153 - 7$
$= -160$

31. $23(2x - 4) - 15(10 - 3x) = 46x - 92 - 150 + 45x = 91x - 242$

33. 

Chapter 9

Graphs of Equations; Data Analysis

Exercise Set 9.1

1. We go to the top of the bar that is above the body weight 200 lb. Then we move horizontally from the top of the bar to the vertical scale listing numbers of drinks. It appears approximately 6 drinks will give a 200-lb person a blood-alcohol level of 0.10%.

3. We see that the bars for weights above 200 lb extend beyond the 6 drink level. Thus, the weight of someone who can consume 6 drinks without reaching a blood-alcohol level of 0.10% is greater than 200 lb.

5. From $3\frac{1}{2}$ on the vertical scale we move horizontally until we reach a bar whose top is above the horizontal line on which we are moving. The first such bar corresponds to a body weight of 120 lb. Thus, we can conclude an individual weighs at least 120 lb if $3\frac{1}{2}$ drinks are consumed without reaching a blood-alcohol level of 0.10%.

7. The section of the pie chart representing housing shows that 32.4% of the expense is for housing.

9. ***Familiarize***. The graph shows that 7.6% of the cost is for child care and education. Let a = the amount spent for child care and education.

Translate. We reword the problem and then translate.

$$\begin{array}{cccccc} \text{What} & \text{is} & 7.6\% & \text{of} & \$136,320? \\ \downarrow & \downarrow & \downarrow & \downarrow & \downarrow \\ a & = & 7.6\% & \cdot & 136,320 \end{array}$$

Solve. We carry out the computation.

$$a = 7.6\% \cdot 136,320 = 0.076 \cdot 136,320 = 10,360.32$$

Check. We repeat the calculation. The answer checks.

State. Child care and education accounted for about \$10,360.32 of the total cost of raising a child to the age of 18.

11. First locate 1991 on the horizontal axis and then move up to the line. Now move across to the vertical scale and read that there were approximately 20,000 alcohol-related deaths in 1991.

13. The lowest point on the graph occurs above 1994. Thus, the lowest number of deaths occurred in 1994.

15. First locate 1994 on the horizontal scale and then move up to the line. Now move across to the vertical scale and read that there were approximately 16,500 alcohol-related deaths in 1994. Similarly, we find that there were approximately 17,000 alcohol-related deaths in 1995. We subtract to find the increase:

$$17,000 - 16,500 = 500$$

Thus, alcohol-related deaths increased by about 500 from 1994 to 1995.

17. $(2, 5)$ is 2 units right and 5 units up.

$(-1, 3)$ is 1 unit left and 3 units up.

$(3, -2)$ is 3 units right and 2 units down.

$(-2, -4)$ is 2 units left and 4 units down.

$(0, 4)$ is 0 units left or right and 4 units up.

$(0, -5)$ is 0 units left or right and 5 units down.

$(5, 0)$ is 5 units right and 0 units up or down.

$(-5, 0)$ is 5 units left and 0 units up or down.

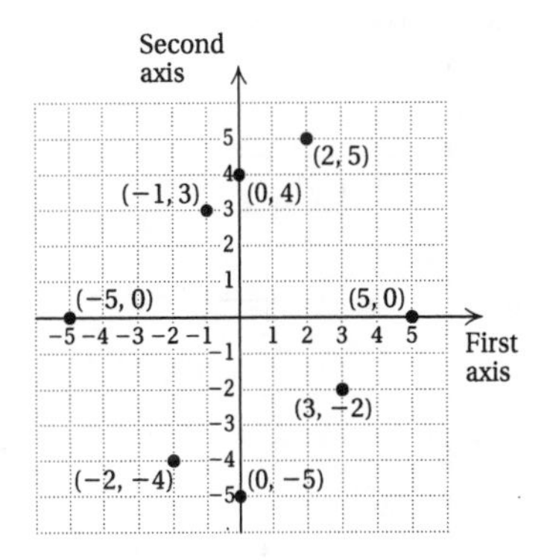

19. Since the first coordinate is negative and the second coordinate positive, the point $(-5, 3)$ is located in quadrant II.

21. Since the first coordinate is positive and the second coordinate negative, the point $(100, -1)$ is in quadrant IV.

23. Since both coordinates are negative, the point $(-6, -29)$ is in quadrant III.

25. Since both coordinates are positive, the point $(3.8, 9.2)$ is in quadrant I.

27. Since the first coordinate is negative and the second coordinate is positive, the point $\left(-\frac{1}{3}, \frac{15}{7}\right)$ is in quadrant II.

29. Since the first coordinate is positive and the second coordinate is negative, the point $\left(12\frac{7}{8}, -1\frac{1}{2}\right)$ is in quadrant IV.

31. In quadrant III, first coordinates are always <u>negative</u> and second coordinates are always <u>negative</u>.

33. In quadrant IV, <u>second</u> coordinates are always negative and <u>first</u> coordinates are always positive.

35.

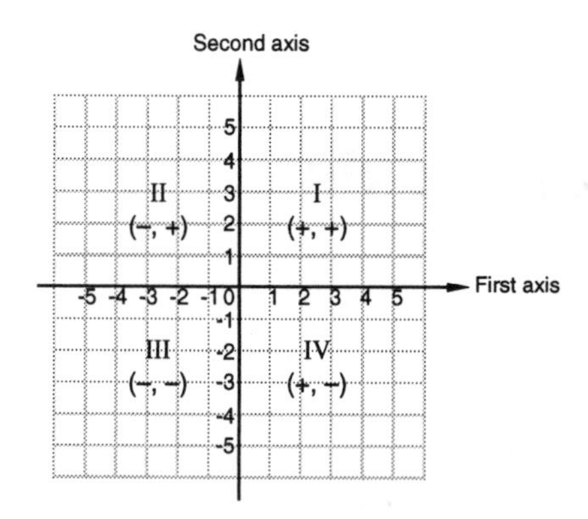

If the first coordinate is positive, then the point must be in either quadrant I or quadrant IV.

37. If the first and second coordinates are equal, they must either be both positive or both negative. The point must be in either quadrant I (both positive) or quadrant III (both negative).

39.

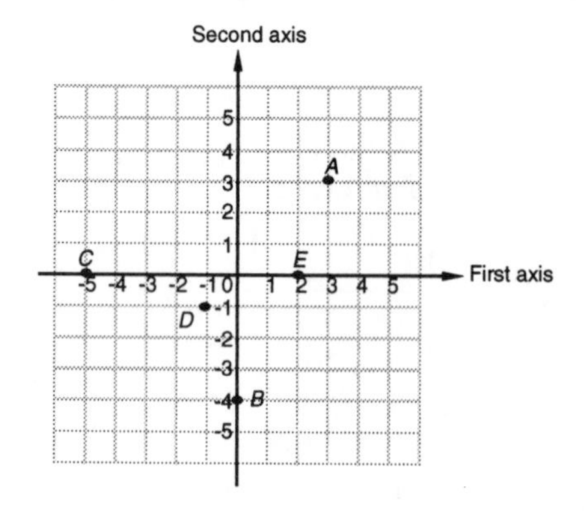

Point A is 3 units right and 3 units up. The coordinates of A are $(3, 3)$.

Point B is 0 units left or right and 4 units down. The coordinates of B are $(0, -4)$.

Point C is 5 units left and 0 units up or down. The coordinates of C are $(-5, 0)$.

Point D is 1 unit left and 1 unit down. The coordinates of D are $(-1, -1)$.

Point E is 2 units right and 0 units up or down. The coordinates of E are $(2, 0)$.

41. The distance of -12 from 0 is 12, so $|-12| = 12$.

43. The distance of 0 from 0 is 0, so $|0| = 0$.

45. ***Familiarize.*** Let x = the amount spent on salaries in 1996, in billions of dollars. Then $17.7\%x$ represents the increase in 1997.

Translate.

$$\underbrace{\text{Amount spent in 1996}}_{x} \quad \underbrace{\text{plus}}_{+} \quad \underbrace{\text{amount of increase}}_{17.7\%x} \quad \underbrace{\text{is}}_{=} \quad \underbrace{\text{amount spent in 1997}}_{1.06}$$

Solve.

$$x + 17.7\%x = 1.06$$
$$1 \cdot x + 0.177x = 1.06$$
$$1.177x = 1.06$$
$$\frac{1.177x}{1.177} = \frac{1.06}{1.177}$$
$$x \approx 0.9$$

Check. We can find 17.7% of 0.9 and add the amount to 0.9:

$17.7\% \times 0.9 = 0.177 \times 0.9 = 0.1593$ and $0.9 + 0.1593 = 1.0593 \approx 1.06$

The answer checks.

State. About \$0.9 billion was spent in 1996.

47. ◈

49.

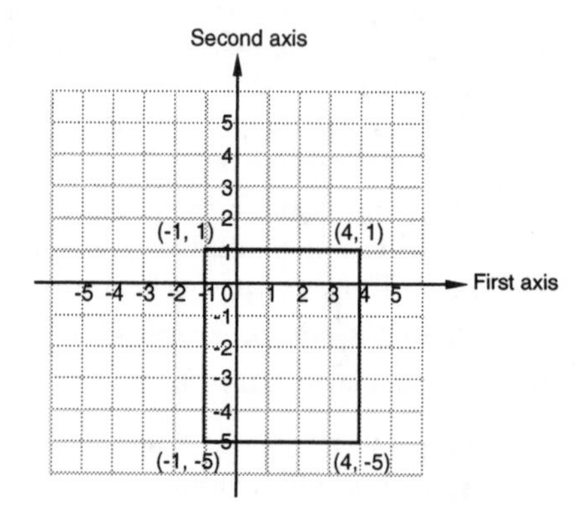

The coordinates of the fourth vertex are $(-1, -5)$.

51. Answers may vary.

We select eight points such that the sum of the coordinates for each point is 6.

$(-1, 7)$	$-1 + 7 = 6$
$(0, 6)$	$0 + 6 = 6$
$(1, 5)$	$1 + 5 = 6$
$(2, 4)$	$2 + 4 = 6$
$(3, 3)$	$3 + 3 = 6$
$(4, 2)$	$4 + 2 = 6$
$(5, 1)$	$5 + 1 = 6$
$(6, 0)$	$6 + 0 = 6$

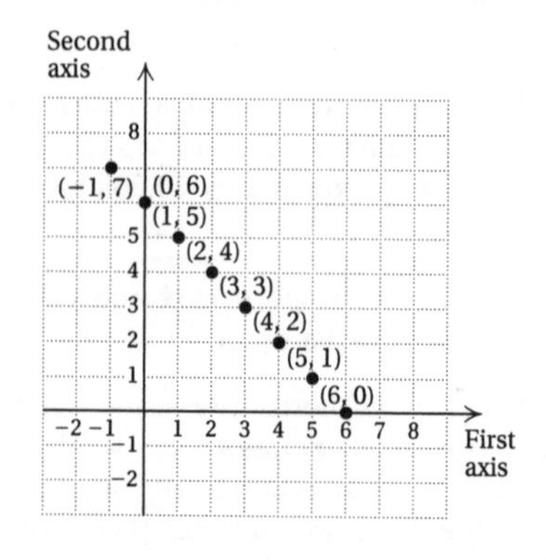

53.

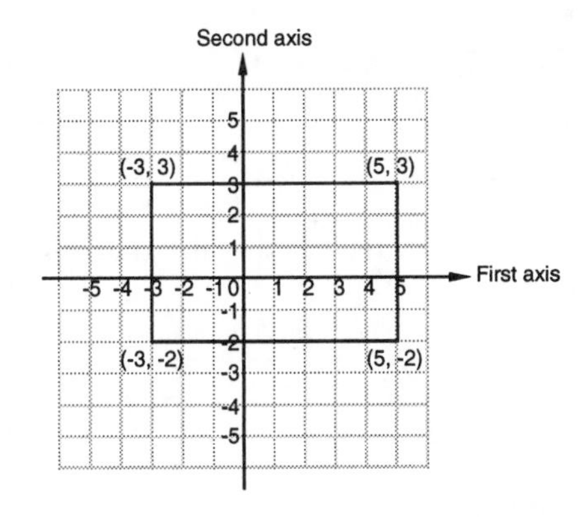

The length is 8, and the width is 5.

$P = 2l + 2w$

$P = 2 \cdot 8 + 2 \cdot 5 = 16 + 10 = 26$

Exercise Set 9.2

1. We substitute 2 for x and 9 for y (alphabetical order of variables).

$$\begin{array}{c|l} \multicolumn{2}{c}{y = 3x - 1} \\ \hline 9 \; ? & 3 \cdot 2 - 1 \\ & 6 - 1 \\ & 5 \qquad \text{FALSE} \end{array}$$

Since $9 = 5$ is false, the pair $(2, 9)$ is not a solution.

3. We substitute 4 for x and 2 for y.

$$\begin{array}{r|l} \multicolumn{2}{c}{2x + 3y = 12} \\ \hline 2 \cdot 4 + 3 \cdot 2 & ? \; 12 \\ 8 + 6 & \\ 14 & \qquad \text{FALSE} \end{array}$$

Since $14 = 12$ is false, the pair (4,2) is not a solution.

5. We substitute 3 for a and -1 for b.

$$\begin{array}{r|l} \multicolumn{2}{c}{3a - 4b = 13} \\ \hline 3 \cdot 3 - 4(-1) & ? \; 13 \\ 9 + 4 & \\ 13 & \qquad \text{TRUE} \end{array}$$

Since $13 = 13$ is true, the pair $(3, -1)$ is a solution.

7. To show that a pair is a solution, we substitute, replacing x with the first coordinate and y with the second coordinate in each pair.

$$\begin{array}{c|l} \multicolumn{2}{c}{y = x - 5} \\ \hline -1 \; ? & 4 - 5 \\ & -1 \qquad \text{TRUE} \end{array} \qquad \begin{array}{c|l} \multicolumn{2}{c}{y = x - 5} \\ \hline -4 \; ? & 1 - 5 \\ & -4 \qquad \text{TRUE} \end{array}$$

In each case the substitution results in a true equation. Thus, $(4, -1)$ and $(1, -4)$ are both solutions of $y = x - 5$. We graph these points and sketch the line passing through them.

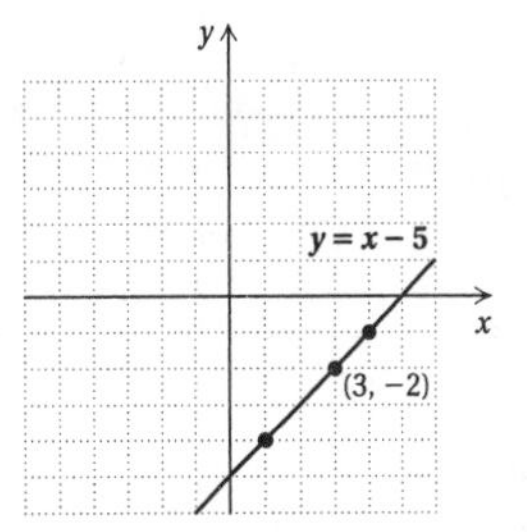

The line appears to pass through $(3, -2)$ also. We check to determine if $(3, -2)$ is a solution of $y = x - 5$.

$$\begin{array}{c|l} \multicolumn{2}{c}{y = x - 5} \\ \hline -2 \; ? & 3 - 5 \\ & -2 \qquad \text{TRUE} \end{array}$$

Thus, $(3, -2)$ is another solution. There are other correct answers, including $(-1, -6)$, $(2, -3)$, $(0, -5)$, $(5, 0)$, and $(6, 1)$.

9. To show that a pair is a solution, we substitute, replacing x with the first coordinate and y with the second coordinate in each pair.

$$\begin{array}{c|l} \multicolumn{2}{c}{y = \frac{1}{2}x + 3} \\ \hline 5 \; ? & \frac{1}{2} \cdot 4 + 3 \\ & 2 + 3 \\ & 5 \qquad \text{TRUE} \end{array} \qquad \begin{array}{c|l} \multicolumn{2}{c}{y = \frac{1}{2}x + 3} \\ \hline 2 \; ? & \frac{1}{2}(-2) + 3 \\ & -1 + 3 \\ & 2 \qquad \text{TRUE} \end{array}$$

In each case the substitution results in a true equation. Thus, $(4, 5)$ and $(-2, 2)$ are both solutions of $y = \frac{1}{2}x + 3$. We graph these points and sketch the line passing through them.

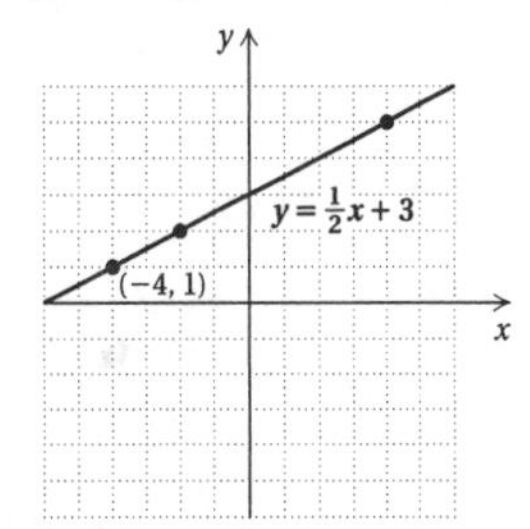

The line appears to pass through $(-4, 1)$ also. We check to determine if $(-4, 1)$ is a solution of $y = \frac{1}{2}x + 3$.

$$\begin{array}{c|l} \multicolumn{2}{c}{y = \frac{1}{2}x + 3} \\ \hline 1 \; ? & \frac{1}{2}(-4) + 3 \\ & -2 + 3 \\ & 1 \qquad \text{TRUE} \end{array}$$

Thus, $(-4, 1)$ is another solution. There are other correct answers, including $(-6, 0)$, $(0, 3)$, $(2, 4)$, and $(6, 6)$.

11. To show that a pair is a solution, we substitute, replacing x with the first coordinate and y with the second coordinate in each pair.

$$\begin{array}{c|c} 4x - 2y = 10 & \\ \hline 4 \cdot 0 - 2(-5) \ ? \ 10 & \\ 10 & \text{TRUE} \end{array}$$

$$\begin{array}{c|c} 4x - 2y = 10 & \\ \hline 4 \cdot 4 - 2 \cdot 3 \ ? \ 10 & \\ 16 - 6 & \\ 10 & \text{TRUE} \end{array}$$

In each case the substitution results in a true equation. Thus, $(0,-5)$ and $(4,3)$ are both solutions of $4x - 2y = 10$. We graph these points and sketch the line passing through them.

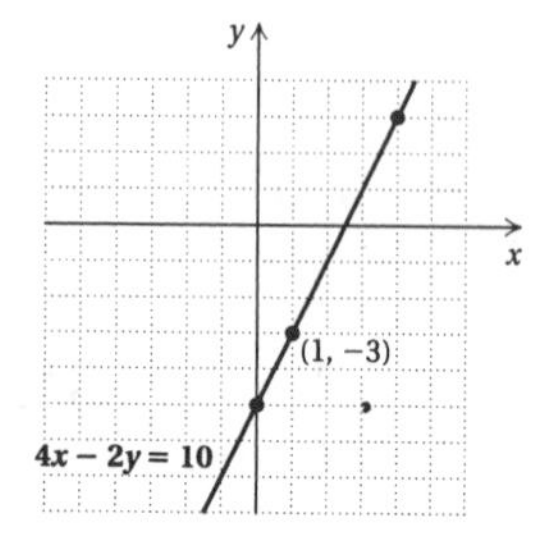

The line appears to pass through $(1,-3)$ also. We check to determine if $(1,-3)$ is a solution of $4x - 2y = 10$.

$$\begin{array}{c|c} 4x - 2y = 10 & \\ \hline 4 \cdot 1 - 2(-3) \ ? \ 10 & \\ 4 + 6 & \\ 10 & \text{TRUE} \end{array}$$

Thus, $(1,-3)$ is another solution. There are other correct answers, including $(2,-1)$, $(3,1)$, and $(5,5)$.

13. $y = x + 1$

The equation is in the form $y = mx + b$. The y-intercept is $(0,1)$. We find two other pairs.

When $x = 3$, $\quad y = 3 + 1 = 4$.
When $x = -5$, $\quad y = -5 + 1 = -4$.

x	y
0	1
3	4
−5	−4

Plot these points, draw the line they determine, and label the graph $y = x + 1$.

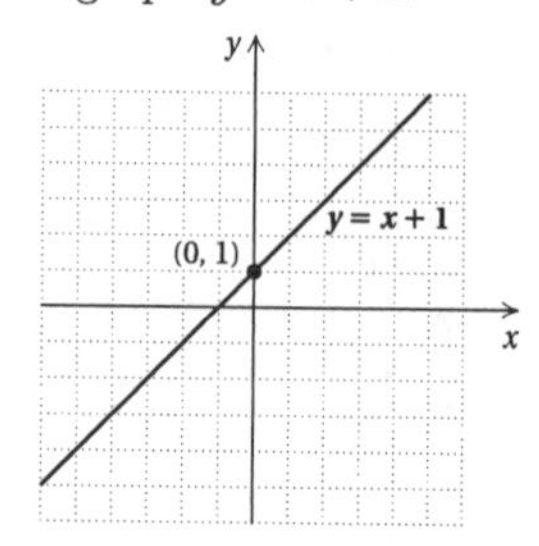

15. $y = x$

The equation is equivalent to $y = x + 0$. The y-intercept is $(0,0)$. We find two other points.

When $x = -2$, $\quad y = -2$.
When $x = 3$, $\quad y = 3$.

x	y
0	0
−2	−2
3	3

Plot these points, draw the line they determine, and label the graph $y = x$.

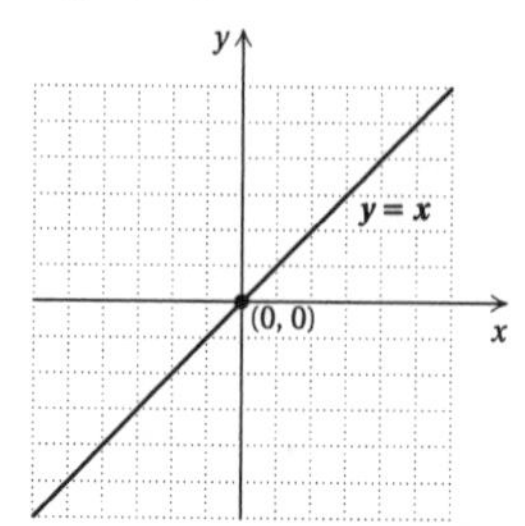

17. $y = \frac{1}{2}x$

The equation is equivalent to $y = \frac{1}{2}x + 0$. The y-intercept is $(0,0)$. We find two other points.

When $x = -4$, $y = \frac{1}{2}(-4) = -2$.

When $x = 4$, $y = \frac{1}{2} \cdot 4 = 2$.

x	y
0	0
−4	−2
4	2

Plot these points, draw the line they determine, and label the graph $y = \frac{1}{2}x$.

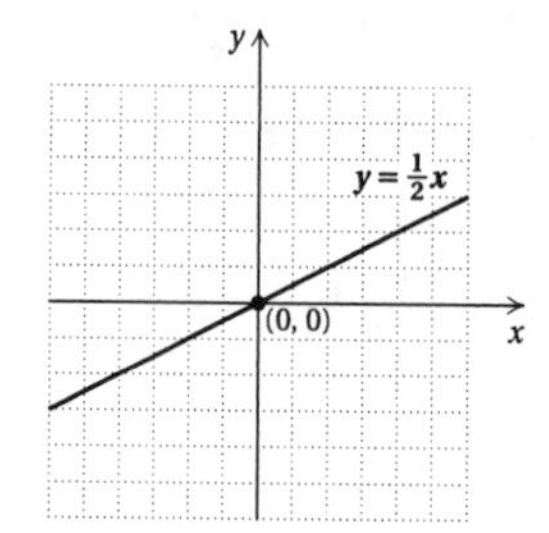

19. $y = x - 3$

The equation is equivalent to $y = x + (-3)$. The y-intercept is $(0,-3)$. We find two other points.

When $x = -2$, $y = -2 - 3 = -5$.

When $x = 4$, $y = 4 - 3 = 1$.

x	y
0	−3
−2	−5
4	1

Plot these points, draw the line they determine, and label the graph $y = x - 3$.

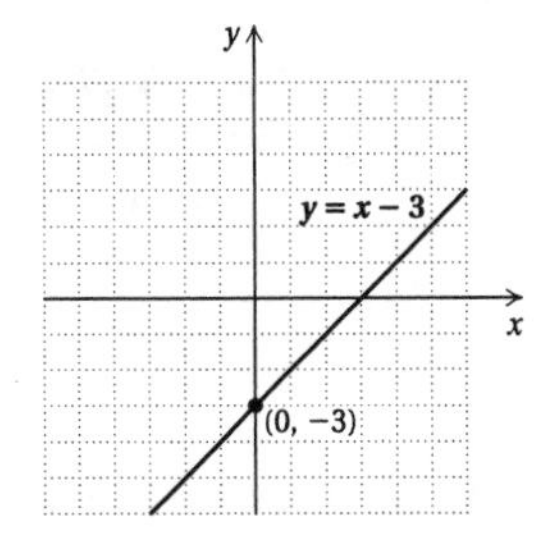

21. $y = 3x - 2 = 3x + (-2)$

The y-intercept is $(0, -2)$. We find two other points.

When $x = -2$, $y = 3(-2) + 2 = -6 + 2 = -4$.

When $x = 1$, $y = 3 \cdot 1 + 2 = 3 + 2 = 5$.

x	y
0	−2
−2	−4
1	5

Plot these points, draw the line they determine, and label the graph $y = 3x + 2$.

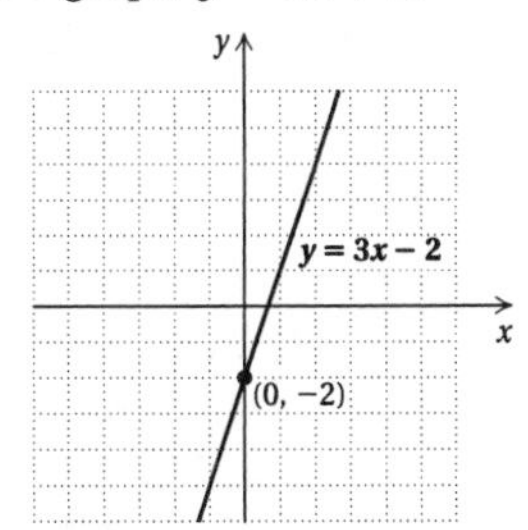

23. $y = \frac{1}{2}x + 1$

The y-intercept is $(0, 1)$. We find two other points using multiples of 2 for x to avoid fractions.

When $x = -4$, $y = \frac{1}{2}(-4) + 1 = -2 + 1 = -1$.

When $x = 4$, $y = \frac{1}{2} \cdot 4 + 1 = 2 + 1 = 3$.

x	y
0	1
−4	−1
4	3

Plot these points, draw the line they determine, and label the graph $y = \frac{1}{2}x + 1$.

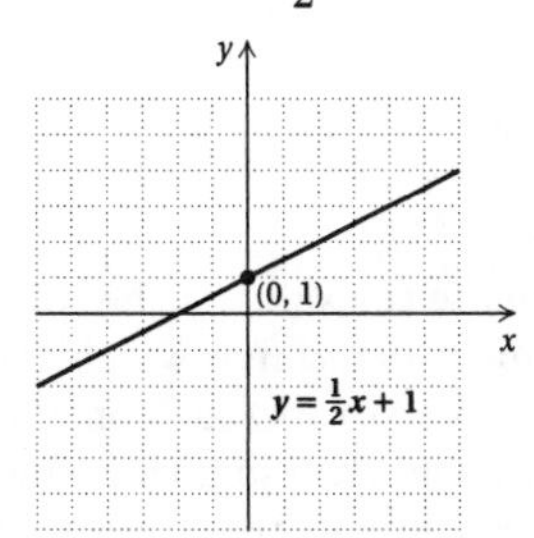

25. $x + y = -5$

$$y = -x - 5$$

$$y = -x + (-5)$$

The y-intercept is $(0, -5)$. We find two other points.

When $x = -4$, $y = -(-4) - 5 = 4 - 5 = -1$.

When $x = -1$, $y = -(-1) - 5 = 1 - 5 = -4$.

x	y
0	−5
−4	−1
−1	−4

Plot these points, draw the line they determine, and label the graph $x + y = -5$.

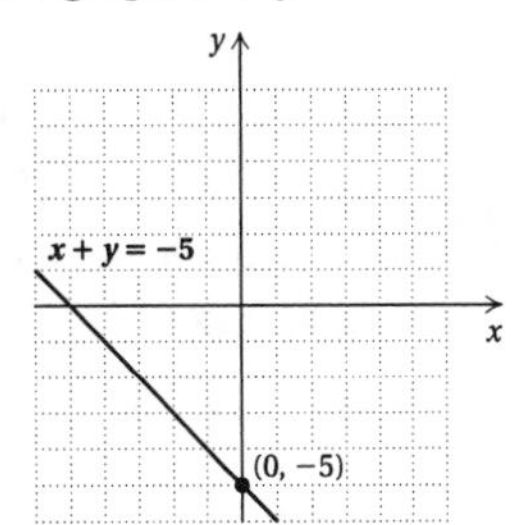

27. $y = \frac{5}{3}x - 2 = \frac{5}{3}x + (-2)$

The y-intercept is $(0, -2)$. We find two other points using multiples of 3 for x to avoid fractions.

When $x = -3$, $y = \frac{5}{3}(-3) - 2 = -5 - 2 = -7$.

When $x = 3$, $y = \frac{5}{3} \cdot 3 - 2 = 5 - 2 = 3$.

x	y
0	−2
−3	−7
3	3

Plot these points, draw the line they determine, and label the graph $y = \frac{5}{3}x - 2$.

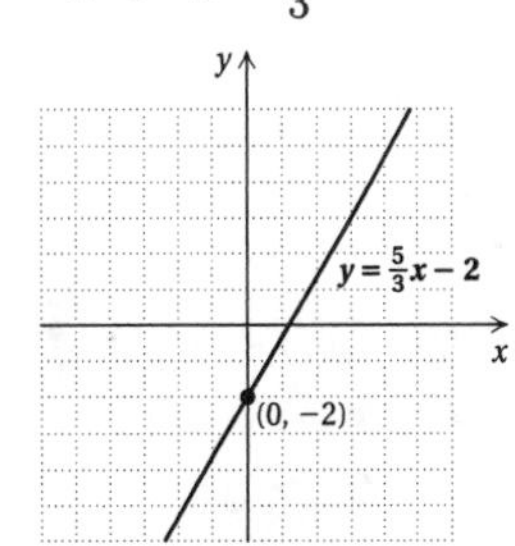

29. $x + 2y = 8$

$$2y = -x + 8$$

$$y = -\frac{1}{2}x + 4$$

The y-intercept is $(0, 4)$. We find two other points using multiples of 2 for x to avoid fractions.

When $x = -2$, $y = -\frac{1}{2}(-2) + 4 = 1 + 4 = 5$.

When $x = 4$, $y = -\frac{1}{2} \cdot 4 + 4 = -2 + 4 = 2$.

x	y
0	4
−2	5
4	2

Plot these points, draw the line they determine, and label the graph $x + 2y = 8$.

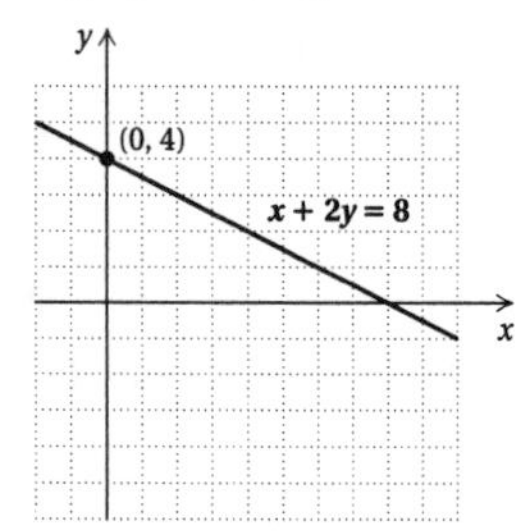

31. $y = \frac{3}{2}x + 1$

The y-intercept is $(0, 1)$. We find two other points using multiples of 2 for x to avoid fractions.

When $x = -4$, $y = \frac{3}{2}(-4) + 1 = -6 + 1 = -5$.

When $x = 2$, $y = \frac{3}{2} \cdot 2 + 1 = 3 + 1 = 4$.

x	y
0	1
-4	-5
2	4

Plot these points, draw the line they determine, and label the graph $y = \frac{3}{2}x + 1$.

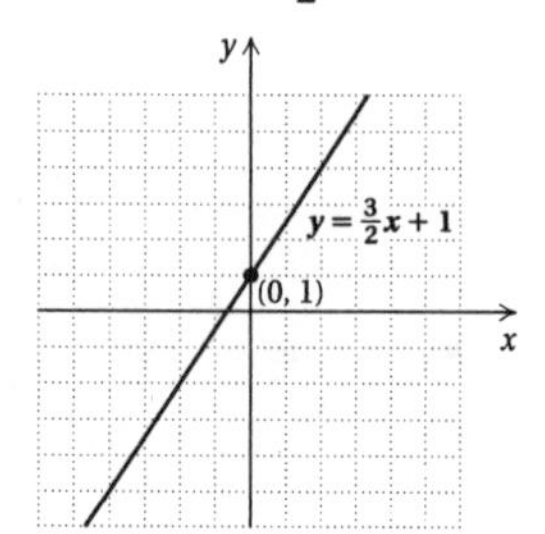

33.
$$\begin{aligned} 8x - 2y &= -10 \\ -2y &= -8x - 10 \\ y &= 4x + 5 \end{aligned}$$

The y-intercept is $(0, 5)$. We find two other points.

When $x = -2$, $y = 4(-2) + 5 = -8 + 5 = -3$.

When $x = -1$, $y = 4(-1) + 5 = -4 + 5 = 1$.

x	y
0	5
-2	-3
-1	1

Plot these points, draw the line they determine, and label the graph $8x - 2y = -10$.

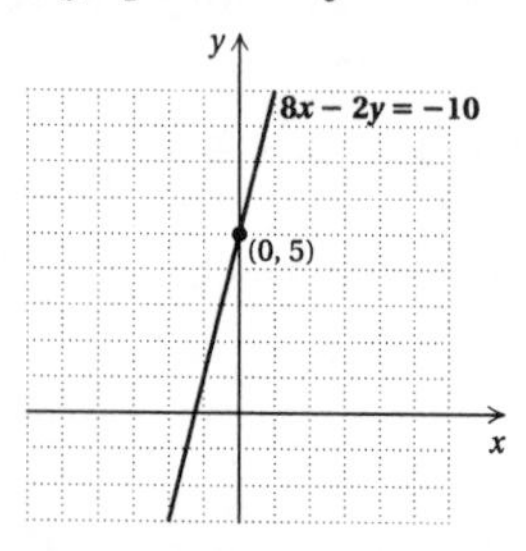

35.
$$\begin{aligned} 8y + 2x &= -4 \\ 8y &= -2x - 4 \\ y &= -\frac{1}{4}x - \frac{1}{2} \\ y &= -\frac{1}{4}x + \left(-\frac{1}{2}\right) \end{aligned}$$

The y-intercept is $\left(0, -\frac{1}{2}\right)$. We find two other points.

When $x = -2$, $y = -\frac{1}{4}(-2) - \frac{1}{2} = \frac{1}{2} - \frac{1}{2} = 0$.

When $x = 2$, $y = -\frac{1}{4} \cdot 2 - \frac{1}{2} = -\frac{1}{2} - \frac{1}{2} = -1$.

x	y
0	$-\frac{1}{2}$
-2	0
2	-1

Plot these points, draw the line they determine, and label the graph $8y + 2x = -4$.

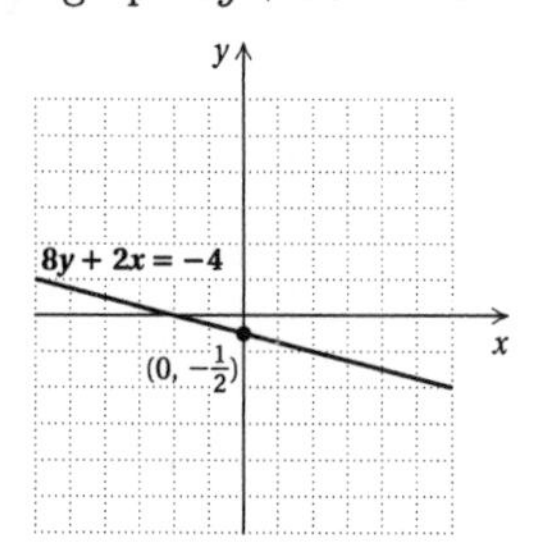

37. a) We substitute 0, 4, and 6 for t and then calculate V.

If $t = 0$, then $V = -50 \cdot 0 + 300 = \$300$.

If $t = 4$, then $V = -50 \cdot 4 + 300 = -200 + 300 = \$100$.

If $t = 6$, then $V = -50 \cdot 6 + 300 = -300 + 300 = \$0$.

b) We plot the three ordered pairs we found in part (a). Note the negative t- and V-values have no meaning in this problem.

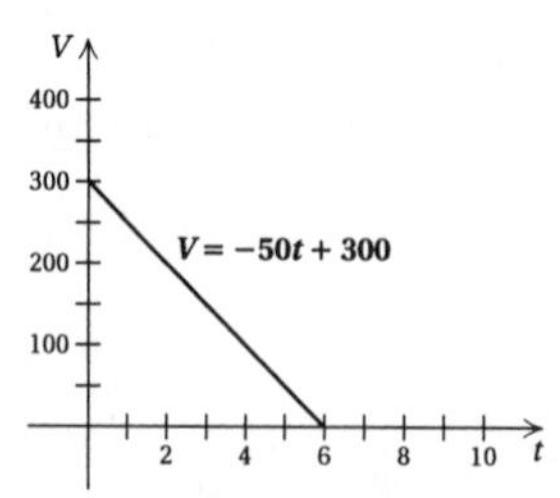

To use the graph to estimate the value of the software after 5 years we must determine which V-value is paired with $t = 5$. We locate 5 on the t-axis, go up to the graph, and then find the value on the V-axis that corresponds to that point. It appears that after 5 years the value of the software is \$50.

c) Substitute 150 for V and then solve for t.

$$
\begin{aligned}
V &= -50t + 300 \\
150 &= -50t + 300 \\
-150 &= -50t \\
3 &= t
\end{aligned}
$$

The value of the software is \$150 after 3 years.

39. a) Substitute 1, 4, 8, and 10 for d and then calculate N.

If $d = 1$, $N = 0.1(1) + 7 = 0.1 + 7 = 7.1$ gal

If $d = 4$, $N = 0.1(4) + 7 = 0.4 + 7 = 7.4$ gal

If $d = 8$, $N = 0.1(8) + 7 = 0.8 + 7 = 7.8$ gal

If $d = 10$, $N = 0.1(10) + 7 = 1 + 7 = 8$ gal

b) Plot the four ordered pairs we found in part (a). Note that negative d- and N-values have no meaning in this problem.

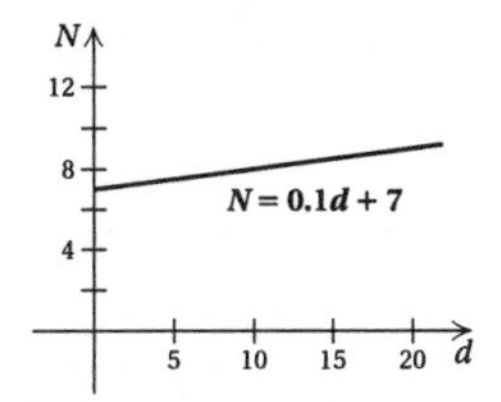

To use the graph to estimate what tea consumption was in 1997 we must determine which N-value is paired with 1997, or with $d = 6$. We locate 6 on the d-axis, go up to the graph, and then find the value on the N-axis that corresponds to that point. It appears that tea consumption was about 7.6 gallons in 1997.

c) Substitute 9 for N and then solve for d.

$$
\begin{aligned}
N &= 0.1d + 7 \\
9 &= 0.1d + 7 \\
2 &= 0.1d \\
20 &= d
\end{aligned}
$$

Tea consumption will be about 9 gallons 20 years after 1991, or in 2011.

41. 2567.03

The digit in the thousands place is 2. The next digit to the right, 5, is 5 or higher, so we round up: 3000.

43. 293.4572

There is no number in the thousands place. The next digit to the right, 2, is less than 5, so we round down: 0.

45. First we find decimal notation for $\frac{7}{8}$.

$$
\begin{array}{r}
0.875 \\
8\,\overline{)\,7.000} \\
\underline{6\,4} \\
60 \\
\underline{56} \\
40 \\
\underline{40} \\
0
\end{array}
$$

Since $\frac{7}{8} = 0.875$, then $-\frac{7}{8} = -0.875$.

47.

$$
\begin{array}{r}
1.828125 \\
64\,\overline{)\,117.000000} \\
\underline{64} \\
53\,0 \\
\underline{51\,2} \\
1\,80 \\
\underline{1\,28} \\
520 \\
\underline{512} \\
80 \\
\underline{64} \\
160 \\
\underline{128} \\
320 \\
\underline{320} \\
0
\end{array}
$$

$\frac{117}{64} = 1.828125$

49.

51. Note that the sum of the coordinates of each point on the graph is 5. Thus, we have $x + y = 5$, or $y = -x + 5$.

53. Note that each y-coordinate is 2 more than the corresponding x-coordinate. Thus, we have $y = x + 2$.

Exercise Set 9.3

1. (a) The graph crosses the y-axis at $(0, 5)$, so the y-intercept is $(0, 5)$.

(b) The graph crosses the x- axis at $(2, 0)$, so the x-intercept is $(2, 0)$.

3. (a) The graph crosses the y-axis at $(0, -4)$, so the y-intercept is $(0, -4)$.

(b) The graph crosses the x-axis at $(3, 0)$, so the x-intercept is $(3, 0)$.

5. $3x + 5y = 15$

(a) To find the y-intercept, let $x = 0$. This is the same as covering up the x-term and then solving.

$$
\begin{aligned}
5y &= 15 \\
y &= 3
\end{aligned}
$$

The y-intercept is $(0, 3)$.

(b) To find the x-intercept, let $y = 0$. This is the same as covering up the y-term and then solving.

$$
\begin{aligned}
3x &= 15 \\
x &= 5
\end{aligned}
$$

The x-intercept is $(5, 0)$.

7. $7x - 2y = 28$

(a) To find the y-intercept, let $x = 0$. This is the same as covering up the x-term and then solving.

$$
\begin{aligned}
-2y &= 28 \\
y &= -14
\end{aligned}
$$

The y−intercept is $(0, -14)$.

(b) To find the x-intercept, let $y = 0$. This is the same as covering up the y-term and then solving.

$$7x = 28$$
$$x = 4$$

The x-intercept is $(4, 0)$.

9. $-4x + 3y = 10$

(a) To find the y-intercept, let $x = 0$. This is the same as covering up the x-term and then solving.

$$3y = 10$$
$$y = \frac{10}{3}$$

The y-intercept is $\left(0, \frac{10}{3}\right)$.

(b) To find the x-intercept, let $y = 0$. This is the same as covering up the y-term and then solving.

$$-4x = 10$$
$$x = -\frac{5}{2}$$

The x-intercept is $\left(-\frac{5}{2}, 0\right)$.

11. $6x - 3 = 9y$

$6x - 9y = 3$ Writing the equation in the form $Ax + By = C$

(a) To find the y-intercept, let $x = 0$. This is the same as covering up the x-term and then solving.

$$-9y = 3$$
$$y = -\frac{1}{3}$$

The y-intercept is $\left(0, -\frac{1}{3}\right)$.

(b) To find the x-intercept, let $y = 0$. This is the same as covering up the y-term and then solving.

$$6x = 3$$
$$x = \frac{1}{2}$$

The x-intercept is $\left(\frac{1}{2}, 0\right)$.

13. $x + 3y = 6$

To find the x-intercept, let $y = 0$. Then solve for x.

$$x + 3y = 6$$
$$x + 3 \cdot 0 = 6$$
$$x = 6$$

Thus, $(6, 0)$ is the x-intercept.

To find the y-intercept, let $x = 0$. Then solve for y.

$$x + 3y = 6$$
$$0 + 3y = 6$$
$$3y = 6$$
$$y = 2$$

Thus, $(0, 2)$ is the y-intercept.

Plot these points and draw the line.

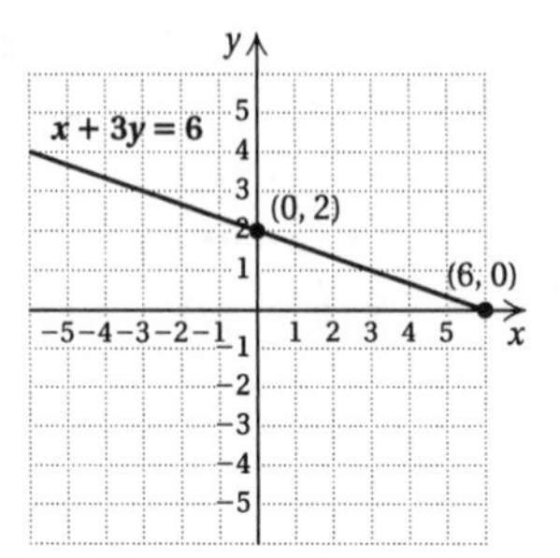

A third point should be used as a check. We substitute any value for x and solve for y.

We let $x = 3$. Then

$$x + 3y = 6$$
$$3 + 3y = 6$$
$$3y = 3$$
$$y = 1$$

The point $(3, 1)$ is on the graph, so the graph is probably correct.

15. $-x + 2y = 4$

To find the x-intercept, let $y = 0$. Then solve for x.

$$-x + 2y = 4$$
$$-x + 2 \cdot 0 = 4$$
$$-x = 4$$
$$x = -4$$

Thus, $(-4, 0)$ is the x-intercept.

To find the y-intercept, let $x = 0$. Then solve for y.

$$-x + 2y = 4$$
$$-0 + 2y = 4$$
$$2y = 4$$
$$y = 2$$

Thus, $(0, 2)$ is the y-intercept.

Plot these points and draw the line.

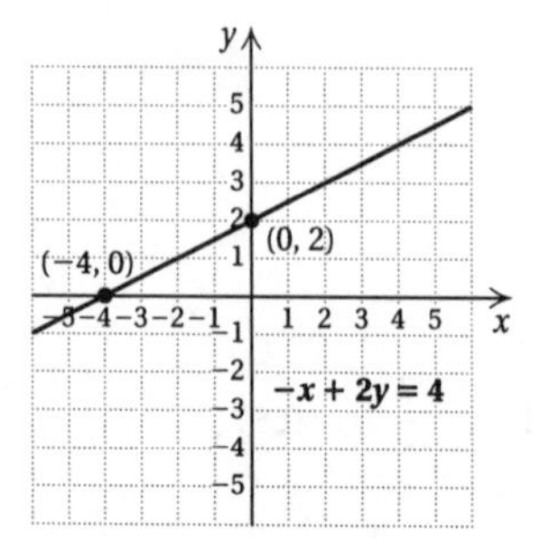

A third point should be used as a check. We substitute any value for x and solve for y.

We let $x = 4$. Then

$$-x + 2y = 4$$
$$-4 + 2y = 4$$
$$2y = 8$$
$$y = 4$$

The point $(4, 4)$ is on the graph, so the graph is probably correct.

17. $3x + y = 6$

To find the x-intercept, let $y = 0$. Then solve for x.

$$3x + y = 6$$
$$3x + 0 = 6$$
$$3x = 6$$
$$x = 2$$

Thus, $(2, 0)$ is the x-intercept.

To find the y-intercept, let $x = 0$. Then solve for y.

$$3x + y = 6$$
$$3 \cdot 0 + y = 6$$
$$y = 6$$

Thus, $(0, 6)$ is the y-intercept.

Plot these points and draw the line.

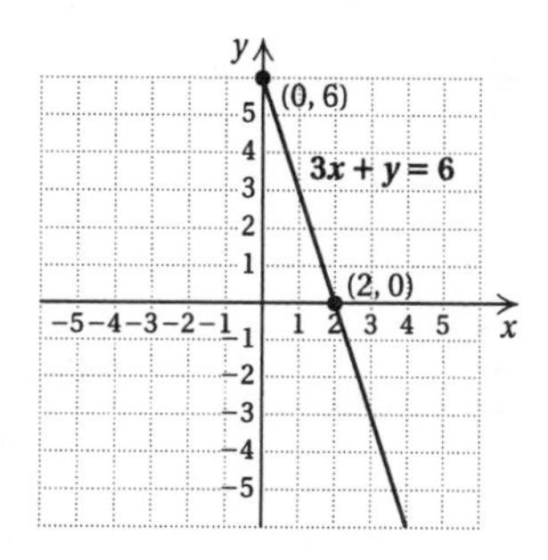

A third point should be used as a check. We substitute any value for x and solve for y.

We let $x = 1$. Then

$$3x + y = 6$$
$$3 \cdot 1 + y = 6$$
$$3 + y = 6$$
$$y = 3$$

The point $(1, 3)$ is on the graph, so the graph is probably correct.

19. $2y - 2 = 6x$

To find the x-intercept, let $y = 0$. Then solve for x.

$$2y - 2 = 6x$$
$$2 \cdot 0 - 2 = 6x$$
$$-2 = 6x$$
$$-\frac{1}{3} = x$$

Thus, $\left(-\frac{1}{3}, 0\right)$ is the x-intercept.

To find the y-intercept, let $x = 0$. Then solve for y.

$$2y - 2 = 6x$$
$$2y - 2 = 6 \cdot 0$$
$$2y - 2 = 0$$
$$2y = 2$$
$$y = 1$$

Thus, $(0, 1)$ is the y-intercept.

It is helpful to plot another point since the intercepts are so close together. This point can also serve as a check.

We let $x = 1$. Then

$$2y - 2 = 6x$$
$$2y - 2 = 6 \cdot 1$$
$$2y - 2 = 6$$
$$2y = 8$$
$$y = 4$$

Plot the point $(1, 4)$ and the intercepts and draw the line.

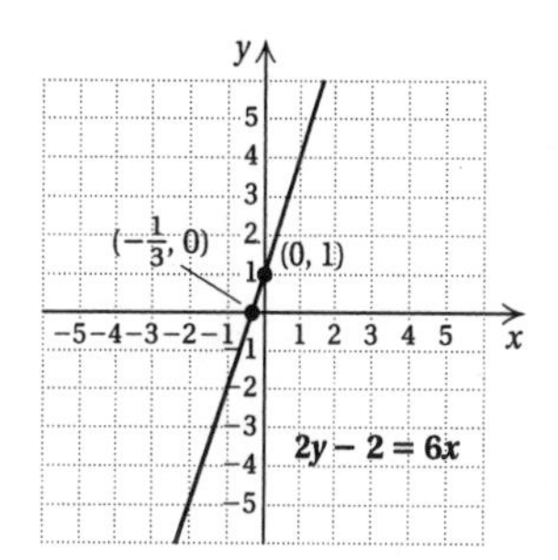

21. $3x - 9 = 3y$

To find the x-intercept, let $y = 0$. Then solve for x.

$$3x - 9 = 3y$$
$$3x - 9 = 3 \cdot 0$$
$$3x - 9 = 0$$
$$3x = 9$$
$$x = 3$$

Thus, $(3, 0)$ is the x-intercept.

To find the y-intercept, let $x = 0$. Then solve for y.

$$3x - 9 = 3y$$
$$3 \cdot 0 - 9 = 3y$$
$$-9 = 3y$$
$$-3 = y$$

Thus, $(0, -3)$ is the y-intercept.

Plot these points and draw the line.

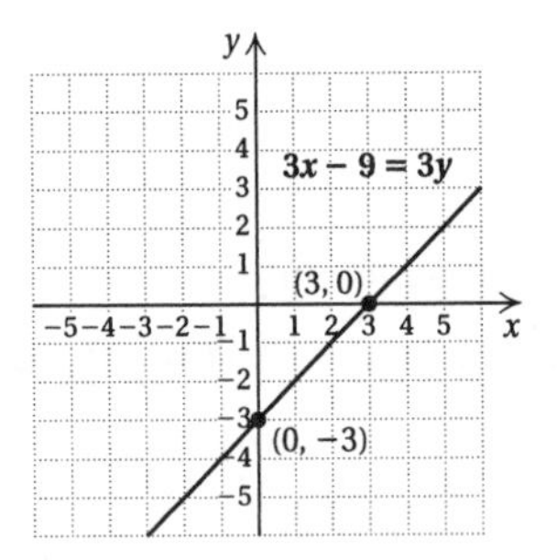

A third point should be used as a check. We substitute any value for x and solve for y.

We let $x = 1$. Then

$$3x - 9 = 3y$$
$$3 \cdot 1 - 9 = 3y$$
$$3 - 9 = 3y$$
$$-6 = 3y$$
$$-2 = y$$

The point $(1, -2)$ is on the graph, so the graph is probably correct.

23. $2x - 3y = 6$

To find the x-intercept, let $y = 0$. Then solve for x.

$$2x - 3y = 6$$
$$2x - 3 \cdot 0 = 6$$
$$2x = 6$$
$$x = 3$$

Thus, $(3, 0)$ is the x-intercept.

To find the y-intercept, let $x = 0$. Then solve for y.

$$2x - 3y = 6$$
$$2 \cdot 0 - 3y = 6$$
$$-3y = 6$$
$$y = -2$$

Thus, $(0, -2)$ is the y-intercept.

Plot these points and draw the line.

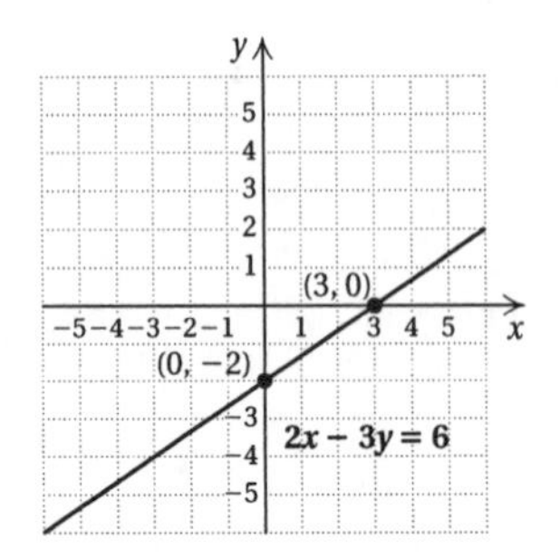

A third point should be used as a check. We substitute any value for x and solve for y.

We let $x = -3$.

$$2x - 3y = 6$$
$$2(-3) - 3y = 6$$
$$-6 - 3y = 6$$
$$-3y = 12$$
$$y = -4$$

The point $(-3, -4)$ is on the graph, so the graph is probably correct.

25. $4x + 5y = 20$

To find the x-intercept, let $y = 0$. Then solve for x.

$$4x + 5y = 20$$
$$4x + 5 \cdot 0 = 20$$
$$4x = 20$$
$$x = 5$$

Thus, $(5, 0)$ is the x-intercept.

To find the y-intercept, let $x = 0$. Then solve for y.

$$4x + 5y = 20$$
$$4 \cdot 0 + 5y = 20$$
$$5y = 20$$
$$y = 4$$

Thus, $(0, 4)$ is the y-intercept.

Plot these points and draw the graph.

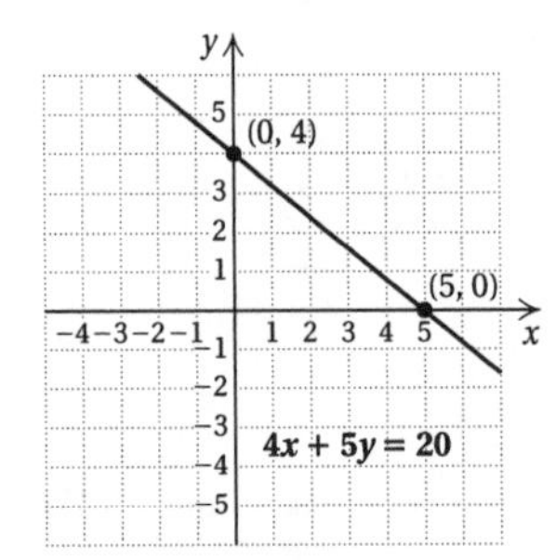

A third point should be used as a check. We substitute any value for x and solve for y.

We let $x = 4$. Then

$$4x + 5y = 20$$
$$4 \cdot 4 + 5y = 20$$
$$16 + 5y = 20$$
$$5y = 4$$
$$y = \frac{4}{5}$$

The point $\left(4, \frac{4}{5}\right)$ is on the graph, so the graph is probably correct.

27. $2x + 3y = 8$

To find the x-intercept, let $y = 0$. Then solve for x.

$$2x + 3y = 8$$
$$2x + 3 \cdot 0 = 8$$
$$2x = 8$$
$$x = 4$$

Thus, $(4, 0)$ is the x-intercept.

To find the y-intercept, let $x = 0$. Then solve for y.

$$2x + 3y = 8$$
$$2 \cdot 0 + 3y = 8$$
$$3y = 8$$
$$y = \frac{8}{3}$$

Thus, $\left(0, \frac{8}{3}\right)$ is the y-intercept.

Plot these points and draw the graph.

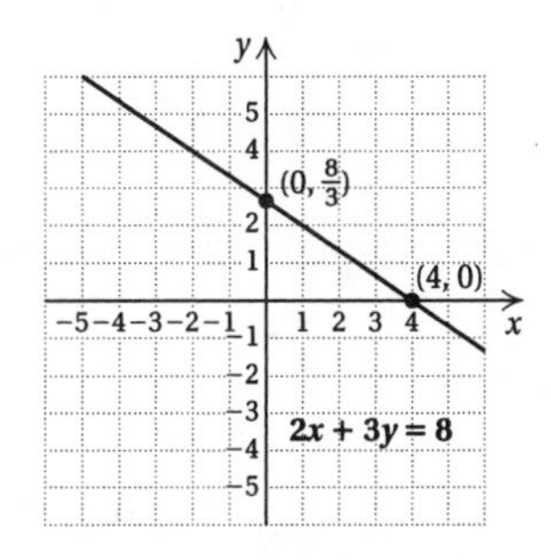

A third point should be used as a check.

We let $x = 1$. Then

$$2x + 3y = 8$$
$$2 \cdot 1 + 3y = 8$$
$$2 + 3y = 8$$
$$3y = 6$$
$$y = 2$$

The point $(1, 2)$ is on the graph, so the graph is probably correct.

29. $x - 3 = y$

To find the x-intercept, let $y = 0$. Then solve for x.

$$x - 3 = y$$
$$x - 3 = 0$$
$$x = 3$$

Thus, $(3, 0)$ is the x-intercept.

To find the y-intercept, let $x = 0$. Then solve for y.

$$x - 3 = y$$
$$0 - 3 = y$$
$$-3 = y$$

Thus, $(0, -3)$ is the y-intercept.

Plot these points and draw the line.

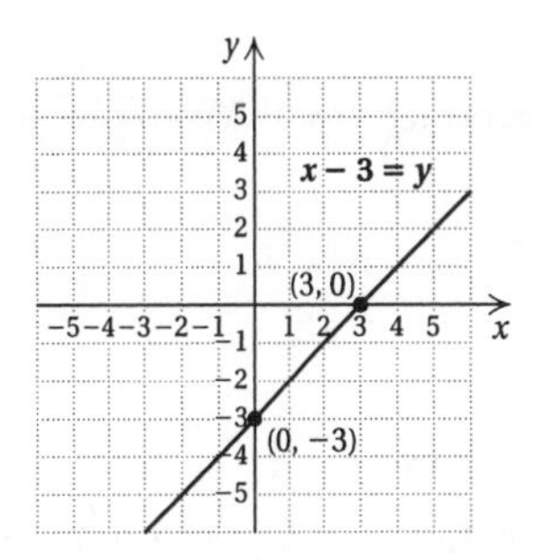

A third point should be used as a check.

We let $x = -2$. Then

$$x - 3 = y$$
$$-2 - 3 = y$$
$$-5 = y$$

The point $(-2, -5)$ is on the graph, so the graph is probably correct.

31. $3x - 2 = y$

To find the x-intercept, let $y = 0$. Then solve for x.

$$3x - 2 = y$$
$$3x - 2 = 0$$
$$3x = 2$$
$$x = \frac{2}{3}$$

Thus, $\left(\frac{2}{3}, 0\right)$ is the x-intercept.

To find the y-intercept, let $x = 0$. Then solve for y.

$$3x - 2 = y$$
$$3 \cdot 0 - 2 = y$$
$$-2 = y$$

Thus, $(0, -2)$ is the y-intercept.

Plot these points and draw the line.

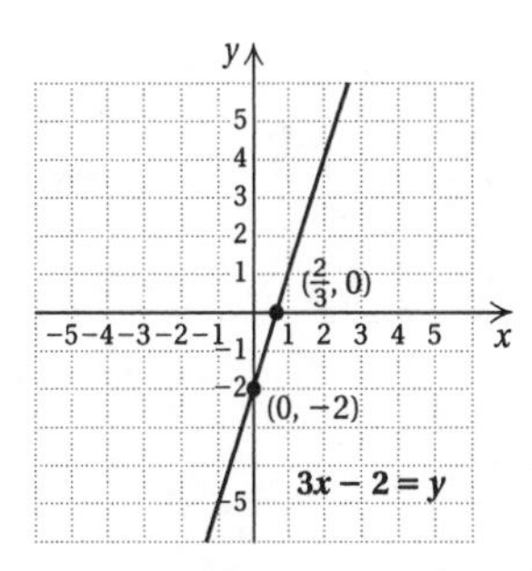

A third point should be used as a check.

We let $x = 2$. Then

$$3x - 2 = y$$
$$3 \cdot 2 - 2 = y$$
$$6 - 2 = y$$
$$4 = y$$

The point $(2, 4)$ is on the graph, so the graph is probably correct.

33. $6x - 2y = 12$

To find the x-intercept, let $y = 0$. Then solve for x.

$$6x - 2y = 12$$
$$6x - 2 \cdot 0 = 12$$
$$6x = 12$$
$$x = 2$$

Thus, $(2, 0)$ is the x-intercept.

To find the y-intercept, let $x = 0$. Then solve for y.

$$6x - 2y = 12$$
$$6 \cdot 0 - 2y = 12$$
$$-2y = 12$$
$$y = -6$$

Thus, $(0, -6)$ is the y-intercept.

Plot these points and draw the line.

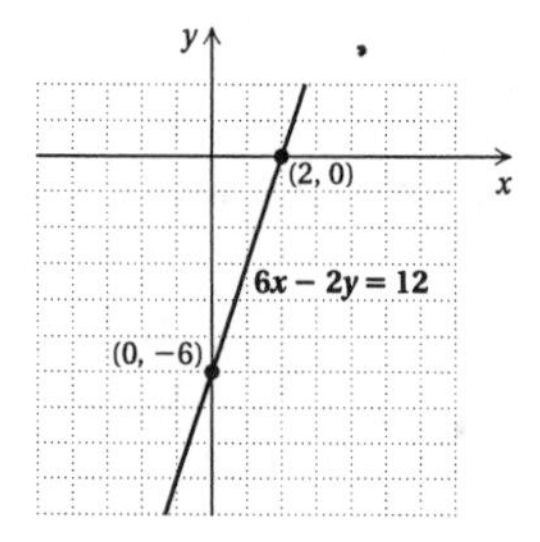

We use a third point as a check.

We let $x = 1$. Then

$$6x - 2y = 12$$
$$6 \cdot 1 - 2y = 12$$
$$6 - 2y = 12$$
$$-2y = 6$$
$$y = -3$$

The point $(1, -3)$ is on the graph, so the graph is probably correct.

35. $3x + 4y = 5$

To find the x-intercept, let $y = 0$. Then solve for x.

$$3x + 4y = 5$$
$$3x + 4 \cdot 0 = 5$$
$$3x = 5$$
$$x = \frac{5}{3}$$

Thus, $\left(\frac{5}{3}, 0\right)$ is the x-intercept.

To find the y-intercept, let $x = 0$. Then solve for y.

$$3x + 4y = 5$$
$$3 \cdot 0 + 4y = 5$$
$$4y = 5$$
$$y = \frac{5}{4}$$

Thus, $\left(0, \frac{5}{4}\right)$ is the y-intercept.

It is helpful to plot another point since the intercepts are so close together. This point can also serve as a check.

We let $x = 3$. Then

$$3x + 4y = 5$$
$$3 \cdot 3 + 4y = 5$$
$$9 + 4y = 5$$
$$4y = -4$$
$$y = -1$$

Plot the point $(3, -1)$ and the intercepts and draw the line.

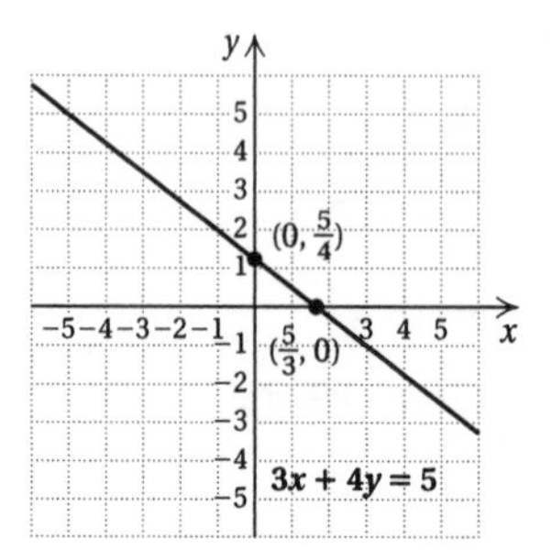

37. $y = -3 - 3x$

To find the x-intercept, let $y = 0$. Then solve for x.

$$y = -3 - 3x$$
$$0 = -3 - 3x$$
$$3x = -3$$
$$x = -1$$

Thus, $(-1, 0)$ is the x-intercept.

To find the y-intercept, let $x = 0$. Then solve for y.

$$y = -3 - 3x$$
$$y = -3 - 3 \cdot 0$$
$$y = -3$$

Thus, $(0, -3)$ is the y-intercept.

Plot these points and draw the graph.

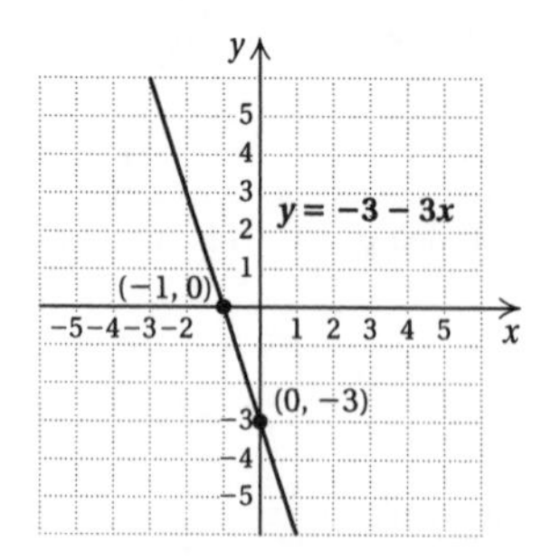

We use a third point as a check.

We let $x = -2$. Then

$$y = -3 - 3x$$
$$y = -3 - 3 \cdot (-2)$$
$$y = -3 + 6$$
$$y = 3$$

The point $(-2, 3)$ is on the graph, so the graph is probably correct.

39. $y - 3x = 0$

To find the x-intercept, let $y = 0$. Then solve for x.

$$0 - 3x = 0$$
$$-3x = 0$$
$$x = 0$$

Thus, $(0, 0)$ is the x-intercept. Note that this is also the y-intercept.

In order to graph the line, we will find a second point.

When $x = 1$, $y - 3 \cdot 1 = 0$

$$y - 3 = 0$$
$$y = 3$$

Plot the points and draw the graph.

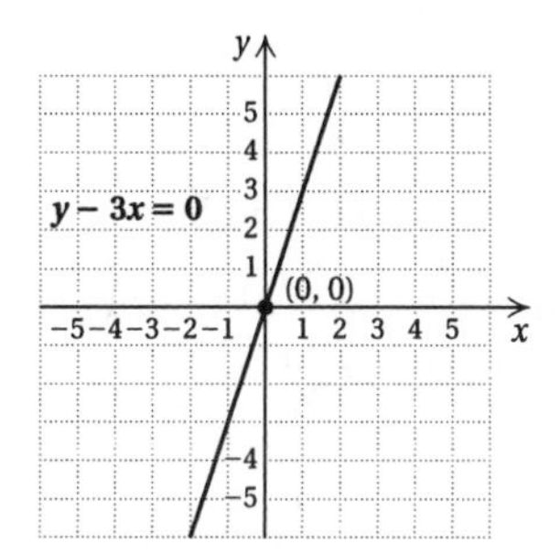

We use a third point as a check.

We let $x = -1$. Then

$$y - 3(-1) = 0$$
$$y + 3 = 0$$
$$y = -3$$

The point $(-1, -3)$ is on the graph, so the graph is probably correct.

41. $x = -2$

Any ordered pair $(-2, y)$ is a solution. The variable x must be -2, but y can be any number we choose. A few solutions are listed below. Plot these points and draw the line.

x	y
-2	-2
-2	0
-2	4

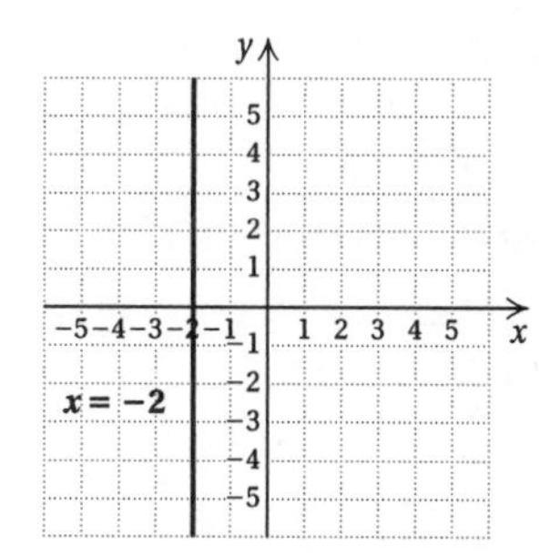

43. $y = 2$

Any ordered pair $(x, 2)$ is a solution. The variable y must be 2, but x can be any number we choose. A few solutions are listed below. Plot these points and draw the line.

x	y
-3	2
0	2
2	2

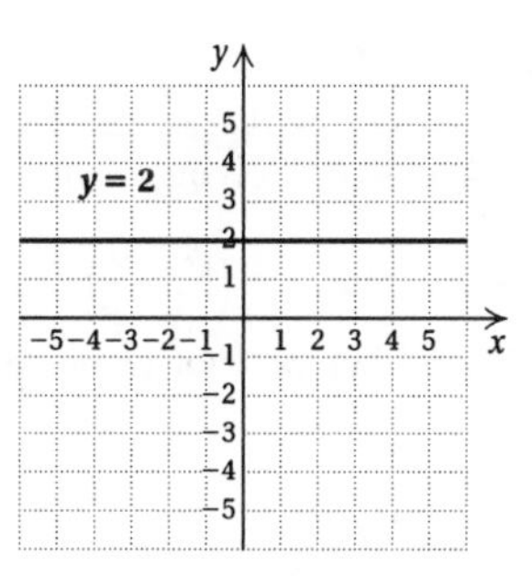

45. $x = 2$

Any ordered pair $(2, y)$ is a solution. The variable x must be 2, but y can be any number we choose. A few solutions are listed below. Plot these points and draw the line.

x	y
2	-1
2	4
2	5

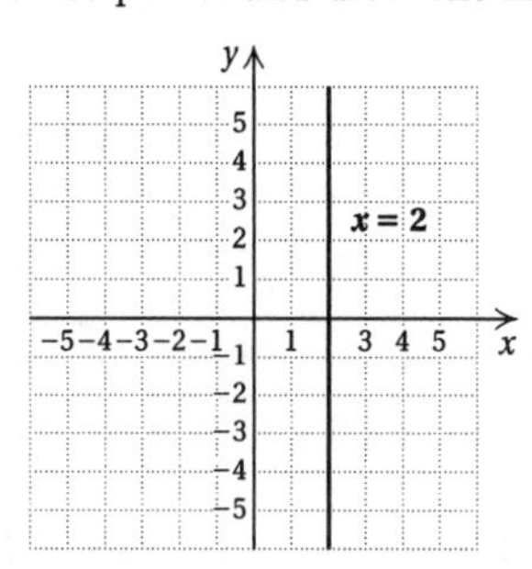

47. $y = 0$

Any ordered pair $(x, 0)$ is a solution. The variable y must be 0, but x can be any number we choose. A few solutions are listed below. Plot these points and draw the line.

x	y
-5	0
-1	0
3	0

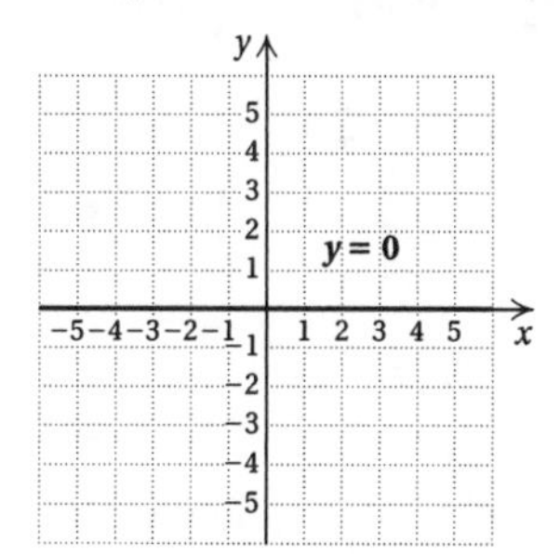

49. $x = \frac{3}{2}$

Any ordered pair $\left(\frac{3}{2}, y\right)$ is a solution. The variable x must be $\frac{3}{2}$, but y can be any number we choose. A few solutions are listed below. Plot these points and draw the line.

x	y
$\frac{3}{2}$	-2
$\frac{3}{2}$	0
$\frac{3}{2}$	4

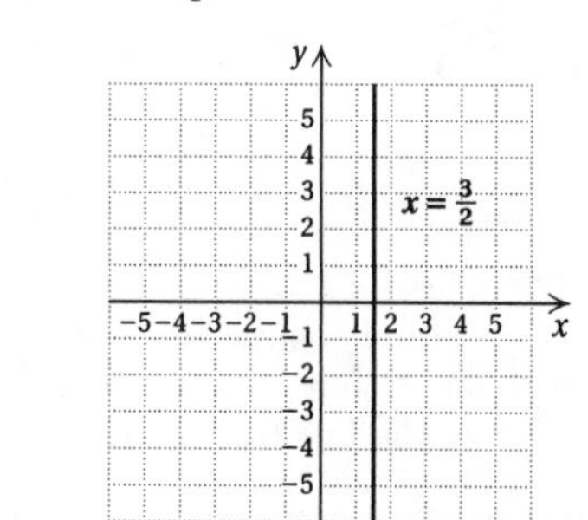

51. $3y = -5$

$y = -\frac{5}{3}$ Solving for y

Any ordered pair $\left(x, -\frac{5}{3}\right)$ is a solution. A few solutions are listed below. Plot these points and draw the line.

x	y
-3	$-\frac{5}{3}$
0	$-\frac{5}{3}$
2	$-\frac{5}{3}$

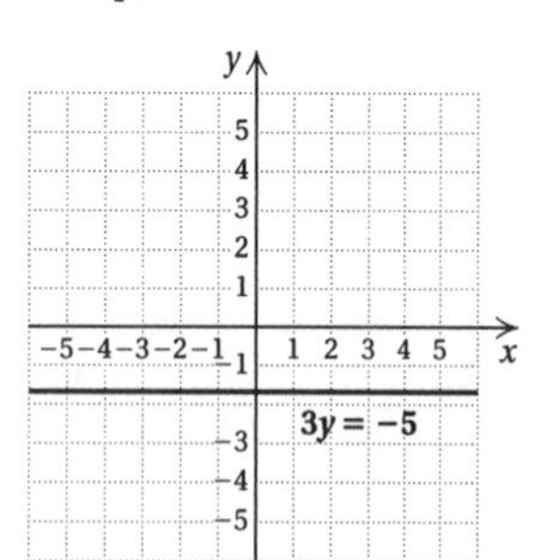

53. $4x + 3 = 0$

$4x = -3$

$x = -\frac{3}{4}$ Solving for x

Any ordered pair $\left(-\frac{3}{4}, y\right)$ is a solution. A few solutions are listed below. Plot these points and draw the line.

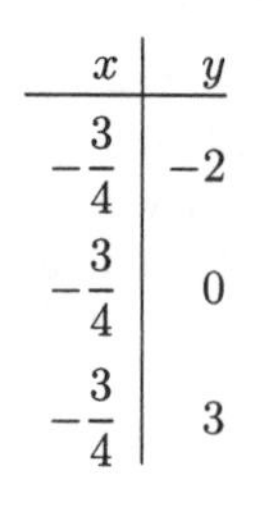

x	y
$-\frac{3}{4}$	-2
$-\frac{3}{4}$	0
$-\frac{3}{4}$	3

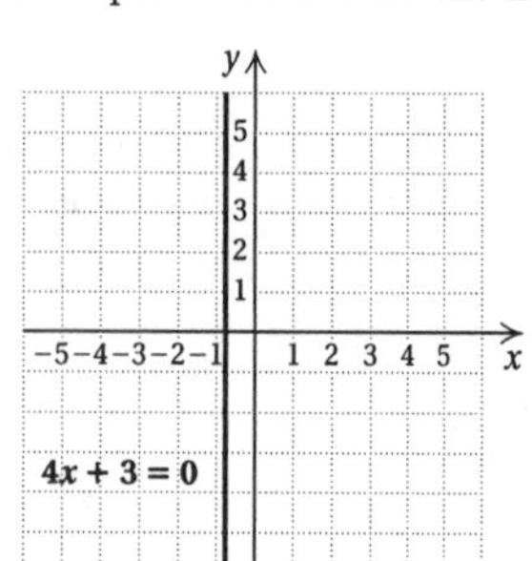

55. $48 - 3y = 0$

$-3y = -48$

$y = 16$ Solving for y

Any ordered pair $(x, 16)$ is a solution. A few solutions are listed below. Plot these points and draw the line.

x	y
-4	16
0	16
2	16

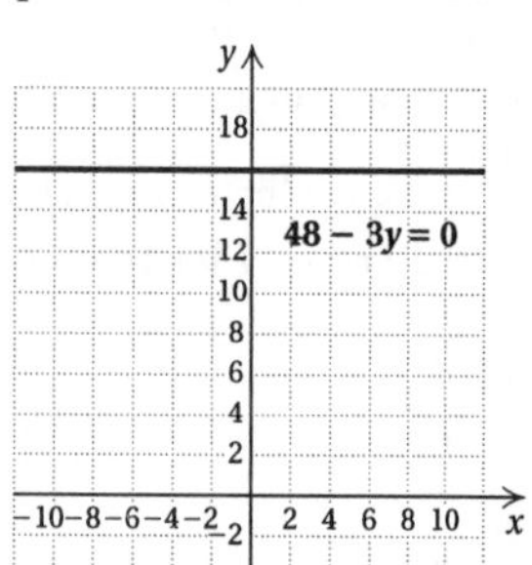

57. Note that every point on the horizontal line passing through $(0, -1)$ has -1 as the y-coordinate. Thus, the equation of the line is $y = -1$.

59. Note that every point on the vertical line passing through $(4, 0)$ has 4 as the x-coordinate. Thus, the equation of the line is $x = 4$.

61. ***Familiarize.*** Let p = the percent of desserts sold that will be pie.

Translate. We reword the problem.

40 is what percent of 250?

$40 = p \cdot 250$

Solve. We solve the equation.

$40 = p \cdot 250$

$\frac{40}{250} = \frac{p \cdot 250}{250}$

$0.16 = p$

$16\% = p$

Check. We can find 16% of 250:

$16\% \cdot 250 = 0.16 \cdot 250 = 40$

The answer checks.

State. 16% of the desserts sold will be pie.

63. ***Familiarize.*** Let c = the cost of the meal before the tip.

Translate. We reword the problem.

\$6.50 is 20% of what?

$6.50 = 20\% \cdot c$

Solve. We solve the equation.

$6.50 = 20\% \cdot c$

$6.50 = 0.2c$

$\frac{6.50}{0.2} = \frac{0.2c}{0.2}$

$32.50 = c$

Check. We can find 20% of 32.50:

$20\% \cdot 32.50 = 0.2 \cdot 32.50 = 6.50$

The answer checks.

State. The cost of the meal before the tip was \$32.50.

65. $-1.6x < 64$

$\frac{-1.6x}{-1.6} > \frac{64}{-1.6}$ Dividing by -1.6 and reversing the inequality symbol

$x > -40$

The solution set is $\{x|x > -40\}$.

67. $x + (x - 1) < (x + 2) - (x + 1)$

$2x - 1 < x + 2 - x - 1$

$2x - 1 < 1$

$2x < 2$

$x < 1$

The solution set is $\{x|x < 1\}$.

69. ◈

71. The y-axis is a vertical line, so it is of the form $x = a$. All points on the y-axis are of the form $(0, y)$, so a must be 0 and the equation is $x = 0$.

73. A line parallel to the x-axis has an equation of the form $y = b$. Since the y-coordinate of one point on the line is -4, then $b = -4$ and the equation is $y = -4$.

75. Substitute -4 for x and 0 for y.

$$3(-4) + k = 5 \cdot 0$$
$$-12 + k = 0$$
$$k = 12$$

Exercise Set 9.4

1. To find the mean we add the numbers and divide by the number of addends, 7.

$$\frac{15 + 40 + 30 + 30 + 45 + 15 + 25}{7} = \frac{200}{7} \approx 28.6$$

To find the median we first list the numbers in order from smallest to largest.

15 15 25 30 30 40 45

The middle number, 30, is the median.

There are two numbers that occur most often, 15 and 30. Thus, the modes are 15 and 30.

3. To find the mean we add the numbers and divide by the number of addends, 6.

$$\frac{81 + 93 + 96 + 98 + 102 + 94}{6} = \frac{564}{6} = 94$$

To find the median we first list the numbers in order from smallest to largest.

81 93 94 96 98 102

There is an even number of numbers. The median is the average of the two middle numbers.

$$\frac{94 + 96}{2} = \frac{190}{2} = 95$$

Each number occurs the same number of times. There is no mode.

5. To find the mean we add the numbers and divide by the number of addends, 5.

$$\frac{23 + 42 + 35 + 37 + 23}{5} = \frac{160}{5} = 32$$

To find the median we first list the numbers in order from smallest to largest.

23 23 35 37 42

The middle number, 35, is the median.

The number 23 occurs most often. It is the mode.

7. To find the mean we add the numbers and divide by the number of addends, 5.

$$\frac{1113 + 610 + 1215 + 798 + 750}{5} = \frac{4486}{5} = 897.2$$

The mean, or average, annual coffee consumption in the given countries is 897.2 cups per person.

To find the median we first list the numbers in order from smallest to largest.

610 750 798 1113 1215

The middle number, 798, is the mean.

Each number occurs the same number of times. There is no mode.

9. To find the mean we add the numbers and divide by the number of addends, 7.

$$\frac{85 + 91 + 90 + 88 + 88 + 85 + 87}{7} = \frac{614}{7} \approx 87.7$$

The mean, or average, height of the 7 tallest men in the NBA is about 87.7 inches.

To find the median we first list the numbers in order from smallest to largest.

85 85 87 88 88 90 91

The middle number, 88, is the median so the median height is 88 inches.

The numbers 85 and 88 occur most often. Thus, the modes are 85 inches and 88 inches.

11. Battery A: Average $= (27.9 + 28.3 + 27.4 + 27.6 + 27.9 + 28.0 + 26.8 + 27.7 + 28.1 + 28.2 + 26.9 + 27.4)/12 = \frac{332.2}{12} \approx 27.7$

Battery B: Average $= (28.3 + 27.6 + 27.8 + 27.4 + 27.9 + 26.9 + 27.8 + 28.1 + 27.9 + 28.7 + 27.6)/11 = \frac{306}{11} \approx 27.8$

Since the average time for Battery B is higher than the average time for Battery A, we conclude that Battery B is better.

13. We interpolate by finding the average of the data values for ages 16 and 18.

$$\frac{162.2 + 162.5}{2} = \frac{324.7}{2} \approx 162.4$$

The missing data value is about 162.4.

We could also have used a graph to find this value as in Example 7.

15. Graph the data and use the graph to extrapolate. Drawing a "representative" line through the data and beyond gives an estimate of about 2008 for the missing data value. Answers will vary depending on the points chosen to determine the "representative" line.

17. Graph the data and use the graph to extrapolate. Drawing a "representative" line through the data and beyond gives an estimate of about 1.39 for the data value corresponding to 1997 and an estimate of about 1.54 for the data value corresponding to 2000. Answers will vary depending on the points chosen to determine the "representative" line.

19. Graph the data and use the graph to extrapolate. Drawing a "representative" line through the data and beyond gives an estimate of about \$1,500,000 for the data value corresponding to 1997 and an estimate of about \$1,700,000 for the data value corresponding to 2000.

21. We interpolate by finding the average of the data values for 17 hours and 19 hours.

$$\frac{81 + 87}{2} = \frac{168}{2} = 84$$

The missing data value is 84.

We could also have used a graph to find this value as in Example 7.

23. $16\% = \frac{16}{100} = \frac{\cancel{4} \cdot 4}{\cancel{4} \cdot 25} = \frac{4}{25}$

25. $37.5\% = \frac{37.5}{100} = \frac{37.5}{100} \cdot \frac{10}{10} = \frac{375}{1000} = \frac{3 \cdot \cancel{125}}{8 \cdot \cancel{125}} = \frac{3}{8}$

27. ***Familiarize.*** Let $p =$ the percent of the cost of the meal represented by the tip.

Translate. We reword the problem.

$$\underbrace{\$8.50}\ \text{is}\ \underbrace{\text{what percent}}\ \text{of}\ \$42.50?$$

$$8.50 = p \cdot 42.50$$

Solve. We solve the equation.

$$8.50 = p \cdot 42.50$$
$$0.2 = p$$
$$20\% = p$$

Check. We can find 20% of 42.50.

$$20\% \cdot 42.50 = 0.2 \cdot 42.50 = 8.50$$

The answer checks.

State. The tip was 20% of the cost of the meal.

29. ***Familiarize.*** Let $c =$ the cost of the meal before the tip was added. Then the tip is $15\% \cdot c$.

Translate. We reword the problem.

$$\underbrace{\text{Cost of meal}}\ \text{plus}\ \text{tip}\ \text{is}\ \underbrace{\text{total cost}}$$

$$c + 15\% \cdot c = 51.92$$

Solve. We solve the equation.

$$c + 15\% \cdot c = 51.92$$
$$1 \cdot c + 0.15c = 51.92$$
$$1.15c = 51.92$$
$$c \approx 45.15$$

Check. We can find 15% of 45.15 and then add this to 45.15.

$15\% \cdot 45.15 = 0.15 \cdot 45.15 \approx 6.77$ and $45.15 + 6.77 = 51.92$

The answer checks.

State. Before the tip the meal cost \$45.15.

31. 

33.

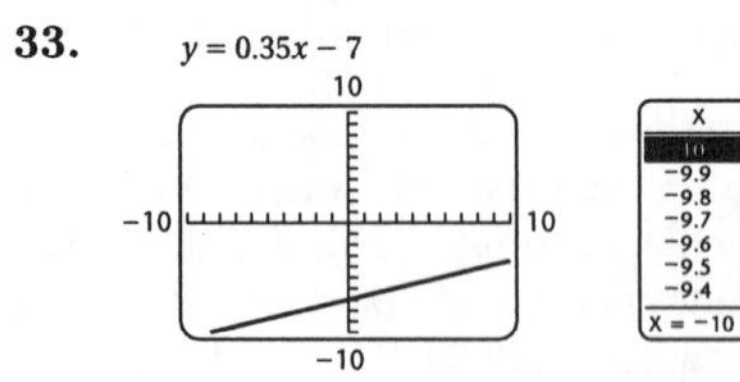

X	Y1
10	−10.5
−9.9	−10.47
−9.8	−10.43
−9.7	−10.4
−9.6	−10.36
−9.5	−10.33
−9.4	−10.29

X = −10

35.

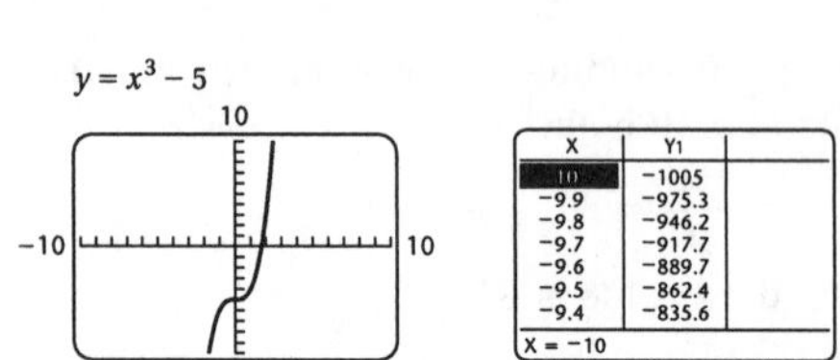

X	Y1
10	−1005
−9.9	−975.3
−9.8	−946.2
−9.7	−917.7
−9.6	−889.7
−9.5	−862.4
−9.4	−835.6

X = −10

Chapter 10

Polynomials: Operations

Exercise Set 10.1

1. 3^4 means $3 \cdot 3 \cdot 3 \cdot 3$.

3. $(1.1)^5$ means $(1.1)(1.1)(1.1)(1.1)(1.1)$.

5. $\left(\frac{2}{3}\right)^4$ means $\left(\frac{2}{3}\right)\left(\frac{2}{3}\right)\left(\frac{2}{3}\right)\left(\frac{2}{3}\right)$.

7. $(7p)^2$ means $(7p)(7p)$.

9. $8k^3$ means $8 \cdot k \cdot k \cdot k$.

11. $a^0 = 1,\ a \neq 0$

13. $b^1 = b$

15. $\left(\frac{2}{3}\right)^0 = 1$

17. $8.38^0 = 1$

19. $(ab)^1 = ab$

21. $ab^1 = a \cdot b^1 = ab$

23. $m^3 = 3^3 = 3 \cdot 3 \cdot 3 = 27$

25. $p^1 = 19^1 = 19$

27. $x^4 = 4^4 = 4 \cdot 4 \cdot 4 \cdot 4 = 256$

29.
$$\begin{aligned} y^2 - 7 &= 10^2 - 7 \\ &= 100 - 7 \quad \text{Evaluating the power} \\ &= 93 \quad \text{Subtracting} \end{aligned}$$

31.
$$\begin{aligned} x^1 + 3 &= 7^1 + 3 \\ &= 7 + 3 \quad (7^1 = 7) \\ &= 10 \end{aligned}$$

$$\begin{aligned} x^0 + 3 &= 7^0 + 3 \\ &= 1 + 3 \quad (7^0 = 1) \\ &= 4 \end{aligned}$$

33.
$$\begin{aligned} A = \pi r^2 &\approx 3.14 \times (34\text{ ft})^2 \\ &\approx 3.14 \times 1156\text{ ft}^2 \quad \text{Evaluating the power} \\ &\approx 3629.84\text{ ft}^2 \end{aligned}$$

35. $3^{-2} = \frac{1}{3^2} = \frac{1}{9}$

37. $10^{-3} = \frac{1}{10^3} = \frac{1}{1000}$

39. $7^{-3} = \frac{1}{7^3} = \frac{1}{343}$

41. $a^{-3} = \frac{1}{a^3}$

43. $\frac{1}{8^{-2}} = 8^2 = 64$

45. $\frac{1}{y^{-4}} = y^4$

47. $\frac{1}{z^{-n}} = z^n$

49. $\frac{1}{4^3} = 4^{-3}$

51. $\frac{1}{x^3} = x^{-3}$

53. $\frac{1}{a^5} = a^{-5}$

55. $2^4 \cdot 2^3 = 2^{4+3} = 2^7$

57. $8^5 \cdot 8^9 = 8^{5+9} = 8^{14}$

59. $x^4 \cdot x^3 = x^{4+3} = x^7$

61. $9^{17} \cdot 9^{21} = 9^{17+21} = 9^{38}$

63. $(3y)^4(3y)^8 = (3y)^{4+8} = (3y)^{12}$

65. $(7y)^1(7y)^{16} = (7y)^{1+16} = (7y)^{17}$

67. $3^{-5} \cdot 3^8 = 3^{-5+8} = 3^3$

69. $x^{-2} \cdot x = x^{-2+1} = x^{-1} = \frac{1}{x}$

71. $x^{14} \cdot x^3 = x^{14+3} = x^{17}$

73. $x^{-7} \cdot x^{-6} = x^{-7+(-6)} = x^{-13} = \frac{1}{x^{13}}$

75. $a^{11} \cdot a^{-3} \cdot a^{-18} = a^{11+(-3)+(-18)} = a^{-10} = \frac{1}{a^{10}}$

77. $t^8 \cdot t^{-8} = t^{8+(-8)} = t^0 = 1$

79. $\frac{7^5}{7^2} = 7^{5-2} = 7^3$

81. $\frac{8^{12}}{8^6} = 8^{12-6} = 8^6$

83. $\frac{y^9}{y^5} = y^{9-5} = y^4$

85. $\frac{16^2}{16^8} = 16^{2-8} = 16^{-6} = \frac{1}{16^6}$

87. $\frac{m^6}{m^{12}} = m^{6-12} = m^{-6} = \frac{1}{m^6}$

89. $\frac{(8x)^6}{(8x)^{10}} = (8x)^{6-10} = (8x)^{-4} = \frac{1}{(8x)^4}$

91. $\dfrac{(2y)^9}{(2y)^9} = (2y)^{9-9} = (2y)^0 = 1$

93. $\dfrac{x}{x^{-1}} = x^{1-(-1)} = x^2$

95. $\dfrac{x^7}{x^{-2}} = x^{7-(-2)} = x^9$

97. $\dfrac{z^{-6}}{z^{-2}} = z^{-6-(-2)} = z^{-4} = \dfrac{1}{z^4}$

99. $\dfrac{x^{-5}}{x^{-8}} = x^{-5-(-8)} = x^3$

101. $\dfrac{m^{-9}}{m^{-9}} = m^{-9-(-9)} = m^0 = 1$

103. $5^2 = 5 \cdot 5 = 25$

$5^{-2} = \dfrac{1}{5^2} = \dfrac{1}{25}$

$\left(\dfrac{1}{5}\right)^2 = \dfrac{1}{5} \cdot \dfrac{1}{5} = \dfrac{1}{25}$

$\left(\dfrac{1}{5}\right)^{-2} = \dfrac{1}{\left(\dfrac{1}{5}\right)^2} = \dfrac{1}{\dfrac{1}{25}} = 1 \cdot \dfrac{25}{1} = 25$

$-5^2 = -(5)(5) = -25$

$(-5)^2 = (-5)(-5) = 25$

105. $64\%t$, or $0.64t$

107.

```
              6 4 .
  2 4.3^ ) 1 5 5 5.2 ^
           1 4 5 8
             9 7 2
             9 7 2
                 0
```

The answer is 64.

109.

$$\begin{aligned}
3x - 4 + 5x - 10x &= x - 8 \\
-2x - 4 &= x - 8 && \text{Collecting like terms} \\
-2x - 4 + 4 &= x - 8 + 4 && \text{Adding 4} \\
-2x &= x - 4 \\
-2x - x &= x - 4 - x && \text{Subtracting } x \\
-3x &= -4 \\
\frac{-3x}{-3} &= \frac{-4}{-3} && \text{Dividing by } -3 \\
x &= \frac{4}{3}
\end{aligned}$$

The solution is $\dfrac{4}{3}$.

111. ***Familiarize.*** Let x = the length of the shorter piece. Then $2x$ = the length of the longer piece.

Translate.

Length of shorter piece	plus	length of longer piece	is	12 in.
↓	↓	↓	↓	↓
x	$+$	$2x$	$=$	12

Solve.

$$\begin{aligned}
x + 2x &= 12 \\
3x &= 12 \\
\frac{3x}{3} &= \frac{12}{3} \\
x &= 4
\end{aligned}$$

If $x = 4$, $2x = 2 \cdot 4 = 8$.

Check. The longer piece, 8 in., is twice as long as the shorter piece, 4 in. Also, 4 in. + 8 in. = 12 in., the total length of the sandwich. The answer checks.

State. The lengths of the pieces are 4 in. and 8 in.

113. ◈

115. Let $y_1 = (x+1)^2$ and $y_2 = x^2+1$. A graph of the equations or a table of values shows that $(x+1)^2 = x^2+1$ is not correct.

117. Let $y_1 = (5x)^0$ and $y_2 = 5x^0$. A graph of the equations or a table of values shows that $(5x)^0 = 5x^0$ is not correct.

119. $(y^{2x})(y^{3x}) = y^{2x+3x} = y^{5x}$

121. $\dfrac{a^{6t}(a^{7t})}{a^{9t}} = \dfrac{a^{6t+7t}}{a^{9t}} = \dfrac{a^{13t}}{a^{9t}} = a^{13t-9t} = a^{4t}$

123. $\dfrac{(0.8)^5}{(0.8)^3(0.8)^2} = \dfrac{(0.8)^5}{(0.8)^{3+2}} = \dfrac{(0.8)^5}{(0.8)^5} = 1$

125. Since the bases are the same, the expression with the larger exponent is larger. Thus, $3^5 > 3^4$.

127. Since the exponents are the same, the expression with the larger base is larger. Thus, $4^3 < 5^3$.

129. Choose any number except 0. For example, let $x = 1$.

$3x^2 = 3 \cdot 1^2 = 3 \cdot 1 = 3$, but

$(3x)^2 = (3 \cdot 1)^2 = 3^2 = 9$.

Exercise Set 10.2

1. $(2^3)^2 = 2^{3 \cdot 2} = 2^6$

3. $(5^2)^{-3} = 5^{2(-3)} = 5^{-6} = \dfrac{1}{5^6}$

5. $(x^{-3})^{-4} = x^{(-3)(-4)} = x^{12}$

7. $(4x^3)^2 = 4^2(x^3)^2$ Raising each factor to the second power

$= 16x^6$

9. $(x^4y^5)^{-3} = (x^4)^{-3}(y^5)^{-3} = x^{4(-3)}y^{5(-3)} = x^{-12}y^{-15} = \dfrac{1}{x^{12}y^{15}}$

11. $(x^{-6}y^{-2})^{-4} = (x^{-6})^{-4}(y^{-2})^{-4} = x^{(-6)(-4)}y^{(-2)(-4)} = x^{24}y^8$

13. $(3x^3y^{-8}z^{-3})^2 = 3^2(x^3)^2(y^{-8})^2(z^{-3})^2 = 9x^6y^{-16}z^{-6} = \dfrac{9x^6}{y^{16}z^6}$

15. $\left(\frac{a^2}{b^3}\right)^4 = \frac{(a^2)^4}{(b^3)^4} = \frac{a^8}{b^{12}}$

17. $\left(\frac{y^3}{2}\right)^2 = \frac{(y^3)^2}{2^2} = \frac{y^6}{4}$

19. $\left(\frac{y^2}{2}\right)^{-3} = \frac{(y^2)^{-3}}{2^{-3}} = \frac{y^{-6}}{2^{-3}} = \frac{\frac{1}{y^6}}{\frac{1}{2^3}} = \frac{1}{y^6} \cdot \frac{2^3}{1} = \frac{8}{y^6}$

21. $\left(\frac{x^2 y}{z}\right)^3 = \frac{(x^2)^3 y^3}{z^3} = \frac{x^6 y^3}{z^3}$

23. $\left(\frac{a^2 b}{cd^3}\right)^{-2} = \frac{(a^2)^{-2} b^{-2}}{c^{-2}(d^3)^{-2}} = \frac{a^{-4} b^{-2}}{c^{-2} d^{-6}} = \frac{\frac{1}{a^4} \cdot \frac{1}{b^2}}{\frac{1}{c^2} \cdot \frac{1}{d^6}} = \frac{\frac{1}{a^4 b^2}}{\frac{1}{c^2 d^6}} =$

$\frac{1}{a^4 b^2} \cdot \frac{c^2 d^6}{1} = \frac{c^2 d^6}{a^4 b^2}$

25. 2.8,000,000,000.

↑ 10 places

Large number, so the exponent is positive.

$28,000,000,000 = 2.8 \times 10^{10}$

27. 9.07,000,000,000,000,000.

↑ 17 places

Large number, so the exponent is positive.

$907,000,000,000,000,000 = 9.07 \times 10^{17}$

29. 0.000003.04

↑ 6 places

Small number, so the exponent is negative.

$0.00000304 = 3.04 \times 10^{-6}$

31. 0.00000001. 8

↑ 8 places

Small number, so the exponent is negative.

$0.000000018 = 1.8 \times 10^{-8}$

33. 1.00,000,000,000.

↑ 11 places

Large number, so the exponent is positive.

$100,000,000,000 = 1.0 \times 10^{11} = 10^{11}$

35. 11.35 million = 11,350,000

1.1,350,000.

↑ 7 places

Large number, so the exponent is positive.

11.35 million $= 1.135 \times 10^7$

37. 8.74×10^7

Positive exponent, so the answer is a large number.

8.7400000.

↑ 7 places

$8.74 \times 10^7 = 87,400,000$

39. 5.704×10^{-8}

Negative exponent, so the answer is a small number.

0.00000005.704

↑ 8 places

$5.704 \times 10^{-8} = 0.00000005704$

41. $10^7 = 1 \times 10^7$

Positive exponent, so the answer is a large number.

1.0000000.

↑ 7 places

$10^7 = 10,000,000$

43. $10^{-5} = 1 \times 10^{-5}$

Negative exponent, so the answer is a small number.

0.00001.

↑ 5 places

$10^{-5} = 0.00001$

45. $(3 \times 10^4)(2 \times 10^5) = (3 \cdot 2) \times (10^4 \cdot 10^5)$
$= 6 \times 10^9$

47. $(5.2 \times 10^5)(6.5 \times 10^{-2}) = (5.2 \cdot 6.5) \times (10^5 \cdot 10^{-2})$
$= 33.8 \times 10^3$

The answer at this stage is 33.8×10^3 but this is not scientific notation since 33.8 is not a number between 1 and 10. We convert 33.8 to scientific notation and simplify.

$33.8 \times 10^3 = (3.38 \times 10^1) \times 10^3 = 3.38 \times (10^1 \times 10^3) = 3.38 \times 10^4$

The answer is 3.38×10^4.

49. $(9.9 \times 10^{-6})(8.23 \times 10^{-8}) = (9.9 \cdot 8.23) \times (10^{-6} \cdot 10^{-8})$
$= 81.477 \times 10^{-14}$

The answer at this stage is 81.477×10^{-14}. We convert 81.477 to scientific notation and simplify.

$81.477 \times 10^{-14} = (8.1477 \times 10^1) \times 10^{-14} = 8.1477 \times (10^1 \times 10^{-14}) = 8.1477 \times 10^{-13}$.

The answer is 8.1477×10^{-13}.

51. $\frac{8.5 \times 10^8}{3.4 \times 10^{-5}} = \frac{8.5}{3.4} \times \frac{10^8}{10^{-5}}$
$= 2.5 \times 10^{8-(-5)}$
$= 2.5 \times 10^{13}$

53. $(3.0 \times 10^6) \div (6.0 \times 10^9) = \frac{3.0 \times 10^6}{6.0 \times 10^9}$
$= \frac{3.0}{6.0} \times \frac{10^6}{10^9}$
$= 0.5 \times 10^{6-9}$
$= 0.5 \times 10^{-3}$

The answer at this stage is 0.5×10^{-3}. We convert 0.5 to scientific notation and simplify.

$0.5 \times 10^{-3} = (5.0 \times 10^{-1}) \times 10^{-3} = 5.0 \times (10^{-1} \times 10^{-3}) = 5.0 \times 10^{-4}$

55. $\dfrac{7.5\times10^{-9}}{2.5\times10^{12}}=\dfrac{7.5}{2.5}\times\dfrac{10^{-9}}{10^{12}}$
$=3.0\times10^{-9-12}$
$=3.0\times10^{-21}$

57. 31.2 million $=31,200,000=3.12\times10^7$

$42,400=4.24\times10^4$

We multiply to find the total income.

$(3.12\times10^7)(4.24\times10^4)=13.2288\times10^{11}$
$=(1.32288\times10)\times10^{11}$
$=1.32288\times10^{12}$

The total income generated by two-person households in 1993 was $\$1.32288\times10^{12}$.

59. 10 billion trillion $=1\times10\times10^9\times10^{12}$
$=1\times10^{22}$

There are 1×10^{22} stars in the known universe.

61. We divide the mass of the sun by the mass of earth.

$\dfrac{1.998\times10^{27}}{6\times10^{21}}=0.333\times10^6$
$=(3.33\times10^{-1})\times10^6$
$=3.33\times10^5$

The mass of the sun is 3.33×10^5 times the mass of earth.

63. First we divide the distance from the earth to the moon by 3 days to find the number of miles per day the space vehicle travels. Note that $240,000=2.4\times10^5$.

$\dfrac{2.4\times10^5}{3}=0.8\times10^5=8\times10^4$

The space vehicle travels 8×10^4 miles per day. Now divide the distance from the earth to Mars by 8×10^4 to find how long it will take the space vehicle to reach Mars. Note that $35,000,000=3.5\times10^7$.

$\dfrac{3.5\times10^7}{8\times10^4}=0.4375\times10^3=4.375\times10^2$

It takes 4.375×10^2 days for the space vehicle to travel from the earth to Mars.

65. $9x-36=9\cdot x-9\cdot4=9(x-4)$

67. $3s+3t+24=3\cdot s+3\cdot t+3\cdot8=3(s+t+8)$

69.
$2x-4-5x+8=x-3$
$-3x+4=x-3$ Collecting like terms
$-3x+4-4=x-3-4$ Subtracting 4
$-3x=x-7$
$-3x-x=x-7-x$ Subtracting x
$-4x=-7$
$\dfrac{-4x}{-4}=\dfrac{-7}{-4}$ Dividing by -4
$x=\dfrac{7}{4}$

The solution is $\dfrac{7}{4}$.

71.
$8(2x+3)-2(x-5)=10$
$16x+24-2x+10=10$ Removing parentheses
$14x+34=10$ Collecting like terms
$14x+34-34=10-34$ Subtracting 34
$14x=-24$
$\dfrac{14x}{14}=\dfrac{-24}{14}$ Dividing by 14
$x=-\dfrac{12}{7}$ Simplifying

The solution is $-\dfrac{12}{7}$.

73. $y=x-5$

The equation is equivalent to $y=x+(-5)$. The y-intercept is $(0,-5)$. We find two other points.

When $x=2$, $y=2-5=-3$.

When $x=4$, $y=4-5=-1$.

x	y
0	-5
2	-3
4	-1

Plot these points, draw the line they determine, and label the graph $y=x-5$.

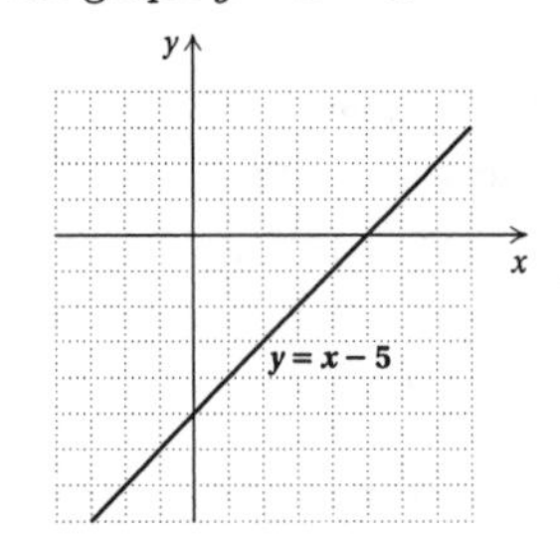

75. ◈

77. $\dfrac{(5.2\times10^6)(6.1\times10^{-11})}{1.28\times10^{-3}}=\dfrac{(5.2\cdot6.1)}{1.28}\times\dfrac{(10^6\cdot10^{-11})}{10^{-3}}$
$=24.78125\times10^{-2}$
$=(2.478125\times10^1)\times10^{-2}$
$=2.478125\times10^{-1}$

79. $\dfrac{(5^{12})^2}{5^{25}}=\dfrac{5^{24}}{5^{25}}=5^{24-25}=5^{-1}=\dfrac{1}{5}$

81. $\dfrac{(3^5)^4}{3^5\cdot3^4}=\dfrac{3^{5\cdot4}}{3^{5+4}}=\dfrac{3^{20}}{3^9}=3^{20-9}=3^{11}$

83. $\left(\dfrac{1}{a}\right)^{-n}=\dfrac{1^{-n}}{a^{-n}}=\dfrac{\frac{1}{1^n}}{\frac{1}{a^n}}=\dfrac{1}{1}\cdot\dfrac{a^n}{1}=a^n$

85. False; let $x=2$, $y=3$, $m=4$, and $n=2$:
$2^4\cdot3^2=16\cdot9=144$, but
$(2\cdot3)^{4\cdot2}=6^8=1,679,616$

87. False; let $x=5$, $y=3$, and $m=2$:
$(5-3)^2=2^2=4$, but
$5^2-3^2=25-9=16$

Exercise Set 10.3

1. $-5x+2=-5\cdot 4+2=-20+2=-18;$

$-5x+2=-5(-1)+2=5+2=7$

3. $2x^2-5x+7=2\cdot 4^2-5\cdot 4+7=2\cdot 16-20+7=32-20+7=19;$

$2x^2-5x+7=2(-1)^2-5(-1)+7=2\cdot 1+5+7=2+5+7=14$

5. $x^3-5x^2+x=4^3-5\cdot 4^2+4=64-5\cdot 16+4=64-80+4=-12;$

$x^3-5x^2+x=(-1)^3-5(-1)^2+(-1)=-1-5\cdot 1-1=-1-5-1=-7$

7. $3x+5=3(-2)+5=-6+5=-1;$

$3x+5=3\cdot 0+5=0+5=5$

9. $x^2-2x+1=(-2)^2-2(-2)+1=4+4+1=9;$

$x^2-2x+1=0^2-2\cdot 0+1=0-0+1=1$

11. $-3x^3+7x^2-3x-2=-3(-2)^3+7(-2)^2-3(-2)-2=-3(-8)+7(4)-3(-2)-2=24+28+6-2=56;$

$-3x^3+7x^2-3x-2=-3\cdot 0^3+7\cdot 0^2-3\cdot 0-2=-3\cdot 0+7\cdot 0-0-2=0+0-0-2=-2$

13. We evaluate the polynomial for $t=10$:

$11.12t^2=11.12(10)^2=11.12(100)=1112$

The skydiver has fallen approximately 1112 ft.

15. We evaluate the polynomial for $x=75$:

$$\begin{aligned}280x-0.4x^2&=280(75)-0.4(75)^2\\&=280(75)-0.4(5625)\\&=21,000-2250\\&=18,750\end{aligned}$$

The total revenue from the sale of 75 TVs is \$18,750.

We evaluate the polynomial for $x=100$:

$$\begin{aligned}280x-0.4x^2&=280(100)-0.4(100)^2\\&=280(100)-0.4(10,000)\\&=28,000-4000\\&=24,000\end{aligned}$$

The total revenue from the sale of 100 TVs is \$24,000.

17. Locate -3 on the x-axis. Then move vertically to the graph and horizontally to the y-axis. It appears that the y-value that is paired with -3 is -4. Thus, the value of $y=5-x^2$ is -4 when $x=-3$.

Locate -1 on the x-axis. Then move vertically to the graph and horizontally to the y-axis. It appears that the y-value that is paired with -1 is 4. Thus, the value of $y=5-x^2$ is 4 when $x=-1$.

Locate 0 on the x-axis. Then move vertically to the graph. We arrive at a point on the y-axis with the y-value 5. Thus, the value of $5-x^2$ is 5 when $x=0$.

Locate 1.5 on the x-axis. Then move vertically to the graph and horizontally to the y-axis. It appears that the y-value that is paired with 1.5 is 2.75. Thus, the value of $y=5-x^2$ is 2.75 when $x=1.5$.

Locate 2 on the x-axis. Then move vertically to the graph and horizontally to the y-axis. It appears that the y-value that is paired with 2 is 1. Thus, the value of $y=5-x^2$ is 1 when $x=2$.

19. Using the polynomial $-0.002d^2+0.8d+6.6$, we get the following results.

For 100 ft: $$\begin{aligned}&-0.002(100)^2+0.8(100)+6.6\\&=-0.002(10,000)+0.8(100)+6.6\\&=-20+80+6.6\\&=66.6\end{aligned}$$

The arrow was about 66.6 ft high after it had traveled 100 ft.

For 200 ft: $$\begin{aligned}&-0.002(200)^2+0.8(200)+6.6\\&=-0.002(40,000)+0.8(200)+6.6\\&=-80+160+6.6\\&=86.6\end{aligned}$$

The arrow was about 86.8 ft high after it had traveled 200 ft.

For 300 ft: $$\begin{aligned}&-0.002(300)^2+0.8(300)+6.6\\&=-0.002(90,000)+0.8(300)+6.6\\&=-180+240+6.6\\&=66.6\end{aligned}$$

The arrow was about 66.6 ft high after it had traveled 300 ft.

For 350 ft: $$\begin{aligned}&-0.002(350)^2+0.8(350)+6.6\\&=-0.002(122,500)+0.8(350)+6.6\\&=-245+280+6.6\\&=41.6\end{aligned}$$

The arrow was about 41.6 ft high after it had traveled 350 ft.

21. $2-3x+x^2=2+(-3x)+x^2$

The terms are 2, $-3x$, and x^2.

23. $5x^3+6x^2-3x^2$

Like terms: $6x^2$ and $-3x^2$ Same variable and exponent

25. $2x^4+5x-7x-3x^4$

Like terms: $2x^4$ and $-3x^4$ Same variable and
Like terms: $5x$ and $-7x$ exponent

27. $3x^5-7x+8+14x^5-2x-9$

Like terms: $3x^5$ and $14x^5$
Like terms: $-7x$ and $-2x$
Like terms: 8 and -9 Constant terms are like terms.

29. $-3x+6$

The coefficient of $-3x$, the first term, is -3.

The coefficient of 6, the second term, is 6.

31. $5x^2+3x+3$

The coefficient of $5x^2$, the first term, is 5.

The coefficient of $3x$, the second term, is 3.

The coefficient of 3, the third term, is 3.

33. $-5x^4 + 6x^3 - 3x^2 + 8x - 2$

The coefficient of $-5x^4$, the first term, is -5.

The coefficient of $6x^3$, the second term, is 6.

The coefficient of $-3x^2$, the third term, is -3.

The coefficient of $8x$, the fourth term, is 8.

The coefficient of -2, the fifth term, is -2.

35. $2x - 5x = (2-5)x = -3x$

37. $x - 9x = 1x - 9x = (1-9)x = -8x$

39. $5x^3 + 6x^3 + 4 = (5+6)x^3 + 4 = 11x^3 + 4$

41. $5x^3 + 6x - 4x^3 - 7x = (5-4)x^3 + (6-7)x =$
$1x^3 + (-1)x = x^3 - x$

43. $6b^5 + 3b^2 - 2b^5 - 3b^2 = (6-2)b^5 + (3-3)b^2 =$
$4b^5 + 0b^2 = 4b^5$

45. $\frac{1}{4}x^5 - 5 + \frac{1}{2}x^5 - 2x - 37 =$
$\left(\frac{1}{4} + \frac{1}{2}\right)x^5 - 2x + (-5 - 37) = \frac{3}{4}x^5 - 2x - 42$

47. $6x^2 + 2x^4 - 2x^2 - x^4 - 4x^2 =$
$6x^2 + 2x^4 - 2x^2 - 1x^4 - 4x^2 =$
$(6-2-4)x^2 + (2-1)x^4 = 0x^2 + 1x^4 =$
$0 + x^4 = x^4$

49. $\frac{1}{4}x^3 - x^2 - \frac{1}{6}x^2 + \frac{3}{8}x^3 + \frac{5}{16}x^3 =$
$\frac{1}{4}x^3 - 1x^2 - \frac{1}{6}x^2 + \frac{3}{8}x^3 + \frac{5}{16}x^3 =$
$\left(\frac{1}{4} + \frac{3}{8} + \frac{5}{16}\right)x^3 + \left(-1 - \frac{1}{6}\right)x^2 =$
$\left(\frac{4}{16} + \frac{6}{16} + \frac{5}{16}\right)x^3 + \left(-\frac{6}{6} - \frac{1}{6}\right)x^2 = \frac{15}{16}x^3 - \frac{7}{6}x^2$

51. $x^5 + x + 6x^3 + 1 + 2x^2 = x^5 + 6x^3 + 2x^2 + x + 1$

53. $5y^3 + 15y^9 + y - y^2 + 7y^8 =$
$15y^9 + 7y^8 + 5y^3 - y^2 + y$

55. $3x^4 - 5x^6 - 2x^4 + 6x^6 = x^4 + x^6 = x^6 + x^4$

57. $-2x + 4x^3 - 7x + 9x^3 + 8 = -9x + 13x^3 + 8 =$
$13x^3 - 9x + 8$

59. $3x + 3x + 3x - x^2 - 4x^2 = 9x - 5x^2 = -5x^2 + 9x$

61. $-x + \frac{3}{4} + 15x^4 - x - \frac{1}{2} - 3x^4 = -2x + \frac{1}{4} + 12x^4 =$
$12x^4 - 2x + \frac{1}{4}$

63. $2x - 4 = 2x^1 - 4x^0$

The degree of $2x$ is 1.

The degree of -4 is 0.

The degree of the polynomial is 1, the largest exponent.

65. $3x^2 - 5x + 2 = 3x^2 - 5x^1 + 2x^0$

The degree of $3x^2$ is 2.

The degree of $-5x$ is 1.

The degree of 2 is 0.

The degree of the polynomial is 2, the largest exponent.

67. $-7x^3 + 6x^2 + 3x + 7 = -7x^3 + 6x^2 + 3x^1 + 7x^0$

The degree of $-7x^3$ is 3.

The degree of $6x^2$ is 2.

The degree of $3x$ is 1.

The degree of 7 is 0.

The degree of the polynomial is 3, the largest exponent.

69. $x^2 - 3x + x^6 - 9x^4 = x^2 - 3x^1 + x^6 - 9x^4$

The degree of x^2 is 2.

The degree of $-3x$ is 1.

The degree of x^6 is 6.

The degree of $-9x^4$ is 4.

The degree of the polynomial is 6, the largest exponent.

71. See the answer section in the text.

73. In the polynomial $x^3 - 27$, there are no x^2 or x terms. The x^2 term (or second-degree term) and the x term (or first-degree term) are missing.

75. In the polynomial $x^4 - x$, there are no x^3, x^2, or x^0 terms. The x^3 term (or third-degree term), the x^2 term (or second-degree term), and the x^0 term (or zero-degree term) are missing.

77. No terms are missing in the polynomial
$2x^3 - 5x^2 + x - 3$.

79. The polynomial $x^2 - 10x + 25$ is a *trinomial* because it has just three terms.

81. The polynomial $x^3 - 7x^2 + 2x - 4$ is *none of these* because it has more than three terms.

83. The polynomial $4x^2 - 25$ is a *binomial* because it has just two terms.

85. The polynomial $40x$ is a *monomial* because it has just one term.

87. ***Familiarize***. Let $a =$ the number of apples the campers had to begin with. Then the first camper ate $\frac{1}{3}a$ apples and $a - \frac{1}{3}a$, or $\frac{2}{3}a$, apples were left. The second camper ate $\frac{1}{3}\left(\frac{2}{3}a\right)$, or $\frac{2}{9}a$, apples, and $\frac{2}{3}a - \frac{2}{9}a$, or $\frac{4}{9}a$, apples were left. The third camper ate $\frac{1}{3}\left(\frac{4}{9}a\right)$, or $\frac{4}{27}a$, apples, and $\frac{4}{9}a - \frac{4}{27}a$, or $\frac{8}{27}a$, apples were left.

Translate. We write an equation for the number of apples left after the third camper eats.

Number of apples left is 8.

$$\underbrace{\qquad\qquad\qquad}_{\downarrow \atop \frac{8}{27}a} \qquad \begin{matrix}\downarrow\downarrow \\ = 8\end{matrix}$$

Solve. We solve the equation.

$$\frac{8}{27}a = 8$$
$$a = \frac{27}{8}\cdot 8$$
$$a = 27$$

Check. If the campers begin with 27 apples, then the first camper eats $\frac{1}{3}\cdot 27$, or 9, and $27-9$, or 18, are left. The second camper then eats $\frac{1}{3}\cdot 18$, or 6 apples and $18-6$, or 12, are left. Finally, the third camper eats $\frac{1}{3}\cdot 12$, or 4 apples and $12-4$, or 8, are left. The answer checks.

State. The campers had 27 apples to begin with.

89. $\frac{1}{8}-\frac{5}{6}=\frac{1}{8}+\left(-\frac{5}{6}\right)$, LCM is 24

$$= \frac{1}{8}\cdot\frac{3}{3}+\left(-\frac{5}{6}\right)\left(\frac{4}{4}\right)$$
$$= \frac{3}{24}+\left(-\frac{20}{24}\right)$$
$$= -\frac{17}{24}$$

91. $5.6-8.2=5.6+(-8.2)=-2.6$

93.

$cx = ab - r$	
$cx + r = ab$	Adding r
$\frac{cx+r}{a} = \frac{ab}{a}$	Dividing by a
$\frac{cx+r}{a} = b$	Simplifying

95. $3x-15y+63=3\cdot x-3\cdot 5y+3\cdot 21=3(x-5y+21)$

97.

99.
$$(3x^2)^3+4x^2\cdot 4x^4-x^4(2x)^2+[(2x)^2]^3-100x^2(x^2)^2$$
$$=27x^6+4x^2\cdot 4x^4-x^4\cdot 4x^2+(2x)^6-100x^2\cdot x^4$$
$$=27x^6+16x^6-4x^6+64x^6-100x^6$$
$$=3x^6$$

101. $(5m^5)^2=5^2m^{5\cdot 2}=25m^{10}$

The degree is 10.

103. Graph $y=5-x^2$. Then use VALUE from the CALC menu to find the y-values that correspond to $x=-3$, $x=-1$, x=0, $x=1.5$, and $x=2$. As before, we find that these values are -4, 4, 5, 2.75, and 1, respectively.

105. Graph $y=-0.002x^2+0.8x+6.6$. Then use VALUE from the CALC menu to find the y-values that correspond to $x=100$, $x=200$, $x=300$, and $x=350$. As before, we find that these values are 66.6 ft, 86.6 ft, 66.6 ft, and 41.6 ft, respectively.

Exercise Set 10.4

1. $(3x+2)+(-4x+3)=(3-4)x+(2+3)=-x+5$

3. $(-6x+2)+(x^2+x-3)=$
$x^2+(-6+1)x+(2-3)=x^2-5x-1$

5. $(x^2-9)+(x^2+9)=(1+1)x^2+(-9+9)=2x^2$

7. $(3x^2-5x+10)+(2x^2+8x-40)=$
$(3+2)x^2+(-5+8)x+(10-40)=5x^2+3x-30$

9. $(1.2x^3+4.5x^2-3.8x)+(-3.4x^3-4.7x^2+23)=$
$(1.2-3.4)x^3+(4.5-4.7)x^2-3.8x+23=$
$-2.2x^3-0.2x^2-3.8x+23$

11. $(1+4x+6x^2+7x^3)+(5-4x+6x^2-7x^3)=$
$(1+5)+(4-4)x+(6+6)x^2+(7-7)x^3=$
$6+0x+12x^2+0x^3=6+12x^2$, or $12x^2+6$

13. $\left(\frac{1}{4}x^4+\frac{2}{3}x^3+\frac{5}{8}x^2+7\right)+\left(-\frac{3}{4}x^4+\frac{3}{8}x^2-7\right)=$
$\left(\frac{1}{4}-\frac{3}{4}\right)x^4+\frac{2}{3}x^3+\left(\frac{5}{8}+\frac{3}{8}\right)x^2+(7-7)=$
$-\frac{2}{4}x^4+\frac{2}{3}x^3+\frac{8}{8}x^2+0=$
$-\frac{1}{2}x^4+\frac{2}{3}x^3+x^2$

15. $(0.02x^5-0.2x^3+x+0.08)+(-0.01x^5+x^4-0.8x-0.02)=$
$(0.02-0.01)x^5+x^4-0.2x^3+(1-0.8)x+(0.08-0.02)=$
$0.01x^5+x^4-0.2x^3+0.2x+0.06$

17. $9x^8-7x^4+2x^2+5)+(8x^7+4x^4-2x)+$
$(-3x^4+6x^2+2x-1)=9x^8+8x^7+(-7+4-3)x^4+$
$(2+6)x^2+(-2+2)x+(5-1)=$
$9x^8+8x^7-6x^4+8x^2+4$

19. Rewrite the problem so the coefficients of like terms have the same number of decimal places.

$$\begin{array}{rrrrr}
0.15x^4 & +\ 0.10x^3 & -\ 0.90x^2 & & \\
 & -\ 0.01x^3 & +\ 0.01x^2 & +\ x & \\
1.25x^4 & & +\ 0.11x^2 & & +\ 0.01 \\
 & 0.27x^3 & & & +\ 0.99 \\
-0.35x^4 & & +\ 15.00x^2 & & -\ 0.03 \\
\hline
1.05x^4 & +\ 0.36x^3 & +\ 14.22x^2 & +\ x & +\ 0.97
\end{array}$$

21. Two equivalent expressions for the additive inverse of $-5x$ are

a) $-(-5x)$ and

b) $5x$. (Changing the sign)

23. Two equivalent expressions for the additive inverse of $-x^2+10x-2$ are

a) $-(-x^2+10x-2)$ and

b) $x^2-10x+2$. (Changing the sign of every term)

25. Two equivalent expressions for the additive inverse of $12x^4 - 3x^3 + 3$ are

a) $-(12x^4 - 3x^3 + 3)$ and

b) $-12x^4 + 3x^3 - 3$. (Changing the sign of every term)

27. We change the sign of every term inside parentheses.

$-(3x - 7) = -3x + 7$

29. We change the sign of every term inside parentheses.

$-(4x^2 - 3x + 2) = -4x^2 + 3x - 2$

31. We change the sign of every term inside parentheses.

$-\left(-4x^4 + 6x^2 + \frac{3}{4}x - 8\right) = 4x^4 - 6x^2 - \frac{3}{4}x + 8$

33. $(3x + 2) - (-4x + 3) = 3x + 2 + 4x - 3$ Changing the sign of every term inside parentheses

$= 7x - 1$

35. $(-6x + 2) - (x^2 + x - 3) = -6x + 2 - x^2 - x + 3$

$= -x^2 - 7x + 5$

37. $(x^2 - 9) - (x^2 + 9) = x^2 - 9 - x^2 - 9 = -18$

39. $(6x^4 + 3x^3 - 1) - (4x^2 - 3x + 3)$

$= 6x^4 + 3x^3 - 1 - 4x^2 + 3x - 3$

$= 6x^4 + 3x^3 - 4x^2 + 3x - 4$

41. $(1.2x^3 + 4.5x^2 - 3.8x) - (-3.4x^3 - 4.7x^2 + 23)$

$= 1.2x^3 + 4.5x^2 - 3.8x + 3.4x^3 + 4.7x^2 - 23$

$= 4.6x^3 + 9.2x^2 - 3.8x - 23$

43. $\frac{5}{8}x^3 - \frac{1}{4}x - \frac{1}{3} - \left(-\frac{1}{8}x^3 + \frac{1}{4}x - \frac{1}{3}\right)$

$= \frac{5}{8}x^3 - \frac{1}{4}x - \frac{1}{3} + \frac{1}{8}x^3 - \frac{1}{4}x + \frac{1}{3}$

$= \frac{6}{8}x^3 - \frac{2}{4}x$

$= \frac{3}{4}x^3 - \frac{1}{2}x$

45. $(0.08x^3 - 0.02x^2 + 0.01x) - (0.02x^3 + 0.03x^2 - 1)$

$= 0.08x^3 - 0.02x^2 + 0.01x - 0.02x^3 - 0.03x^2 + 1$

$= 0.06x^3 - 0.05x^2 + 0.01x + 1$

47.
$$\begin{array}{l} x^2 + 5x + 6 \\ \underline{x^2 + 2x} \end{array}$$

$$\begin{array}{ll} x^2 + 5x + 6 & \\ \underline{-x^2 - 2x} & \text{Changing signs} \\ 3x + 6 & \text{Adding} \end{array}$$

49.
$$\begin{array}{l} 5x^4 + 6x^3 - 9x^2 \\ \underline{-6x^4 - 6x^3 \qquad + 8x + 9} \end{array}$$

$$\begin{array}{ll} 5x^4 + 6x^3 - 9x^2 & \\ \underline{6x^4 + 6x^3 \qquad - 8x - 9} & \text{Changing signs} \\ 11x^4 + 12x^3 - 9x^2 - 8x - 9 & \text{Adding} \end{array}$$

51.
$$\begin{array}{l} x^5 \qquad\qquad\qquad - 1 \\ \underline{x^5 - x^4 + x^3 - x^2 + x - 1} \end{array}$$

$$\begin{array}{ll} x^5 \qquad\qquad\qquad - 1 & \\ \underline{-x^5 + x^4 - x^3 + x^2 - x + 1} & \text{Changing signs} \\ x^4 - x^3 + x^2 - x & \text{Adding} \end{array}$$

53.

A (3x by x), B (x by x), C (x by x), D (4 by x)

The area of a rectangle is the product of the length and width. The sum of the areas is found as follows:

$$\begin{array}{ccccccccc} & \text{Area of } A & + & \text{Area of } B & + & \text{Area of } C & + & \text{Area of } D \\ = & 3x \cdot x & + & x \cdot x & + & x \cdot x & + & 4 \cdot x \\ = & 3x^2 & + & x^2 & + & x^2 & + & 4x \\ = & 5x^2 & + & 4x & & & & \end{array}$$

A polynomial for the sum of the areas is $5x^2 + 4x$.

55. We add the lengths of the sides:

$4a + 7 + a + \frac{1}{2}a + 3 + a + 2a + 3a$

$= \left(4 + 1 + \frac{1}{2} + 1 + 2 + 3\right)a + (7 + 3)$

$= 11\frac{1}{2}a + 10$, or $\frac{23}{2}a + 10$

57. $8x + 3x = 66$

$11x = 66$ Collecting like terms

$\frac{11x}{11} = \frac{66}{11}$ Dividing by 11

$x = 6$

The solution is 6.

59. $\frac{3}{8}x + \frac{1}{4} - \frac{3}{4}x = \frac{11}{16} + x$, LCM is 16

$16\left(\frac{3}{8}x + \frac{1}{4} - \frac{3}{4}x\right) = 16\left(\frac{11}{16} + x\right)$ Clearing fractions

$6x + 4 - 12x = 11 + 16x$

$-6x + 4 = 11 + 16x$ Collecting like terms

$-6x + 4 - 4 = 11 + 16x - 4$ Subtracting 4

$-6x = 7 + 16x$

$-6x - 16x = 7 + 16x - 16x$ Subtracting $16x$

$-22x = 7$

$\frac{-22x}{-22} = \frac{7}{-22}$ Dividing by -22

$x = -\frac{7}{22}$

The solution is $-\frac{7}{22}$.

61.
$$\begin{aligned} 1.5x - 2.7x &= 22 - 5.6x \\ 10(1.5x - 2.7x) &= 10(22 - 5.6x) && \text{Clearing decimals} \\ 15x - 27x &= 220 - 56x \\ -12x &= 220 - 56x && \text{Collecting like terms} \\ 44x &= 220 && \text{Adding } 56x \\ x &= \frac{220}{44} && \text{Dividing by 44} \\ x &= 5 && \text{Simplifying} \end{aligned}$$

The solution is 5.

63.
$$\begin{aligned} 6(y-3) - 8 &= 4(y+2) + 5 \\ 6y - 18 - 8 &= 4y + 8 + 5 && \text{Removing parentheses} \\ 6y - 26 &= 4y + 13 && \text{Collecting like terms} \\ 6y - 26 + 26 &= 4y + 13 + 26 && \text{Adding 26} \\ 6y &= 4y + 39 \\ 6y - 4y &= 4y + 39 - 4y && \text{Subtracting } 4y \\ 2y &= 39 \\ \frac{2y}{2} &= \frac{39}{2} && \text{Dividing by 2} \\ y &= \frac{39}{2} \end{aligned}$$

The solution is $\frac{39}{2}$.

65.
$$\begin{aligned} 3x - 7 &\le 5x + 13 \\ -2x - 7 &\le 13 && \text{Subtracting } 5x \\ -2x &\le 20 && \text{Adding 7} \\ x &\ge -10 && \text{Dividing by } -2 \text{ and reversing the inequality symbol} \end{aligned}$$

The solution set is $\{x|x \ge -10\}$.

67. ◈

69.

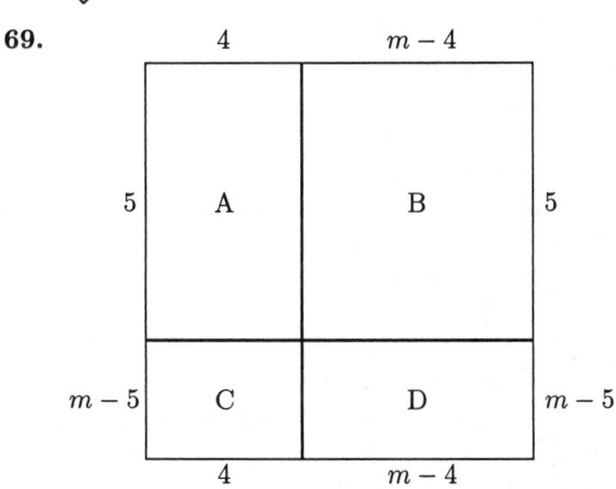

We can add the areas of the four rectangles A, B, C, and D.

$5 \cdot 4 + 5(m-4) + 4(m-5) + (m-4)(m-5)$, or

$20 + 5(m-4) + 4(m-5) + (m-4)(m-5)$

The length and width of the figure can also be expressed as $4 + (m-4)$ and $5 + (m-5)$, or m and m. Then the area can be expressed as $m \cdot m$, or m^2.

71.

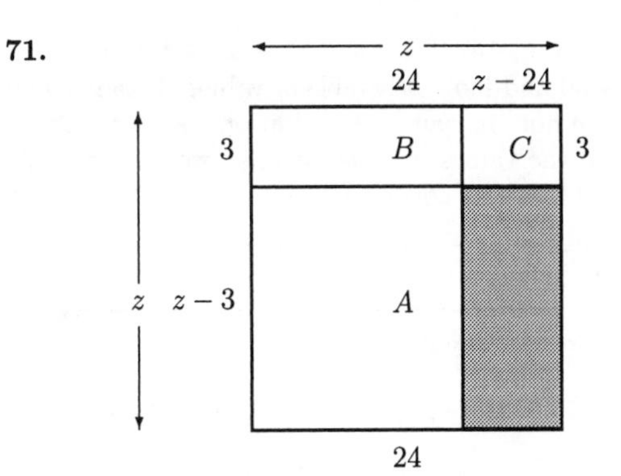

We label the sides A, B, and C with additional information. The area of the square is $z \cdot z$, or z^2. The area of the shaded section is z^2 minus the areas of sections A, B, and C.

Area of shaded section = Area of square − Area of A − Area of B − Area of C

$$\begin{aligned} \text{Area of shaded section} &= z \cdot z - 24(z-3) - 3 \cdot 24 - 3(z-24) \\ &= z^2 - 24z + 72 - 72 - 3z + 72 \\ &= z^2 - 27z + 72 \end{aligned}$$

A polynomial for the shaded area is $z^2 - 27z + 72$.

73.

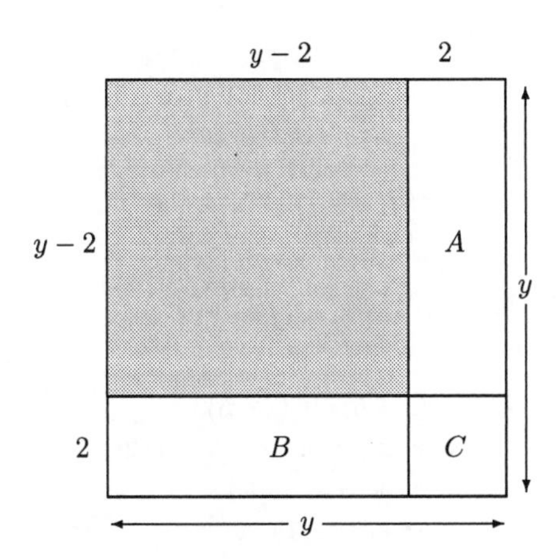

The shaded area is $(y-2)^2$. We find it as follows:

Shaded area = Area of square − Area of A − Area of B − Area of C

$$\begin{aligned} (y-2)^2 &= y^2 - 2(y-2) - 2(y-2) - 2 \cdot 2 \\ (y-2)^2 &= y^2 - 2y + 4 - 2y + 4 - 4 \\ (y-2)^2 &= y^2 - 4y + 4 \end{aligned}$$

75.
$$\begin{aligned} &(3x^2 - 4x + 6) - (-2x^2 + 4) + (-5x - 3) \\ &= 3x^2 - 4x + 6 + 2x^2 - 4 - 5x - 3 \\ &= 5x^2 - 9x - 1 \end{aligned}$$

77.
$$\begin{aligned} &(-4 + x^2 + 2x^3) - (-6 - x + 3x^3) - (-x^2 - 5x^3) \\ &= -4 + x^2 + 2x^3 + 6 + x - 3x^3 + x^2 + 5x^3 \\ &= 4x^3 + 2x^2 + x + 2 \end{aligned}$$

79. Enter $y_1 = (3x^2 - 5x + 10) + (2x^2 + 8x - 40)$ and $y_2 = 5x^2 + 3x - 30$ and look at a table of values. If the y_1-and y_2-values are not the same, the addition was not correct. In this case, the values are the same, so we can conclude that the addition was correct.

Exercise Set 10.5

1. $(8x^2)(5) = (8 \cdot 5)x^2 = 40x^2$

3. $(-x^2)(-x) = (-1x^2)(-1x) = (-1)(-1)(x^2 \cdot x) = x^3$

5. $(8x^5)(4x^3) = (8 \cdot 4)(x^5 \cdot x^3) = 32x^8$

7. $(0.1x^6)(0.3x^5) = (0.1)(0.3)(x^6 \cdot x^5) = 0.03x^{11}$

9. $\left(-\frac{1}{5}x^3\right)\left(-\frac{1}{3}x\right) = \left(-\frac{1}{5}\right)\left(-\frac{1}{3}\right)(x^3 \cdot x) = \frac{1}{15}x^4$

11. $(-4x^2)(0) = 0$ Any number multiplied by 0 is 0.

13. $(3x^2)(-4x^3)(2x^6) = (3)(-4)(2)(x^2 \cdot x^3 \cdot x^6) = -24x^{11}$

15. $$\begin{aligned} 2x(-x+5) &= 2x(-x) + 2x(5) \\ &= -2x^2 + 10x \end{aligned}$$

17. $$\begin{aligned} -5x(x-1) &= -5x(x) - 5x(-1) \\ &= -5x^2 + 5x \end{aligned}$$

19. $$\begin{aligned} x^2(x^3+1) &= x^2(x^3) + x^2(1) \\ &= x^5 + x^2 \end{aligned}$$

21. $$\begin{aligned} 3x(2x^2 - 6x + 1) &= 3x(2x^2) + 3x(-6x) + 3x(1) \\ &= 6x^3 - 18x^2 + 3x \end{aligned}$$

23. $$\begin{aligned} -6x^2(x^2 + x) &= -6x^2(x^2) - 6x^2(x) \\ &= -6x^4 - 6x^3 \end{aligned}$$

25. $$\begin{aligned} 3y^2(6y^4 + 8y^3) &= 3y^2(6y^4) + 3y^2(8y^3) \\ &= 18y^6 + 24y^5 \end{aligned}$$

27. $$\begin{aligned} (x+6)(x+3) &= (x+6)x + (x+6)3 \\ &= x \cdot x + 6 \cdot x + x \cdot 3 + 6 \cdot 3 \\ &= x^2 + 6x + 3x + 18 \\ &= x^2 + 9x + 18 \end{aligned}$$

29. $$\begin{aligned} (x+5)(x-2) &= (x+5)x + (x+5)(-2) \\ &= x \cdot x + 5 \cdot x + x(-2) + 5(-2) \\ &= x^2 + 5x - 2x - 10 \\ &= x^2 + 3x - 10 \end{aligned}$$

31. $$\begin{aligned} (x-4)(x-3) &= (x-4)x + (x-4)(-3) \\ &= x \cdot x - 4 \cdot x + x(-3) - 4(-3) \\ &= x^2 - 4x - 3x + 12 \\ &= x^2 - 7x + 12 \end{aligned}$$

33. $$\begin{aligned} (x+3)(x-3) &= (x+3)x + (x+3)(-3) \\ &= x \cdot x + 3 \cdot x + x(-3) + 3(-3) \\ &= x^2 + 3x - 3x - 9 \\ &= x^2 - 9 \end{aligned}$$

35. $$\begin{aligned} (5-x)(5-2x) &= (5-x)5 + (5-x)(-2x) \\ &= 5 \cdot 5 - x \cdot 5 + 5(-2x) - x(-2x) \\ &= 25 - 5x - 10x + 2x^2 \\ &= 25 - 15x + 2x^2 \end{aligned}$$

37. $$\begin{aligned} (2x+5)(2x+5) &= (2x+5)2x + (2x+5)5 \\ &= 2x \cdot 2x + 5 \cdot 2x + 2x \cdot 5 + 5 \cdot 5 \\ &= 4x^2 + 10x + 10x + 25 \\ &= 4x^2 + 20x + 25 \end{aligned}$$

39. $$\begin{aligned} \left(x - \frac{5}{2}\right)\left(x + \frac{2}{5}\right) &= \left(x - \frac{5}{2}\right)x + \left(x - \frac{5}{2}\right)\frac{2}{5} \\ &= x \cdot x - \frac{5}{2} \cdot x + x \cdot \frac{2}{5} - \frac{5}{2} \cdot \frac{2}{5} \\ &= x^2 - \frac{5}{2}x + \frac{2}{5}x - 1 \\ &= x^2 - \frac{25}{10}x + \frac{4}{10}x - 1 \\ &= x^2 - \frac{21}{10}x - 1 \end{aligned}$$

41. $$\begin{aligned} (x-2.3)(x+4.7) &= (x-2.3)x + (x-2.3)4.7 \\ &= x \cdot x - 2.3 \cdot x + x \cdot 4.7 - 2.3(4.7) \\ &= x^2 - 2.3x + 4.7x - 10.81 \\ &= x^2 + 2.4x - 10.81 \end{aligned}$$

43. $$\begin{aligned} &(x^2 + x + 1)(x - 1) \\ &= (x^2 + x + 1)x + (x^2 + x + 1)(-1) \\ &= x^2 \cdot x + x \cdot x + 1 \cdot x + x^2(-1) + x(-1) + 1(-1) \\ &= x^3 + x^2 + x - x^2 - x - 1 \\ &= x^3 - 1 \end{aligned}$$

45. $$\begin{aligned} &(2x+1)(2x^2 + 6x + 1) \\ &= 2x(2x^2 + 6x + 1) + 1(2x^2 + 6x + 1) \\ &= 2x \cdot 2x^2 + 2x \cdot 6x + 2x \cdot 1 + 1 \cdot 2x^2 + 1 \cdot 6x + 1 \cdot 1 \\ &= 4x^3 + 12x^2 + 2x + 2x^2 + 6x + 1 \\ &= 4x^3 + 14x^2 + 8x + 1 \end{aligned}$$

47. $$\begin{aligned} &(y^2 - 3)(3y^2 - 6y + 2) \\ &= y^2(3y^2 - 6y + 2) - 3(3y^2 - 6y + 2) \\ &= y^2 \cdot 3y^2 + y^2(-6y) + y^2 \cdot 2 - 3 \cdot 3y^2 - 3(-6y) - 3 \cdot 2 \\ &= 3y^4 - 6y^3 + 2y^2 - 9y^2 + 18y - 6 \\ &= 3y^4 - 6y^3 - 7y^2 + 18y - 6 \end{aligned}$$

49. $$\begin{aligned} &(x^3 + x^2)(x^3 + x^2 - x) \\ &= x^3(x^3 + x^2 - x) + x^2(x^3 + x^2 - x) \\ &= x^3 \cdot x^3 + x^3 \cdot x^2 + x^3(-x) + x^2 \cdot x^3 + x^2 \cdot x^2 + x^2(-x) \\ &= x^6 + x^5 - x^4 + x^5 + x^4 - x^3 \\ &= x^6 + 2x^5 - x^3 \end{aligned}$$

51. $$\begin{aligned} &(-5x^3 - 7x^2 + 1)(2x^2 - x) \\ &= (-5x^3 - 7x^2 + 1)2x^2 + (-5x^3 - 7x^2 + 1)(-x) \\ &= -5x^3 \cdot 2x^2 - 7x^2 \cdot 2x^2 + 1 \cdot 2x^2 - 5x^3(-x) - 7x^2(-x) + \\ &\quad 1(-x) \\ &= -10x^5 - 14x^4 + 2x^2 + 5x^4 + 7x^3 - x \\ &= -10x^5 - 9x^4 + 7x^3 + 2x^2 - x \end{aligned}$$

53.
$$\begin{array}{rl}
1+x+x^2 & \text{Line up like terms} \\
-1-x+x^2 & \text{in columns} \\
\hline
x^2+x^3+x^4 & \text{Multiplying the top row by } x^2 \\
-x-x^2-x^3 & \text{Multiplying by } -x \\
-1-x-x^2 & \text{Multiplying by } -1 \\
\hline
-1-2x-x^2 \qquad +x^4 &
\end{array}$$

55.
$$\begin{array}{rl}
2t^2-t-4 & \\
3t^2+2t-1 & \\
\hline
-2t^2+t+4 & \text{Multiplying by } -1 \\
4t^3-2t^2-8t & \text{Multiplying by } 2t \\
6t^4-3t^3-12t^2 & \text{Multiplying by } 3t^2 \\
\hline
6t^4+t^3-16t^2-7t+4 &
\end{array}$$

57.
$$\begin{array}{rl}
x \quad -x^3 \quad +x^5 & \\
-1+x^2 \quad +x^4 & \text{Rewriting in ascending order} \\
\hline
x^5-x^7+x^9 & \text{Multiplying by } x^4 \\
x^3-x^5+x^7 & \text{Multiplying by } x^2 \\
-x+x^3-x^5 & \text{Multiplying by } -1 \\
\hline
-x+2x^3-x^5 \qquad +x^9 &
\end{array}$$

59.
$$\begin{array}{r}
x^3+x^2+x+1 \\
x-1 \\
\hline
-x^3-x^2-x-1 \\
x^4+x^3+x^2+x \\
\hline
x^4 \qquad\qquad -1
\end{array}$$

61. $-\frac{1}{4}-\frac{1}{2}=-\frac{1}{4}-\frac{1}{2}\cdot\frac{2}{2}=-\frac{1}{4}-\frac{2}{4}=-\frac{3}{4}$

63. $(10-2)(10+2)=8\cdot 12=96$

65. $15x-18y+12=3\cdot 5x-3\cdot 6y+3\cdot 4=$
$3(5x-6y+4)$

67. $-9x-45y+15=-3\cdot 3x-3\cdot 15y-3(-5)=$
$-3(3x+15y-5)$

69. $y=\frac{1}{2}x-3$

The equation is equivalent to $y=\frac{1}{2}x+(-3)$. The y-intercept is $(0,-3)$. We find two other points, using multiples of 2 for x to avoid fractions.

When $x=-2$, $y=\frac{1}{2}(-2)-3=-1-3=-4$.

When $x=4$, $y=\frac{1}{2}\cdot 4-3=2-3=-1$.

x	y
0	-3
-2	-4
4	-1

Plot these points, draw the line they determine, and label the graph $y=\frac{1}{2}x-3$.

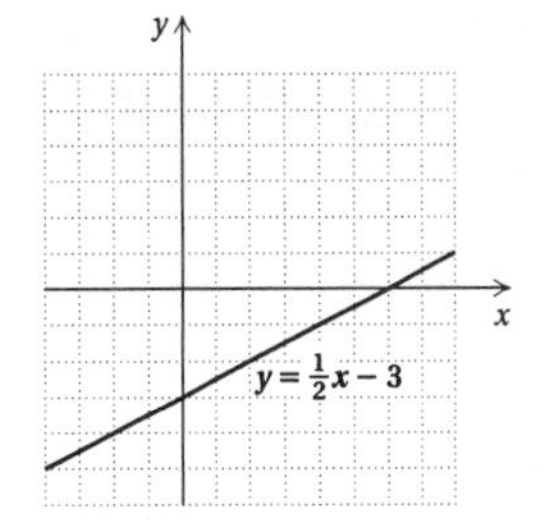

71.

73. The shaded area is the area of the large rectangle less the area of the small rectangle:

$$\begin{aligned}4t(21t+8)-2t(3t-4)&=84t^2+32t-6t^2+8t\\&=78t^2+40t\end{aligned}$$

75. Let b = the length of the base. The $b+4$ = the height. Let A represent the area.

$$\begin{aligned}
\text{Area} &= \frac{1}{2}\times\text{ base }\times\text{ height}\\
A &= \frac{1}{2}\cdot b\cdot(b+4)\\
A &= \frac{1}{2}b(b+4)\\
A &= \frac{1}{2}b^2+2b
\end{aligned}$$

77.
$$\begin{aligned}
&(x-2)(x-7)-(x-2)(x-7)\\
&=x(x-7)-2(x-7)-[x(x-7)-2(x-7)]\\
&=x^2-7x-2x+14-(x^2-7x-2x+14)\\
&=x^2-9x+14-(x^2-9x+14)\\
&=x^2-9x+14-x^2+9x-14\\
&=0
\end{aligned}$$

Exercise Set 10.6

1. $(x+1)(x^2+3)$

F O I L

$=x\cdot x^2+x\cdot 3+1\cdot x^2+1\cdot 3$

$=x^3+3x+x^2+3$

3. $(x^3+2)(x+1)$

F O I L

$=x^3\cdot x+x^3\cdot 1+2\cdot x+2\cdot 1$

$=x^4+x^3+2x+2$

5. $(y+2)(y-3)$

F O I L

$=y\cdot y+y\cdot(-3)+2\cdot y+2\cdot(-3)$

$=y^2-3y+2y-6$

$=y^2-y-6$

7. $(3x+2)(3x+2)$

F O I L

$=3x\cdot 3x+3x\cdot 2+2\cdot 3x+2\cdot 2$

$=9x^2+6x+6x+4$

$=9x^2+12x+4$

9. $(5x-6)(x+2)$
F O I L
$= 5x \cdot x + 5x \cdot 2 + (-6) \cdot x + (-6) \cdot 2$
$= 5x^2 + 10x - 6x - 12$
$= 5x^2 + 4x - 12$

11. $(3t-1)(3t+1)$
F O I L
$= 3t \cdot 3t + 3t \cdot 1 + (-1) \cdot 3t + (-1) \cdot 1$
$= 9t^2 + 3t - 3t - 1$
$= 9t^2 - 1$

13. $(4x-2)(x-1)$
F O I L
$= 4x \cdot x + 4x \cdot (-1) + (-2) \cdot x + (-2) \cdot (-1)$
$= 4x^2 - 4x - 2x + 2$
$= 4x^2 - 6x + 2$

15. $\left(p-\frac{1}{4}\right)\left(p+\frac{1}{4}\right)$
F O I L
$= p \cdot p + p \cdot \frac{1}{4} + \left(-\frac{1}{4}\right) \cdot p + \left(-\frac{1}{4}\right) \cdot \frac{1}{4}$
$= p^2 + \frac{1}{4}p - \frac{1}{4}p - \frac{1}{16}$
$= p^2 - \frac{1}{16}$

17. $(x-0.1)(x+0.1)$
F O I L
$= x \cdot x + x \cdot (0.1) + (-0.1) \cdot x + (-0.1)(0.1)$
$= x^2 + 0.1x - 0.1x - 0.01$
$= x^2 - 0.01$

19. $(2x^2+6)(x+1)$
F O I L
$= 2x^3 + 2x^2 + 6x + 6$

21. $(-2x+1)(x+6)$
F O I L
$= -2x^2 - 12x + x + 6$
$= -2x^2 - 11x + 6$

23. $(a+7)(a+7)$
F O I L
$= a^2 + 7a + 7a + 49$
$= a^2 + 14a + 49$

25. $(1+2x)(1-3x)$
F O I L
$= 1 - 3x + 2x - 6x^2$
$= 1 - x - 6x^2$

27. $(x^2+3)(x^3-1)$
F O I L
$= x^5 - x^2 + 3x^3 - 3$

29. $(3x^2-2)(x^4-2)$
F O I L
$= 3x^6 - 6x^2 - 2x^4 + 4$

31. $(2.8x-1.5)(4.7x+9.3)$
F O I L
$= 2.8x(4.7x) + 2.8x(9.3) - 1.5(4.7x) - 1.5(9.3)$
$= 13.16x^2 + 26.04x - 7.05x - 13.95$
$= 13.16x^2 + 18.99x - 13.95$

33. $(3x^5+2)(2x^2+6)$
F O I L
$= 6x^7 + 18x^5 + 4x^2 + 12$

35. $(8x^3+1)(x^3+8)$
F O I L
$= 8x^6 + 64x^3 + x^3 + 8$
$= 8x^6 + 65x^3 + 8$

37. $(4x^2+3)(x-3)$
F O I L
$= 4x^3 - 12x^2 + 3x - 9$

39. $(4y^4+y^2)(y^2+y)$
F O I L
$= 4y^6 + 4y^5 + y^4 + y^3$

41. $(x+4)(x-4)$ Product of sum and difference of two terms
$= x^2 - 4^2$
$= x^2 - 16$

43. $(2x+1)(2x-1)$ Product of sum and difference of two terms
$= (2x)^2 - 1^2$
$= 4x^2 - 1$

45. $(5m-2)(5m+2)$ Product of sum and difference of two terms
$= (5m)^2 - 2^2$
$= 25m^2 - 4$

47. $(2x^2+3)(2x^2-3)$ Product of sum and difference of two terms
$= (2x^2)^2 - 3^2$
$= 4x^4 - 9$

49. $(3x^4-4)(3x^4+4)$
$= (3x^4)^2 - 4^2$
$= 9x^8 - 16$

51. $(x^6-x^2)(x^6+x^2)$
$= (x^6)^2 - (x^2)^2$
$= x^{12} - x^4$

53. $(x^4+3x)(x^4-3x)$
$= (x^4)^2 - (3x)^2$
$= x^8 - 9x^2$

55. $(x^{12}-3)(x^{12}+3)$
$= (x^{12})^2 - 3^2$
$= x^{24} - 9$

57. $(2y^8+3)(2y^8-3)$
$= (2y^8)^2 - 3^2$
$= 4y^{16} - 9$

59. $\left(\frac{5}{8}x - 4.3\right)\left(\frac{5}{8}x + 4.3\right)$
$= \left(\frac{5}{8}x\right)^2 - (4.3)^2$
$= \frac{25}{64}x^2 - 18.49$

61. $(x+2)^2 = x^2 + 2\cdot x\cdot 2 + 2^2$ Square of a binomial sum
$= x^2 + 4x + 4$

63. $(3x^2+1)$ Square of a binomial sum
$= (3x^2)^2 + 2\cdot 3x^2\cdot 1 + 1^2$
$= 9x^4 + 6x^2 + 1$

65. $\left(a - \frac{1}{2}\right)^2$ Square of a binomial sum
$= a^2 - 2\cdot a\cdot \frac{1}{2} + \left(\frac{1}{2}\right)^2$
$= a^2 - a + \frac{1}{4}$

67. $(3+x)^2 = 3^2 + 2\cdot 3\cdot x + x^2$
$= 9 + 6x + x^2$

69. $(x^2+1)^2 = (x^2)^2 + 2\cdot x^2\cdot 1 + 1^2$
$= x^4 + 2x^2 + 1$

71. $(2-3x^4)^2 = 2^2 - 2\cdot 2\cdot 3x^4 + (3x^4)^2$
$= 4 - 12x^4 + 9x^8$

73. $(5+6t^2)^2 = 5^2 + 2\cdot 5\cdot 6t^2 + (6t^2)^2$
$= 25 + 60t^2 + 36t^4$

75. $\left(x - \frac{5}{8}\right)^2 = x^2 - 2\cdot x\cdot \frac{5}{8} + \left(\frac{5}{8}\right)^2$
$= x^2 - \frac{5}{4}x + \frac{25}{64}$

77. $(3-2x^3)^2 = 3^2 - 2\cdot 3\cdot 2x^3 + (2x^3)^2$
$= 9 - 12x^3 + 4x^6$

79. $4x(x^2+6x-3)$ Product of a monomial and a trinomial
$= 4x\cdot x^2 + 4x\cdot 6x + 4x(-3)$
$= 4x^3 + 24x^2 - 12x$

81. $\left(2x^2 - \frac{1}{2}\right)\left(2x^2 - \frac{1}{2}\right)$ Square of a binomial difference
$= (2x^2)^2 - 2\cdot 2x^2\cdot \frac{1}{2} + \left(\frac{1}{2}\right)^2$
$= 4x^4 - 2x^2 + \frac{1}{4}$

83. $(-1+3p)(1+3p)$
$= (3p-1)(3p+1)$ Product of the sum and difference of two terms
$= (3p)^2 - 1^2$
$= 9p^2 - 1$

85. $3t^2(5t^3 - t^2 + t)$ Product of a monomial and a trinomial
$= 3t^2\cdot 5t^3 + 3t^2(-t^2) + 3t^2\cdot t$
$= 15t^5 - 3t^4 + 3t^3$

87. $(6x^4+4)^2$ Square of a binomial sum
$= (6x^4)^2 + 2\cdot 6x^4\cdot 4 + 4^2$
$= 36x^8 + 48x^4 + 16$

89. $(3x+2)(4x^2+5)$ Product of two binomials; use FOIL
$= 3x\cdot 4x^2 + 3x\cdot 5 + 2\cdot 4x^2 + 2\cdot 5$
$= 12x^3 + 15x + 8x^2 + 10$

91. $(8-6x^4)^2$ Square of a binomial difference
$= 8^2 - 2\cdot 8\cdot 6x^4 + (6x^4)^2$
$= 64 - 96x^4 + 36x^8$

93.

$$\begin{array}{r} t^2+t+1 \\ t-1 \\ \hline -t^2-t-1 \\ t^3+t^2+t \quad \\ \hline t^3 \qquad\quad -1 \end{array}$$

95. $3^2 + 4^2 = 9 + 16 = 25$
$(3+4)^2 = 7^2 = 49$

97. $9^2 - 5^2 = 81 - 25 = 56$
$(9-5)^2 = 4^2 = 16$

99. ***Familiarize***. Let t = the number of watts used by the television set. Then $10t$ = the number of watts used by the lamps, and $40t$ = the number of watts used by the air conditioner.

Translate.

Lamp watts	+	Air conditioner watts	+	Television watts	=	Total watts
↓	↓	↓	↓	↓	↓	↓
$10t$	+	$40t$	+	t	=	2550

Solve. We solve the equation.

$$10t + 40t + t = 2550$$
$$51t = 2550$$
$$t = 50$$

The possible solution is:

Television, t: 50 watts

Lamps, $10t$: $10 \cdot 50$, or 500 watts

Air conditioner, $40t$: $40 \cdot 50$, or 2000 watts

Check. The number of watts used by the lamps, 500, is 10 times 50, the number used by the television. The number of watts used by the air conditioner, 2000, is 40 times 50, the number used by the television. Also, $50+500+2000 = 2550$, the total wattage used.

State. The television uses 50 watts, the lamps use 500 watts, and the air conditioner uses 2000 watts.

101.
$$\begin{aligned} 3(x-2) &= 5(2x+7) \\ 3x-6 &= 10x+35 && \text{Removing parentheses} \\ 3x-6+6 &= 10x+35+6 && \text{Adding 6} \\ 3x &= 10x+41 \\ 3x-10x &= 10x+41-10x && \text{Subtracting } 10x \\ -7x &= 41 \\ \frac{-7x}{-7} &= \frac{41}{-7} && \text{Dividing by } -7 \\ x &= -\frac{41}{7} \end{aligned}$$

The solution is $-\frac{41}{7}$.

103.
$$\begin{aligned} 3x-2y &= 12 \\ -2y &= -3x+12 && \text{Subtracting } 3x \\ \frac{-2y}{-2} &= \frac{-3x+12}{-2} && \text{Dividing by } -2 \\ y &= \frac{3x-12}{2}, \text{ or} \\ y &= \frac{3}{2}x-6 \end{aligned}$$

105. ◈

107.
$$\begin{aligned} & 5x(3x-1)(2x+3) \\ &= 5x(6x^2+7x-3) && \text{Using FOIL} \\ &= 30x^3+35x^2-15x \end{aligned}$$

109.
$$\begin{aligned} & [(a-5)(a+5)]^2 \\ &= (a^2-25)^2 && \text{Finding the product of a sum and difference of same two terms} \\ &= a^4-50a^2+625 && \text{Squaring a binomial} \end{aligned}$$

111.
$$\begin{aligned} & (3t^4-2)^2 1(3t^4+2)^2 \\ &= [(3t^4-2)(3t^4+2)]^2 \\ &= (9t^8-4)^2 \\ &= 81t^{16}-72t^8+16 \end{aligned}$$

113.
$$\begin{aligned} (x+2)(x-5) &= (x+1)(x-3) \\ x^2-5x+2x-10 &= x^2-3x+x-3 \\ x^2-3x-10 &= x^2-2x-3 \\ -3x-10 &= -2x-3 && \text{Adding } -x^2 \\ -3x+2x &= 10-3 && \text{Adding } 2x \text{ and } 10 \\ -x &= 7 \\ x &= -7 \end{aligned}$$

The solution is -7.

115. See the answer section in the text.

117.

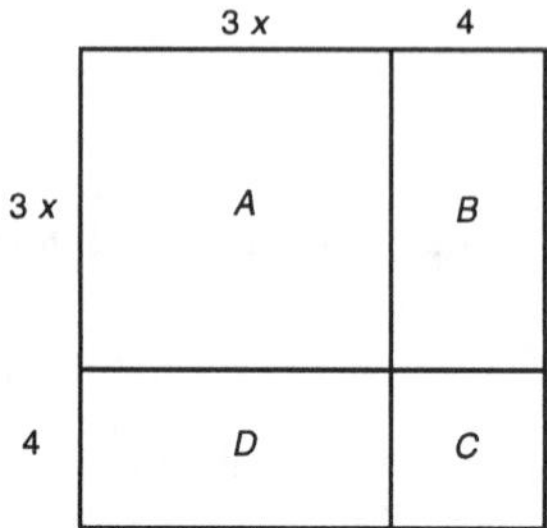

We can find the shaded area in two ways.

Method 1: The figure is a square with side $3x+4$, so the area is $(3x+4)^2 = 9x^2+24x+16$.

Method 2: We add the areas of A, B, C, and D.

$3x \cdot 3x + 3x \cdot 4 + 4 \cdot 4 + 3x \cdot 4 = 9x^2+12x+16+12x = 9x^2+24x+16$.

Either way, we find that the total shaded area is $9x^2+24x+16$.

119. Enter $y_1 = (x-1)^2$ and $y_2 = x^2-2x+1$. Then compare the graphs or the y_1-and y_2-values in a table. It appears that the graphs are the same and that the y_1-and y_2-values are the same, so $(x-1)^2 = x^2-2x+1$ is correct.

121. Enter $y_1 = (x-3)(x+3)$ and $y_2 = x^2-6$. Then compare the graphs or the y_1-and y_2-values in a table. The graphs are not the same nor are the y_1-and y_2-values, so $(x-3)(x+3) = x^2-6$ is not correct.

Exercise Set 10.7

1. We replace x by 3 and y by -2.

$x^2-y^2+xy = 3^2-(-2)^2+3(-2) = 9-4-6 = -1$

3. We replace x by 3 and y by -2.

$x^2-3y^2+2xy = 3^2-3(-2)^2+2\cdot 3(-2) = 9-3\cdot 4+2\cdot 3(-2) = 9-12-12 = -15$

5. We replace x by 3, y by -2, and z by -5.

$8xyz = 8\cdot 3\cdot(-2)\cdot(-5) = 240$

7. We replace x by 3, y by -2, and z by -5.

$xyz^2-z = 3(-2)(-5)^2-(-5) = 3(-2)(25)-(-5) = -150+5 = -145$

9. We replace h by 165 and A by 20.

$$\begin{aligned} & 0.041h - 0.018A - 2.69 \\ &= 0.041(165) - 0.018(20) - 2.69 \\ &= 6.765 - 0.36 - 2.69 \\ &= 6.405 - 2.69 \\ &= 3.715 \end{aligned}$$

The lung capacity of a 20-year-old woman who is 165 cm tall is 3.715 liters.

11. Evaluate the polynomial for $h = 50$, $v = 40$, and $t = 2$.

$$\begin{aligned} & h + vt - 4.9t^2 \\ &= 50 + 40 \cdot 2 - 4.9(2)^2 \\ &= 50 + 80 - 19.6 \\ &= 110.4 \end{aligned}$$

The rocket will be 110.4 m above the ground 2 seconds after blast off.

13. Replace h by 6.3, r by 1.2, and π by 3.14.

$$\begin{aligned} 2\pi rh + 2\pi r^2 &\approx 2(3.14)(1.2)(6.3) + 2(3.14)(1.2)^2 \\ &\approx 2(3.14)(1.2)(6.3) + 2(3.14)(1.44) \\ &\approx 47.4768 + 9.0432 \\ &\approx 56.52 \end{aligned}$$

The surface area of the can is about 56.52 in^2.

15. $x^3y - 2xy + 3x^2 - 5$

Term	Coefficient	Degree	
x^3y	1	4	(Think: $x^3y = x^3y^1$)
$-2xy$	-2	2	(Think: $-2xy = -2x^1y^1$)
$3x^2$	3	2	
-5	-5	0	(Think: $-5 = -5x^0$)

The degree of the polynomial is the degree of the term of highest degree.

The term of highest degree is x^3y. Its degree is 4. The degree of the polynomial is 4.

17. $17x^2y^3 - 3x^3yz - 7$

Term	Coefficient	Degree	
$17x^2y^3$	17	5	
$-3x^3yz$	-3	5	(Think: $-3x^3yz = -3x^3y^1z^1$)
-7	-7	0	(Think: $-7 = -7x^0$)

The terms of highest degree are $17x^2y^3$ and $-3x^3yz$. Each has degree 5. The degree of the polynomial is 5.

19. $a + b - 2a - 3b = (1-2)a + (1-3)b = -a - 2b$

21. $3x^2y - 2xy^2 + x^2$

There are *no* like terms, so none of the terms can be collected.

23.

$$\begin{aligned} & 6au + 3av + 14au + 7av \\ &= (6+14)au + (3+7)av \\ &= 20au + 10av \end{aligned}$$

25.

$$\begin{aligned} & 2u^2v - 3uv^2 + 6u^2v - 2uv^2 \\ &= (2+6)u^2v + (-3-2)uv^2 \\ &= 8u^2v - 5uv^2 \end{aligned}$$

27.

$$\begin{aligned} & (2x^2 - xy + y^2) + (-x^2 - 3xy + 2y^2) \\ &= (2-1)x^2 + (-1-3)xy + (1+2)y^2 \\ &= x^2 - 4xy + 3y^2 \end{aligned}$$

29.

$$\begin{aligned} & (r - 2s + 3) + (2r + s) + (s + 4) \\ &= (1+2)r + (-2+1+1)s + (3+4) \\ &= 3r + 0s + 7 \\ &= 3r + 7 \end{aligned}$$

31.

$$\begin{aligned} & (b^3a^2 - 2b^2a^3 + 3ba + 4) + (b^2a^3 - 4b^3a^2 + 2ba - 1) \\ &= (1-4)b^3a^2 + (-2+1)b^2a^3 + (3+2)ba + (4-1) \\ &= -3b^3a^2 - b^2a^3 + 5ba + 3, \text{ or} \\ & -a^3b^2 - 3a^2b^3 + 5ab + 3 \end{aligned}$$

33.

$$\begin{aligned} & (a^3 + b^3) - (a^2b - ab^2 + b^3 + a^3) \\ &= a^3 + b^3 - a^2b + ab^2 - b^3 - a^3 \\ &= (1-1)a^3 - a^2b + ab^2 + (1-1)b^3 \\ &= -a^2b + ab^2 \end{aligned}$$

35.

$$\begin{aligned} & (xy - ab - 8) - (xy - 3ab - 6) \\ &= xy - ab - 8 - xy + 3ab + 6 \\ &= (1-1)xy + (-1+3)ab + (-8+6) \\ &= 2ab - 2 \end{aligned}$$

37.

$$\begin{aligned} & (-2a + 7b - c) - (-3b + 4c - 8d) \\ &= -2a + 7b - c + 3b - 4c + 8d \\ &= -2a + (7+3)b + (-1-4)c + 8d \\ &= -2a + 10b - 5c + 8d \end{aligned}$$

39.

F O I L

$$\begin{aligned} (3z - u)(2z + 3u) &= 6z^2 + 9zu - 2uz - 3u^2 \\ &= 6z^2 + 7zu - 3u^2 \end{aligned}$$

41.

F O I L

$$\begin{aligned} (a^2b - 2)(a^2b - 5) &= a^4b^2 - 5a^2b - 2a^2b + 10 \\ &= a^4b^2 - 7a^2b + 10 \end{aligned}$$

43.

$$\begin{aligned} (a^3 + bc)(a^3 - bc) &= (a^3)^2 - (bc)^2 \\ & \quad [(A+B)(A-B) = A^2 - B^2] \\ &= a^6 - b^2c^2 \end{aligned}$$

45.

$$\begin{array}{r} y^4x + y^2 + 1 \\ y^2 + 1 \\ \hline y^4x + y^2 + 1 \\ y^6x + y^4 \quad\quad + y^2 \\ \hline y^6x + y^4 + y^4x + 2y^2 + 1 \end{array}$$

47. $(3xy - 1)(4xy + 2)$

F O I L

$$\begin{aligned} &= 12x^2y^2 + 6xy - 4xy - 2 \\ &= 12x^2y^2 + 2xy - 2 \end{aligned}$$

49. $(3 - c^2d^2)(4 + c^2d^2)$

F O I L

$$\begin{aligned} &= 12 + 3c^2d^2 - 4c^2d^2 - c^4d^4 \\ &= 12 - c^2d^2 - c^4d^4 \end{aligned}$$

51. $(m^2 - n^2)(m + n)$

F O I L

$$= m^3 + m^2n - mn^2 - n^3$$

53. $(xy + x^5y^5)(x^4y^4 - xy)$
F O I L
$= x^5y^5 - x^2y^2 + x^9y^9 - x^6y^6$
$= x^9y^9 - x^6y^6 + x^5y^5 - x^2y^2$

55. $(x + h)^2$
$= x^2 + 2xh + h^2 \quad [(A+B)^2 = A^2 + 2AB + B^2]$

57. $(r^3t^2 - 4)^2$
$= (r^3t^2)^2 - 2 \cdot r^3t^2 \cdot 4 + 4^2$
$[(A-B)^2 = A^2 - 2AB + B^2]$
$= r^6t^4 - 8r^3t^2 + 16$

59. $(p^4 + m^2n^2)^2$
$= (p^4)^2 + 2 \cdot p^4 \cdot m^2n^2 + (m^2n^2)^2$
$[(A+B)^2 = A^2 + 2AB + B^2]$
$= p^8 + 2p^4m^2n^2 + m^4n^4$

61. $\left(2a^3 - \frac{1}{2}b^3\right)^2$
$= (2a^3)^2 - 2 \cdot 2a^3 \cdot \frac{1}{2}b^3 + \left(\frac{1}{2}b^3\right)^2$
$[(A-B)^2 = A^2 - 2AB + B^2]$
$= 4a^6 - 2a^3b^3 + \frac{1}{4}b^6$

63. $3a(a - 2b)^2 = 3a(a^2 - 4ab + 4b^2)$
$= 3a^3 - 12a^2b + 12ab^2$

65. $(2a - b)(2a + b) = (2a)^2 - b^2 = 4a^2 - b^2$

67. $(c^2 - d)(c^2 + d) = (c^2)^2 - d^2$
$= c^4 - d^2$

69. $(ab + cd^2)(ab - cd^2) = (ab)^2 - (cd^2)^2$
$= a^2b^2 - c^2d^4$

71. $(x + y - 3)(x + y + 3)$
$= [(x + y) - 3][(x + y) + 3]$
$= (x + y)^2 - 3^2$
$= x^2 + 2xy + y^2 - 9$

73. $[x + y + z][x - (y + z)]$
$= [x + (y + z)][x - (y + z)]$
$= x^2 - (y + z)^2$
$= x^2 - (y^2 + 2yz + z^2)$
$= x^2 - y^2 - 2yz - z^2$

75. $(a + b + c)(a - b - c)$
$= [a + (b + c)][a - (b + c)]$
$= a^2 - (b + c)^2$
$= a^2 - (b^2 + 2bc + c^2)$
$= a^2 - b^2 - 2bc - c^2$

77. The first coordinate is positive and the second coordinate is negative, so $(2, -5)$ is in quadrant IV.

79. Both coordinates are positive, so $(16, 23)$ is in quadrant I.

81. $2x = -10$
$x = -5$

Any ordered pair $(-5, y)$ is a solution. The variable x must be -5, but y can be any number we choose. A few solutions are listed below. Plot these points and draw the line.

x	y
-5	-3
-5	0
-5	4

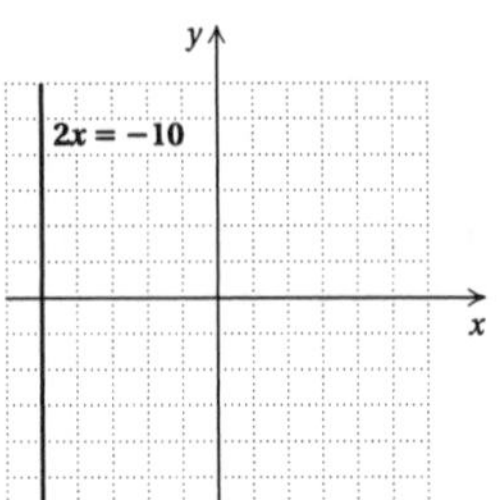

83. $8y - 16 = 0$
$8y = 16$
$y = 2$

Any ordered pair $(x, 2)$ is a solution. The variable y must be 2, but x can be any number we choose. A few solutions are listed below. Plot these points and draw the line.

x	y
-4	2
0	2
3	2

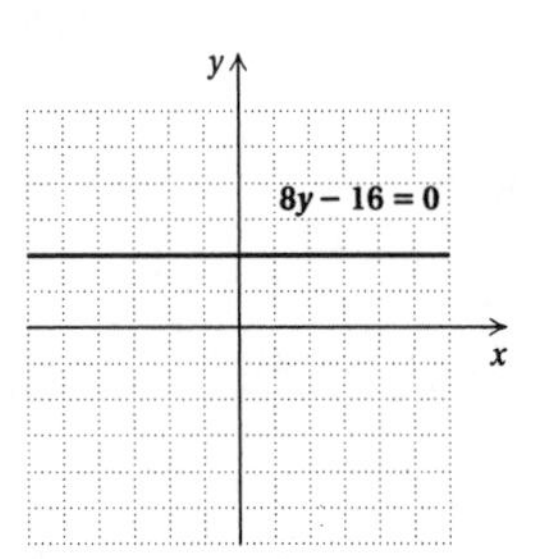

85. To find the mean we add the numbers and divide by the number of addends, 7.

$$\frac{23 + 31 + 24 + 31 + 25 + 28 + 31}{7} = \frac{193}{7} \approx 27.57$$

To find the median, we first list the numbers from smallest to largest.

23, 24, 25, 28, 31, 31, 31

The middle number, 28, is the median.

The number 31 occurs most often, so it is the mode.

87. ◈

89. It is helpful to add additional labels to the figure.

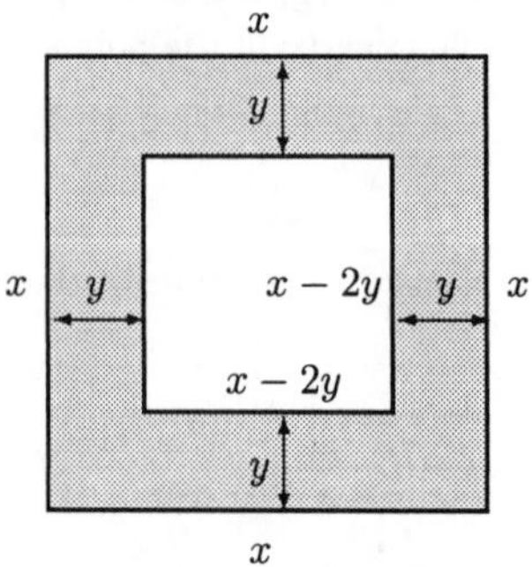

The area of the large square is $x \cdot x$, or x^2. The area of the small square is $(x - 2y)(x - 2y)$, or $(x - 2y)^2$.

$$\begin{array}{ccccc} \text{Area of shaded region} & = & \text{Area of large square} & - & \text{Area of small square} \\ \text{Area of shaded region} & = & x^2 & - & (x-2y)^2 \end{array}$$

$$= x^2 - (x^2 - 4xy + 4y^2)$$
$$= x^2 - x^2 + 4xy - 4y^2$$
$$= 4xy - 4y^2$$

91. It is helpful to add additional labels to the figure.

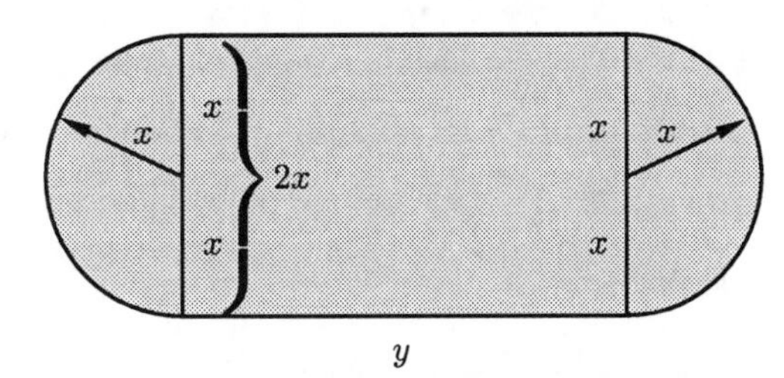

The two semicircles make a circle with radius x. The area of that circle is πx^2. The area of the rectangle is $2x \cdot y$. The sum of the two regions, $\pi x^2 + 2xy$, is the area of the shaded region.

93. The lateral surface area of the outer portion of the solid is the lateral surface area of a right circular cylinder with radius n and height h. The lateral surface area of the inner portion is the lateral surface area of a right circular cylinder with radius m and height h. Recall that the formula for the lateral surface area of a right circular cylinder with radius r and height h is $2\pi rh$.

The surface area of the top is the area of a circle with radius n less the area of a circle with radius m. The surface area of the bottom is the same as the surface area of the top.

Thus, the surface area of the solid is

$$2\pi nh + 2\pi mh + 2\pi n^2 - 2\pi m^2.$$

Exercise Set 10.8

1. $\dfrac{24x^4}{8} = \dfrac{24}{8} \cdot x^4 = 3x^4$

Check: We multiply.

$3x^4 \cdot 8 = 24x^4$

3. $\dfrac{25x^3}{5x^2} = \dfrac{25}{5} \cdot \dfrac{x^3}{x^2} = 5x^{3-2} = 5x$

Check: We multiply.

$5x \cdot 5x^2 = 25x^3$

5. $\dfrac{-54x^{11}}{-3x^8} = \dfrac{-54}{-3} \cdot \dfrac{x^{11}}{x^8} = 18x^{11-8} = 18x^3$

Check: We multiply.

$18x^3(-3x^8) = -54x^{11}$

7. $\dfrac{64a^5b^4}{16a^2b^3} = \dfrac{64}{16} \cdot \dfrac{a^5}{a^2} \cdot \dfrac{b^4}{b^3} = 4a^{5-2}b^{4-3} = 4a^3b$

Check: We multiply.

$(4a^3b)(16a^2b^3) = 64a^5b^4$

9. $\dfrac{24x^4 - 4x^3 + x^2 - 16}{8}$

$= \dfrac{24x^4}{8} - \dfrac{4x^3}{8} + \dfrac{x^2}{8} - \dfrac{16}{8}$

$= 3x^4 - \dfrac{1}{2}x^3 + \dfrac{1}{8}x^2 - 2$

Check: We multiply.

$$\begin{array}{r} 3x^4 - \frac{1}{2}x^3 + \frac{1}{8}x^2 - 2 \\ 8 \\ \hline 24x^4 - 4x^3 + x^2 - 16 \end{array}$$

11. $\dfrac{u - 2u^2 - u^5}{u}$

$= \dfrac{u}{u} - \dfrac{2u^2}{u} - \dfrac{u^5}{u}$

$= 1 - 2u - u^4$

Check: We multiply.

$$\begin{array}{r} 1 - 2u - u^4 \\ u \\ \hline u - 2u^2 - u^5 \end{array}$$

13. $(15t^3 + 24t^2 - 6t) \div (3t)$

$= \dfrac{15t^3 + 24t^2 - 6t}{3t}$

$= \dfrac{15t^3}{3t} + \dfrac{24t^2}{3t} - \dfrac{6t}{3t}$

$= 5t^2 + 8t - 2$

Check: We multiply.

$$\begin{array}{r} 5t^2 + 8t - 2 \\ 3t \\ \hline 15t^3 + 24t^2 - 6t \end{array}$$

15. $(20x^6 - 20x^4 - 5x^2) \div (-5x^2)$

$= \dfrac{20x^6 - 20x^4 - 5x^2}{-5x^2}$

$= \dfrac{20x^6}{-5x^2} - \dfrac{20x^4}{-5x^2} - \dfrac{5x^2}{-5x^2}$

$= -4x^4 - (-4x^2) - (-1)$

$= -4x^4 + 4x^2 + 1$

Check: We multiply.

$$\begin{array}{r} -4x^4 + 4x^2 + 1 \\ -5x^2 \\ \hline 20x^6 - 20x^4 - 5x^2 \end{array}$$

17. $(24x^5 - 40x^4 + 6x^3) \div (4x^3)$

$= \dfrac{24x^5 - 40x^4 + 6x^3}{4x^3}$

$= \dfrac{24x^5}{4x^3} - \dfrac{40x^4}{4x^3} + \dfrac{6x^3}{4x^3}$

$= 6x^2 - 10x + \dfrac{3}{2}$

Check: We multiply.

$$\begin{array}{r} 6x^2 - 10x + \frac{3}{2} \\ 4x^3 \\ \hline 24x^5 - 40x^4 + 6x^3 \end{array}$$

19. $\frac{18x^2 - 5x + 2}{2}$

$= \frac{18x^2}{2} - \frac{5x}{2} + \frac{2}{2}$

$= 9x^2 - \frac{5}{2}x + 1$

Check: We multiply.

$$\begin{array}{r} 9x^2 - \frac{5}{2}x + 1 \\ 2 \\ \hline 18x^2 - 5x + 2 \end{array}$$

21. $\frac{12x^3 + 26x^2 + 8x}{2x}$

$= \frac{12x^3}{2x} + \frac{26x^2}{2x} + \frac{8x}{2x}$

$= 6x^2 + 13x + 4$

Check: We multiply.

$$\begin{array}{r} 6x^2 + 13x + 4 \\ 2x \\ \hline 12x^3 + 26x^2 + 8x \end{array}$$

23. $\frac{9r^2s^2 + 3r^2s - 6rs^2}{3rs}$

$= \frac{9r^2s^2}{3rs} + \frac{3r^2s}{3rs} - \frac{6rs^2}{3rs}$

$= 3rs + r - 2s$

Check: We multiply.

$$\begin{array}{r} 3rs + r - 2s \\ 3rs \\ \hline 9r^2s^2 + 3r^2s - 6rs^2 \end{array}$$

25.

$$\begin{array}{rl} x + 2 \\ x + 2 \,\overline{)\, x^2 + 4x + 4} \\ x^2 + 2x \\ \hline 2x + 4 & \leftarrow (x^2 + 4x) - (x^2 + 2x) \\ 2x + 4 \\ \hline 0 & \leftarrow (2x + 4) - (2x + 4) \end{array}$$

The answer is $x + 2$.

27.

$$\begin{array}{rl} x - 5 \\ x - 5 \,\overline{)\, x^2 - 10x - 25} \\ x^2 - 5x \\ \hline -5x - 25 & \leftarrow (x^2 - 10x) - (x^2 - 5x) \\ -5x + 25 \\ \hline -50 & \leftarrow (-5x - 25) - (-5x + 25) \end{array}$$

The answer is $x - 5 + \frac{-50}{x - 5}$.

29.

$$\begin{array}{rl} x - 2 \\ x + 6 \,\overline{)\, x^2 + 4x - 14} \\ x^2 + 6x \\ \hline -2x - 14 & \leftarrow (x^2 + 4x) - (x^2 + 6x) \\ -2x - 12 \\ \hline -2 & \leftarrow (-2x - 14) - (-2x - 12) \end{array}$$

The answer is $x - 2 + \frac{-2}{x + 6}$.

31.

$$\begin{array}{rl} x - 3 \\ x + 3 \,\overline{)\, x^2 + 0x - 9} & \leftarrow \text{Filling in the missing term} \\ x^2 + 3x \\ \hline -3x - 9 & \leftarrow x^2 - (x^2 + 3x) \\ -3x - 9 \\ \hline 0 & \leftarrow (-3x - 9) - (-3x - 9) \end{array}$$

The answer is $x - 3$.

33.

$$\begin{array}{rl} x^4 - x^3 + x^2 - x + 1 \\ x + 1 \,\overline{)\, x^5 + 0x^4 + 0x^3 + 0x^2 + 0x + 1} & \leftarrow \text{Filling in missing terms} \\ x^5 + x^4 \\ \hline -x^4 & \leftarrow x^5 - (x^5 + x^4) \\ -x^4 - x^3 \\ \hline x^3 & \leftarrow -x^4 - (-x^4 - x^3) \\ x^3 + x^2 \\ \hline -x^2 & \leftarrow x^3 - (x^3 + x^2) \\ -x^2 - x \\ \hline x + 1 & \leftarrow -x^2 - (-x^2 - x) \\ x + 1 \\ \hline 0 & \leftarrow (x + 1) - (x + 1) \end{array}$$

The answer is $x^4 - x^3 + x^2 - x + 1$.

35.

$$\begin{array}{rl} 2x^2 - 7x + 4 \\ 4x + 3 \,\overline{)\, 8x^3 - 22x^2 - 5x + 12} \\ 8x^3 + 6x^2 \\ \hline -28x^2 - 5x & \leftarrow (8x^3 - 22x^2) - (8x^3 + 6x^2) \\ -28x^2 - 21x \\ \hline 16x + 12 & \leftarrow (-28x^2 - 5x) - (-28x^2 - 21x) \\ 16x + 12 \\ \hline 0 & \leftarrow (16x + 12) - (16x + 12) \end{array}$$

The answer is $2x^2 - 7x + 4$.

37.

$$\begin{array}{rl} x^3 - 6 \\ x^3 - 7 \,\overline{)\, x^6 - 13x^3 + 42} \\ x^6 - 7x^3 \\ \hline -6x^3 + 42 & \leftarrow (x^6 - 13x^3) - (x^6 - 7x^3) \\ -6x^3 + 42 \\ \hline 0 & \leftarrow (-6x^3 + 42) - (-6x^3 + 42) \end{array}$$

The answer is $x^3 - 6$.

39.
$$\begin{array}{r l}
 & x^3+2x^2+4x+8 \\ \cline{2-2}
x-2 \,\big) & x^4+0x^3+0x^2+0x-16 \\
 & \underline{x^4-2x^3} \\
 & \quad 2x^3 \quad \leftarrow x^4-(x^4-2x^3) \\
 & \quad \underline{2x^3-4x^2} \\
 & \qquad 4x^2 \quad \leftarrow 2x^3-(2x^3-4x^2) \\
 & \qquad \underline{4x^2-8x} \\
 & \qquad\quad 8x-16 \leftarrow 4x^2-(4x^2-8x) \\
 & \qquad\quad \underline{8x-16} \\
 & \qquad\qquad 0 \leftarrow (8x-16)-(8x-16)
\end{array}$$

The answer is x^3+2x^2+4x+8.

41.
$$\begin{array}{r l}
 & t^2+1 \\ \cline{2-2}
t-1 \,\big) & t^3-t^2+t-1 \\
 & \underline{t^3-t^2} \qquad \leftarrow (t^3-t^2)-(t^3-t^2) \\
 & \quad 0+t-1 \\
 & \quad \underline{t-1} \leftarrow (t-1)-(t-1) \\
 & \qquad 0
\end{array}$$

The answer is t^2+1.

43. $17-45=17+(-45)=-28$

45. $-2.3-(-9.1)=-2.3+9.1=6.8$

47. ***Familiarize***. Let w = the width. Then $w+15$ = the length. We draw a picture.

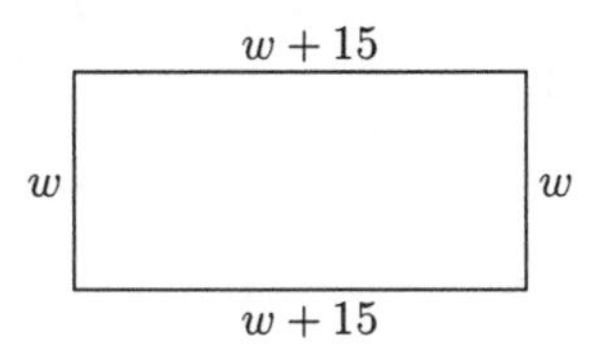

We will use the fact that the perimeter is 640 ft to find w (the width). Then we can find $w+15$ (the length) and multiply the length and the width to find the area.

Translate.

Width+Width+ Length + Length =Perimeter

$w \;+\; w \;+(w+15)+(w+15)= \;640$

Solve.

$$\begin{aligned}
w+w+(w+15)+(w+15) &= 640 \\
4w+30 &= 640 \\
4w &= 610 \\
w &= 152.5
\end{aligned}$$

If the width is 152.5, then the length is 152.5+15, or 167.5. The area is (167.5)(152.5), or 25,543.75 ft^2.

Check. The length, 167.5 ft, is 15 ft greater than the width, 152.5 ft. The perimeter is $152.5+152.5+167.5+167.5$, or 640 ft. We should also recheck the computation we used to find the area. The answer checks.

State. The area is 25,543.75 ft^2.

49.
$$\begin{aligned}
-6(2-x)+10(5x-7) &= 10 \\
-12+6x+50x-70 &= 10 \\
56x-82 &= 10 \quad \text{Collecting like terms} \\
56x-82+82 &= 10+82 \quad \text{Adding 82} \\
56x &= 92 \\
\frac{56x}{56} &= \frac{92}{56} \quad \text{Dividing by 56} \\
x &= \frac{23}{14}
\end{aligned}$$

The solution is $\frac{23}{14}$.

51. $4x-12+24y=4\cdot x-4\cdot 3+4\cdot 6y=4(x-3+6y)$

53. ◈

55.
$$\begin{array}{r l}
 & x^2+5 \\ \cline{2-2}
x^2+4 \,\big) & x^4+9x^2+20 \\
 & \underline{x^4+4x^2} \\
 & \qquad 5x^2+20 \\
 & \qquad \underline{5x^2+20} \\
 & \qquad\qquad 0
\end{array}$$

The answer is x^2+5.

57.
$$\begin{array}{r l}
 & a+3 \\ \cline{2-2}
5a^2-7a-2 \,\big) & 5a^3+8a^2-23a-1 \\
 & \underline{5a^3-7a^2-2a} \\
 & \quad 15a^2-21a-1 \\
 & \quad \underline{15a^2-21a-6} \\
 & \qquad\qquad 5
\end{array}$$

The answer is $a+3+\frac{5}{5a^2-7a-2}$.

59. We rewrite the dividend in descending order.

$$\begin{array}{r l}
 & 2x^2+x-3 \\ \cline{2-2}
3x^3-2x-1 \,\big) & 6x^5+3x^4-13x^3-4x^2+5x+3 \\
 & \underline{6x^5 \qquad -4x^3-2x^2} \\
 & \quad 3x^4-9x^3-2x^2+5x \\
 & \quad \underline{3x^4 \qquad -2x^2-x} \\
 & \qquad -9x^3 \qquad +6x+3 \\
 & \qquad \underline{-9x^3 \qquad +6x+3} \\
 & \qquad\qquad\qquad 0
\end{array}$$

The answer is $2x^2+x-3$.

61.

$$\begin{array}{r|l} & a^5+a^4b+a^3b^2+a^2b^3+ab^4+b^5 \\ \hline a-b & a^6+0a^5b+0a^4b^2+0a^3b^3+0a^2b^4+0ab^5-b^6 \\ & \underline{a^6-a^5b} \\ & \quad a^5b \\ & \quad \underline{a^5b-a^4b^2} \\ & \qquad a^4b^2 \\ & \qquad \underline{a^4b^2-a^3b^3} \\ & \qquad\quad a^3b^3 \\ & \qquad\quad \underline{a^3b^3-a^2b^4} \\ & \qquad\qquad a^2b^4 \\ & \qquad\qquad \underline{a^2b^4-ab^5} \\ & \qquad\qquad\quad ab^5-b^6 \\ & \qquad\qquad\quad \underline{ab^5-b^6} \\ & \qquad\qquad\qquad\quad 0 \end{array}$$

The answer is $a^5+a^4b+a^3b^2+a^2b^3+ab^4+b^5$.

63.

$$\begin{array}{r|l} & x+5 \\ \hline x-1 & x^2+4x+c \\ & \underline{x^2-x} \\ & \quad 5x+c \\ & \quad \underline{5x-5} \\ & \qquad c+5 \end{array}$$

We set the remainder equal to 0.

$c+5=0$

$c=-5$

Thus, c must be -5.

65.

$$\begin{array}{r|l} & c^2x+(-2c+c^2) \\ \hline x-1 & c^2x^2-2cx+1 \\ & \underline{c^2x^2-c^2x} \\ & (-2c+c^2)x+1 \\ & \underline{(-2c+c^2)x-(-2c+c^2)} \\ & \qquad 1+(-2c+c^2) \end{array}$$

We set the remainder equal to 0.

$c^2-2c+1=0$

$(c-1)^2=0$

$c=1$

Thus, c must be 1.

Chapter 11

Polynomials: Factoring

Exercise Set 11.1

1. Answers may vary. $8x^3 = (4x^2)(2x) = (-8)(-x^3) = (2x^2)(4x)$

3. Answers may vary. $-10a^6 = (-5a^5)(2a) = (10a^3)(-a^3) = (-2a^2)(5a^4)$

5. Answers may vary. $24x^4 = (6x)(4x^3) = (-3x^2)(-8x^2) = (2x^3)(12x)$

7. $x^2 - 6x = x \cdot x - x \cdot 6$ Factoring each term

$= x(x-6)$ Factoring out the common factor x

9. $2x^2 + 6x = 2x \cdot x + 2x \cdot 3$ Factoring each term

$= 2x(x+3)$ Factoring out the common factor $2x$

11. $x^3 + 6x^2 = x^2 \cdot x + x^2 \cdot 6$ Factoring each term

$= x^2(x+6)$ Factoring out x^2

13. $8x^4 - 24x^2 = 8x^2 \cdot x^2 - 8x^2 \cdot 3$

$= 8x^2(x^2 - 3)$ Factoring out $8x^2$

15. $2x^2 + 2x - 8 = 2 \cdot x^2 + 2 \cdot x - 2 \cdot 4$

$= 2(x^2 + x - 4)$ Factoring out 2

17. $17x^5y^3 + 34x^3y^2 + 51xy$

$= 17xy \cdot x^4y^2 + 17xy \cdot 2x^2y + 17xy \cdot 3$

$= 17xy(x^4y^2 + 2x^2y + 3)$

19. $6x^4 - 10x^3 + 3x^2 = x^2 \cdot 6x^2 - x^2 \cdot 10x + x^2 \cdot 3$

$= x^2(6x^2 - 10x + 3)$

21. $x^5y^5 + x^4y^3 + x^3y^3 - x^2y^2$

$= x^2y^2 \cdot x^3y^3 + x^2y^2 \cdot x^2y + x^2y^2 \cdot xy + x^2y^2(-1)$

$= x^2y^2(x^3y^3 + x^2y + xy - 1)$

23. $2x^7 - 2x^6 - 64x^5 + 4x^3$

$= 2x^3 \cdot x^4 - 2x^3 \cdot x^3 - 2x^3 \cdot 32x^2 + 2x^3 \cdot 2$

$= 2x^3(x^4 - x^3 - 32x^2 + 2)$

25. $1.6x^4 - 2.4x^3 + 3.2x^2 + 6.4x$

$= 0.8x(2x^3) - 0.8x(3x^2) + 0.8x(4x) + 0.8x(8)$

$= 0.8x(2x^3 - 3x^2 + 4x + 8)$

27. $\frac{5}{3}x^6 + \frac{4}{3}x^5 + \frac{1}{3}x^4 + \frac{1}{3}x^3$

$= \frac{1}{3}x^3(5x^3) + \frac{1}{3}x^3(4x^2) + \frac{1}{3}x^3(x) + \frac{1}{3}x^3(1)$

$= \frac{1}{3}x^3(5x^3 + 4x^2 + x + 1)$

29. Factor: $x^2(x+3) + 2(x+3)$

The binomial $x+3$ is common to both terms:

$x^2(x+3) + 2(x+3) = (x^2+2)(x+3)$

31. $5a^3(2a-7) - (2a-7)$

$= 5a^3(2a-7) - 1(2a-7)$

$= (5a^3 - 1)(2a-7)$

33. $x^3 + 3x^2 + 2x + 6$

$= (x^3 + 3x^2) + (2x + 6)$

$= x^2(x+3) + 2(x+3)$ Factoring each binomial

$= (x^2+2)(x+3)$ Factoring out the common factor $x+3$

35. $2x^3 + 6x^2 + x + 3$

$= (2x^3 + 6x^2) + (x+3)$

$= 2x^2(x+3) + 1(x+3)$ Factoring each binomial

$= (2x^2+1)(x+3)$

37. $8x^3 - 12x^2 + 6x - 9 = 4x^2(2x-3) + 3(2x-3)$

$= (4x^2+3)(2x-3)$

39. $12x^3 - 16x^2 + 3x - 4$

$= 4x^2(3x-4) + 1(3x-4)$ Factoring 1 out of the second binomial

$= (4x^2+1)(3x-4)$

41. $5x^3 - 5x^2 - x + 1$

$= (5x^3 - 5x^2) + (-x+1)$

$= 5x^2(x-1) - 1(x-1)$ Check: $-1(x-1) = -x+1$

$= (5x^2-1)(x-1)$

43. $x^3 + 8x^2 - 3x - 24 = x^2(x+8) - 3(x+8)$

$= (x^2-3)(x+8)$

45. $2x^3 - 8x^2 - 9x + 36 = 2x^2(x-4) - 9(x-4)$

$= (2x^2-9)(x-4)$

47. $-2x < 48$

$x > -24$ Dividing by -2 and reversing the inequality symbol

The solution set is $\{x|x > -24\}$.

49. $\frac{-108}{-4} = 27$ (The quotient of two negative numbers is positive.)

51. $(y+5)(y+7) = y^2 + 7y + 5y + 35$ Using FOIL

$= y^2 + 12y + 35$

53. $(y+7)(y-7) = y^2 - 7^2 = y^2 - 49$

$[(A+B))(A-B) = A^2 - B^2]$

55. $x + y = 4$

To find the x-intercept, let $y = 0$. Then solve for x.

$$x + y = 4$$
$$x + 0 = 4$$
$$x = 4$$

The x-intercept is $(4, 0)$.

To find the y-intercept, let $x = 0$. Then solve for y.

$$x + y = 4$$
$$0 + y = 4$$
$$y = 4$$

The y-intercept is $(0, 4)$.

Plot these points and draw the line.

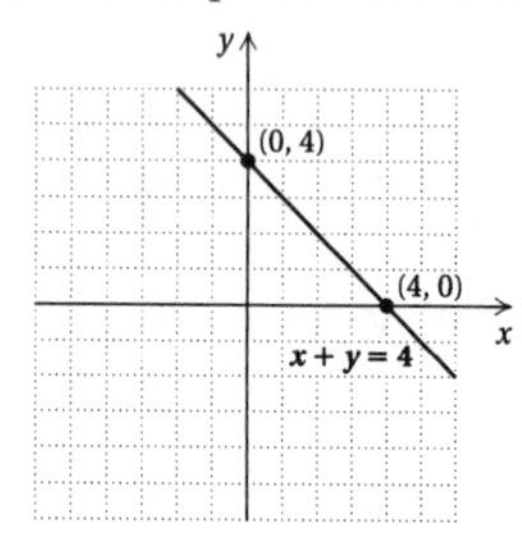

A third point should be used as a check. We substitute any value for x and solve for y. We let $x = 2$. Then

$$x + y = 4$$
$$2 + y = 4$$
$$y = 2$$

The point $(2, 2)$ is on the graph, so the graph is probably correct.

57. $5x - 3y = 15$

To find the x-intercept, let $y = 0$. Then solve for x.

$$5x - 3y = 15$$
$$5x - 3 \cdot 0 = 15$$
$$5x = 15$$
$$x = 3$$

The x-intercept is $(3, 0)$.

To find the y-intercept, let $x = 0$. Then solve for y.

$$5x - 3y = 15$$
$$5 \cdot 0 - 3y = 15$$
$$-3y = 15$$
$$y = -5$$

The y-intercept is $(0, -5)$.

Plot these points and draw the line.

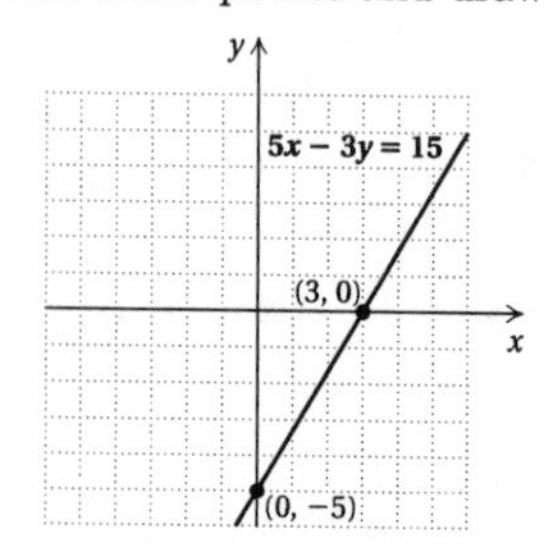

A third point should be used as a check. We substitute any value for x and solve for y. We let $x = 6$. Then

$$5x - 3y = 15$$
$$5 \cdot 6 - 3y = 15$$
$$30 - 3y = 15$$
$$-3y = -15$$
$$y = 5$$

The point $(6, 5)$ is on the graph, so the graph is probably correct.

59. ◈

61. $4x^5 + 6x^3 + 6x^2 + 9 = 2x^3(2x^2 + 3) + 3(2x^2 + 3)$
$= (2x^3 + 3)(2x^2 + 3)$

63. $x^{12} + x^7 + x^5 + 1 = x^7(x^5 + 1) + (x^5 + 1)$
$= (x^7 + 1)(x^5 + 1)$

65. $p^3 + p^2 - 3p + 10 = p^2(p + 1) - (3p - 10)$

This polynomial is not factorable using factoring by grouping.

Exercise Set 11.2

1. $x^2 + 8x + 15$

Since the constant term and coefficient of the middle term are both positive, we look for a factorization of 15 in which both factors are positive. Their sum must be 8.

Pairs of factors	Sums of factors
1, 15	16
3, 5	8

The numbers we want are 3 and 5.

$x^2 + 8x + 15 = (x + 3)(x + 5)$.

3. $x^2 + 7x + 12$

Since the constant term is positive and the coefficient of the middle term is positive, we look for a factorization of 12 in which both factors are positive. Their sum must be 7.

Pairs of factors	Sums of factors
1, 12	13
2, 6	8
3, 4	7

The numbers we want are 3 and 4.

$x^2 + 7x + 12 = (x + 3)(x + 4)$.

5. $x^2 - 6x + 9$

Since the constant term is positive and the coefficient of the middle term is negative, we look for a factorization of 9 in which both factors are negative. Their sum must be -6.

Pairs of factors	Sums of factors
−1, −9	−10
−3, −3	−6

The numbers we want are -3 and -3.

$x^2 - 6x + 9 = (x-3)(x-3)$, or $(x-3)^2$.

7. $x^2 + 9x + 14$

Since the constant term is positive and the coefficient of the middle term is positive, we look for a factorization of 14 in which both factors are positive. Their sum must be 9.

Pairs of factors	Sums of factors
1, 14	15
2, 7	9

The numbers we want are 2 and 7.

$x^2 + 9x + 14 = (x+2)(x+7)$.

9. $b^2 + 5b + 4$

Since the constant term is positive and the coefficient of the middle term is positive, we look for a factorization of 4 in which both factors are positive. Their sum must be 5.

Pairs of factors	Sums of factors
1, 4	5
2, 2	4

The numbers we want are 1 and 4.

$b^2 + 5b + 4 = (b+1)(b+4)$.

11. $x^2 + \frac{2}{3}x + \frac{1}{9}$

Since the constant term is positive and the coefficient of the middle term is positive, we look for a factorization of $\frac{1}{9}$ in which both factors are positive. Their sum must be $\frac{2}{3}$.

Pairs of factors	Sums of factors
$1, \frac{1}{9}$	$\frac{10}{9}$
$\frac{1}{3}, \frac{1}{3}$	$\frac{2}{3}$

The numbers we want are $\frac{1}{3}$ and $\frac{1}{3}$.

$x^2 + \frac{2}{3}x + \frac{1}{9} = \left(x + \frac{1}{3}\right)\left(x + \frac{1}{3}\right)$, or $\left(x + \frac{1}{3}\right)^2$.

13. $d^2 - 7d + 10$

Since the constant term is positive and the coefficient of the middle term is negative, we look for a factorization of 10 in which both factors are negative. Their sum must be -7.

Pairs of factors	Sums of factors
$-1, -10$	-11
$-2, -5$	-7

The numbers we want are -2 and -5.

$d^2 - 7d + 10 = (d-2)(d-5)$.

15. $y^2 - 11y + 10$

Since the constant term is positive and the coefficient of the middle term is negative, we look for a factorization of 10 in which both factors are negative. Their sum must be -11.

Pairs of factors	Sums of factors
$-1, -10$	-11
$-2, -5$	-7

The numbers we want are -1 and -10.

$y^2 - 11y + 10 = (y-1)(y-10)$.

17. $x^2 + x - 42$

Since the constant term is negative, we look for a factorization of -42 in which one factor is positive and one factor is negative. Their sum must be 1, the coefficient of the middle term.

Pairs of factors	Sums of factors
$-1, 42$	41
$1, -42$	-41
$-2, 21$	19
$2, -21$	-19
$-3, 14$	11
$3, -14$	-11
$-6, 7$	1
$6, -7$	-1

The numbers we want are -6 and 7.

$x^2 + x - 42 = (x-6)(x+7)$.

19. $x^2 - 7x - 18$

Since the constant term is negative, we look for a factorization of -18 in which one factor is positive and one factor is negative. Their sum must be -7, the coefficient of the middle term.

Pairs of factors	Sums of factors
$-1, 18$	17
$1, -18$	-17
$-2, 9$	7
$2, -9$	-7
$-3, 6$	3
$3, -6$	-3

The numbers we want are 2 and -9.

$x^2 - 7x - 18 = (x+2)(x-9)$.

21. $x^3 - 6x^2 - 16x = x(x^2 - 6x - 16)$

After factoring out the common factor, x, we consider $x^2 - 6x - 16$. Since the constant term is negative, we look for a factorization of -16 in which one factor is positive and one factor is negative. Their sum must be -6, the coefficient of the middle term.

Pairs of factors	Sums of factors
$-1,\ 16$	15
$1,\ -16$	-15
$-2,\ 8$	6
$2,\ -8$	-6
$-4,\ 4$	0

The numbers we want are 2 and -8.

Then $x^2 - 6x - 16 = (x+2)(x-8)$, so $x^3 - 6x^2 - 16x = x(x+2)(x-8)$.

23. $y^3 - 4y^2 - 45y = y(y^2 - 4y - 45)$

After factoring out the common factor, y, we consider $y^2 - 4y - 45$. Since the constant term is negative, we look for a factorization of -45 in which one factor is positive and one factor is negative. Their sum must be -4, the coefficient of the middle term.

Pairs of factors	Sums of factors
$-1,\ 45$	44
$1,\ -45$	-44
$-3,\ 15$	12
$3,\ -15$	-12
$-5,\ 9$	4
$5,\ -9$	-4

The numbers we want are 5 and -9.

Then $y^2 - 4y - 45 = (y+5)(y-9)$, so $y^3 - 4y^2 - 45y = y(y+5)(y-9)$.

25. $-2x - 99 + x^2 = x^2 - 2x - 99$

Since the constant term is negative, we look for a factorization of -99 in which one factor is positive and one factor is negative. Their sum must be -2, the coefficient of the middle term.

Pairs of factors	Sums of factors
$-1,\ 99$	98
$1,\ -99$	-98
$-3,\ 33$	30
$3,\ -33$	-30
$-9,\ 11$	2
$9,\ -11$	-2

The numbers we want are 9 and -11.

$-2x - 99 + x^2 = (x+9)(x-11)$.

27. $c^4 + c^2 - 56$

Consider this trinomial as $(c^2)^2 + c^2 - 56$. We look for numbers p and q such that $c^4 + c^2 - 56 = (c^2 + p)(c^2 + q)$. Since the constant term is negative, we look for a factorization of -56 in which one factor is positive and one factor is negative. Their sum must be 1.

Pairs of factors	Sums of factors
$-1,\ 56$	55
$1,\ -56$	-55
$-2,\ 28$	26
$2,\ -28$	-26
$-4,\ 14$	12
$4,\ -14$	-12
$-7,\ 8$	1
$7,\ -8$	-1

The numbers we want are -7 and 8.

$c^4 + c^2 - 56 = (c^2 - 7)(c^2 + 8)$.

29. $a^4 + 2a^2 - 35$

Consider this trinomial as $(a^2)^2 + 2a^2 - 35$. We look for numbers p and q such that $a^4 + 2a^2 - 35 = (a^2 + p)(a^2 + q)$. Since the constant term is negative, we look for a factorization of -35 in which one factor is positive and one factor is negative. Their sum must be 2.

Pairs of factors	Sums of factors
$-1,\ 35$	34
$1,\ -35$	-34
$-5,\ 7$	2
$5,\ -7$	-2

The numbers we want are -5 and 7.

$a^4 + 2a^2 - 35 = (a^2 - 5)(a^2 + 7)$.

31. $x^2 + x + 1$

Since the constant term and the coefficient of the middle term are both positive, we look for a factorization of 1 in which both factors are positive. The sum must be 1. The only possible pair of factors is 1 and 1, but their sum is not 1. Thus, this polynomial is not factorable into binomials.

33. $7 - 2p + p^2 = p^2 - 2p + 7$

Since the constant term is positive and the coefficient of the middle term is negative, we look for a factorization of 7 in which both factors are negative. The sum must be -2. The only possible pair of factors is -1 and -7, but their sum is not -2. Thus, this polynomial is not factorable into binomials.

35. $x^2 + 20x + 100$

We look for two factors, both positive, whose product is 100 and whose sum is 20.

They are 10 and 10. $10 \cdot 10 = 100$ and $10 + 10 = 20$.

$x^2 + 20x + 100 = (x+10)(x+10)$, or $(x+10)^2$.

37. $x^4 - 21x^3 - 100x^2 = x^2(x^2 - 21x - 100)$

After factoring out the common factor, x^2, we consider $x^2 - 21x - 100$. We look for two factors, one positive and one negative, whose product is -100 and whose sum is -21.

They are 4 and -25. $4 \cdot (-25) = -100$ and $4 + (-25) = -21$.

Then $x^2 - 21x - 100 = (x+4)(x-25)$, so $x^4 - 21x^3 - 100x^2 = x^2(x+4)(x-25)$.

39. $x^2 - 21x - 72$

We look for two factors, one positive and one negative, whose product is -72 and whose sum is -21. They are 3 and -24.

$x^2 - 21x - 72 = (x+3)(x-24)$.

41. $x^2 - 25x + 144$

We look for two factors, both negative, whose product is 144 and whose sum is -25. They are -9 and -16.

$x^2 - 25x + 144 = (x-9)(x-16)$.

43. $a^2 + a - 132$

We look for two factors, one positive and one negative, whose product is -132 and whose sum is 1. They are -11 and 12.

$a^2 + a - 132 = (a-11)(a+12)$.

45. $120 - 23x + x^2 = x^2 - 23x + 120$

We look for two factors, both negative, whose product is 120 and whose sum is -23. They are -8 and -15.

$x^2 - 23x + 120 = (x-8)(x-15)$.

47. First write the polynomial in descending order and factor out -1.

$108 - 3x - x^2 = -x^2 - 3x + 108 = -1(x^2 + 3x - 108)$

Now we factor the polynomial $x^2 + 3x - 108$. We look for two factors, one positive and one negative, whose product is -108 and whose sum is 3. They are -9 and 12.

$x^2 + 3x - 108 = (x-9)(x+12)$

The final answer must include -1 which was factored out above.

$-x^2 - 3x + 108 = -1(x-9)(x+12)$, or $-(x-9)(x+12)$

Using the distributive law to find $-1(x-9)$, we see that $-1(x-9)(x+12)$ can also be expressed as $(-x+9)(x+12)$, or $(9-x)(12+x)$.

49. $y^2 - 0.2y - 0.08$

We look for two factors, one positive and one negative, whose product is -0.08 and whose sum is-0.2. They are -0.4 and 0.2.

$y^2 - 0.2y - 0.08 = (y-0.4)(y+0.2)$.

51. $p^2 + 3pq - 10q^2 = p^2 + 3pq - 10q^2$

Think of $3q$ as a "coefficient" of p. Then we look for factors of $-10q^2$ whose sum is $3q$. They are $5q$ and $-2q$.

$p^2 + 3pq - 10q^2 = (p+5q)(p-2q)$.

53. $m^2 + 5mn + 4n^2 = m^2 + 5nm + 4n^2$

We look for factors of $4n^2$ whose sum is $5n$. They are $4n$ and n.

$m^2 + 5mn + 4n^2 = (m+4n)(m+n)$

55. $s^2 - 2st - 15t^2 = s^2 - 2ts - 15t^2$

We look for factors of $-15t^2$ whose sum is $-2t$. They are $-5t$ and $3t$.

$s^2 - 2st - 15t^2 = (s-5t)(s+3t)$

57. $8x(2x^2 - 6x + 1) = 8x \cdot 2x^2 - 8x \cdot 6x + 8x \cdot 1 = 16x^3 - 48x^2 + 8x$

59. $(7w+6)^2 = (7w)^2 + 2 \cdot 7w \cdot 6 + 6^2 = 49w^2 + 84w + 36$

61. $(4w-11)(4w+11) = (4w)^2 - (11)^2 = 16w^2 - 121$

63. $3x - 8 = 0$

$3x = 8$ Adding 8 on both sides

$x = \dfrac{8}{3}$ Dividing by 3 on both sides

The solution is $\dfrac{8}{3}$.

65. ***Familiarize.*** Let n = the number of people arrested the year before.

Translate. We reword the problem.

Number arrested the year before	less	1.2%	of	that number	is	29,200.
↓	↓	↓	↓	↓	↓	↓
n	$-$	1.2%	$\cdot$	n	$=$	29,200

Carry out. We solve the equation.

$$n - 1.2\% \cdot n = 29,200$$
$$1 \cdot n - 0.012n = 29,200$$
$$0.988n = 29,200$$
$$n \approx 29,555 \quad \text{Rounding}$$

Check. 1.2% of 29,555 is $0.012(29,555) \approx 355$ and $29,555 - 355 = 29,200$. The answer checks.

State. Approximately 29,555 people were arrested the year before.

67.

69. $y^2 + my + 50$

We look for pairs of factors whose product is 50. The sum of each pair is represented by m.

Pairs of factors whose product is -50	Sums of factors
1, 50	51
-1, -50	-51
2, 25	27
-2, -25	-27
5, 10	15
-5, -10	-15

The polynomial $y^2 + my + 50$ can be factored if m is 51, -51, 27, -27, 15, or -15.

71. $x^2 - \dfrac{1}{2}x - \dfrac{3}{16}$

We look for two factors, one positive and one negative, whose product is $-\dfrac{3}{16}$ and whose sum is $-\dfrac{1}{2}$.

They are $-\dfrac{3}{4}$ and $\dfrac{1}{4}$.

$-\frac{3}{4}\cdot\frac{1}{4}=-\frac{3}{16}$ and $-\frac{3}{4}+\frac{1}{4}=-\frac{2}{4}=-\frac{1}{2}$.

$x^2-\frac{1}{2}x-\frac{3}{16}=\left(x-\frac{3}{4}\right)\left(x+\frac{1}{4}\right)$

73. $x^2+\frac{30}{7}x-\frac{25}{7}$

We look for two factors, one positive and one negative, whose product is $-\frac{25}{7}$ and whose sum is $\frac{30}{7}$.

They are 5 and $-\frac{5}{7}$.

$5\cdot\left(-\frac{5}{7}\right)=-\frac{25}{7}$ and $5+\left(-\frac{5}{7}\right)=\frac{35}{7}+\left(-\frac{5}{7}\right)=\frac{30}{7}$.

$x^2+\frac{30}{7}x-\frac{25}{7}=(x+5)\left(x-\frac{5}{7}\right)$

75. $b^{2n}+7b^n+10$

Consider this trinomial as $(b^n)^2+7b^n+10$. We look for numbers p and q such that $b^{2n}+7b^n+10=(b^n+p)(b^n+q)$. We find two factors, both positive, whose product is 10 and whose sum is 7. They are 5 and 2.

$b^{2n}+7b^n+10=(b^n+5)(b^n+2)$

77. We first label the drawing with additional information.

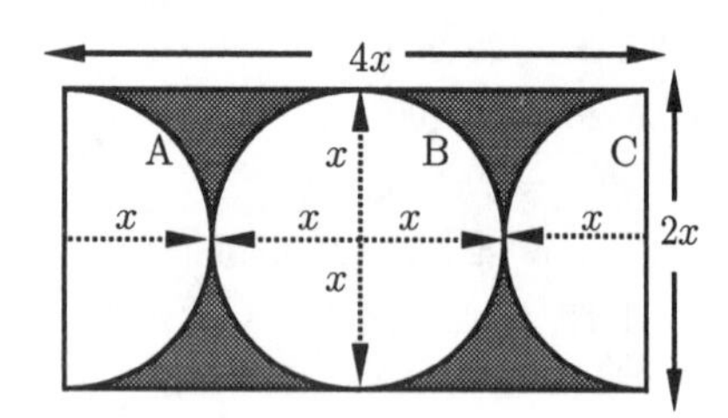

$4x$ represents the length of the rectangle and $2x$ the width. The area of the rectangle is $4x\cdot 2x$, or $8x^2$.

The area of semicircle A is $\frac{1}{2}\pi x^2$.

The area of circle B is πx^2.

The area of semicircle C is $\frac{1}{2}\pi x^2$.

$$\begin{aligned}\text{Area of shaded region} &= \text{Area of rectangle} - \text{Area of } A - \text{Area of } B - \text{Area of } C\\ \text{Area of shaded region} &= 8x^2 - \frac{1}{2}\pi x^2 - \pi x^2 - \frac{1}{2}\pi x^2\\ &= 8x^2-2\pi x^2\\ &= 2x^2(4-\pi)\end{aligned}$$

The shaded area can be represented by $2x^2(4-\pi)$.

Exercise Set 11.3

1. $2x^2-7x-4$

(1) Look for a common factor. There is none (other than 1 or -1).

(2) Factor the first term, $2x^2$. The only possibility is $2x$, x. The desired factorization is of the form:

$(2x+\quad)(x+\quad)$

(3) Factor the last term, -4, which is negative. The possibilities are -4, 1 and 4, -1 and 2, -2.

(4) Look for combinations of factors from steps (2) and (3) such that the sum of their products is the middle term, $-7x$. We try some possibilities:

$(2x-4)(x+1)=2x^2-2x-4$

$(2x+4)(x-1)=2x^2+2x-4$

$(2x+2)(x-2)=2x^2-2x-4$

$(2x+1)(x-4)=2x^2-7x-4$

The factorization is $(2x+1)(x-4)$.

3. $5x^2-x-18$

(1) There is no common factor (other than 1 or -1).

(2) Factor the first term, $5x^2$. The only possibility is $5x$, x. The desired factorization is of the form:

$(5x+\quad)(x+\quad)$

(3) Factor the last term, -18. The possibilities are -18, 1 and 18, -1 and -9, 2 and 9, -2 and -6, 3 and 6, -3.

(4) Look for combinations of factors from steps (2) and (3) such that the sum of their products is the middle term, x. We try some possibilities:

$(5x-18)(x+1)=5x^2-13x-18$

$(5x+18)(x-1)=5x^2+13x-18$

$(5x+9)(x-2)=5x^2-x-18$

The factorization is $(5x+9)(x-2)$.

5. $6x^2+23x+7$

(1) There is no common factor (other than 1 or -1).

(2) Factor the first term, $6x^2$. The possibilities are $6x$, x and $3x$, $2x$. We have these as possibilities for factorizations:

$(6x+\quad)(x+\quad)$ and $(3x+\quad)(2x+\quad)$

(3) Factor the last term, 7. The possibilities are 7, 1 and -7, -1.

(4) Look for combinations of factors from steps (2) and (3) such that the sum of their products is the middle term, $23x$. Since all signs are positive, we need consider only plus signs. We try some possibilities:

$(6x+7)(x+1)=6x^2+13x+7$

$(3x+7)(2x+1)=6x^2+17x+7$

$(6x+1)(x+7)=6x^2+43x+7$

$(3x+1)(2x+7)=6x^2+23x+7$

The factorization is $(3x+1)(2x+7)$.

7. $3x^2 + 4x + 1$

(1) There is no common factor (other than 1 or -1).

(2) Factor the first term, $3x^2$. The only possibility is $3x$, x. The desired factorization is of the form:

$$(3x+ \quad)(x+ \quad)$$

(3) Factor the last term, 1. The possibilities are 1, 1 and -1, -1.

(4) Look for combinations of factors from steps (2) and (3) such that the sum of their products is the middle term, $4x$. Since all signs are positive, we need consider only plus signs. There is only one such possibility:

$(3x + 1)(x + 1) = 3x^2 + 4x + 1$

The factorization is $(3x + 1)(x + 1)$.

9. $4x^2 + 4x - 15$

(1) There is no common factor (other than 1 or -1).

(2) Factor the first term, $4x^2$. The possibilities are $4x$, x and $2x$, $2x$. We have these as possibilities for factorizations:

$$(4x+ \quad)(x+ \quad) \text{ and } (2x+ \quad)(2x+ \quad)$$

(3) Factor the last term, -15. The possibilities are 15, -1 and -15, 1 and 5, -3 and -5, 3.

(4) We try some possibilities:

$(4x + 15)(x - 1) = 4x^2 + 11x - 15$

$(2x + 15)(2x - 1) = 4x^2 + 28x - 15$

$(4x - 15)(x + 1) = 4x^2 - 11x - 15$

$(2x - 15)(2x + 1) = 4x^2 - 28x - 15$

$(4x + 5)(x - 3) = 4x^2 - 7x - 15$

$(2x + 5)(2x - 3) = 4x^2 + 4x - 15$

The factorization is $(2x + 5)(2x - 3)$.

11. $2x^2 - x - 1$

(1) There is no common factor (other than 1 or -1).

(2) Factor the first term, $2x^2$. The only possibility is $2x$, x. The desired factorization is of the form:

$$(2x+ \quad)(x+ \quad)$$

(3) Factor the last term, -1. The only possibility is -1, 1.

(4) We try the possibilities:

$(2x - 1)(x + 1) = 2x^2 + x - 1$

$(2x + 1)(x - 1) = 2x^2 - x - 1$

The factorization is $(2x + 1)(x - 1)$.

13. $9x^2 + 18x - 16$

(1) There is no common factor (other than 1 or -1).

(2) Factor the first term, $9x^2$. The possibilities are $9x$, x and $3x$, $3x$. We have these as possibilities for factorizations:

$$(9x+ \quad)(x+ \quad) \text{ and } (3x+ \quad)(3x+ \quad)$$

(3) Factor the last term, -16. The possibilities are 16, -1 and -16, 1 and 8, -2 and -8, 2 and 4, -4.

(4) We try some possibilities:

$(9x + 16)(x - 1) = 9x^2 + 7x - 16$

$(3x + 16)(3x - 1) = 9x^2 + 45x - 16$

$(9x - 16)(x + 1) = 9x^2 - 7x - 16$

$(3x - 16)(3x + 1) = 9x^2 - 45x - 16$

$(9x + 8)(x - 2) = 9x^2 - 10x - 16$

$(3x + 8)(3x - 2) = 9x^2 + 18x - 16$

The factorization is $(3x + 8)(3x - 2)$.

15. $3x^2 - 5x - 2$

(1) There is no common factor (other than 1 or -1).

(2) Factor the first term, $3x^2$. The only possibility is $3x$, x. The desired factorization is of the form:

$$(3x+ \quad)(x+ \quad)$$

(3) Factor the last term, -2. The possibilities are 2, -1 and -2 and 1.

(4) We try some possibilities:

$(3x + 2)(x - 1) = 3x^2 - x - 2$

$(3x - 2)(x + 1) = 3x^2 + x - 2$

$(3x - 1)(x + 2) = 3x^2 + 5x - 2$

$(3x + 1)(x - 2) = 3x^2 - 5x - 2$

The factorization is $(3x + 1)(x - 2)$.

17. $12x^2 + 31x + 20$

(1) There is no common factor (other than 1 or -1).

(2) Factor the first term, $12x^2$. The possibilities are $12x$, x and $6x$, $2x$ and $4x$, $3x$. We have these as possibilities for factorizations:

$(12x+ \quad)(x+ \quad)$ and $(6x+ \quad)(2x+ \quad)$ and $(4x+ \quad)(3x+ \quad)$

(3) Factor the last term, 20. Since all signs are positive, we need consider only positive pairs of factors. Those factor pairs are 20, 1 and 10, 2 and 5, 4.

(4) We can immediately reject all possibilities in which either factor has a common factor, such as $(12x + 20)$ or $(6x + 4)$, because we determined at the outset that there are no common factors. We try some of the remaining possibilities:

$$(12x + 1)(x + 20) = 12x^2 + 241x + 20$$
$$(12x + 5)(x + 4) = 12x^2 + 53x + 20$$
$$(6x + 1)(2x + 20) = 12x^2 + 122x + 20$$
$$(4x + 5)(3x + 4) = 12x^2 + 31x + 20$$

The factorization is $(4x + 5)(3x + 4)$.

19. $14x^2 + 19x - 3$

(1) There is no common factor (other than 1 or -1).

(2) Factor the first term, $14x^2$. The possibilities are $14x$, x and $7x$, $2x$. We have these as possibilities for factorizations:

$(14x+\quad)(x+\quad)$ and $(7x+\quad)(2x+\quad)$

(3) Factor the last term, -3. The possibilities are -1, 3 and 3, -1.

(4) We try some possibilities:

$$(14x - 1)(x + 3) = 14x^2 + 41x - 3$$
$$(7x - 1)(2x + 3) = 7x^2 + 19x - 3$$

The factorization is $(7x - 1)(2x + 3)$.

21. $9x^2 + 18x + 8$

(1) There is no common factor (other than 1 or -1).

(2) Factor the first term, $9x^2$. The possibilities are $9x$, x and $3x$, $3x$. We have these as possibilities for factorizations:

$(9x+\quad)(x+\quad)$ and $(3x+\quad)(3x+\quad)$

(3) Factor the last term, 8. Since all signs are positive, we need consider only positive pairs of factors. Those factor pairs are 8, 1 and 4, 2.

(4) We try some possibilities:

$$(9x + 8)(x + 1) = 9x^2 + 17x + 8$$
$$(3x + 8)(3x + 1) = 9x^2 + 27x + 8$$
$$(9x + 4)(x + 2) = 9x^2 + 22x + 8$$
$$(3x + 4)(3x + 2) = 9x^2 + 18x + 8$$

The factorization is $(3x + 4)(3x + 2)$.

23. $49 - 42x + 9x^2 = 9x^2 - 42x + 49$

(1) There is no common factor (other than 1 or -1).

(2) Factor the first term, $9x^2$. The possibilities are $9x$, x and $3x$, $3x$. We have these as possibilities for factorizations:

$(9x+\quad)(x+\quad)$ and $(3x+\quad)(3x+\quad)$

(3) Factor 49. Since 49 is positive and the middle term is negative, we need consider only negative pairs of factors. Those factor pairs are -49, -1 and -7, -7.

(4) We try some possibilities:

$$(9x - 49)(x - 1) = 9x^2 - 58x + 49$$
$$(3x - 49)(3x - 1) = 9x^2 - 150x + 49$$
$$(9x - 7)(x - 7) = 9x^2 - 70x + 49$$
$$(3x - 7)(3x - 7) = 9x^2 - 42x + 49$$

The factorization is $(3x - 7)(3x - 7)$, or $(3x - 7)^2$. This can also be expressed as follows:

$(3x - 7)^2 = (-1)^2(3x - 7)^2 = [-1 \cdot (3x - 7)]^2 = (-3x + 7)^2$, or $(7 - 3x)^2$

25. $24x^2 + 47x - 2$

(1) There is no common factor (other than 1 or -1).

(2) Factor the first term, $24x^2$. The possibilities are $24x$, x and $12x$, $2x$ and $6x$, $4x$ and $3x$, $8x$. We have these as possibilities for factorizations:

$(24x+\quad)(x+\quad)$ and $(12x+\quad)(2x+\quad)$ and $(6x+\quad)(4x+\quad)$ and $(3x+\quad)(8x+\quad)$

(3) Factor the last term, -2. The possibilities are 2, -1 and -2, 1.

(4) We can immediately reject all possibilities in which either factor has a common factor, such as $(24x + 2)$ or $(12x - 2)$, because we determined at the outset that there are no common factors. We try some of the remaining possibilities:

$$(24x - 1)(x + 2) = 24x^2 + 47x - 2$$

The factorization is $(24x - 1)(x + 2)$.

27. $35x^2 - 57x - 44$

(1) There is no common factor (other than 1 or -1).

(2) Factor the first term, $35x^2$. The possibilities are $35x$, x and $7x$, $5x$. We have these as possibilities for factorizations:

$(35x+\quad)(x+\quad)$ and $(7x+\quad)(5x+\quad)$

(3) Factor the last term, -44. The possibilities are 1, -44 and -1, 44 and 2, -22 and -2, 22 and 4, -11, and -4, 11.

(4) We try some possibilities:

$(35x + 1)(x - 44) = 35x^2 - 1539x - 44$

$(7x + 1)(5x - 44) = 35x^2 - 303x - 44$

$(35x + 2)(x - 22) = 35x^2 - 768x - 44$

$(7x + 2)(5x - 22) = 35x^2 - 144x - 44$

$(35x + 4)(x - 11) = 35x^2 - 381x - 44$

$(7x + 4)(5x - 11) = 35x^2 - 57x - 44$

The factorization is $(7x + 4)(5x - 11)$.

29. $20 + 6x - 2x^2$

(1) We factor out the common factor, 2:
$2(10 + 3x - x^2)$

Then we factor the trinomial $10 + 3x - x^2$.

(2) Factor 10. The possibilities are 10, 1 and 5, 2. We have these as possibilities for factorizations:

$(10+\quad)(1+\quad)$ and $(5+\quad)(2+\quad)$

Note that the second term of each factor is an x-term.

(3) Factor $-x^2$. The only possibility is x, $-x$.

(4) We try some possibilities:

$(10 + x)(1 - x) = 10 - 9x - x^2$

$(5 + x)(2 - x) = 10 - 3x - x^2$

$(5 - x)(2 + x) = 10 + 3x - x^2$

The factorization of $10+3x-x^2$ is $(5-x)(2+x)$. We must include the common factor in order to get a factorization of the original trinomial.

$20 + 6x - 2x^2 = 2(5 - x)(2 + x)$

31. $12x^2 + 28x - 24$

(1) We factor out the common factor, 4:
$4(3x^2 + 7x - 6)$

Then we factor the trinomial $3x^2 + 7x - 6$.

(2) Factor $3x^2$. The only possibility is $3x$, x. The desired factorization is of the form:

$(3x+\quad)(x+\quad)$

(3) Factor -6. The possibilities are 6, -1 and -6, 1 and 3, -2 and -3, 2.

(4) We can immediately reject all possibilities in which either factor has a common factor, such as $(3x + 6)$ or $(3x - 3)$, because we factored out the largest common factor at the outset. We try some of the remaining possibilities:

$(3x - 1)(x + 6) = 3x^2 + 17x - 6$

$(3x - 2)(x + 3) = 3x^2 + 7x - 6$

The factorization of $3x^2+7x-6$ is $(3x-2)(x+3)$. We must include the common factor in order to get a factorization of the original trinomial.

$12x^2 + 28x - 24 = 4(3x - 2)(x + 3)$

33. $30x^2 - 24x - 54$

(1) We factor out the common factor, 6:
$6(5x^2 - 4x - 9)$

Then we factor the trinomial $5x^2 - 4x - 9$.

(2) Factor $5x^2$. The only possibility is $5x$, x. The desired factorization is of the form:

$(5x+\quad)(x+\quad)$

(3) Factor -9. The possibilities are 9, -1 and -9, 1 and 3, -3.

(4) We try some possibilities:

$(5x + 9)(x - 1) = 5x^2 + 4x - 9$

$(5x - 9)(x + 1) = 5x^2 - 4x - 9$

The factorization of $5x^2-4x-9$ is $(5x-9)(x+1)$. We must include the common factor in order to get a factorization of the original trinomial.

$30x^2 - 24x - 54 = 6(5x - 9)(x + 1)$

35. $4y + 6y^2 - 10 = 6y^2 + 4y - 10$

(1) We factor out the common factor, 2:
$2(3y^2 + 2y - 5)$

Then we factor the trinomial $3y^2 + 2y - 5$.

(2) Factor $3y^2$. The only possibility is $3y$, y. The desired factorization is of the form:

$(3y+\quad)(y+\quad)$

(3) Factor -5. The possibilities are 5, -1 and -5, 1.

(4) We try some possibilities:

$(3y + 5)(y - 1) = 3y^2 + 2y - 5$

Then $3y^2 + 2y - 5 = (3y + 5)(y - 1)$, so $6y^2 + 4y - 10 = 2(3y + 5)(y - 1)$.

37. $3x^2 - 4x + 1$

(1) There is no common factor (other than 1 or -1).

(2) Factor the first term, $3x^2$. The only possibility is $3x$, x. The desired factorization is of the form:

$(3x+\quad)(x+\quad)$

(3) Factor the last term, 1. Since 1 is positive and the middle term is negative, we need consider only negative factor pairs. The only such pair is -1, -1.

(4) There is only one possibility:

$(3x - 1)(x - 1) = 3x^2 - 4x + 1$

The factorization is $(3x - 1)(x - 1)$.

39. $12x^2 - 28x - 24$

(1) We factor out the common factor, 4:
$4(3x^2 - 7x - 6)$

Then we factor the trinomial $3x^2 - 7x - 6$.

(2) Factor $3x^2$. The only possibility is $3x$, x. The desired factorization is of the form:

$$(3x+ \quad)(x+ \quad)$$

(3) Factor -6. The possibilities are $6, -1$ and $-6, 1$ and $3, -2$ and $-3, 2$.

(4) We can immediately reject all possibilities in which either factor has a common factor, such as $(3x - 6)$ or $(3x + 3)$, because we factored out the largest common factor at the outset. We try some of the remaining possibilities:

$$(3x - 1)(x + 6) = 3x^2 + 17x - 6$$
$$(3x - 2)(x + 3) = 3x^2 + 7x - 6$$
$$(3x + 2)(x - 3) = 3x^2 - 7x - 6$$

Then $3x^2 - 7x - 6 = (3x + 2)(x - 3)$, so $12x^2 - 28x - 24 = 4(3x + 2)(x - 3)$.

41. $-1 + 2x^2 - x = 2x^2 - x - 1$

(1) There is no common factor (other than 1 or -1).

(2) Factor the first term, $2x^2$. The only possibility is $2x$, x. The desired factorization is of the form:

$$(2x+ \quad)(x+ \quad)$$

(3) Factor -1. The only possibility is $1, -1$.

(4) We try some possibilities:

$$(2x + 1)(x - 1) = 2x^2 - x - 1$$

The factorization is $(2x + 1)(x - 1)$.

43. $9x^2 - 18x - 16$

(1) There is no common factor (other than 1 or -1).

(2) Factor the first term, $9x^2$. The possibilities are $9x$, x and $3x$, $3x$. We have these as possibilities for factorizations:

$$(9x+ \quad)(x+ \quad) \text{ and } (3x+ \quad)(3x+ \quad)$$

(3) Factor the last term, -16. The possibilities are $16, -1$ and $-16, 1$ and $8, -2$ and $-8, 2$ and $4, -4$.

(4) We try some possibilities:

$$(9x + 16)(x - 1) = 9x^2 + 7x - 16$$
$$(3x + 16)(3x - 1) = 9x^2 + 45x - 16$$
$$(9x + 8)(x - 2) = 9x^2 - 10x - 16$$
$$(3x + 8)(3x - 2) = 9x^2 + 18x - 16$$
$$(3x - 8)(3x + 2) = 9x^2 - 18x - 16$$

The factorization is $(3x - 8)(3x + 2)$.

45. $15x^2 - 25x - 10$

(1) Factor out the common factor, 5:

$5(3x^2 - 5x - 2)$

Then we factor the trinomial $3x^2 - 5x - 2$. This was done in Exercise 15. We know that $3x^2 - 5x - 2 = (3x + 1)(x - 2)$, so $15x^2 - 25x - 10 = 5(3x + 1)(x - 2)$.

47. $12p^3 + 31p^2 + 20p$

(1) We factor out the common factor, p:

$p(12p^2 + 31p + 20)$

Then we factor the trinomial $12p^2 + 31p + 20$. This was done in Exercise 17 although the variable is x in that exercise. We know that $12p^2 + 31p + 20 = (3p + 4)(4p + 5)$, so $12p^3 + 31p^2 + 20p = p(3p + 4)(4p + 5)$.

49. $14x^4 + 19x^3 - 3x^2$

(1) Factor out the common factor, x^2:

$x^2(14x^2 + 19x - 3)$

Then we factor the trinomial $14x^2 + 19x - 3$. This was done in Exercise 19. We know that $14x^2 + 19x - 3 = (7x - 1)(2x + 3)$, so $14x^4 + 19x^3 - 3x^2 = x^2(7x - 1)(2x + 3)$.

51. $168x^3 - 45x^2 + 3x$

(1) Factor out the common factor, $3x$:

$3x(56x^2 - 15x + 1)$

Then we factor the trinomial $56x^2 - 15x + 1$.

(2) Factor $56x^2$. The possibilities are $56x$, x and $28x$, $2x$ and $14x$, $4x$ and $7x$, $8x$. We have these as possibilities for factorizations:

$(56x+ \quad)(x+ \quad)$ and $(28x+ \quad)(2x+ \quad)$ and

$(14x+ \quad)(4x+ \quad)$ and $(7x+ \quad)(8x+ \quad)$

(3) Factor 1. Since 1 is positive and the middle term is negative we need consider only the negative factor pair $-1, -1$.

(4) We try some possibilities:

$$(56x - 1)(x - 1) = 56x^2 - 57x + 1$$
$$(28x - 1)(2x - 1) = 56x^2 - 30x + 1$$
$$(14x - 1)(4x - 1) = 56x^2 - 18x + 1$$
$$(7x - 1)(8x - 1) = 56x^2 - 15x + 1$$

Then $56x^2 - 15x + 1 = (7x - 1)(8x - 1)$, so $168x^3 - 45x^2 + 3x = 3x(7x - 1)(8x - 1)$.

53. $15x^4 - 19x^2 + 6 = 15(x^2)^2 - 19x^2 + 6$

(1) There is no common factor (other than 1 or -1).

(2) Factor the first term, $15x^4$. The possibilities are $15x^2$, x^2 and $5x^2$, $3x^2$. We have these as possibilities for factorizations:

$$(15x^2+ \quad)(x^2+ \quad) \text{ and } (5x^2+ \quad)(3x^2+ \quad)$$

(3) Factor 6. Since 6 is positive and the middle term is negative, we need consider only negative factor pairs. Those pairs are $-6, -1$ and $-3, -2$.

(4) We can immediately reject all possibilities in which either factor has a common factor, such as $(15x^2 - 6)$ or $(3x^2 - 3)$, because we determined at the outset that there is no common factor. We try some of the remaining possibilities:

$$(15x^2 - 1)(x^2 - 6) = 15x^4 - 91x^2 + 6$$
$$(15x^2 - 2)(x^2 - 3) = 15x^4 - 47x^2 + 6$$
$$(5x^2 - 6)(3x^2 - 1) = 15x^4 - 23x^2 + 6$$
$$(5x^2 - 3)(3x^2 - 2) = 15x^4 - 19x^2 + 6$$

The factorization is $(5x^2 - 3)(3x^2 - 2)$.

55. $25t^2 + 80t + 64$

(1) There is no common factor (other than 1 or -1).

(2) Factor the first term, $25t^2$. The possibilities are $25t$, t and $5t$, $5t$. We have these as possibilities for factorizations:

$(25t+ \quad)(t+ \quad)$ and $(5t+ \quad)(5t+ \quad)$

(3) Factor the last term, 64. Since all signs are positive, we need consider only positive pairs of factors. Those factor pairs are 64, 1 and 32, 2 and 16, 4 and 8, 8.

(4) We try some possibilities:

$$(25t + 64)(t + 1) = 25t^2 + 89t + 64$$
$$(5t + 32)(5t + 2) = 25t^2 + 170t + 64$$
$$(25t + 16)(t + 4) = 25t^2 + 116t + 64$$
$$(5t + 8)(5t + 8) = 25t^2 + 80t + 64$$

The factorization is $(5t + 8)(5t + 8)$ or $(5t + 8)^2$.

57. $6x^3 + 4x^2 - 10x$

(1) Factor out the common factor, $2x$:

$2x(3x^2 + 2x - 5)$

Then we factor the trinomial $3x^2 + 2x - 5$. We did this in Exercise 35 (after we factored 2 out of the original trinomial). We know that $3x^2 + 2x - 5 = (3x + 5)(x - 1)$, so $6x^3 + 4x^2 - 10x = 2x(3x + 5)(x - 1)$.

59. $25x^2 + 79x + 64$

We follow the same procedure as in Exercise 55. None of the possibilities works. Thus, $25x^2 + 79x + 64$ is not factorable.

61. $6x^2 - 19x - 5$

We follow the same procedure as in Exercise 43. None of the possibilities works. Thus, $6x^2 - 19x - 5$ is not factorable.

63. $12m^2 - mn - 20n^2$

(1) There is no common factor (other than 1 or -1).

(2) Factor the first term, $12m^2$. The possibilities are $12m$, m and $6m$, $2m$ and $3m$, $4m$. We have these as possibilities for factorizations:

$(12m+ \quad)(m+ \quad)$ and $(6m+ \quad)(2m+ \quad)$ and $(3m+ \quad)(4m+ \quad)$

(3) Factor the last term, $-20n^2$. The possibilities are $20n$, $-n$ and $-20n$, n and $10n$, $-2n$ and $-10n$, $2n$ and $5n$, $-4n$ and $-5n$, $4n$.

(4) We can immediately reject all possibilities in which either factor has a common factor, such as $(12m+20n)$ or $(4m-2n)$, because we determined at the outset that there is no common factor. We try some of the remaining possibilities:

$$(12m - n)(m + 20n) = 12m^2 + 239mn - 20n^2$$
$$(12m + 5n)(m - 4n) = 12m^2 - 43mn - 20n^2$$
$$(3m - 20n)(4m + n) = 12m^2 - 77mn - 20n^2$$
$$(3m - 4n)(4m + 5n) = 12m^2 - mn - 20n^2$$

The factorization is $(3m - 4n)(4m + 5n)$.

65. $6a^2 - ab - 15b^2$

(1) There is no common factor (other than 1 or -1).

(2) Factor the first term, $6a^2$. The possibilities are $6a$, a and $3a$, $2a$. We have these as possibilities for factorizations:

$(6a+ \quad)(a+ \quad)$ and $(3a+ \quad)(2a+ \quad)$

(3) Factor the last term, $-15b^2$. The possibilities are $15b$, $-b$ and $-15b$, b and $5b$, $-3b$ and $-5b$, $3b$.

(4) We can immediately reject all possibilities in which either factor has a common factor, such as $(6a + 15b)$ or $(3a - 3b)$, because we determined at the outset that there is no common factor. We try some of the remaining possibilities:

$$(6a - b)(a + 15b) = 6a^2 + 89ab - 15b^2$$
$$(3a - b)(2a + 15b) = 6a^2 + 43ab - 15b^2$$
$$(6a + 5b)(a - 3b) = 6a^2 - 13ab - 15b^2$$
$$(3a + 5b)(2a - 3b) = 6a^2 + ab - 15b^2$$
$$(3a - 5b)(2a + 3b) = 6a^2 - ab - 15b^2$$

The factorization is $(3a - 5b)(2a + 3b)$.

67. $9a^2 + 18ab + 8b^2$

(1) There is no common factor (other than 1 or -1).

(2) Factor the first term, $9a^2$. The possibilities are $9a$, a and $3a$, $3a$. We have these as possibilities for factorizations:

$$(9a+\quad)(a+\quad) \text{ and } (3a+\quad)(3a+\quad)$$

(3) Factor $8b^2$. Since all signs are positive, we need consider only pairs of factors with positive coefficients. Those factor pairs are $8b$, b and $4b$, $2b$.

(4) We try some possibilities:

$$(9a+8b)(a+b) = 9a^2+17ab+8b^2$$
$$(3a+8b)(3a+b) = 9a^2+27ab+8b^2$$
$$(9a+4b)(a+2b) = 9a^2+22ab+8b^2$$
$$(3a+4b)(3a+2b) = 9a^2+18ab+8b^2$$

The factorization is $(3a+4b)(3a+2b)$.

69. $35p^2+34pq+8q^2$

(1) There is no common factor (other than 1 or -1).

(2) Factor the first term, $35p^2$. The possibilities are $35p$, p and $7p$, $5p$. We have these as possibilities for factorizations:

$$(35p+\quad)(p+\quad) \text{ and } (7p+\quad)(5p+\quad)$$

(3) Factor $8q^2$. Since all signs are positive, we need consider only pairs of factors with positive coefficients. Those factor pairs are $8q$, q and $4q$, $2q$.

(4) We try some possibilities:

$$(35p+8q)(p+q) = 35p^2+43pq+8q^2$$
$$(7p+8q)(5p+q) = 35p^2+47pq+8q^2$$
$$(35p+4q)(p+2q) = 35p^2+74pq+8q^2$$
$$(7p+4q)(5p+2q) = 35p^2+34pq+8p^2$$

The factorization is $(7p+4q)(5p+2q)$.

71. $18x^2-6xy-24y^2$

(1) Factor out the common factor, 6:

$6(3x^2-xy-4y^2)$

Then we factor the trinomial $3x^2-xy-4y^2$.

(2) Factor $3x^2$. The only possibility is $3x$, x. The desired factorization is of the form:

$$(3x+\quad)(x+\quad)$$

(3) Factor $-4y^2$. The possibilities are $4y$, $-y$ and $-4y$, y and $2y$, $-2y$.

(4) We try some possibilities:

$$(3x+4y)(x-y) = 3x^2+xy-4y^2$$
$$(3x-4y)(x+y) = 3x^2-xy-4y^2$$

Then $3x^2-xy-4y^2 = (3x-4y)(x+y)$, so $18x^2-6xy-24y^2 = 6(3x-4y)(x+y)$.

73.

$A = pq-7$

$A+7 = pq$ Adding 7

$\dfrac{A+7}{p} = q$ Dividing by p

75. $3x+2y=6$

$2y = 6-3x$ Subtracting $3x$

$y = \dfrac{6-3x}{2}$ Dividing by 2

77. $5-4x < -11$

$-4x < -16$ Subtracting 5

$x > 4$ Dividing by -4 and reversing the inequality symbol

The solution set is $\{x|x>4\}$.

79. Graph: $y = \dfrac{2}{5}x-1$

Because the equation is in the form $y=mx+b$, we know the y-intercept is $(0,-1)$. We find two other points on the line, substituting multiples of 5 for x to avoid fractions.

When $x=-5$, $y = \dfrac{2}{5}(-5)-1 = -2-1 = -3$.

When $x=5$, $y = \dfrac{2}{5}(5)-1 = 2-1 = 1$.

x	y
0	-1
-5	-3
5	1

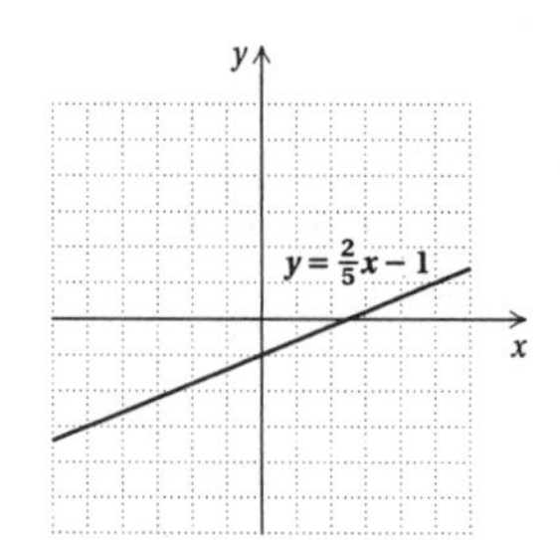

81. $(3x-5)(3x+5) = (3x)^2-5^2 = 9x^2-25$

83. ◈

85. $20x^{2n}+16x^n+3 = 20(x^n)^2+16x^n+3$

(1) There is no common factor (other than 1 and -1).

(2) Factor the first term, $20x^{2n}$. The possibilities are $20x^n$, x^n and $10x^n$, $2x^n$ and $5x^n$, $4x^n$. We have these as possibilities for factorizations:

$(20x^n+\quad)(x^n+\quad)$ and $(10x^n+\quad)(2x^n+\quad)$ and $(5x^n+\quad)(4x^n+\quad)$

(3) Factor the last term, 3. Since all signs are positive, we need consider only the positive factor pair 3, 1.

(4) We try some possibilities:

$$(20x^n+3)(x^n+1) = 20x^{2n}+23x^n+3$$
$$(10x^n+3)(2x^n+1) = 20x^{2n}+16x^n+3$$

The factorization is $(10x^n+3)(2x^n+1)$.

87. $3x^{6a} - 2x^{3a} - 1 = 3(x^{3a})^2 - 2x^{3a} - 1$

(1) There is no common factor (other than 1 or -1).

(2) Factor the first term, $3x^{6a}$. The only possibility is $3x^{3a}$, x^{3a}. The desired factorization is of the form:

$$(3x^{3a}+ \quad)(x^{3a}+ \quad)$$

(3) Factor the last term, -1. The only possibility is -1, 1.

(4) We try the possibilities:

$$(3x^{3a} - 1)(x^{3a} + 1) = 3x^{6a} + 2x^{3a} - 1$$
$$(3x^{3a} + 1)(x^{3a} - 1) = 3x^{6a} - 2x^{3a} - 1$$

The factorization is $(3x^{3a} + 1)(x^{3a} - 1)$.

Exercise Set 11.4

1. $x^2 + 2x + 7x + 14 = x(x + 2) + 7(x + 2)$
$= (x + 7)(x + 2)$

3. $x^2 - 4x - x + 4 = x(x - 4) - 1(x - 4)$
$= (x - 1)(x - 4)$

5. $6x^2 + 4x + 9x + 6 = 2x(3x + 2) + 3(3x + 2)$
$= (2x + 3)(3x + 2)$

7. $3x^2 - 4x - 12x + 16 = x(3x - 4) - 4(3x - 4)$
$= (x - 4)(3x - 4)$

9. $35x^2 - 40x + 21x - 24 = 5x(7x - 8) + 3(7x - 8)$
$= (5x + 3)(7x - 8)$

11. $4x^2 + 6x - 6x - 9 = 2x(2x + 3) - 3(2x + 3)$
$= (2x - 3)(2x + 3)$

13. $2x^4 + 6x^2 + 5x^2 + 15 = 2x^2(x^2 + 3) + 5(x^2 + 3)$
$= (2x^2 + 5)(x^2 + 3)$

15. $2x^2 - 7x - 4$

(a) First factor out a common factor, if any. There is none (other than 1 or -1).

(b) Multiply the leading coefficient, 2 and the constant, -4: $2(-4) = -8$.

(c) Look for a factorization of -8 in which the sum of the factors is the coefficient of the middle term, -7.

Pairs of factors	Sums of factors
-1, 8	7
1, -8	-7
-2, 4	2
2, -4	-2

(d) Split the middle term: $-7x = 1x - 8x$

(e) Factor by grouping:

$2x^2 - 7x - 4 = 2x^2 + x - 8x - 4$
$= x(2x + 1) - 4(2x + 1)$
$= (x - 4)(2x + 1)$

17. $3x^2 + 4x - 15$

(a) First factor out a common factor, if any. There is none (other than 1 or -1).

(b) Multiply the leading coefficient, 3, and the constant, -15: $3(-15) = -45$.

(c) Look for a factorization of -45 in which the sum of the factors is the coefficient of the middle term, 4.

Pairs of factors	Sums of factors
-1, 45	44
1, -45	-44
-3, 15	12
3, -15	-12
-5, 9	4
5, -9	-4

(d) Split the middle term: $4x = -5x + 9x$

(e) Factor by grouping:

$3x^2 + 4x - 15 = 3x^2 - 5x + 9x - 15$
$= x(3x - 5) + 3(3x - 5)$
$= (x + 3)(3x - 5)$

19. $6x^2 + 23x + 7$

(a) First factor out a common factor, if any. There is none (other than 1 or -1).

(b) Multiply the leading coefficient, 6, and the constant, 7: $6 \cdot 7 = 42$.

(c) Look for a factorization of 42 in which the sum of the factors is the coefficient of the middle term, 23. We only need to consider positive factors.

Pairs of factors	Sums of factors
1, 42	43
2, 21	23
3, 14	17
6, 7	13

(d) Split the middle term: $23x = 2x + 21x$

(e) Factor by grouping:

$6x^2 + 23x + 7 = 6x^2 + 2x + 21x + 7$
$= 2x(3x + 1) + 7(3x + 1)$
$= (2x + 7)(3x + 1)$

21. $3x^2 + 4x + 1$

(a) First factor out a common factor, if any. There is none (other than 1 or -1).

(b) Multiply the leading coefficient, 3, and the constant, 1: $3 \cdot 1 = 3$.

(c) Look for a factorization of 3 in which the sum of the factors is the coefficient of the middle term, 4. The numbers we want are 1 and 3: $1 \cdot 3 = 3$ and $1 + 3 = 4$.

(d) Split the middle term: $4x = 1x + 3x$

(e) Factor by grouping:

$$\begin{aligned} 3x^2 + 4x + 1 &= 3x^2 + x + 3x + 1 \\ &= x(3x+1) + 1(3x+1) \\ &= (x+1)(3x+1) \end{aligned}$$

23. $4x^2 + 4x - 15$

(a) First factor out a common factor, if any. There is none (other than 1 or -1).

(b) Multiply the leading coefficient, 4, and the constant, -15: $4(-15) = -60$.

(c) Look for a factorization of -60 in which the sum of the factors is the coefficient of the middle term, 4.

Pairs of factors	Sums of factors
$-1,\ 60$	59
$1, -60$	-59
$-2,\ 30$	28
$2, -30$	-28
$-3,\ 20$	17
$3, -20$	-17
$-4,\ 15$	11
$4, -15$	-11
$-5,\ 12$	7
$5, -12$	-7
$-6,\ 10$	4
$6, -10$	-4

(d) Split the middle term: $4x = -6x + 10x$

(e) Factor by grouping:

$$\begin{aligned} 4x^2 + 4x - 15 &= 4x^2 - 6x + 10x - 15 \\ &= 2x(2x-3) + 5(2x-3) \\ &= (2x+5)(2x-3) \end{aligned}$$

25. $2x^2 + x - 1$

(a) First factor out a common factor, if any. There is none (other than 1 or -1).

(b) Multiply the leading coefficient, 2, and the constant, -1: $2(-1) = -2$.

(c) Look for a factorization of -2 in which the sum of the factors is the coefficient of the middle term, 1. The numbers we wand are 2 and -1: $2(-1) = -2$ and $2 - 1 = 1$.

(d) Split the middle term: $x = 2x - 1x$

(e) Factor by grouping:

$$\begin{aligned} 2x^2 + x - 1 &= 2x^2 + 2x - x - 1 \\ &= 2x(x+1) - 1(x+1) \\ &= (2x-1)(x+1) \end{aligned}$$

27. $9x^2 - 18x - 16$

(a) First factor out a common factor, if any. There is none (other than 1 or -1).

(b) Multiply the leading coefficient, 9, and the constant, -16: $9(-16) = -144$.

(c) Look for a factorization of -144, so the sum of the factors is the coefficient of the middle term, -18.

Pairs of factors	Sums of factors
$-1,\ 144$	143
$1, -144$	-143
$-2,\ 72$	70
$2, -72$	-70
$-3,\ 48$	45
$3, -48$	-45
$-4,\ 36$	32
$4, -36$	-32
$-6,\ 24$	18
$6, -24$	-18
$-8,\ 18$	10
$8, -18$	-10
$-9,\ 16$	7
$9, -16$	-7
$-12,\ 12$	0

(d) Split the middle term: $-18x = 6x - 24x$

(e) Factor by grouping:

$$\begin{aligned} 9x^2 - 18x - 16 &= 9x^2 + 6x - 24x - 16 \\ &= 3x(3x+2) - 8(3x+2) \\ &= (3x-8)(3x+2) \end{aligned}$$

29. $3x^2 + 5x - 2$

(a) First factor out a common factor, if any. There is none (other than 1 or -1).

(b) Multiply the leading coefficient, 3, and the constant, -2: $3(-2) = -6$.

(c) Look for a factorization of -6 in which the sum of the factors is the coefficient of the middle term, 5. The numbers we wand are -1 and 6: $-1(6) = -6$ and $-1 + 6 = 5$.

(d) Split the middle term: $5x = -1x + 6x$

(e) Factor by grouping:

$$\begin{aligned} 3x^2 + 5x - 2 &= 3x^2 - x + 6x - 2 \\ &= x(3x-1) + 2(3x-1) \\ &= (x+2)(3x-1) \end{aligned}$$

31. $12x^2 - 31x + 20$

(a) First factor out a common factor, if any. There is none (other than 1 or -1).

(b) Multiply the leading coefficient, 12, and the constant, 20: $12 \cdot 20 = 240$.

(c) Look for a factorization of 240 in which the sum of the factors is the coefficient of the middle term, −31. We only need to consider negative factors.

Pairs of factors	Sums of factors
−1, −240	−241
−2, −120	−122
−3, −8	−83
−4, −60	−64
−5, −48	−53
−6, −40	−46
−8, −30	−38
−10, −24	−34
−12, −20	−32
−15, −16	−31

(d) Split the middle term: $-31x = -15x - 16x$

(e) Factor by grouping:

$$\begin{aligned}12x^2 - 31x + 20 &= 12x^2 - 15x - 16x + 20\\ &= 3x(4x-5) - 4(4x-5)\\ &= (3x-4)(4x-5)\end{aligned}$$

33. $14x^2 + 19x - 3$

(a) First factor out a common factor, if any. There is none (other than 1 or −1).

(b) Multiply the leading coefficient, 14, and the constant, −3: $14(-3) = -42$.

(c) Look for a factorization of −42 so that the sum of the factors is the coefficient of the middle term, 19.

Pairs of factors	Sums of factors
−1, 42	41
1, −42	−41
−2, 21	19
2, −21	−19
−3, 14	11
3, −14	−11
−6, 7	1
6, −7	−1

(d) Split the middle term: $19x = -2x + 21x$

(e) Factor by grouping:

$$\begin{aligned}14x^2 + 19x - 3 &= 14x^2 - 2x + 21x - 3\\ &= 2x(7x-1) + 3(7x-1)\\ &= (2x+3)(7x-1)\end{aligned}$$

35. $9x^2 + 18x + 8$

(a) First factor out a common factor, if any. There is none (other than 1 or −1).

(b) Multiply the leading coefficient, 9, and the constant, 8: $9 \cdot 8 = 72$.

(c) Look for a factorization of 72 in which the sum of the factors is the coefficient of the middle term, 18. We only need to consider positive factors.

Pairs of factors	Sums of factors
1, 72	73
2, 36	38
3, 24	27
4, 18	22
6, 12	18
8, 9	17

(d) Split the middle term: $18x = 6x + 12x$

(e) Factor by grouping:

$$\begin{aligned}9x^2 + 18x + 8 &= 9x^2 + 6x + 12x + 8\\ &= 3x(3x+2) + 4(3x+2)\\ &= (3x+4)(3x+2)\end{aligned}$$

37. $49 - 42x + 9x^2 = 9x^2 - 42x + 49$

(a) First factor out a common factor, if any. There is none (other than 1 or −1).

(b) Multiply the leading coefficient, 9, and the constant, 49: $9 \cdot 49 = 441$.

(c) Look for a factorization of 441 in which the sum of the factors is the coefficient of the middle term, −42. We only need to consider negative factors.

Pairs of factors	Sums of factors
−1, −441	−442
−3, −147	−150
−7, −63	−70
−9, −49	−58
−21, −21	−42

(d) Split the middle term: $-42x = -21x - 21x$

(e) Factor by grouping:

$$\begin{aligned}9x^2 - 42x + 49 &= 9x^2 - 21x - 21x + 49\\ &= 3x(3x-7) - 7(3x-7)\\ &= (3x-7)(3x-7), \text{ or}\\ &\quad (3x-7)^2\end{aligned}$$

39. $24x^2 + 47x - 2$

(a) First factor out a common factor, if any. There is none (other than 1 or −1).

(b) Multiply the leading coefficient, 24, and the constant, −2: $24(-2) = -48$.

(c) Look for a factorization of −48 in which the sum of the factors is the coefficient of the middle term, 47. The numbers we want are 48 and −1: $48(-1) = -48$ and $48 + (-1) = 47$.

(d) Split the middle term: $47x = 48x - 1x$

(e) Factor by grouping:

$$24x^2 + 47x - 2 = 24x^2 + 48x - x - 2$$
$$= 24x(x+2) - 1(x+2)$$
$$= (24x-1)(x+2)$$

41. $35x^5 - 57x^4 - 44x^3$

(a) We first factor out the common factor, x^3.

$x^3(35x^2 - 57x - 44)$

(b) Now we factor the trinomial $35x^2 - 57x - 44$. Multiply the leading coefficient, 35, and the constant, -44: $35(-44) = -1540$.

(c) Look for a factorization of -1540 in which the sum of the factors is the coefficient of the middle term, -57.

Pairs of factors	Sums of factors
7, −220	−213
10, −154	−144
11, −140	−129
14, −110	−96
20, −77	−57

(d) Split the middle term: $-57x = 20x - 77x$

(e) Factor by grouping:

$$35x^2 - 57x - 44 = 35x^2 + 20x - 77x - 44$$
$$= 5x(7x+4) - 11(7x+4)$$
$$= (5x-11)(7x+4)$$

We must include the common factor to get a factorization of the original trinomial.

$35x^5 - 57x^4 - 44x^3 = x^3(5x-11)(7x+4)$

43. $60x + 18x^2 - 6x^3$

(a) We first factor out the common factor, $6x$.

$60x + 18x^2 - 6x^3 = 6x(10 + 3x - x^2)$

(b) Now we factor the trinomial $10+3x-x^2$. Multiply the leading coefficient, -1, and the constant, 10: $-1(10) = -10$.

(c) Look for a factorization of -10 in which the sum of the factors is the coefficient of the middle term, 3. The numbers we want are 5 and -2: $5(-2) = -10$ and $5 + (-2) = 3$.

(d) Split the middle term: $3x = 5x - 2x$

(e) Factor by grouping:

$$10 + 3x - x^2 = 10 + 5x - 2x - x^2$$
$$= 5(2+x) - x(2+x)$$
$$= (5-x)(2+x)$$

We must include the common factor to get a factorization of the original trinomial.

$60x + 18x^2 - 6x^3 = 6x(5-x)(2+x)$

45. $15x^3 + 33x^4 + 6x^5 = 6x^5 + 33x^4 + 15x^3$

(a) First factor out the common factor, $3x^3$.

$3x^3(2x^2 + 11x + 5)$

(b) Now we factor the trinomial $2x^2 + 11x + 5$. Multiply the leading coefficient, 2, and the constant, 5: $2 \cdot 5 = 10$.

(c) Look for a factorization of 10 in which the sum of the factors is the coefficient of the middle term, 11. We only need to consider positive factors.

Pairs of factors	Sums of factors
2, 5	7
1, 10	11

(d) Split the middle term: $11x = x + 10x$

(e) Factor by grouping:

$$2x^2 + 11x + 5 = 2x^2 + x + 10x + 5$$
$$= x(2x+1) + 5(2x+1)$$
$$= (x+5)(2x+1)$$

We must include the common factor to get a factorization of the original trinomial.

$15x^3 + 33x^4 + 6x^5 = 3x^3(x+5)(2x+1)$, or $3x^3(5+x)(1+2x)$

47. $-10x > 1000$

$\dfrac{-10x}{-10} < \dfrac{1000}{-10}$ Dividing by -10 and reversing the inequality symbol

$x < -100$

The solution set is $\{x|x < -100\}$.

49. $6 - 3x \geq -18$

$-3x \geq -24$ Subtracting 6

$x \leq 8$ Dividing by -3 and reversing the inequality symbol

The solution set is $\{x|x \leq 8\}$.

51. $\frac{1}{2}x - 6x + 10 \leq x - 5x$

$2\left(\frac{1}{2}x - 6x + 10\right) \leq 2(x - 5x)$ Multiplying by 2 to clear the fraction

$x - 12x + 20 \leq 2x - 10x$

$-11x + 20 \leq -8x$ Collecting like terms

$20 \leq 3x$ Adding $11x$

$\frac{20}{3} \leq x$ Dividing by 3

The solution set is $\left\{x|x \geq \frac{20}{3}\right\}$.

53. $3x - 6x + 2(x-4) > 2(9-4x)$

$3x - 6x + 2x - 8 > 18 - 8x$ Removing parentheses

$-x - 8 > 18 - 8x$ Collecting like terms

$7x > 26$ Adding $8x$ and 8

$x > \frac{26}{7}$ Dividing by 7

The solution set is $\left\{x|x > \frac{26}{7}\right\}$.

55. 

57. $9x^{10} - 12x^5 + 4$

(a) First factor out a common factor, if any. There is none (other than 1 or -1).

(b) Multiply the leading coefficient, 9, and the constant, 4: $9 \cdot 4 = 36$.

(c) Look for a factorization of 36 in which the sum of the factors is the coefficient of the middle term, -12. The factors we want are -6 and -6.

(d) Split the middle term: $-12x^5 = -6x^5 - 6x^5$

(e) Factor by grouping:

$$\begin{aligned} 9x^{10} - 12x^5 + 4 &= 9x^{10} - 6x^5 - 6x^5 + 4 \\ &= 3x^5(3x^5 - 2) - 2(3x^5 - 2) \\ &= (3x^5 - 2)(3x^5 - 2), \text{ or} \\ &= (3x^5 - 2)^2 \end{aligned}$$

59. $16x^{10} + 8x^5 + 1$

(a) First factor out a common factor, if any. There is none (other than 1 or -1).

(b) Multiply the leading coefficient, 16, and the constant, 1: $16 \cdot 1 = 16$.

(c) Look for a factorization of 16 in which the sum of the factors is the coefficient of the middle term, 8. The factors we want are 4 and 4.

(d) Split the middle term: $8x^5 = 4x^5 + 4x^5$

(e) Factor by grouping:

$$\begin{aligned} 16x^{10} + 8x^5 + 1 &= 16x^{10} + 4x^5 + 4x^5 + 1 \\ &= 4x^5(4x^5 + 1) + 1(4x^5 + 1) \\ &= (4x^5 + 1)(4x^5 + 1), \text{ or} \\ &= (4x^5 + 1)^2 \end{aligned}$$

61.-69.

Exercise Set 11.5

1. $x^2 - 14x + 49$

(a) We know that x^2 and 49 are squares.

(b) There is no minus sign before either x^2 or 49.

(c) If we multiply the square roots, x and 7, and double the product, we get $2 \cdot x \cdot 7 = 14x$. This is the opposite of the remaining term, $-14x$.

Thus, $x^2 - 14x + 49$ is a trinomial square.

3. $x^2 + 16x - 64$

Both x^2 and 64 are squares, but there is a minus sign before 64. Thus, $x^2 + 16x - 64$ is not a trinomial square.

5. $x^2 - 2x + 4$

(a) Both x^2 and 4 are squares.

(b) There is no minus sign before either x^2 or 4.

(c) If we multiply the square roots, x and 2, and double the product, we get $2 \cdot x \cdot 2 = 4x$. This is neither the remaining term nor its opposite.

Thus, $x^2 - 2x + 4$ is not a trinomial square.

7. $9x^2 - 36x + 24$

Only one term is a square. Thus, $9x^2 - 36x + 24$ is not a trinomial square.

9.
$$\begin{aligned} x^2 - 14x + 49 &= x^2 - 2 \cdot x \cdot 7 + 7^2 = (x-7)^2 \\ &\quad\uparrow \quad \uparrow \quad \uparrow \quad \uparrow \quad \uparrow \\ &= A^2 - 2\ A\ B + B^2 = (A-B)^2 \end{aligned}$$

11.
$$\begin{aligned} x^2 + 16x + 64 &= x^2 + 2 \cdot x \cdot 8 + 8^2 = (x+8)^2 \\ &\quad\uparrow \quad \uparrow \quad \uparrow \quad \uparrow \quad \uparrow \\ &= A^2 + 2\ A\ B + B^2 = (A+B)^2 \end{aligned}$$

13. $x^2 - 2x + 1 = x^2 - 2 \cdot x \cdot 1 + 1^2 = (x-1)^2$

15.
$$\begin{aligned} 4 + 4x + x^2 &= x^2 + 4x + 4 \qquad \text{Changing the order} \\ &= x^2 + 2 \cdot x \cdot 2 + 2^2 \\ &= (x+2)^2 \end{aligned}$$

17. $q^4 - 6q^2 + 9 = (q^2)^2 - 2 \cdot q^2 \cdot 3 + 3^2 = (q^2 - 3)^2$

19.
$$\begin{aligned} 49 + 56y + 16y^2 &= 16y^2 + 56y + 49 \\ &= (4y)^2 + 2 \cdot 4y \cdot 7 + 7^2 \\ &= (4y + 7)^2 \end{aligned}$$

21.
$$\begin{aligned} 2x^2 - 4x + 2 &= 2(x^2 - 2x + 1) \\ &= 2(x^2 - 2 \cdot x \cdot 1 + 1^2) \\ &= 2(x-1)^2 \end{aligned}$$

23.
$$\begin{aligned} x^3 - 18x^2 + 81x &= x(x^2 - 18x + 81) \\ &= x(x^2 - 2 \cdot x \cdot 9 + 9^2) \\ &= x(x-9)^2 \end{aligned}$$

25.
$$\begin{aligned} 12q^2 - 36q + 27 &= 3(4q^2 - 12q + 9) \\ &= 3[(2q)^2 - 2 \cdot 2q \cdot 3 + 3^2] \\ &= 3(2q-3)^2 \end{aligned}$$

27.
$$\begin{aligned} 49 - 42x + 9x^2 &= 7^2 - 2 \cdot 7 \cdot 3x + (3x)^2 \\ &= (7 - 3x)^2 \end{aligned}$$

29.
$$\begin{aligned} 5y^4 + 10y^2 + 5 &= 5(y^4 + 2y^2 + 1) \\ &= 5[(y^2)^2 + 2 \cdot y^2 \cdot 1 + 1^2] \\ &= 5(y^2 + 1)^2 \end{aligned}$$

31.
$$\begin{aligned} 1 + 4x^4 + 4x^2 &= 1^2 + 2 \cdot 1 \cdot 2x^2 + (2x^2)^2 \\ &= (1 + 2x^2)^2 \end{aligned}$$

33.
$$\begin{aligned} 4p^2 + 12pq + 9q^2 &= (2p)^2 + 2 \cdot 2p \cdot 3q + (3q)^2 \\ &= (2p + 3q)^2 \end{aligned}$$

35.
$$\begin{aligned} a^2 - 6ab + 9b^2 &= a^2 - 2 \cdot a \cdot 3b + (3b)^2 \\ &= (a - 3b)^2 \end{aligned}$$

37. $81a^2 - 18ab + b^2 = (9a)^2 - 2 \cdot 9a \cdot b + b^2$
$= (9a - b)^2$

39. $36a^2 + 96ab + 64b^2 = 4(9a^2 + 24ab + 16b^2)$
$= 4[(3a)^2 + 2 \cdot 3a \cdot 4b + (4b)^2]$
$= 4(3a + 4b)^2$

41. $x^2 - 4$

(a) The first expression is a square: x^2

The second expression is a square: $4 = 2^2$

(b) The terms have different signs.

$x^2 - 4$ is a difference of squares.

43. $x^2 + 25$

The terms do not have different signs.

$x^2 + 25$ is not a difference of squares.

45. $x^2 - 45$

The number 45 is not a square.

$x^2 - 45$ is not a difference of squares.

47. $16x^2 - 25y^2$

(a) The first expression is a square: $16x^2 = (4x)^2$

The second expression is a square: $25y^2 = (5y)^2$

(b) The terms have different signs.

$16x^2 - 25y^2$ is a difference of squares.

49. $y^2 - 4 = y^2 - 2^2 = (y + 2)(y - 2)$

51. $p^2 - 9 = p^2 - 3^2 = (p + 3)(p - 3)$

53. $-49 + t^2 = t^2 - 49 = t^2 - 7^2 = (t + 7)(t - 7)$

55. $a^2 - b^2 = (a + b)(a - b)$

57. $25t^2 - m^2 = (5t)^2 - m^2 = (5t + m)(5t - m)$

59. $100 - k^2 = 10^2 - k^2 = (10 + k)(10 - k)$

61. $16a^2 - 9 = (4a)^2 - 3^2 = (4a + 3)(4a - 3)$

63. $4x^2 - 25y^2 = (2x)^2 - (5y)^2 = (2x + 5y)(2x - 5y)$

65. $8x^2 - 98 = 2(4x^2 - 49) = 2[(2x)^2 - 7^2] = 2(2x + 7)(2x - 7)$

67. $36x - 49x^3 = x(36 - 49x^2) = x[6^2 - (7x)^2] = x(6 + 7x)(6 - 7x)$

69. $49a^4 - 81 = (7a^2)^2 - 9^2 = (7a^2 + 9)(7a^2 - 9)$

71. $a^4 - 16$
$= (a^2)^2 - 4^2$
$= (a^2 + 4)(a^2 - 4)$ Factoring a difference of squares
$= (a^2 + 4)(a + 2)(a - 2)$ Factoring further: $a^2 - 4$ is a difference of squares.

73. $5x^4 - 405$
$5(x^4 - 81)$
$= 5[(x^2)^2 - 9^2]$
$= 5(x^2 + 9)(x^2 - 9)$
$= 5(x^2 + 9)(x + 3)(x - 3)$ Factoring $x^2 - 9$

75. $1 - y^8$
$= 1^2 - (y^4)^2$
$= (1 + y^4)(1 - y^4)$
$= (1 + y^4)(1 + y^2)(1 - y^2)$ Factoring $1 - y^4$
$= (1 + y^4)(1 + y^2)(1 + y)(1 - y)$ Factoring $1 - y^2$

77. $x^{12} - 16$
$= (x^6)^2 - 4^2$
$= (x^6 + 4)(x^6 - 4)$
$= (x^6 + 4)(x^3 + 2)(x^3 - 2)$ Factoring $x^6 - 4$

79. $y^2 - \frac{1}{16} = y^2 - \left(\frac{1}{4}\right)^2$
$= \left(y + \frac{1}{4}\right)\left(y - \frac{1}{4}\right)$

81. $25 - \frac{1}{49}x^2 = 5^2 - \left(\frac{1}{7}x\right)^2$
$= \left(5 + \frac{1}{7}x\right)\left(5 - \frac{1}{7}x\right)$

83. $16m^4 - t^4$
$= (4m^2)^2 - (t^2)^2$
$= (4m^2 + t^2)(4m^2 - t^2)$
$= (4m^2 + t^2)(2m + t)(2m - t)$ Factoring $4m^2 - t^2$

85. $-110 \div 10$ The quotient of a negative number and a positive number is negative.

$-110 \div 10 = -11$

87. $-\frac{2}{3} \div \frac{4}{5} = -\frac{2}{3} \cdot \frac{5}{4} = -\frac{10}{12} = -\frac{2 \cdot 5}{2 \cdot 6} = -\frac{\cancel{2} \cdot 5}{\cancel{2} \cdot 6} = -\frac{5}{6}$

89. $-64 \div (-32)$ The quotient of two negative numbers is a positive number.

$-64 \div (-32) = 2$

91. The shaded region is a square with sides of length $x - y - y$, or $x - 2y$. Its area is $(x - 2y)(x - 2y)$, or $(x - 2y)^2$. Multiplying, we get the polynomial $x^2 - 4xy + 4y^2$.

93. $y^5 \cdot y^7 = y^{5+7} = y^{12}$

95. $y - 6x = 6$

To find the x-intercept, let $y = 0$. Then solve for x.

$y - 6x = 6$
$0 - 6x = 6$
$-6x = 6$
$x = -1$

The x-intercept is $(-1, 0)$.

To find the y-intercept, let $x = 0$. Then solve for y.

$$y - 6x = 6$$
$$y - 6 \cdot 0 = 6$$
$$y = 6$$

The y-intercept is $(0, 6)$.

Plot these points and draw the line.

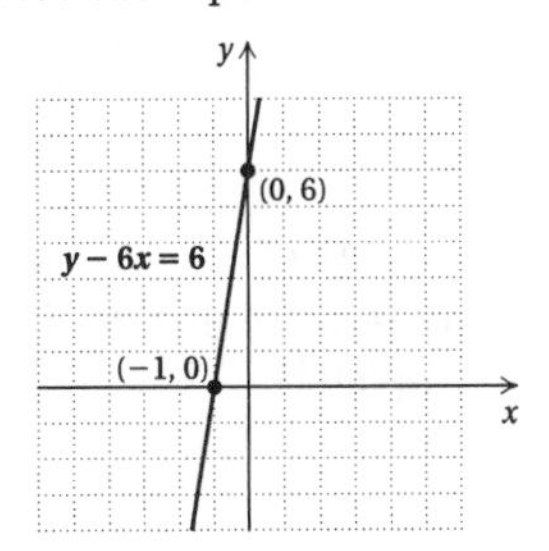

A third point should be used as a check. We substitute any value for x and solve for y. We let $x = -2$. Then

$$y - 6x = 6$$
$$y - 6(-2) = 6$$
$$y + 12 = 6$$
$$y = -6$$

The point $(-2, -6)$ is on the graph, so the graph is probably correct.

97. ◈

99. $49x^2 - 216$

There is no common factor. Also, $49x^2$ is a square, but 216 is not so this expression is not a difference of squares. It is not factorable.

101. $x^2 + 22x + 121 = x^2 + 2 \cdot x \cdot 11 + 11^2$
$= (x + 11)^2$

103. $18x^3 + 12x^2 + 2x = 2x(9x^2 + 6x + 1)$
$= 2x[(3x)^2 + 2 \cdot 3x \cdot 1 + 1^2]$
$= 2x(3x + 1)^2$

105. $x^8 - 2^8$
$= (x^4 + 2^4)(x^4 - 2^4)$
$= (x^4 + 2^4)(x^2 + 2^2)(x^2 - 2^2)$
$= (x^4 + 2^4)(x^2 + 2^2)(x + 2)(x - 2)$, or
$= (x^4 + 16)(x^2 + 4)(x + 2)(x - 2)$

107. $3x^5 - 12x^3 = 3x^3(x^2 - 4) = 3x^3(x + 2)(x - 2)$

109. $18x^3 - \dfrac{8}{25}x = 2x\left(9x^2 - \dfrac{4}{25}\right) = 2x\left(3x + \dfrac{2}{5}\right)\left(3x - \dfrac{2}{5}\right)$

111. $0.49p - p^3 = p(0.49 - p^2) = p(0.7 + p)(0.7 - p)$

113. $0.64x^2 - 1.21 = (0.8x)^2 - (1.1)^2 = (0.8x + 1.1)(0.8x - 1.1)$

115. $(x+3)^2 - 9 = [(x+3)+3][(x+3)-3] = (x+6)x$, or $x(x+6)$

117. $x^2 - \left(\dfrac{1}{x}\right)^2 = \left(x + \dfrac{1}{x}\right)\left(x - \dfrac{1}{x}\right)$

119. $81 - b^{4k} = 9^2 - (b^{2k})^2$
$= (9 + b^{2k})(9 - b^{2k})$
$= (9 + b^{2k})[3^2 - (b^k)^2]$
$= (9 + b^{2k})(3 + b^k)(3 - b^k)$

121. $9b^{2n} + 12b^n + 4 = (3b^n)^2 + 2 \cdot 3b^n \cdot 2 + 2^2 =$
$(3b^n + 2)^2$

123. $(y + 3)^2 + 2(y + 3) + 1$
$= (y + 3)^2 + 2 \cdot (y + 3) \cdot 1 + 1^2$
$= [(y + 3) + 1]^2$
$= (y + 4)^2$

125. If $cy^2 + 6y + 1$ is the square of a binomial, then $2 \cdot a \cdot 1 = 6$ where $a^2 = c$. Then $a = 3$, so $c = a^2 = 3^2 = 9$. (The polynomial is $9y^2 + 6y + 1$.)

127. Enter $y_1 = x^2 + 9$ and $y_2 = (x + 3)(x + 3)$ and look at a table of values. The y_1-and y_2-values are not the same, so the factorization is not correct.

129. Enter $y_1 = x^2 + 9$ and $y_2 = (x + 3)^2$ and look at a table of values. The y_1-and y_2-values are not the same, so the factorization is not correct.

Exercise Set 11.6

1. $3x^2 - 192 = 3(x^2 - 64)$ — 3 is a common factor
$= 3(x^2 - 8^2)$ — Difference of squares
$= 3(x + 8)(x - 8)$

3. $a^2 + 25 - 10a = a^2 - 10a + 25$
$= a^2 - 2 \cdot a \cdot 5 + 5^2$ — Trinomial square
$= (a - 5)^2$

5. $2x^2 - 11x + 12$

There is no common factor (other than 1). This polynomial has three terms, but it is not a trinomial square. Multiply the leading coefficient and the constant, 2 and 12: $2 \cdot 12 = 24$. Try to factor 24 so that the sum of the factors is -11. The numbers we want are -3 and -8: $-3(-8) = 24$ and $-3 + (-8) = -11$. Split the middle term and factor by grouping.

$2x^2 - 11x + 12 = 2x^2 - 3x - 8x + 12$
$= x(2x - 3) - 4(2x - 3)$
$= (x - 4)(2x - 3)$

7. $x^3 + 24x^2 + 144x$
$= x(x^2 + 24x + 144)$ — x is a common factor
$= x(x^2 + 2 \cdot x \cdot 12 + 12^2)$ — Trinomial square
$= x(x + 12)^2$

9. $x^3 + 3x^2 - 4x - 12$
$= x^2(x + 3) - 4(x + 3)$ — Factoring by grouping
$= (x^2 - 4)(x + 3)$
$= (x + 2)(x - 2)(x + 3)$ — Factoring the difference of squares

11. $48x^2 - 3 = 3(16x^2 - 1)$ 3 is a common factor
$= 3[(4x)^2 - 1^2]$ Difference of squares
$= 3(4x+1)(4x-1)$

13. $9x^3 + 12x^2 - 45x$
$= 3x(3x^2 + 4x - 15)$ $3x$ is a common factor
$= 3x(3x-5)(x+3)$ Factoring the trinomial

15. $x^2 + 4$ is a *11sum* of squares with no common factor. It cannot be factored.

17. $x^4 + 7x^2 - 3x^3 - 21x = x(x^3 + 7x - 3x^2 - 21)$
$= x[x(x^2+7) - 3(x^2+7)]$
$= x[(x-3)(x^2+7)]$
$= x(x-3)(x^2+7)$

19. $x^5 - 14x^4 + 49x^3$
$= x^3(x^2 - 14x + 49)$ x^3 is a common factor
$= x^3(x^2 - 2 \cdot x \cdot 7 + 7^2)$ Trinomial square
$= x^3(x-7)^2$

21. $20 - 6x - 2x^2$
$= -2(-10 + 3x + x^2)$ -2 is a common factor
$= -2(x^2 + 3x - 10)$ Writing in descending order
$= -2(x+5)(x-2)$, Using trial and error
or $2(5+x)(2-x)$

23. $x^2 - 6x + 1$

There is no common factor (other than 1 or -1). This is not a trinomial square, because $-6x \neq 2 \cdot x \cdot 1$ and $-6x \neq -2 \cdot x \cdot 1$. We try factoring using the refined trial and error procedure. We look for two factors of 1 whose sum is -6. There are none. The polynomial cannot be factored.

25. $4x^4 - 64$
$= 4(x^4 - 16)$ 4 is a common factor
$= 4[(x^2)^2 - 4^2]$ Difference of squares
$= 4(x^2+4)(x^2-4)$ Difference of squares
$= 4(x^2+4)(x+2)(x-2)$

27. $1 - y^8$ Difference of squares
$= (1+y^4)(1-y^4)$ Difference of squares
$= (1+y^4)(1+y^2)(1-y^2)$ Difference of squares
$= (1+y^4)(1+y^2)(1+y)(1-y)$

29. $x^5 - 4x^4 + 3x^3$
$= x^3(x^2 - 4x + 3)$ x^3 is a common factor
$= x^3(x-3)(x-1)$ Factoring the trinomial using trial and error

31. $\frac{1}{81}x^6 - \frac{8}{27}x^3 + \frac{16}{9}$
$= \frac{1}{9}\left(\frac{1}{9}x^6 - \frac{8}{3}x^3 + 16\right)$ $\frac{1}{9}$ is a common factor
$= \frac{1}{9}\left[\left(\frac{1}{3}x^3\right)^2 - 2 \cdot \frac{1}{3}x^3 \cdot 4 + 4^2\right]$ Trinomial square
$= \frac{1}{9}\left(\frac{1}{3}x^3 - 4\right)^2$

33. $mx^2 + my^2$
$= m(x^2 + y^2)$ m is a common factor

The factor with more than one term cannot be factored further, so we have factored completely.

35. $9x^2y^2 - 36xy = 9xy(xy - 4)$

37. $2\pi rh + 2\pi r^2 = 2\pi r(h + r)$

39. $(a+b)(x-3) + (a+b)(x+4)$
$= (a+b)[(x-3) + (x+4)]$ $(a+b)$ is a common factor
$= (a+b)(2x+1)$

41. $(x-1)(x+1) - y(x+1) = (x+1)(x-1-y)$
$(x+1)$ is a common factor

43. $n^2 + 2n + np + 2p$
$= n(n+2) + p(n+2)$ Factoring by grouping
$= (n+p)(n+2)$

45. $6q^2 - 3q + 2pq - p$
$= 3q(2q-1) + p(2q-1)$ Factoring by grouping
$= (3q+p)(2q-1)$

47. $4b^2 + a^2 - 4ab$
$= a^2 - 4ab + 4b^2$ Rearranging
$= a^2 - 2 \cdot a \cdot 2b + (2b)^2$ Trinomial square
$= (a - 2b)^2$

(Note that if we had rewritten the polynomial as $4b^2 - 4ab + a^2$, we might have written the result as $(2b-a)^2$. The two factorizations are equivalent.)

49. $16x^2 + 24xy + 9y^2$
$= (4x)^2 + 2 \cdot 4x \cdot 3y + (3y)^2$ Trinomial square
$= (4x + 3y)^2$

51. $49m^4 - 112m^2n + 64n^2$
$= (7m^2)^2 - 2 \cdot 7m^2 \cdot 8n + (8n)^2$ Trinomial square
$= (7m^2 - 8n)^2$

53. $y^4 + 10y^2z^2 + 25z^4$
$= (y^2)^2 + 2 \cdot y^2 \cdot 5z^2 + (5z^2)^2$ Trinomial square
$= (y^2 + 5z^2)^2$

55. $\frac{1}{4}a^2 + \frac{1}{3}ab + \frac{1}{9}b^2$
$= \left(\frac{1}{2}a\right)^2 + 2 \cdot \frac{1}{2}a \cdot \frac{1}{3}b + \left(\frac{1}{3}b\right)^2$
$= \left(\frac{1}{2}a + \frac{1}{3}b\right)^2$

57. $a^2 - ab - 2b^2 = (a - 2b)(a + b)$ Using trial and error

59. $2mn - 360n^2 + m^2$

$= m^2 + 2mn - 360n^2$ Rewriting

$= (m + 20n)(m - 18n)$ Using trial and error

61. $m^2n^2 - 4mn - 32 = (mn - 8)(mn + 4)$ Using trial and error

63. $a^2b^6 + 4ab^5 - 32b^4$

$= b^4(a^2b^2 + 4ab - 32)$ b^4 is a common factor

$= b^4(ab + 8)(ab - 4)$ Using trial and error

65. $a^5 + 4a^4b - 5a^3b^2$

$= a^3(a^2 + 4ab - 5b^2)$ a^3 is a common factor

$= a^3(a + 5b)(a - b)$ Factoring the trinomial

67. $a^2 - \frac{1}{25}b^2$

$= a^2 - \left(\frac{1}{5}b\right)^2$ Difference of squares

$= \left(a + \frac{1}{5}b\right)\left(a - \frac{1}{5}b\right)$

69. $x^2 - y^2 = (x + y)(x - y)$ Difference of squares

71. $16 - p^4q^4$

$= 4^2 - (p^2q^2)^2$ Difference of squares

$= (4 + p^2q^2)(4 - p^2q^2)$ $4 - p^2q^2$ is a difference of squares

$= (4 + p^2q^2)(2 + pq)(2 - pq)$

73. $1 - 16x^{12}y^{12}$

$= 1^2 - (4x^6y^6)^2$ Difference of squares

$= (1 + 4x^6y^6)(1 - 4x^6y^6)$ $1 - 4x^6y^6$ is a difference of squares

$= (1 + 4x^6y^6)(1 + 2x^3y^3)(1 - 2x^3y^3)$

75. $q^3 + 8q^2 - q - 8$

$= q^2(q + 8) - (q + 8)$ Factoring by grouping

$= (q^2 - 1)(q + 8)$

$= (q + 1)(q - 1)(q + 8)$ Factoring the difference of squares

77. $112xy + 49x^2 + 64y^2$

$= 49x^2 + 112xy + 64y^2$ Rearranging

$= (7x)^2 + 2 \cdot 7x \cdot 8y + (8y)^2$ Trinomial square

$= (7x + 8y)^2$

79. The highest point on the graph lies above 1990, so sports-car sales were highest in 1990.

81. We locate 68,000 on the vertical axis and then go across horizontally to the graph until we reach a dot on the graph. There are two dots that correspond to 68,000. One is above 1992 and the other is above 1996, so about 68,000 sports-cars were sold in 1992 and in 1996.

83. From the graph we see that in 1995 about 57,000 cars were sold and in 1997 about 75,000 cars were sold. We subtract to find the increase.

$$75,000 - 57,000 = 18,000$$

Sales increased by about 18,000 from 1995 to 1997.

85. $\frac{7}{5} \div \left(-\frac{11}{10}\right)$

$= \frac{7}{5} \cdot \left(-\frac{10}{11}\right)$ Multiplying by the reciprocal of the divisor

$= -\frac{7 \cdot 10}{5 \cdot 11}$

$= -\frac{7 \cdot 5 \cdot 2}{5 \cdot 11} = -\frac{7 \cdot 2}{11} \cdot \frac{5}{5}$

$= -\frac{14}{11}$

87. $A = aX + bX - 7$

$A + 7 = aX + bX$

$A + 7 = X(a + b)$

$\frac{A + 7}{a + b} = X$

89. ◈

91. $a^4 - 2a^2 + 1 = (a^2)^2 - 2 \cdot a^2 \cdot 1 + 1^2$

$= (a^2 - 1)^2$

$= [(a + 1)(a - 1)]^2$

$= (a + 1)^2(a - 1)^2$

93. $12.25x^2 - 7x + 1 = (3.5x)^2 - 2 \cdot (3.5x) \cdot 1 + 1^2$

$= (3.5x - 1)^2$

95. $5x^2 + 13x + 7.2$

Multiply the leading coefficient and the constant, 5 and 7.2: $5(7.2) = 36$. Try to factor 36 so that the sum of the factors is 13. The numbers we want are 9 and 4. Split the middle term and factor by grouping:

$5x^2 + 13x + 7.2 = 5x^2 + 9x + 4x + 7.2$

$= 5x(x + 1.8) + 4(x + 1.8)$

$= (5x + 4)(x + 1.8)$

97. $18 + y^3 - 9y - 2y^2$

$= y^3 - 2y^2 - 9y + 18$

$= y^2(y - 2) - 9(y - 2)$

$= (y^2 - 9)(y - 2)$

$= (y + 3)(y - 3)(y - 2)$

99. $a^3 + 4a^2 + a + 4 = a^2(a + 4) + 1(a + 4)$

$= (a^2 + 1)(a + 4)$

101. $x^4 - 7x^2 - 18 = (x^2 - 9)(x^2 + 2)$

$= (x + 3)(x - 3)(x^2 + 2)$

103. $x^3 - x^2 - 4x + 4 = x^2(x - 1) - 4(x - 1)$

$= (x^2 - 4)(x - 1)$

$= (x + 2)(x - 2)(x - 1)$

105. $y^2(y-1)-2y(y-1)+(y-1)$
$= (y-1)(y^2-2y+1)$
$= (y-1)(y-1)^2$
$= (y-1)^3$

107. $(y+4)^2+2x(y+4)+x^2$
$= (y+4)^2+2\cdot(y+4)\cdot x+x^2$ Trinomial square
$= (y+4+x)^2$

Exercise Set 11.7

1. $(x+4)(x+9)=0$

$x+4=0$ or $x+9=0$ Using the principle of zero products

$x=-4$ or $x=-9$ Solving the two equations separately

Check:

For -4

$$\begin{array}{c|c} (x+4)(x+9) & 0 \\ \hline (-4+4)(-4+9) \; ? & 0 \\ 0\cdot 5 & \\ 0 & \text{TRUE} \end{array}$$

For -9

$$\begin{array}{c|c} (x+4)(x+9) & 0 \\ \hline (9+4)(-9+9) \; ? & 0 \\ 13\cdot 0 & \\ 0 & \text{TRUE} \end{array}$$

The solutions are -4 and -9.

3. $(x+3)(x-8)=0$

$x+3=0$ or $x-8=0$ Using the principle of zero products

$x=-3$ or $x=8$

Check:

For -3

$$\begin{array}{c|c} (x+3)(x-8) & 0 \\ \hline (-3+3)(-3-8) \; ? & 0 \\ 0(-11) & \\ 0 & \text{TRUE} \end{array}$$

For 8

$$\begin{array}{c|c} (x+3)(x-8) & 0 \\ \hline (8+3)(8-8) \; ? & 0 \\ 11\cdot 0 & \\ 0 & \text{TRUE} \end{array}$$

The solutions are -3 and 8.

5. $(x+12)(x-11)=0$

$x+12=0$ or $x-11=0$

$x=-12$ or $x=11$

The solutions are -12 and 11.

7. $x(x+3)=0$

$x=0$ or $x+3=0$

$x=0$ or $x=-3$

The solutions are 0 and -3.

9. $0=y(y+18)$

$y=0$ or $y+18=0$

$y=0$ or $y=-18$

The solutions are 0 and -18.

11. $(2x+5)(x+4)=0$

$2x+5=0$ or $x+4=0$

$2x=-5$ or $x=-4$

$x=-\frac{5}{2}$ or $x=-4$

The solutions are $-\frac{5}{2}$ and -4.

13. $(5x+1)(4x-12)=0$

$5x+1=0$ or $4x-12=0$

$5x=-1$ or $4x=12$

$x=-\frac{1}{5}$ or $x=3$

The solutions are $-\frac{1}{5}$ and 3.

15. $(7x-28)(28x-7)=0$

$7x-28=0$ or $28x-7=0$

$7x=28$ or $28x=7$

$x=4$ or $x=\frac{7}{28}=\frac{1}{4}$

The solutions are 4 and $\frac{1}{4}$.

17. $2x(3x-2)=0$

$2x=0$ or $3x-2=0$

$x=0$ or $3x=2$

$x=0$ or $x=\frac{2}{3}$

The solutions are 0 and $\frac{2}{3}$.

19. $\left(\frac{1}{5}+2x\right)\left(\frac{1}{9}-3x\right)=0$

$\frac{1}{5}+2x=0$ or $\frac{1}{9}-3x=0$

$2x=-\frac{1}{5}$ or $-3x=-\frac{1}{9}$

$x=-\frac{1}{10}$ or $x=\frac{1}{27}$

The solutions are $-\frac{1}{10}$ and $\frac{1}{27}$.

21. $(0.3x-0.1)(0.05x+1)=0$

$0.3x - 0.1 = 0$ or $0.05x + 1 = 0$

$0.3x = 0.1$ or $0.05x = -1$

$x = \frac{0.1}{0.3}$ or $x = -\frac{1}{0.05}$

$x = \frac{1}{3}$ or $x = -20$

The solutions are $\frac{1}{3}$ and -20.

23. $9x(3x - 2)(2x - 1) = 0$

$9x = 0$ or $3x - 2 = 0$ or $2x - 1 = 0$

$x = 0$ or $3x = 2$ or $2x = 1$

$x = 0$ or $x = \frac{2}{3}$ or $x = \frac{1}{2}$

The solutions are 0, $\frac{2}{3}$, and $\frac{1}{2}$.

25. $x^2 + 6x + 5 = 0$

$(x + 5)(x + 1) = 0$ Factoring

$x + 5 = 0$ or $x + 1 = 0$ Using the principle of zero products

$x = -5$ or $x = -1$

The solutions are -5 and -1.

27. $x^2 + 7x - 18 = 0$

$(x + 9)(x - 2) = 0$ Factoring

$x + 9 = 0$ or $x - 2 = 0$ Using the principle of zero products

$x = -9$ or $x = 2$

The solutions are -9 and 2.

29. $x^2 - 8x + 15 = 0$

$(x - 5)(x - 3) = 0$

$x - 5 = 0$ or $x - 3 = 0$

$x = 5$ or $x = 3$

The solutions are 5 and 3.

31. $x^2 - 8x = 0$

$x(x - 8) = 0$

$x = 0$ or $x - 8 = 0$

$x = 0$ or $x = 8$

The solutions are 0 and 8.

33. $x^2 + 18x = 0$

$x(x + 18) = 0$

$x = 0$ or $x + 18 = 0$

$x = 0$ or $x = -18$

The solutions are 0 and -18.

35. $x^2 = 16$

$x^2 - 16 = 0$ Subtracting 16

$(x - 4)(x + 4) = 0$

$x - 4 = 0$ or $x + 4 = 0$

$x = 4$ or $x = -4$

The solutions are 4 and -4.

37. $9x^2 - 4 = 0$

$(3x - 2)(3x + 2) = 0$

$3x - 2 = 0$ or $3x + 2 = 0$

$3x = 2$ or $3x = -2$

$x = \frac{2}{3}$ or $x = -\frac{2}{3}$

The solutions are $\frac{2}{3}$ and $-\frac{2}{3}$.

39. $0 = 6x + x^2 + 9$

$0 = x^2 + 6x + 9$ Writing in descending order

$0 = (x + 3)(x + 3)$

$x + 3 = 0$ or $x + 3 = 0$

$x = -3$ or $x = -3$

There is only one solution, -3.

41. $x^2 + 16 = 8x$

$x^2 - 8x + 16 = 0$ Subtracting $8x$

$(x - 4)(x - 4) = 0$

$x - 4 = 0$ or $x - 4 = 0$

$x = 4$ or $x = 4$

There is only one solution, 4.

43. $5x^2 = 6x$

$5x^2 - 6x = 0$

$x(5x - 6) = 0$

$x = 0$ or $5x - 6 = 0$

$x = 0$ or $5x = 6$

$x = 0$ or $x = \frac{6}{5}$

The solutions are 0 and $\frac{6}{5}$.

45. $6x^2 - 4x = 10$

$6x^2 - 4x - 10 = 0$

$2(3x^2 - 2x - 5) = 0$

$2(3x - 5)(x + 1) = 0$

$3x - 5 = 0$ or $x + 1 = 0$

$3x = 5$ or $x = -1$

$x = \frac{5}{3}$ or $x = -1$

The solutions are $\frac{5}{3}$ and -1.

47. $12y^2 - 5y = 2$

$12y^2 - 5y - 2 = 0$

$(4y + 1)(3y - 2) = 0$

$4y + 1 = 0$ or $3y - 2 = 0$

$4y = -1$ or $3y = 2$

$y = -\frac{1}{4}$ or $y = \frac{2}{3}$

The solutions are $-\frac{1}{4}$ and $\frac{2}{3}$.

49.
$$t(3t+1)=2$$
$$3t^2+t=2 \quad \text{Multiplying on the left}$$
$$3t^2+t-2=0 \quad \text{Subtracting 2}$$
$$(3t-2)(t+1)=0$$
$$3t-2=0 \quad \text{or} \quad t+1=0$$
$$3t=2 \quad \text{or} \quad t=-1$$
$$t=\frac{2}{3} \quad \text{or} \quad t=-1$$
The solutions are $\frac{2}{3}$ and -1.

51.
$$100y^2=49$$
$$100y^2-49=0$$
$$(10y+7)(10y-7)=0$$
$$10y+7=0 \quad \text{or} \quad 10y-7=0$$
$$10y=-7 \quad \text{or} \quad 10y=7$$
$$y=-\frac{7}{10} \quad \text{or} \quad y=\frac{7}{10}$$
The solutions are $-\frac{7}{10}$ and $\frac{7}{10}$.

53.
$$x^2-5x=18+2x$$
$$x^2-5x-18-2x=0 \quad \text{Subtracting 18 and } 2x$$
$$x^2-7x-18=0$$
$$(x-9)(x+2)=0$$
$$x-9=0 \quad \text{or} \quad x+2=0$$
$$x=9 \quad \text{or} \quad x=-2$$
The solutions are 9 and -2.

55. $10x^2-23x+12=0$
$$(5x-4)(2x-3)=0$$
$$5x-4=0 \quad \text{or} \quad 2x-3=0$$
$$5x=4 \quad \text{or} \quad 2x=3$$
$$x=\frac{4}{5} \quad \text{or} \quad x=\frac{3}{2}$$
The solutions are $\frac{4}{5}$ and $\frac{3}{2}$.

57. We let $y=0$ and solve for x.
$$0=x^2+3x-4$$
$$0=(x+4)(x-1)$$
$$x+4=0 \quad \text{or} \quad x-1=0$$
$$x=-4 \quad \text{or} \quad x=1$$
The x-intercepts are $(-4,0)$ and $(1,0)$.

59. We let $y=0$ and solve for x
$$0=2x^2+x-10$$
$$0=(2x+5)(x-2)$$
$$2x+5=0 \quad \text{or} \quad x-2=0$$
$$2x=-5 \quad \text{or} \quad x=2$$
$$x=-\frac{5}{2} \quad \text{or} \quad x=2$$
The x-intercepts are $\left(-\frac{5}{2},0\right)$ and $(2,0)$.

61. $(a+b)^2$

63. $144 \div -9 = -16$

The two numbers have different signs, so their quotient is negative.

65.
$$-\frac{5}{8}\div\frac{3}{16}=-\frac{5}{8}\cdot\frac{16}{3}$$
$$=-\frac{5\cdot 16}{8\cdot 3}$$
$$=-\frac{5\cdot \not{8}\cdot 2}{\not{8}\cdot 3}$$
$$=-\frac{10}{3}$$

67. ◈

69.
$$b(b+9)=4(5+2b)$$
$$b^2+9b=20+8b$$
$$b^2+9b-8b-20=0$$
$$b^2+b-20=0$$
$$(b+5)(b-4)=0$$
$$b+5=0 \quad \text{or} \quad b-4=0$$
$$b=-5 \quad \text{or} \quad b=4$$
The solutions are -5 and 4.

71.
$$(t-3)^2=36$$
$$t^2-6t+9=36$$
$$t^2-6t-27=0$$
$$(t-9)(t+3)=0$$
$$t-9=0 \quad \text{or} \quad t+3=0$$
$$t=9 \quad \text{or} \quad t=-3$$
The solutions are 9 and -3.

73.
$$x^2-\frac{1}{64}=0$$
$$\left(x-\frac{1}{8}\right)\left(x+\frac{1}{8}\right)=0$$
$$x-\frac{1}{8}=0 \quad \text{or} \quad x+\frac{1}{8}=0$$
$$x=\frac{1}{8} \quad \text{or} \quad x=-\frac{1}{8}$$
The solutions are $\frac{1}{8}$ and $-\frac{1}{8}$.

75.
$$\frac{5}{16}x^2=5$$
$$\frac{5}{16}x^2-5=0$$
$$5\left(\frac{1}{16}x^2-1\right)=0$$
$$5\left(\frac{1}{4}x-1\right)\left(\frac{1}{4}x+1\right)=0$$
$$\frac{1}{4}x-1=0 \quad \text{or} \quad \frac{1}{4}x+1=0$$
$$\frac{1}{4}x=1 \quad \text{or} \quad \frac{1}{4}x=-1$$
$$x=4 \quad \text{or} \quad x=-4$$
The solutions are 4 and -4.

77. (a)

$$\begin{aligned} x = -3 \quad &\text{or} \quad x = 4 \\ x+3=0 \quad &\text{or} \quad x-4=0 \\ (x+3)(x-4) &= 0 \quad \text{Principle of zero products} \\ x^2 - x - 12 &= 0 \quad \text{Multiplying} \end{aligned}$$

(b)

$$\begin{aligned} x = -3 \quad &\text{or} \quad x = -4 \\ x+3=0 \quad &\text{or} \quad x+4=0 \\ (x+3)(x+4) &= 0 \\ x^2 + 7x + 12 &= 0 \end{aligned}$$

(c)

$$\begin{aligned} x = \frac{1}{2} \quad &\text{or} \quad x = \frac{1}{2} \\ x - \frac{1}{2} = 0 \quad &\text{or} \quad x - \frac{1}{2} = 0 \\ \left(x-\frac{1}{2}\right)\left(x-\frac{1}{2}\right) &= 0 \\ x^2 - x + \frac{1}{4} &= 0, \quad \text{or} \\ 4x^2 - 4x + 1 &= 0 \quad \text{Multiplying by 4} \end{aligned}$$

(d)

$$\begin{aligned} (x-5)(x+5) &= 0 \\ x^2 - 25 &= 0 \end{aligned}$$

(e)

$$\begin{aligned} (x-0)(x-0.1)\left(x-\frac{1}{4}\right) &= 0 \\ x\left(x-\frac{1}{10}\right)\left(x-\frac{1}{4}\right) &= 0 \\ x\left(x^2 - \frac{7}{20}x + \frac{1}{40}\right) &= 0 \\ x^3 - \frac{7}{20}x^2 + \frac{1}{40}x &= 0, \quad \text{or} \\ 40x^3 - 14x^2 + x &= 0 \quad \text{Multiplying by 40} \end{aligned}$$

Exercise Set 11.8

1. ***Familiarize.*** Let x = the number, or numbers.

Translate. We reword the problem.

7 plus the square of a number is 32.

$$7 + x^2 = 32$$

Solve. We solve the equation.

$$\begin{aligned} 7 + x^2 &= 32 \\ 7 + x^2 - 32 &= 32 - 32 \quad \text{Subtracting 32} \\ x^2 - 25 &= 0 \quad \text{Simplifying} \\ (x+5)(x-5) &= 0 \quad \text{Factoring} \\ x+5=0 \quad \text{or} \quad x-5 &= 0 \quad \text{Using the principle of zero products} \\ x = -5 \quad \text{or} \quad x &= 5 \end{aligned}$$

Check. The square of both -5 and 5 is 25, and 7 added to 25 is 32. Both numbers check.

State. There are two such numbers, -5 and 5.

3. ***Familiarize.*** Let x = the number, or numbers.

Translate. We reword the problem.

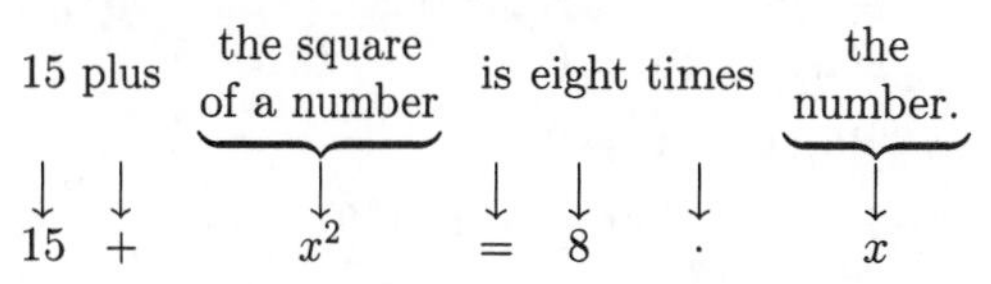

Solve. We solve the equation.

$$\begin{aligned} 15 + x^2 &= 8x \\ 15 + x^2 - 8x &= 8x - 8x \quad \text{Subtracting } 8x \\ 15 + x^2 - 8x &= 0 \quad \text{Simplifying} \\ x^2 - 8x + 15 &= 0 \quad \text{Writing in descending order} \\ (x-5)(x-3) &= 0 \quad \text{Factoring} \\ x-5=0 \quad \text{or} \quad x-3 &= 0 \quad \text{Using the principle of zero products} \\ x = 5 \quad \text{or} \quad x &= 3 \end{aligned}$$

Check. The square of 5 is 25. Fifteen more than 25 is 40 and $8 \cdot 5 = 40$, so 5 checks. The square of 3 is 9. Fifteen more than 9 is 24 and $8 \cdot 3 = 24$, so 3 also checks.

State. There are two such numbers, 5 and 3.

5. ***Familiarize.*** Using the labels shown on the drawing in the text, we let w = the width and $w + 5$ = the length. Recall that the area of a rectangle is length times width.

Translate. We reword the problem.

Length times width is area.

$$(w+5) \cdot w = 84$$

Solve. We solve the equation.

$$\begin{aligned} (w+5)w &= 84 \\ w^2 + 5w &= 84 \\ w^2 + 5w - 84 &= 84 - 84 \\ w^2 + 5w - 84 &= 0 \\ (w+12)(w-7) &= 0 \\ w+12=0 \quad \text{or} \quad w-7 &= 0 \\ w = -12 \quad \text{or} \quad w &= 7 \end{aligned}$$

Check. The width of a rectangle cannot have a negative measure, so -12 cannot be a solution. Suppose the width is 7 cm. The length is 5 cm greater than the width, so the length is 12 cm and the area is $12 \cdot 7$, or 84 cm^2. The numbers check in the original problem.

State. The length is 12 cm, and the width is 7 cm.

7. ***Familiarize.*** The page numbers on facing pages are consecutive integers. Let x = the smaller integer. Then $x + 1$ = the larger integer.

Translate. We reword the problem.

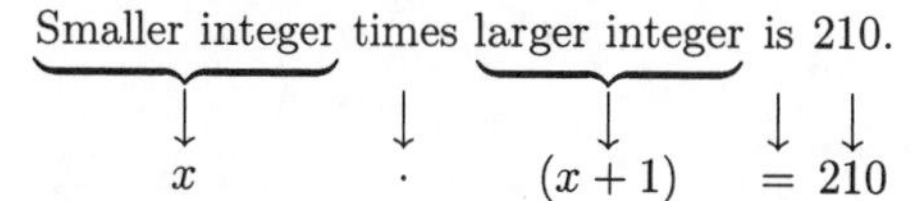

Solve. We solve the equation.

$$x(x+1) = 210$$
$$x^2 + x = 210$$
$$x^2 + x - 210 = 0$$
$$(x+15)(x-14) = 0$$
$$x + 15 = 0 \quad \text{or} \quad x - 14 = 0$$
$$x = -15 \quad \text{or} \quad x = 14$$

Check. The solutions of the equation are -15 and 14. Since a page number cannot be negative, -15 cannot be a solution of the original problem. We only need to check 14. When $x = 14$, then $x+1 = 15$, and $14 \cdot 15 = 210$. This checks.

State. The page numbers are 14 and 15.

9. ***Familiarize.*** Let x = the smaller even integer. Then $x + 2$ = the larger even integer.

Translate. We reword the problem.

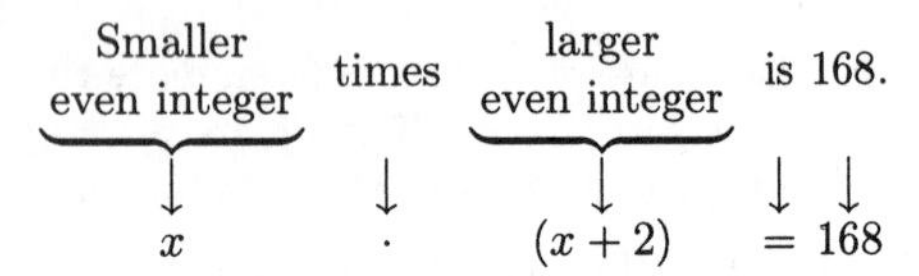

Solve.

$$x(x+2) = 168$$
$$x^2 + 2x = 168$$
$$x^2 + 2x - 168 = 0$$
$$(x+14)(x-12) = 0$$
$$x + 14 = 0 \quad \text{or} \quad x - 12 = 0$$
$$x = -14 \quad \text{or} \quad x = 12$$

Check. The solutions of the equation are -14 and 12. When x is -14, then $x + 2$ is -12 and $-14(-12) = 168$. The numbers -14 and -12 are consecutive even integers which are solutions of the problem. When x is 12, then $x + 2$ is 14 and $12 \cdot 14 = 168$. The numbers 12 and 14 are also consecutive even integers which are solutions of the problem.

State. We have two solutions, each of which consists of a pair of numbers: -14 and -12, and 12 and 14.

11. ***Familiarize.*** Let x = the smaller odd integer. Then $x + 2$ = the larger odd integer.

Translate. We reword the problem.

$$\underbrace{\text{Smaller odd integer}}_{x} \quad \underbrace{\text{times}}_{\cdot} \quad \underbrace{\text{larger odd integer}}_{(x+2)} \quad \underbrace{\text{is 255.}}_{= 255}$$

Solve.

$$x(x+2) = 255$$
$$x^2 + 2x = 255$$
$$x^2 + 2x - 255 = 0$$
$$(x-15)(x+17) = 0$$
$$x - 15 = 0 \quad \text{or} \quad x + 17 = 0$$
$$x = 15 \quad \text{or} \quad x = -17$$

Check. The solutions of the equation are 15 and -17. When x is 15, then $x + 2$ is 17 and $15 \cdot 17 = 255$. The numbers 15 and 17 are consecutive odd integers which are solutions to the problem. When x is -17, then $x + 2$ is -15 and $-17(-15) = 255$. The numbers -17 and -15 are also consecutive odd integers which are solutions to the problem.

State. We have two solutions, each of which consists of a pair of numbers: 15 and 17, and -17 and -15.

13. ***Familiarize.*** First draw a picture. Let x = the length of a side of the square.

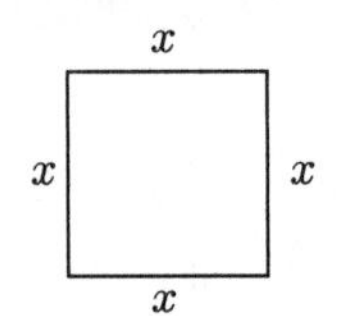

The area of the square is $x \cdot x$, or x^2. The perimeter of the square is $x + x + x + x$, or $4x$.

Translate.

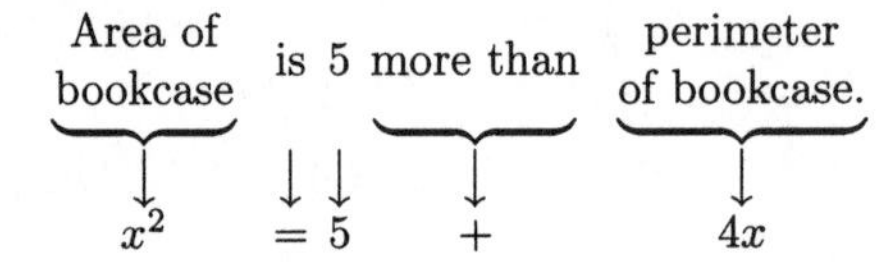

Solve.

$$x^2 = 5 + 4x$$
$$x^2 - 4x - 5 = 0$$
$$(x-5)(x+1) = 0$$
$$x - 5 = 0 \quad \text{or} \quad x + 1 = 0$$
$$x = 5 \quad \text{or} \quad x = -1$$

Check. The solutions of the equation are 5 and -1. The length of a side cannot be negative, so we only check 5. The area is $5 \cdot 5$, or 25. The perimeter is $5 + 5 + 5 + 5$, or 20. The area, 25, is 5 more than the perimeter, 20. This checks.

State. The length of a side is 5.

15. ***Familiarize.*** Using the labels shown on the drawing in the text, we let b = the base and $b+1$ = the height. Recall that the formula for the area of a triangle is $\frac{1}{2} \cdot (\text{base}) \cdot (\text{height})$.

Translate.

$$\underbrace{\frac{1}{2}}_{\frac{1}{2}} \quad \underbrace{\cdot}_{\cdot} \quad \underbrace{\text{base}}_{b} \quad \underbrace{\cdot}_{\cdot} \quad \underbrace{\text{height}}_{(b+1)} \quad \underbrace{\text{is}}_{=} \quad \underbrace{\text{area.}}_{15}$$

Solve. We solve the equation.

$$\frac{1}{2}b(b+1) = 15$$

$$2 \cdot \frac{1}{2}b(b+1) = 2 \cdot 15 \quad \text{Multiplying by 2}$$

$$b(b+1) = 30 \quad \text{Simplifying}$$

$$b^2 + b = 30$$

$$b^2 + b - 30 = 0$$

$$(b+6)(b-5) = 0$$

$$b+6 = 0 \quad \text{or} \quad b-5 = 0$$

$$b = -6 \quad \text{or} \quad b = 5$$

Check. The base of a triangle cannot have a negative length, so -6 cannot be a solution. Suppose the base is 5 cm. The height is 1 cm greater than the base, so the height is 6 cm and the area is $\frac{1}{2} \cdot 5 \cdot 6$, or 15 cm^2. These numbers check.

State. The height is 6 cm, and the base is 5 cm.

17. ***Familiarize.*** We make a drawing. Let x = the length of a side of the original square. Then $x+3$ = the length of a side of the enlarged square.

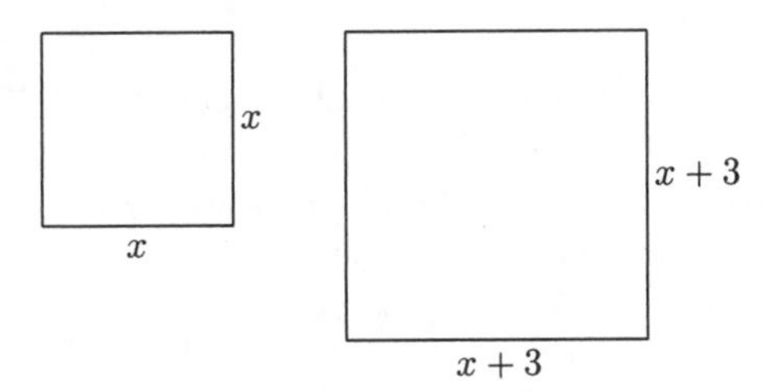

Recall that the area of a square is found by squaring the length of a side.

Translate.

Area of enlarged square is the square of the lengthened side.

$$81 = (x+3)^2$$

Solve.

$$81 = (x+3)^2$$

$$81 = x^2 + 6x + 9$$

$$0 = x^2 + 6x - 72$$

$$0 = (x+12)(x-6)$$

$$x+12 = 0 \quad \text{or} \quad x-6 = 0$$

$$x = -12 \quad \text{or} \quad x = 6$$

Check. The solutions of the equation are -12 and 6. The length of a side cannot be negative, so -12 cannot be a solution. Suppose the length of a side of the original square is 6 km. Then the length of a side of the new square is $6+3$, or 9 km. Its area is 9^2, or 81 km^2. The numbers check.

State. The length of a side of the original square is 6 km.

19. ***Familiarize.*** We will use the formula $h = 180t - 16t^2$.

Translate. Substitute 464 for h.

$$464 = 180t - 16t^2$$

Solve. We solve the equation.

$$464 = 180t - 16t^2$$

$$16t^2 - 180t + 464 = 0$$

$$4(4t^2 - 45t + 116) = 0$$

$$4(4t-29)(t-4) = 0$$

$$4t - 29 = 0 \quad \text{or} \quad t-4 = 0$$

$$4t = 29 \quad \text{or} \quad t = 4$$

$$t = \frac{29}{4} \quad \text{or} \quad t = 4$$

Check. The solutions of the equation are $\frac{29}{4}$, or $7\frac{1}{4}$, and 4. Since we want to find how many seconds it takes the rocket to *first* reach a height of 464 ft, we check the smaller number, 4. We substitute 4 for t in the formula.

$$h = 180t - 16t^2$$

$$h = 180 \cdot 4 - 16(4)^2$$

$$h = 180 \cdot 4 - 16 \cdot 16$$

$$h = 720 - 256$$

$$h = 464$$

The answer checks.

State. The rocket will first reach a height of 464 ft after 4 seconds.

21. ***Familiarize.*** Let x = the smaller odd positive integer. Then $x+2$ = the larger odd positive integer.

Translate.

Square of the smaller odd positive integer + Square of the larger odd positive integer is 74

$$x^2 + (x+2)^2 = 74$$

Solve.

$$x^2 + (x+2)^2 = 74$$

$$x^2 + x^2 + 4x + 4 = 74$$

$$2x^2 + 4x - 70 = 0$$

$$2(x^2 + 2x - 35) = 0$$

$$2(x+7)(x-5) = 0$$

$$x+7 = 0 \quad \text{or} \quad x-5 = 0$$

$$x = -7 \quad \text{or} \quad x = 5$$

Check. The solutions of the equation are -7 and 5. The problem asks for odd positive integers, so -7 cannot be a solution. When x is 5, $x+2$ is 7. The numbers 5 and 7 are consecutive odd positive integers. The sum of their squares, $25+49$, is 74. The numbers check.

State. The integers are 5 and 7.

23. ***Familiarize.*** Reread Example 3 in Section 4.3.

Translate. Substitute 14 for n.

$$14^2 - 14 = N$$

Solve. We do the computation on the left.

$$14^2 - 14 = N$$
$$196 - 14 = N$$
$$182 = N$$

Check. We can redo the computation, or we can solve the equation $n^2 - n = 182$. The answer checks.

State. 182 games will be played.

25. ***Familiarize***. Reread Example 3 in Section 4.3.

Translate. Substitute 132 for N.

$$n^2 - n = 132$$

Solve.

$$n^2 - n = 132$$
$$n^2 - n - 132 = 0$$
$$(n - 12)(n + 11) = 0$$
$$n - 12 = 0 \quad \text{or} \quad n + 11 = 0$$
$$n = 12 \quad \text{or} \quad n = -11$$

Check. The solutions of the equation are 12 and -11. Since the number of teams cannot be negative, -11 cannot be a solution. But 12 checks since $12^2 - 12 = 144 - 12 = 132$.

State. There are 12 teams in the league.

27. ***Familiarize***. We will use the formula $N = \frac{1}{2}(n^2 - n)$.

Translate. Substitute 100 for n.

$$N = \frac{1}{2}(100^2 - 100)$$

Solve. We do the computation on the right.

$$N = \frac{1}{2}(10,000 - 100)$$
$$N = \frac{1}{2}(9900)$$
$$N = 4950$$

Check. We can redo the computation, or we can solve the equation $4950 = \frac{1}{2}(n^2 - n)$. The answer checks.

State. 4950 handshakes are possible.

29. ***Familiarize***. We will use the formula $N = \frac{1}{2}(n^2 - n)$.

Translate. Substitute 300 for N.

$$300 = \frac{1}{2}(n^2 - n)$$

Solve. We solve the equation.

$$2 \cdot 300 = 2 \cdot \frac{1}{2}(n^2 - n) \quad \text{Multiplying by 2}$$
$$600 = n^2 - n$$
$$0 = n^2 - n - 600$$
$$0 = (n + 24)(n - 25)$$
$$n + 24 = 0 \quad \text{or} \quad n - 25 = 0$$
$$n = -24 \quad \text{or} \quad n = 25$$

Check. The number of people at a meeting cannot be negative, so -24 cannot be a solution. But 25 checks since $\frac{1}{2}(25^2 - 25) = \frac{1}{2}(625 - 25) = \frac{1}{2} \cdot 600 = 300$.

State. There were 25 people at the party.

31. ***Familiarize***. We make a drawing. Let x = the length of the unknown leg. Then $x + 2$ = the length of the hypotenuse.

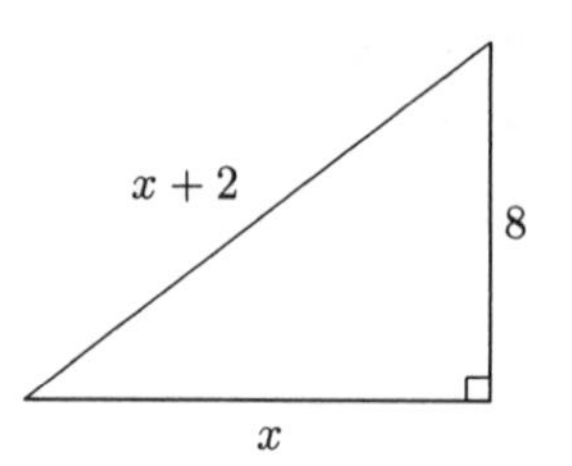

Translate. Use the Pythagorean theorem.

$$a^2 + b^2 = c^2$$
$$8^2 + x^2 = (x + 2)^2$$

Solve. We solve the equation.

$$8^2 + x^2 = (x + 2)^2$$
$$64 + x^2 = x^2 + 4x + 4$$
$$60 = 4x \quad \text{Subtracting } x^2 \text{ and } 4$$
$$15 = x$$

Check. When $x = 15$, then $x+2 = 17$ and $8^2 + 15^2 = 17^2$. Thus, 15 and 17 check.

State. The lengths of the hypotenuse and the other leg are 17 ft and 15 ft, respectively.

33. $(3x - 5y)(3x + 5y) = (3x)^2 - (5y)^2 = 9x^2 - 25y^2$

35. $(3x+5y)^2 = (3x)^2 + 2 \cdot 3x \cdot 5y + (5y)^2 = 9x^2 + 30xy + 25y^2$

37. $4x - 16y = 64$

To find the x-intercept, let $y = 0$ and solve for x.

$$4x - 16y = 64$$
$$4x - 16 \cdot 0 = 64$$
$$4x = 64$$
$$x = 16$$

The x-intercept is $(16, 0)$.

To find the y-intercept, let $x = 0$ and solve for y.

$$4x - 16y = 64$$
$$4 \cdot 0 - 16y = 64$$
$$-16y = 64$$
$$y = -4$$

The y-intercept is $(0, -4)$.

39. $x - 1.3y = 6.5$

To find the x-intercept, let $y = 0$ and solve for x.

$$x - 1.3y = 6.5$$
$$x - 1.3(0) = 6.5$$
$$x = 6.5$$

The x-intercept is $(6.5, 0)$.

To find the y-intercept, let $x = 0$ and solve for y.

$$x - 1.3y = 6.5$$
$$0 - 1.3y = 6.5$$
$$-1.3y = 6.5$$
$$y = -5$$

The y-intercept is $(0, -5)$.

41. 

43. ***Familiarize***. Using the labels shown on the drawing in the text, we let x = the width of the walk. Then the length and width of the rectangle formed by the pool and walk together are $40 + 2x$ and $20 + 2x$, respectively.

Translate.

$$\underbrace{\text{Area}}\ \text{is}\ \underbrace{\text{length}}\ \text{times}\ \underbrace{\text{width.}}$$
$$1500 = (40 + 2x) \cdot (20 + 2x)$$

Solve. We solve the equation.

$1500 = (40 + 2x)(20 + 2x)$

$1500 = 2(20 + x) \cdot 2(10 + x)$ Factoring 2 out of each factor on the right

$1500 = 4 \cdot (20 + x)(10 + x)$

$375 = (20 + x)(10 + x)$ Dividing by 4

$375 = 200 + 30x + x^2$

$0 = x^2 + 30x - 175$

$0 = (x + 35)(x - 5)$

$x + 35 = 0$ or $x - 5 = 0$

$x = -35$ or $x = 5$

Check. The solutions of the equation are -35 and 5. Since the width of the walk cannot be negative, -35 is not a solution. When $x = 5$, $40 + 2x = 40 + 2 \cdot 5$, or 50 and $20 + 2x = 20 + 2 \cdot 5$, or 30. The total area of the pool and walk is $50 \cdot 30$, or 1500 ft^2. This checks.

State. The width of the walk is 5 ft.

45. ***Familiarize***. Let y = the ten's digit. Then $y + 4$ = the one's digit and $10y + y + 4$, or $11y + 4$, represents the number.

Translate.

$$\underbrace{\text{The number}}\ \text{plus}\ \underbrace{\text{the product of the digits}}\ \text{is 58.}$$
$$11y + 4 + y(y + 4) = 58$$

Solve. We solve the equation.

$11y + 4 + y(y + 4) = 58$

$11y + 4 + y^2 + 4y = 58$

$y^2 + 15y + 4 = 58$

$y^2 + 15y - 54 = 0$

$(y + 18)(y - 3) = 0$

$y + 18 = 0$ or $y - 3 = 0$

$y = -18$ or $y = 3$

Check. Since -18 cannot be a digit of the number, we only need to check 3. When $y = 3$, then $y + 4 = 7$ and the number is 37. We see that $37 + 3 \cdot 7 = 37 + 21$, or 58. The result checks.

State. The number is 37.

47. ***Familiarize***. We make a drawing. Let w = the width of the piece of cardboard. Then $2w$ = the length.

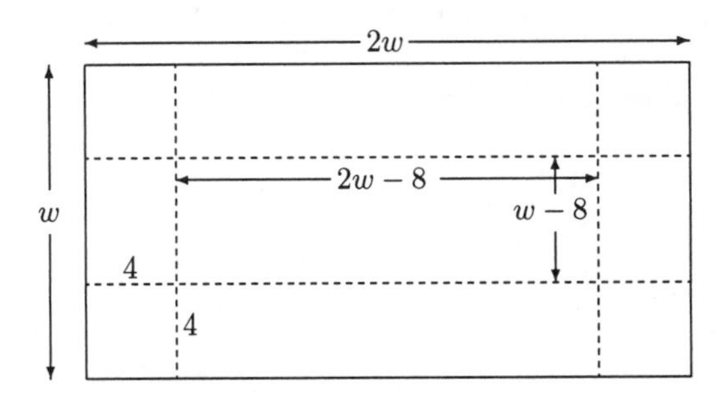

The box will have length $2w - 8$, width $w - 8$, and height 4. Recall that the formula for volume is V = length $\times$ width $\times$ height.

Translate.

$$\underbrace{\text{The volume}}\ \text{is}\ \underbrace{616\text{cm}^3.}$$
$$(2w - 8)(w - 8)(4) = 616$$

Solve. We solve the equation.

$(2w - 8)(w - 8)(4) = 616$

$(2w^2 - 24w + 64)(4) = 616$

$8w^2 - 96w + 256 = 616$

$8w^2 - 96w - 360 = 0$

$8(w^2 - 12w - 45) = 0$

$w^2 - 12w - 45 = 0$ Dividing by 8

$(w - 15)(w + 3) = 0$

$w - 15 = 0$ or $w + 3 = 0$

$w = 15$ or $w = -3$

Check. The width cannot be negative, so we only need to check 15. When $w = 15$, then $2w = 30$ and the dimensions of the box are $30 - 8$ by $15 - 8$ by 4, or 22 by 7 by 4. The volume is $22 \cdot 7 \cdot 4$, or 616.

State. The cardboard is 30 cm by 15 cm.

Chapter 12

Rational Expressions and Equations

Exercise Set 12.1

1. $\dfrac{-3}{2x}$

To determine which numbers make the rational expression undefined, we set the denominator equal to 0 and solve:

$$2x = 0$$
$$x = 0$$

The expression is undefined for the replacement number 0.

3. $\dfrac{5}{x-8}$

To determine which numbers make the rational expression undefined, we set the denominator equal to 0 and solve:

$$x - 8 = 0$$
$$x = 8$$

The expression is undefined for the replacement number 8.

5. $\dfrac{3}{2y+5}$

Set the denominator equal to 0 and solve:

$$2y + 5 = 0$$
$$2y = -5$$
$$y = -\frac{5}{2}$$

The expression is undefined for the replacement number $-\dfrac{5}{2}$.

7. $\dfrac{x^2+11}{x^2-3x-28}$

Set the denominator equal to 0 and solve:

$$x^2 - 3x - 28 = 0$$
$$(x-7)(x+4) = 0$$
$$x - 7 = 0 \quad \text{or} \quad x + 4 = 0$$
$$x = 7 \quad \text{or} \quad x = -4$$

The expression is undefined for the replacement numbers 7 and -4.

9. $\dfrac{m^3-2m}{m^2-25}$

Set the denominator equal to 0 and solve:

$$m^2 - 25 = 0$$
$$(m+5)(m-5) = 0$$
$$m + 5 = 0 \quad \text{or} \quad m - 5 = 0$$
$$m = -5 \quad \text{or} \quad m = 5$$

The expression is undefined for the replacement numbers -5 and 5.

11. $\dfrac{x-4}{3}$

Since the denominator is the constant 3, there are no replacement numbers for which the expression is undefined.

13. $\dfrac{4x}{4x}\cdot\dfrac{3x^2}{5y} = \dfrac{(4x)(3x^2)}{(4x)(5y)}$ Multiplying the numerators and the denominators

15. $\dfrac{2x}{2x}\cdot\dfrac{x-1}{x+4} = \dfrac{2x(x-1)}{2x(x+4)}$ Multiplying the numerators and the denominators

17. $\dfrac{3-x}{4-x}\cdot\dfrac{-1}{-1} = \dfrac{(3-x)(-1)}{(4-x)(-1)}$, or $\dfrac{-1(3-x)}{-1(4-x)}$

19. $\dfrac{y+6}{y+6}\cdot\dfrac{y-7}{y+2} = \dfrac{(y+6)(y-7)}{(y+6)(y+2)}$

21.

$$\frac{8x^3}{32x} = \frac{8\cdot x\cdot x^2}{8\cdot 4\cdot x} \quad \text{Factoring numerator and denominator}$$
$$= \frac{8x}{8x}\cdot\frac{x^2}{4} \quad \text{Factoring the rational expression}$$
$$= 1\cdot\frac{x^2}{4} \quad \left(\frac{8x}{8x}=1\right)$$
$$= \frac{x^2}{4} \quad \text{We removed a factor of 1.}$$

23.

$$\frac{48p^7q^5}{18p^5q^4} = \frac{8\cdot 6\cdot p^5\cdot p^2\cdot q^4\cdot q}{6\cdot 3\cdot p^5\cdot q^4} \quad \text{Factoring numerator and denominator}$$
$$= \frac{6p^5q^4}{6p^5q^4}\cdot\frac{8p^2q}{3} \quad \text{Factoring the rational expression}$$
$$= 1\cdot\frac{8p^2q}{3} \quad \left(\frac{6p^5q^4}{6p^5q^4}=1\right)$$
$$= \frac{8p^2q}{3} \quad \text{Removing a factor of 1}$$

25.

$$\frac{4x-12}{4x} = \frac{4(x-3)}{4\cdot x}$$
$$= \frac{4}{4}\cdot\frac{x-3}{x}$$
$$= 1\cdot\frac{x-3}{x}$$
$$= \frac{x-3}{x}$$

27.

$$\frac{3m^2+3m}{6m^2+9m} = \frac{3m(m+1)}{3m(2m+3)}$$
$$= \frac{3m}{3m}\cdot\frac{m+1}{2m+3}$$
$$= 1\cdot\frac{m+1}{2m+3}$$
$$= \frac{m+1}{2m+3}$$

29. $\dfrac{a^2-9}{a^2+5a+6} = \dfrac{(a-3)(a+3)}{(a+2)(a+3)}$
$= \dfrac{a-3}{a+2} \cdot \dfrac{a+3}{a+3}$
$= \dfrac{a-3}{a+2} \cdot 1$
$= \dfrac{a-3}{a+2}$

31. $\dfrac{a^2-10a+21}{a^2-11a+28} = \dfrac{(a-7)(a-3)}{(a-7)(a-4)}$
$= \dfrac{a-7}{a-7} \cdot \dfrac{a-3}{a-4}$
$= 1 \cdot \dfrac{a-3}{a-4}$
$= \dfrac{a-3}{a-4}$

33. $\dfrac{x^2-25}{x^2-10x+25} = \dfrac{(x-5)(x+5)}{(x-5)(x-5)}$
$= \dfrac{x-5}{x-5} \cdot \dfrac{x+5}{x-5}$
$= 1 \cdot \dfrac{x+5}{x-5}$
$= \dfrac{x+5}{x-5}$

35. $\dfrac{a^2-1}{a-1} = \dfrac{(a-1)(a+1)}{a-1}$
$= \dfrac{a-1}{a-1} \cdot \dfrac{a+1}{1}$
$= 1 \cdot \dfrac{a+1}{1}$
$= a+1$

37. $\dfrac{x^2+1}{x+1}$ cannot be simplified.

Neither the numerator nor the denominator can be factored.

39. $\dfrac{6x^2-54}{4x^2-36} = \dfrac{2 \cdot 3(x^2-9)}{2 \cdot 2(x^2-9)}$
$= \dfrac{2(x^2-9)}{2(x^2-9)} \cdot \dfrac{3}{2}$
$= 1 \cdot \dfrac{3}{2}$
$= \dfrac{3}{2}$

41. $\dfrac{6t+12}{t^2-t-6} = \dfrac{6(t+2)}{(t-3)(t+2)}$
$= \dfrac{6}{t-3} \cdot \dfrac{t+2}{t+2}$
$= \dfrac{6}{t-3} \cdot 1$
$= \dfrac{6}{t-3}$

43. $\dfrac{2t^2+6t+4}{4t^2-12t-16} = \dfrac{2(t^2+3t+2}{4(t^2-3t-4)}$
$= \dfrac{2(t+2)(t+1)}{2 \cdot 2(t-4)(t+1)}$
$= \dfrac{2(t+1)}{2(t+1)} \cdot \dfrac{t+2}{2(t-4)}$
$= 1 \cdot \dfrac{t+2}{2(t-4)}$
$= \dfrac{t+2}{2(t-4)}$

45. $\dfrac{t^2-4}{(t+2)^2} = \dfrac{(t-2)(t+2)}{(t+2)(t+2)}$
$= \dfrac{t-2}{t+2} \cdot \dfrac{t+2}{t+2}$
$= \dfrac{t-2}{t+2} \cdot 1$
$= \dfrac{t-2}{t+2}$

47. $\dfrac{6-x}{x-6} = \dfrac{-(-6+x)}{x-6}$
$= \dfrac{-1(x-6)}{x-6}$
$= -1 \cdot \dfrac{x-6}{x-6}$
$= -1 \cdot 1$
$= -1$

49. $\dfrac{a-b}{b-a} = \dfrac{-(-a+b)}{b-a}$
$= \dfrac{-1(b-a)}{b-a}$
$= -1 \cdot \dfrac{b-a}{b-a}$
$= -1 \cdot 1$
$= -1$

51. $\dfrac{6t-12}{2-t} = \dfrac{-6(-t+2)}{2-t}$
$= \dfrac{-6(2-t)}{2-t}$
$= \dfrac{-6(2-t)}{2-t}$
$= -6$

53. $\dfrac{x^2-1}{1-x} = \dfrac{(x+1)(x-1)}{-1(-1+x)}$
$= \dfrac{(x+1)(x-1)}{-1(x-1)}$
$= \dfrac{(x+1)(x-1)}{-1(x-1)}$
$= -(x+1)$
$= -x-1$

55. $\dfrac{4x^3}{3x}\cdot\dfrac{14}{x}=\dfrac{4x^3\cdot 14}{3x\cdot x}$ Multiplying the numerators and the denominators

$=\dfrac{4\cdot x\cdot x\cdot x\cdot 14}{3\cdot x\cdot x}$ Factoring the numerator and the denominator

$=\dfrac{4\cdot\cancel{x}\cdot\cancel{x}\cdot x\cdot 14}{3\cdot\cancel{x}\cdot\cancel{x}}$ Removing a factor of 1

$=\dfrac{56x}{3}$ Simplifying

57. $\dfrac{3c}{d^2}\cdot\dfrac{4d}{6c^3}=\dfrac{3c\cdot 4d}{d^2\cdot 6c^3}$ Multiplying the numerators and the denominators

$=\dfrac{3\cdot c\cdot 2\cdot 2\cdot d}{d\cdot d\cdot 3\cdot 2\cdot c\cdot c\cdot c}$ Factoring the numerator and the denominator

$=\dfrac{\cancel{3}\cdot\cancel{c}\cdot\cancel{2}\cdot 2\cdot\cancel{d}}{\cancel{d}\cdot d\cdot\cancel{3}\cdot\cancel{2}\cdot\cancel{c}\cdot c\cdot c}$

$=\dfrac{2}{dc^2}$

59. $\dfrac{x^2-3x-10}{(x-2)^2}\cdot\dfrac{x-2}{x-5}=\dfrac{(x^2-3x-10)(x-2)}{(x-2)^2(x-5)}$

$=\dfrac{(x-5)(x+2)(x-2)}{(x-2)(x-2)(x-5)}$

$=\dfrac{\cancel{(x-5)}(x+2)\cancel{(x-2)}}{\cancel{(x-2)}(x-2)\cancel{(x-5)}}$

$=\dfrac{x+2}{x-2}$

61. $\dfrac{a^2-9}{a^2}\cdot\dfrac{a^2-3a}{a^2+a-12}=\dfrac{(a-3)(a+3)(a)(a-3)}{a\cdot a(a+4)(a-3)}$

$=\dfrac{\cancel{(a-3)}(a+3)(\cancel{a})(a-3)}{\cancel{a}\cdot a(a+4)\cancel{(a-3)}}$

$=\dfrac{(a-3)(a+3)}{a(a+4)}$

63. $\dfrac{4a^2}{3a^2-12a+12}\cdot\dfrac{3a-6}{2a}=\dfrac{4a^2(3a-6)}{(3a^2-12a+12)2a}$

$=\dfrac{2\cdot 2\cdot a\cdot a\cdot 3\cdot(a-2)}{3\cdot(a-2)\cdot(a-2)\cdot 2\cdot a}$

$=\dfrac{\cancel{2}\cdot 2\cdot\cancel{a}\cdot a\cdot\cancel{3}\cdot\cancel{(a-2)}}{\cancel{3}\cdot\cancel{(a-2)}\cdot(a-2)\cdot\cancel{2}\cdot\cancel{a}}$

$=\dfrac{2a}{a-2}$

65. $\dfrac{t^4-16}{t^4-1}\cdot\dfrac{t^2+1}{t^2+4}$

$=\dfrac{(t^4-16)(t^2+1)}{(t^4-1)(t^2+4)}$

$=\dfrac{(t^2+4)(t+2)(t-2)(t^2+1)}{(t^2+1)(t+1)(t-1)(t^2+4)}$

$=\dfrac{\cancel{(t^2+4)}(t+2)(t-2)\cancel{(t^2+1)}}{\cancel{(t^2+1)}(t+1)(t-1)\cancel{(t^2+4)}}$

$=\dfrac{(t+2)(t-2)}{(t+1)(t-1)}$

67. $\dfrac{(x+4)^3}{(x+2)^3}\cdot\dfrac{x^2+4x+4}{x^2+8x+16}$

$=\dfrac{(x+4)^3(x^2+4x+4)}{(x+2)^3(x^2+8x+16)}$

$=\dfrac{(x+4)(x+4)(x+4)(x+2)(x+2)}{(x+2)(x+2)(x+2)(x+4)(x+4)}$

$=\dfrac{\cancel{(x+4)}\cancel{(x+4)}(x+4)\cancel{(x+2)}\cancel{(x+2)}}{\cancel{(x+2)}\cancel{(x+2)}(x+2)\cancel{(x+4)}\cancel{(x+4)}}$

$=\dfrac{x+4}{x+2}$

69. $\dfrac{5a^2-180}{10a^2-10}\cdot\dfrac{20a+20}{2a-12}=\dfrac{(5a^2-180)(20a+20)}{(10a^2-10)(2a-12)}$

$=\dfrac{5(a+6)(a-6)(2)(10)(a+1)}{10(a+1)(a-1)(2)(a-6)}$

$=\dfrac{5(a+6)\cancel{(a-6)}\cancel{(2)}\cancel{(10)}\cancel{(a+1)}}{\cancel{10}\cancel{(a+1)}(a-1)\cancel{(2)}\cancel{(a-6)}}$

$=\dfrac{5(a+6)}{a-1}$

71. ***Familiarize.*** Let $x=$ the smaller even integer. Then $x+2=$ the larger even integer.

Translate. We reword the problem.

Smaller even integer	times	larger even integer	is	360.
$\downarrow$	$\downarrow$	$\downarrow$	$\downarrow$	$\downarrow$
x	$\cdot$	$(x+2)$	$=$	360

Solve.

$$x(x+2)=360$$
$$x^2+2x=360$$
$$x^2+2x-360=0$$
$$(x+20)(x-18)=0$$
$$x+20=0 \quad\text{or}\quad x-18=0$$
$$x=-20 \quad\text{or}\quad x=18$$

Check. The solutions of the equation are -20 and 18. When $x=-20$, then $x+2=-18$ and $-20(-18)=360$. The numbers -20 and -18 are consecutive even integers which are solutions to the problem. When $x=18$, then $x+2=20$ and $18\cdot 20=360$. The numbers 18 and 20 are also consecutive even integers which are solutions to the problem.

State. We have two solutions, each of which consists of a pair of numbers: -20 and -18, and 18 and 20.

73. x^2-x-56

We look for a pair of numbers whose product is -56 and whose sum is -1. The numbers are -8 and 7.

$$x^2-x-56=(x-8)(x+7)$$

75. $x^5-2x^4-35x^3=x^3(x^2-2x-35)=x^3(x-7)(x+5)$

77. $16-t^4=4^2-(t^2)^2$ Difference of squares

$=(4+t^2)(4-t^2)$

$=(4+t^2)(2^2-t^2)$ Difference of squares

$=(4+t^2)(2+t)(2-t)$

79. $x^2 - 9x + 14$

We look for a pair of numbers whose product is 14 and whose sum is -9. The numbers are -2 and -7.

$$x^2 - 9x + 14 = (x-2)(x-7)$$

81.
$$16x^2 - 40xy + 25y^2$$
$$= (4x)^2 - 2\cdot 4x\cdot 5y + (5y)^2 \quad \text{Trinomial square}$$
$$= (4x-5y)^2$$

83. ◈

85.
$$\frac{x^4 - 16y^2}{(x^2+4y^2)(x-2y)}$$
$$= \frac{(x^2+4y^2)(x+2y)(x-2y)}{(x^2+4y^2)(x-2y)}$$
$$= \frac{\cancel{(x^2+4y^2)}(x+2y)\cancel{(x-2y)}}{\cancel{(x^2+4y^2)}\cancel{(x-2y)}(1)}$$
$$= x+2y$$

87.
$$\frac{t^4-1}{t^4-81}\cdot\frac{t^2-9}{t^2+1}\cdot\frac{(t-9)^2}{(t+1)^2}$$
$$= \frac{(t^2+1)(t+1)(t-1)(t+3)(t-3)(t-9)(t-9)}{(t^2+9)(t+3)(t-3)(t^2+1)(t+1)(t+1)}$$
$$= \frac{\cancel{(t^2+1)}\cancel{(t+1)}(t-1)\cancel{(t+3)}\cancel{(t-3)}(t-9)(t-9)}{(t^2+9)\cancel{(t+3)}\cancel{(t-3)}\cancel{(t^2+1)}\cancel{(t+1)}(t+1)}$$
$$= \frac{(t-1)(t-9)(t-9)}{(t^2+9)(t+1)}, \text{ or } \frac{(t-1)(t-9)^2}{(t^2+9)(t+1)}$$

89.
$$\frac{x^2-y^2}{(x-y)^2}\cdot\frac{x^2-2xy+y^2}{x^2-4xy-5y^2}$$
$$= \frac{(x+y)(x-y)(x-y)(x-y)}{(x-y)(x-y)(x-5y)(x+y)}$$
$$= \frac{\cancel{(x+y)}\cancel{(x-y)}\cancel{(x-y)}(x-y)}{\cancel{(x-y)}\cancel{(x-y)}(x-5y)\cancel{(x+y)}}$$
$$= \frac{x-y}{x-5y}$$

Exercise Set 12.2

1. The reciprocal of $\frac{4}{x}$ is $\frac{x}{4}$ because $\frac{4}{x}\cdot\frac{x}{4} = 1$.

3. The reciprocal of x^2-y^2 is $\frac{1}{x^2-y^2}$ because $\frac{x^2-y^2}{1}\cdot\frac{1}{x^2-y^2} = 1$.

5. The reciprocal of $\frac{1}{a+b}$ is $a+b$ because $\frac{1}{a+b}\cdot(a+b) = 1$.

7. The reciprocal of $\frac{x^2+2x-5}{x^2-4x+7}$ is $\frac{x^2-4x+7}{x^2+2x-5}$ because $\frac{x^2+2x-5}{x^2-4x+7}\cdot\frac{x^2-4x+7}{x^2+2x-5} = 1$.

9.
$$\frac{2}{5}\div\frac{4}{3} = \frac{2}{5}\cdot\frac{3}{4} \quad \text{Multiplying by the reciprocal of the divisor}$$
$$= \frac{2\cdot 3}{5\cdot 4}$$
$$= \frac{2\cdot 3}{5\cdot 2\cdot 2} \quad \text{Factoring the denominator}$$
$$= \frac{\cancel{2}\cdot 3}{5\cdot\cancel{2}\cdot 2} \quad \text{Removing a factor of 1}$$
$$= \frac{3}{10} \quad \text{Simplifying}$$

11.
$$\frac{2}{x}\div\frac{8}{x} = \frac{2}{x}\cdot\frac{x}{8} \quad \text{Multiplying by the reciprocal of the divisor}$$
$$= \frac{2\cdot x}{x\cdot 8}$$
$$= \frac{2\cdot x\cdot 1}{x\cdot 2\cdot 4} \quad \text{Factoring the numerator and the denominator}$$
$$= \frac{\cancel{2}\cdot\cancel{x}\cdot 1}{\cancel{x}\cdot\cancel{2}\cdot 4} \quad \text{Removing a factor of 1}$$
$$= \frac{1}{4} \quad \text{Simplifying}$$

13.
$$\frac{a}{b^2}\div\frac{a^2}{b^3} = \frac{a}{b^2}\cdot\frac{b^3}{a^2} \quad \text{Multiplying by the reciprocal of the divisor}$$
$$= \frac{a\cdot b^3}{b^2\cdot a^2}$$
$$= \frac{a\cdot b^2\cdot b}{b^2\cdot a\cdot a}$$
$$= \frac{\cancel{a}\cdot\cancel{b^2}\cdot b}{\cancel{b^2}\cdot\cancel{a}\cdot a}$$
$$= \frac{b}{a}$$

15.
$$\frac{a+2}{a-3}\div\frac{a-1}{a+3} = \frac{a+2}{a-3}\cdot\frac{a+3}{a-1}$$
$$= \frac{(a+2)(a+3)}{(a-3)(a-1)}$$

17.
$$\frac{x^2-1}{x}\div\frac{x+1}{x-1} = \frac{x^2-1}{x}\cdot\frac{x-1}{x+1}$$
$$= \frac{(x^2-1)(x-1)}{x(x+1)}$$
$$= \frac{(x-1)(x+1)(x-1)}{x(x+1)}$$
$$= \frac{(x-1)\cancel{(x+1)}(x-1)}{x\cancel{(x+1)}}$$
$$= \frac{(x-1)^2}{x}$$

19. $\dfrac{x+1}{6} \div \dfrac{x+1}{3} = \dfrac{x+1}{6} \cdot \dfrac{3}{x+1}$

$= \dfrac{(x+1)\cdot 3}{6(x+1)}$

$= \dfrac{3(x+1)}{2\cdot 3(x+1)}$

$= \dfrac{1\cdot \cancel{3}\cancel{(x+1)}}{2\cdot \cancel{3}\cancel{(x+1)}}$

$= \dfrac{1}{2}$

21. $\dfrac{5x-5}{16} \div \dfrac{x-1}{6} = \dfrac{5x-5}{16} \cdot \dfrac{6}{x-1}$

$= \dfrac{(5x-5)\cdot 6}{16(x-1)}$

$= \dfrac{5(x-1)\cdot 2\cdot 3}{2\cdot 8(x-1)}$

$= \dfrac{5\cancel{(x-1)}\cdot \cancel{2}\cdot 3}{\cancel{2}\cdot 8\cancel{(x-1)}}$

$= \dfrac{15}{8}$

23. $\dfrac{-6+3x}{5} \div \dfrac{4x-8}{25} = \dfrac{-6+3x}{5} \cdot \dfrac{25}{4x-8}$

$= \dfrac{(-6+3x)\cdot 25}{5(4x-8)}$

$= \dfrac{3(x-2)\cdot 5\cdot 5}{5\cdot 4(x-2)}$

$= \dfrac{3\cancel{(x-2)}\cdot \cancel{5}\cdot 5}{\cancel{5}\cdot 4\cancel{(x-2)}}$

$= \dfrac{15}{4}$

25. $\dfrac{a+2}{a-1} \div \dfrac{3a+6}{a-5} = \dfrac{a+2}{a-1} \cdot \dfrac{a-5}{3a+6}$

$= \dfrac{(a+2)(a-5)}{(a-1)(3a+6)}$

$= \dfrac{(a+2)(a-5)}{(a-1)\cdot 3\cdot (a+2)}$

$= \dfrac{\cancel{(a+2)}(a-5)}{(a-1)\cdot 3\cdot \cancel{(a+2)}}$

$= \dfrac{a-5}{3(a-1)}$

27. $\dfrac{x^2-4}{x} \div \dfrac{x-2}{x+2} = \dfrac{x^2-4}{x} \cdot \dfrac{x+2}{x-2}$

$= \dfrac{(x^2-4)(x+2)}{x(x-2)}$

$= \dfrac{(x-2)(x+2)(x+2)}{x(x-2)}$

$= \dfrac{\cancel{(x-2)}(x+2)(x+2)}{x\cancel{(x-2)}}$

$= \dfrac{(x+2)^2}{x}$

29. $\dfrac{x^2-9}{4x+12} \div \dfrac{x-3}{6} = \dfrac{x^2-9}{4x+12} \cdot \dfrac{6}{x-3}$

$= \dfrac{(x^2-9)\cdot 6}{(4x+12)(x-3)}$

$= \dfrac{(x-3)(x+3)\cdot 3\cdot 2}{2\cdot 2(x+3)(x-3)}$

$= \dfrac{\cancel{(x-3)}\cancel{(x+3)}\cdot 3\cdot \cancel{2}}{\cancel{2}\cdot 2\cancel{(x+3)}\cancel{(x-3)}}$

$= \dfrac{3}{2}$

31. $\dfrac{c^2+3c}{c^2+2c-3} \div \dfrac{c}{c+1} = \dfrac{c^2+3c}{c^2+2c-3} \cdot \dfrac{c+1}{c}$

$= \dfrac{(c^2+3c)(c+1)}{(c^2+2c-3)c}$

$= \dfrac{c(c+3)(c+1)}{(c+3)(c-1)c}$

$= \dfrac{\cancel{c}\cancel{(c+3)}(c+1)}{\cancel{(c+3)}(c-1)\cancel{c}}$

$= \dfrac{c+1}{c-1}$

33. $\dfrac{2y^2-7y+3}{2y^2+3y-2} \div \dfrac{6y^2-5y+1}{3y^2+5y-2}$

$= \dfrac{2y^2-7y+3}{2y^2+3y-2} \cdot \dfrac{3y^2+5y-2}{6y^2-5y+1}$

$= \dfrac{(2y^2-7y+3)(3y^2+5y-2)}{(2y^2+3y-2)(6y^2-5y+1)}$

$= \dfrac{(2y-1)(y-3)(3y-1)(y+2)}{(2y-1)(y+2)(3y-1)(2y-1)}$

$= \dfrac{\cancel{(2y-1)}(y-3)\cancel{(3y-1)}\cancel{(y+2)}}{\cancel{(2y-1)}\cancel{(y+2)}\cancel{(3y-1)}(2y-1)}$

$= \dfrac{y-3}{2y-1}$

35. $\dfrac{x^2-1}{4x+4} \div \dfrac{2x^2-4x+2}{8x+8} = \dfrac{x^2-1}{4x+4} \cdot \dfrac{8x+8}{2x^2-4x+2}$

$= \dfrac{(x^2-1)(8x+8)}{(4x+4)(2x^2-4x+2)}$

$= \dfrac{(x+1)(x-1)(2)(4)(x+1)}{4(x+1)(2)(x-1)(x-1)}$

$= \dfrac{\cancel{(x+1)}\cancel{(x-1)}\cancel{(2)}\cancel{(4)}(x+1)}{\cancel{4}\cancel{(x+1)}\cancel{(2)}(x-1)\cancel{(x-1)}}$

$= \dfrac{x+1}{x-1}$

37. ***Familiarize***. Let s = Bonnie's score on the last test.

Translate. The average of the four scores must be at least 90. This means it must be greater than or equal to 90. We translate.

$$\frac{96+98+89+s}{4} \geq 90$$

Solve. We solve the inequality. First we multiply by 4 to clear the fraction.

$$4\left(\frac{96+98+89+s}{4}\right) \geq 4\cdot 90$$
$$96+98+89+s \geq 360$$
$$283+s \geq 360$$
$$s \geq 77 \quad \text{Subtracting 283}$$

Check. We can do a partial check by substituting a value for s less than 77 and a value for s greater than 77.

For $s=76$: $\dfrac{96+98+89+76}{4} = 89.75 < 90$

For $s=78$: $\dfrac{96+98+89+78}{4} = 90.25 \leq 90$

Since the average is less than 90 for a value of s less than 77 and greater than or equal to 90 for a value greater than or equal to 77, the answer is probably correct.

State. The scores on the last test that will earn Bonnie an A are $\{s|s \geq 77\}$.

39. $(8x^3-3x^2+7)-(8x^2+3x-5) =$
$8x^3-3x^2+7-8x^2-3x+5 =$
$8x^3-11x^2-3x+12$

41. $(2x^{-3}y^4)^2 = 2^2(x^{-3})^2(y^4)^2$

$= 2^2x^{-6}y^8$ Multiplying exponents

$= 4x^{-6}y^8$ $(2^2=4)$

$= \dfrac{4y^8}{x^6}$ $\left(x^{-6}=\dfrac{1}{x^6}\right)$

43. $\left(\dfrac{2x^3}{y^5}\right)^2 = \dfrac{2^2(x^3)^2}{(y^5)^2}$

$= \dfrac{2^2x^6}{y^{10}}$ Multiplying exponents

$= \dfrac{4x^6}{y^{10}}$ $(2^2=4)$

45. ◈

47. $\dfrac{3a^2-5ab-12b^2}{3ab+4b^2} \div (3b^2-ab)$

$= \dfrac{3a^2-5ab-12b^2}{3ab+4b^2}\cdot\dfrac{1}{3b^2-ab}$

$= \dfrac{(3a+4b)(a-3b)}{b(3a+4b)\cdot b(3b-a)}$

$= \dfrac{(3a+4b)(-1)(3b-a)}{b(3a+4b)\cdot b(3b-a)}$

$= \dfrac{\cancel{(3a+4b)}(-1)\cancel{(3b-a)}}{b\cancel{(3a+4b)}\cdot b\cancel{(3b-a)}}$

$= -\dfrac{1}{b^2}$

49. The volume V of a rectangular solid is given by the formula $V = l\cdot w\cdot h$, where l = the length, w = the width, and h = the height. We substitute in the formula and solve for h.

$$V = l\cdot w\cdot h$$
$$x-3 = \frac{x-3}{x-7}\cdot\frac{x+y}{x-7}\cdot h$$
$$\frac{x-7}{\cancel{x-3}}\cdot\frac{x-7}{x+y}\cdot\cancel{(x-3)} = \frac{\cancel{x-7}}{\cancel{x-3}}\cdot\frac{\cancel{x-7}}{\cancel{x+y}}\cdot\frac{\cancel{x-3}}{\cancel{x-7}}\cdot\frac{\cancel{x+y}}{\cancel{x-7}}\cdot h$$
$$\frac{(x-7)^2}{x+y} = h$$

The height is $\dfrac{(x-7)^2}{x+y}$.

Exercise Set 12.3

1. $12 = 2\cdot2\cdot3$

$27 = 3\cdot3\cdot3$

$\text{LCM} = 2\cdot2\cdot3\cdot3\cdot3$, or 108

3. $8 = 2\cdot2\cdot2$

$9 = 3\cdot3$

$\text{LCM} = 2\cdot2\cdot2\cdot3\cdot3$, or 72

5. $6 = 2\cdot3$

$9 = 3\cdot3$

$21 = 3\cdot7$

$\text{LCM} = 2\cdot3\cdot3\cdot7$, or 126

7. $24 = 2\cdot2\cdot2\cdot3$

$36 = 2\cdot2\cdot3\cdot3$

$40 = 2\cdot2\cdot2\cdot5$

$\text{LCM} = 2\cdot2\cdot2\cdot3\cdot3\cdot5$, or 360

9. $10 = 2\cdot5$

$100 = 2\cdot2\cdot5\cdot5$

$500 = 2\cdot2\cdot5\cdot5\cdot5$

$\text{LCM} = 2\cdot2\cdot5\cdot5\cdot5$, or 500

(We might have observed at the outset that both 10 and 100 are factors of 500, so the LCM is 500.)

11. $24 = 2\cdot2\cdot2\cdot3$

$18 = 2\cdot3\cdot3$

$\text{LCD} = 2\cdot2\cdot2\cdot3\cdot3$, or 72

$$\frac{7}{24}+\frac{11}{18} = \frac{7}{2\cdot2\cdot2\cdot3}\cdot\frac{3}{3}+\frac{11}{2\cdot3\cdot3}\cdot\frac{2\cdot2}{2\cdot2}$$
$$= \frac{21}{2\cdot2\cdot2\cdot3\cdot3}+\frac{44}{2\cdot2\cdot2\cdot3\cdot3}$$
$$= \frac{65}{72}$$

13. $\dfrac{1}{6}+\dfrac{3}{40}$

$=\dfrac{1}{2\cdot 3}+\dfrac{3}{2\cdot 2\cdot 2\cdot 5}$

LCD is $2\cdot 2\cdot 2\cdot 3\cdot 5$, or 120

$=\dfrac{1}{2\cdot 3}\cdot\dfrac{2\cdot 2\cdot 5}{2\cdot 2\cdot 5}+\dfrac{3}{2\cdot 2\cdot 2\cdot 5}\cdot\dfrac{3}{3}$

$=\dfrac{20+9}{2\cdot 2\cdot 2\cdot 3\cdot 5}$

$=\dfrac{29}{120}$

15. $\dfrac{1}{20}+\dfrac{1}{30}+\dfrac{2}{45}$

$=\dfrac{1}{2\cdot 2\cdot 5}+\dfrac{1}{2\cdot 3\cdot 5}+\dfrac{2}{3\cdot 3\cdot 5}$

LCD is $2\cdot 2\cdot 3\cdot 3\cdot 5$, or 180

$=\dfrac{1}{2\cdot 2\cdot 5}\cdot\dfrac{3\cdot 3}{3\cdot 3}+\dfrac{1}{2\cdot 3\cdot 5}\cdot\dfrac{2\cdot 3}{2\cdot 3}+\dfrac{2}{3\cdot 3\cdot 5}\cdot\dfrac{2\cdot 2}{2\cdot 2}$

$=\dfrac{9+6+8}{2\cdot 2\cdot 3\cdot 3\cdot 5}$

$=\dfrac{23}{180}$

17. $6x^2=2\cdot 3\cdot x\cdot x$

$12x^3=2\cdot 2\cdot 3\cdot x\cdot x\cdot x$

$\text{LCM}=2\cdot 2\cdot 3\cdot x\cdot x\cdot x$, or $12x^3$

19. $2x^2=2\cdot x\cdot x$

$6xy=2\cdot 3\cdot x\cdot y$

$18y^2=2\cdot 3\cdot 3\cdot y\cdot y$

$\text{LCM}=2\cdot 3\cdot 3\cdot x\cdot x\cdot y\cdot y$, or $18x^2y^2$

21. $2(y-3)=2\cdot(y-3)$

$6(y-3)=2\cdot 3\cdot(y-3)$

$\text{LCM}=2\cdot 3\cdot(y-3)$, or $6(y-3)$

23. t, $t+2$, $t-2$

The expressions are not factorable, so the LCM is their product:

$\text{LCM}=t(t+2)(t-2)$

25. $x^2-4=(x+2)(x-2)$

$x^2+5x+6=(x+3)(x+2)$

$\text{LCM}=(x+2)(x-2)(x+3)$

27. $t^3+4t^2+4t=t(t^2+4t+4)=t(t+2)(t+2)$

$t^2-4t=t(t-4)$

$\text{LCM}=t(t+2)(t+2)(t-4)=t(t+2)^2(t-4)$

29. $a+1=a+1$

$(a-1)^2=(a-1)(a-1)$

$a^2-1=(a+1)(a-1)$

$\text{LCM}=(a+1)(a-1)(a-1)=(a+1)(a-1)^2$

31. $m^2-5m+6=(m-3)(m-2)$

$m^2-4m+4=(m-2)(m-2)$

$\text{LCM}=(m-3)(m-2)(m-2)=(m-3)(m-2)^2$

33. $2+3x=2+3x$

$4-9x^2=(2+3x)(2-3x)$

$2-3x=2-3x$

$\text{LCM}=(2+3x)(2-3x)$

35. $10v^2+30v=10v(v+3)=2\cdot 5\cdot v(v+3)$

$5v^2+35v+60=5(v^2+7v+12)$

$=5(v+4)(v+3)$

$\text{LCM}=2\cdot 5\cdot v(v+3)(v+4)=10v(v+3)(v+4)$

37. $9x^3-9x^2-18x=9x(x^2-x-2)$

$=3\cdot 3\cdot x(x-2)(x+1)$

$6x^5-24x^4+24x^3=6x^3(x^2-4x+4)$

$=2\cdot 3\cdot x\cdot x\cdot x(x-2)(x-2)$

$\text{LCM}=2\cdot 3\cdot 3\cdot x\cdot x\cdot x(x-2)(x-2)(x+1)=$
$18x^3(x-2)^2(x+1)$

39. $x^5+4x^4+4x^3=x^3(x^2+4x+4)$

$=x\cdot x\cdot x(x+2)(x+2)$

$3x^2-12=3(x^2-4)=3(x+2)(x-2)$

$2x+4=2(x+2)$

$\text{LCM}=2\cdot 3\cdot x\cdot x\cdot x(x+2)(x+2)(x-2)$

$=6x^3(x+2)^2(x-2)$

41. $x^2-6x+9=x^2-2\cdot x\cdot 3+3^2$ Trinomial square

$=(x-3)^2$

43. $x^2-9=x^2-3^2$ Difference of squares

$=(x+3)(x-3)$

45. $x^2+6x+9=x^2+2\cdot x\cdot 3+3^2$ Trinomial square

$=(x+3)^2$

47. Locate 1970 on the horizontal axis, go up to the graph, and then go over to the corresponding point on the vertical axis. We read that about 54% of those married in 1970 will divorce.

49. Locate 1990 on the horizontal axis, go up to the graph, and then go over to the corresponding point on the vertical axis. We read that about 74% of those married in 1990 will divorce.

51. Locate 50 on the vertical axis, go across to the graph, and then go down to the corresponding point on the horizontal axis. We read that the divorce percentage was about 50% in 1965.

53. ◈

Exercise Set 12.4

1. $\frac{5}{8}+\frac{3}{8}+\frac{5+3}{8}=\frac{8}{8}=1$

3. $\frac{1}{3+x}+\frac{5}{3+x}=\frac{1+5}{3+x}=\frac{6}{3+x}$

5. $\frac{x^2+7x}{x^2-5x}+\frac{x^2-4x}{x^2-5x}=\frac{(x^2+7x)+(x^2-4x)}{x^2-5x}$
$=\frac{2x^2+3x}{x^2-5x}$
$=\frac{x(2x+3)}{x(x-5)}$
$=\frac{\not{x}(2x+3)}{\not{x}(x-5)}$
$=\frac{2x+3}{x-5}$

7. $\frac{7}{8}+\frac{5}{-8}=\frac{7}{8}+\frac{5}{-8}\cdot\frac{-1}{-1}$
$=\frac{7}{8}+\frac{-5}{8}$
$=\frac{7+(-5)}{8}$
$=\frac{2}{8}=\frac{\not{2}\cdot 1}{4\cdot\not{2}}$
$=\frac{1}{4}$

9. $\frac{3}{t}+\frac{4}{-t}=\frac{3}{t}+\frac{4}{-t}\cdot\frac{-1}{-1}$
$=\frac{3}{t}+\frac{-4}{t}$
$=\frac{3+(-4)}{t}$
$=\frac{-1}{t}$
$=-\frac{1}{t}$

11. $\frac{2x+7}{x-6}+\frac{3x}{6-x}=\frac{2x+7}{x-6}+\frac{3x}{6-x}\cdot\frac{-1}{-1}$
$=\frac{2x+7}{x-6}+\frac{-3x}{x-6}$
$=\frac{(2x+7)+(-3x)}{x-6}$
$=\frac{-x+7}{x-6}$

13. $\frac{y^2}{y-3}+\frac{9}{3-y}=\frac{y^2}{y-3}+\frac{9}{3-y}\cdot\frac{-1}{-1}$
$=\frac{y^2}{y-3}+\frac{-9}{y-3}$
$=\frac{y^2+(-9)}{y-3}$
$=\frac{y^2-9}{y-3}$
$=\frac{(y+3)(y-3)}{y-3}$
$=\frac{(y+3)\not{(y-3)}}{1\not{(y-3)}}$
$=y+3$

15. $\frac{b-7}{b^2-16}+\frac{7-b}{16-b^2}=\frac{b-7}{b^2-16}+\frac{7-b}{16-b^2}\cdot\frac{-1}{-1}$
$=\frac{b-7}{b^2-16}+\frac{b-7}{b^2-16}$
$=\frac{(b-7)+(b-7)}{b^2-16}$
$=\frac{2b-14}{b^2-16}$

17. $\frac{a^2}{a-b}+\frac{b^2}{b-a}=\frac{a^2}{a-b}+\frac{b^2}{b-a}\cdot\frac{-1}{-1}$
$=\frac{a^2}{a-b}+\frac{-b^2}{a-b}$
$=\frac{a^2+(-b^2)}{a-b}$
$=\frac{a^2-b^2}{a-b}$
$=\frac{(a+b)(a-b)}{a-b}$
$=\frac{(a+b)\not{(a-b)}}{1\not{(a-b)}}$
$=a+b$

19. $\frac{x+3}{x-5}+\frac{2x-1}{5-x}+\frac{2(3x-1)}{x-5}$
$=\frac{x+3}{x-5}+\frac{2x-1}{5-x}\cdot\frac{-1}{-1}+\frac{2(3x-1)}{x-5}$
$=\frac{x+3}{x-5}+\frac{1-2x}{x-5}+\frac{2(3x-1)}{x-5}$
$=\frac{(x+3)+(1-2x)+(6x-2)}{x-5}$
$=\frac{5x+2}{x-5}$

21. $\dfrac{2(4x+1)}{5x-7}+\dfrac{3(x-2)}{7-5x}+\dfrac{-10x-1}{5x-7}$

$=\dfrac{2(4x+1)}{5x-7}+\dfrac{3(x-2)}{7-5x}\cdot\dfrac{-1}{-1}+\dfrac{-10x-1}{5x-7}$

$=\dfrac{2(4x+1)}{5x-7}+\dfrac{-3(x-2)}{5x-7}+\dfrac{-10x-1}{5x-7}$

$=\dfrac{(8x+2)+(-3x+6)+(-10x-1)}{5x-7}$

$=\dfrac{-5x+7}{5x-7}$

$=\dfrac{-1(5x-7)}{5x-7}$

$=\dfrac{-1(\cancel{5x-7})}{\cancel{5x-7}}$

$=-1$

23. $\dfrac{x+1}{(x+3)(x-3)}+\dfrac{4(x-3)}{(x-3)(x+3)}+\dfrac{(x-1)(x-3)}{(3-x)(x+3)}$

$=\dfrac{x+1}{(x+3)(x-3)}+\dfrac{4(x-3)}{(x-3)(x+3)}+\dfrac{(x-1)(x-3)}{(3-x)(x+3)}\cdot\dfrac{-1}{-1}$

$=\dfrac{x+1}{(x+3)(x-3)}+\dfrac{4(x-3)}{(x-3)(x+3)}+\dfrac{-1(x^2-4x+3)}{(x-3)(x+3)}$

$=\dfrac{(x+1)+(4x-12)+(-x^2+4x-3)}{(x+3)(x-3)}$

$=\dfrac{-x^2+9x-14}{(x+3)(x-3)}$

25. $\dfrac{2}{x}+\dfrac{5}{x^2}=\dfrac{2}{x}+\dfrac{5}{x\cdot x}$ LCD $=x\cdot x$, or x^2

$=\dfrac{2}{x}\cdot\dfrac{x}{x}+\dfrac{5}{x\cdot x}$

$=\dfrac{2x+5}{x^2}$

27. $\left.\begin{array}{l}6r=2\cdot3\cdot r\\8r=2\cdot2\cdot2\cdot r\end{array}\right\}\text{LCD}=2\cdot2\cdot2\cdot3\cdot r\text{, or }24r$

$\dfrac{5}{6r}+\dfrac{7}{8r}=\dfrac{5}{6r}\cdot\dfrac{4}{4}+\dfrac{7}{8r}\cdot\dfrac{3}{3}$

$=\dfrac{20+21}{24r}$

$=\dfrac{41}{24r}$

29. $\left.\begin{array}{l}xy^2=x\cdot y\cdot y\\x^2y=x\cdot x\cdot y\end{array}\right\}\text{LCD}=x\cdot x\cdot y\cdot y\text{, or }x^2y^2$

$\dfrac{4}{xy^2}+\dfrac{6}{x^2y}=\dfrac{4}{xy^2}\cdot\dfrac{x}{x}+\dfrac{6}{x^2y}\cdot\dfrac{y}{y}$

$=\dfrac{4x+6y}{x^2y^2}$

31. $\left.\begin{array}{l}9t^3=3\cdot3\cdot t\cdot t\cdot t\\6t^2=2\cdot3\cdot t\cdot t\end{array}\right\}\text{LCD}=2\cdot3\cdot3\cdot t\cdot t\cdot t\text{, or }18t^3$

$\dfrac{2}{9t^3}+\dfrac{1}{6t^2}=\dfrac{2}{9t^3}\cdot\dfrac{2}{2}+\dfrac{1}{6t^2}\cdot\dfrac{3t}{3t}$

$=\dfrac{4+3t}{18t^3}$

33. LCD $=x^2y^2$ (See Exercise 29.)

$\dfrac{x+y}{xy^2}+\dfrac{3x+y}{x^2y}=\dfrac{x+y}{xy^2}\cdot\dfrac{x}{x}+\dfrac{3x+y}{x^2y}\cdot\dfrac{y}{y}$

$=\dfrac{x(x+y)+y(3x+y)}{x^2y^2}$

$=\dfrac{x^2+xy+3xy+y^2}{x^2y^2}$

$=\dfrac{x^2+4xy+y^2}{x^2y^2}$

35. The denominators do not factor, so the LCD is their product, $(x-2)(x+2)$.

$\dfrac{3}{x-2}+\dfrac{3}{x+2}=\dfrac{3}{x-2}\cdot\dfrac{x+2}{x+2}+\dfrac{3}{x+2}\cdot\dfrac{x-2}{x-2}$

$=\dfrac{3(x+2)+3(x-2)}{(x-2)(x+2)}$

$=\dfrac{3x+6+3x-6}{(x-2)(x+2)}$

$=\dfrac{6x}{(x-2)(x+2)}$

37. $\left.\begin{array}{l}3x=3\cdot x\\x+1=x+1\end{array}\right\}\text{LCD}=3x(x+1)$

$\dfrac{3}{x+1}+\dfrac{2}{3x}=\dfrac{3}{x+1}\cdot\dfrac{3x}{3x}+\dfrac{2}{3x}\cdot\dfrac{x+1}{x+1}$

$=\dfrac{9x+2(x+1)}{3x(x+1)}$

$=\dfrac{9x+2x+2}{3x(x+1)}$

$=\dfrac{11x+2}{3x(x+1)}$

39. $\left.\begin{array}{l}x^2-16=(x+4)(x-4)\\x-4=x-4\end{array}\right\}\text{LCD}=(x+4)(x-4)$

$\dfrac{2x}{x^2-16}+\dfrac{x}{x-4}=\dfrac{2x}{(x+4)(x-4)}+\dfrac{x}{x-4}\cdot\dfrac{x+4}{x+4}$

$=\dfrac{2x+x(x+4)}{(x+4)(x-4)}$

$=\dfrac{2x+x^2+4x}{(x+4)(x-4)}$

$=\dfrac{x^2+6x}{(x+4)(x-4)}$

41. $\dfrac{5}{z+4}+\dfrac{3}{3z+12}=\dfrac{5}{z+4}+\dfrac{3}{3(z+4)}$ LCD $=3(z+4)$

$=\dfrac{5}{z+4}\cdot\dfrac{3}{3}+\dfrac{3}{3(z+4)}$

$=\dfrac{15+3}{3(z+4)}=\dfrac{18}{3(z+4)}$

$=\dfrac{3\cdot6}{3(z+4)}=\dfrac{\cancel{3}\cdot6}{\cancel{3}(z+4)}$

$=\dfrac{6}{z+4}$

43. $\dfrac{3}{x-1}+\dfrac{2}{(x-1)^2}$ LCD $=(x-1)^2$

$=\dfrac{3}{x-1}\cdot\dfrac{x-1}{x-1}+\dfrac{2}{(x-1)^2}$

$=\dfrac{3(x-1)+2}{(x-1)^2}$

$=\dfrac{3x-3+2}{(x-1)^2}$

$=\dfrac{3x-1}{(x-1)^2}$

45. $\dfrac{4a}{5a-10}+\dfrac{3a}{10a-20}=\dfrac{4a}{5(a-2)}+\dfrac{3a}{2\cdot5(a-2)}$

LCD $=2\cdot5(a-2)$

$=\dfrac{4a}{5(a-2)}\cdot\dfrac{2}{2}+\dfrac{3a}{2\cdot5(a-2)}$

$=\dfrac{8a+3a}{10(a-2)}$

$=\dfrac{11a}{10(a-2)}$

47. $\dfrac{x+4}{x}+\dfrac{x}{x+4}$ LCD $=x(x+4)$

$=\dfrac{x+4}{x}\cdot\dfrac{x+4}{x+4}+\dfrac{x}{x+4}\cdot\dfrac{x}{x}$

$=\dfrac{(x+4)^2+x^2}{x(x+4)}$

$=\dfrac{x^2+8x+16+x^2}{x(x+4)}$

$=\dfrac{2x^2+8x+16}{x(x+4)}$

49. $\dfrac{4}{a^2-a-2}+\dfrac{3}{a^2+4a+3}$

$=\dfrac{4}{(a-2)(a+1)}+\dfrac{3}{(a+3)(a+1)}$

LCD $=(a-2)(a+1)(a+3)$

$=\dfrac{4}{(a-2)(a+1)}\cdot\dfrac{a+3}{a+3}+\dfrac{3}{(a+3)(a+1)}\cdot\dfrac{a-2}{a-2}$

$=\dfrac{4(a+3)+3(a-2)}{(a-2)(a+1)(a+3)}$

$=\dfrac{4a+12+3a-6}{(a-2)(a+1)(a+3)}$

$=\dfrac{7a+6}{(a-2)(a+1)(a+3)}$

51. $\dfrac{x+3}{x-5}+\dfrac{x-5}{x+3}$ LCD $=(x-5)(x+3)$

$=\dfrac{x+3}{x-5}\cdot\dfrac{x+3}{x+3}+\dfrac{x-5}{x+3}\cdot\dfrac{x-5}{x-5}$

$=\dfrac{(x+3)^2+(x-5)^2}{(x-5)(x+3)}$

$=\dfrac{x^2+6x+9+x^2-10x+25}{(x-5)(x+3)}$

$=\dfrac{2x^2-4x+34}{(x-5)(x+3)}$

53. $\dfrac{a}{a^2-1}+\dfrac{2a}{a^2-a}$

$=\dfrac{a}{(a+1)(a-1)}+\dfrac{2a}{a(a-1)}$

LCD $=a(a+1)(a-1)$

$=\dfrac{a}{(a+1)(a-1)}\cdot\dfrac{a}{a}+\dfrac{2a}{a(a-1)}\cdot\dfrac{a+1}{a+1}$

$=\dfrac{a^2+2a(a+1)}{a(a+1)(a-1)}=\dfrac{a^2+2a^2+2a}{a(a+1)(a-1)}$

$=\dfrac{3a^2+2a}{a(a+1)(a-1)}=\dfrac{a(3a+2)}{a(a+1)(a-1)}$

$=\dfrac{\not{a}(3a+2)}{\not{a}(a+1)(a-1)}=\dfrac{3a+2}{(a+1)(a-1)}$

55. $\dfrac{6}{x-y}+\dfrac{4x}{y^2-x^2}$

$=\dfrac{6}{x-y}+\dfrac{4x}{(y-x)(y+x)}$

$=\dfrac{6}{x-y}+\dfrac{4x}{(y-x)(y+x)}\cdot\dfrac{-1}{-1}$

$=\dfrac{6}{x-y}+\dfrac{-4x}{(x-y)(x+y)}$

$[-1(y-x)=x-y; y+x=x+y]$

LCD $=(x-y)(x+y)$

$=\dfrac{6}{x-y}\cdot\dfrac{x+y}{x+y}+\dfrac{-4x}{(x-y)(x+y)}$

$=\dfrac{6(x+y)-4x}{(x-y)(x+y)}$

$=\dfrac{6x+6y-4x}{(x-y)(x+y)}$

$=\dfrac{2x+6y}{(x-y)(x+y)}$

57. $\dfrac{4-a}{25-a^2}+\dfrac{a+1}{a-5}$

$=\dfrac{4-a}{25-a^2}\cdot\dfrac{-1}{-1}+\dfrac{a+1}{a-5}$

$=\dfrac{a-4}{a^2-25}+\dfrac{a+1}{a-5}$

$=\dfrac{a-4}{(a+5)(a-5)}+\dfrac{a+1}{a-5}$

LCD $=(a+5)(a-5)$

$=\dfrac{a-4}{(a+5)(a-5)}+\dfrac{a+1}{a-5}\cdot\dfrac{a+5}{a+5}$

$=\dfrac{a-4}{(a+5)(a-5)}+\dfrac{(a+1)(a+5)}{(a+5)(a-5)}$

$=\dfrac{(a-4)+(a+1)(a+5)}{(a+5)(a-5)}$

$=\dfrac{a-4+a^2+6a+5}{(a+5)(a-5)}$

$=\dfrac{a^2+7a+1}{(a+5)(a-5)}$

59. $\dfrac{2}{t^2+t-6}+\dfrac{3}{t^2-9}$

$=\dfrac{2}{(t+3)(t-2)}+\dfrac{3}{(t+3)(t-3)}$

LCD $=(t+3)(t-2)(t-3)$

$=\dfrac{2}{(t+3)(t-2)}\cdot\dfrac{t-3}{t-3}+\dfrac{3}{(t+3)(t-3)}\cdot\dfrac{t-2}{t-2}$

$=\dfrac{2(t-3)+3(t-2)}{(t+3)(t-2)(t-3)}$

$=\dfrac{2t-6+3t-6}{(t+3)(t-2)(t-3)}$

$=\dfrac{5t-12}{(t+3)(t-2)(t-3)}$

61. $(x^2+x)-(x+1)=x^2+x-x-1=x^2-1$

63. $(2x^4y^3)^{-3}=\dfrac{1}{(2x^4y^3)^3}=\dfrac{1}{2^3(x^4)^3(y^3)^3}=\dfrac{1}{8x^{12}y^9}$

65. $\left(\dfrac{x^{-4}}{y^7}\right)^3=\dfrac{(x^{-4})^3}{(y^7)^3}=\dfrac{x^{-12}}{y^{21}}=\dfrac{1}{x^{12}y^{21}}$

67. $y=\dfrac{1}{2}x-5=\dfrac{1}{2}x+(-5)$

The y-intercept is $(0,-5)$. We find two other pairs.

When $x=2$, $y=\dfrac{1}{2}\cdot 2-5=1-5=-4$.

When $x=4$, $y=\dfrac{1}{2}\cdot 4-5=2-5=-3$.

x	y
0	-5
2	-4
4	-3

Plot these points, draw the line they determine, and label the graph $y=\dfrac{1}{2}x-5$.

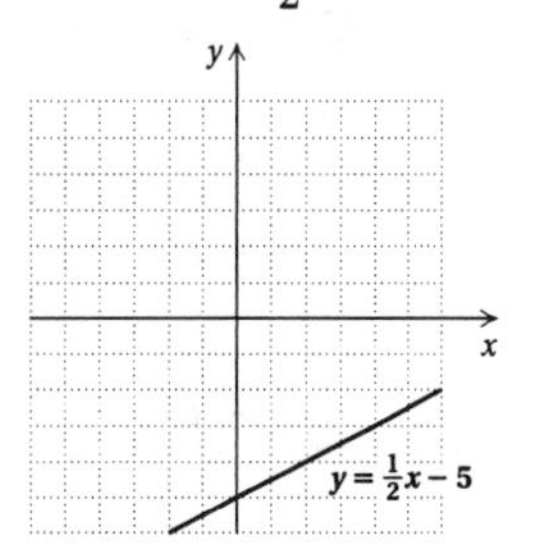

69. $y=3$

Any ordered pair $(x,3)$ is a solution. The variable y must be 3, but x can be any number we choose. A few solutions are listed below. Plot these points and draw the line.

x	y
-4	3
0	3
3	3

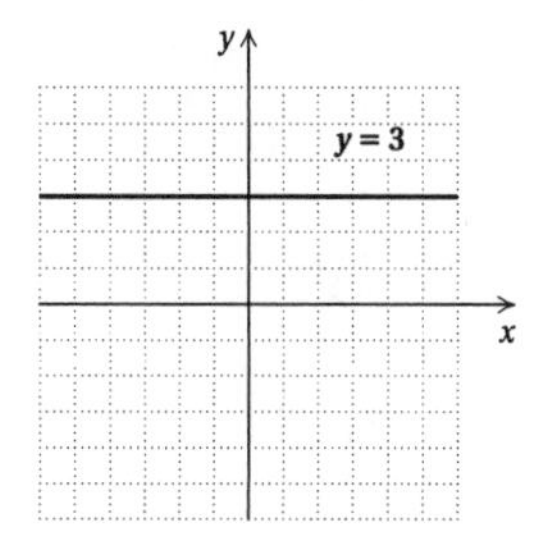

71. $3x-7=5x+9$

$-2x-7=9$ Subtracting $5x$

$-2x=16$ Adding 7

$x=-8$ Dividing by -2

The solution is -8.

73. $x^2-8x+15=0$

$(x-3)(x-5)=0$

$x-3=0$ or $x-5=0$ Principle of zero products

$x=3$ or $x=5$

The solutions are 3 and 5.

75. 

77. To find the perimeter we add the lengths of the sides:

$\dfrac{y+4}{3}+\dfrac{y+4}{3}+\dfrac{y-2}{5}+\dfrac{y-2}{5}$ LCD $=3\cdot 5$

$=\dfrac{y+4}{3}\cdot\dfrac{5}{5}+\dfrac{y+4}{3}\cdot\dfrac{5}{5}+\dfrac{y-2}{5}\cdot\dfrac{3}{3}+\dfrac{y-2}{5}\cdot\dfrac{3}{3}$

$=\dfrac{5y+20+5y+20+3y-6+3y-6}{3\cdot 5}$

$=\dfrac{16y+28}{15}$

To find the area we multiply the length and the width:

$\left(\dfrac{y+4}{3}\right)\left(\dfrac{y-2}{5}\right)=\dfrac{(y+4)(y-2)}{3\cdot 5}=\dfrac{y^2+2y-8}{15}$

79. $\dfrac{5}{z+2}+\dfrac{4z}{z^2-4}+2$

$=\dfrac{5}{z+2}+\dfrac{4z}{(z+2)(z-2)}+\dfrac{2}{1}$ LCD $=(z+2)(z-2)$

$=\dfrac{5}{z+2}\cdot\dfrac{z-2}{z-2}+\dfrac{4z}{(z+2)(z-2)}+\dfrac{2}{1}\cdot\dfrac{(z+2)(z-2)}{(z+2)(z-2)}$

$=\dfrac{5z-10+4z+2(z^2-4)}{(z+2)(z-2)}$

$=\dfrac{5z-10+4z+2z^2-8}{(z+2)(z-2)}=\dfrac{2z^2+9z-18}{(z+2)(z-2)}$

$=\dfrac{(2z-3)(z+6)}{(z+2)(z-2)}$

81. $\dfrac{3z^2}{z^4-4}+\dfrac{5z^2-3}{2z^4+z^2-6}$

$=\dfrac{3z^2}{(z^2+2)(z^2-2)}+\dfrac{5z^2-3}{(2z^2-3)(z^2+2)}$

LCD $=(z^2+2)(z^2-2)(2z^2-3)$

$=\dfrac{3z^2}{(z^2+2)(z^2-2)}\cdot\dfrac{2z^2-3}{2z^2-3}+$

$\dfrac{5z^2-3}{(2z^2-3)(z^2+2)}\cdot\dfrac{z^2-2}{z^2-2}$

$=\dfrac{6z^4-9z^2+5z^4-13z^2+6}{(z^2+2)(z^2-2)(2z^2-3)}$

$=\dfrac{11z^4-22z^2+6}{(z^2+2)(z^2-2)(2z^2-3)}$

83.-87.

Exercise Set 12.5

1. $\dfrac{7}{x}-\dfrac{3}{x}=\dfrac{7-3}{x}=\dfrac{4}{x}$

3. $\dfrac{y}{y-4}-\dfrac{4}{y-4}=\dfrac{y-4}{y-4}=1$

5. $\dfrac{2x-3}{x^2+3x-4}-\dfrac{x-7}{x^2+3x-4}$

$=\dfrac{2x-3-(x-7)}{x^2+3x-4}$

$=\dfrac{2x-3-x+7}{x^2+3x-4}$

$=\dfrac{x+4}{x^2+3x-4}$

$=\dfrac{x+4}{(x+4)(x-1)}$

$=\dfrac{(x+4)\cdot 1}{(x+4)(x-1)}$

$=\dfrac{1}{x-1}$

7. $\dfrac{11}{6}-\dfrac{5}{-6}=\dfrac{11}{6}-\dfrac{5}{-6}\cdot\dfrac{-1}{-1}$

$=\dfrac{11}{6}-\dfrac{-5}{6}$

$=\dfrac{11-(-5)}{6}$

$=\dfrac{11+5}{6}$

$=\dfrac{16}{6}$

$=\dfrac{8}{3}$

9. $\dfrac{5}{a}-\dfrac{8}{-a}=\dfrac{5}{a}-\dfrac{8}{-a}\cdot\dfrac{-1}{-1}$

$=\dfrac{5}{a}-\dfrac{-8}{a}$

$=\dfrac{5-(-8)}{a}$

$=\dfrac{5+8}{a}$

$=\dfrac{13}{a}$

11. $\dfrac{4}{y-1}-\dfrac{4}{1-y}=\dfrac{4}{y-1}-\dfrac{4}{1-y}\cdot\dfrac{-1}{-1}$

$=\dfrac{4}{y-1}-\dfrac{4(-1)}{(1-y)(-1)}$

$=\dfrac{4}{y-1}-\dfrac{-4}{y-1}$

$=\dfrac{4-(-4)}{y-1}$

$=\dfrac{4+4}{y-1}$

$=\dfrac{8}{y-1}$

13. $\dfrac{3-x}{x-7}-\dfrac{2x-5}{7-x}=\dfrac{3-x}{x-7}-\dfrac{2x-5}{7-x}\cdot\dfrac{-1}{-1}$

$=\dfrac{3-x}{x-7}-\dfrac{(2x-5)(-1)}{(7-x)(-1)}$

$=\dfrac{3-x}{x-7}-\dfrac{5-2x}{x-7}$

$=\dfrac{(3-x)-(5-2x)}{x-7}$

$=\dfrac{3-x-5+2x}{x-7}$

$=\dfrac{x-2}{x-7}$

15. $\dfrac{a-2}{a^2-25}-\dfrac{6-a}{25-a^2}=\dfrac{a-2}{a^2-25}-\dfrac{6-a}{25-a^2}\cdot\dfrac{-1}{-1}$

$=\dfrac{a-2}{a^2-25}-\dfrac{(6-a)(-1)}{(25-a^2)(-1)}$

$=\dfrac{a-2}{a^2-25}-\dfrac{a-6}{a^2-25}$

$=\dfrac{(a-2)-(a-6)}{a^2-25}$

$=\dfrac{a-2-a+6}{a^2-25}$

$=\dfrac{4}{a^2-25}$

17. $\dfrac{4-x}{x-9}-\dfrac{3x-8}{9-x}=\dfrac{4-x}{x-9}-\dfrac{3x-8}{9-x}\cdot\dfrac{-1}{-1}$

$=\dfrac{4-x}{x-9}-\dfrac{8-3x}{x-9}$

$=\dfrac{(4-x)-(8-3x)}{x-9}$

$=\dfrac{4-x-8+3x}{x-9}$

$=\dfrac{2x-4}{x-9}$

19. $\dfrac{2(x-1)}{2x-3}-\dfrac{3(x+2)}{2x-3}-\dfrac{x-1}{3-2x}$

$=\dfrac{2(x-1)}{2x-3}-\dfrac{3(x+2)}{2x-3}-\dfrac{x-1}{3-2x}\cdot\dfrac{-1}{-1}$

$=\dfrac{2(x-1)}{2x-3}-\dfrac{3(x+2)}{2x-3}-\dfrac{1-x}{2x-3}$

$=\dfrac{(2x-2)-(3x+6)-(1-x)}{2x-3}$

$=\dfrac{2x-2-3x-6-1+x}{2x-3}$

$=\dfrac{-9}{2x-3}$

21. $\dfrac{a-2}{10}-\dfrac{a+1}{5}=\dfrac{a-2}{10}-\dfrac{a+1}{5}\cdot\dfrac{2}{2}$ LCD = 10

$=\dfrac{a-2}{10}-\dfrac{2(a+1)}{10}$

$=\dfrac{(a-2)-2(a+1)}{10}$

$=\dfrac{a-2-2a-2}{10}$

$=\dfrac{-a-4}{10}$

23. $\dfrac{4z-9}{3z}-\dfrac{3z-8}{4z}=\dfrac{4z-9}{3z}\cdot\dfrac{4}{4}-\dfrac{3z-8}{4z}\cdot\dfrac{3}{3}$

LCD $=3\cdot 4\cdot z$, or $12z$

$=\dfrac{16z-36}{12z}-\dfrac{9z-24}{12z}$

$=\dfrac{16z-36-(9z-24)}{12z}$

$=\dfrac{16z-36-9z+24}{12z}$

$=\dfrac{7z-12}{12z}$

25. $\dfrac{4x+2t}{3xt^2}-\dfrac{5x-3t}{x^2t}$ LCD $=3x^2t^2$

$=\dfrac{4x+2t}{3xt^2}\cdot\dfrac{x}{x}-\dfrac{5x-3t}{x^2t}\cdot\dfrac{3t}{3t}$

$=\dfrac{4x^2+2tx}{3x^2t^2}-\dfrac{15xt-9t^2}{3x^2t^2}$

$=\dfrac{4x^2+2tx-(15xt-9t^2)}{3x^2t^2}$

$=\dfrac{4x^2+2tx-15xt+9t^2}{3x^2t^2}$

$=\dfrac{4x^2-13xt+9t^2}{3x^2t^2}$

27. $\dfrac{5}{x+5}-\dfrac{3}{x-5}$ LCD $=(x+5)(x-5)$

$=\dfrac{5}{x+5}\cdot\dfrac{x-5}{x-5}-\dfrac{3}{x-5}\cdot\dfrac{x+5}{x+5}$

$=\dfrac{5x-25}{(x+5)(x-5)}-\dfrac{3x+15}{(x+5)(x-5)}$

$=\dfrac{5x-25-(3x+15)}{(x+5)(x-5)}$

$=\dfrac{5x-25-3x-15}{(x+5)(x-5)}$

$=\dfrac{2x-40}{(x+5)(x-5)}$

29. $\dfrac{3}{2t^2-2t}-\dfrac{5}{2t-2}$

$=\dfrac{3}{2t(t-1)}-\dfrac{5}{2(t-1)}$ LCD $=2t(t-1)$

$=\dfrac{3}{2t(t-1)}-\dfrac{5}{2(t-1)}\cdot\dfrac{t}{t}$

$=\dfrac{3}{2t(t-1)}-\dfrac{5t}{2t(t-1)}$

$=\dfrac{3-5t}{2t(t-1)}$

31. $\dfrac{2s}{t^2-s^2}-\dfrac{s}{t-s}$ LCD $=(t-s)(t+s)$

$$=\frac{2s}{(t-s)(t+s)}-\frac{s}{t-s}\cdot\frac{t+s}{t+s}$$
$$=\frac{2s}{(t-s)(t+s)}-\frac{st+s^2}{(t-s)(t+s)}$$
$$=\frac{2s-(st+s^2)}{(t-s)(t+s)}$$
$$=\frac{2s-st-s^2}{(t-s)(t+s)}$$

33. $\dfrac{y-5}{y}-\dfrac{3y-1}{4y}=\dfrac{y-5}{y}\cdot\dfrac{4}{4}-\dfrac{3y-1}{4y}$ LCD $=4y$

$$=\frac{4y-20}{4y}-\frac{3y-1}{4y}$$
$$=\frac{4y-20-(3y-1)}{4y}$$
$$=\frac{4y-20-3y+1}{4y}$$
$$=\frac{y-19}{4y}$$

35. $\dfrac{a}{x+a}-\dfrac{a}{x-a}$ LCD $=(x+a)(x-a)$

$$=\frac{a}{x+a}\cdot\frac{x-a}{x-a}-\frac{a}{x-a}\cdot\frac{x+a}{x+a}$$
$$=\frac{ax-a^2}{(x+a)(x-a)}-\frac{ax+a^2}{(x+a)(x-a)}$$
$$=\frac{ax-a^2-(ax+a^2)}{(x+a)(x-a)}$$
$$=\frac{ax-a^2-ax-a^2}{(x+a)(x-a)}$$
$$=\frac{-2a^2}{(x+a)(x-a)}$$

37. $\dfrac{5x}{x^2-9}-\dfrac{4}{3-x}$

$$=\frac{5x}{(x+3)(x-3)}-\frac{4}{3-x}$$ $x-3$ and $3-x$ are opposites

$$=\frac{5x}{(x+3)(x-3)}-\frac{4}{3-x}\cdot\frac{-1}{-1}$$
$$=\frac{5x}{(x+3)(x-3)}-\frac{-4}{x-3}$$ LCD $=(x+3)(x-3)$

$$=\frac{5x}{(x+3)(x-3)}-\frac{-4}{x-3}\cdot\frac{x+3}{x+3}$$
$$=\frac{5x}{(x+3)(x-3)}-\frac{-4x-12}{(x+3)(x-3)}$$
$$=\frac{5x-(-4x-12)}{(x+3)(x-3)}$$
$$=\frac{5x+4x+12}{(x+3)(x-3)}$$
$$=\frac{9x+12}{(x+3)(x-3)}$$

39. $\dfrac{t^2}{2t^2-2t}-\dfrac{1}{2t-2}$

$$=\frac{t^2}{2t(t-1)}-\frac{1}{2(t-1)}$$ LCD $=2t(t-1)$

$$=\frac{t^2}{2t(t-1)}-\frac{1}{2(t-1)}\cdot\frac{t}{t}$$
$$=\frac{t^2}{2t(t-1)}-\frac{t}{2t(t-1)}$$
$$=\frac{t^2-t}{2t(t-1)}$$
$$=\frac{t(t-1)}{2t(t-1)}$$
$$=\frac{\cancel{t}(\cancel{t-1})(1)}{2\cancel{t}(\cancel{t-1})}$$
$$=\frac{1}{2}$$

41. $\dfrac{x}{x^2+5x+6}-\dfrac{2}{x^2+3x+2}$

$$=\frac{x}{(x+3)(x+2)}-\frac{2}{(x+2)(x+1)}$$
LCD $=(x+3)(x+2)(x+1)$

$$=\frac{x}{(x+3)(x+2)}\cdot\frac{x+1}{x+1}-\frac{2}{(x+2)(x+1)}\cdot\frac{x+3}{x+3}$$
$$=\frac{x^2+x}{(x+3)(x+2)(x+1)}-\frac{2x+6}{(x+3)(x+2)(x+1)}$$
$$=\frac{x^2+x-(2x+6)}{(x+3)(x+2)(x+1)}$$
$$=\frac{x^2+x-2x-6}{(x+3)(x+2)(x+1)}$$
$$=\frac{x^2-x-6}{(x+3)(x+2)(x+1)}$$
$$=\frac{(x-3)(x+2)}{(x+3)(x+2)(x+1)}$$
$$=\frac{(x-3)(\cancel{x+2})}{(x+3)(\cancel{x+2})(x+1)}$$
$$=\frac{x-3}{(x+3)(x+1)}$$

43. $\dfrac{3(2x+5)}{x-1}-\dfrac{3(2x-3)}{1-x}+\dfrac{6x+1}{x-1}$

$$=\frac{3(2x+5)}{x-1}-\frac{3(2x-3)}{1-x}\cdot\frac{-1}{-1}+\frac{6x-1}{x-1}$$
$$=\frac{3(2x+5)}{x-1}-\frac{-3(2x-3)}{x-1}+\frac{6x-1}{x-1}$$
$$=\frac{(6x+15)-(-6x+9)+(6x-1)}{x-1}$$
$$=\frac{6x+15+6x-9+6x-1}{x-1}$$
$$=\frac{18x+5}{x-1}$$

45. $\dfrac{x-y}{x^2-y^2}+\dfrac{x+y}{x^2-y^2}-\dfrac{2x}{x^2-y^2}$

$=\dfrac{x-y+x+y-2x}{x^2-y^2}$

$=\dfrac{0}{x^2-y^2}$

$=0$

47. $\dfrac{10}{2y-1}-\dfrac{6}{1-2y}+\dfrac{y}{2y-1}+\dfrac{y-4}{1-2y}$

$=\dfrac{10}{2y-1}-\dfrac{6}{1-2y}\cdot\dfrac{-1}{-1}+\dfrac{y}{2y-1}+\dfrac{y-4}{1-2y}\cdot\dfrac{-1}{-1}$

$=\dfrac{10}{2y-1}-\dfrac{-6}{2y-1}+\dfrac{y}{2y-1}+\dfrac{4-y}{2y-1}$

$=\dfrac{10-(-6)+y+4-y}{2y-1}$

$=\dfrac{10+6+y+4-y}{2y-1}$

$=\dfrac{20}{2y-1}$

49. $\dfrac{a+6}{4-a^2}-\dfrac{a+3}{a+2}+\dfrac{a-3}{2-a}$

$=\dfrac{a+6}{(2+a)(2-a)}-\dfrac{a+3}{2+a}+\dfrac{a-3}{2-a}$

$a+2=2+a$; LCD $=(2+a)(2-a)$

$=\dfrac{a+6}{(2+a)(2-a)}-\dfrac{a+3}{2+a}\cdot\dfrac{2-a}{2-a}+\dfrac{a-3}{2-a}\cdot\dfrac{2+a}{2+a}$

$=\dfrac{(a+6)-(a+3)(2-a)+(a-3)(2+a)}{(2+a)(2-a)}$

$=\dfrac{a+6-(-a^2-a+6)+(a^2-a-6)}{(2+a)(2-a)}$

$=\dfrac{a+6+a^2+a-6+a^2-a-6}{(2+a)(2-a)}$

$=\dfrac{2a^2+a-6}{(2+a)(2-a)}$

$=\dfrac{(2a-3)(a+2)}{(2+a)(2-a)}$

$=\dfrac{(2a-3)\cancel{(2+a)}}{\cancel{(2+a)}(2-a)}$

$=\dfrac{2a-3}{2-a}$

51. $\dfrac{2z}{1-2z}+\dfrac{3z}{2z+1}-\dfrac{3}{4z^2-1}$

$=\dfrac{2z}{1-2z}\cdot\dfrac{-1}{-1}+\dfrac{3z}{2z+1}-\dfrac{3}{4z^2-1}$

$=\dfrac{-2z}{2z-1}+\dfrac{3z}{2z+1}-\dfrac{3}{(2z-1)(2z+1)}$

LCD $=(2z-1)(2z+1)$

$=\dfrac{-2z}{2z-1}\cdot\dfrac{2z+1}{2z+1}+\dfrac{3z}{2z+1}\cdot\dfrac{2z-1}{2z-1}-\dfrac{3}{(2z-1)(2z+1)}$

$=\dfrac{(-4z^2-2z)+(6z^2-3z)-3}{(2z-1)(2z+1)}$

$=\dfrac{2z^2-5z-3}{(2z-1)(2z+1)}$

$=\dfrac{(z-3)(2z+1)}{(2z-1)(2z+1)}$

$=\dfrac{(z-3)\cancel{(2z+1)}}{(2z-1)\cancel{(2z+1)}}$

$=\dfrac{z-3}{2z-1}$

53. $\dfrac{1}{x+y}-\dfrac{1}{x-y}+\dfrac{2x}{x^2-y^2}$

$=\dfrac{1}{x+y}-\dfrac{1}{x-y}+\dfrac{2x}{(x+y)(x-y)}$

LCD $=(x+y)(x-y)$

$=\dfrac{1}{x+y}\cdot\dfrac{x-y}{x-y}-\dfrac{1}{x-y}\cdot\dfrac{x+y}{x+y}+\dfrac{2x}{(x+y)(x-y)}$

$=\dfrac{x-y-(x+y)+2x}{(x+y)(x-y)}$

$=\dfrac{x-y-x-y+2x}{(x+y)(x-y)}$

$=\dfrac{2x-2y}{(x+y)(x-y)}$

$=\dfrac{2(x-y)}{(x+y)(x-y)}$

$=\dfrac{2\cancel{(x-y)}}{(x+y)\cancel{(x-y)}}$

$=\dfrac{2}{x+y}$

55. $\dfrac{x^8}{x^3}=x^{8-3}=x^5$

57. $(a^2b^{-5})^{-4}=a^{2(-4)}b^{-5(-4)}=a^{-8}b^{20}=\dfrac{b^{20}}{a^8}$

59. $\dfrac{66x^2}{11x^5}=\dfrac{6\cdot\cancel{11}\cdot\cancel{x^2}}{\cancel{11}\cdot\cancel{x^2}\cdot x^3}=\dfrac{6}{x^3}$

61. The shaded area has dimensions $x-6$ by $x-3$. Then the area is $(x-6)(x-3)$, or $x^2-9x+18$.

63. ◈

65. $\dfrac{2x+11}{x-3}\cdot\dfrac{3}{x+4}+\dfrac{2x+1}{4+x}\cdot\dfrac{3}{3-x}$

$$= \frac{6x+33}{(x-3)(x+4)}+\frac{6x+3}{(4+x)(3-x)}$$
$$= \frac{6x+33}{(x-3)(x+4)}+\frac{6x+3}{(4+x)(3-x)}\cdot\frac{-1}{-1}$$
$$= \frac{6x+33}{(x-3)(x+4)}+\frac{-6x-3}{(x+4)(x-3)}$$
$$= \frac{6x+33-6x-3}{(x-3)(x+4)}$$
$$= \frac{30}{(x-3)(x+4)}$$

67. $\dfrac{x}{x^4-y^4}-\left(\dfrac{1}{x+y}\right)^2$

$$= \frac{x}{(x^2+y^2)(x+y)(x-y)}-\frac{1}{(x+y)^2}$$
$$\text{LCD } = (x^2+y^2)(x+y)^2(x-y)$$
$$= \frac{x}{(x^2+y^2)(x+y)(x-y)}\cdot\frac{x+y}{x+y}-\frac{1}{(x+y)^2}\cdot\frac{(x^2+y^2)(x-y)}{(x^2+y^2)(x-y)}$$
$$= \frac{x(x+y)-(x^2+y^2)(x-y)}{(x^2+y^2)(x+y)^2(x-y)}$$
$$= \frac{x^2+xy-(x^3-x^2y+xy^2-y^3)}{(x^2+y^2)(x+y)^2(x-y)}$$
$$= \frac{x^2+xy-x^3+x^2y-xy^2+y^3}{(x^2+y^2)(x+y)^2(x-y)}$$

69. Let $l =$ the length of the missing side.

$$\frac{a^2-5a-9}{a-6}+\frac{a^2-6}{a-6}+l = 2a+5$$
$$\frac{2a^2-5a-15}{a-6}+l = 2a+5$$
$$l = 2a+5-\frac{2a^2-5a-15}{a-6}$$
$$l = (2a+5)\cdot\frac{a-6}{a-6}-\frac{2a^2-5a-15}{a-6}$$
$$l = \frac{2a^2-7a-30}{a-6}-\frac{2a^2-5a-15}{a-6}$$
$$l = \frac{2a^2-7a-30-(2a^2-5a-15)}{a-6}$$
$$l = \frac{2a^2-7a-30-2a^2+5a+15}{a-6}$$
$$l = \frac{-2a-15}{a-6}$$

The length of the missing side is $\dfrac{-2a-15}{a-6}$.

Now find the area.

$$A = \frac{1}{2}\cdot b\cdot h$$
$$A = \frac{1}{2}\left(\frac{-2a-15}{a-6}\right)\left(\frac{a^2-6}{a-6}\right)$$
$$A = \frac{(-2a-15)(a^2-6)}{2(a-6)^2}, \text{ or}$$
$$A = \frac{-2a^3-15a^2+12a+90}{2a^2-24a+72}$$

71. Enter $y_1 = \dfrac{3}{2t^2-2t}-\dfrac{5}{2t-2}$ and $y_2 = \dfrac{3-5t}{2t(t-1)}$ and look at a table of values. If corresponding y_1- and y_2-values are the same, the subtraction is correct.

73. Enter $y_1 = \dfrac{2s}{t^2-s^2}-\dfrac{s}{t-s}$ and $y_2 = \dfrac{2s-st-s^2}{(t-s)(t+s)}$ and look at a table of values. If corresponding y_1- and y_2-values are the same, the subtraction is correct.

75. Enter $y_1 = \dfrac{y-5}{y}-\dfrac{3y-1}{4y}$ and $y_2 = \dfrac{y-19}{4y}$ and look at a table of values. If corresponding y_1- and y_2-values are the same, the subtraction is correct.

Exercise Set 12.6

1. $\dfrac{4}{5}-\dfrac{2}{3}=\dfrac{x}{9}$, LCM = 45

$$45\left(\frac{4}{5}-\frac{2}{3}\right) = 45\cdot\frac{x}{9}$$
$$45\cdot\frac{4}{5}-45\cdot\frac{2}{3} = 45\cdot\frac{x}{9}$$
$$36-30 = 5x$$
$$6 = 5x$$
$$\frac{6}{5} = x$$

Check:

$\frac{4}{5}-\frac{2}{3}$	$=\ \frac{x}{9}$
$\frac{4}{5}-\frac{2}{3}$?	$\frac{\frac{6}{5}}{9}$
$\frac{12}{15}-\frac{10}{15}$	$\frac{6}{5}\cdot\frac{1}{9}$
$\frac{2}{15}$	$\frac{2}{15}$ TRUE

This checks, so the solution is $\dfrac{6}{5}$.

3. $\dfrac{3}{5}+\dfrac{1}{8}=\dfrac{1}{x}$, LCM = $40x$

$$40x\left(\frac{3}{5}+\frac{1}{8}\right) = 40x\cdot\frac{1}{x}$$
$$40x\cdot\frac{3}{5}+40x\cdot\frac{1}{8} = 40x\cdot\frac{1}{x}$$
$$24x+5x = 40$$
$$29x = 40$$
$$x = \frac{40}{29}$$

Check:

$$\begin{array}{c|c}
\multicolumn{2}{c}{\frac{3}{5}+\frac{1}{8}=\frac{1}{x}} \\ \hline
\frac{3}{5}+\frac{1}{8} \;?\; & \frac{1}{\frac{40}{29}} \\
\frac{24}{40}+\frac{5}{40} & 1\cdot\frac{29}{40} \\
\frac{29}{40} & \frac{29}{40} \quad \text{TRUE}
\end{array}$$

This checks, so the solution is $\frac{40}{29}$.

5.

$$\frac{3}{8}+\frac{4}{5}=\frac{x}{20}, \text{ LCM} = 40$$

$$40\left(\frac{3}{8}+\frac{4}{5}\right)=40\cdot\frac{x}{20}$$

$$40\cdot\frac{3}{8}+40\cdot\frac{4}{5}=40\cdot\frac{x}{20}$$

$$15+32=2x$$

$$47=2x$$

$$\frac{47}{2}=x$$

Check:

$$\begin{array}{c|c}
\multicolumn{2}{c}{\frac{3}{8}+\frac{4}{5}=\frac{x}{20}} \\ \hline
\frac{3}{8}+\frac{4}{5} \;?\; & \frac{\frac{47}{2}}{20} \\
\frac{15}{40}+\frac{32}{40} & \frac{47}{2}\cdot\frac{1}{20} \\
\frac{47}{40} & \frac{47}{40} \quad \text{TRUE}
\end{array}$$

This checks, so the solution is $\frac{47}{2}$.

7.

$$\frac{1}{x}=\frac{2}{3}-\frac{5}{6}, \text{ LCM} = 6x$$

$$6x\cdot\frac{1}{x}=6x\left(\frac{2}{3}-\frac{5}{6}\right)$$

$$6x\cdot\frac{1}{x}=6x\cdot\frac{2}{3}-6x\cdot\frac{5}{6}$$

$$6=4x-5x$$

$$6=-x$$

$$-6=x$$

Check:

$$\begin{array}{c|c}
\multicolumn{2}{c}{\frac{1}{x}=\frac{2}{3}-\frac{5}{6}} \\ \hline
\frac{1}{-6} \;?\; & \frac{2}{3}-\frac{5}{6} \\
-\frac{1}{6} & \frac{4}{6}-\frac{5}{6} \\
 & -\frac{1}{6} \quad \text{TRUE}
\end{array}$$

This checks, so the solution is -6.

9.

$$\frac{1}{6}+\frac{1}{8}=\frac{1}{t}, \text{ LCM} = 24t$$

$$24t\left(\frac{1}{6}+\frac{1}{8}\right)=24t\cdot\frac{1}{t}$$

$$24t\cdot\frac{1}{6}+24t\cdot\frac{1}{8}=24t\cdot\frac{1}{t}$$

$$4t+3t=24$$

$$7t=24$$

$$t=\frac{24}{7}$$

Check:

$$\begin{array}{c|c}
\multicolumn{2}{c}{\frac{1}{6}+\frac{1}{8}=\frac{1}{t}} \\ \hline
\frac{1}{6}+\frac{1}{8} \;?\; & \frac{1}{24/7} \\
\frac{4}{24}+\frac{3}{24} & 1\cdot\frac{7}{24} \\
\frac{7}{24} & \frac{7}{24} \quad \text{TRUE}
\end{array}$$

This checks, so the solution is $\frac{24}{7}$.

11.

$$x+\frac{4}{x}=-5, \text{ LCM} = x$$

$$x\left(x+\frac{4}{x}\right)=x(-5)$$

$$x\cdot x+x\cdot\frac{4}{x}=x(-5)$$

$$x^2+4=-5x$$

$$x^2+5x+4=0$$

$$(x+4)(x+1)=0$$

$$x+4=0 \quad\text{or}\quad x+1=0$$

$$x=-4 \quad\text{or}\quad x=-1$$

Check:

$$\begin{array}{c|c}
\multicolumn{2}{c}{x+\frac{4}{x}=-5} \\ \hline
-4+\frac{4}{-4} \;?\; & -5 \\
-4-1 & \\
-5 & \quad \text{TRUE}
\end{array}
\qquad
\begin{array}{c|c}
\multicolumn{2}{c}{x+\frac{4}{x}=-5} \\ \hline
-1+\frac{4}{-1} \;?\; & -5 \\
-1-4 & \\
-5 & \quad \text{TRUE}
\end{array}$$

Both of these check, so the two solutions are -4 and -1.

13.

$$\frac{x}{4}-\frac{4}{x}=0, \text{ LCM} = 4x$$

$$4x\left(\frac{x}{4}-\frac{4}{x}\right)=4x\cdot 0$$

$$4x\cdot\frac{x}{4}-4x\cdot\frac{4}{x}=4x\cdot 0$$

$$x^2-16=0$$

$$(x+4)(x-4)=0$$

$$x+4=0 \quad\text{or}\quad x-4=0$$

$$x=-4 \quad\text{or}\quad x=4$$

Check:

$$\begin{array}{c|c} \frac{x}{4}-\frac{4}{x}=0 & \\ \hline \frac{-4}{4}-\frac{4}{-4} & ?\ 0 \\ -1-(-1) & \\ -1+1 & \\ 0 & \text{TRUE} \end{array} \qquad \begin{array}{c|c} \frac{x}{4}-\frac{4}{x}=0 & \\ \hline \frac{4}{4}-\frac{4}{4} & ?\ 0 \\ 1-1 & \\ 0 & \text{TRUE} \end{array}$$

Both of these check, so the two solutions are -4 and 4.

15.
$$\frac{5}{x}=\frac{6}{x}-\frac{1}{3},\ \text{LCM}=3x$$
$$3x\cdot\frac{5}{x}=3x\left(\frac{6}{x}-\frac{1}{3}\right)$$
$$3x\cdot\frac{5}{x}=3x\cdot\frac{6}{x}-3x\cdot\frac{1}{3}$$
$$15=18-x$$
$$-3=-x$$
$$3=x$$

Check:

$$\begin{array}{c|c} \frac{5}{x} & \frac{6}{x}-\frac{1}{3} \\ \hline \frac{5}{3} & ?\ \frac{6}{3}-\frac{1}{3} \\ & \frac{5}{3} \quad \text{TRUE} \end{array}$$

This checks, so the solution is 3.

17.
$$\frac{5}{3x}+\frac{3}{x}=1,\ \text{LCM}=3x$$
$$3x\left(\frac{5}{3x}+\frac{3}{x}\right)=3x\cdot 1$$
$$3x\cdot\frac{5}{3x}+3x\cdot\frac{3}{x}=3x\cdot 1$$
$$5+9=3x$$
$$14=3x$$
$$\frac{14}{3}=x$$

Check:

$$\begin{array}{c|c} \frac{5}{3x}+\frac{3}{x}=1 & \\ \hline \frac{5}{3\cdot(14/3)}+\frac{3}{(14/3)} & ?\ 1 \\ \frac{5}{14}+\frac{9}{14} & \\ \frac{14}{14} & \\ 1 & \text{TRUE} \end{array}$$

This checks, so the solution is $\frac{14}{3}$.

19.
$$\frac{t-2}{t+3}=\frac{3}{8},\ \text{LCM}=8(t+3)$$
$$8(t+3)\left(\frac{t-2}{t+3}\right)=8(t+3)\left(\frac{3}{8}\right)$$
$$8(t-2)=3(t+3)$$
$$8t-16=3t+9$$
$$5t=25$$
$$t=5$$

Check:

$$\begin{array}{c|c} \frac{t-2}{t+3} & \frac{3}{8} \\ \hline \frac{5-2}{5+3} & ?\ \frac{3}{8} \\ \frac{3}{8} & \quad \text{TRUE} \end{array}$$

This checks, so the solution is 5.

21.
$$\frac{2}{x+1}=\frac{1}{x-2},\ \text{LCM}=(x+1)(x-2)$$
$$(x+1)(x-2)\cdot\frac{2}{x+1}=(x+1)(x-2)\cdot\frac{1}{x-2}$$
$$2(x-2)=x+1$$
$$2x-4=x+1$$
$$x=5$$

This checks, so the solution is 5.

23.
$$\frac{x}{6}-\frac{x}{10}=\frac{1}{6},\ \text{LCM}=30$$
$$30\left(\frac{x}{6}-\frac{x}{10}\right)=30\cdot\frac{1}{6}$$
$$30\cdot\frac{x}{6}-30\cdot\frac{x}{10}=30\cdot\frac{1}{6}$$
$$5x-3x=5$$
$$2x=5$$
$$x=\frac{5}{2}$$

This checks, so the solution is $\frac{5}{2}$.

25.
$$\frac{t+2}{5}-\frac{t-2}{4}=1,\ \text{LCM}=20$$
$$20\left(\frac{t+2}{5}-\frac{t-2}{4}\right)=20\cdot 1$$
$$20\left(\frac{t+2}{5}\right)-20\left(\frac{t-2}{4}\right)=20\cdot 1$$
$$4(t+2)-5(t-2)=20$$
$$4t+8-5t+10=20$$
$$-t+18=20$$
$$-t=2$$
$$t=-2$$

This checks, so the solution is -2.

27.
$$\frac{5}{x-1} = \frac{3}{x+2},$$
$$\text{LCD} = (x-1)(x+2)$$
$$(x-1)(x+2)\cdot\frac{5}{x-1} = (x-1)(x+2)\cdot\frac{3}{x+2}$$
$$5(x+2) = 3(x-1)$$
$$5x+10 = 3x-3$$
$$2x = -13$$
$$x = -\frac{13}{2}$$

This checks, so the solution is $-\frac{13}{2}$.

29.
$$\frac{a-3}{3a+2} = \frac{1}{5},\ \text{LCM} = 5(3a+2)$$
$$5(3a+2)\cdot\frac{a-3}{3a+2} = 5(3a+2)\cdot\frac{1}{5}$$
$$5(a-3) = 3a+2$$
$$5a-15 = 3a+2$$
$$2a = 17$$
$$a = \frac{17}{2}$$

This checks, so the solution is $\frac{17}{2}$.

31.
$$\frac{x-1}{x-5} = \frac{4}{x-5},\ \text{LCM} = x-5$$
$$(x-5)\cdot\frac{x-1}{x-5} = (x-5)\cdot\frac{4}{x-5}$$
$$x-1 = 4$$
$$x = 5$$

The number 5 is not a solution because it makes a denominator zero. Thus, there is no solution.

33.
$$\frac{2}{x+3} = \frac{5}{x},\ \text{LCM} = x(x+3)$$
$$x(x+3)\cdot\frac{2}{x+3} = x(x+3)\cdot\frac{5}{x}$$
$$2x = 5(x+3)$$
$$2x = 5x+15$$
$$-15 = 3x$$
$$-5 = x$$

This checks, so the solution is -5.

35.
$$\frac{x-2}{x-3} = \frac{x-1}{x+1},\ \text{LCM} = (x-3)(x+1)$$
$$(x-3)(x+1)\cdot\frac{x-2}{x-3} = (x-3)(x+1)\cdot\frac{x-1}{x+1}$$
$$(x+1)(x-2) = (x-3)(x-1)$$
$$x^2-x-2 = x^2-4x+3$$
$$-x-2 = -4x+3$$
$$3x = 5$$
$$x = \frac{5}{3}$$

This checks, so the solution is $\frac{5}{3}$.

37.
$$\frac{1}{x+3} + \frac{1}{x-3} = \frac{1}{x^2-9},$$
$$\text{LCM} = (x+3)(x-3)$$
$$(x+3)(x-3)\left(\frac{1}{x+3}+\frac{1}{x-3}\right) = (x+3)(x-3)\cdot\frac{1}{(x+3)(x-3)}$$
$$(x-3)+(x+3) = 1$$
$$2x = 1$$
$$x = \frac{1}{2}$$

This checks, so the solution is $\frac{1}{2}$.

39.
$$\frac{x}{x+4} - \frac{4}{x-4} = \frac{x^2+16}{x^2-16},$$
$$\text{LCM} = (x+4)(x-4)$$
$$(x+4)(x-4)\left(\frac{x}{x+4}-\frac{x}{x-4}\right) = (x+4)(x-4)\cdot\frac{x^2+16}{(x+4)(x-4)}$$
$$x(x-4)-4(x+4) = x^2+16$$
$$x^2-4x-4x-16 = x^2+16$$
$$x^2-8x-16 = x^2+16$$
$$-8x-16 = 16$$
$$-8x = 32$$
$$x = -4$$

The number -4 is not a solution because it makes a denominator zero. Thus, there is no solution.

41.
$$\frac{4-a}{8-a} = \frac{4}{a-8}$$ $8-a$ and $a-8$ are opposites
$$\frac{4-a}{8-a}\cdot\frac{-1}{-1} = \frac{4}{a-8}$$
$$\frac{a-4}{a-8} = \frac{4}{a-8},\ \text{LCM} = a-8$$
$$(a-8)\left(\frac{a-4}{a-8}\right) = (a-8)\left(\frac{4}{a-8}\right)$$
$$a-4 = 4$$
$$a = 8$$

The number 8 is not a solution because it makes a denominator zero. Thus, there is no solution.

43.
$$2-\frac{a-2}{a+3}=\frac{a^2-4}{a+3},\ \text{LCM}=a+3$$
$$(a+3)\left(2-\frac{a-2}{a+3}\right)=(a+3)\cdot\frac{a^2-4}{a+3}$$
$$2(a+3)-(a-2)=a^2-4$$
$$2a+6-a+2=a^2-4$$
$$0=a^2-a-12$$
$$0=(a-4)(a+3)$$
$$a-4=0 \text{ or } a+3=0$$
$$a=4 \text{ or } a=-3$$
Only 4 checks, so the solution is 4.

45. $(a^2b^5)^{-3}=\dfrac{1}{(a^2b^5)^3}=\dfrac{1}{(a^2)^3(b^5)^3}=\dfrac{1}{a^6b^{15}}$

47. $\left(\dfrac{2x}{t^2}\right)^4=\dfrac{(2x)^4}{(t^2)^4}=\dfrac{2^4x^4}{t^8}=\dfrac{16x^4}{t^8}$

49. $4x^{-5}\cdot 8x^{11}=4\cdot 8x^{-5+11}=32x^6$

51. $5x+10y=20$

To find the x-intercept, let $y=0$. Then solve for x.
$$5x+10y=20$$
$$5x+10\cdot 0=20$$
$$5x=20$$
$$x=4$$
The x-intercept is $(4,0)$.

To find the y-intercept, let $x=0$. Then solve for y.
$$5x+10y=20$$
$$5\cdot 0+10y=20$$
$$10y=20$$
$$y=2$$
The y-intercept is $(0,2)$.

Plot these points and draw the line.

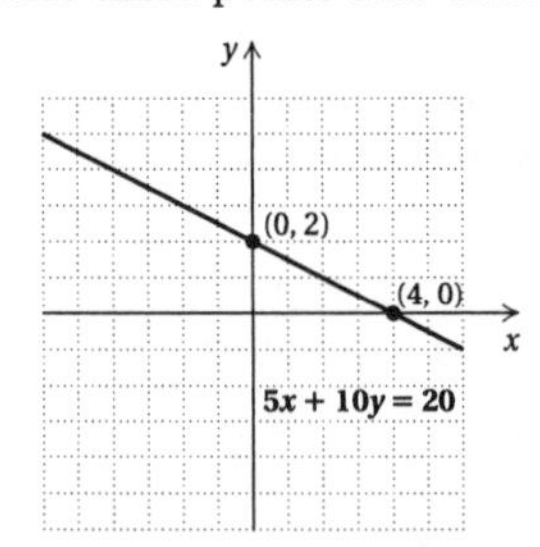

A third point should be used as a check. We substitute any value for x and solve for y. We let $x=-4$. Then
$$5x+10y=20$$
$$5(-4)+10y=20$$
$$-20+10y=20$$
$$10y=40$$
$$y=4.$$
The point $(-4,4)$ is on the graph, so the graph is probably correct.

53. $10y-4x=-20$

To find the x-intercept, let $y=0$. Then solve for x.
$$10y-4x=-20$$
$$10\cdot 0-4x=-20$$
$$-4x=-20$$
$$x=5$$
The x-intercept is $(5,0)$.

To find the y-intercept, let $x=0$. Then solve for y.
$$10y-4x=-20$$
$$10y-4\cdot 0=-20$$
$$10y=-20$$
$$y=-2$$
The y-intercept is $(0,-2)$.

Plot these points and draw the line.

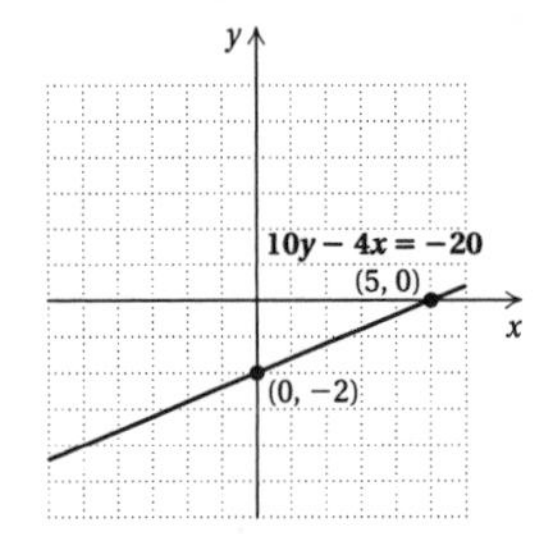

A third point should be used as a check. We substitute any value for x and solve for y. We let $x=-5$. Then
$$10y-4x=-20$$
$$10y-4(-5)=-20$$
$$10y+20=-20$$
$$10y=-40$$
$$y=-4.$$
The point $(-5,-4)$ is on the graph, so the graph is probably correct.

55. ◈

57.
$$\frac{4}{y-2}-\frac{2y-3}{y^2-4}=\frac{5}{y+2},$$
$$\text{LCM}=(y+2)(y-2)$$
$$(y+2)(y-2)\left(\frac{4}{y-2}-\frac{2y-3}{(y+2)(y-2)}\right)=(y+2)(y-2)\cdot\frac{5}{y+2}$$
$$4(y+2)-(2y-3)=5(y-2)$$
$$4y+8-2y+3=5y-10$$
$$2y+11=5y-10$$
$$21=3y$$
$$7=y$$
This checks, so the solution is 7.

59. $$\frac{x+1}{x+2} = \frac{x+3}{x+4},$$
LCM $= (x+2)(x+4)$

$$(x+2)(x+4)\left(\frac{x+1}{x+2}\right) = (x+2)(x+4)\left(\frac{x+3}{x+4}\right)$$
$$(x+4)(x+1) = (x+2)(x+3)$$
$$x^2+5x+4 = x^2+5x+6$$
$$4 = 6 \quad \text{Subtracting } x^2 \text{ and } 5x$$

We get a false equation, so the original equation has no solution.

61. $$4a-3 = \frac{a+13}{a+1}, \text{ LCM} = a+1$$
$$(a+1)(4a-3) = (a+1)\cdot\frac{a+13}{a+1}$$
$$4a^2+a-3 = a+13$$
$$4a^2-16 = 0$$
$$4(a+2)(a-2) = 0$$
$$a+2=0 \quad \text{or} \quad a-2=0$$
$$a=-2 \quad \text{or} \quad a=2$$

Both of these check, so the two solutions are -2 and 2.

63. $$\frac{y^2-4}{y+3} = 2-\frac{y-2}{y+3}, \text{ LCM} = y+3$$
$$(y+3)\cdot\frac{y^2-4}{y+3} = (y+3)\left(2-\frac{y-2}{y+3}\right)$$
$$y^2-4 = 2(y+3)-(y-2)$$
$$y^2-4 = 2y+6-y+2$$
$$y^2-4 = y+8$$
$$y^2-y-12 = 0$$
$$(y-4)(y+3) = 0$$
$$y-4=0 \quad \text{or} \quad y+3=0$$
$$y=4 \quad \text{or} \quad y=-3$$

The number 4 is a solution, but -3 is not because it makes a denominator zero.

65. 

Exercise Set 12.7

1. ***Familiarize***. Let x = the number. Then $\frac{1}{x}$ is the reciprocal of the number.

Translate.

The reciprocal of 6 plus the reciprocal of 8 is the reciprocal of the number.

$$\frac{1}{6} + \frac{1}{8} = \frac{1}{x}$$

Solve. We solve the equation.

$$\frac{1}{6}+\frac{1}{8} = \frac{1}{x}, \text{ LCM} = 24x$$
$$24x\left(\frac{1}{6}+\frac{1}{8}\right) = 24x\cdot\frac{1}{x}$$
$$24x\cdot\frac{1}{6}+24x\cdot\frac{1}{8} = 24x\cdot\frac{1}{x}$$
$$4x+3x = 24$$
$$7x = 24$$
$$x = \frac{24}{7}$$

Check. The reciprocal of $\frac{24}{7}$ is $\frac{7}{24}$. Also, $\frac{1}{6}+\frac{1}{8} = \frac{4}{24}+\frac{3}{24} = \frac{7}{24}$, so the value checks.

State. The number is $\frac{24}{7}$.

3. ***Familiarize***. Let x = the smaller number. Then $x+5$ = the larger number.

Translate.

The larger number divided by the smaller number is $\frac{4}{3}$.

$$(x+5) \div x = \frac{4}{3}$$

Solve. We solve the equation.

$$\frac{x+5}{x} = \frac{4}{3}, \text{ LCM} = 3x$$
$$3x\left(\frac{x+5}{x}\right) = 3x\cdot\frac{4}{3}$$
$$3(x+5) = 4x$$
$$3x+15 = 4x$$
$$15 = x$$

Check. If the smaller number is 15, then the larger is $15+5$, or 20. The quotient of 20 divided by 15 is $\frac{20}{15}$, or $\frac{4}{3}$. The values check.

State. The numbers are 15 and 20.

5. ***Familiarize***. We complete the table shown in the text.

$$d = r \cdot t$$

	Distance	Speed	Time	
Car	150	r	t	$\rightarrow 150 = r(t)$
Truck	350	$r+40$	t	$\rightarrow 350 = (r+40)t$

Translate. We apply the formula $d = rt$ along the rows of the table to obtain two equations:

$$150 = rt,$$
$$350 = (r+40)t$$

Then we solve each equation for t and set the results equal:

Solving $150 = rt$ for t: $t = \frac{150}{r}$

Solving $350 = (r+40)t$ for t: $t = \frac{350}{r+40}$

Thus, we have

$$\frac{150}{r} = \frac{350}{r+40}.$$

Solve. We multiply by the LCM, $r(r+40)$.

$$r(r+40)\cdot\frac{150}{r} = r(r+40)\cdot\frac{350}{r+40}$$
$$150(r+40) = 350r$$
$$150r + 6000 = 350r$$
$$6000 = 200r$$
$$30 = r$$

Check. If r is 30 km/h, then $r+40$ is 70 km/h. The time for the car is 150/30, or 5 hr. The time for the truck is 350/70, or 5 hr. The times are the same. The values check.

State. The speed of the car is 30 km/h, and the speed of the truck is 70 km/h.

7. ***Familiarize.*** We complete the table shown in the text.

$$d \quad = \quad r \quad \cdot \quad t$$

	Distance	Speed	Time
Freight	330	$r-14$	t
Passenger	400	r	t

Translate. From the rows of the table we have two equations:

$$330 = (r-14)t,$$
$$400 = rt$$

We solve each equation for t and set the results equal:

Solving $330 = (r-14)t$ for t: $t = \frac{330}{r-14}$

Solving $400 = rt$ for t: $t = \frac{400}{r}$

Thus, we have

$$\frac{330}{r-14} = \frac{400}{r}.$$

Solve. We multiply by the LCM, $r(r-14)$.

$$r(r-14)\cdot\frac{330}{r-14} = r(r-14)\cdot\frac{400}{r}$$
$$330r = 400(r-14)$$
$$330r = 400r - 5600$$
$$-70r = -5600$$
$$r = 80$$

Then substitute 80 for r in either equation to find t:

$$t = \frac{400}{r}$$
$$t = \frac{400}{80} \quad \text{Substituting 80 for } r$$
$$t = 5$$

Check. If $r = 80$, then $r - 14 = 66$. In 5 hr the freight train travels $66\cdot 5$, or 330 mi, and the passenger train travels $80\cdot 5$, or 400 mi. The values check.

State. The speed of the passenger train is 80 mph. The speed of the freight train is 66 mph.

9. ***Familiarize.*** We let r represent the speed going. Then $2r$ is the speed returning. We let t represent the time going. Then $t-3$ represents the time returning. We organize the information in a table.

$$d \quad = \quad r \quad \cdot \quad t$$

	Distance	Speed	Time
Going	120	r	t
Returning	120	$2r$	$t-3$

Translate. The rows of the table give us two equations:

$$120 = rt,$$
$$120 = 2r(t-3)$$

We can solve each equation for r and set the results equal:

Solving $120 = rt$ for r: $r = \frac{120}{t}$

Solving $120 = 2r(t-3)$ for r: $r = \frac{120}{2(t-3)}$, or

$$r = \frac{60}{t-3}$$

Then $\frac{120}{t} = \frac{60}{t-3}$.

Solve. We multiply on both sides by the LCM, $t(t-3)$.

$$t(t-3)\cdot\frac{120}{t} = t(t-3)\cdot\frac{60}{t-3}$$
$$120(t-3) = 60t$$
$$120t - 360 = 60t$$
$$-360 = -60t$$
$$6 = t$$

Then substitute 6 for t in either equation to find r, the speed going:

$$r = \frac{120}{t}$$
$$r = \frac{120}{6} \quad \text{Substituting 6 for } t$$
$$r = 20$$

Check. If $r = 20$ and $t = 6$, then $2r = 2\cdot 20$, or 40 mph and $t - 3 = 6 - 3$, or 3 hr. The distance going is $6\cdot 20$, or 120 mi. The distance returning is $40\cdot 3$, or 120 mi. The numbers check.

State. The speed going is 20 mph.

11. ***Familiarize.*** It takes the Brother machine 10 min working alone and the Xerox machine 8 min working alone. Then in 1 min the Brother machine does $\frac{1}{10}$ of the job, and the Xerox machine does $\frac{1}{8}$ of the job. Working together they can do $\frac{1}{10}+\frac{1}{8}$, or $\frac{9}{40}$ of the job in 1 min. In 2 min the Brother machine does $2\left(\frac{1}{10}\right)$ of the job, and the Xerox machine does $2\left(\frac{1}{8}\right)$ of the job. Working together they can do $2\left(\frac{1}{10}\right)+2\left(\frac{1}{8}\right)$, or $\frac{9}{20}$ of the job in 2 min. Continuing this reasoning, we find that they do $4\left(\frac{1}{10}\right)+4\left(\frac{1}{8}\right)$, or $\frac{9}{10}$ of the job in 4 min working together and $5\left(\frac{1}{10}\right)+5\left(\frac{1}{8}\right)$, or $1\frac{1}{8}$ of the job in 5 min working together. Since $1\frac{1}{8}$ is more of the job than needs to be done, we see that the answer is somewhere between 4 min and 5 min.

Translate. Using the work principle, we see that we want some number t such that

$$\frac{t}{10}+\frac{t}{8}=1.$$

Solve. We solve the equation.

$$\frac{t}{10}+\frac{t}{8}=1,\ \text{LCM}=40$$
$$40\left(\frac{t}{10}+\frac{t}{8}\right)=40\cdot 1$$
$$4t+5t=40$$
$$9t=40$$
$$t=\frac{40}{9},\ \text{or}\ 4\frac{4}{9}$$

Check. Note that in $\frac{40}{9}$ min, the portion of the job done is $\frac{1}{10}\cdot\frac{40}{9}+\frac{1}{8}\cdot\frac{40}{9}=\frac{4}{9}+\frac{5}{9}=1$ job.

State. It would take $4\frac{4}{9}$ min for the two machines to fax the report, working together.

13. ***Familiarize***. The job takes Rory 12 hours working alone and Mira 9 hours working alone. Then in 1 hour Rory does $\frac{1}{12}$ of the job and Mira does $\frac{1}{9}$ of the job. Working together they can do $\frac{1}{12}+\frac{1}{9}$, or $\frac{7}{36}$ of the job in 1 hour. In two hours, Rory does $2\left(\frac{1}{12}\right)$ of the job and Mira does $2\left(\frac{1}{9}\right)$ of the job. Working together they can do $2\left(\frac{1}{12}\right)+2\left(\frac{1}{9}\right)$, or $\frac{7}{18}$ of the job in two hours. In 3 hours they can do $3\left(\frac{1}{12}\right)+3\left(\frac{1}{9}\right)$, or $\frac{7}{12}$ of the job. In 4 hours, they can do $\frac{7}{9}$, and in 5 hours they can do $\frac{35}{36}$ of the job. In 6 hours, they can do $\frac{7}{6}$, or $1\frac{1}{6}$ of the job which is more of the job than needs to be done. The answer is somewhere between 5 hr and 6 hr.

Translate. Using the work principle, we see that we want some number t such that

$$t\left(\frac{1}{12}\right)+t\left(\frac{1}{9}\right)=1.$$

Solve. We solve the equation.

$$\frac{t}{12}+\frac{t}{9}=1,\ \text{LCM=36}$$
$$36\left(\frac{t}{12}+\frac{t}{9}\right)=36\cdot 1$$
$$3t+4t=36$$
$$7t=36$$
$$t=\frac{36}{7},\ \text{or}\ 5\frac{1}{7}$$

Check. The check can be done by recalculating. We also have another check. In the familiarization step we learned the time must be between 5 hr and 6 hr. The answer, $5\frac{1}{7}$ hr, is between 5 hr and 6 hr and is less than 9 hours, the time it takes Mira alone.

State. Working together, it takes them $5\frac{1}{7}$ hr to fit the kitchen.

15. $\frac{54\text{ days}}{6\text{ days}}=9$

17. $\frac{4.6\text{ km}}{2\text{ hr}}=2.3\text{ km/h}$

19. ***Familiarize***. A 120-lb person should eat at least 44 g of protein each day, and we wish to find the minimum protein required for a 180-lb person. We can set up ratios. We let p = the minimum number of grams of protein a 180-lb person should eat each day.

Translate. If we assume the rates of protein intake are the same, the ratios are the same and we have an equation.

$$\begin{matrix}\text{Protein}\rightarrow\\ \text{Weight}\rightarrow\end{matrix}\frac{44}{120}=\frac{p}{180}\begin{matrix}\leftarrow\text{Protein}\\ \leftarrow\text{Weight}\end{matrix}$$

Solve. We solve the equation.

$$360\cdot\frac{44}{120}=360\cdot\frac{p}{180}\quad\text{Multiplying by the LCM, 360}$$
$$3\cdot 44=2\cdot p$$
$$132=2p$$
$$66=p$$

Check. $\frac{44}{120}=\frac{4\cdot 11}{4\cdot 30}=\frac{\cancel{4}\cdot 11}{\cancel{4}\cdot 30}=\frac{11}{30}$ and

$\frac{66}{180}=\frac{6\cdot 11}{6\cdot 30}=\frac{\cancel{6}\cdot 11}{\cancel{6}\cdot 30}=\frac{11}{30}$. The ratios are the same.

State. A 180-lb person should eat a minimum of 66 g of protein each day.

21. ***Familiarize***. A student travels 234 kilometers in 14 days, and we wish to find how far the student would travel in 42 days. We can set up ratios. We let K = the number of kilometers the student would travel in 42 days.

Translate. Assuming the rates are the same, we can translate to a proportion.

$$\begin{matrix}\text{Kilometers}\rightarrow\\ \text{Days}\rightarrow\end{matrix}\frac{K}{42}=\frac{234}{14}\begin{matrix}\leftarrow\text{Kilometers}\\ \leftarrow\text{Days}\end{matrix}$$

Solve. We solve the equation. We multiply by 42 to get K alone.

$$42 \cdot \frac{K}{42} = 42 \cdot \frac{234}{14}$$
$$K = \frac{9828}{14}$$
$$K = 702$$

Check.

$$\frac{702}{42} \approx 16.7 \text{ and } \frac{234}{14} \approx 16.7.$$

The ratios are the same.

State. The student would travel 702 kilometers in 42 days.

23. ***Familiarize***. A sample of 144 firecrackers contained 9 duds, and we wish to find how many duds could be expected in a sample of 3200 firecrackers. We can set up ratios, letting $d =$ the number of duds expected in a sample of 3200 firecrackers.

Translate. Assuming the rates of occurrence of duds are the same, we can translate to a proportion.

$$\begin{array}{r} \text{Duds} \rightarrow \\ \text{Sample size} \rightarrow \end{array} \frac{9}{144} = \frac{d}{3200} \begin{array}{l} \leftarrow \text{Duds} \\ \leftarrow \text{Sample size} \end{array}$$

Solve. We solve the equation. We multiply by 3200 to get d alone.

$$3200 \cdot \frac{9}{144} = 3200 \cdot \frac{d}{3200}$$
$$\frac{28,800}{144} = d$$
$$200 = d$$

Check.

$$\frac{9}{144} = 0.0625 \text{ and } \frac{200}{3200} = 0.0625$$

The ratios are the same.

State. You would expect 200 duds in a sample of 3200 firecrackers.

25. a) ***Familiarize***. We assume that Martinez's rate of hitting 17 home runs in 44 games will continue for the 162 game season. We let $H =$ the number of home runs Martinez can hit in 162 games.

Translate. Since we assume the rates are the same, we can translate to a proportion.

$$\begin{array}{r} \text{Number of} \\ \text{home runs} \longrightarrow \\ \text{Number of} \longrightarrow \\ \text{games} \end{array} \frac{17}{44} = \frac{H}{162} \begin{array}{l} \text{Number of} \\ \longleftarrow \text{home runs} \\ \longleftarrow \text{Number of} \\ \text{games} \end{array}$$

Solve. We solve the equation.

$$\frac{17}{44} = \frac{H}{162}$$
$$17 \cdot 162 = 44H \quad \text{Equating cross products}$$
$$\frac{17 \cdot 162}{44} = \frac{44H}{44} \quad \text{Dividing by 44}$$
$$62.59 \approx H$$

Check.

$$\frac{17}{44} \approx 0.386 \text{ and } \frac{62.59}{162} \approx 0.386$$

The ratios are the same.

State. Martinez could hit about 63 home runs.

b) It could be predicted that Martinez would break Maris' record.

27. ***Familiarize***. The ratio of blue whales tagged to the total blue whale population, P, is $\frac{500}{P}$. Of the 400 blue whales checked later, 20 were tagged. The ratio of blue whales tagged to blue whales checked is $\frac{20}{400}$.

Translate. Assuming the two ratios are the same, we can translate to a proportion.

$$\begin{array}{c} \text{Whales tagged} \\ \text{originally} \\ \text{Whale} \\ \text{population} \end{array} \begin{array}{c} \longrightarrow \\ \longrightarrow \end{array} \frac{500}{P} = \frac{20}{400} \begin{array}{c} \longleftarrow \\ \longleftarrow \end{array} \begin{array}{c} \text{Tagged whales} \\ \text{caught later} \\ \text{Whales caught} \\ \text{later} \end{array}$$

Solve. We solve the equation.

$$400P \cdot \frac{500}{P} = 400P \cdot \frac{20}{400} \quad \text{Multiplying by the LCM, } 400P$$
$$400 \cdot 500 = P \cdot 20$$
$$200,000 = 20P$$
$$10,000 = P$$

Check.

$$\frac{500}{10,000} = \frac{1}{20} \text{ and } \frac{20}{400} = \frac{1}{20}.$$

The ratios are the same.

State. The blue whale population is about 10,000.

29. ***Familiarize***. The ratio of the weight of an object on Mars to the weight of an object on earth is 0.4 to 1.

a) We wish to find how much a 12-ton rocket would weigh on Mars.

b) We wish to find how much a 120-lb astronaut would weigh on Mars.

We can set up ratios. We let $r =$ the weight of a 12-ton rocket and $a =$ the weight of a 120-lb astronaut on Mars.

Translate. Assuming the ratios are the same, we can translate to proportions.

a)

$$\begin{array}{r} \text{Weight} \\ \text{on Mars} \rightarrow \\ \text{Weight} \rightarrow \\ \text{on earth} \end{array} \frac{0.4}{1} = \frac{r}{12} \begin{array}{l} \text{Weight} \\ \leftarrow \text{on Mars} \\ \leftarrow \text{Weight} \\ \text{on earth} \end{array}$$

b)
$$\begin{array}{lcl}\text{Weight on Mars} \rightarrow & \dfrac{0.4}{1} = \dfrac{a}{120} & \leftarrow \text{Weight on Mars}\\ \text{Weight on earth} \rightarrow & & \leftarrow \text{Weight on earth}\end{array}$$

Solve. We solve each proportion.

a) $\dfrac{0.4}{1} = \dfrac{r}{12}$ b) $\dfrac{0.4}{1} = \dfrac{1}{120}$

$12(0.4) = r$ $120(0.4) = a$

$4.8 = r$ $48 = a$

Check. $\dfrac{0.4}{1} = 0.4$, $\dfrac{4.8}{12} = 0.4$, and $\dfrac{48}{120} = 0.4$.

The ratios are the same.

State. a) A 12-ton rocket would weigh 4.8 tons on Mars.

b) A 120-lb astronaut would weigh 48 lb on Mars.

31. ***Familiarize.*** Let g = the number of additional games the team must win in order to finish with a 0.750 record. Then the total number of wins will be $25 + g$, and there will be a total of $36 + 12$, or 48 games in the season. We can set up ratios.

Translate. Assuming the ratios are the same, we can set up a proportion.

$$\begin{array}{lcl}\text{Games won} \rightarrow & \dfrac{25+g}{48} = \dfrac{0.750}{1} & \leftarrow \text{Games won}\\ \text{Games played} \rightarrow & & \leftarrow \text{Games played}\end{array}$$

Solve. We solve the proportion.

$$48\left(\frac{25+g}{48}\right) = 48 \cdot \frac{0.750}{1}$$
$$25 + g = 36$$
$$g = 11$$

Check. $\dfrac{25+11}{48} = \dfrac{36}{48} = 0.75$ and $\dfrac{0.750}{1} = 0.75$.

The ratios are the same.

State. The team must win 11 more games.

33. We write a proportion and then solve it.

$$\frac{b}{6} = \frac{7}{4}$$
$$b = \frac{7}{4} \cdot 6 \quad \text{Multiplying by 6}$$
$$b = \frac{42}{4}$$
$$b = \frac{21}{2}, \text{ or } 10.5$$

(Note that the proportions $\dfrac{6}{b} = \dfrac{4}{7}$, $\dfrac{b}{7} = \dfrac{6}{4}$, or $\dfrac{7}{b} = \dfrac{4}{6}$ could also be used.)

35. We write a proportion and then solve it.

$$\frac{4}{f} = \frac{6}{4}$$
$$4f \cdot \frac{4}{f} = 4f \cdot \frac{6}{4}$$
$$16 = 6f$$
$$\frac{8}{3} = f \quad \text{Simplifying}$$

(One of the following proportions could also be used: $\dfrac{f}{4} = \dfrac{4}{6}$, $\dfrac{4}{f} = \dfrac{9}{6}$, $\dfrac{f}{4} = \dfrac{6}{9}$, $\dfrac{4}{9} = \dfrac{f}{6}$, $\dfrac{9}{4} = \dfrac{6}{f}$)

37. We write a proportion and then solve it.

$$\frac{h}{7} = \frac{10}{6}$$
$$h = \frac{10}{6} \cdot 7 \quad \text{Multiplying by 7}$$
$$h = \frac{70}{6}$$
$$h = \frac{35}{3} \quad \text{Simplifying}$$

(Note that the proportions $\dfrac{7}{h} = \dfrac{6}{10}$, $\dfrac{h}{10} = \dfrac{7}{6}$, or $\dfrac{10}{h} = \dfrac{6}{7}$ could also be used.)

39. $x^5 \cdot x^6 = x^{5+6} = x^{11}$

41. $x^{-5} \cdot x^{-6} = x^{-5+(-6)} = x^{-11} = \dfrac{1}{x^{11}}$

43. ◈

45. ***Familiarize.*** The job takes Larry 8 days working alone and Moe 10 days working alone. Let x represent the number of days it would take Curly working alone. Then in 1 day Larry does $\dfrac{1}{8}$ of the job, Moe does $\dfrac{1}{10}$ of the job, and Curly does $\dfrac{1}{x}$ of the job. In 1 day they would complete $\dfrac{1}{8} + \dfrac{1}{10} + \dfrac{1}{x}$ of the job, and in 3 days they would complete $3\left(\dfrac{1}{8} + \dfrac{1}{10} + \dfrac{1}{x}\right)$, or $\dfrac{3}{8} + \dfrac{3}{10} + \dfrac{3}{x}$.

Translate. The amount done in 3 days is one entire job, so we have

$$\frac{3}{8} + \frac{3}{10} + \frac{3}{x} = 1.$$

Solve. We solve the equation.

$$\frac{3}{8} + \frac{3}{10} + \frac{3}{x} = 1, \text{ LCM} = 40x$$
$$40x\left(\frac{3}{8} + \frac{3}{10} + \frac{3}{x}\right) = 40x \cdot 1$$
$$40x \cdot \frac{3}{8} + 40x \cdot \frac{3}{10} + 40x \cdot \frac{3}{x} = 40x$$
$$15x + 12x + 120 = 40x$$
$$120 = 13x$$
$$\frac{120}{13} = x$$

Check. If it takes Curly $\frac{120}{13}$, or $9\frac{3}{13}$ days, to complete the job, then in one day Curly does $\frac{1}{\frac{120}{13}}$, or $\frac{13}{120}$, of the job, and in 3 days he does $3\left(\frac{13}{120}\right)$, or $\frac{13}{40}$, of the job. The portion of the job done by Larry, Moe, and Curly in 3 days is $\frac{3}{8}+\frac{3}{10}+\frac{13}{40}=\frac{15}{40}+\frac{12}{40}+\frac{13}{40}=\frac{40}{40}=1$ entire job. The answer checks.

State. It will take Curly $9\frac{3}{13}$ days to complete the job working alone.

47. ***Familiarize***. Let x represent the numerator and $x+1$ represent the denominator of the original fraction. The fraction is $\frac{x}{x+1}$. If 2 is subtracted from the numerator and the denominator, the resulting fraction is $\frac{x-2}{x+1-2}$, or $\frac{x-2}{x-1}$.

Translate.

$$\underbrace{\text{The resulting fraction}}_{\downarrow} \text{ is } \frac{1}{2}.$$

$$\frac{x-2}{x-1} = \frac{1}{2}$$

Solve. We solve the equation.

$$\frac{x-2}{x-1}=\frac{1}{2},\ \text{LCM}=2(x-1)$$

$$2(x-1)\cdot\frac{x-2}{x-1}=2(x-1)\cdot\frac{1}{2}$$

$$2(x-2)=x-1$$

$$2x-4=x-1$$

$$x=3$$

Check. If $x=3$, then $x+1=4$ and the original fraction is $\frac{3}{4}$. If 2 is subtracted from both numerator and denominator, the resulting fraction is $\frac{3-2}{4-2}$, or $\frac{1}{2}$. The value checks.

State. The original fraction was $\frac{3}{4}$.

49. ***Familiarize***. Let $t=$ the number of minutes after 5:00 at which the hands of the clock will first be together. While the minute hand moves through t minutes, the hour hand moves through $t/12$ minutes. At 5:00 the hour hand is on the 25-minute mark. We wish to find when a move of the minute hand through t minutes is equal to $25+t/12$ minutes.

Translate. We use the last sentence of the familiarization step to write an equation.

$$t=25+\frac{t}{12}$$

Solve. We solve the equation.

$$t=25+\frac{t}{12}$$

$$12\cdot t=12\left(25+\frac{t}{12}\right)$$

$$12t=300+t \quad \text{Multiplying by 12}$$

$$11t=300$$

$$t=\frac{300}{11} \text{ or } 27\frac{3}{11}$$

Check. At $27\frac{3}{11}$ minutes after 5:00, the minute hand is at the $27\frac{3}{11}$-minutes mark and the hour hand is at the $25+\frac{27\frac{3}{11}}{12}$-minute mark. Simplifying $25+\frac{27\frac{3}{11}}{12}$, we get $25+\frac{\frac{300}{11}}{12}=25+\frac{300}{11}\cdot\frac{1}{12}=25+\frac{25}{11}=25+2\frac{3}{11}=27\frac{3}{11}$. Thus, the hands are together.

State. The hands are first together $27\frac{3}{11}$ minutes after 5:00.

Exercise Set 12.8

1.

$$S=2\pi rh$$

$$\frac{S}{2\pi h}=\frac{2\pi rh}{2\pi h} \quad \text{Dividing by } 2\pi h$$

$$\frac{S}{2\pi h}=r \quad \text{Simplifying}$$

3.

$$A=\frac{1}{2}bh$$

$$2\cdot A=2\cdot\frac{1}{2}bh \quad \text{Multiplying by 2}$$

$$2A=bh \quad \text{Simplifying}$$

$$\frac{2A}{h}=\frac{bh}{h} \quad \text{Dividing by } h$$

$$\frac{2a}{h}=b \quad \text{Simplifying}$$

5.

$$S=180(n-2)$$

$$S=180n-360 \quad \text{Removing parentheses}$$

$$S+360=180n \quad \text{Adding 360}$$

$$\frac{S+360}{180}=n \quad \text{Dividing by 180}$$

7. $V = \frac{1}{3}k(B+b+4M)$

$3 \cdot V = 3 \cdot \frac{1}{3}k(B+b+4M)$ Multiplying by 3

$3V = k(B+b+4M)$ Simplifying

$3V = kB + kb + 4kM$ Removing parentheses

$3V - kB - 4kM = kb$ Subtracting kB and $4kM$

$\frac{3V - kB - 4kM}{k} = b$ Dividing by k

9. $S(r-1) = rl - a$

$Sr - S = rl - a$ Removing parentheses

$Sr - rl = S - a$ Adding S and subtracting rl

$r(S-l) = S - a$ Factoring out r

$r = \frac{S-a}{S-l}$ Dividing by $S - l$

11. $A = \frac{1}{2}h(b_1 + b_2)$

$2A = h(b_1 + b_2)$ Multiplying by 2

$\frac{2A}{b_1 + b_2} = h$ Dividing by $b_1 + b_2$

13. $\frac{A-B}{AB} = Q$

$AB \cdot \frac{A-B}{AB} = AB \cdot Q$ Multiplying by AB

$A - B = ABQ$ Simplifying

$A = ABQ + B$ Adding B

$A = B(AQ+1)$ Factoring out B

$\frac{A}{AQ+1} = B$ Dividing by $AQ + 1$

15. $\frac{1}{p} + \frac{1}{q} = \frac{1}{f}$, LCM $= pqf$

$pqf\left(\frac{1}{p} + \frac{1}{q}\right) = pqf \cdot \frac{1}{f}$ Multiplying by pqf

$pqf \cdot \frac{1}{p} + pqf \cdot \frac{1}{q} = pqf \cdot \frac{1}{f}$ Removing parentheses

$qf + pf = pq$ Simplifying

$qf = pq - pf$ Subtracting pf

$qf = p(q-f)$ Factoring out p

$\frac{qf}{q-f} = p$ Dividing by $q - f$

17. $\frac{A}{P} = 1 + r$

$P \cdot \frac{A}{P} = P \cdot (1+r)$ Multiplying by P

$A = P(1+r)$ Simplifying

19. $\frac{1}{R} = \frac{1}{r_1} + \frac{1}{r_2}$, LCM $= Rr_1r_2$

$Rr_1r_2 \cdot \frac{1}{R} = Rr_1r_2\left(\frac{1}{r_1} + \frac{1}{r_2}\right)$ Multiplying by Rr_1r_2

$Rr_1r_2 \cdot \frac{1}{R} = Rr_1r_2 \cdot \frac{1}{r_1} + Rr_1r_2 \cdot \frac{1}{r_2}$ Removing parentheses

$r_1r_2 = Rr_2 + Rr_1$ Simplifying

$r_1r_2 = R(r_2 + r_1)$ Factoring out R

$\frac{r_1r_2}{r_2 + r_1} = R$ Dividing by $r_2 + r_1$

21. $\frac{A}{B} = \frac{C}{D}$, LCM $= BD$

$BD \cdot \frac{A}{B} = BD \cdot \frac{C}{D}$ Multiplying by BD

$DA = BC$ Simplifying

$D = \frac{BC}{A}$ Dividing by A

23. $h_1 = q\left(1 + \frac{h_2}{p}\right)$

$h_1 = q + \frac{qh_2}{p}$ Removing parentheses

$h_1 - q = \frac{qh_2}{p}$ Subtracting q

$p(h_1 - q) = qh_2$ Multiplying by p

$\frac{p(h_1 - q)}{q} = h_2$ Dividing by q

25. $C = \frac{Ka - b}{a}$

$a \cdot C = a \cdot \left(\frac{Ka - b}{a}\right)$ Multiplying by a

$aC = Ka - b$ Simplifying

$b = Ka - aC$ Adding b and subtracting aC

$b = a(K - C)$ Factoring out a

$\frac{b}{K-C} = a$ Dividing by $K - C$

27. $(5x^3 - 7x^2 + 9) - (8x^3 - 2x^2 + 4)$

$= 5x^3 - 7x^2 + 9 - 8x^3 + 2x^2 - 4$

$= (5-8)x^3 + (-7+2)x^2 + (9-4)$

$= -3x^3 - 5x^2 + 5$

29. $x^2 - 4 = x^2 - 2^2$ Difference of squares

$= (x+2)(x-2)$

31. $49m^2 - 112mn + 64n^2$

$= (7m)^2 - 2 \cdot 7m \cdot 8n + (8n)^2$ Trinomial square

$= (7m - 8n)^2$

33.
$$\begin{aligned}
&y^4-1\\
&=(y^2)^2-1^2 \quad \text{Difference of squares}\\
&=(y^2+1)(y^2-1)\\
&=(y^2+1)(y^2-1^2) \quad \text{Difference of squares}\\
&=(y^2+1)(y+1)(y-1)
\end{aligned}$$

35.
$$\begin{array}{r}
x^2+\ 2x+\ 8\\
x-2\,\overline{)\,x^3+0x^2+4x-\ 4}\\
\underline{x^3-2x^2}\\
2x^2+4x\\
\underline{2x^2-4x}\\
8x-\ 4\\
\underline{8x-16}\\
12
\end{array}$$

The answer is $x^2+2x+8+\dfrac{12}{x-2}$.

To check we multiply x^2+2x+8 by $x-2$ and then add the remainder, 12.

$$\begin{array}{r}
x^2+2x+\ 8\\
\underline{x-\ 2}\\
\underline{-2x^2-4x-16}\\
\underline{x^3+\ 2x^2+8x}\\
x^3+4x-16
\end{array}$$

$(x^3+4x-16)+12=x^3+4x-4$

37.

39.
$$\begin{aligned}
u&=-F\Big(E-\frac{P}{T}\Big)\\
u&=-EF+\frac{FP}{T} &&\text{Removing parentheses}\\
T\cdot u&=T\Big(-EF+\frac{FP}{T}\Big) &&\text{Multiplying by } T\\
Tu&=-EFT+FP\\
Tu+EFT&=FP &&\text{Adding } EFT\\
T(u+EF)&=FP &&\text{Factoring out } T\\
T&=\frac{FP}{u+EF} &&\text{Dividing by } u+EF
\end{aligned}$$

41.
$$\begin{aligned}
N&=\frac{(b+d)f_1-v}{(b-v)f_2}\\
N&=\frac{bf_1+df_1-v}{bf_2-vf_2}\\
N(bf_2-vf_2)&=bf_1+df_1-v\\
Nbf_2-Nvf_2&=bf_1+df_1-v\\
v-Nvf_2&=bf_1+df_1-Nbf_2\\
v(1-Nf_2)&=bf_1+df_1-Nbf_2\\
v&=\frac{bf_1+df_1-Nbf_2}{1-Nf_2},\ \text{or}\\
v&=\frac{Nbf_2-bf_1-df_1}{Nf_2-1}
\end{aligned}$$

Exercise Set 12.9

1.
$$\begin{aligned}
&\frac{1+\frac{9}{16}}{1-\frac{3}{4}} &&\text{LCM of the denominators is 16.}\\
&=\frac{1+\frac{9}{16}}{1-\frac{3}{4}}\cdot\frac{16}{16} &&\text{Multiplying by 1 using } \frac{16}{16}\\
&=\frac{\Big(1+\frac{9}{16}\Big)16}{\Big(1-\frac{3}{4}\Big)16} &&\text{Multiplying numerator and denominator by 16}\\
&=\frac{1(16)+\frac{9}{16}(16)}{1(16)-\frac{3}{4}(16)}\\
&=\frac{16+9}{16-12}\\
&=\frac{25}{4}
\end{aligned}$$

3.
$$\begin{aligned}
&\frac{1-\frac{3}{5}}{1+\frac{1}{5}}\\
&=\frac{1\cdot\frac{5}{5}-\frac{3}{5}}{1\cdot\frac{5}{5}+\frac{1}{5}} &&\text{Getting a common denominator in numerator and in denominator}\\
&=\frac{\frac{5}{5}-\frac{3}{5}}{\frac{5}{5}+\frac{1}{5}}\\
&=\frac{\frac{2}{5}}{\frac{6}{5}} &&\text{Subtracting in numerator; adding in denominator}\\
&=\frac{2}{5}\cdot\frac{5}{6} &&\text{Multiplying by the reciprocal of the divisor}\\
&=\frac{2\cdot5}{5\cdot2\cdot3}\\
&=\frac{\cancel{2}\cdot\cancel{5}\cdot1}{\cancel{5}\cdot\cancel{2}\cdot3}\\
&=\frac{1}{3}
\end{aligned}$$

5. $$\frac{\frac{1}{2}+\frac{3}{4}}{\frac{5}{8}-\frac{5}{6}} = \frac{\frac{1}{2}\cdot\frac{2}{2}+\frac{3}{4}}{\frac{5}{8}\cdot\frac{3}{3}-\frac{5}{6}\cdot\frac{4}{4}}$$ Getting a common denominator in numerator and denominator

$$= \frac{\frac{2}{4}+\frac{3}{4}}{\frac{15}{24}-\frac{20}{24}}$$

$$= \frac{\frac{5}{4}}{\frac{-5}{24}}$$ Adding in numerator; subtracting in denominator

$$= \frac{5}{4}\cdot\frac{24}{-5}$$ Multiplying by the reciprocal of the divisor

$$= \frac{5\cdot 4\cdot 6}{4\cdot(-1)\cdot 5}$$

$$= \frac{\cancel{5}\cdot\cancel{4}\cdot 6}{\cancel{4}\cdot(-1)\cdot\cancel{5}}$$

$$= -6$$

7. $$\frac{\frac{1}{x}+3}{\frac{1}{x}-5}$$ LCM of the denominators is x.

$$= \frac{\frac{1}{x}+3}{\frac{1}{x}-5}\cdot\frac{x}{x}$$ Multiplying by 1 using $\frac{x}{x}$

$$= \frac{\left(\frac{1}{x}+3\right)x}{\left(\frac{1}{x}-5\right)x}$$

$$= \frac{\frac{1}{x}\cdot x+3\cdot x}{\frac{1}{x}\cdot x-5\cdot x}$$

$$= \frac{1+3x}{1-5x}$$

9. $$\frac{4-\frac{1}{x^2}}{2-\frac{1}{x}}$$ LCM of the denominators is x^2.

$$= \frac{4-\frac{1}{x^2}}{2-\frac{1}{x}}\cdot\frac{x^2}{x^2}$$

$$= \frac{\left(4-\frac{1}{x^2}\right)x^2}{\left(2-\frac{1}{x}\right)x^2}$$

$$= \frac{4\cdot x^2-\frac{1}{x^2}\cdot x^2}{2\cdot x^2-\frac{1}{x}\cdot x^2}$$

$$= \frac{4x^2-1}{2x^2-x}$$

$$= \frac{(2x+1)(2x-1)}{x(2x-1)}$$ Factoring numerator and denominator

$$= \frac{(2x+1)\cancel{(2x-1)}}{x\cancel{(2x-1)}}$$

$$= \frac{2x+1}{x}$$

11. $$\frac{8+\frac{8}{d}}{1+\frac{1}{d}} = \frac{8\cdot\frac{d}{d}+\frac{8}{d}}{1\cdot\frac{d}{d}+\frac{1}{d}}$$

$$= \frac{\frac{8d+8}{d}}{\frac{d+1}{d}}$$

$$= \frac{8d+8}{d}\cdot\frac{d}{d+1}$$

$$= \frac{8(d+1)(d)}{d(d+1)}$$

$$= \frac{8\cancel{(d+1)}\cancel{(d)}}{\cancel{d}\cancel{(d+1)}(1)}$$

$$= 8$$

13. $\dfrac{\frac{x}{8}-\frac{8}{x}}{\frac{1}{8}+\frac{1}{x}}$ LCM of the denominators is $8x$.

$$= \frac{\frac{x}{8}-\frac{8}{x}}{\frac{1}{8}+\frac{1}{x}} \cdot \frac{8x}{8x}$$

$$= \frac{\left(\frac{x}{8}-\frac{8}{x}\right)8x}{\left(\frac{1}{8}+\frac{1}{x}\right)8x}$$

$$= \frac{\frac{x}{8}(8x)-\frac{8}{x}(8x)}{\frac{1}{8}(8x)+\frac{1}{x}(8x)}$$

$$= \frac{x^2-64}{x+8}$$

$$= \frac{(x+8)(x-8)}{x+8}$$

$$= \frac{\cancel{(x+8)}(x-8)}{1\cancel{(x+8)}}$$

$$= x-8$$

15. $\dfrac{1+\frac{1}{y}}{1-\frac{1}{y^2}} = \dfrac{1\cdot\frac{y}{y}+\frac{1}{y}}{1\cdot\frac{y^2}{y^2}-\frac{1}{y^2}}$

$$= \frac{\frac{y+1}{y}}{\frac{y^2-1}{y^2}}$$

$$= \frac{y+1}{y} \cdot \frac{y^2}{y^2-1}$$

$$= \frac{(y+1)y\cdot y}{y(y+1)(y-1)}$$

$$= \frac{\cancel{(y+1)}\cancel{y}\cdot y}{\cancel{y}\cancel{(y+1)}(y-1)}$$

$$= \frac{y}{y-1}$$

17. $\dfrac{\frac{1}{5}-\frac{1}{a}}{\frac{5-a}{5}}$ LCM of the denominators is $5a$.

$$= \frac{\frac{1}{5}-\frac{1}{a}}{\frac{5-a}{5}} \cdot \frac{5a}{5a}$$

$$= \frac{\left(\frac{1}{5}-\frac{1}{a}\right)5a}{\left(\frac{5-a}{5}\right)5a}$$

$$= \frac{\frac{1}{5}(5a)-\frac{1}{a}(5a)}{a(5-a)}$$

$$= \frac{a-5}{5a-a^2}$$

$$= \frac{a-5}{-a(-5+a)}$$

$$= \frac{1\cancel{(a-5)}}{-a\cancel{(a-5)}}$$

$$= -\frac{1}{a}$$

19. $\dfrac{\frac{1}{a}+\frac{1}{b}}{\frac{1}{a^2}-\frac{1}{b^2}}$ LCM of the denominators is a^2b^2.

$$= \frac{\frac{1}{a}+\frac{1}{b}}{\frac{1}{a^2}-\frac{1}{b^2}} \cdot \frac{a^2b^2}{a^2b^2}$$

$$= \frac{\left(\frac{1}{a}+\frac{1}{b}\right)\cdot a^2b^2}{\left(\frac{1}{a^2}-\frac{1}{b^2}\right)\cdot a^2b^2}$$

$$= \frac{\frac{1}{a}\cdot a^2b^2+\frac{1}{b}\cdot a^2b^2}{\frac{1}{a^2}\cdot a^2b^2-\frac{1}{b^2}\cdot a^2b^2}$$

$$= \frac{ab^2+a^2b}{b^2-a^2}$$

$$= \frac{ab(b+a)}{(b+a)(b-a)}$$

$$= \frac{ab\cancel{(b+a)}}{\cancel{(b+a)}(b-a)}$$

$$= \frac{ab}{b-a}$$

21. $\dfrac{\dfrac{p}{q}+\dfrac{q}{p}}{\dfrac{1}{p}+\dfrac{1}{q}}$ LCM of the denominators is pq.

$$= \frac{\left(\frac{p}{q}+\frac{q}{p}\right)\cdot pq}{\left(\frac{1}{p}+\frac{1}{q}\right)\cdot pq}$$

$$= \frac{\frac{p}{q}\cdot pq+\frac{q}{p}\cdot pq}{\frac{1}{p}\cdot pq+\frac{1}{q}\cdot pq}$$

$$= \frac{p^2+q^2}{q+p}$$

23. $(2x^3-4x^2+x-7)+(4x^4+x^3+4x^2+x)$

$= 4x^4+3x^3+2x-7$

25. $p^2-10p+25 = p^2-2\cdot p\cdot 5+5^2$ Trinomial square

$= (p-5)^2$

27. $50p^2-100 = 50(p^2-2)$ Factoring out the common factor

Since p^2-2 cannot be factored, we have factored completely.

29. ***Familiarize***. Let w = the width of the rectangle. Then $w+3$ = the length. Recall that the formula for the area of a rectangle is $A = lw$ and the formula for the perimeter of a rectangle is $P = 2l+2w$.

Translate. We substitute in the formula for area.

$10 = lw$

$10 = (w+3)w$

Solve.

$10 = (w+3)w$

$10 = w^2+3w$

$0 = w^2+3w-10$

$0 = (w+5)(w-2)$

$w+5=0$ or $w-2=0$

$w=-5$ or $w=2$

Check. Since the width cannot be negative, we only check 2. If $w=2$, then $w+3 = 2+3$, or 5. Since $2\cdot 5 = 10$, the given area, the answer checks. Now we find the perimeter:

$P = 2l+2w$

$P = 2\cdot 5+2\cdot 2$

$P = 10+4$

$P = 14$

We can check this by repeating the calculation.

State. The perimeter is 14 yd.

31. ◈

33. $$\frac{1}{\frac{2}{x-1}-\frac{1}{3x-2}}$$

$$= \frac{1}{\frac{2}{x-1}-\frac{1}{3x-2}}\cdot\frac{(x-1)(3x-2)}{(x-1)(3x-2)}$$

$$= \frac{(x-1)(3x-2)}{\left(\frac{2}{x-1}-\frac{1}{3x-2}\right)(x-1)(3x-2)}$$

$$= \frac{(x-1)(3x-2)}{\frac{2}{x-1}(x-1)(3x-2)-\frac{1}{3x-2}(x-1)(3x-2)}$$

$$= \frac{(x-1)(3x-2)}{2(3x-2)-(x-1)}$$

$$= \frac{(x-1)(3x-2)}{6x-4-x+1}$$

$$= \frac{(x-1)(3x-2)}{5x-3}$$

35. $$\frac{\frac{a}{b}-\frac{c}{d}}{\frac{b}{a}-\frac{d}{c}} = \frac{\frac{a}{b}\cdot\frac{d}{d}-\frac{c}{d}\cdot\frac{b}{b}}{\frac{b}{a}\cdot\frac{c}{c}-\frac{d}{c}\cdot\frac{a}{a}}$$

$$= \frac{\frac{ad-bc}{bd}}{\frac{bc-ad}{ac}}$$

$$= \frac{ad-bc}{bd}\cdot\frac{ac}{bc-ad}$$

$$= \frac{-1(bc-ad)(ac)}{bd(bc-ad)}$$

$$= \frac{-1\cancel{(bc-ad)}(ac)}{bd\cancel{(bc-ad)}}$$

$$= -\frac{ac}{bd}$$

37. $$1+\cfrac{1}{1+\cfrac{1}{1+\cfrac{1}{1+\cfrac{1}{x}}}} = 1+\cfrac{1}{1+\cfrac{1}{1+\cfrac{1}{\cfrac{x+1}{x}}}}$$

$$= 1+\cfrac{1}{1+\cfrac{1}{1+\cfrac{x}{x+1}}}$$

$$= 1+\cfrac{1}{1+\cfrac{1}{\cfrac{x+1+x}{x+1}}}$$

$$= 1+\cfrac{1}{1+\cfrac{1}{\cfrac{2x+1}{x+1}}}$$

$$= 1+\cfrac{1}{1+\cfrac{x+1}{2x+1}}$$

$$= 1+\cfrac{1}{\cfrac{2x+1+x+1}{2x+1}}$$

$$= 1+\cfrac{1}{\cfrac{3x+2}{2x+1}}$$

$$= 1+\frac{2x+1}{3x+2}$$

$$= \frac{3x+2+2x+1}{3x+2}$$

$$= \frac{5x+3}{3x+2}$$

Chapter 13

Graphs, Slope, and Applications

Exercise Set 13.1

1. We consider (x_1, y_1) to be $(-3, 5)$ and (x_2, y_2) to be $(4, 2)$.

$$m = \frac{y_2 - y_1}{x_2 - x_1} = \frac{2-5}{4-(-3)} = \frac{-3}{7} = -\frac{3}{7}$$

3. We consider (x_1, y_1) to be $(-4, -2)$ and (x_2, y_2) to be $(3, -2)$.

$$m = \frac{y_2 - y_1}{x_2 - x_1} = \frac{-2-(-2)}{3-(-4)} = \frac{0}{7} = 0$$

5. We plot $(-2, 4)$ and $(3, 0)$ and draw the line containing these points.

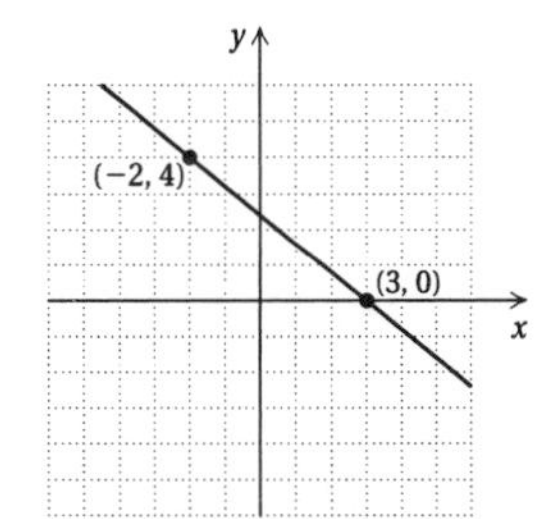

To find the slope, consider (x_1, y_1) to be $(-2, 4)$ and (x_2, y_2) to be $(3, 0)$.

$$m = \frac{y_2 - y_1}{x_2 - x_1} = \frac{0-4}{3-(-2)} = \frac{-4}{5} = -\frac{4}{5}$$

7. We plot $(-4, 0)$ and $(-5, -3)$ and draw the line containing these points.

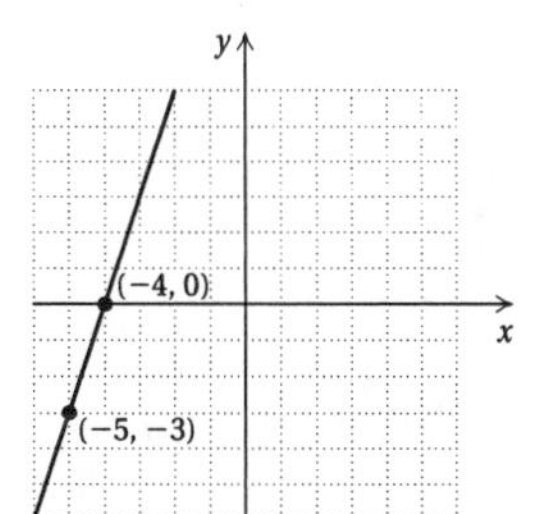

To find the slope, consider (x_1, y_1) to be $(-4, 0)$ and (x_2, y_2) to be $(-5, -3)$.

$$m = \frac{y_2 - y_1}{x_2 - x_1} = \frac{-3-0}{-5-(-4)} = \frac{-3}{-1} = 3$$

9. $m = \dfrac{-\frac{1}{2} - \frac{3}{2}}{2-5} = \dfrac{-2}{-3} = \dfrac{2}{3}$

11. $m = \dfrac{-2-3}{4-4} = \dfrac{-5}{0}$

Since division by 0 is undefined, the slope is undefined.

13. $y = -10x + 7$

The equation is in the form $y = mx + b$, where $m = -10$. Thus, the slope is -10.

15. $y = 3.78x - 4$

The equation is in the form $y = mx + b$, where $m = 3.78$. Thus, the slope is 3.78.

17. We solve for y, obtaining an equation of the form $y = mx + b$.

$$\begin{aligned} 3x - y &= 4 \\ -y &= -3x + 4 \\ -1(-y) &= -1(-3x + 4) \\ y &= 3x - 4 \end{aligned}$$

The slope is 3.

19. We solve for y, obtaining an equation of the form $y = mx + b$.

$$\begin{aligned} x + 5y &= 10 \\ 5y &= -x + 10 \\ y &= \frac{1}{5}(-x + 10) \\ y &= -\frac{1}{5}x + 2 \end{aligned}$$

The slope is $-\dfrac{1}{5}$.

21. We solve for y, obtaining an equation of the form $y = mx + b$.

$$\begin{aligned} 3x + 2y &= 6 \\ 2y &= -3x + 6 \\ y &= \frac{1}{2}(-3x + 6) \\ y &= -\frac{3}{2}x + 3 \end{aligned}$$

The slope is $-\dfrac{3}{2}$.

23. We solve for y, obtaining an equation of the form $y = mx + b$.

$$\begin{aligned} 5x - 7y &= 14 \\ -7y &= -5x + 14 \\ y &= -\frac{1}{7}(-5x + 14) \\ y &= \frac{5}{7}x - 2 \end{aligned}$$

The slope is $\dfrac{5}{7}$.

25. $y = -2.74x$

The equation is in the form $y = mx + b$, where $m = -2.74$. Thus, the slope is -2.74.

27. We solve for y, obtaining an equation of the form $y = mx + b$.

$$9x = 3y + 5$$
$$9x - 5 = 3y$$
$$\frac{1}{3}(9x - 5) = y$$
$$3x - \frac{5}{3} = y$$

The slope is 3.

29. We solve for y, obtaining an equation of the form $y = mx + b$.

$$5x - 4y + 12 = 0$$
$$5x + 12 = 4y$$
$$\frac{1}{4}(5x + 12) = y$$
$$\frac{5}{4}x + 3 = y$$

The slope is $\frac{5}{4}$.

31. $y = 4$

The equation can be thought of as $y = 0 \cdot x + 4$, so the slope is 0.

33. $m = \frac{\text{rise}}{\text{run}} = \frac{2.4}{8.2} = \frac{2.4}{8.2} \cdot \frac{10}{10} = \frac{24}{82}$

$= \frac{\not{2} \cdot 12}{\not{2} \cdot 41} = \frac{12}{41}$

35. $m = \frac{\text{rise}}{\text{run}} = \frac{56}{258} = \frac{\not{2} \cdot 28}{\not{2} \cdot 129} = \frac{28}{129}$

37. The rate (or slope) can be found using any two points on the graph. We use $(2, 30)$ and $(10, 150)$.

$$\text{Rate} = \frac{\text{change in calories burned}}{\text{corresponding change in time}}$$
$$= \frac{150 - 30}{10 - 2}$$
$$= \frac{120}{8}$$
$$= 15 \text{ calories per minute}$$

39. Long's Peak rises 14,255 ft $-$ 9600 ft $=$ 4655 ft.

$$\text{Grade} = \frac{4655}{15,840} \approx 0.294 \approx 29.4\%$$

41. We multiply and divide in order from left to right.

$$11 \cdot 6 \div 3 \cdot 2 \div 7 = 66 \div 3 \cdot 2 \div 7$$
$$= 22 \cdot 2 \div 7$$
$$= 44 \div 7$$
$$= \frac{44}{7}$$

43. $[10 - 3(7 - 2)]$

$= [10 - 3 \cdot 5]$ Subtracting inside the parentheses

$= [10 - 15]$ Multiplying

$= -5$ Subtracting

45. $\frac{4^2 + 2^2}{5^3 - 4^2}$

$= \frac{64 + 4}{125 - 16}$ Evaluating exponential expressions

$= \frac{68}{109}$ Adding the numerator; subtracting in the denominator

47. We carry out the divisions in order from left to right.

$$1000 \div 100 \div 10 \div 2 = 10 \div 10 \div 2$$
$$= 1 \div 2$$
$$= \frac{1}{2}$$

49.

Exercise Set 13.2

1. $y = -4x - 9$

The equation is already in the form $y = mx + b$. The slope is -4 and the y-intercept is $(0, -9)$.

3. $y = 1.8x$

We can think of $y = 1.8x$ as $y = 1.8x + 0$. The slope is 1.8 and the y-intercept is $(0, 0)$.

5. We solve for y.

$$-8x - 7y = 21$$
$$-7y = 8x + 21$$
$$y = -\frac{1}{7}(8x + 21)$$
$$y = -\frac{8}{7}x - 3$$

The slope is $-\frac{8}{7}$ and the y-intercept is $(0, -3)$.

7. We solve for y.

$$4x = 9y + 7$$
$$4x - 7 = 9y$$
$$\frac{1}{9}(4x - 7) = y$$
$$\frac{4}{9}x - \frac{7}{9} = y$$

The slope is $\frac{4}{9}$ and the y-intercept is $\left(0, -\frac{7}{9}\right)$.

9. We solve for y.

$$-6x = 4y + 2$$
$$-6x - 2 = 4y$$
$$\frac{1}{4}(-6x - 2) = y$$
$$-\frac{3}{2}x - \frac{1}{2} = y$$

The slope is $-\frac{3}{2}$ and the y-intercept is $\left(0, -\frac{1}{2}\right)$.

11. $y = -17$

We can think of $y = -17$ as $y = 0x - 17$. The slope is 0 and the y-intercept is $(0, -17)$.

13. We substitute -7 for m and -13 for b in the equation $y = mx + b$.

$$y = -7x - 13$$

15. We substitute 1.01 for m and -2.6 for b in the equation $y = mx + b$.

$$y = 1.01x - 2.6$$

17. We know the slope is -2, so the equation is $y = -2x + b$. Using the point $(-3, 0)$, we substitute -3 for x and 0 for y in $y = -2x + b$. Then we solve for b.

$$\begin{aligned} y &= -2x + b \\ 0 &= -2(-3) + b \\ 0 &= 6 + b \\ -6 &= b \end{aligned}$$

Thus, we have the equation $y = -2x - 6$.

19. We know the slope is $\frac{3}{4}$, so the equation is $y = \frac{3}{4}x + b$. Using the point $(2, 4)$, we substitute 2 for x and 4 for y in $y = \frac{3}{4}x + b$. Then we solve for b.

$$\begin{aligned} y &= \frac{3}{4}x + b \\ 4 &= \frac{3}{4} \cdot 2 + b \\ 4 &= \frac{3}{2} + b \\ \frac{5}{2} &= b \end{aligned}$$

Thus, we have the equation $y = \frac{3}{4}x + \frac{5}{2}$.

21. We know the slope is 1, so the equation is $y = 1 \cdot x + b$, or $y = x + b$. Using the point $(2, -6)$, we substitute 2 for x and -6 for y in $y = x + b$. Then we solve for y.

$$\begin{aligned} y &= x + b \\ -6 &= 2 + b \\ -8 &= b \end{aligned}$$

Thus, we have the equation $y = x - 8$.

23. We substitute -3 for m and 3 for b in the equation $y = mx + b$.

$$y = -3x + 3$$

25. $(12, 16)$ and $(1, 5)$

First we find the slope.

$$m = \frac{16 - 5}{12 - 1} = \frac{11}{11} = 1$$

Thus, $y = 1 \cdot x + b$, or $y = x + b$. We can use either point to find b. We choose $(1, 5)$. Substitute 1 for x and 5 for y in $y = x + b$.

$$\begin{aligned} y &= x + b \\ 5 &= 1 + b \\ 4 &= b \end{aligned}$$

Thus, the equation is $y = x + 4$.

27. $(0, 4)$ and $(4, 2)$

First we find the slope.

$$m = \frac{4 - 2}{0 - 4} = \frac{2}{-4} = -\frac{1}{2}$$

Thus, $y = -\frac{1}{2}x + b$. One of the given points is the y-intercept $(0, 4)$. Thus, we substitute 4 for b in $y = -\frac{1}{2}x + b$. The equation is $y = -\frac{1}{2}x + 4$.

29. $(3, 2)$ and $(1, 5)$

First we find the slope.

$$m = \frac{2 - 5}{3 - 1} = \frac{-3}{2} = -\frac{3}{2}$$

Thus, $y = -\frac{3}{2}x + b$. We can use either point to find b. We choose $(3, 2)$. Substitute 3 for x and 2 for y in $y = -\frac{3}{2}x + b$.

$$\begin{aligned} y &= -\frac{3}{2}x + b \\ 2 &= -\frac{3}{2} \cdot 3 + b \\ 2 &= -\frac{9}{2} + b \\ \frac{13}{2} &= b \end{aligned}$$

Thus, the equation is $y = -\frac{3}{2}x + \frac{13}{2}$.

31. $(-4, 5)$ and $(-2, -3)$

First we find the slope.

$$m = \frac{5 - (-3)}{-4 - (-2)} = \frac{8}{-2} = -4$$

Thus, $y = -4x + b$. We can use either point to find b. We choose $(-4, 5)$. Substitute -4 for x and 5 for y in $y = -4x + b$.

$$\begin{aligned} y &= -4x + b \\ 5 &= -4(-4) + b \\ 5 &= 16 + b \\ -11 &= b \end{aligned}$$

Thus, the equation is $y = -4x - 11$.

33. a) First we find the slope.

$$m = \frac{150 - 105}{20 - 80} = \frac{45}{-60} = -0.75$$

Thus, $T = -0.75a + b$. We can use either point to find b. We choose $(20, 150)$. Substitute 20 for a and 150 for T in $T = -0.75a + b$.

$$\begin{aligned} 150 &= -0.75(20) + b \\ 150 &= -15 + b \\ 165 &= b \end{aligned}$$

Thus, the equation is $T = -0.75a + 165$.

b) The rate of change is the slope, -0.75 beats per minute per year.

c) Substitute 50 for a and calculate T.

$$T = -0.75a + 165$$
$$T = -0.75(50) + 165$$
$$T = -37.5 + 165$$
$$T = 127.5$$

The target heart rate for a 50 year old person is 127.5 beats per minute.

35. $2x^2 + 6x = 0$

$2x(x+3) = 0$

$x = 0 \quad or \quad x + 3 = 0$

$x = 0 \quad or \quad x = -3$

The solutions are 0 and -3.

37. $x^2 - x - 6 = 0$

$(x-3)(x+2) = 0$

$x - 3 = 0 \quad or \quad x + 2 = 0$

$x = 3 \quad or \quad x = -2$

The solutions are 3 and -2.

39. $2x^2 + 11x = 21$

$2x^2 + 11x - 21 = 0$

$(2x-3)(x+7) = 0$

$2x - 3 = 0 \quad or \quad x + 7 = 0$

$2x = 3 \quad or \quad x = -7$

$x = \frac{3}{2} \quad or \quad x = -7$

The solutions are $\frac{3}{2}$ and -7.

41. $x^2 + 5x - 14 = 0$

$(x+7)(x-2) = 0$

$x + 7 = 0 \quad or \quad x - 2 = 0$

$x = -7 \quad or \quad x = 2$

The solutions are -7 and 2.

43. $3x - 4(9-x) = 17$

$3x - 36 + 4x = 17$

$7x - 36 = 17$

$7x = 53$

$x = \frac{53}{7}$

The solution is $\frac{53}{7}$.

45. $40(2x-7) = 50(4-6x)$

$80x - 280 = 200 - 300x$

$380x - 280 = 200$

$380x = 480$

$x = \frac{480}{380}$

$x = \frac{24}{19}$

The solution is $\frac{24}{19}$.

47. ◈

49. First find the slope of $3x - y + 4 = 0$.

$3x - y + 4 = 0$

$3x + 4 = y$

The slope is 3.

Thus, $y = 3x + b$. Using the point $(2, -3)$, we substitute 2 for x and -3 for y in $y = 3x + b$. Then we solve for b.

$y = 3x + b$

$-3 = 3 \cdot 2 + b$

$-3 = 6 + b$

$-9 = b$

Thus, the equation is $y = 3x - 9$.

51. First find the slope of $3x - 2y = 8$.

$3x - 2y = 8$

$-2y = -3x + 8$

$y = \frac{3}{2}x - 4$

The slope is $\frac{3}{2}$.

Then find the y-intercept of $2y + 3x = -4$.

$2y + 3x = -4$

$2y = -3x - 4$

$y = -\frac{3}{2}x - 2$

The y-intercept is $(0, -2)$.

Finally, write the equation of the line with slope $\frac{3}{2}$ and y-intercept $(0, -2)$.

$y = mx + b$

$y = \frac{3}{2}x + (-2)$

$y = \frac{3}{2}x - 2$

Exercise Set 13.3

1. 1. The first equation is already solved for y:

$y = x + 4$

2. We solve the second equation for y:

$y - x = -3$

$y = x - 3$

The slope of each line is 1. The y-intercepts, $(0, 4)$ and $(0, -3)$, are different. The lines are parallel.

3. We solve each equation for y:

1. $y + 3 = 6x$

$y = 6x - 3$

2. $-6x - y = 2$

$-y = 6x + 2$

$y = -6x - 2$

The slope of the first line is 6 and of the second is -6. Since the slopes are different, the lines are not parallel.

5. We solve each equation for y:

1. $10y + 32x = 16.4$ 2. $y + 3.5 = 0.3125x$

$10y = -32x + 16.4$ $y = 0.3125x - 3.5$

$y = -3.2x + 1.64$

The slope of the first line is -3.2 and of the second is 0.3125. Since the slopes are different, the lines are not parallel.

7. 1. The first equation is already solved for y:

$y = 2x + 7$

2. We solve the second equation for y:

$$5y + 10x = 20$$
$$5y = -10x + 20$$
$$y = -2x + 4$$

The slope of the first line is 2 and of the second is -2. Since the slopes are different, the lines are not parallel.

9. We solve each equation for y:

1. $3x - y = -9$ 2. $2y - 6x = -2$

$3x + 9 = y$ $2y = 6x - 2$

$y = 3x - 1$

The slope of each line is 3. The y-intercepts, $(0, 9)$ and $(0, -1)$ are different. The lines are parallel.

11. $x = 3$,

$x = 4$

These are vertical lines with equations of the form $x = p$ and $x = q$, where $p \neq q$. Thus, they are parallel.

13. 1. The first equation is already solved for y:

$y = -4x + 3$

2. We solve the second equation for y:

$$4y + x = -1$$
$$4y = -x - 1$$
$$y = -\frac{1}{4}x - \frac{1}{4}$$

The slopes are -4 and $-\frac{1}{4}$. Their product is $-4\left(-\frac{1}{4}\right) = 1$. Since the product of the slopes is not -1, the lines are not perpendicular.

15. We solve each equation for y:

1. $x + y = 6$ 2. $4y - 4x = 12$

$y = -x + 6$ $4y = 4x + 12$

$y = x + 3$

The slopes are -1 and 1. Their product is $-1 \cdot 1 = -1$. The lines are perpendicular.

17. 1. The first equation is already solved for y:

$y = -0.3125x + 11$

2. We solve the second equation for y:

$$y - 3.2x = -14$$
$$y = 3.2x - 14$$

The slopes are -0.3125 and 3.2. Their product is $-0.3125(3.2) = -1$. The lines are perpendicular.

19. 1. The first equation is already solved for y:

$y = -x + 8$

2. We solve the second equation for y:

$$x - y = -1$$
$$x + 1 = y$$

The slopes are -1 and 1. Their product is $-1 \cdot 1 = -1$. The lines are perpendicular.

21. We solve each equation for y:

1.
$$\frac{3}{8}x - \frac{y}{2} = 1$$
$$8\left(\frac{3}{8}x - \frac{y}{2}\right) = 8 \cdot 1$$
$$8 \cdot \frac{3}{8}x - 8 \cdot \frac{y}{2} = 8$$
$$3x - 4y = 8$$
$$-4y = -3x + 8$$
$$y = \frac{3}{4}x - 2$$

2.
$$\frac{4}{3}x - y + 1 = 0$$
$$\frac{4}{3}x + 1 = y$$

The slopes are $\frac{3}{4}$ and $\frac{4}{3}$. Their product is $\frac{3}{4}\left(\frac{4}{3}\right) = 1$. Since the product of the slopes is not -1, the lines are not perpendicular.

23. $x = 0$,

$y = -2$

The first line is vertical and the second is horizontal, so the lines are perpendicular.

25. ***Familiarize.*** We let $t =$ the time the second train travels before it overtakes the first train. Then the first train travels for $t + 2$ hr before it is overtaken. The trains travel the same distance. We organize the information in a table.

	Distance	Speed	Time
First train	d	70	$t + 2$
Second train	d	90	t

Translate. From the rows of the table we get two equations:

$$d = 70(t + 2)$$
$$d = 90t$$

Since the right sides of both equations are equal to d, we set them equal to each other.

$$70(t + 2) = 90t$$

Solve. We solve the equation.

$$70(t+2) = 90t$$
$$70t + 140 = 90t$$
$$140 = 20t$$
$$7 = t$$

Check. In 7 hr the second train travels $90 \cdot 7$, or 630 mi. In $7+2$, or 9 hr, the first train travels $70 \cdot 9$, or 630 mi. Since the distances are the same, the result checks.

State. The second train will overtake the first train 7 hr after the second train leaves the station (or 9 hr after the first train leaves the station).

27. $\frac{x^2}{x+4} = \frac{16}{x+4}$, LCM is $x+4$

$$(x+4) \cdot \frac{x^2}{x+4} = (x+4) \cdot \frac{16}{x+4}$$
$$x^2 = 16$$
$$x^2 - 16 = 0$$
$$(x+4)(x-4) = 0$$
$$x+4 = 0 \quad or \quad x-4 = 0$$
$$x = -4 \quad or \quad x = 4$$

Check: For $x = -4$:

$$\frac{x^2}{x+4} = \frac{16}{x+4}$$

$$\frac{(-4)^2}{-4+4} \; ? \; \frac{16}{-4+4}$$

$$\frac{16}{0} \;\Big|\; \frac{16}{0} \quad \text{UNDEFINED}$$

For $x = 4$:

$$\frac{x^2}{x+4} = \frac{16}{x+4}$$

$$\frac{4^2}{4+4} \; ? \; \frac{16}{4+4}$$

$$\frac{16}{8} \;\Big|\; \frac{16}{8} \quad \text{TRUE}$$

The number 4 checks, but -4 does not. The solution is 4.

29. $\frac{t}{3} + \frac{t}{10} = 1$, LCM is 30

$$30\left(\frac{t}{3} + \frac{t}{10}\right) = 30 \cdot 1$$
$$30 \cdot \frac{t}{3} + 30 \cdot \frac{t}{10} = 30$$
$$10t + 3t = 30$$
$$13t = 30$$
$$t = \frac{30}{13}$$

The solution is $\frac{30}{13}$.

31.

$$\frac{4}{x-2} + \frac{7}{x-3} = \frac{10}{x^2-5x+6}$$
$$\frac{4}{x-2} + \frac{7}{x-3} = \frac{10}{(x-2)(x-3)},$$

LCM is $(x-2)(x-3)$

$$(x-2)(x-3)\left(\frac{4}{x-2} + \frac{7}{x-3}\right) = (x-2)(x-3) \cdot \frac{10}{(x-2)(x-3)}$$
$$(x-2)(x-3) \cdot \frac{4}{x-2} + (x-2)(x-3) \cdot \frac{7}{x-3} = 10$$
$$4(x-3) + 7(x-2) = 10$$
$$4x - 12 + 7x - 14 = 10$$
$$11x - 26 = 10$$
$$11x = 36$$
$$x = \frac{36}{11}$$

The number $\frac{36}{11}$ checks and is the solution.

33.

35.-39.

41.-45.

47. First we find the slope of the given line:

$$y - 3x = 4$$
$$y = 3x + 4$$

The slope is 3.

Then we use the slope-intercept equation to write the equation of a line with slope 3 and y-intercept $(0, 6)$:

$$y = mx + b$$
$$y = 3x + 6 \quad \text{Substituting 3 for } m \text{ and 6 for } b$$

49. First we find the slope of the given line:

$$3y - x = 0$$
$$3y = x$$
$$y = \frac{1}{3}x$$

The slope is $\frac{1}{3}$.

We can find the slope of the line perpendicular to the given line by taking the reciprocal of $\frac{1}{3}$ and changing the sign. We get -3.

Then we use the slope-intercept equation to write the equation of a line with slope -3 and y-intercept $(0, 2)$:

$$y = mx + b$$
$$y = -3x + 2 \quad \text{Substituting } -3 \text{ for } m \text{ and 2 for } b$$

51. First we find the slope of the given line:

$$4x - 8y = 12$$
$$-8y = -4x + 12$$
$$y = \frac{1}{2}x - \frac{3}{2}$$

The slope is $\frac{1}{2}$, so the equation is $y = \frac{1}{2}x + b$. Substitute -2 for x and 0 for y and solve for b.

$$y = \frac{1}{2}x + b$$
$$0 = \frac{1}{2}(-2) + b$$
$$0 = -1 + b$$
$$1 = b$$

Thus, the equation is $y = \frac{1}{2}x + 1$.

53. We find the slope of each line:

1. $4y = kx - 6$

$$y = \frac{k}{4}x - \frac{3}{2}$$

2. $5x + 20y = 12$

$$20y = -5x + 12$$
$$y = -\frac{1}{4}x + \frac{3}{5}$$

The slopes are $\frac{k}{4}$ and $-\frac{1}{4}$. If the lines are perpendicular, the product of their slopes is -1.

$$\frac{k}{4}\left(-\frac{1}{4}\right) = -1$$
$$-\frac{k}{16} = -1$$
$$k = 16$$

55. First we find the equation of A, a line containing the points $(1, -1)$ and $(4, 3)$:

The slope is $\frac{3-(-1)}{4-1} = \frac{4}{3}$, so the equation is $y = \frac{4}{3}x + b$. Use either point to find b. We choose $(1, -1)$.

$$y = \frac{4}{3}x + b$$
$$-1 = \frac{4}{3} \cdot 1 + b$$
$$-1 = \frac{4}{3} + b$$
$$-\frac{7}{3} = b$$

Thus, the equation is $y = \frac{4}{3}x - \frac{7}{3}$.

The slope of A is $\frac{4}{3}$. Since A and B are perpendicular we find the slope of B by taking the reciprocal of $\frac{4}{3}$ and changing the sign. We get $-\frac{3}{4}$, so the equation is $y = -\frac{3}{4}x + b$. We use the point $(1, -1)$ to find b.

$$y = -\frac{3}{4}x + b$$
$$-1 = -\frac{3}{4} \cdot 1 + b$$
$$-1 = -\frac{3}{4} + b$$
$$-\frac{1}{4} = b$$

Thus, the equation is $y = -\frac{3}{4}x - \frac{1}{4}$.

Exercise Set 13.4

1. We use alphabetical order to replace x by -3 and y by -5.

$$\begin{array}{r|l} -x - 3y & < 18 \\ \hline -(-3) - 3(-5) & ?\ 18 \\ 3 + 15 & \\ 18 & \quad \text{FALSE} \end{array}$$

Since $18 < 18$ is false, $(-3, -5)$ is not a solution.

3. We use alphabetical order to replace x by $\frac{1}{2}$ and y by $-\frac{1}{4}$.

$$\begin{array}{r|l} 7y - 9x & \le -3 \\ \hline 7\left(-\frac{1}{4}\right) - 9 \cdot \frac{1}{2} & ?\ -3 \\ -\frac{7}{4} - \frac{9}{2} & \\ -\frac{7}{4} - \frac{18}{4} & \\ -\frac{25}{4} & \\ -6\frac{1}{4} & \quad \text{TRUE} \end{array}$$

Since $-6\frac{1}{4} \le -3$ is true, $\left(\frac{1}{2}, -\frac{1}{4}\right)$ is a solution.

5. Graph $x > 2y$.

First graph the line $x = 2y$, or $y = \frac{1}{2}x$. Two points on the line are $(0, 0)$ and $(4, 2)$. We draw a dashed line since the inequality symbol is $>$. Then we pick a test point that is not on the line. We try $(-2, 1)$.

$$\begin{array}{r|l} x & > 2y \\ \hline -2 & ?\ 2 \cdot 1 \\ & \ 2 \quad \text{FALSE} \end{array}$$

We see that $(-2, 1)$ is not a solution of the inequality, so we shade the points in the region that does not contain $(-2, 1)$.

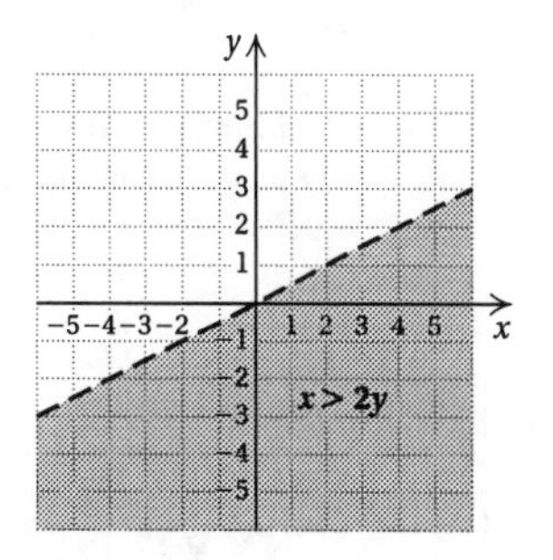

7. Graph $y \le x - 3$.

First graph the line $y = x - 3$. The intercepts are $(0, -3)$ and $(3, 0)$. We draw a solid line since the inequality symbol is $\leq$. Then we pick a test point that is not on the line. We try $(0,0)$.

$$\begin{array}{c|c} \multicolumn{2}{c}{y \leq x - 3} \\ \hline 0 \; ? & 0 - 3 \\ & -3 \quad \text{FALSE} \end{array}$$

We see that $(0,0)$ is not a solution of the inequality, so we shade the region that does not contain $(0,0)$.

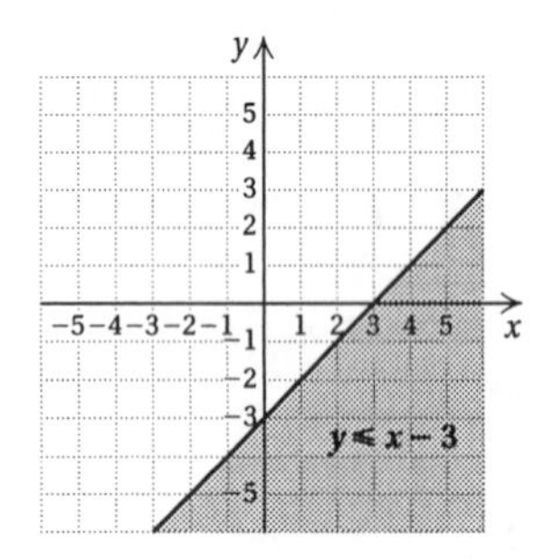

9. Graph $y < x + 1$.

First graph the line $y = x + 1$. The intercepts are $(0,1)$ and $(-1,0)$. We draw a dashed line since the inequality symbol is $<$. Then we pick a test point that is not on the line. We try $(0,0)$.

$$\begin{array}{c|c} \multicolumn{2}{c}{y < x + 1} \\ \hline 0 \; ? & 0 + 1 \\ & 1 \quad \text{TRUE} \end{array}$$

Since $(0,0)$ is a solution of the inequality, we shade the region that contains $(0,0)$.

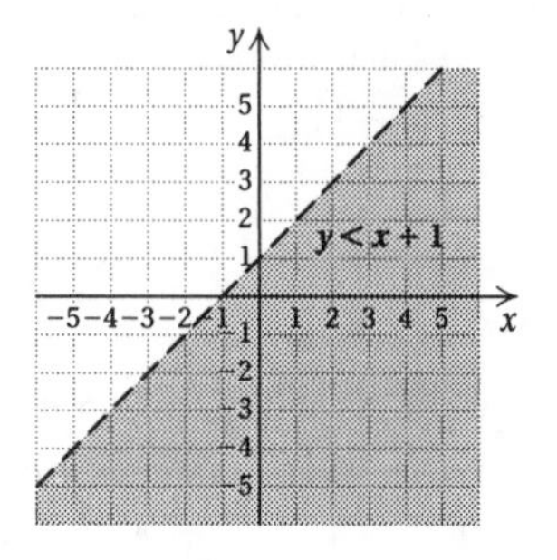

11. Graph $y \geq x - 2$.

First graph the line $y = x - 2$. The intercepts are $(0,-2)$ and $(2,0)$. We draw a solid line since the inequality symbol is $\geq$. Then we test the point $(0,0)$.

$$\begin{array}{c|c} \multicolumn{2}{c}{y \geq x - 2} \\ \hline 0 \; ? & 0 - 2 \\ & -2 \quad \text{TRUE} \end{array}$$

Since $(0,0)$ is a solution of the inequality, we shade the region containing $(0,0)$.

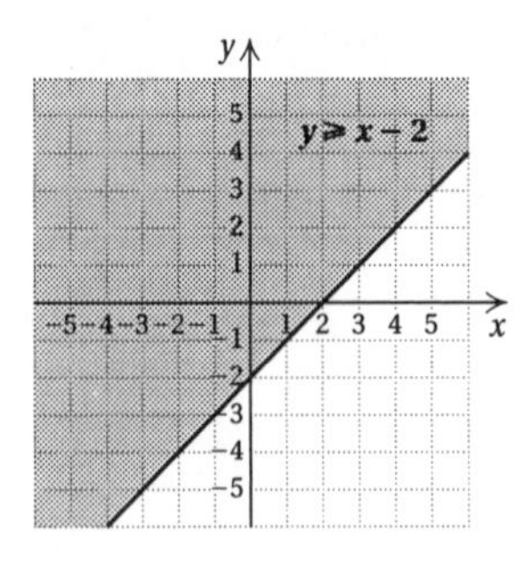

13. Graph $y \leq 2x - 1$.

First graph the line $y = 2x - 1$. The intercepts are $(0,-1)$ and $\left(\frac{1}{2}, 0\right)$. We draw a solid line since the inequality symbol

is $\leq$. Then we test the point $(0,0)$.

$$\begin{array}{c|c} \multicolumn{2}{c}{y \leq 2x - 1} \\ \hline 0 \; ? & 2 \cdot 0 - 1 \\ & -1 \quad \text{FALSE} \end{array}$$

Since $(0,0)$ is not a solution of the inequality, we shade the region that does not contain $(0,0)$.

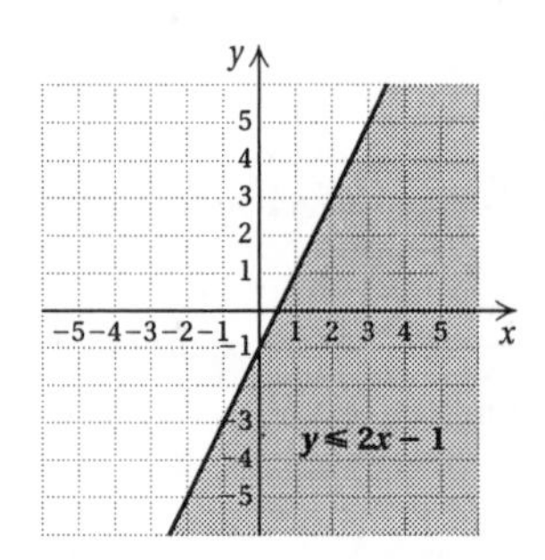

15. Graph $x + y \leq 3$.

First graph the line $x + y = 3$. The intercepts are $(0,3)$ and $(3,0)$. We draw a solid line since the inequality symbol is $\leq$. Then we test the point $(0,0)$.

$$\begin{array}{c|c} \multicolumn{2}{c}{x + y \leq 3} \\ \hline 0 + 0 & ? \; 3 \\ 0 & \quad \text{TRUE} \end{array}$$

Since $(0,0)$ is a solution of the inequality, we shade the region that contains $(0,0)$.

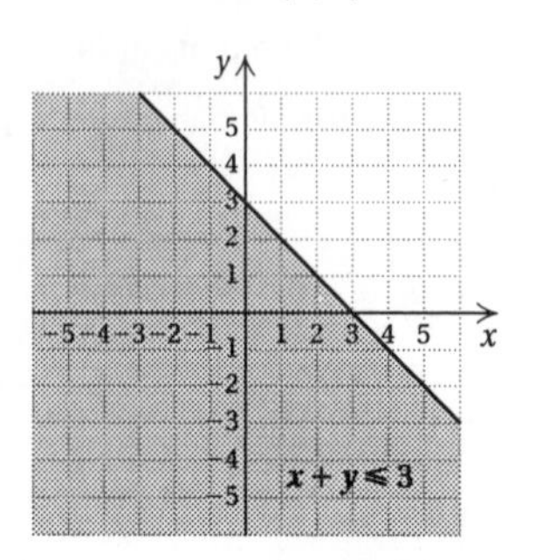

17. Graph $x - y > 7$.

First graph the line $x - y = 7$. The intercepts are $(0, -7)$ and $(7, 0)$. We draw a dashed line since the inequality symbol is $>$. Then we test the point $(0, 0)$.

$$\begin{array}{c|c} x - y > 7 & \\ \hline 0 - 0 \ ? \ 7 & \\ 0 & \text{FALSE} \end{array}$$

Since $(0, 0)$ is not a solution of the inequality, we shade the region that does not contain $(0, 0)$.

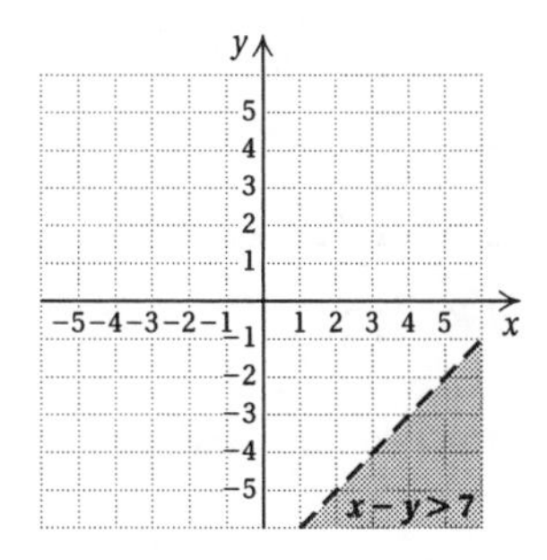

19. Graph $2x + 3y \leq 12$.

First graph the line $2x + 3y = 12$. The intercepts are $(0, 4)$ and $(6, 0)$. We draw a solid line since the inequality symbol is $\leq$. Then we test the point $(0, 0)$.

$$\begin{array}{c|c} 2x + 3y \leq 12 & \\ \hline 2 \cdot 0 + 3 \cdot 0 \ ? \ 12 & \\ 0 & \text{TRUE} \end{array}$$

Since $(0, 0)$ is a solution of the inequality, we shade the region containing $(0, 0)$.

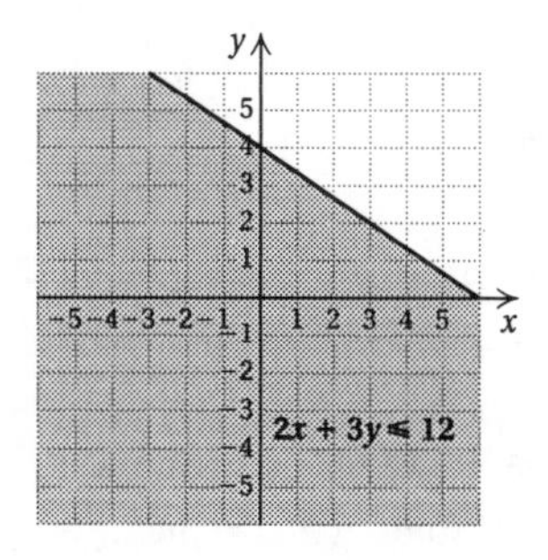

21. Graph $y \geq 1 - 2x$.

First graph the line $y = 1 - 2x$. The intercepts are $(0, 1)$ and $\left(\frac{1}{2}, 0\right)$. We draw a solid line since the inequality symbol is $\geq$. Then we test the point $(0, 0)$.

$$\begin{array}{c|c} y \geq 1 - 2x & \\ \hline 0 \ ? \ 1 - 2 \cdot 0 & \\ & 1 \quad \text{FALSE} \end{array}$$

Since $(0, 0)$ is not a solution of the inequality, we shade the region that does not contain $(0, 0)$.

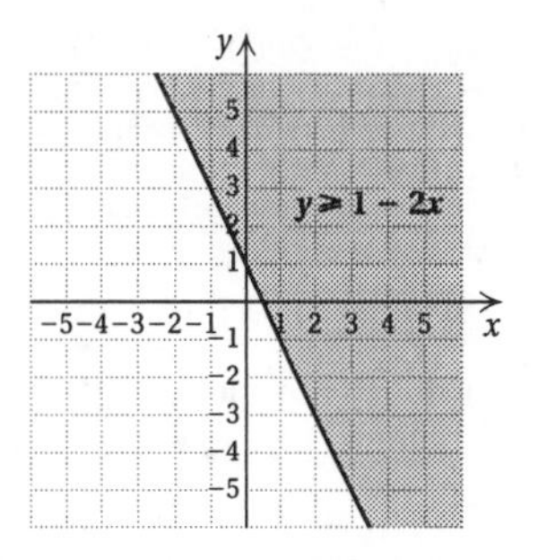

23. Graph $2x - 3y > 6$.

First graph the line $2x - 3y = 6$. The intercepts are $(0, -2)$ and $(3, 0)$. We draw a dashed line since the inequality symbol is $>$. Then we test the point $(0, 0)$.

$$\begin{array}{c|c} 2x - 3y > 6 & \\ \hline 2 \cdot 0 - 3 \cdot 0 \ ? \ 6 & \\ 0 & \text{FALSE} \end{array}$$

Since $(0, 0)$ is not a solution of the inequality, we shade the region that does not contain $(0, 0)$.

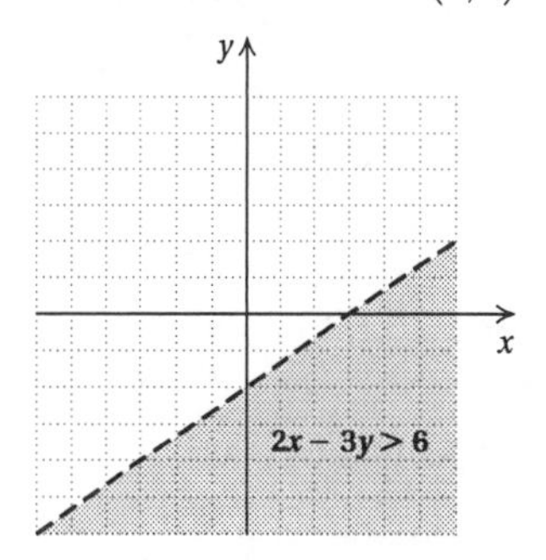

25. Graph $y \leq 3$.

First graph the line $y = 3$ using a solid line since the inequality symbol is $\leq$. Then pick a test point that is not on the line. We choose $(1, -2)$. We can write the inequality as $0x + y \leq 3$.

$$\begin{array}{c|c} 0x + y \leq 3 & \\ \hline 0 \cdot 1 + (-2) \ ? \ 3 & \\ -2 & \text{TRUE} \end{array}$$

Since $(1, -2)$ is a solution of the inequality, we shade the region containing $(1, -2)$.

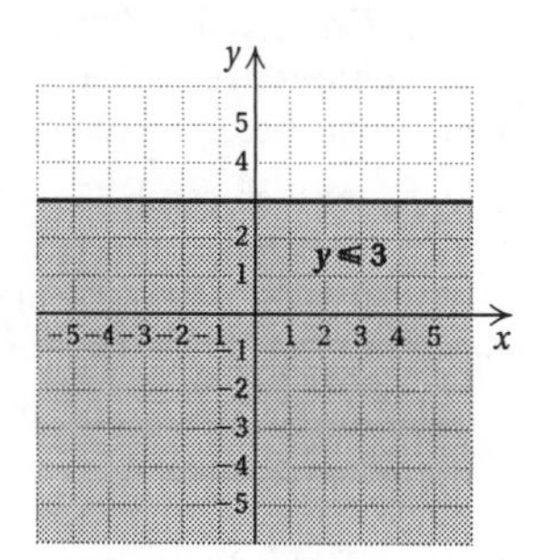

27. Graph $x \geq -1$.

Graph the line $x = 1$ using a solid line since the inequality symbol is $\geq$. Then pick a test point that is not on the line. We choose $(2,3)$. We can write the inequality as $x + 0y \geq -1$.

$$\begin{array}{c|c} x + 0y \geq -1 & \\ \hline 2 + 0 \cdot 3 \ ? \ -1 & \\ 2 & \text{TRUE} \end{array}$$

Since $(2,3)$ is a solution of the inequality, we shade the region containing $(2,3)$.

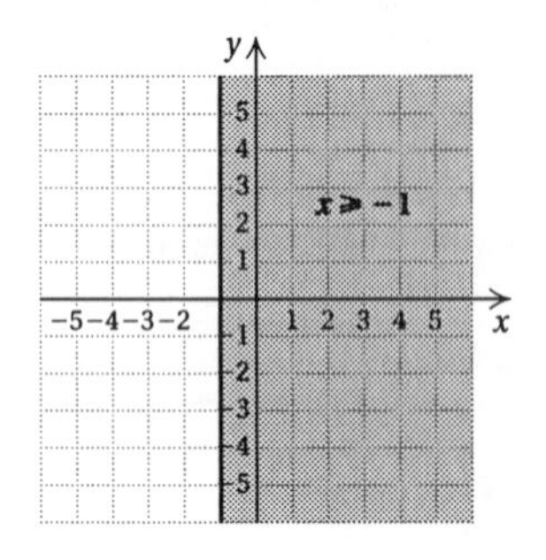

29.
$$\frac{12}{x} = \frac{48}{x+9}, \text{ LCM is } x(x+9)$$
$$x(x+9) \cdot \frac{12}{x} = x(x+9) \cdot \frac{48}{x+9}$$
$$12(x+9) = 48x$$
$$12x + 108 = 48x$$
$$108 = 36x$$
$$3 = x$$

The number 3 checks and is the solution.

31.
$$x^2 + 16 = 8x$$
$$x^2 - 8x + 16 = 0$$
$$(x-4)(x-4) = 0$$
$$x - 4 = 0 \quad \text{or} \quad x - 4 = 0$$
$$x = 4 \quad \text{or} \quad x = 4$$

The solution is 4.

33. 

35. The c children weigh $35c$ kg, and the a adults weigh $75a$ kg. Together, the children and adults weigh $35c + 75a$ kg. When this total is more than 1000 kg the elevator is overloaded, so we have $35c + 75a > 1000$. (Of course, c and a would also have to be nonnegative, but we will not deal with nonnegativity constraints here.)

To graph $35c+75a > 1000$, we first graph $35c+75a = 1000$ using a dashed line. Two points on the line are $(4,20)$ and $(11,5)$. (We are using alphabetical order of variables.) Then we test the point $(0,0)$.

$$\begin{array}{c|c} 35c + 75a > 1000 & \\ \hline 35 \cdot 0 + 75 \cdot 0 \ ? \ 1000 & \\ 0 & \text{FALSE} \end{array}$$

Since $(0,0)$ is not a solution of the inequality, we shade the region that does not contain $(0,0)$.

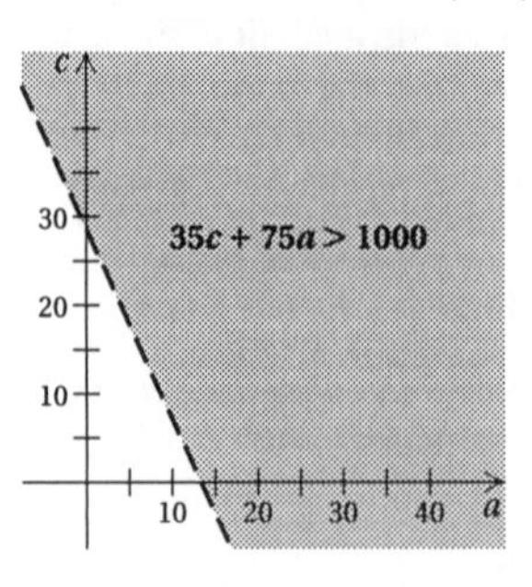

Exercise Set 13.5

1. We substitute to find k.

$y = kx$

$36 = k \cdot 9$ Substituting 36 for y and 9 for x

$\frac{36}{9} = k$

$4 = k$ k is the variation constant.

The equation of the variation is $y = 4x$.

3. We substitute to find k.

$y = kx$

$0.8 = k \cdot 0.5$ Substituting 0.8 for y and 0.5 for x

$\frac{0.8}{0.5} = k$

$\frac{8}{5} = k$ k is the variation constant.

The equation of the variation is $y = \frac{8}{5}x$.

5. We substitute to find k.

$y = kx$

$630 = k \cdot 175$ Substituting 630 for y and 175 for x

$\frac{630}{175} = k$

$3.6 = k$ k is the variation constant.

The equation of the variation is $y = 3.6x$.

7. We substitute to find k.

$y = kx$

$500 = k \cdot 60$ Substituting 500 for y and 60 for x

$\frac{500}{60} = k$

$\frac{25}{3} = k$ k is the variation constant.

The equation of the variation is $y = \frac{25}{3}x$.

9. ***Familiarize and Translate.*** The problem states that we have direct variation between the variables P and H. Thus, an equation $P = kH$, $k > 0$, applies. As the number of hours increases, the paycheck increases.

Solve.

a) First find an equation of variation.

$P = kH$

$84 = k \cdot 15$ Substituting 78.75 for P and 15 for H

$\frac{84}{15} = k$

$5.6 = k$

The equation of variation is $P = 5.6H$.

b) Use the equation to find the pay for 35 hours work.

$P = 5.6H$

$P = 5.6(35)$ Substituting 35 for H

$P = 196$

Check. This check might be done by repeating the computations. We might also do some reasoning about the answer. The paycheck increased from \$84 to \$196. Similarly, the hours increased from 15 to 35.

State. For 35 hours work, the paycheck is \$196.

11. ***Familiarize and Translate.*** The problem states that we have direct variation between the variables C and S. Thus, an equation $C = kS$, $k > 0$, applies. As the depth increases, the cost increases.

Solve.

a) First find an equation of variation.

$C = kS$

$75 = k \cdot 6$ Substituting 75 for C and 6 for S

$\frac{75}{6} = k$

$12.5 = k$

The equation of variation is $C = 12.5S$.

b) Use the equation to find the cost of filling the sandbox to a depth of 8 inches.

$C = 12.5S$

$C = 12.5(8)$ Substituting 8 for S

$C = 100$

Check. In addition to repeating the computations, we can also do some reasoning. The depth increased from 6 inches to 8 inches. Similarly, the cost increased from \$75 to \$100.

State. The sand will cost \$100.

13. ***Familiarize and Translate.*** This problem states that we have direct variation between the variables S and W. Thus, an equation $S = kW$, $k > 0$, applies. As the weight increases, the number of servings increases.

Solve.

a) First find an equation of variation.

$S = kW$

$40 = k \cdot 14$ Substituting 40 for S and 14 for W

$\frac{40}{14} = k$

$\frac{20}{7} = k$

The equation of variation is $S = \frac{20}{7}W$.

b) Use the equation to find the number of servings from an 8-kg turkey.

$S = \frac{20}{7}W$

$S = \frac{20}{7} \cdot 8$ Substituting 8 for W

$S = \frac{160}{7}$, or $22\frac{6}{7}$

Check. A check can always be done by repeating the computations. We can also do some reasoning about the answer. The number of servings decreased from 40 to $22\frac{6}{7}$. Similarly, the weight decreased from 14 kg to 8 kg.

State. $22\frac{6}{7}$ servings can be obtained from an 8-kg turkey.

15. ***Familiarize and Translate.*** The problem states that we have direct variation between the variables M and E. Thus, an equation $M = kE$, $k > 0$, applies. As the weight on earth increases, the weight on the moon increases.

Solve.

a) First find an equation of variation.

$M = kE$

$28.6 = k \cdot 171.6$ Substituting 28.6 for M and 171.6 for E

$286 = 1716k$ Clearing decimals

$\frac{286}{1716} = k$

$\frac{1}{6} = k$

The equation of variation is $M = \frac{1}{6}E$.

b) Use the equation to find how much a 220-lb person would weigh on the moon.

$M = \frac{1}{6}E$

$M = \frac{1}{6} \cdot 220$ Substituting 220 for E

$M = \frac{220}{6}$, or $36.\overline{6}$

Check. In addition to repeating the computations we can do some reasoning. The weight on the earth increased from 171.6 lb to 220 lb. Similarly, the weight on the moon increased from 28.6 lb to $36.\overline{6}$ lb.

State. A 220-lb person would weigh $36.\overline{6}$ lb on the moon.

17. ***Familiarize and Translate***. The problem states that we have direct variation between the variables N and S. Thus, an equation $N = kS$, $k > 0$, applies. As the speed of the internal processor increases, the number of instructions increases.

Solve.

a) First find an equation of variation.

$$N = kS$$

$$2,000,000 = k \cdot 25 \quad \text{Substituting 2,000,000 for } N \text{ and 25 for } S$$

$$\frac{2,000,000}{25} = k$$

$$80,000 = k$$

The equation of variation is $N = 80,000S$.

b) Use the equation to find how many instructions the processor will perform at a speed of 200 megahertz.

$$N = 80,000S$$

$$N = 80,000 \cdot 200 \quad \text{Substituting 200 for } S$$

$$N = 16,000,000$$

Check. In addition to repeating the computations we can do some reasoning. The speed of the processor increased from 25 to 200 megahertz. Similarly, the number of instructions increased from 2,000,000 to 16,000,000.

State. The processor will perform 16,000,000 instructions running at a speed of 200 megahertz.

19. We substitute to find k.

$$y = \frac{k}{x}$$

$$3 = \frac{k}{25} \quad \text{Substituting 3 for } y \text{ and 25 for } x$$

$$25 \cdot 3 = k$$

$$75 = k$$

The equation of variation is $y = \frac{75}{x}$.

21. We substitute to find k.

$$y = \frac{k}{x}$$

$$10 = \frac{k}{8} \quad \text{Substituting 10 for } y \text{ and 8 for } x$$

$$8 \cdot 10 = k$$

$$80 = k$$

The equation of variation is $y = \frac{80}{x}$.

23. We substitute to find k.

$$y = \frac{k}{x}$$

$$6.25 = \frac{k}{0.16} \quad \text{Substituting 6.25 for } y \text{ and 0.16 for } x$$

$$0.16(6.25) = k$$

$$1 = k$$

The equation of variation is $y = \frac{1}{x}$.

25. We substitute to find k.

$$y = \frac{k}{x}$$

$$50 = \frac{k}{42} \quad \text{Substituting 50 for } y \text{ and 42 for } x$$

$$42 \cdot 50 = k$$

$$2100 = k$$

The equation of variation is $y = \frac{2100}{x}$.

27. We substitute to find k.

$$y = \frac{k}{x}$$

$$0.2 = \frac{k}{0.3} \quad \text{Substituting 0.2 for } y \text{ and 0.3 for } x$$

$$0.06 = k$$

The equation of variation is $y = \frac{0.06}{x}$.

29. a) It seems reasonable that, as the number of hours of production increases, the number of compact-disc players produced will increase, so direct variation might apply.

b) ***Familiarize***. Let H = the number of hours the production line is working, and let P = the number of compact-disc players produced. An equation $P = kH$, $k > 0$, applies. (See part (a)).

Translate. We write an equation of variation.

Number of players produced varies directly as hours of production. This translates to $P = kH$.

Solve.

a) First we find an equation of variation.

$$P = kH$$

$$15 = k \cdot 8 \quad \text{Substituting 8 for } H \text{ and 15 for } P$$

$$\frac{15}{8} = k$$

The equation of variation is $P = \frac{15}{8}H$.

b) Use the equation to find the number of players produced in 37 hr.

$$P = \frac{15}{8}H$$

$$P = \frac{15}{8} \cdot 37 \qquad \text{Substituting 37 for } H$$

$$P = \frac{555}{8} = 69\frac{3}{8}$$

Check. In addition to repeating the computations, we can do some reasoning. The number of hours increased from 8 to 37. Similarly, the number of compact disc players produced increased from 15 to $69\frac{3}{8}$.

State. About $69\frac{3}{8}$ compact-disc players can be produced in 37 hr.

31. a) It seems reasonable that, as the number of workers increases, the number of hours required to do the job decreases, so inverse variation might apply.

b) ***Familiarize.*** Let T = the time required to cook the meal and N = the number of cooks. An equation $T = k/N$, $k > 0$, applies. (See part (a)).

Translate. We write an equation of variation. Time varies inversely as the number of cooks. This translates to $T = \frac{k}{N}$.

Solve.

a) First find the equation of variation.

$$T = \frac{k}{N}$$

$$4 = \frac{k}{9} \qquad \text{Substituting 4 for } T \text{ and 9 for } N$$

$$36 = k$$

The equation of variation is $T = \frac{36}{N}$.

b) Use the equation to find the amount of time it takes 8 cooks to prepare the dinner.

$$T = \frac{36}{N}$$

$$T = \frac{36}{8} \qquad \text{Substituting 8 for } N$$

$$T = 4.5$$

Check. The check might be done by repeating the computation. We might also analyze the results. The number of cooks decreased from 9 to 8, and the time increased from 4 hr to 4.5 hr. This is what we would expect with inverse variation.

State. It will take 8 cooks 4.5 hr to prepare the dinner.

33. ***Familiarize.*** The problem states that we have inverse variation between the variables N and P. Thus, an equation $N = k/P$, $k > 0$, applies. As the miles per gallon rating increases, the number of gallons required to travel the fixed distance decreases.

Translate. We write an equation of variation. Number of gallons varies inversely as miles per gallon rating. This translates to $N = \frac{k}{P}$.

Solve.

a) First find an equation of variation.

$$N = \frac{k}{P}$$

$$20 = \frac{k}{14} \qquad \text{Substituting 20 for } N \text{ and 14 for } P$$

$$280 = k$$

The equation is $N = \frac{280}{P}$.

b) Use the equation to find the number of gallons of gasoline needed for a car that gets 28 mpg.

$$N = \frac{k}{P}$$

$$N = \frac{280}{28} \qquad \text{Substituting 28 for } P$$

$$N = 10$$

Check. In addition to repeating the computations, we can analyze the results. The number of miles per gallon increased from 14 to 28, and the number of gallons required decreased from 20 to 10. This is what we would expect with inverse variation.

State. A car that gets 28 mpg will need 10 gallons of gasoline to travel the fixed distance.

35. ***Familiarize.*** The problem states that we have inverse variation between the variables I and R. Thus, an equation $I = k/R$, $k > 0$, applies. As the resistance increases, the current decreases.

Translate. We write an equation of variation. Current varies inversely as resistance. This translates to $I = \frac{k}{R}$.

Solve.

a) First find an equation of variation.

$$I = \frac{k}{R}$$

$$96 = \frac{k}{20} \qquad \text{Substituting 96 for } I \text{ and 20 for } R$$

$$1920 = k$$

The equation of variation is $I = \frac{1920}{R}$.

b) Use the equation to find the current when the resistance is 60 ohms.

$$I = \frac{1920}{R}$$

$$I = \frac{1920}{60} \qquad \text{Substituting 60 for } R$$

$$I = 32$$

Check. The check might be done by repeating the computations. We might also analyze the results. The resistance increased from 20 ohms to 60 ohms, and the current decreased from 96 amperes to 32 amperes. This is what we would expect with inverse variation.

State. The current is 32 amperes when the resistance is 60 ohms.

37. ***Familiarize.*** Let $S =$ the size of the files. The problem states that we have inverse variation between the variables N and S. Thus, an equation $N = k/S$, $k > 0$, applies. As the size of the files increases, the number of files that can be held decreases.

Translate. We write an equation of variation. Number of files varies inversely as the size of the files. This translates to $N = \frac{k}{S}$.

Solve.

a) First find an equation of variation.

$$N = \frac{k}{S}$$

$$1600 = \frac{k}{50,000} \quad \text{Substituting 1600 for } N \text{ and 50,000 for } S$$

$$80,000,000 = k$$

The equation of variation is $N = \frac{80,000,000}{S}$.

b) Use the equation to find the number of files the disk will hold if each is 125,000 bytes.

$$N = \frac{80,000,000}{S}$$

$$N = \frac{80,000,000}{125,000} \quad \text{Substituting 125,000 for } S$$

$$N = 640$$

Check. The check might be done by repeating the computations. We might also analyze the results. The size of each file increased from 50,000 to 125,000 bytes and the number of files decreased from 1600 to 640. This is what we would expect with inverse variation.

State. The disk will hold 640 files when each is 125,000 bytes.

39. ***Familiarize.*** The problem states that we have inverse variation between the variables A and d. Thus, an equation $A = k/d$, $k > 0$, applies. As the distance increases, the apparent size decreases.

Translate. We write an equation of variation. Apparent size varies inversely as the distance. This translates to $A = \frac{k}{d}$.

Solve.

a) First find an equation of variation.

$$A = \frac{k}{d}$$

$$27.5 = \frac{k}{30} \quad \text{Substituting 27.5 for } A \text{ and 30 for } d$$

$$825 = k$$

The equation of variation is $A = \frac{825}{d}$.

b) Use the equation to find the apparent size when the distance is 100 ft.

$$A = \frac{825}{d}$$

$$A = \frac{825}{100} \quad \text{Substituting 100 for } d$$

$$A = 8.25$$

Check. The check might be done by repeating the computations. We might also analyze the results. The distance increased from 30 ft to 100 ft, and the apparent size decreased from 27.5 ft to 8.25 ft. This is what we would expect with inverse variation.

State. The flagpole will appear to be 8.25 ft tall when it is 100 ft from the observer.

41.

$$\frac{x+2}{x+5} = \frac{x-4}{x-6}, \quad \text{LCM is } (x+5)(x-6)$$

$$(x+5)(x-6) \cdot \frac{x+2}{x+5} = (x+5)(x-6) \cdot \frac{x-4}{x-6}$$

$$(x-6)(x+2) = (x+5)(x-4)$$

$$x^2 - 4x - 12 = x^2 + x - 20$$

$$-4x - 12 = x - 20 \quad \text{Subtracting } x^2$$

$$-5x = -8 \quad \text{Subtracting } x \text{ and adding 12}$$

$$x = \frac{8}{5}$$

The number $\frac{8}{5}$ checks and is the solution.

43.

$$x^2 - 25x + 144 = 0$$

$$(x-9)(x-16) = 0$$

$$x - 9 = 0 \quad or \quad x - 16 = 0$$

$$x = 9 \quad or \quad x = 16$$

The solutions are 9 and 16.

45.

$$35x^2 + 8 = 34x$$

$$35x^2 - 34x + 8 = 0$$

$$(7x-4)(5x-2) = 0$$

$$7x - 4 = 0 \quad or \quad 5x - 2 = 0$$

$$7x = 4 \quad or \quad 5x = 2$$

$$x = \frac{4}{7} \quad or \quad x = \frac{2}{5}$$

The solutions are $\frac{4}{7}$ and $\frac{2}{5}$.

47. We do the divisions in order from left to right.

$$3^7 \div 3^4 \div 3^3 \div 3 = 3^3 \div 3^3 \div 3$$

$$= 1 \div 3$$

$$= \frac{1}{3}$$

49. ◈

51. ◈

53. The y-values get larger.

55. $P^2 = kt$

57. $P = kV^3$

Chapter 14

Systems of Equations

Exercise Set 14.1

1. We check by substituting alphabetically 1 for x and 5 for y.

$$\begin{array}{c|c} \multicolumn{2}{c}{5x - 2y = -5} \\ \hline 5 \cdot 1 - 2 \cdot 5 \ ? \ -5 & \\ 5 - 10 & \\ -5 & \text{TRUE} \end{array} \qquad \begin{array}{c|c} \multicolumn{2}{c}{3x - 7y = -32} \\ \hline 3 \cdot 1 - 7 \cdot 5 \ ? \ -32 & \\ 3 - 35 & \\ -32 & \text{TRUE} \end{array}$$

The ordered pair $(1, 5)$ is a solution of both equations, so it is a solution of the system of equations.

3. We check by substituting alphabetically 4 for a and 2 for b.

$$\begin{array}{c|c} \multicolumn{2}{c}{3b - 2a = -2} \\ \hline 3 \cdot 2 - 2 \cdot 4 \ ? \ -2 & \\ 6 - 8 & \\ -2 & \text{TRUE} \end{array} \qquad \begin{array}{c|c} \multicolumn{2}{c}{b + 2a = 8} \\ \hline 2 + 2 \cdot 4 \ ? \ 8 & \\ 2 + 8 & \\ 10 & \text{FALSE} \end{array}$$

The ordered pair $(4, 2)$ is not a solution of $b + 2a = 8$, so it is not a solution of the system of equations.

5. We check by substituting alphabetically 15 for x and 20 for y.

$$\begin{array}{c|c} \multicolumn{2}{c}{3x - 2y = 5} \\ \hline 3 \cdot 15 - 2 \cdot 20 \ ? \ 5 & \\ 45 - 40 & \\ 5 & \text{TRUE} \end{array} \qquad \begin{array}{c|c} \multicolumn{2}{c}{6x - 5y = -10} \\ \hline 6 \cdot 15 - 5 \cdot 20 \ ? \ -10 & \\ 90 - 100 & \\ -10 & \text{TRUE} \end{array}$$

The ordered pair $(15, 20)$ is a solution of both equations, so it is a solution of the system of equations.

7. We check by substituting alphabetically -1 for x and 1 for y.

$$\begin{array}{c|c} \multicolumn{2}{c}{x = -1} \\ \hline -1 \ ? \ -1 & \text{TRUE} \end{array} \qquad \begin{array}{c|c} \multicolumn{2}{c}{x - y = -2} \\ \hline -1 - 1 \ ? \ -2 & \\ -2 & \text{TRUE} \end{array}$$

The ordered pair $(-1, 1)$ is a solution of both equations, so it is a solution of the system of equations.

9. We check by substituting alphabetically 18 for x and 3 for y.

$$\begin{array}{c|c} \multicolumn{2}{c}{y = \frac{1}{6}x} \\ \hline 3 \ ? \ \frac{1}{6} \cdot 18 & \\ & 3 \quad \text{TRUE} \end{array} \qquad \begin{array}{c|c} \multicolumn{2}{c}{2x - y = 33} \\ \hline 2 \cdot 18 \ ? \ 33 & \\ 36 - 3 & \\ 33 & \text{TRUE} \end{array}$$

The ordered pair $(18, 3)$ is a solution of both equations, so it is a solution of the system of equations.

11. We graph the equations.

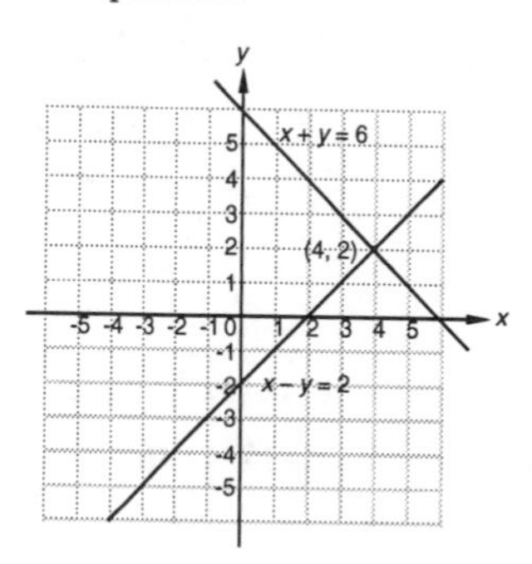

The point of intersection looks as if it has coordinates $(4, 2)$.

Check:

$$\begin{array}{c|c} \multicolumn{2}{c}{x - y = 2} \\ \hline 4 - 2 \ ? \ 2 & \\ 2 & \text{TRUE} \end{array} \qquad \begin{array}{c|c} \multicolumn{2}{c}{x + y = 6} \\ \hline 4 + 2 \ ? \ 6 & \\ 6 & \text{TRUE} \end{array}$$

The solution is $(4, 2)$.

13. We graph the equations.

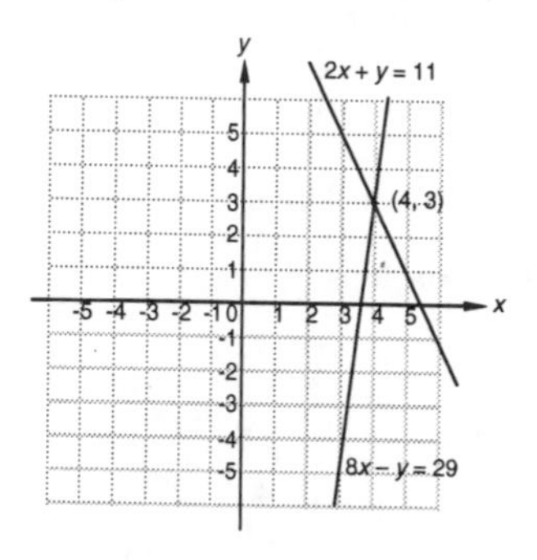

The point of intersection looks as if it has coordinates $(4, 3)$.

Check:

$$\begin{array}{c|c} \multicolumn{2}{c}{8x - y = 29} \\ \hline 8 \cdot 4 - 3 \ ? \ 29 & \\ 32 - 3 & \\ 29 & \text{TRUE} \end{array} \qquad \begin{array}{c|c} \multicolumn{2}{c}{2x + y = 11} \\ \hline 2 \cdot 4 + 3 \ ? \ 11 & \\ 8 + 3 & \\ 11 & \text{TRUE} \end{array}$$

The solution is $(4, 3)$.

15. We graph the equations.

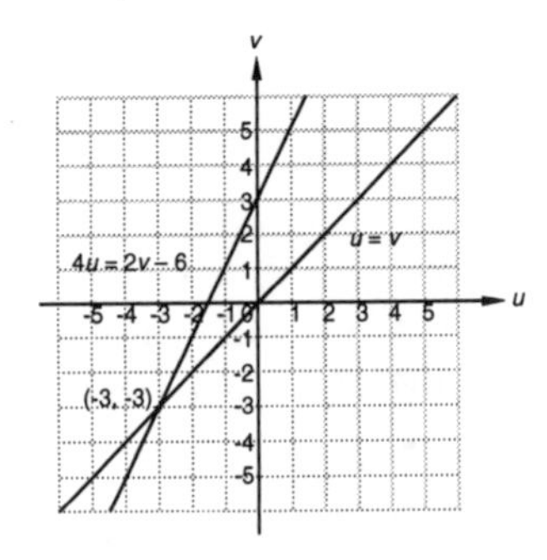

The point of intersection looks as if it has coordinates $(-3,-3)$.

Check:

$u = v$	
$-3 \ ? \ -3$	TRUE

$4u = 2v - 6$		
$4(-3)$	$? \ 2(-3) - 6$	
-12	$-6 - 6$	
	-12	TRUE

The solution is $(-3,-3)$.

17. We graph the equations.

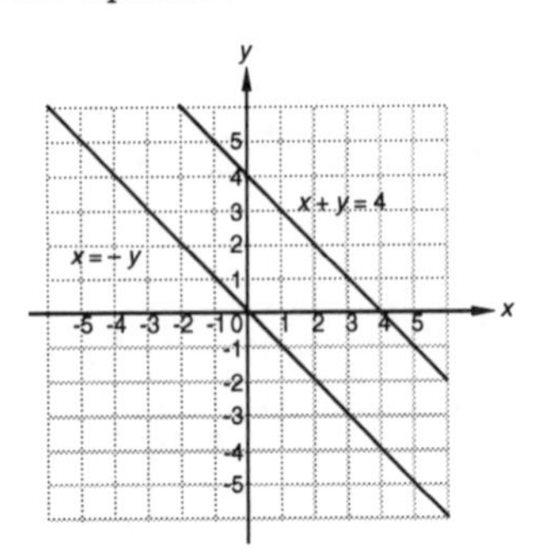

The lines are parallel. There is no solution.

19. We graph the equations.

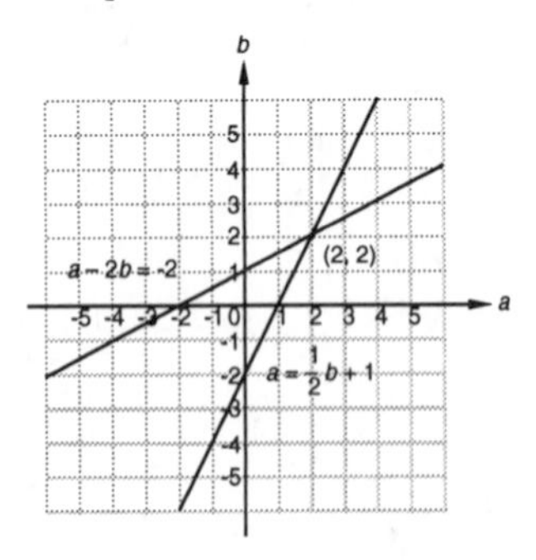

The point of intersection looks as if it has coordinates $(2,2)$.

Check:

$a = \frac{1}{2}b + 1$		
$2 \ ? \ \frac{1}{2} \cdot 2 + 1$		
	$1 + 1$	
	2	TRUE

$a - 2b = -2$		
$2 - 2 \cdot 2$	$? \ -2$	
$2 - 4$		
-2		TRUE

The solution is $(2,2)$.

21. We graph the equations.

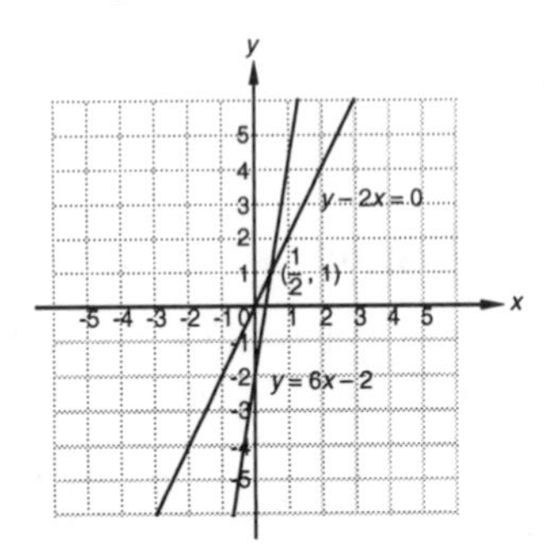

The point of intersection looks as if it has coordinates $\left(\frac{1}{2}, 1\right)$.

Check:

$y - 2x = 0$		
$1 - 2 \cdot \frac{1}{2}$	$? \ 0$	
$1 - 1$		
0		TRUE

$y = 6x - 2$		
1	$? \ 6 \cdot \frac{1}{2} - 2$	
	$3 - 2$	
	1	TRUE

The solution is $\left(\frac{1}{2}, 1\right)$.

23. We graph the equations.

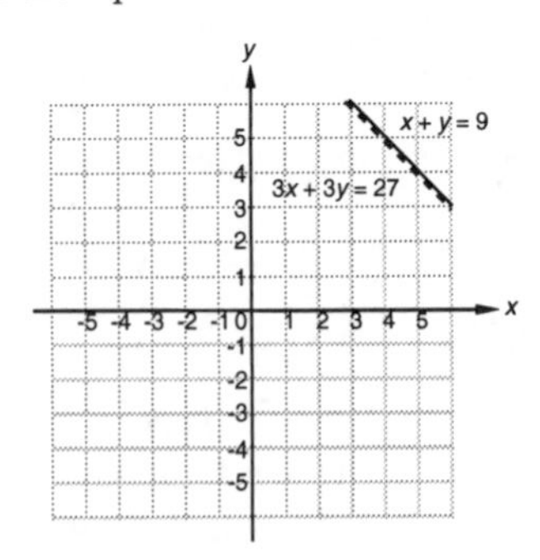

The lines coincide. The system has an infinite number of solutions.

25. We graph the equations.

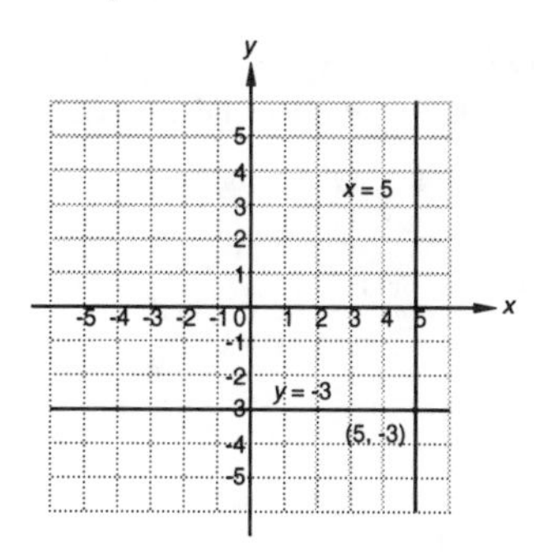

The point of intersection looks as if it has coordinates $(5, -3)$.

Check:

$x = 5$		$y = -3$	
$5 \; ? \; 5$	TRUE	$-3 \; ? \; -3$	TRUE

The solution is $(5, -3)$.

27.
$$\begin{aligned}(9x^{-5})(12x^{-8}) &= 9 \cdot 12 \cdot x^{-5} \cdot x^{-8}\\ &= 108x^{-5+(-8)}\\ &= 108x^{-13}\\ &= \frac{108}{x^{13}}\end{aligned}$$

29.
$$\frac{1}{x} - \frac{1}{x^2} + \frac{1}{x+1}, \text{ LCM is } x^2(x+1)$$
$$= \frac{1}{x} \cdot \frac{x(x+1)}{x(x+1)} - \frac{1}{x^2} \cdot \frac{x+1}{x+1} + \frac{1}{x+1} \cdot \frac{x^2}{x^2}$$
$$= \frac{x(x+1) - (x+1) + x^2}{x^2(x+1)}$$
$$= \frac{x^2 + x - x - 1 + x^2}{x^2(x+1)}$$
$$= \frac{2x^2 - 1}{x^2(x+1)}$$

31.
$$\frac{x+2}{x-4} - \frac{x+1}{x+4}, \text{ LCM is } (x-4)(x+4)$$
$$= \frac{x+2}{x-4} \cdot \frac{x+4}{x+4} - \frac{x+1}{x+4} \cdot \frac{x-4}{x-4}$$
$$= \frac{(x+2)(x+4) - (x+1)(x-4)}{(x-4)(x+4)}$$
$$= \frac{x^2 + 6x + 8 - (x^2 - 3x - 4)}{(x-4)(x+4)}$$
$$= \frac{x^2 + 6x + 8 - x^2 + 3x + 4}{(x-4)(x+4)}$$
$$= \frac{9x + 12}{(x-4)(x+4)}$$

33. The polynomial has exactly three terms, so it is a trinomial.

35. The polynomial has exactly one term, so it is a monomial.

37. 

39. $(2, -3)$ is a solution of $Ax - 3y = 13$. Substitute 2 for x and -3 for y and solve for A.

$$\begin{aligned}Ax - 3y &= 13\\ A \cdot 2 - 3(-3) &= 13\\ 2A + 9 &= 13\\ 2A &= 4\\ A &= 2\end{aligned}$$

$(2, -3)$ is a solution of $x - By = 8$. Substitute 2 for x and -3 for y and solve for B.

$$\begin{aligned}x - By &= 8\\ 2 - B(-3) &= 8\\ 2 + 3B &= 8\\ 3B &= 6\\ B &= 2\end{aligned}$$

41. Answers may vary. Any two equations with a solution of $(6, -2)$ will do. One possibility is

$x + y = 4,$

$x - y = 8.$

43.-49.

Exercise Set 14.2

1. $x + y = 10, \quad (1)$

$y = x + 8 \quad (2)$

We substitute $x + 8$ for y in Equation (1) and solve for x.

$x + y = 10$	Equation (1)
$x + (x + 8) = 10$	Substituting
$2x + 8 = 10$	Collecting like terms
$2x = 2$	Subtracting 8
$x = 1$	Dividing by 2

Next we substitute 1 for x in either equation of the original system and solve for y. We choose Equation (2) since it has y alone on one side.

$y = x + 8$	Equation (2)
$y = 1 + 8$	Substituting
$y = 9$	

We check the ordered pair $(1, 9)$.

$x + y = 10$		$y = x + 8$	
$1 + 9 \; ? \; 10$		$9 \; ? \; 1 + 8$	
10	TRUE	9	TRUE

Since $(1, 9)$ checks in both equations, it is the solution.

3. $y = x - 6,$ (1)

$x + y = -2$ (2)

We substitute $x - 6$ for y in Equation (2) and solve for x.

$x + y = -2$ Equation (2)

$x + (x - 6) = -2$ Substituting

$2x - 6 = -2$ Collecting like terms

$2x = 4$ Adding 6

$x = 2$ Dividing by 2

Next we substitute 2 for x in either equation of the original system and solve for y. We choose Equation (1) since it has y alone on one side.

$y = x - 6$ Equation (1)

$y = 2 - 6$ Substituting

$y = -4$

We check the ordered pair $(2, -4)$.

$y = x - 6$		$x + y = -2$	
$-4 \ ? \ 2 - 6$		$2 + (-4) \ ? \ -2$	
-4	TRUE	-2	TRUE

Since $(2, -4)$ checks in both equations, it is the solution.

5. $y = 2x - 5,$ (1)

$3y - x = 5$ (2)

We substitute $2x - 5$ for y in Equation (2) and solve for x.

$3y - x = 5$ Equation (2)

$3(2x - 5) - x = 5$ Substituting

$6x - 15 - x = 5$ Removing parentheses

$5x - 15 = 5$ Collecting like terms

$5x = 20$ Adding 15

$x = 4$ Dividing by 5

Next we substitute 4 for x in either equation of the original system and solve for y.

$y = 2x - 5$ Equation (1)

$y = 2 \cdot 4 - 5$ Substituting

$y = 8 - 5$

$y = 3$

We check the ordered pair $(4, 3)$.

$y = 2x - 5$		$3y - x = 5$	
$3 \ ? \ 2 \cdot 4 - 5$		$3 \cdot 3 - 4 \ ? \ 5$	
$8 - 5$		$9 - 4$	
3	TRUE	5	TRUE

Since $(4, 3)$ checks in both equations, it is the solution.

7. $x = -2y,$ (1)

$x + 4y = 2$ (2)

We substitute $-2y$ for x in Equation (2) and solve for y.

$x + 4y = 2$ Equation (2)

$-2y + 4y = 2$ Substituting

$2y = 2$ Collecting like terms

$y = 1$ Dividing by 2

Next we substitute 1 for y in either equation of the original system and solve for x.

$x = -2y$ Equation (1)

$x = -2 \cdot 1$

$x = -2$

We check the ordered pair $(-2, 1)$.

$x = -2y$		$3y - x = 5$	
$-2 \ ? \ -2 \cdot 1$		$3 \cdot 1 - (-2) \ ? \ 5$	
-2	TRUE	$3 + 2$	
		5	TRUE

Since $(-2, 1)$ checks in both equations, it is the solution.

9. $x - y = 6,$ (1)

$x + y = -2$ (2)

We solve Equation (1) for x.

$x - y = 6$ Equation (1)

$x = y + 6$ Adding y (3)

We substitute $y + 6$ for x in Equation (2) and solve for y.

$x + y = -2$ Equation (2)

$(y + 6) + y = -2$ Substituting

$2y + 6 = -2$ Collecting like terms

$2y = -8$ Subtracting 6

$y = -4$ Dividing by 2

Now we substitute -4 for y in Equation (3) and compute x.

$x = y + 6 = -4 + 6 = 2$

The ordered pair $(2, -4)$ checks in both equations. It is the solution.

11. $y - 2x = -6,$ (1)

$2y - x = 5$ (2)

We solve Equation (1) for y.

$y - 2x = -6$ Equation (1)

$y = 2x - 6$ (3)

We substitute $2x - 6$ for y in Equation (2) and solve for x.

$2y - x = 5$ Equation (2)

$2(2x - 6) - x = 5$ Substituting

$4x - 12 - x = 5$ Removing parentheses

$3x - 12 = 5$ Collecting like terms

$3x = 17$ Adding 12

$x = \frac{17}{3}$ Dividing by 3

We substitute $\frac{17}{3}$ for x in Equation (3) and compute y.

$$y = 2x - 6 = 2\left(\frac{17}{3}\right) - 6 = \frac{34}{3} - \frac{18}{3} = \frac{16}{3}$$

The ordered pair $\left(\frac{17}{3}, \frac{16}{3}\right)$ checks in both equations. It is the solution.

13. $2x + 3y = -2,\quad (1)$

$2x - y = 9\quad (2)$

We solve Equation (2) for y.

$2x - y = 9$ Equation (2)

$2x = 9 + y$ Adding y

$2x - 9 = y$ Subtracting 9 (3)

We substitute $2x - 9$ for y in Equation (1) and solve for x.

$2x + 3y = -2$ Equation (1)

$2x + 3(2x - 9) = -2$ Substituting

$2x + 6x - 27 = -2$ Removing parentheses

$8x - 27 = -2$ Collecting like terms

$8x = 25$ Adding 27

$x = \frac{25}{8}$ Dividing by 8

Now we substitute $\frac{25}{8}$ for x in Equation (3) and compute y.

$$y = 2x - 9 = 2\left(\frac{25}{8}\right) - 9 = \frac{25}{4} - \frac{36}{4} = -\frac{11}{4}$$

The ordered pair $\left(\frac{25}{8}, -\frac{11}{4}\right)$ checks in both equations. It is the solution.

15. $x - y = -3,\quad (1)$

$2x + 3y = -6\quad (2)$

We solve Equation (1) for x.

$x - y = -3$ Equation (1)

$x = y - 3$ (3)

We substitute $y - 3$ for x in Equation (2) and solve for y.

$2x + 3y = -6$ Equation (2)

$2(y - 3) + 3y = -6$ Substituting

$2y - 6 + 3y = -6$ Removing parentheses

$5y - 6 = -6$ Collecting like terms

$5y = 0$ Adding 6

$y = 0$ Dividing by 5

Now we substitute 0 for y in Equation (3) and compute x.

$x = y - 3 = 0 - 3 = -3$

The ordered pair $(-3, 0)$ checks in both equations. It is the solution.

17. $r - 2s = 0,\quad (1)$

$4r - 3s = 15\quad (2)$

We solve Equation (1) for r.

$r - 2s = 0$ Equation (1)

$r = 2s$ (3)

We substitute $2s$ for r in Equation (2) and solve for s.

$4r - 3s = 15$ Equation (2)

$4(2s) - 3s = 15$ Substituting

$8s - 3s = 15$ Removing parentheses

$5s = 15$ Collecting like terms

$s = 3$ Dividing by 5

Now we substitute 3 for s in Equation (3) and compute r.

$r = 2s = 2 \cdot 3 = 6$

The ordered pair $(6, 3)$ checks in both equations. It is the solution.

19. ***Familiarize.*** We let x = the larger number and y = the smaller number.

Translate. We translate the first statement.

The sum of two numbers is 37.

$x + y = 37$

Now we translate the second statement.

One number is 5 more than the other.

$x = 5 + y$

The resulting system is

$x + y = 37,\quad (1)$

$x = 5 + y.\quad (2)$

Solve. We solve the system of equations. We substitute $5 + y$ for x in Equation (1) and solve for y.

$x + y = 37$ Equation (1)

$(5 + y) + y = 37$ Substituting

$5 + 2y = 37$ Collecting like terms

$2y = 32$ Subtracting 5

$y = 16$ Dividing by 2

We go back to the original equations and substitute 16 for y. We use Equation (2).

$x = 5 + y$ Equation (2)

$x = 5 + 16$ Substituting

$x = 21$

Check. The sum of 21 and 16 is 37. The number 21 is 5 more than the number 16. These numbers check.

State. The numbers are 21 and 16.

21. ***Familiarize.*** Let x = one number and y = the other.

Translate. We reword and translate.

The sum of two numbers is 52.

$x + y = 52$

The difference of two numbers is 28.

$x - y = 28$

(The second statement could also be translated as $y - x = 28$.)

The resulting system is

$$x + y = 52, \quad (1)$$
$$x - y = 28. \quad (2)$$

Solve. We solve the system. First we solve Equation (2) for x.

$$x - y = 28 \qquad \text{Equation (2)}$$
$$x = y + 28 \qquad \text{Adding } y \qquad (3)$$

We substitute $y + 28$ for x in Equation (1) and solve for y.

$$x + y = 52 \qquad \text{Equation (1)}$$
$$(y + 28) + y = 52 \qquad \text{Substituting}$$
$$2y + 28 = 52 \qquad \text{Collecting like terms}$$
$$2y = 24 \qquad \text{Subtracting 28}$$
$$y = 12 \qquad \text{Dividing by 2}$$

Now we substitute 12 for y in Equation (3) and compute x.

$$x = y + 28 = 12 + 28 = 40$$

Check. The sum of 40 and 12 is 52, and their difference is 28. These numbers check.

State. The numbers are 40 and 12.

23. ***Familiarize.*** We let x = the larger number and y = the smaller number.

Translate. We translate the first statement.

The difference between two numbers is 12.

$$x - y = 12$$

Now we translate the second statement.

Two times the larger number is five times the smaller.

$$2x = 5y$$

The resulting system is

$$x - y = 12, \quad (1)$$
$$2x = 5y. \quad (2)$$

Solve. We solve the system. First we solve Equation (1) for x.

$$x - y = 12 \qquad \text{Equation (1)}$$
$$x = y + 12 \qquad \text{Adding } y \qquad (3)$$

We substitute $y + 12$ for x in Equation (2) and solve for y.

$$2x = 5y \qquad \text{Equation (2)}$$
$$2(y + 12) = 5y \qquad \text{Substituting}$$
$$2y + 24 = 5y \qquad \text{Removing parentheses}$$
$$24 = 3y \qquad \text{Subtracting 2y}$$
$$8 = y \qquad \text{Dividing by 3}$$

Now we substitute 8 for y in Equation (3) and compute x.

$$x = y + 12 = 8 + 12 = 20$$

Check. The difference between 20 and 8 is 12. Two times 20, or 40, is five times 8. These numbers check.

State. The numbers are 20 and 8.

25. ***Familiarize.*** From the drawing in the text we see that we have a rectangle with length l and width w.

Translate. The perimeter of a rectangle is $2l + 2w$. We translate the first statement.

The perimeter is 1300 mi.

$$2l + 2w = 1300$$

Then we reword and translate the second statement.

The width is the length less 110 mi.

$$w = l - 110$$

The resulting system is

$$2l + 2w = 1300, \quad (1)$$
$$w = l - 110. \quad (2)$$

Solve. We solve the system. We substitute $l - 110$ for w in Equation (1) and solve for l.

$$2l + 2w = 1300 \qquad \text{Equation (1)}$$
$$2l + 2(l - 110) = 1300 \qquad \text{Substituting}$$
$$2l + 2l - 220 = 1300 \qquad \text{Removing parentheses}$$
$$4l - 220 = 1300 \qquad \text{Collecting like terms}$$
$$4l = 1520 \qquad \text{Adding 220}$$
$$l = 380 \qquad \text{Dividing by 4}$$

Now we substitute 380 for l in Equation (2) and solve for w.

$$w = l - 110 \qquad \text{Equation (2)}$$
$$w = 380 - 110 \qquad \text{Substituting}$$
$$w = 270$$

Check. A possible solution is a length of 380 mi and a width of 270 mi. The perimeter would be $2 \cdot 380 + 2 \cdot 270$, or $760 + 540$, or 1300. Also, the width is 110 mi less than the length. These numbers check.

State. The length is 380 mi, and the width is 270 mi.

27. ***Familiarize.*** We make a drawing. We let l = the length and w = the width.

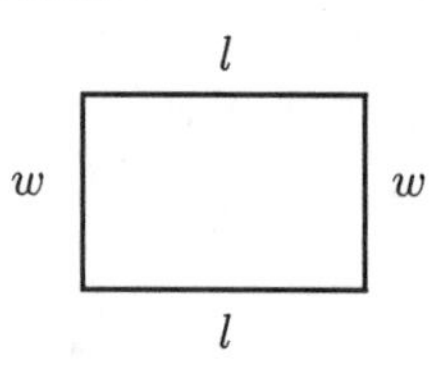

Translate. The perimeter is $2l + 2w$. We translate the first statement.

The perimeter is $10\frac{1}{2}$ in.

$$2l + 2w = 10\frac{1}{2}$$

We translate the second statement.

$$\underbrace{\text{The length}}_{\downarrow \atop l} \ \underset{\downarrow \atop =}{\text{is}} \ \underbrace{\text{twice the width.}}_{\downarrow \atop 2w}$$

The resulting system is

$$2l + 2w = 10\frac{1}{2}, \quad (1)$$
$$l = 2w. \quad (2)$$

Solve. We solve the system. We substitute $2w$ for l in Equation (1) and solve for w. We also express $10\frac{1}{2}$ as $\frac{21}{2}$.

$$2l + 2w = 10\frac{1}{2} \quad \text{Equation (1)}$$
$$2(2w) + 2w = \frac{21}{2} \quad \text{Substituting}$$
$$4w + 2w = \frac{21}{2} \quad \text{Removing parentheses}$$
$$6w = \frac{21}{2} \quad \text{Collecting like terms}$$
$$w = \frac{1}{6} \cdot \frac{21}{2} \quad \text{Multiplying by } \frac{1}{6}$$
$$w = \frac{1 \cdot 21}{6 \cdot 2}$$
$$w = \frac{1 \cdot \not{3} \cdot 7}{\not{3} \cdot 2 \cdot 2}$$
$$w = \frac{7}{4}, \text{ or } 1\frac{3}{4}$$

Now we substitute $\frac{7}{4}$ for w in Equation (2) and solve for l.

$$l = 2w \quad \text{Equation (2)}$$
$$l = 2 \cdot \frac{7}{4} \quad \text{Substituting}$$
$$l = \frac{2 \cdot 7}{4}$$
$$l = \frac{\not{2} \cdot 7}{\not{2} \cdot 2}$$
$$l = \frac{7}{2}, \text{ or } 3\frac{1}{2}$$

Check. A possible solution is a length of $\frac{7}{2}$, or $3\frac{1}{2}$ in. and a width of $\frac{7}{4}$, or $1\frac{3}{4}$ in. The perimeter would be $2 \cdot \frac{7}{2} + 2 \cdot \frac{7}{4}$, or $7 + \frac{7}{2}$, or $10\frac{1}{2}$ in. Also, twice the width is $2 \cdot \frac{7}{4}$, or $\frac{7}{2}$, which is the length. These numbers check.

State. The length is $3\frac{1}{2}$ in., and the width is $1\frac{3}{4}$ in.

29. Graph: $2x - 3y = 6$

To find the x-intercept, let $y = 0$. Then solve for x.

$$2x - 3 \cdot 0 = 6$$
$$2x = 6$$
$$x = 3$$

The x-intercept is $(3, 0)$.

To find the y-intercept, let $x = 0$. Then solve for y.

$$2 \cdot 0 - 3y = 6$$
$$-3y = 6$$
$$y = -2$$

The y-intercept is $(0, -2)$.

We plot these points and draw the line.

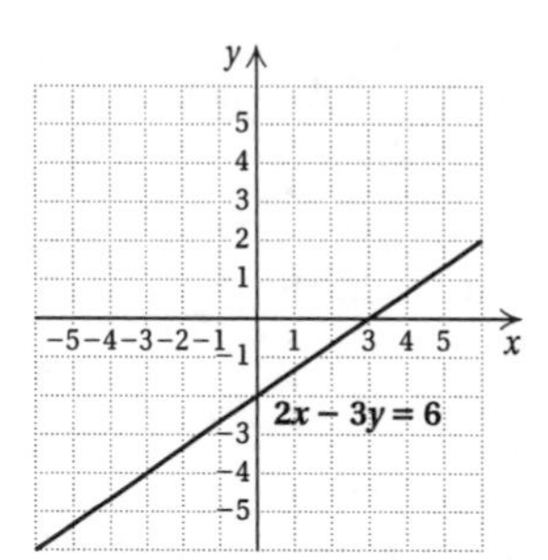

A third point should be used as a check. We let $x = -3$:

$$2(-3) - 3y = 6$$
$$-6 - 3y = 6$$
$$-3y = 12$$
$$y = -4$$

The point $(-3, -4)$ is on the graph, so our graph is probably correct.

31. Graph: $y = 2x - 5$

We select several values for x and compute the corresponding y-values.

When $x = 0$, $y = 2 \cdot 0 - 5 = 0 - 5 = -5$.

When $x = 2$, $y = 2 \cdot 2 - 5 = 4 - 5 = -1$.

When $x = 4$, $y = 2 \cdot 4 - 5 = 8 - 5 = 3$.

x	y	(x, y)
0	−5	$(0, -5)$
2	−1	$(2, -1)$
4	3	$(4, 3)$

We plot these points and draw the line connecting them.

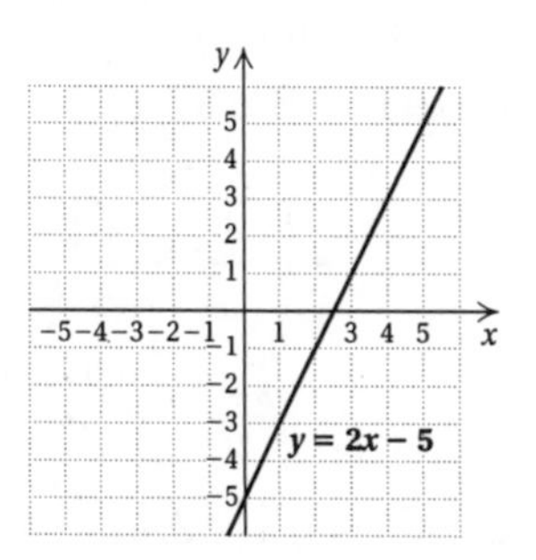

33. $6x^2 - 13x + 6$

The possibilities are $(x+\quad)(6x+\quad)$ and $(2x+\quad)(3x+\quad)$. We look for a pair of factors of the last term, 6, which produces the correct middle term. Since the last term is positive and the middle term is negative, we need only consider negative pairs. The factorization is $(2x-3)(3x-2)$.

35. $4x^2+3x+2$

The possibilities are $(x+\quad)(4x+\quad)$ and $(2x+\quad)(2x+\quad)$. We look for a pair of factors of the last term, 2, which produce the correct middle term. Since the last term and the middle term are both positive, we need only consider positive pairs. We find that there is no possibility that works. The trinomial cannot be factored.

37. 

39. First put the equations in "$y=$" form by solving for y. We get

$$y_1 = x-5,$$
$$y_2 = (-1/2)x + 7/2.$$

Then graph these equations in the standard window and use the CALC-INTERSECT feature to find the coordinates of the point of intersection of the graphs. The solution is $(5.\overline{6}, 0.\overline{6})$.

41. First put the equations in "$y=$" form by solving for y. We get

$$y_1 = 2.35x - 5.97,$$
$$y_2 = (1/2.14)x + (4.88/2.14).$$

Then graph these equations in the standard window and use the CALC-INTERSECT feature to find the coordinates of the point of intersection of the graphs. The solution is approximately $(4.38, 4.33)$.

Exercise Set 14.3

1.
$$x - y = 7 \quad (1)$$
$$x + y = 5 \quad (2)$$
$$2x = 12 \quad \text{Adding}$$
$$x = 6 \quad \text{Dividing by 2}$$

Substitute 6 for x in either of the original equations and solve for y.

$$x + y = 5 \quad \text{Equation (2)}$$
$$6 + y = 5 \quad \text{Substituting}$$
$$y = -1 \quad \text{Subtracting 6}$$

Check:

$x - y = 7$			$x + y = 5$		
$6-(-1)$? 7			$6+(-1)$? 5		
$6+1$			5	TRUE	
7	TRUE				

Since $(6,-1)$ checks, it is the solution.

3.
$$x + y = 8 \quad (1)$$
$$-x + 2y = 7 \quad (2)$$
$$3y = 15 \quad \text{Adding}$$
$$y = 5 \quad \text{Dividing by 3}$$

Substitute 5 for y in either of the original equations and solve for x.

$$x + y = 8 \quad \text{Equation (1)}$$
$$x + 5 = 8 \quad \text{Substituting}$$
$$x = 3$$

Check:

$x + y = 8$		$-x + 2y = 7$	
$3+5$? 8		$-3+2\cdot 5$? 7	
8	TRUE	$-3+10$	
		7	TRUE

Since $(3,5)$ checks, it is the solution.

5.
$$5x - y = 5 \quad (1)$$
$$3x + y = 11 \quad (2)$$
$$8x = 16 \quad \text{Adding}$$
$$x = 2 \quad \text{Dividing by 8}$$

Substitute 2 for x in either of the original equations and solve for y.

$$3x + y = 11 \quad \text{Equation (2)}$$
$$3\cdot 2 + y = 11 \quad \text{Substituting}$$
$$6 + y = 11$$
$$y = 5$$

Check:

$5x - y = 5$		$3x + y = 11$	
$5\cdot 2-5$? 5		$3\cdot 2+5$? 11	
$10-5$		$6+5$	
5	TRUE	11	TRUE

Since $(2,5)$ checks, it is the solution.

7.
$$4a + 3b = 7 \quad (1)$$
$$-4a + b = 5 \quad (2)$$
$$4b = 12 \quad \text{Adding}$$
$$b = 3$$

Substitute 3 for b in either of the original equations and solve for a.

$$4a + 3b = 7 \quad \text{Equation (1)}$$
$$4a + 3\cdot 3 = 7 \quad \text{Substituting}$$
$$4a + 9 = 7$$
$$4a = -2$$
$$a = -\frac{1}{2}$$

Check:

$4a + 3b = 7$		$-4a + b = 5$	
$4\left(-\frac{1}{2}\right)+3\cdot 3$? 7		$-4\left(-\frac{1}{2}\right)+3$? 5	
$-2+9$		$2+3$	
7	TRUE	5	TRUE

Since $\left(-\frac{1}{2}, 3\right)$ checks, it is the solution.

9. $8x - 5y = -9$ (1)

$3x + 5y = -2$ (2)

$11x = -11$ Adding

$x = -1$

Substitute -1 for x in either of the original equations and solve for y.

$3x + 5y = -2$ Equation (2)

$3(-1) + 5y = -2$ Substituting

$-3 + 5y = -2$

$5y = 1$

$y = \frac{1}{5}$

Check:

$$\begin{array}{c|c} 8x - 5y = -9 & \\ \hline 8(-1) - 5\left(\frac{1}{5}\right) \ ? \ -9 & \\ -8 - 1 & \\ -9 & \text{TRUE} \end{array} \qquad \begin{array}{c|c} 3x + 5y = -2 & \\ \hline 3(-1) + 5\left(\frac{1}{5}\right) \ ? \ -2 & \\ -3 + 1 & \\ -2 & \text{TRUE} \end{array}$$

Since $\left(-1, \frac{1}{5}\right)$ checks, it is the solution.

11. $4x - 5y = 7$

$-4x + 5y = 7$

$0 = 14$ Adding

We obtain a false equation, $0 = 14$, so there is no solution.

13. $x + y = -7,$ (1)

$3x + y = -9$ (2)

We multiply on both sides of Equation (1) by -1 and then add.

$-x - y = 7$ Multiplying by -1

$3x + y = -9$ Equation (2)

$2x = -2$ Adding

$x = -1$

Substitute -1 for x in one of the original equations and solve for y.

$x + y = -7$ Equation (1)

$-1 + y = -7$ Substituting

$y = -6$

Check:

$$\begin{array}{c|c} x + y = -7 & \\ \hline -1 + (-6) \ ? \ -7 & \\ -7 & \text{TRUE} \end{array} \qquad \begin{array}{c|c} 3x + y = -9 & \\ \hline 3(-1) + (-6) \ ? \ -9 & \\ -3 - 6 & \\ -9 & \text{TRUE} \end{array}$$

Since $(-1, -6)$ checks, it is the solution.

15. $3x - y = 8,$ (1)

$x + 2y = 5$ (2)

We multiply on both sides of Equation (1) by 2 and then add.

$6x - 2y = 16$ Multiplying by 2

$x + 2y = 5$ Equation (2)

$7x = 21$ Adding

$x = 3$

Substitute 3 for x in one of the original equations and solve for y.

$x + 2y = 5$ Equation (2)

$3 + 2y = 5$ Substituting

$2y = 2$

$y = 1$

Check:

$$\begin{array}{c|c} 3x - y = 8 & \\ \hline 3 \cdot 3 - 1 \ ? \ 8 & \\ 9 - 1 & \\ 8 & \text{TRUE} \end{array} \qquad \begin{array}{c|c} x + 2y = 5 & \\ \hline 3 + 2 \cdot 1 \ ? \ 5 & \\ 3 + 2 & \\ 5 & \text{TRUE} \end{array}$$

Since $(3, 1)$ checks, it is the solution.

17. $x - y = 5,$ (1)

$4x - 5y = 17$ (2)

We multiply on both sides of Equation (1) by -4 and then add.

$-4x + 4y = -20$ Multiplying by -4

$4x - 5y = 17$ Equation (2)

$-y = -3$ Adding

$y = 3$

Substitute 3 for y in one of the original equations and solve for x.

$x - y = 5$ Equation (1)

$x - 3 = 5$ Substituting

$x = 8$

Check:

$$\begin{array}{c|c} x - y = 5 & \\ \hline 8 - 3 \ ? \ 5 & \\ 5 & \text{TRUE} \end{array} \qquad \begin{array}{c|c} 4x - 5y = 17 & \\ \hline 4 \cdot 8 - 5 \cdot 3 \ ? \ 17 & \\ 32 - 15 & \\ 17 & \text{TRUE} \end{array}$$

Since $(8, 3)$ checks, it is the solution.

19. $2w - 3z = -1,$ (1)

$3w + 4z = 24$ (2)

We use the multiplication principle with both equations and then add.

$8w - 12z = -4$ Multiplying (1) by 4

$9w + 12z = 72$ Multiplying (2) by 3

$17w = 68$ Adding

$w = 4$

Substitute 4 for w in one of the original equations and solve for z.

$$\begin{aligned} 3w + 4z &= 24 && \text{Equation (2)} \\ 3 \cdot 4 + 4z &= 24 && \text{Substituting} \\ 12 + 4z &= 24 \\ 4z &= 12 \\ z &= 3 \end{aligned}$$

Check:

$$\begin{array}{c|l} 2w - 3z = -1 & \\ \hline 2 \cdot 4 - 3 \cdot 3 \ ? \ -1 & \\ 8 - 9 & \\ -1 & \text{TRUE} \end{array} \qquad \begin{array}{c|l} 3w + 4z = 24 & \\ \hline 3 \cdot 4 + 4 \cdot 3 \ ? \ 24 & \\ 12 + 12 & \\ 24 & \text{TRUE} \end{array}$$

Since $(4, 3)$ checks, it is the solution.

21. $2a + 3b = -1, \quad (1)$

$3a + 5b = -2 \quad (2)$

We use the multiplication principle with both equations and then add.

$$\begin{aligned} -10a - 15b &= 5 && \text{Multiplying (1) by } -5 \\ 9a + 15b &= -6 && \text{Multiplying (2) by 3} \\ \hline -a \qquad &= -1 && \text{Adding} \\ a &= 1 \end{aligned}$$

Substitute 1 for a in one of the original equations and solve for b.

$$\begin{aligned} 2a + 3b &= -1 && \text{Equation (1)} \\ 2 \cdot 1 + 3b &= -1 && \text{Substituting} \\ 2 + 3b &= -1 \\ 3b &= -3 \\ b &= -1 \end{aligned}$$

Check:

$$\begin{array}{c|l} 2a + 3b = -1 & \\ \hline 2 \cdot 1 + 3(-1) \ ? \ -1 & \\ 2 - 3 & \\ -1 & \text{TRUE} \end{array} \qquad \begin{array}{c|l} 3a + 5b = -2 & \\ \hline 3 \cdot 1 + 5(-1) \ ? \ -2 & \\ 3 - 5 & \\ -2 & \text{TRUE} \end{array}$$

Since $(1, -1)$ checks, it is the solution.

23. $x = 3y, \quad (1)$

$5x + 14 = y \quad (2)$

We first get each equation in the form $Ax + By = C$.

$$\begin{aligned} x - 3y &= 0, && \text{(1a)} && \text{Adding } -3y \\ 5x - y &= -14 && \text{(2a)} && \text{Adding } -y - 14 \end{aligned}$$

We multiply by -5 on both sides of Equation (1a) and add.

$$\begin{aligned} -5x + 15y &= 0 && \text{Multiplying by } -5 \\ 5x - \quad y &= -14 \\ \hline 14y &= -14 && \text{Adding} \\ y &= -1 \end{aligned}$$

Substitute -1 for y in Equation (1) and solve for x.

$$\begin{aligned} x - 3y &= 0 \\ x - 3(-1) &= 0 && \text{Substituting} \\ x + 3 &= 0 \\ x &= -3 \end{aligned}$$

Check:

$$\begin{array}{c|l} x - 3y = 0 & \\ \hline -3 - 3(-1) \ ? \ 0 & \\ -3 + 3 & \\ 0 & \text{TRUE} \end{array} \qquad \begin{array}{c|l} 5x - y = -14 & \\ \hline 5(-3) - (-1) \ ? \ -14 & \\ -15 + 1 & \\ -14 & \text{TRUE} \end{array}$$

Since $(-3, -1)$ checks, it is the solution.

25. $2x + 5y = 16, \quad (1)$

$3x - 2y = 5 \quad (2)$

We use the multiplication principle with both equations and then add.

$$\begin{aligned} 4x + 10y &= 32 && \text{Multiplying (1) by 2} \\ 15x - 10y &= 25 && \text{Multiplying (2) by 5} \\ \hline 19x \qquad &= 57 \\ x &= 3 \end{aligned}$$

Substitute 3 for x in one of the original equations and solve for y.

$$\begin{aligned} 2x + 5y &= 16 && \text{Equation (1)} \\ 2 \cdot 3 + 5y &= 16 && \text{Substituting} \\ 6 + 5y &= 16 \\ 5y &= 10 \\ y &= 2 \end{aligned}$$

Check:

$$\begin{array}{c|l} 2x + 5y = 16 & \\ \hline 2 \cdot 3 + 5 \cdot 2 \ ? \ 16 & \\ 6 + 10 & \\ 16 & \text{TRUE} \end{array} \qquad \begin{array}{c|l} 3x - 2y = 5 & \\ \hline 3 \cdot 3 - 2 \cdot 2 \ ? \ 5 & \\ 9 - 4 & \\ 5 & \text{TRUE} \end{array}$$

Since $(3, 2)$ checks, it is the solution.

27. $p = 32 + q, \quad (1)$

$3p = 8q + 6 \quad (2)$

First we write each equation in the form $Ap + Bq = C$.

$$\begin{aligned} p - q &= 32, && \text{(1a)} && \text{Subtracting } q \\ 3p - 8q &= 6 && \text{(2a)} && \text{Subtracting } 8q \end{aligned}$$

Now we multiply both sides of Equation (1a) by -3 and then add.

$$\begin{aligned} -3p + \quad 3q &= -96 && \text{Multiplying by } -3 \\ 3p - \quad 8q &= 6 && \text{Equation (2a)} \\ \hline -5q &= -90 && \text{Adding} \\ q &= 18 \end{aligned}$$

Substitute 18 for q in Equation (1) and solve for p.

$$\begin{aligned} p &= 32 + q \\ p &= 32 + 18 && \text{Substituting} \\ p &= 50 \end{aligned}$$

Check:

$$\begin{array}{c|c} p - q = 32 & \\ \hline 50 - 18 \ ? \ 32 & \\ 32 & \text{TRUE} \end{array} \qquad \begin{array}{c|c} 3p - 8q = 6 & \\ \hline 3 \cdot 50 - 8 \cdot 18 \ ? \ 6 & \\ 150 - 144 & \\ 6 & \text{TRUE} \end{array}$$

Since $(50, 18)$ checks, it is the solution.

29. $3x - 2y = 10$, (1)

$-6x + 4y = -20$ (2)

We multiply by 2 on both sides of Equation (1) and add.

$$\begin{aligned} 6x - 4y &= 20 \\ -6x + 4y &= -20 \\ \hline 0 &= 0 \end{aligned}$$

We get an obviously true equation, so the system has an infinite number of solutions.

31. $0.06x + 0.05y = 0.07$,

$0.04x - 0.03y = 0.11$

We first multiply each equation by 100 to clear the decimals.

$6x + 5y = 7$, (1)

$4x - 3y = 11$ (2)

We use the multiplication principle with both equations of the resulting system.

$$\begin{aligned} 18x + 15y &= 21 \quad \text{Multiplying (1) by 3} \\ 20x - 15y &= 55 \quad \text{Multiplying (2) by 5} \\ \hline 38x \qquad &= 76 \quad \text{Adding} \\ x &= 2 \end{aligned}$$

Substitute 2 for x in Equation (1) and solve for y.

$$\begin{aligned} 6x + 5y &= 7 \\ 6 \cdot 2 + 5y &= 7 \\ 12 + 5y &= 7 \\ 5y &= -5 \\ y &= -1 \end{aligned}$$

Check:

$$\begin{array}{c|c} 0.06x + 0.05y = 0.07 & \\ \hline 0.06(2) + 0.05(-1) \ ? \ 0.07 & \\ 0.12 - 0.05 & \\ 0.07 & \text{TRUE} \end{array}$$

$$\begin{array}{c|c} 0.04x - 0.03y = 0.11 & \\ \hline 0.04(2) - 0.03(-1) \ ? \ 0.11 & \\ 0.08 + 0.03 & \\ 0.11 & \text{TRUE} \end{array}$$

Since $(2, -1)$ checks, it is the solution.

33. $\frac{1}{3}x + \frac{3}{2}y = \frac{5}{4}$,

$\frac{3}{4}x - \frac{5}{6}y = \frac{3}{8}$

First we clear the fractions. We multiply on both sides of the first equation by 12 and on both sides of the second equation by 24.

$$\begin{aligned} 12\left(\frac{1}{3}x + \frac{3}{2}y\right) &= 12 \cdot \frac{5}{4} \\ 12 \cdot \frac{1}{3}x + 12 \cdot \frac{3}{2}y &= 15 \\ 4x + 18y &= 15 \end{aligned}$$

$$\begin{aligned} 24\left(\frac{3}{4}x - \frac{5}{6}y\right) &= 24 \cdot \frac{3}{8} \\ 24 \cdot \frac{3}{4}x - 24 \cdot \frac{5}{6}y &= 9 \\ 18x - 20y &= 9 \end{aligned}$$

The resulting system is

$4x + 18y = 15$, (1)

$18x - 20y = 9$. (2)

We use the multiplication principle with both equations.

$$\begin{aligned} 72x + 324y &= 270 \quad \text{Multiplying (1) by 18} \\ -72x + 80y &= -36 \quad \text{Multiplying (2) by } -4 \\ \hline 404y &= 234 \\ y &= \frac{234}{404}, \text{ or } \frac{117}{202} \end{aligned}$$

Substitute $\frac{117}{202}$ for y in Equation (1) and solve for x.

$$\begin{aligned} 4x + 18\left(\frac{117}{202}\right) &= 15 \\ 4x + \frac{1053}{101} &= 15 \\ 4x &= \frac{462}{101} \\ x &= \frac{1}{4} \cdot \frac{462}{101} \\ x &= \frac{231}{202} \end{aligned}$$

The ordered pair $\left(\frac{231}{202}, \frac{117}{202}\right)$ checks in both equations. It is the solution.

35. $-4.5x + 7.5y = 6$,

$-x + 1.5y = 5$

First we clear the decimals by multiplying by 10 on both sides of each equation.

$$\begin{aligned} 10(-4.5x + 7.5y) &= 10 \cdot 6 \\ -45x + 75y &= 60 \end{aligned}$$

$$\begin{aligned} 10(-x + 1.5y) &= 10 \cdot 5 \\ -10x + 15y &= 50 \end{aligned}$$

The resulting system is

$-45x + 75y = 60$, (1)

$-10x + 15y = 50$. (2)

We multiply both sides of Equation (2) by -5 and then add.

$$\begin{array}{rll} -45x + 75y &= 60 & \text{Equation (1)} \\ 50x - 75y &= -250 & \text{Multiplying by } -5 \\ \hline 5x \qquad &= -190 & \text{Adding} \\ x &= -38 & \end{array}$$

Substitute -38 for x in Equation (2) and solve for y.

$$\begin{aligned} -10x + 15y &= 50 \\ -10(-38) + 15y &= 50 \\ 380 + 15y &= 50 \\ 15y &= -330 \\ y &= -22 \end{aligned}$$

The ordered pair $(-38, -22)$ checks in both equations. It is the solution.

37. ***Familiarize***. We let M = the number of miles driven and C = the total cost of the rental.

Translate. We reword and translate the first statement, using \$0.39 for 39 cents.

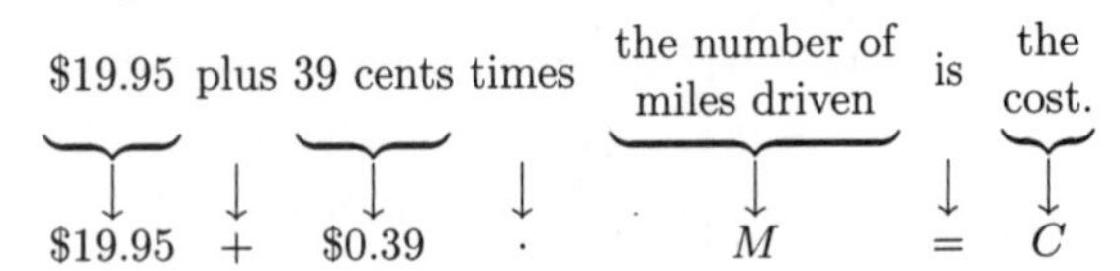

We reword and translate the second statement using \$0.29 for 29 cents.

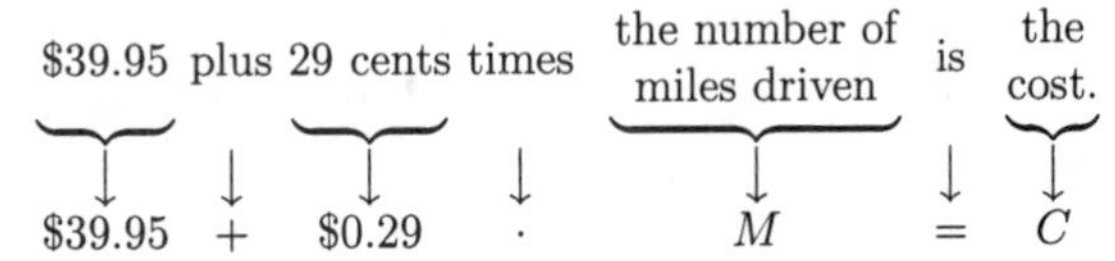

The resulting system is

$$19.95 + 0.39M = C,$$
$$39.95 + 0.29M = C.$$

Solve. We solve the system of equations. We clear the decimals by multiplying on both sides of each equation by 100.

$$\begin{array}{rl} 1995 + 39M = 100C, & (1) \\ 3995 + 29M = 100C & (2) \end{array}$$

We multiply Equation (2) by -1 and add.

$$\begin{array}{rl} 1995 + 39M &= 100C \\ -3995 - 29M &= -100C \\ \hline -2000 + 10M &= 0 \\ 10M &= 2000 \\ M &= 200 \end{array}$$

Check. For 200 mi, the cost of the Quick-Haul van is $\$19.95 + \$0.39(200)$, or $\$19.95 + \$78 = \$97.95$. The cost of the other company's van is $\$39.95 + \$0.29(200)$, or $\$39.95 + \$58$, or \$97.95. Thus the costs are the same when the mileage is 200.

State. When the moving vans are driven 200 miles, the costs will be the same.

39. ***Familiarize***. Let x = the smaller angle and y = the larger angle.

Translate. We reword the problem.

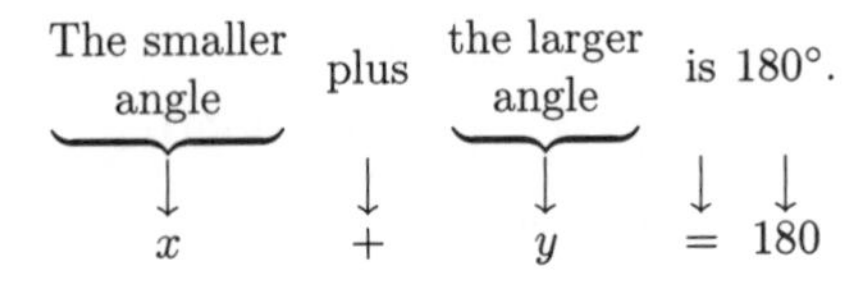

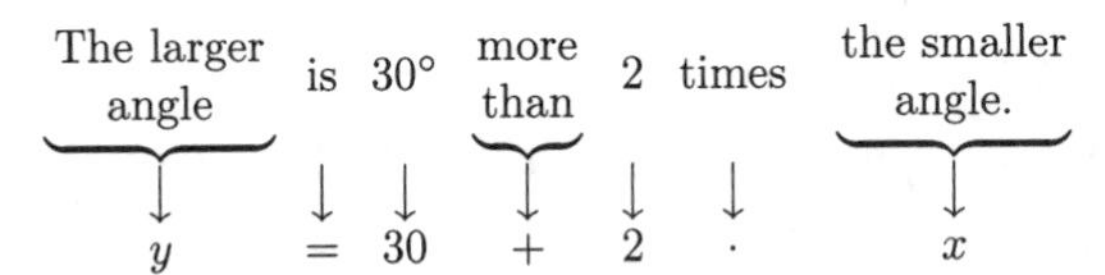

The resulting system is

$$x + y = 180,$$
$$y = 30 + 2x.$$

Solve. We solve the system. We will use the elimination method although we could also easily use the substitution method. First we get the second equation in the form $Ax + By = C$.

$$\begin{array}{rl} x + y = 180 & (1) \\ -2x + y = 30 & (2) \text{ Adding } -2x \end{array}$$

Now we multiply Equation (2) by -1 and add.

$$\begin{array}{rl} x + y &= 180 \\ 2x - y &= -30 \\ \hline 3x \qquad &= 150 \\ x &= 50 \end{array}$$

Then we substitute 50 for x in Equation (1) and solve for y.

$$\begin{array}{rll} x + y &= 180 & \text{Equation (1)} \\ 50 + y &= 180 & \text{Substituting} \\ y &= 130 & \end{array}$$

Check. The sum of the angles is $50° + 130°$, or $180°$, so the angles are supplementary. Also, $30°$ more than two times the $50°$ angle is $30° + 2 \cdot 50°$, or $30° + 100°$, or $130°$, the other angle. These numbers check.

State. The angles are 50° and 130°.

41. ***Familiarize***. We let x = the larger angle and y = the smaller angle.

Translate. We reword and translate the first statement.

$$\underbrace{\text{The sum of two angles}}_{x + y} \ \underset{=}{\text{is}} \ \underset{90}{90°}.$$

We reword and translate the second statement.

$$\underbrace{\text{The difference of two angles}}_{x - y} \ \underset{=}{\text{is}} \ \underset{34}{34°}.$$

The resulting system is

$$x + y = 90,$$
$$x - y = 34.$$

Solve. We solve the system.

$$\begin{array}{rl} x + y = 90, & (1) \\ \underline{x - y = 34} & (2) \\ 2x \quad = 124 & \text{Adding} \\ x = 62 & \end{array}$$

Now we substitute 62 for x in Equation (1) and solve for y.

$$\begin{array}{rl} x + y = 90 & \text{Equation (1)} \\ 62 + y = 90 & \text{Substituting} \\ y = 28 & \end{array}$$

Check. The sum of the angles is $62° + 28°$, or $90°$, so the angles are complementary. The difference of the angles is $62° - 28°$, or $34°$. These numbers check.

State. The angles are 62° and 28°.

43. ***Familiarize.*** We let x = the number of hectares of hay that should be planted and y = the number of hectares of oats that should be planted.

Translate. We reword and translate the first statement.

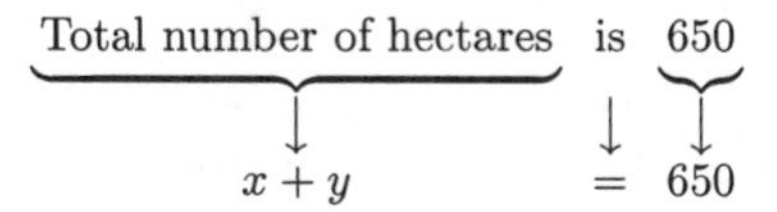

Now we reword and translate the second statement.

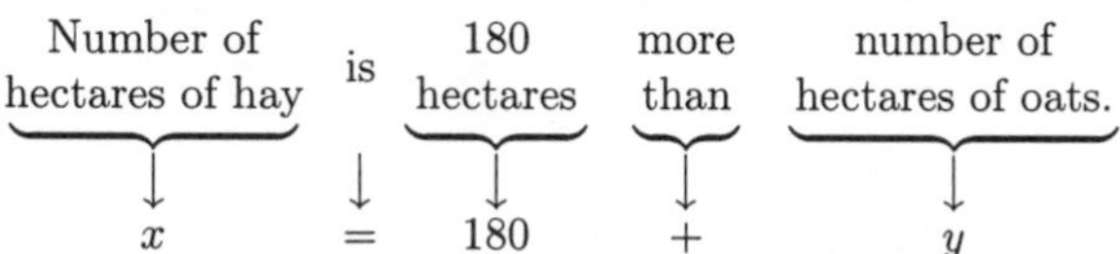

The resulting system is

$$x + y = 650,$$
$$x = 180 + y$$

Solve. We solve the system. We will use the elimination method, although we could also easily use the substitution method. First we get the second equation in the form $Ax + By = C$. Then we add the equations.

$$\begin{array}{rll} x + y = 650 & (1) & \\ \underline{x - y = 180} & (2) & \text{Subtracting } y \\ 2x \quad = 830 & \text{Adding} & \\ x = 415 & & \end{array}$$

Now we substitute 415 for x in Equation (1) and solve for y.

$$\begin{array}{rl} x + y = 650 & \text{Equation (1)} \\ 415 + y = 650 & \text{Substituting} \\ y = 235 & \end{array}$$

Check. The total number of hectares is $415 + 235$, or 650. Also, the number of hectares of hay is 180 more than the number of hectares of oats. These numbers check.

State. The owners should plant 415 hectares of hay and 235 hectares of oats.

45. $x^{-2} \cdot x^{-5} = x^{-2+(-5)} = x^{-7} = \dfrac{1}{x^7}$

47. $x^2 \cdot x^{-5} = x^{2+(-5)} = x^{-3} = \dfrac{1}{x^3}$

49. $\dfrac{x^{-2}}{x^{-5}} = x^{-2-(-5)} = x^3$

51. $(a^2b^{-3})(a^5b^{-6}) = a^{2+5}b^{-3+(-6)} = a^7b^{-9} = \dfrac{a^7}{b^9}$

53.
$$\begin{aligned} \frac{x^2 - 5x + 6}{x^2 - 4} &= \frac{(x-3)(x-2)}{(x+2)(x-2)} \\ &= \frac{(x-3)\cancel{(x-2)}}{(x+2)\cancel{(x-2)}} \\ &= \frac{x-3}{x+2} \end{aligned}$$

55.
$$\begin{aligned} &\frac{x-2}{x+3} - \frac{2x-5}{x-4} \quad \text{LCD is } (x+3)(x-4) \\ &= \frac{x-2}{x+3} \cdot \frac{x-4}{x-4} - \frac{2x-5}{x-4} \cdot \frac{x+3}{x+3} \\ &= \frac{(x-2)(x-4)}{(x+3)(x-4)} - \frac{(2x-5)(x+3)}{(x-4)(x+3)} \\ &= \frac{x^2 - 6x + 8}{(x+3)(x-4)} - \frac{2x^2 + x - 15}{(x-4)(x+3)} \\ &= \frac{x^2 - 6x + 8 - (2x^2 + x - 15)}{(x+3)(x-4)} \\ &= \frac{x^2 - 6x + 8 - 2x^2 - x + 15}{(x+3)(x-4)} \\ &= \frac{-x^2 - 7x + 23}{(x+3)(x-4)} \end{aligned}$$

57.

59.-67.

69.-77.

79. ***Familiarize.*** We let x = Will's age now and y = his father's age now. In 20 years their ages will be $x + 20$ and $y + 20$.

Translate. We translate the first statement.

Will's age is 20% of his father's age.

$$\begin{array}{ccccc} x & = & 20\% & \cdot & y, \text{ or} \\ x & = & 0.2 & \cdot & y \end{array}$$

We reword and translate the second statement.

Will's age in 20 years will be 52% of his father's age in 20 years.

$$\begin{array}{ccccc} x + 20 & = & 52\% & \cdot & (y + 20), \text{ or} \\ x + 20 & = & 0.52 & \cdot & (y + 20) \end{array}$$

The resulting system is

$$x = 0.2y,$$
$$x + 20 = 0.52(y + 20).$$

Solve. We solve the system of equations. We will use the elimination method, although we could also easily use the substitution method. First we clear decimals by multiplying on both sides of the first equation by 10 and on both sides of the second equation by 100.

$$10x = 2y,$$
$$100x + 2000 = 52(y + 20)$$

Now we write each equation in the form $Ax + By = C$.

$$10x - 2y = 0, \quad (1)$$
$$100x - 52y = -960 \quad (2)$$

We multiply Equation (1) by -10 and then add.

$$-100x + 20y = 0$$
$$100x - 52y = -960$$
$$-32y = -960$$
$$y = 30$$

Substitute 30 for y in Equation (1) and solve for x.

$$10x - 2y = 0$$
$$10x - 2 \cdot 30 = 0$$
$$10x - 60 = 0$$
$$10x = 60$$
$$x = 6$$

Check. Will's age now, 6, is 20% of 30, his father's age now. Will's age 20 years from now, 26, is 52% of 50, his father's age 20 years from now. These numbers check.

State. Will is 6 years old now, and his father is 30 years old.

81. $3(x - y) = 9,$

$x + y = 7$

First we remove parentheses in the first equation.

$$3x - 3y = 9, \quad (1)$$
$$x + y = 7 \quad (2)$$

Then we multiply Equation (2) by 3 and add.

$$3x - 3y = 9$$
$$3x + 3y = 21$$
$$6x = 30$$
$$x = 5$$

Now we substitute 5 for x in Equation (2) and solve for y.

$$x + y = 7$$
$$5 + y = 7$$
$$y = 2$$

The ordered pair $(5, 2)$ checks and is the solution.

83. $2(5a - 5b) = 10,$

$-5(6a + 2b) = 10$

First we remove parentheses.

$$10a - 10b = 10, \quad (1)$$
$$-30a - 10b = 10 \quad (2)$$

Then we multiply Equation (2) by -1 and add.

$$10a - 10b = 10$$
$$30a + 10b = -10$$
$$40a = 0$$
$$a = 0$$

Substitute 0 for a in Equation (1) and solve for b.

$$10 \cdot 0 - 10b = 10$$
$$-10b = 10$$
$$b = -1$$

The ordered pair $(0, -1)$ checks and is the solution.

85. ***Familiarize***. Let x = the number of rabbits and y = the number of pheasants. The x rabbits have x heads and $4x$ feet, and the y pheasants have y heads and $2y$ feet.

Translate. We reword and translate.

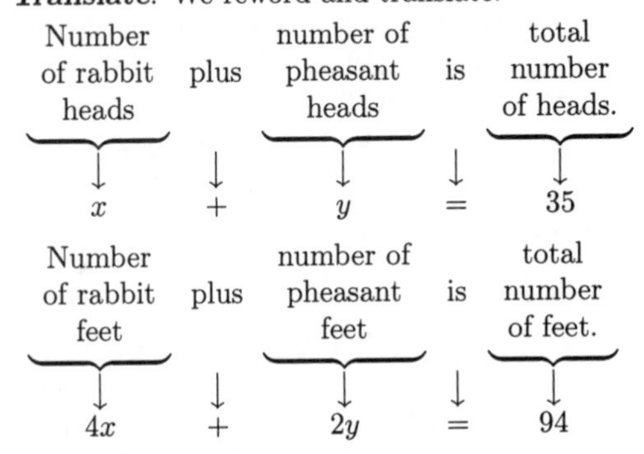

The resulting system is

$$x + y = 35, \quad (1)$$
$$4x + 2y = 94. \quad (2)$$

Solve. We multiply the first equation by -2 and then add.

$$-2x - 2y = -70$$
$$4x + 2y = 94$$
$$2x = 24$$
$$x = 12$$

Substitute 12 for x in Equation (1) and solve for y.

$$x + y = 35$$
$$12 + y = 35$$
$$y = 23$$

Check. If there are 12 rabbits and 23 pheasants, then there are $12+23$, or 35 heads, and $4 \cdot 12 + 2 \cdot 23$, or $48 + 46$, or 94 feet. The answer checks.

State. There are 12 rabbits and 23 pheasants.

Exercise Set 14.4

1. ***Familiarize***. Let x = the number of two-pointers and y = the number of foul shots Shaquille O'Neill made. Then he scored $2x$ points from two-pointers and y points from foul shots.

Translate. We reword and translate.

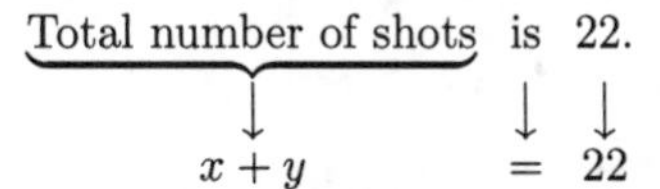

Total number of points is 36.

$2x+y = 36$

The resulting system is

$x+y=22,\quad (1)$

$2x+y=36.\quad (2)$

Solve. We use the elimination method. Multiply the first equation by -1 and then add.

$$\begin{array}{r} -x - y = -22 \\ 2x + y = 36 \\ \hline x \quad = 14 \end{array}$$

We find y by substituting in Equation (1).

$x+y=22$

$14+y=22$

$y=8$

Check. The total number of shots is $14+8$, or 22. The total number of points is $2\cdot 14+8$, or $28+8$, or 36. The answer checks.

State. Shaquille O'Neill made 14 two-pointers and 8 foul shots.

3. ***Familiarize***. Let k = the age of the Kuyatt's house now and m = the age of the Marconi's house now. Eight years ago the houses' ages were $k-8$ and $m-8$.

Translate. We reword and translate.

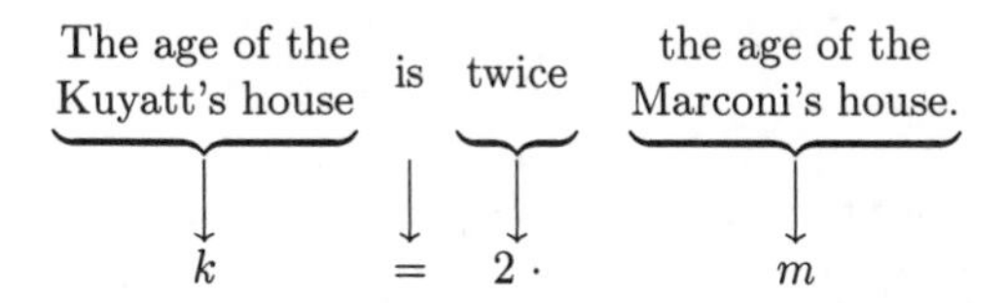

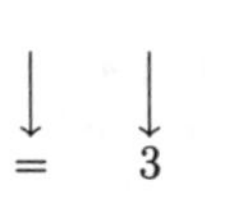

was three times the age of the Marconi's house 8 years ago.

$k-8 = 3 \cdot m-8$

The resulting system is

$k=2m,\quad (1)$

$k-8=3(m-8).\quad (2)$

Solve. We use the substitution method. We substitute $2m$ for k in Equation (2) and solve for m.

$k-8=3(m-8)$

$2m-8=3(m-8)$

$2m-8=3m-24$

$-8=m-24$

$16=m$

We find k by substituting 16 for m in Equation (1).

$k=2m$

$k=2\cdot 16$

$k=32$

Check. The age of the Kuyatt's house, 32 years, is twice the age of the Marconi's house, 16 years. Eight years ago, when the Kuyatt's house was 24 years old and the Marconi's house was 8 years old, the Kuyatt's house was three times as old as the Marconi's house. These numbers check.

State. The Kuyatt's house is 32 years old, and the Marconi's house is 16 years old.

5. ***Familiarize***. Let R = Randy's age now and M = Mandy's age now. In twelve years their ages will be $R+12$ and $M+12$.

Translate. We reword and translate.

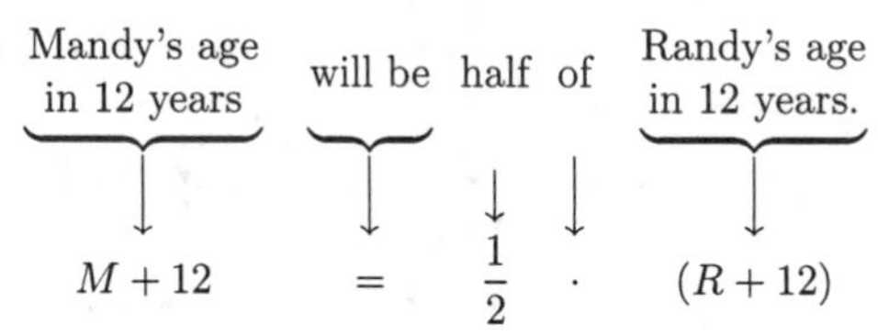

Mandy's age in 12 years will be half of Randy's age in 12 years.

$M+12 = \frac{1}{2} \cdot (R+12)$

The resulting system is

$R=4M,\quad (1)$

$M+12=\frac{1}{2}(R+12).\quad (2)$

Solve. We use the substitution method. We substitute $4M$ for R in Equation (2) and solve for M.

$M+12=\frac{1}{2}(R+12)$

$M+12=\frac{1}{2}(4M+12)$

$M+12=2M+6$

$12=M+6$

$6=M$

We find R by substituting 6 for M in Equation (1).

$R=4M$

$R=4\cdot 6$

$R=24$

Check. Randy's age now, 24, is 4 times 6, Mandy's age. In 12 yr, when Randy will be 36 and Mandy 18, Mandy's age will be half of Randy's age. These numbers check.

State. Randy is 24 years old now, and Mandy is 6.

7. ***Familiarize***. We complete the table in the text. Note that x represents the number of pounds of Brazilian coffee to be used and y represents the number of pounds of Turkish coffee.

Type of coffee	Brazilian	Turkish	Mixture
Cost of coffee	\$19	\$22	\$20
Amount (in pounds)	x	y	300
Mixture	$19x$	$22y$	\$20(300), or \$6000

Equation from second row: $x + y = 300$

Equation from third row: $19x + 22y = 6000$

Translate. The second and third rows of the table give us two equations. Since the total amount of the mixture is 300 lb, we have

$$x + y = 300.$$

The value of the Brazilian coffee is $19x$ (x lb at \$19 per pound), the value of the Turkish coffee is $22y$ (y lb at \$22 per pound), and the value of the mixture is \$20(300) or \$6000. Thus we have

$$19x + 22y = 6000.$$

The resulting system is

$$x + y = 300, \quad (1)$$
$$19x + 22y = 6000. \quad (2)$$

Solve. We use the elimination method. We multiply on both sides of Equation (1) by -19 and then add.

$$-19x - 19y = -5700 \quad \text{Multiplying by } -19$$
$$\underline{19x + 22y = 6000}$$
$$3y = 300$$
$$y = 100$$

We go back to Equation (1) and substitute 100 for y.

$$x + y = 300$$
$$x + 100 = 300$$
$$x = 200$$

Check. The sum of 200 and 100 is 300. The value of the mixture is \$19(200)+\$22(100), or \$3800+\$2200, or \$6000. These values check.

State. 200 lb of Brazilian coffee and 100 lb of Turkish coffee should be used.

9. ***Familiarize***. Let d represent the number of dimes and q the number of quarters. Then, $10d$ represents the value of the dimes in cents, and $25q$ represents the value of the quarters in cents. The total value is \$15.25, or 1525¢. The total number of coins is 103.

Translate.

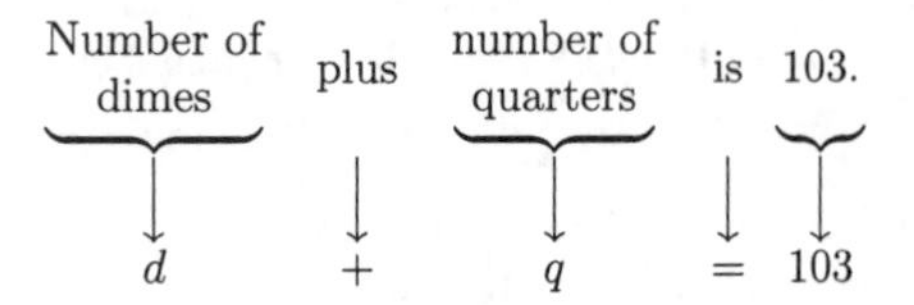

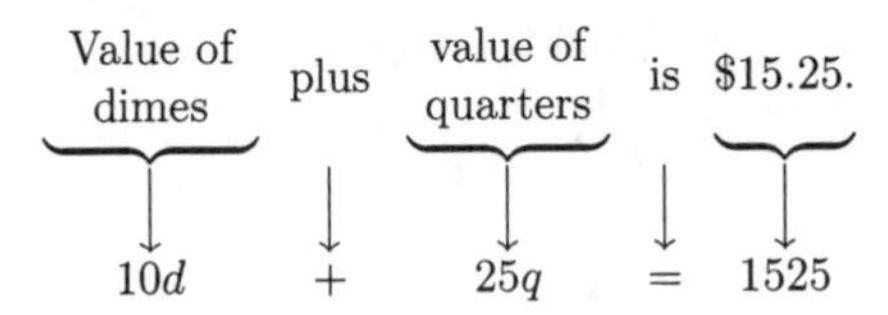

The resulting system is

$$d + q = 103, \quad (1)$$
$$10d + 25q = 1525. \quad (2)$$

Solve. We use the addition method. We multiply Equation (1) by -10 and then add.

$$-10d - 10q = -1030 \quad \text{Multiplying by } -10$$
$$\underline{10d + 25q = 1525}$$
$$15q = 495 \quad \text{Adding}$$
$$q = 33$$

Now we substitute 33 for q in one of the original equations and solve for d.

$$d + q = 103 \quad (1)$$
$$d + 33 = 103 \quad \text{Substituting}$$
$$d = 70$$

Check. The number of dimes plus the number of quarters is 70+33, or 103. The total value in cents is $10 \cdot 70 + 25 \cdot 33$, or 700+825, or 1525. This is equal to \$15.25. This checks.

State. There are 70 dimes and 33 quarters.

11. ***Familiarize***. Let p = the cost of one slice of pizza and s = the cost of one soda.

Translate.

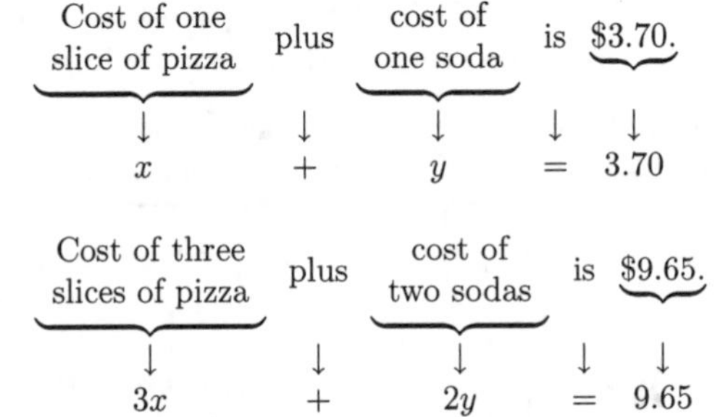

We multiply both equations by 100 to clear decimals. (We could also multiply the first equation by 10 and the second by 100.) The resulting system is

$$100x + 100y = 370, \quad (1)$$
$$300x + 200y = 965. \quad (2)$$

Solve. We use the elimination method. We multiply Equation (1) by -2 and then add.

$$-200x - 200y = -740 \quad \text{Multiplying by } -2$$
$$\underline{300x + 200y = 965}$$
$$100x = 225$$
$$x = 2.25$$

We go back to Equation (1) and substitute 2.25 for x.

$$100x + 100y = 370$$
$$100(2.25) + 100y = 370$$
$$225 + 100y = 370$$
$$100y = 145$$
$$y = 1.45$$

Check. If one slice of pizza costs \$2.25 and one soda costs \$1.45, then they cost \$2.25+\$1.45, or \$3.70 together. Also, three slices of pizza and two sodas cost 3(\$2.25)+2(\$1.45), or \$6.75 + \$2.90, or \$9.65. These numbers check.

State. One slice of pizza costs \$2.25, and one soda costs \$1.45.

13. ***Familiarize***. Let x = the number of cardholders tickets that were sold and y = the number of non-cardholders tickets. We arrange the information in a table.

	Card-holders	Non-card-holders	Total
Price	\$2.25	\$3	
Number sold	x	y	203
Money taken in	$2.25x$	$3y$	\$513

Translate. The last two rows of the table give us two equations. The total number of tickets sold was 203, so we have

$$x + y = 203.$$

The total amount of money collected was \$513, so we have

$$2.25x + 3y = 513.$$

We can multiply the second equation on both sides by 100 to clear decimals. The resulting system is

$$x + y = 203, \quad (1)$$
$$225x + 300y = 51,300. \quad (2)$$

Solve. We use the elimination method. We multiply on both sides of Equation (1) by −225 and then add.

$$-225x - 225y = -46,675 \quad \text{Multiplying by } -225$$
$$225x + 300y = 51,300$$
$$75y = 5625$$
$$y = 75$$

We go back to Equation (1) and substitute 75 for y.

$$x + y = 203$$
$$x + 75 = 203$$
$$x = 128$$

Check. The number of tickets sold was 128 + 75, or 203. The money collected was \$2.25(128) + \$3(75), or \$288 + \$225, or \$513. These numbers check.

State. 128 cardholders tickets and 75 non-cardholders tickets were sold.

15. ***Familiarize***. Let b = the number of Upper Box tickets and r = the number of Lower Reserved tickets that were bought. We arrange the information in a table.

	Upper Box	Lower Reserved	Total
Price	\$10	\$9.50	
Number bought	b	r	29
Cost	$12b$	$9.5r$	\$318

Translate. The last two rows of the table give us two equations. The total number of tickets bought was 29, so we have

$$b + r = 29.$$

The total cost of the tickets was \$318, so we have

$$12b + 9.5r = 318.$$

We multiply the second equation on both sides by 10 to clear decimals. The resulting system is

$$b + r = 29, \quad (1)$$
$$120b + 95r = 3180. \quad (2)$$

Solve. We use the elimination method. We multiply on both sides of Equation (1) by −95 and then add.

$$-95b - 95r = -2755 \quad \text{Multiplying by } -95$$
$$120b + 95r = 3180$$
$$25b = 425$$
$$b = 17$$

We go back to Equation (1) and substitute 17 for b.

$$b + r = 29$$
$$17 + r = 29$$
$$r = 12$$

Check. The number of tickets sold was 17 + 12, or 29. The total cost was \$12(17) + \$9.5(12), or \$204 + \$114, or \$318. These numbers check.

State. They bought 17 Upper Box and 12 Lower Reserved tickets.

17. ***Familiarize***. We complete the table in the text. Note that x represents the number of liters of solution A to be used and y represents the number of liters of solution B.

Type of solution	A	B	Mixture
Amount of solution	x	y	100 L
Percent of acid	50%	80%	68%
Amount of acid in solution	$50\%x$	$80\%y$	68% × 100, or 68 L

Equation from first row: $x + y = 100$

Equation from second row: $50\%x + 80\%y = 68$

Translate. The first and third rows of the table give us two equations. Since the total amount of solution is 100 liters, we have

$x + y = 100.$

The amount of acid in the mixture is to be 68% of 100, or 68 liters. The amounts of acid from the two solutions are 50%x and 80%y. Thus

$$50\%x + 80\%y = 68,$$
or $$0.5x + 0.8y = 68,$$
or $$5x + 8y = 680 \quad \text{Clearing decimals}$$

The resulting system is

$$x + y = 100, \quad (1)$$
$$5x + 8y = 680. \quad (2)$$

Solve. We use the elimination method. We multiply on both sides of Equation (1) by -5 and then add.

$$-5x - 5y = -500 \quad \text{Multiplying by } -5$$
$$5x + 8y = 680$$
$$3y = 180$$
$$y = 60$$

We go back to Equation (1) and substitute 60 for y.

$$x + y = 100$$
$$x + 60 = 100$$
$$x = 40$$

Check. We consider $x = 40$ and $y = 60$. The sum is 100. Now 50% of 40 is 20 and 80% of 60 is 48. These add up to 68. The numbers check.

State. 40 liters of solution A and 60 liters of solution B should be used.

19. ***Familiarize***. We let $x =$ the number of pounds of hay and $y =$ the number of pounds of grain that should be fed to the horse each day. We arrange the information in a table.

Type of feed	Hay	Grain	Mixture
Amount of feed	x	y	15
Percent of protein	6%	12%	8%
Amount of protein in mixture	6%x	12%y	8% × 15, or 1.2 lb

Translate. The first and last rows of the table give us two equations. The total amount of feed is 15 lb, so we have

$x + y = 15.$

The amount of protein in the mixture is to be 8% of 15 lb, or 1.2 lb. The amounts of protein from the two feeds are 6%x and 12%y. Thus

$$6\%x + 12\%y = 1.2, \quad \text{or}$$
$$0.06x + 0.12y = 1.2, \quad \text{or}$$
$$6x + 12y = 120 \quad \text{Clearing decimals}$$

The resulting system is

$$x + y = 15, \quad (1)$$
$$6x + 12y = 120. \quad (2)$$

Solve. We use the elimination method. Multiply on both sides of Equation (1) by -6 and then add.

$$-6x - 6y = -90$$
$$6x + 12y = 120$$
$$6y = 30$$
$$y = 5$$

We go back to Equation (1) and substitute 5 for y.

$$x + y = 15$$
$$x + 5 = 15$$
$$x = 10$$

Check. The sum of 10 and 5 is 15. Also, 6% of 10 is 0.6 and 12% of 5 is 0.6, and $0.6 + 0.6 = 1.2$. These numbers check.

State. Irene should feed her horse 10 lb of hay and 5 lb of grain each day.

21. ***Familiarize***. We arrange the information in a table. Let $x =$ the amount of seed A and $y =$ the amount of seed B to be used.

Type of seed	A	B	Mixture
Cost of seed	\$2.50	\$1.75	\$2.14
Amount (in pounds)	x	y	75
Mixture	$2.50x$	$1.75y$	\$2.14(75), or \$160.50

Translate. The last two rows of the table give us two equations.

Since the total amount of grass seed is 75 lb, we have

$x + y = 75.$

The value of seed A is $2.50x$ (x lb at \$2.50 per pound), and the value of seed B is $1.75y$ (y lb at \$1.75 per pound). The value of the mixture is \$2.14(75), or \$160.50, so we have

$$2.50x + 1.75y = 160.50, \quad \text{or}$$
$$250x + 175y = 16,050 \quad \text{Clearing decimals}$$

The resulting system is

$$x + y = 75, \quad (1)$$
$$250x + 175y = 16,050. \quad (2)$$

Solve. We use the elimination method. We multiply Equation (1) by -250 and then add.

$$-250x - 250y = -18,750$$
$$250x + 175y = 16,050$$
$$-75y = -2700$$
$$y = 36$$

Next we substitute 36 for y in one of the original equations and solve for x.

$$x + y = 75 \quad (1)$$
$$x + 36 = 75$$
$$x = 39$$

Check. We consider $x = 39$ lb and $y = 36$ lb. The sum is 75 lb. The value of the mixture is \$2.50(39) + \$1.75(36), or \$97.50 + \$63.00, or \$160.50. These values check.

State. 39 lb of seed A and 36 lb of seed B should be used.

23. ***Familiarize.*** We arrange the information in a table. Let a = the number of type A questions and b = the number of type B questions.

Type of question	A	B	Mixture (Test)
Number	a	b	16
Time	3 min	6 min	
Value	10 points	15 points	
Mixture (Test)	$3a$ min, $10a$ points	$6b$ min, $15b$ points	60 min, 180 points

Translate. The table actually gives us three equations. Since the total number of questions is 16, we have

$$a + b = 16.$$

The total time is 60 min, so we have

$$3a + 6b = 60.$$

The total number of points is 180, so we have

$$10a + 15b = 180.$$

The resulting system is

$$a + b = 16, \quad (1)$$
$$3a + 6b = 60, \quad (2)$$
$$10a + 15b = 180. \quad (3)$$

Solve. We will solve the system composed of Equations (1) and (2) and then check to see that this solution also satisfies Equation (3). We multiply equation (1) by -3 and add.

$$-3a - 3b = -48$$
$$3a + 6b = 60$$
$$3b = 12$$
$$b = 4$$

Now we substitute 4 for b in Equation (1) and solve for a.

$$a + b = 16$$
$$a + 4 = 16$$
$$a = 12$$

Check. We consider $a = 12$ questions and $b = 4$ questions. The total number of questions is 16. The time required is $3 \cdot 12 + 6 \cdot 4$, or $36 + 24$, or 60 min. The total points are $10 \cdot 12 + 15 \cdot 4$, or $120 + 60$, or 180. These values check.

State. 12 questions of type A and 4 questions of type B were answered correctly.

25. ***Familiarize.*** We let x = the number of pages in large type and y = the number of pages in small type. We arrange the information in a table.

Size of type	Large	Small	Mixture (Book)
Words per page	1300	1850	
Number of pages	x	y	12
Number of words	$1300x$	$1850y$	18,526

Translate. The last two rows of the table give us two equations. The total number of pages in the document is 12, so we have

$$x + y = 12.$$

The number of words on the pages with large type is $1300x$ (x pages with 1300 words per page), and the number of words on the pages with small type is $1850y$ (y pages with 1850 words per page). The total number of words is 18,526, so we have

$$1300x + 1850y = 18,526.$$

The resulting system is

$$x + y = 12, \quad (1)$$
$$1300x + 1850y = 18,526. \quad (2)$$

Solve. We use the elimination method. We multiply on both sides of Equation (1) by -1300 and then add.

$$-1300x - 1300y = -15,600 \quad \text{Multiplying by } -1300$$
$$1300x + 1850y = 18,526$$
$$550y = 2926$$
$$y = 5.32$$

We go back to Equation (1) and substitute 5.32 for y.

$$x + y = 12$$
$$x + 5.32 = 12$$
$$x = 6.68$$

Check. The sum of 6.68 and 5.32 is 12. The number of words in large type is 1300(6.68), or 8684, and the number of words in small type is 1850(5.32), or 9842. Then the total number of words is 8684 + 9842, or 18,526. These numbers check.

State. There were 6.68, or $6\frac{17}{25}$ pages, in large type and 5.32, or $5\frac{8}{25}$ pages, in small type.

27. $25x^2 - 81 = (5x)^2 - 9^2$
$= (5x + 9)(5x - 9)$

29. $4x^2 + 100 = 4(x^2 + 25)$

31. $y = -2x - 3$

The equation is in the form $y = mx + b$, so the y-intercept is $(0, -3)$.

To find the x-intercept, we let $y = 0$ and solve for x.

$$0 = -2x - 3$$
$$2x = -3$$
$$x = -\frac{3}{2}$$

The x-intercept is $\left(-\frac{3}{2}, 0\right)$.

We plot the intercepts and draw the line.

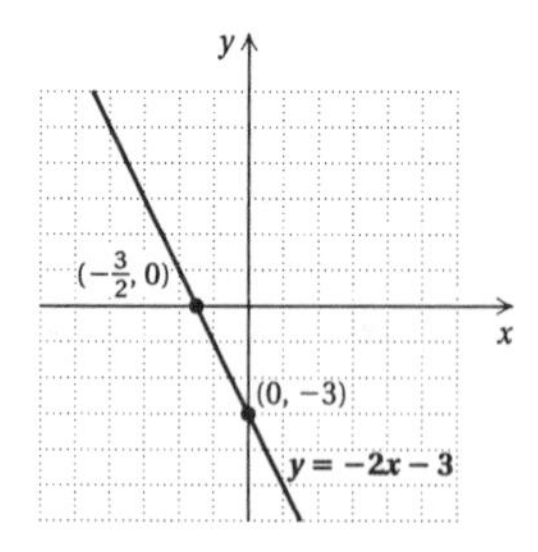

A third point should be used as a check.

For example, let $x = -3$. Then

$$y = -2(-3) - 3 = 6 - 3 = 3.$$

It appears that the point $(-3, 3)$ is on the graph, so the graph is probably correct.

33. $5x - 2y = -10$

To find the y-intercept, let $x = 0$ and solve for y.

$$5 \cdot 0 - 2y = -10$$
$$-2y = -10$$
$$y = 5$$

The y-intercept is $(0, 5)$.

To find the x-intercept, let $y = 0$ and solve for x.

$$5x - 2 \cdot 0 = -10$$
$$5x = -10$$
$$x = -2$$

The x-intercept is $(-2, 0)$.

We plot the intercepts and draw the line.

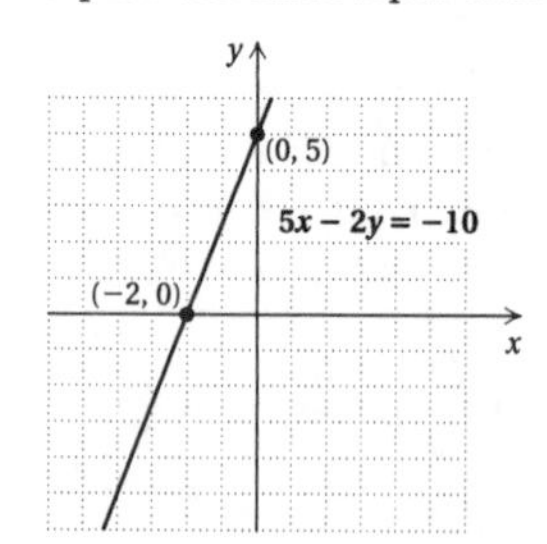

A third point should be used as a check. For example, let $x = -4$. then

$$5(-4) - 2y = -10$$
$$-20 - 2y = -10$$
$$-2y = 10$$
$$y = -5$$

It appears that the point $(-4, -5)$ is on the graph, so the graph is probably correct.

35. ◈

37. ***Familiarize***. Let x = the number of gallons of 87-octane gas and y = the number of gallons of 93-octane gas that should be used. We arrange the information in a table.

Type of gas	87-octane	93-octane	Mixture
Amount of gas	x	y	18
Octane rating	87	93	89
Mixture	$87x$	$93y$	$18 \cdot 89$, or 1602

Translate. The first and last rows of the table give us a system of equations.

$$x + y = 18, \quad (1)$$
$$87x + 93y = 1602 \quad (2)$$

Solve. We multiply Equation (1) by -87 and then add.

$$-87x - 87y = -1566$$
$$87x + 93y = 1602$$
$$\overline{\quad 6y = 36 \quad}$$
$$y = 6$$

Then substitute 6 for y in Equation (1) and solve for x.

$$x + y = 18$$
$$x + 6 = 18$$
$$x = 12$$

Check. The total amount of gas is 12 gal + 6 gal, or 18 gal. Also $87(12) + 93(6) = 1044 + 558 = 1602$. The answer checks.

State. 12 gal of 87-octane gas and 6 gal of 93-octane gas should be blended.

39. ***Familiarize***. In a table we arrange the information regarding the solution after some of the 30% solution is drained and replaced with pure antifreeze. We let x represent the amount of the original (30%) solution remaining, and we let y represent the amount of the 30% mixture that is drained and replaced with pure antifreeze.

Type of solution	Original (30%)	Pure antifreeze	Mixture
Amount of solution	x	y	16
Percent of antifreeze	30%	100%	50%
Amount of antifreeze in solution	$0.3x$	$1 \cdot y$, or y	$0.5(16)$, or 8

Translate. The table gives us two equations.

Amount of solution: $x + y = 16$

Amount of antifreeze in solution: $0.3x + y = 8$, or $3x + 10y = 80$

The resulting system is

$$x + y = 16, \quad (1)$$
$$3x + 10y = 80. \quad (2)$$

Solve. We multiply Equation (1) by -3 and then add.

$$\begin{array}{r} -3x - 3y = -48 \\ 3x + 10y = 80 \\ \hline 7y = 32 \\ y = \frac{32}{7}, \text{ or } 4\frac{4}{7} \end{array}$$

Then we substitute $4\frac{4}{7}$ for y in Equation (1) and solve for x.

$$x + y = 16$$
$$x + 4\frac{4}{7} = 16$$
$$x = 11\frac{3}{7}$$

Check. When $x = 11\frac{3}{7}$ L and $y = 4\frac{4}{7}$ L, the total is 16 L. The amount of antifreeze in the mixture is $0.3\left(11\frac{3}{7}\right) + 4\frac{4}{7}$, or $\frac{3}{10} \cdot \frac{80}{7} + \frac{32}{7}$, or $\frac{24}{7} + \frac{32}{7} = \frac{56}{7}$, or 8 L. This is 50% of 16 L, so the numbers check.

State. $4\frac{4}{7}$ of the original mixture should be drained and replaced with pure antifreeze.

41. ***Familiarize.*** We arrange the information in a table. Let x = the number of liters of skim milk and y = the number of liters of 3.2% milk.

Type of milk	4.6%	Skim	3.2% (Mixture)
Amount of milk	100 L	x	y
Percent of butterfat	4.6%	0%	3.2%
Amount of butterfat in milk	$4.6\% \times 100$, or 4.6 L	$0\% \cdot x$, or 0 L	$3.2\%y$

Translate. The first and third rows of the table give us two equations.

Amount of milk: $100 + x = y$

Amount of butterfat: $4.6 + 0 = 3.2\%y$, or $4.6 = 0.032y$.

The resulting system is

$$100 + x = y,$$
$$4.6 = 0.032y.$$

Solve. We solve the second equation for y.

$$4.6 = 0.032y$$
$$\frac{4.6}{0.032} = y$$
$$143.75 = y$$

We substitute 143.75 for y in the first equation and solve for x.

$$100 + x = y$$
$$100 + x = 143.75$$
$$x = 43.75$$

Check. We consider $x = 43.75$ L and $y = 143.75$ L. The difference between 143.75 L and 43.75 L is 100 L. There is no butterfat in the skim milk. There are 4.6 liters of butterfat in the 100 liters of the 4.6% milk. Thus there are 4.6 liters of butterfat in the mixture. This checks because 3.2% of 143.75 is 4.6.

State. 43.75 L of skim milk should be used.

43. ***Familiarize.*** Let l = the length and w = the width of the first frame. Then the length and width of the second frame are $l + 2$ and $w + 1$, respectively. The perimeter of the second frame is $2(l+2) + 2(w+1)$ and the cost of that frame is $\$0.40[2(l+2) + 2(w+1)]$.

Translate.

Length of first frame is 3 more than width.

$$l = 3 + w$$

Cost of second frame is \$22.40.

$$0.40[2(l+2) + 2(w+1)] = 22.40$$

We simplify the second equation and write the resulting system.

$$l = 3 + w,$$
$$0.8l + 0.8w = 20$$

Solve. First we multiply by 10 on both sides of the second equation to clear decimals. We have:

$$l = 3 + w, \quad (1)$$
$$8l + 8w = 200 \quad (2)$$

Now we substitute $3 + w$ for l in Equation (2) and solve for w.

$$8l + 8w = 200$$
$$8(3 + w) + 8w + 200$$
$$24 + 8w + 8w = 200$$
$$24 + 16w = 200$$
$$16w = 176$$
$$w = 11$$

Substitute 11 for w in Equation (1) and solve for l.

$$l = 3 + w$$
$$l = 3 + 11$$
$$l = 14$$

Check. The length, 14 in., is 3 in. longer than the width, 11 in. The dimensions of the second frame are $14 + 2$ by $11 + 1$, or 16 in. by 12 in. The cost of this frame is $\$0.40(2\cdot16+2\cdot12) = \$0.40(32+24) = \$0.40(56) = \$22.40$. The answer checks.

State. The dimensions of the first frame are 14 in. by 11 in.

45. ***Familiarize.*** Let x represent the part invested at 6% and y represent the part invested at 6.5%. The interest earned from the 6% investment is $6\% \cdot x$. The interest earned from the 6.5% investment is $6.5\% \cdot y$. The total investment is \$54,000, and the total interest earned is \$3385.

Translate.

$x + y = 54,000,$

$6\%x + 6.5\%y = \$3385,$ or $60x + 65y = 3,385,000$

Solve. Multiply the first equation by -60 and add.

$$\begin{aligned} -60x - 60y &= -3,240,000 \\ 60x + 65y &= 3,385,000 \\ \hline 5y &= 145,000 \\ y &= 29,000 \end{aligned}$$

Substitute 29,000 for y in the first equation and solve for x.

$$\begin{aligned} x + y &= 54,000 \\ x + 29,000 &= 54,000 \\ x &= 25,000 \end{aligned}$$

Check. We consider \$25,000 invested at 6% and \$29,000 invested at 6.5%. The sum of the investments is \$54,000. The interest earned is $6\% \cdot 25,000 + 6.5\% \cdot 29,000$, or $1500 + 1885$, or \$3385. These numbers check.

State. \$25,000 was invested at 6%, and \$29,000 was invested at 6.5%.

Exercise Set 14.5

1. ***Familiarize.*** We first make a drawing.

	30 mph	
Slow car	t hours	d miles

	46 mph	
Fast car	t hours	$d + 72$ miles

We let d = the distance the slow car travels. Then $d+72$ = the distance the fast car travels. We call the time t. We complete the table in the text, filling in the distances as well as the other information.

$d = r \cdot t$

	Distance	Speed	Time
Slow car	d	30	t
Fast car	$d + 72$	46	t

Translate. We get an equation $d = rt$ from each row of the table. Thus we have

$d = 30t,$ (1)

$d + 72 = 46t.$ (2)

Solve. We use the substitution method. We substitute $30t$ for d in Equation (2).

$$\begin{aligned} d + 72 &= 46t \\ 30t + 72 &= 46t \quad \text{Substituting} \\ 72 &= 16t \quad \text{Subtracting } 30t \\ 4.5 &= t \quad \text{Dividing by 16} \end{aligned}$$

Check. In 4.5 hr the slow car travels 30(4.5), or 135 mi, and the fast car travels 46(4.5), or 207 mi. Since 207 is 72 more than 135, our result checks.

State. The trains will be 72 mi apart in 4.5 hr.

3. ***Familiarize.*** First make a drawing.

Station	72 mph	
Slow train	$t + 3$ hours	d miles

Station	120 mph	
Fast train	t hours	d miles

Trains meet here.

From the drawing we see that the distances are the same. Let's call the distance d. Let t represent the time for the faster train and $t+3$ represent the time for the slower train. We complete the table in the text.

$d = r \cdot t$

	Distance	Speed	Time
Slow train	d	72	$t + 3$
Fast train	d	120	t

Equation from first row: $d = 72(t + 3)$

Equation from second row: $d = 120t$

Translate. Using $d = rt$ in each row of the table, we get the following system of equations:

$d = 72(t + 3),$ (1)

$d = 120t.$ (2)

Solve. Substitute $120t$ for d in Equation (1) and solve for t.

$$\begin{aligned} d &= 72(t + 3) \\ 120t &= 72(t + 3) \quad \text{Substituting} \\ 120t &= 72t + 216 \\ 48t &= 216 \\ t &= \frac{216}{48} \\ t &= 4.5 \end{aligned}$$

Check. When $t = 4.5$ hours, the faster train will travel 120(4.5), or 540 mi, and the slower train will travel 72(7.5), or 540 mi. In both cases we get the distance 540 mi.

State. In 4.5 hours after the second train leaves, the second train will overtake the first train. We can also state the answer as 7.5 hours after the first train leaves.

5. ***Familiarize.*** We first make a drawing.

With the current	$r + 6$
4 hours	d kilometers

Against the current	$r - 6$
10 hours	d kilometers

From the drawing we see that the distances are the same. Let d represent the distance. Let r represent the speed of the canoe in still water. Then, when the canoe is traveling with the current, its speed is $r + 6$. When it is traveling against the current, its speed is $r - 6$. We complete the table in the text.

$$d = r \cdot t$$

	Distance	Speed	Time
With current	d	$r+6$	4
Against current	d	$r-6$	10

Equation from first row: $d = (r+6)4$

Equation from second row: $d = (r-6)10$

Translate. Using $d = rt$ in each row of the table, we get the following system of equations:

$$d = (r+6)4, \quad (1)$$
$$d = (r-6)10 \quad (2)$$

Solve. Substitute $(r+6)4$ for d in Equation (2) and solve for r.

$$d = (r-6)10$$
$$(r+6)4 = (r-6)10 \quad \text{Substituting}$$
$$4r + 24 = 10r - 60$$
$$84 = 6r$$
$$14 = r$$

Check. When $r = 14$, $r + 6 = 20$ and $20 \cdot 4 = 80$, the distance. When $r = 14$, $r - 6 = 8$ and $8 \cdot 10 = 80$. In both cases, we get the same distance.

State. The speed of the canoe in still water is 14 km/h.

7. ***Familiarize.*** First make a drawing.

Passenger 96 km/h

$t - 2$ hours d kilometers

Freight 64 km/h

t hours d kilometers

Central City Clear Creek

From the drawing we see that the distances are the same. Let d represent the distance. Let t represent the time for the freight train. Then the time for the passenger train is $t - 2$. We organize the information in a table.

$$d = r \cdot t$$

	Distance	Speed	Time
Passenger	d	96	$t-2$
Freight	d	64	t

Translate. From each row of the table we get an equation.

$$d = 96(t-2), \quad (1)$$
$$d = 64t \quad (2)$$

Solve. Substitute $64t$ for d in Equation (1) and solve for t.

$$d = 96(t-2)$$
$$64t = 96(t-2) \quad \text{Substituting}$$
$$64t = 96t - 192$$
$$192 = 32t$$
$$6 = t$$

Next we substitute 6 for t in one of the original equations and solve for d.

$$d = 64t \quad \text{Equation (2)}$$
$$d = 64 \cdot 6 \quad \text{Substituting}$$
$$d = 384$$

Check. If the time is 6 hr, then the distance the passenger train travels is $96(6-2)$, or 384 km. The freight train travels $64(6)$, or 384 km. The distances are the same.

State. It is 384 km from Central City to Clear Creek.

9. ***Familiarize.*** We first make a drawing.

Downstream $r+6$

3 hours d miles

Upstream $r-6$

5 hours d miles

We let r represent the speed of the boat in still water and d represent the distance Antoine traveled downstream before he turned back. We organize the information in a table.

$$d = r \cdot t$$

	Distance	Speed	Time
Downstream	d	$r+6$	3
Upstream	d	$r-6$	5

Translate. Using $d = rt$ in each row of the table, we get the following system of equations:

$$d = (r+6)3, \quad (1)$$
$$d = (r-6)5 \quad (2)$$

Solve. Substitute $(r+6)3$ for d in Equation (2) and solve for r.

$$d = (r-6)5$$
$$(r+6)3 = (r-6)5 \quad \text{Substituting}$$
$$3r + 18 = 5r - 30$$
$$48 = 2r$$
$$24 = r$$

If $r = 24$, then $d = (r+6)3 = (24+6)3 = 30 \cdot 3 = 90$.

Check. If $r = 24$, then $r + 6 = 24 + 6 = 30$ and $r - 6 = 24 - 6 = 18$. If Antoine travels for 3 hr at 30 mph, then he travels $3 \cdot 30$, or 90 mi, downstream. If he travels for 5 hr at 18 mph, then he also travels $5 \cdot 18$, or 90 mi, upstream. Since the distances are the same, the answer checks.

State. (a) Antoine must travel at a speed of 24 mph.

(b) Antoine traveled 90 mi downstream before he turned back.

11. ***Familiarize***. We first make a drawing.

230 ft/min

Toddler $t+1$ min d ft

660 ft/min

Mother t min d ft

They meet here.

From the drawing we see that the distances are the same. Let's call the distance d. Let $t =$ the time the mother runs. Then $t+1 =$ the time the toddler runs. We arrange the information in a table.

$d = r \cdot t$

	Distance	Speed	Time
Toddler	d	230	$t+1$
Mother	d	660	t

Translate. Using $d = rt$ in each row of the table we get two equations.

$$d = 230(t+1), \quad (1)$$
$$d = 660t \quad (2)$$

Solve. Substitute $660t$ for d in Equation (1) and solve for t.

$$d = 230(t+1)$$
$$660t = 230(t+1) \quad \text{Substituting}$$
$$660t = 230t + 230$$
$$430t = 230$$
$$t = \frac{230}{430}, \text{ or } \frac{23}{43}$$

Check. When $t = \frac{23}{43}$ the toddler will travel $230\left(1\frac{23}{43}\right)$, or $230 \cdot \frac{66}{43}$, or $\frac{15,180}{43}$ ft and the mother will travel $660 \cdot \frac{23}{43}$, or $\frac{15,180}{43}$ ft. Since the distances are the same, our result checks.

State. The mother will overtake the toddler $\frac{23}{43}$ min after she starts running. We can also state the answer as $1\frac{23}{43}$ min after the toddler starts running.

13. ***Familiarize***. First make a drawing.

Home t hr 45 mph | $(2-t)$ hr 6 mph Work

Motorcycle distance | Walking distance

25 miles

Let t represent the time the motorcycle was driven. Then $2-t$ represents the time the rider walked. We organize the information in a table.

$d = r \cdot t$

	Distance	Speed	Time
Motorcycling	Motorcycle distance	45	t
Walking	Walking distance	6	$2-t$
Total	25		

Translate. From the drawing we see that

Motorcycle distance + Walking distance = 25

Then using $d = rt$ in each row of the table we get

$$45t + 6(2-t) = 25$$

Solve. We solve this equation for t.

$$45t + 12 - 6t = 25$$
$$39t + 12 = 25$$
$$39t = 13$$
$$t = \frac{13}{39}$$
$$t = \frac{1}{3}$$

Check. The problem asks us to find how far the motorcycle went before it broke down. If $t = \frac{1}{3}$, then $45t$ (the distance the motorcycle traveled) $= 45 \cdot \frac{1}{3}$, or 15 and $6(2-t)$ (the distance walked) $= 6\left(2 - \frac{1}{3}\right) = 6 \cdot \frac{5}{3}$, or 10. The total of these distances is 25, so $\frac{1}{3}$ checks.

State. The motorcycle went 15 miles before it broke down.

15. $$\frac{8x^2}{24x} = \frac{8}{24} \cdot \frac{x^2}{x} = \frac{1}{3} \cdot x^{2-1} = \frac{x}{3}$$

17. $$\frac{5a+15}{10} = \frac{5(a+3)}{5 \cdot 2} = \frac{\cancel{5}(a+3)}{\cancel{5} \cdot 2} = \frac{a+3}{2}$$

19. $$\frac{2x^2-50}{x^2-25} = \frac{2(x^2-25)}{x^2-25} = \frac{2}{1} \cdot \frac{x^2-25}{x^2-25} = 2$$

21. $$\frac{x^2-3x-10}{x^2-2x-15} = \frac{(x-5)(x+2)}{(x-5)(x+3)} = \frac{\cancel{(x-5)}(x+2)}{\cancel{(x-5)}(x+3)} = \frac{x+2}{x+3}$$

23. $$\frac{(x^2+6x+9)(x-2)}{(x^2-4)(x+3)} = \frac{(x+3)(x+3)(x-2)}{(x+2)(x-2)(x+3)} = \frac{\cancel{(x+3)}(x+3)\cancel{(x-2)}}{(x+2)\cancel{(x-2)}\cancel{(x+3)}} = \frac{x+3}{x+2}$$

25. $$\frac{6x^2+18x+12}{6x^2-6} = \frac{6(x^2+3x+2)}{6(x^2-1)} = \frac{6(x+1)(x+2)}{6(x+1)(x-1)} = \frac{\cancel{6}(\cancel{x+1})(x+2)}{\cancel{6}(\cancel{x+1})(x-1)} = \frac{x+2}{x-1}$$

27. ◈

29. ***Familiarize***. We arrange the information in a table. Let d = the length of the route and t = Lindbergh's time. Note that 16 hr and 57 min = $16\frac{57}{60}$ hr = 16.95 hr.

$d = r \cdot t$

	Distance	Speed	Time
Lindbergh	d	107.4	t
Hughes	d	217.1	$t - 16.95$

Translate. From the rows of the table we get two equations.

$$d = 107.4t, \quad (1)$$
$$d = 217.1(t - 16.95) \quad (2)$$

Solve. We substitute $107.4t$ for d in Equation (2) and solve for t.

$$d = 217.1(t - 16.95)$$
$$107.4t = 217.1(t - 16.95)$$
$$107.4t = 217.1t - 3679.845$$
$$-109.7t = -3679.845$$
$$t \approx 33.54$$

Now we go back to Equation (1) and substitute 33.54 for t.

$$d = 107.4t$$
$$d = 107.4(33.54)$$
$$d \approx 3602$$

Check. When $t \approx 33.54$, Lindbergh traveled $107.4(33.54) \approx 3602$ mi, and Hughes traveled $217.1(16.59) \approx 3602$ mi. Since the distances are the same, our result checks.

State. The route was 3602 mi long. (Answers may vary slightly due to rounding differences.)

31. ***Familiarize***. We arrange the information in a table. Let's call the distance d. When the riverboat is traveling upstream its speed is $12 - 4$, or 8 mph. Its speed traveling downstream is $12 + 4$, or 16 mph.

$d = r \cdot t$

	Distance	Speed	Time
Upstream	d	8	Time upstream
Downstream	d	16	Time downstream
Total			1

Translate. From the table we see that (Time upstream) + (Time downstream) = 1. Then using $d = rt$, in the form $\frac{d}{r} = t$, in each row of the table we get

$$\frac{d}{8} + \frac{d}{16} = 1.$$

Solve. We solve the equation. The LCM is 16.

$$\frac{d}{8} + \frac{d}{16} = 1$$
$$16\left(\frac{d}{8} + \frac{d}{16}\right) = 16 \cdot 1$$
$$16 \cdot \frac{d}{8} + 16 \cdot \frac{d}{16} = 16$$
$$2d + d = 16$$
$$3d = 16$$
$$d = \frac{16}{3}, \text{ or } 5\frac{1}{3}$$

Check. When $d = \frac{16}{3}$,

$$\text{(Time upstream)} + \text{(Time downstream)} = \frac{\frac{16}{3}}{8} + \frac{\frac{16}{3}}{16} = \frac{16}{3} \cdot \frac{1}{8} + \frac{16}{3} \cdot \frac{1}{16} = \frac{2}{3} + \frac{1}{3} = 1 \text{ hr}$$

Thus the distance of $\frac{16}{3}$ mi, or $5\frac{1}{3}$ mi checks.

State. The pilot should travel $5\frac{1}{3}$ mi upstream before turning around.

Chapter 15

Radical Expressions and Equations

Exercise Set 15.1

1. The square roots of 4 are 2 and -2, because $2^2 = 4$ and $(-2)^2 = 4$.

3. The square roots of 9 are 3 and -3, because $3^2 = 9$ and $(-3)^2 = 9$.

5. The square roots of 100 are 10 and -10, because $10^2 = 100$ and $(-10)^2 = 100$.

7. The square roots of 169 are 13 and -13, because $13^2 = 169$ and $(-13)^2 = 169$.

9. The square roots of 256 are 16 and -16, because $16^2 = 256$ and $(-16)^2 = 256$.

11. $\sqrt{4} = 2$, taking the principal square root.

13. $\sqrt{9} = 3$, so $-\sqrt{9} = -3$.

15. $\sqrt{36} = 6$, so $-\sqrt{36} = -6$.

17. $\sqrt{225} = 15$, so $-\sqrt{225} = -15$.

19. $\sqrt{361} = 19$, taking the principal square root.

21. 2.236

23. 20.785

25. $\sqrt{347.7} \approx 18.647$, so $-\sqrt{347.7} \approx -18.647$.

27. 2.779

29. 120

31. a) We substitute 25 into the formula:

$N = 2.5\sqrt{25} = 2.5(5) = 12.5 \approx 13$

b) We substitute 89 into the formula and use a calculator to find an approximation.

$N = 2.5\sqrt{89} \approx 2.5(9.434) = 23.585 \approx 24$

33. The radicand is the expression under the radical, 200.

35. The radicand is the expression under the radical, $a - 4$.

37. The radicand is the expression under the radical, $t^2 + 1$.

39. The radicand is the expression under the radical, $\dfrac{3}{x+2}$.

41. No, because the radicand is negative

43. Yes, because the radicand is nonnegative

45. $\sqrt{c^2} = c$ Since c is assumed to be nonnegative

47. $\sqrt{9x^2} = \sqrt{(3x)^2} = 3x$ Since $3x$ is assumed to be nonnegative

49. $\sqrt{(8p)^2} = 8p$ Since $8p$ is assumed to be nonnegative

51. $\sqrt{(ab)^2} = ab$

53. $\sqrt{(34d)^2} = 34d$

55. $\sqrt{(x+3)^2} = x + 3$

57. $\sqrt{a^2 - 10a + 25} = \sqrt{(a-5)^2} = a - 5$

59. $\sqrt{4a^2 - 20a + 25} = \sqrt{(2a-5)^2} = 2a - 5$

61. ***Familiarize.*** This problem states that we have direct variation between F and I. Thus, an equation $F = kI$, $k > 0$, applies. As the income increases, the amount spent on food increases.

Translate. We write an equation of variation.

Amount spent on food varies directly as the income.

This translates to $F = kI$.

Solve.

a) First find an equation of variation.

$$F = kI$$

$$10,192 = k \cdot 39,200 \quad \text{Substituting 10,192 for } F \text{ and 39,200 for } I$$

$$\frac{10,192}{39,200} = k$$

$$0.26 = k$$

The equation of variation is $F = 0.26I$.

b) We use the equation to find how much a family spends on food when their income is \$41,000.

$$F = 0.26I$$

$$F = 0.26(\$41,000) \quad \text{Substituting \$41,000 for } I$$

$$F = \$10,660$$

Check. Let us do some reasoning about the answer. The income increased from \$39,200 to \$41,000. Similarly, the amount spend on food increased from \$10,192 to \$10,660. This is what we would expect with direct variation.

State. The amount spent on food is \$10,660.

63.

$$\frac{x^2 + 10x - 11}{x^2 - 1} \div \frac{x+11}{x+1}$$

$$= \frac{x^2 + 10x - 11}{x^2 - 1} \cdot \frac{x+1}{x+11}$$

$$= \frac{(x^2 + 10x - 11)(x+1)}{(x^2-1)(x+11)}$$

$$= \frac{(x+11)(x-1)(x+1)}{(x+1)(x-1)(x+11)}$$

$$= 1$$

65. ◈

67. To approximate $\sqrt{3}$, locate 3 on the x-axis, move up vertically to the graph, and then move left horizontally to the y-axis to read the approximation.

$\sqrt{3} \approx 1.7$ (Answers may vary.)

To approximate $\sqrt{5}$, locate 5 on the x-axis, move up vertically to the graph, and then move left horizontally to the y-axis to read the approximation.

$\sqrt{5} \approx 2.2$ (Answers may vary.)

To approximate $\sqrt{7}$, locate 7 on the x-axis, move up vertically to the graph, and then move left horizontally to the y-axis to read the approximation.

$\sqrt{7} \approx 2.6$ (Answers may vary.)

69. $-\sqrt{36} < -\sqrt{33} < -\sqrt{25}$, or $-6 < -\sqrt{33} < -5$; $-\sqrt{33}$ is between -6 and -5.

71. $\sqrt{y^2} = -7$ has no solution, because the principal square root is nonnegative.

73. ***Familiarize.*** Let s represent the length of a side of the square. Then the area is $s \cdot s$, or s^2.

Translate.

$$\underbrace{\text{The area of a square}}_{s^2} \ \underset{=}{\text{is}} \ \underset{3}{3.}$$

Solve. We solve the equation.

$$s^2 = 3$$

$$s = \sqrt{3} \quad \text{or} \quad s = -\sqrt{3}$$

(The solutions s of the equation are the square roots of 3, since $s^2 = 3$.)

Check. Since length cannot be negative, we only need to check $\sqrt{3}$. If the length of a side of a square is $\sqrt{3}$, then the area is $\sqrt{3} \cdot \sqrt{3}$, or $(\sqrt{3})^2$, or 3. The answer checks.

State. The length of a side of the square is $\sqrt{3}$.

Exercise Set 15.2

1. $\sqrt{12} = \sqrt{4 \cdot 3}$ 4 is a perfect square.

$= \sqrt{4}\,\sqrt{3}$ Factoring into a product of radicals

$= 2\sqrt{3}$ Taking the square root

3. $\sqrt{75} = \sqrt{25 \cdot 3}$ 25 is a perfect square.

$= \sqrt{25}\,\sqrt{3}$ Factoring into a product of radicals

$= 5\sqrt{3}$ Taking the square root

5. $\sqrt{20} = \sqrt{4 \cdot 5}$ 4 is a perfect square.

$= \sqrt{4}\,\sqrt{5}$ Factoring into a product of radicals

$= 2\sqrt{5}$ Taking the square root

7. $\sqrt{600} = \sqrt{100 \cdot 6}$ 100 is a perfect square.

$= \sqrt{100} \cdot \sqrt{6}$ Factoring into a product of radicals

$= 10\sqrt{6}$ Taking the square root

9. $\sqrt{486} = \sqrt{81 \cdot 6}$ 81 is a perfect square.

$= \sqrt{81} \cdot \sqrt{6}$ Factoring into a product of radicals

$= 9\sqrt{6}$ Taking the square root

11. $\sqrt{9x} = \sqrt{9 \cdot x} = \sqrt{9}\,\sqrt{x} = 3\sqrt{x}$

13. $\sqrt{48x} = \sqrt{16 \cdot 3 \cdot x} = \sqrt{16}\,\sqrt{3x} = 4\sqrt{3x}$

15. $\sqrt{16a} = \sqrt{16 \cdot a} = \sqrt{16}\,\sqrt{a} = 4\sqrt{a}$

17. $\sqrt{64y^2} = \sqrt{64}\,\sqrt{y^2} = 8y$, or
$\sqrt{64y^2} = \sqrt{(8y)^2} = 8y$

19. $\sqrt{13x^2} = \sqrt{13}\,\sqrt{x^2} = \sqrt{13} \cdot x$, or $x\sqrt{13}$

21. $\sqrt{8t^2} = \sqrt{2 \cdot 4 \cdot t^2} = \sqrt{4}\,\sqrt{t^2}\,\sqrt{2} = 2t\sqrt{2}$

23. $\sqrt{180} = \sqrt{36 \cdot 5} = 6\sqrt{5}$

25. $\sqrt{288y} = \sqrt{144 \cdot 2 \cdot y} = \sqrt{144}\,\sqrt{2y} = 12\sqrt{2y}$

27. $\sqrt{28x^2} = \sqrt{4 \cdot 7 \cdot x^2} = \sqrt{4}\,\sqrt{x^2}\,\sqrt{7} = 2x\sqrt{7}$

29. $\sqrt{x^2 - 6x + 9} = \sqrt{(x-3)^2} = x - 3$

31. $\sqrt{8x^2 + 8x + 2} = \sqrt{2(4x^2 + 4x + 1)} =$
$\sqrt{2(2x+1)^2} = \sqrt{2}\,\sqrt{(2x+1)^2} = \sqrt{2}\,(2x+1)$

33. $\sqrt{36y + 12y^2 + y^3} = \sqrt{y(36 + 12y + y^2)} =$
$\sqrt{y(6+y)^2} = \sqrt{y}\,\sqrt{(6+y)^2} = \sqrt{y}\,(6+y)$

35. $\sqrt{x^6} = \sqrt{(x^3)^2} = x^3$

37. $\sqrt{x^{12}} = \sqrt{(x^6)^2} = x^6$

39. $\sqrt{x^5} = \sqrt{x^4 \cdot x}$ One factor is a perfect square

$= \sqrt{x^4}\,\sqrt{x}$

$= \sqrt{(x^2)^2}\,\sqrt{x}$

$= x^2\sqrt{x}$

41. $\sqrt{t^{19}} = \sqrt{t^{18} \cdot t} = \sqrt{t^{18}}\,\sqrt{t} = \sqrt{(t^9)^2}\,\sqrt{t} = t^9\sqrt{t}$

43. $\sqrt{(y-2)^8} = \sqrt{[(y-2)^4]^2} = (y-2)^4$

45. $\sqrt{4(x+5)^{10}} = \sqrt{4[(x+5)^5]^2} = \sqrt{4}\,\sqrt{[(x+5)^5]^2} =$
$2(x+5)^5$

47. $\sqrt{36m^3} = \sqrt{36 \cdot m^2 \cdot m} = \sqrt{36}\,\sqrt{m^2}\,\sqrt{m} = 6m\sqrt{m}$

49. $\sqrt{8a^5} = \sqrt{2 \cdot 4 \cdot a^4 \cdot a} = \sqrt{2 \cdot 4 \cdot (a^2)^2 \cdot a} =$
$\sqrt{4}\,\sqrt{(a^2)^2}\,\sqrt{2a} = 2a^2\sqrt{2a}$

51. $\sqrt{104p^{17}} = \sqrt{4 \cdot 26 \cdot p^{16} \cdot p} = \sqrt{4 \cdot 26 \cdot (p^8)^2 \cdot p} =$
$\sqrt{4}\,\sqrt{(p^8)^2}\,\sqrt{26p} = 2p^8\sqrt{26p}$

53. $\sqrt{448x^6y^3} = \sqrt{64 \cdot 7 \cdot x^6 \cdot y^2 \cdot y} =$
$\sqrt{64 \cdot 7 \cdot (x^3)^2 \cdot y^2 \cdot y} =$
$\sqrt{64}\,\sqrt{(x^3)^2}\,\sqrt{y^2}\,\sqrt{7y} = 8x^3y\,\sqrt{7y}$

55. $\sqrt{3}\,\sqrt{18} = \sqrt{3 \cdot 18}$ Multiplying
$= \sqrt{3 \cdot 3 \cdot 6}$ Looking for perfect-square factors or pairs of factors
$= \sqrt{3 \cdot 3}\,\sqrt{6}$
$= 3\sqrt{6}$

57. $\sqrt{15}\,\sqrt{6} = \sqrt{15 \cdot 6}$ Multiplying
$= \sqrt{5 \cdot 3 \cdot 3 \cdot 2}$ Looking for perfect-square factors or pairs of factors
$= \sqrt{3 \cdot 3}\,\sqrt{5 \cdot 2}$
$= 3\sqrt{10}$

59. $\sqrt{18}\,\sqrt{14x} = \sqrt{18 \cdot 14x} = \sqrt{3 \cdot 3 \cdot 2 \cdot 2 \cdot 7 \cdot x} =$
$\sqrt{3 \cdot 3}\,\sqrt{2 \cdot 2}\,\sqrt{7x} = 3 \cdot 2\sqrt{7x} = 6\sqrt{7x}$

61. $\sqrt{3x}\,\sqrt{12y} = \sqrt{3x \cdot 12y} = \sqrt{3 \cdot x \cdot 3 \cdot 4 \cdot y} =$
$\sqrt{3 \cdot 3 \cdot 4 \cdot x \cdot y} = \sqrt{3 \cdot 3}\,\sqrt{4}\,\sqrt{x \cdot y} = 3 \cdot 2\sqrt{xy} = 6\sqrt{xy}$

63. $\sqrt{13}\,\sqrt{13} = \sqrt{13 \cdot 13} = 13$

65. $\sqrt{5b}\,\sqrt{15b} = \sqrt{5b \cdot 15b} = \sqrt{5 \cdot b \cdot 5 \cdot 3 \cdot b} =$
$\sqrt{5 \cdot 5 \cdot b \cdot b \cdot 3} = \sqrt{5 \cdot 5}\,\sqrt{b \cdot b}\,\sqrt{3} = 5b\sqrt{3}$

67. $\sqrt{2t}\,\sqrt{2t} = \sqrt{2t \cdot 2t} = 2t$

69. $\sqrt{ab}\,\sqrt{ac} = \sqrt{ab \cdot ac} = \sqrt{a \cdot a \cdot b \cdot c} = \sqrt{a \cdot a}\,\sqrt{b \cdot c} =$
$a\sqrt{bc}$

71. $\sqrt{2x^2y}\,\sqrt{4xy^2} = \sqrt{2x^2y \cdot 4xy^2} = \sqrt{2 \cdot x^2 \cdot y \cdot 4 \cdot x \cdot y^2} =$
$\sqrt{4}\,\sqrt{x^2}\,\sqrt{y^2}\,\sqrt{2xy} = 2xy\sqrt{2xy}$

73. $\sqrt{18}\,\sqrt{18} = \sqrt{18 \cdot 18} = 18$

75. $\sqrt{5}\,\sqrt{2x-1} = \sqrt{5(2x-1)} = \sqrt{10x-5}$

77. $\sqrt{x+2}\,\sqrt{x+2} = \sqrt{(x+2)^2} = x+2$

79. $\sqrt{18x^2y^3}\,\sqrt{6xy^4} = \sqrt{18x^2y^3 \cdot 6xy^4} =$
$\sqrt{3 \cdot 6 \cdot x^2 \cdot y^2 \cdot y \cdot 6 \cdot x \cdot y^4} = \sqrt{6 \cdot 6 \cdot x^2 \cdot y^6 \cdot 3 \cdot x \cdot y} =$
$\sqrt{6 \cdot 6}\,\sqrt{x^2}\,\sqrt{y^6}\,\sqrt{3xy} = 6xy^3\sqrt{3xy}$

81. $\sqrt{50x^4y^6}\,\sqrt{10xy} = \sqrt{50x^4y^6 \cdot 10xy} =$
$\sqrt{5 \cdot 10 \cdot x^4 \cdot y^6 \cdot 10 \cdot x \cdot y} = \sqrt{10 \cdot 10 \cdot x^4 \cdot y^6 \cdot 5 \cdot x \cdot y} =$
$\sqrt{10 \cdot 10}\,\sqrt{x^4}\,\sqrt{y^6}\,\sqrt{5xy} = 10x^2y^3\sqrt{5xy}$

83.
$x - y = -6$ (1)
$x + y = 2$ (2)
$2x = -4$ Adding
$x = -2$

Now we substitute -2 for x in one of the original equations and solve for y.

$x + y = 2$ Equation (2)
$-2 + y = 2$ Substituting
$y = 4$

Since $(-2, 4)$ checks in both equations, it is the solution.

85. $3x - 2y = 4$, (1)
$2x + 5y = 9$ (2)

We will us the elimination method. We multiply on both sides of Equation (1) by 5 and on both sides of Equation (2) by 2. Then we add

$15x - 10y = 20$
$4x + 10y = 18$
$19x = 38$
$x = 2$

Now we substitute 2 for x in one of the original equations and solve for y.

$2x + 5y = 9$ Equation (2)
$2 \cdot 2 + 5y = 9$
$4 + 5y = 9$
$5y = 5$
$y = 1$

Since $(2, 1)$ checks, it is the solution.

87. ***Familiarize***. We let l = the length of the rectangle and w = the width. Recall that the perimeter of a rectangle is $2l + 2w$, and the area is lw.

Translate. We translate the first statement.

The perimeter of a rectangle is 84 ft.
$2l + 2w = 84$

Now we translate the second statement.

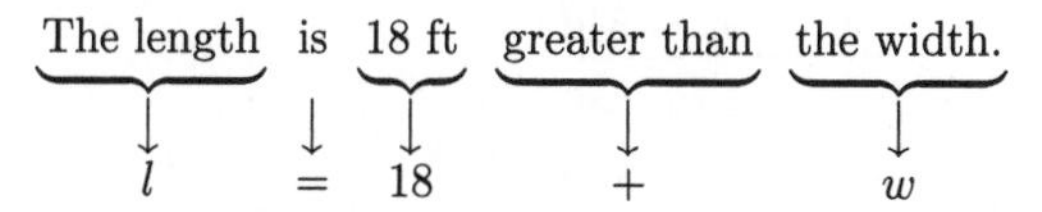

The resulting system is

$2l + 2w = 84$, (1)
$l = 18 + w$. (2)

Solve. We use the substitution method. We substitute $15 + w$ for l in Equation (1) and solve for w.

$2l + 2w = 84$ Equation (1)
$2(18 + w) + 2w = 84$ Substituting
$36 + 2w + 2w = 84$
$36 + 4w = 84$
$4w = 48$
$w = 12$

Substitute 12 for w in Equation (2) and compute l.

$l = 18 + w = 18 + 12 = 30$

When $l = 30$ ft and $w = 12$ ft, the area is 30 ft $\cdot$ 12 ft, or 360 ft^2.

Check. When $l = 30$ ft and $w = 12$ ft, the perimeter is $2 \cdot 30$ ft $+ 2 \cdot 12$ ft, or 60 ft $+ 24$ ft, or 84 ft. The length, 30 ft, is 18 ft greater than 12 ft, the width. We recheck the computation of the area.

State. The area of the rectangle is 360 ft^2.

89. ***Familiarize***. Let $x =$ the number of liters of 30% solution and $y =$ the number of liters of 50% solution to be used. We organize the information in a table.

Type of solution	30% insecticide	50% insecticide	Mixture
Amount of solution	x	y	200 L
Percent of insecticide	30%	50%	42%
Amount of insecticide in solution	$30\%x$	$50\%y$	$42\% \times 200$, or 84 L

Translate. The first and last rows of the table give us two equations. Since the total amount of solution is 200 L, we have

$$x + y = 200.$$

The amount of insecticide in the mixture is to be 42% of 200, or 84 L. The amounts of insecticide from the two solutions are $30\%x$ and $50\%y$. Thus

$$30\%x + 50\%y = 84,$$

or $\quad 0.3x + 0.5y = 84,$

or $\quad 3x + 5y = 840$ Clearing decimals

The resulting system is

$$x + y = 200, \quad (1)$$
$$3x + 5y = 840. \quad (2)$$

Solve. We use the elimination method. Multiply by -3 on both sides of Equation (1) and then add.

$$-3x - 3y = -600$$
$$3x + 5y = 840$$
$$2y = 240$$
$$y = 120$$

We go back to Equation (1) and substitute 120 for y.

$$x + y = 200$$
$$x + 120 = 200$$
$$x = 80$$

Check. The sum of 80 L and 120 L is 200 L. Also, 30% of 80 is 24 and 50% of 120 is 60 and $24 + 60 = 84$. The answer checks.

State. 80 L of 80% solution and 120 L of 50% solution should be used.

91. ***Familiarize***. We first make a drawing.

Downstream $r + 2$
2 hours d miles

Upstream $r - 2$
3 hours d miles

We let r represent the speed at which Greg and Beth were paddling the canoe and d represent the distance they traveled downriver. We organize the information in a table.

$$d \;=\; r \;\cdot\; t$$

	Distance	Speed	Time
Downstream	d	$r+2$	2
Upstream	d	$r-2$	3

Translate. Using $d = rt$ in each row of the table, we get the following system of equations:

$$d = (r+2)2, \quad (1)$$
$$d = (r-2)3 \quad (2)$$

Solve. Substitute $(r+2)2$ for d in Equation (2) and solve for r.

$$d = (r-2)3$$
$$(r+2)2 = (r-2)3 \quad \text{Substituting}$$
$$2r + 4 = 3r - 6$$
$$10 = r$$

Check. If $r = 10$, then $r + 2 = 10 + 2 = 12$ and $r - 2 = 10 - 2 = 8$. If Greg and Beth travel at 12 mph for 2 hr, they travel $12 \cdot 2$, or 24 mi. If they travel at 8 mph for 3 hr, they travel $8 \cdot 3$, or 24 mi. Since the distances are the same, the answer checks.

State. Greg and Beth were paddling the canoe at a speed of 10 mph.

93. ◈

95. $\sqrt{x^2 - x - 2} = \sqrt{(x-2)(x+1)} = \sqrt{x-2}\,\sqrt{x+1}$

97. $\sqrt{2x^2 - 5x - 12} = \sqrt{(2x+3)(x-4)} =$
$\sqrt{2x+3}\,\sqrt{x-4}$

99. $\sqrt{a^2 - b^2} = \sqrt{(a+b)(a-b)} = \sqrt{a+b}\,\sqrt{a-b}$

101. $\sqrt{0.25} = \sqrt{(0.5)^2} = 0.5$

103. $\sqrt{9a^6} = \sqrt{(3a^3)^2} = 3a^3$

105. $\sqrt{2y}\,\sqrt{3}\,\sqrt{8y} = \sqrt{2y \cdot 3 \cdot 8y} = \sqrt{2 \cdot y \cdot 3 \cdot 2 \cdot 4 \cdot y} =$
$\sqrt{2 \cdot 2 \cdot 4 \cdot y \cdot y \cdot 3} = \sqrt{2 \cdot 2}\,\sqrt{4}\,\sqrt{y \cdot y}\,\sqrt{3} =$
$2 \cdot 2 \cdot y\sqrt{3} = 4y\sqrt{3}$

107. $\sqrt{27(x+1)}\,\sqrt{12y(x+1)^2}$
$\sqrt{27(x+1) \cdot 12y(x+1)^2} =$
$\sqrt{9 \cdot 3 \cdot (x+1) \cdot 4 \cdot 3 \cdot y(x+1)^2} =$
$\sqrt{9 \cdot 3 \cdot 3 \cdot 4 \cdot (x+1)^2 \cdot (x+1)y} =$
$\sqrt{9}\,\sqrt{3 \cdot 3}\,\sqrt{4}\,\sqrt{(x+1)^2}\,\sqrt{(x+1)y}$
$3 \cdot 3 \cdot 2(x+1)\sqrt{(x+1)y} = 18(x+1)\sqrt{(x+1)y}$

109. $\sqrt{x}\,\sqrt{2x}\,\sqrt{10x^5} = \sqrt{x \cdot 2x \cdot 10x^5} =$
$\sqrt{x \cdot 2 \cdot x \cdot 2 \cdot 5 \cdot x^4 \cdot x} = \sqrt{x \cdot x \cdot 2 \cdot 2 \cdot x^4 \cdot 5 \cdot x} =$
$\sqrt{x \cdot x}\,\sqrt{2 \cdot 2}\,\sqrt{x^4}\,\sqrt{5x} = x \cdot 2 \cdot x^2\sqrt{5x} = 2x^3\sqrt{5x}$

Exercise Set 15.3

1. $\dfrac{\sqrt{18}}{\sqrt{2}}=\sqrt{\dfrac{18}{2}}=\sqrt{9}=3$

3. $\dfrac{\sqrt{108}}{\sqrt{3}}=\sqrt{\dfrac{108}{3}}=\sqrt{36}=6$

5. $\dfrac{\sqrt{65}}{\sqrt{13}}=\sqrt{\dfrac{65}{13}}=\sqrt{5}$

7. $\dfrac{\sqrt{3}}{\sqrt{75}}=\sqrt{\dfrac{3}{75}}=\sqrt{\dfrac{1}{25}}=\dfrac{1}{5}$

9. $\dfrac{\sqrt{12}}{\sqrt{75}}=\sqrt{\dfrac{12}{75}}=\sqrt{\dfrac{4}{25}}=\dfrac{2}{5}$

11. $\dfrac{\sqrt{8x}}{\sqrt{2x}}=\sqrt{\dfrac{8x}{2x}}=\sqrt{4}=2$

13. $\dfrac{\sqrt{63y^3}}{\sqrt{7y}}=\sqrt{\dfrac{63y^3}{7y}}=\sqrt{9y^2}=3y$

15. $\sqrt{\dfrac{16}{49}}=\dfrac{\sqrt{16}}{\sqrt{49}}=\dfrac{4}{7}$

17. $\sqrt{\dfrac{1}{36}}=\dfrac{\sqrt{1}}{\sqrt{36}}=\dfrac{1}{6}$

19. $-\sqrt{\dfrac{16}{81}}=-\dfrac{\sqrt{16}}{\sqrt{81}}=-\dfrac{4}{9}$

21. $\sqrt{\dfrac{64}{289}}=\dfrac{\sqrt{64}}{\sqrt{289}}=\dfrac{8}{17}$

23. $\sqrt{\dfrac{1690}{1960}}=\sqrt{\dfrac{169\cdot 10}{196\cdot 10}}=\sqrt{\dfrac{169}{196}\cdot\dfrac{10}{10}}=\sqrt{\dfrac{169}{196}}\cdot 1=$
$\sqrt{\dfrac{169}{196}}=\dfrac{\sqrt{169}}{\sqrt{196}}=\dfrac{13}{14}$

25. $\sqrt{\dfrac{25}{x^2}}=\dfrac{\sqrt{25}}{\sqrt{x^2}}=\dfrac{5}{x}$

27. $\sqrt{\dfrac{9a^2}{625}}=\dfrac{\sqrt{9a^2}}{\sqrt{625}}=\dfrac{3a}{25}$

29. $\sqrt{\dfrac{2}{5}}=\sqrt{\dfrac{2}{5}\cdot\dfrac{5}{5}}=\sqrt{\dfrac{10}{25}}=\dfrac{\sqrt{10}}{\sqrt{25}}=\dfrac{\sqrt{10}}{5}$

31. $\sqrt{\dfrac{7}{8}}=\sqrt{\dfrac{7}{8}\cdot\dfrac{2}{2}}=\sqrt{\dfrac{14}{16}}=\dfrac{\sqrt{14}}{\sqrt{16}}=\dfrac{\sqrt{14}}{4}$

33. $\sqrt{\dfrac{1}{12}}=\sqrt{\dfrac{1}{12}\cdot\dfrac{3}{3}}=\sqrt{\dfrac{3}{36}}=\dfrac{\sqrt{3}}{\sqrt{36}}=\dfrac{\sqrt{3}}{6}$

35. $\sqrt{\dfrac{5}{18}}=\sqrt{\dfrac{5}{18}\cdot\dfrac{2}{2}}=\sqrt{\dfrac{10}{36}}=\dfrac{\sqrt{10}}{\sqrt{36}}=\dfrac{\sqrt{10}}{6}$

37. $\dfrac{3}{\sqrt{5}}=\dfrac{3}{\sqrt{5}}\cdot\dfrac{\sqrt{5}}{\sqrt{5}}=\dfrac{3\sqrt{5}}{5}$

39. $\sqrt{\dfrac{8}{3}}=\sqrt{\dfrac{8}{3}\cdot\dfrac{3}{3}}=\sqrt{\dfrac{24}{9}}=\dfrac{\sqrt{4\cdot 6}}{\sqrt{9}}=\dfrac{\sqrt{4}\,\sqrt{6}}{\sqrt{9}}=\dfrac{2\sqrt{6}}{3}$

41. $\sqrt{\dfrac{3}{x}}=\sqrt{\dfrac{3}{x}\cdot\dfrac{x}{x}}=\sqrt{\dfrac{3x}{x^2}}=\dfrac{\sqrt{3x}}{\sqrt{x^2}}=\dfrac{\sqrt{3x}}{x}$

43. $\sqrt{\dfrac{x}{y}}=\sqrt{\dfrac{x}{y}\cdot\dfrac{y}{y}}=\sqrt{\dfrac{xy}{y^2}}=\dfrac{\sqrt{xy}}{\sqrt{y^2}}=\dfrac{\sqrt{xy}}{y}$

45. $\sqrt{\dfrac{x^2}{20}}=\sqrt{\dfrac{x^2}{20}\cdot\dfrac{5}{5}}=\sqrt{\dfrac{5x^2}{100}}=\dfrac{\sqrt{x^2\cdot 5}}{\sqrt{100}}=\dfrac{\sqrt{x^2}\,\sqrt{5}}{\sqrt{100}}=\dfrac{x\sqrt{5}}{10}$

47. $\dfrac{\sqrt{7}}{\sqrt{2}}=\dfrac{\sqrt{7}}{\sqrt{2}}\cdot\dfrac{\sqrt{2}}{\sqrt{2}}=\dfrac{\sqrt{14}}{2}$

49. $\dfrac{\sqrt{9}}{\sqrt{8}}=\dfrac{\sqrt{9}}{\sqrt{8}}\cdot\dfrac{\sqrt{2}}{\sqrt{2}}=\dfrac{\sqrt{9\cdot 2}}{\sqrt{16}}=\dfrac{3\sqrt{2}}{4}$

51. $\dfrac{\sqrt{3}}{\sqrt{2}}=\dfrac{\sqrt{3}}{\sqrt{2}}\cdot\dfrac{\sqrt{2}}{\sqrt{2}}=\dfrac{\sqrt{6}}{2}$

53. $\dfrac{2}{\sqrt{2}}=\dfrac{2}{\sqrt{2}}\cdot\dfrac{\sqrt{2}}{\sqrt{2}}=\dfrac{2\sqrt{2}}{2}=\sqrt{2}$

55. $\dfrac{\sqrt{5}}{\sqrt{11}}=\dfrac{\sqrt{5}}{\sqrt{11}}\cdot\dfrac{\sqrt{11}}{\sqrt{11}}=\dfrac{\sqrt{55}}{11}$

57. $\dfrac{\sqrt{7}}{\sqrt{12}}=\dfrac{\sqrt{7}}{\sqrt{12}}\cdot\dfrac{\sqrt{3}}{\sqrt{3}}=\dfrac{\sqrt{21}}{\sqrt{36}}=\dfrac{\sqrt{21}}{6}$

59. $\dfrac{\sqrt{48}}{\sqrt{32}}=\sqrt{\dfrac{48}{32}}=\sqrt{\dfrac{3}{2}}=\sqrt{\dfrac{3}{2}\cdot\dfrac{2}{2}}=\sqrt{\dfrac{6}{4}}=\dfrac{\sqrt{6}}{\sqrt{4}}=\dfrac{\sqrt{6}}{2}$

61. $\dfrac{\sqrt{450}}{\sqrt{18}}=\sqrt{\dfrac{450}{18}}=\sqrt{25}=5$

63. $\dfrac{\sqrt{3}}{\sqrt{x}}=\dfrac{\sqrt{3}}{\sqrt{x}}\cdot\dfrac{\sqrt{x}}{\sqrt{x}}=\dfrac{\sqrt{3x}}{x}$

65. $\dfrac{4y}{\sqrt{5}}=\dfrac{4y}{\sqrt{5}}\cdot\dfrac{\sqrt{5}}{\sqrt{5}}=\dfrac{4y\sqrt{5}}{5}$

67. $\dfrac{\sqrt{a^3}}{\sqrt{8}}=\dfrac{\sqrt{a^3}}{\sqrt{8}}\cdot\dfrac{\sqrt{2}}{\sqrt{2}}=\dfrac{\sqrt{2a^3}}{\sqrt{16}}=\dfrac{\sqrt{a^2\cdot 2a}}{\sqrt{16}}=\dfrac{a\sqrt{2a}}{4}$

69. $\dfrac{\sqrt{56}}{\sqrt{12x}}=\sqrt{\dfrac{56}{12x}}=\sqrt{\dfrac{14}{3x}}=\sqrt{\dfrac{14}{3x}\cdot\dfrac{3x}{3x}}=\sqrt{\dfrac{42x}{3x\cdot 3x}}=$
$\dfrac{\sqrt{42x}}{3x}$

71. $\dfrac{\sqrt{27c}}{\sqrt{32c^3}}=\sqrt{\dfrac{27c}{32c^2}}=\sqrt{\dfrac{27}{32c^2}}=\sqrt{\dfrac{27}{32c^2}\cdot\dfrac{2}{2}}=\sqrt{\dfrac{54}{64c^2}}=$
$\sqrt{\dfrac{9\cdot 6}{64c^2}}=\dfrac{3\sqrt{6}}{8c}$

73. $\dfrac{\sqrt{y^5}}{\sqrt{xy^2}}=\sqrt{\dfrac{y^5}{xy^2}}=\sqrt{\dfrac{y^3}{x}}=\sqrt{\dfrac{y^3}{x}\cdot\dfrac{x}{x}}=\sqrt{\dfrac{xy^3}{x^2}}=$
$\sqrt{\dfrac{y^2\cdot xy}{x^2}}=\dfrac{y\sqrt{xy}}{x}$

75. $\frac{\sqrt{45mn^2}}{\sqrt{32m}} = \sqrt{\frac{45mn^2}{32m}} = \sqrt{\frac{45n^2}{32}} = \sqrt{\frac{45n^2}{32} \cdot \frac{2}{2}} =$

$\sqrt{\frac{90n^2}{64}} = \frac{\sqrt{90n^2}}{\sqrt{64}} = \frac{\sqrt{9 \cdot n^2 \cdot 10}}{8} = \frac{3n\sqrt{10}}{8}$

77. $x = y + 2,\quad (1)$

$x + y = 6 \quad (2)$

We substitute $y + 2$ for x in Equation (2) and solve for y.

$$(y+2) + y = 6$$
$$2y + 2 = 6$$
$$2y = 4$$
$$y = 2$$

Substitute 2 for y in Equation (1) to find x.

$$x = 2 + 2 = 4$$

The ordered pair $(4, 2)$ checks in both equations. It is the solution.

79. $2x - 3y = 7 \quad (1)$
$2x - 3y = 9 \quad (2)$

We multiply Equation (2) by -1 and add.

$$\begin{array}{r} 2x - 3y = 7 \\ -2x + 3y = -9 \\ \hline 0 = -2 \end{array}$$

We get a false equation. The system of equations has no solution.

81.
$$\begin{array}{ll} x + y = -7 & (1) \\ x - y = 2 & (2) \\ \hline 2x \quad = -5 & \text{Adding} \\ x = -\frac{5}{2} & \end{array}$$

Substitute $-\frac{5}{2}$ for x in Equation (1) to find y.

$$x + y = -7 \quad \text{Equation (1)}$$
$$-\frac{5}{2} + y = -7 \quad \text{Substituting}$$
$$y = -\frac{9}{2}$$

The ordered pair $\left(-\frac{5}{2}, -\frac{9}{2}\right)$ checks in both equations. It is the solution.

83. $(3x - 7)(3x + 7) = (3x)^2 - 7^2 = 9x^2 - 49$

85. $9x - 5y + 12x - 4y$
$= (9x + 12x) + (-5y - 4y)$
$= 21x - 9y$

87. ◈

89. 2 ft: $T \approx 2(3.14)\sqrt{\frac{2}{32}} \approx 6.28\sqrt{\frac{1}{16}} \approx 6.28\left(\frac{1}{4}\right) \approx$ 1.57 sec

8 ft: $T \approx 2(3.14)\sqrt{\frac{8}{32}} \approx 6.28\sqrt{\frac{1}{4}} \approx 6.28\left(\frac{1}{2}\right) \approx$ 3.14 sec

64 ft: $T \approx 2(3.14)\sqrt{\frac{64}{32}} \approx 6.28\sqrt{2} \approx$ $(6.28)(1.414) \approx 8.88$ sec

100 ft: $T \approx 2(3.14)\sqrt{\frac{100}{32}} \approx 6.28\sqrt{\frac{50}{16}} \approx \frac{6.28\sqrt{50}}{4} \approx$ $\frac{6.28(7.071)}{4} \approx 11.10$ sec

91. $T = 2\pi\sqrt{\frac{\frac{32}{\pi^2}}{32}} = 2\pi\sqrt{\frac{32}{\pi^2} \cdot \frac{1}{32}} = 2\pi\sqrt{\frac{1}{\pi^2}} = 2\pi\left(\frac{1}{\pi}\right) = 2$ sec

The time it takes the pendulum to swing from one side to the other and back is 2 sec, so it takes 1 sec to swing from one side to the other.

93. $\sqrt{\frac{5}{1600}} = \frac{\sqrt{5}}{\sqrt{1600}} = \frac{\sqrt{5}}{40}$

95. $\sqrt{\frac{1}{5x^3}} = \sqrt{\frac{1}{5x^3} \cdot \frac{5x}{5x}} = \sqrt{\frac{5x}{25x^4}} = \frac{\sqrt{5x}}{\sqrt{25x^4}} = \frac{\sqrt{5x}}{5x^2}$

97. $\sqrt{\frac{3a}{b}} = \sqrt{\frac{3a}{b} \cdot \frac{b}{b}} = \sqrt{\frac{3ab}{b^2}} = \frac{\sqrt{3ab}}{\sqrt{b^2}} = \frac{\sqrt{3ab}}{b}$

99. $\sqrt{0.009} = \sqrt{\frac{9}{1000}} = \sqrt{\frac{9}{1000} \cdot \frac{10}{10}} = \sqrt{\frac{90}{10,000}} =$

$\frac{\sqrt{90}}{\sqrt{10,000}} = \frac{\sqrt{9 \cdot 10}}{100} = \frac{\sqrt{9}\,\sqrt{10}}{100} = \frac{3\sqrt{10}}{100}$

101. $\sqrt{\frac{1}{x^2} - \frac{2}{xy} + \frac{1}{y^2}}$, LCD is x^2y^2

$= \sqrt{\frac{1}{x^2} \cdot \frac{y^2}{y^2} - \frac{2}{xy} \cdot \frac{xy}{xy} + \frac{1}{y^2} \cdot \frac{x^2}{x^2}}$

$= \sqrt{\frac{y^2 - 2xy + x^2}{x^2y^2}}$

$= \sqrt{\frac{(y-x)^2}{x^2y^2}}$

$= \frac{\sqrt{(y-x)^2}}{\sqrt{x^2y^2}}$

$= \frac{y-x}{xy}$

Exercise Set 15.4

1. $7\sqrt{3} + 9\sqrt{3} = (7+9)\sqrt{3}$
$= 16\sqrt{3}$

3. $7\sqrt{5} - 3\sqrt{5} = (7-3)\sqrt{5}$
$= 4\sqrt{5}$

5. $6\sqrt{x} + 7\sqrt{x} = (6+7)\sqrt{x}$
$= 13\sqrt{x}$

7. $4\sqrt{d} - 13\sqrt{d} = (4-13)\sqrt{d}$
$= -9\sqrt{d}$

9. $5\sqrt{8}+15\sqrt{2} = 5\sqrt{4\cdot 2}+15\sqrt{2}$
$= 5\cdot 2\sqrt{2}+15\sqrt{2}$
$= 10\sqrt{2}+15\sqrt{2}$
$= 25\sqrt{2}$

11. $\sqrt{27}-2\sqrt{3} = \sqrt{9\cdot 3}-2\sqrt{3}$
$= 3\sqrt{3}-2\sqrt{3}$
$= (3-2)\sqrt{3}$
$= 1\sqrt{3}$
$= \sqrt{3}$

13. $\sqrt{45}-\sqrt{20} = \sqrt{9\cdot 5}-\sqrt{4\cdot 5}$
$= 3\sqrt{5}-2\sqrt{5}$
$= (3-2)\sqrt{5}$
$= 1\sqrt{5}$
$= \sqrt{5}$

15. $\sqrt{72}+\sqrt{98} = \sqrt{36\cdot 2}+\sqrt{49\cdot 2}$
$= 6\sqrt{2}+7\sqrt{2}$
$= (6+7)\sqrt{2}$
$= 13\sqrt{2}$

17. $2\sqrt{12}+\sqrt{27}-\sqrt{48} = 2\sqrt{4\cdot 3}+\sqrt{9\cdot 3}-\sqrt{16\cdot 3}$
$= 2\cdot 2\sqrt{3}+3\sqrt{3}-4\sqrt{3}$
$= 4\sqrt{3}+3\sqrt{3}-4\sqrt{3}$
$= (4+3-4)\sqrt{3}$
$= 3\sqrt{3}$

19. $\sqrt{18}-3\sqrt{8}+\sqrt{50} = \sqrt{9\cdot 2}-3\sqrt{4\cdot 2}+\sqrt{25\cdot 2}$
$= 3\sqrt{2}-3\cdot 2\sqrt{2}+5\sqrt{2}$
$= 3\sqrt{2}-6\sqrt{2}+5\sqrt{2}$
$= (3-6+5)\sqrt{2}$
$= 2\sqrt{2}$

21. $2\sqrt{27}-3\sqrt{48}+3\sqrt{12} = 2\sqrt{9\cdot 3}-3\sqrt{16\cdot 3}+3\sqrt{4\cdot 3}$
$= 2\cdot 3\sqrt{3}-3\cdot 4\sqrt{3}+3\cdot 2\sqrt{3}$
$= 6\sqrt{3}-12\sqrt{3}+6\sqrt{3}$
$= (6-12+6)\sqrt{3}$
$= 0\sqrt{3}$
$= 0$

23. $\sqrt{4x}+\sqrt{81x^3} = \sqrt{4\cdot x}+\sqrt{81\cdot x^2\cdot x}$
$= 2\sqrt{x}+9x\sqrt{x}$
$= (2+9x)\sqrt{x}$

25. $\sqrt{27}-\sqrt{12x^2} = \sqrt{9\cdot 3}-\sqrt{4\cdot 3\cdot x^2}$
$= 3\sqrt{3}-2x\sqrt{3}$
$= (3-2x)\sqrt{3}$

27. $\sqrt{8x+8}+\sqrt{2x+2} = \sqrt{4(2x+2)}+\sqrt{2x+2}$
$= 2\sqrt{2x+2}+1\sqrt{2x+2}$
$= (2+1)\sqrt{2x+2}$
$= 3\sqrt{2x+2}$

29. $\sqrt{x^5-x^2}+\sqrt{9x^3-9} = \sqrt{x^2(x^3-1)}+\sqrt{9(x^3-1)}$
$= x\sqrt{x^3-1}+3\sqrt{x^3-1}$
$= (x+3)\sqrt{x^3-1}$

31. $4a\sqrt{a^2b}+a\sqrt{a^2b^3}-5\sqrt{b^3}$
$= 4a\sqrt{a^2\cdot b}+a\sqrt{a^2\cdot b^2\cdot b}-5\sqrt{b^2\cdot b}$
$= 4a\cdot a\sqrt{b}+a\cdot a\cdot b\sqrt{b}-5\cdot b\sqrt{b}$
$= 4a^2\sqrt{b}+a^2b\sqrt{b}-5b\sqrt{b}$
$= (4a^2+a^2b-5b)\sqrt{b}$

33. $\sqrt{3}-\sqrt{\frac{1}{3}} = \sqrt{3}-\sqrt{\frac{1}{3}\cdot\frac{3}{3}}$
$= \sqrt{3}-\frac{\sqrt{3}}{3}$
$= \left(1-\frac{1}{3}\right)\sqrt{3}$
$= \frac{2}{3}\sqrt{3}$, or $\frac{2\sqrt{3}}{3}$

35. $5\sqrt{2}+3\sqrt{\frac{1}{2}} = 5\sqrt{2}+3\sqrt{\frac{1}{2}\cdot\frac{2}{2}}$
$= 5\sqrt{2}+\frac{3}{2}\sqrt{2}$
$= \left(5+\frac{3}{2}\right)\sqrt{2}$
$= \frac{13}{2}\sqrt{2}$, or $\frac{13\sqrt{2}}{2}$

37. $\sqrt{\frac{2}{3}}-\sqrt{\frac{1}{6}} = \sqrt{\frac{2}{3}\cdot\frac{3}{3}}-\sqrt{\frac{1}{6}\cdot\frac{6}{6}}$
$= \frac{\sqrt{6}}{3}-\frac{\sqrt{6}}{6}$
$= \left(\frac{1}{3}-\frac{1}{6}\right)\sqrt{6}$
$= \frac{1}{6}\sqrt{6}$, or $\frac{\sqrt{6}}{6}$

39. $\sqrt{3}(\sqrt{5}-1) = \sqrt{3}\,\sqrt{5}-\sqrt{3}\cdot 1$
$= \sqrt{15}-\sqrt{3}$

41. $(2+\sqrt{3})(5-\sqrt{7})$
$= 2\cdot 5-2\sqrt{7}+\sqrt{3}\cdot 5-\sqrt{3}\sqrt{7}$ Using FOIL
$= 10-2\sqrt{7}+5\sqrt{3}-\sqrt{21}$

43. $(2-\sqrt{5})^2$
$= 2^2-2\cdot 2\cdot\sqrt{5}+(\sqrt{5})^2$
Using $(A-B)^2 = A^2-2AB+B^2$
$= 4-4\sqrt{5}+5$
$= 9-4\sqrt{5}$

45. $(\sqrt{2}+8)(\sqrt{2}-8)$

$= (\sqrt{2})^2 - 8^2$ Using $(A+B)(A-B) = A^2 - B^2$

$= 2 - 64$

$= -62$

47. $(\sqrt{6}-\sqrt{5})(\sqrt{6}+\sqrt{5})$

$= (\sqrt{6})^2 - (\sqrt{5})^2$ Using $(A+B)(A-B) = A^2 - B^2$

$= 6 - 5$

$= 1$

49. $(3\sqrt{5}-2)(\sqrt{5}+1)$

$= 3\sqrt{5}\,\sqrt{5} + 3\sqrt{5} - 2\sqrt{5} - 2$ Using FOIL

$= 3\cdot 5 + 3\sqrt{5} - 2\sqrt{5} - 2$

$= 15 + \sqrt{5} - 2$

$= 13 + \sqrt{5}$

51. $(\sqrt{x}-\sqrt{y})^2 = (\sqrt{x})^2 - 2\sqrt{x}\,\sqrt{y} + (\sqrt{y})^2$

Using $(A-B)^2 = A^2 - 2AB + B^2$

$= x - 2\sqrt{xy} + y$

53. We multiply by 1 using the conjugate of $\sqrt{3}-\sqrt{5}$, which is $\sqrt{3}+\sqrt{5}$, as the numerator and denominator.

$$\frac{2}{\sqrt{3}-\sqrt{5}} = \frac{2}{\sqrt{3}-\sqrt{5}}\cdot\frac{\sqrt{3}+\sqrt{5}}{\sqrt{3}+\sqrt{5}} \quad \text{Multiplying by 1}$$

$$= \frac{2(\sqrt{3}+\sqrt{5})}{(\sqrt{3}-\sqrt{5})(\sqrt{3}+\sqrt{5})} \quad \text{Multiplying}$$

$$= \frac{2\sqrt{3}+2\sqrt{5}}{(\sqrt{3})^2-(\sqrt{5})^2} = \frac{2\sqrt{3}+2\sqrt{5}}{3-5}$$

$$= \frac{2\sqrt{3}+2\sqrt{5}}{-2} = \frac{2(\sqrt{3}+\sqrt{5})}{-2}$$

$$= -(\sqrt{3}+\sqrt{5}) = -\sqrt{3}-\sqrt{5}$$

55. We multiply by 1 using the conjugate of $\sqrt{3}+\sqrt{2}$, which is $\sqrt{3}-\sqrt{2}$, as the numerator and denominator.

$$\frac{\sqrt{3}-\sqrt{2}}{\sqrt{3}+\sqrt{2}} = \frac{\sqrt{3}-\sqrt{2}}{\sqrt{3}+\sqrt{2}}\cdot\frac{\sqrt{3}-\sqrt{2}}{\sqrt{3}-\sqrt{2}} \quad \text{Multiplying by 1}$$

$$= \frac{(\sqrt{3}-\sqrt{2})^2}{(\sqrt{3}+\sqrt{2})(\sqrt{3}-\sqrt{2})}$$

$$= \frac{(\sqrt{3})^2 - 2\sqrt{3}\,\sqrt{2} + (\sqrt{2})^2}{(\sqrt{3})^2-(\sqrt{2})^2}$$

$$= \frac{3-2\sqrt{6}+2}{3-2} = \frac{5-2\sqrt{6}}{1}$$

$$= 5-2\sqrt{6}$$

57. We multiply by 1 using the conjugate of $\sqrt{10}+1$, which is $\sqrt{10}-1$, as the numerator and denominator.

$$\frac{4}{\sqrt{10}+1} = \frac{4}{\sqrt{10}+1}\cdot\frac{\sqrt{10}-1}{\sqrt{10}-1}$$

$$= \frac{4(\sqrt{10}-1)}{(\sqrt{10}+1)(\sqrt{10}-1)}$$

$$= \frac{4\sqrt{10}-4}{(\sqrt{10})^2-1^2} = \frac{4\sqrt{10}-4}{10-1}$$

$$= \frac{4\sqrt{10}-4}{9}$$

59. We multiply by 1 using the conjugate of $3+\sqrt{7}$, which is $3-\sqrt{7}$, as the numerator and denominator.

$$\frac{1-\sqrt{7}}{3+\sqrt{7}} = \frac{1-\sqrt{7}}{3+\sqrt{7}}\cdot\frac{3-\sqrt{7}}{3-\sqrt{7}}$$

$$= \frac{(1-\sqrt{7})(3-\sqrt{7})}{(3+\sqrt{7})(3-\sqrt{7})}$$

$$= \frac{3-\sqrt{7}-3\sqrt{7}+\sqrt{7}\,\sqrt{7}}{3^2-(\sqrt{7})^2}$$

$$= \frac{3-\sqrt{7}-3\sqrt{7}+7}{9-7} = \frac{10-4\sqrt{7}}{2}$$

$$= \frac{2(5-2\sqrt{7})}{2} = 5-2\sqrt{7}$$

61. We multiply by 1 using the conjugate of $4+\sqrt{x}$, which is $4-\sqrt{x}$, as the numerator and denominator.

$$\frac{3}{4+\sqrt{x}} = \frac{3}{4+\sqrt{x}}\cdot\frac{4-\sqrt{x}}{4-\sqrt{x}}$$

$$= \frac{3(4-\sqrt{x})}{(4+\sqrt{x})(4-\sqrt{x})}$$

$$= \frac{12-3\sqrt{x}}{4^2-(\sqrt{x})^2}$$

$$= \frac{12-3\sqrt{x}}{16-x}$$

63. We multiply by 1 using the conjugate of $8-\sqrt{x}$, which is $8+\sqrt{x}$, as the numerator and denominator.

$$\frac{3+\sqrt{2}}{8-\sqrt{x}} = \frac{3+\sqrt{2}}{8-\sqrt{x}}\cdot\frac{8+\sqrt{x}}{8+\sqrt{x}}$$

$$= \frac{(3+\sqrt{2})(8+\sqrt{x})}{(8-\sqrt{x})(8+\sqrt{x})}$$

$$= \frac{3\cdot 8+3\cdot\sqrt{x}+\sqrt{2}\cdot 8+\sqrt{2}\cdot\sqrt{x}}{8^2-(\sqrt{x})^2}$$

$$= \frac{24+3\sqrt{x}+8\sqrt{2}+\sqrt{2x}}{64-x}$$

65.
$$\begin{aligned} 3x+5+2(x-3) &= 4-6x \\ 3x+5+2x-6 &= 4-6x \\ 5x-1 &= 4-6x \\ 11x-1 &= 4 \\ 11x &= 5 \\ x &= \frac{5}{11} \end{aligned}$$

The solution is $\frac{5}{11}$.

67.
$$\begin{aligned} x^2-5x &= 6 \\ x^2-5x-6 &= 0 \\ (x+1)(x-6) &= 0 \end{aligned}$$
$$x+1=0 \quad or \quad x-6=0$$
$$x=-1 \quad or \quad x=6$$

The solutions are -1 and 6.

69. ***Familiarize***. Let x = the number of liters of Jolly Juice and y = the number of liters of Real Squeeze in the mixture. We organize the given information in a table.

	Jolly Juice	Real Squeeze	Mixture
Amount	x	y	8
Percent real fruit juice	3%	6%	5.4%
Amount of real fruit juice	$0.03x$	$0.06y$	0.054(8), or 0.432

Translate. We get two equation from the first and third rows of the table.

$$x + y = 8,$$
$$0.03x + 0.06y = 0.432$$

Clearing decimals gives

$$x + y = 8, \quad (1)$$
$$30x + 60y = 432. \quad (2)$$

Carry out. We use elimination. Multiply Equation (1) by -30 and add.

$$\begin{aligned} -30x - 30y &= -240 \\ 30x + 60y &= 432 \\ \hline 30y &= 192 \\ y &= 6.4 \end{aligned}$$

Now substitute 6.4 for y in Equation (1) and solve for x.

$$\begin{aligned} x+y &= 8 \\ x+6.4 &= 8 \\ x &= 1.6 \end{aligned}$$

Check. The sum of 1.6 and 6.4 is 8. The amount of real fruit juice in this mixture is $0.03(1.6) + 0.06(6.4)$, or $0.048 + 0.384$, or 0.432 L. The answer checks.

State. 1.6 L of Jolly Juice and 6.4 L of Real Squeeze should be used.

71. For $x=-1$, $y=(-1)^3-5(-1)^2+(-1)-2=-1-5-1-2=-9$.
For $x=0$, $y=0^3-5\cdot 0^2+0-2=0-0+0-2=-2$.
For $x=1$, $y=1^3-5\cdot 1^2+1-2=1-5+1-2=-5$.
For $x=3$, $y=3^3-5\cdot 3^2+3-2=27-45+3-2=-17$.
For $x=4.85$, $y=(4.85)^3-5(4.85)^2+4.85-2=$
$114.084125-117.6125+4.85-2=-0.678375$

These values could have been estimated using the graph also.

73. ◈

75. Since $\sqrt{a^2+b^2} \neq \sqrt{a^2}+\sqrt{b^2}$ for $a=2$ and $b=3$, the two expressions are not equivalent.

77. Enter $y_1=\sqrt{x^2+4}$ and $y_2=\sqrt{x}+2$ and look at the graphs or a table of values. Since the graphs do not coincide and the values for y_1 and y_2 are different, the given statement is not correct.

79.
$$\begin{aligned} &\frac{1}{3}\sqrt{27}+\sqrt{8}+\sqrt{300}-\sqrt{18}-\sqrt{162} \\ &= \frac{1}{3}\sqrt{9\cdot 3}+\sqrt{4\cdot 2}+\sqrt{100\cdot 3}-\sqrt{9\cdot 2}-\sqrt{81\cdot 2} \\ &= \frac{1}{3}\cdot 3\sqrt{3}+2\sqrt{2}+10\sqrt{3}-3\sqrt{2}-9\sqrt{2} \\ &= \sqrt{3}+2\sqrt{2}+10\sqrt{3}-3\sqrt{2}-9\sqrt{2} \\ &= (1+10)\sqrt{3}+(2-3-9)\sqrt{2} \\ &= 11\sqrt{3}-10\sqrt{2} \end{aligned}$$

81.
$$\begin{aligned} (3\sqrt{x+2})^2 &= 3^2(\sqrt{x+2})^2 \\ &= 9(x+2) \end{aligned}$$

The statement is true.

Exercise Set 15.5

1.
$$\begin{aligned} \sqrt{x} &= 6 \\ (\sqrt{x})^2 &= 6^2 \quad \text{Squaring both sides} \\ x &= 36 \quad \text{Simplifying} \end{aligned}$$

Check: $\sqrt{x}=6$
$\sqrt{36}\ ?\ 6$
$6 \mid$ TRUE

The solution is 36.

3.
$$\begin{aligned} \sqrt{x} &= 4.3 \\ (\sqrt{x})^2 &= (4.3)^2 \quad \text{Squaring both sides} \\ x &= 18.49 \quad \text{Simplifying} \end{aligned}$$

Check: $\sqrt{x}=4.3$
$\sqrt{18.49}\ ?\ 4.3$
$4.3 \mid$ TRUE

The solution is 18.49.

5. $\sqrt{y+4} = 13$

$(\sqrt{y+4})^2 = 13^2$ Squaring both sides

$y + 4 = 169$ Simplifying

$y = 165$ Subtracting 4

Check: $\sqrt{y+4} = 13$

$\sqrt{165+4} \; ? \; 13$

$\sqrt{169}$

13 TRUE

The solution is 165.

7. $\sqrt{2x+4} = 25$

$(\sqrt{2x+4})^2 = 25^2$ Squaring both sides

$2x + 4 = 625$ Simplifying

$2x = 621$ Subtracting 4

$x = \frac{621}{2}$ Dividing by 2

Check: $\sqrt{2x+4} = 25$

$\sqrt{2 \cdot \frac{621}{2} + 4} \; ? \; 25$

$\sqrt{621+4}$

$\sqrt{625}$

25 TRUE

The solution is $\frac{621}{2}$.

9. $3 + \sqrt{x-1} = 5$

$\sqrt{x-1} = 2$ Subtracting 3

$(\sqrt{x-1})^2 = 2^2$ Squaring both sides

$x - 1 = 4$

$x = 5$

Check: $3 + \sqrt{x-1} = 5$

$3 + \sqrt{5-1} \; ? \; 5$

$3 + \sqrt{4}$

$3 + 2$

5 TRUE

The solution is 5.

11. $6 - 2\sqrt{3n} = 0$

$6 = 2\sqrt{3n}$ Adding $2\sqrt{3n}$

$6^2 = (2\sqrt{3n})^2$ Squaring both sides

$36 = 4 \cdot 3n$

$36 = 12n$

$3 = n$

Check: $6 - 2\sqrt{3n} = 0$

$6 - 2\sqrt{3 \cdot 3} \; ? \; 0$

$6 - 2 \cdot 3$

$6 - 6$

0 TRUE

The solution is 3.

13. $\sqrt{5x-7} = \sqrt{x+10}$

$(\sqrt{5x-7})^2 = (\sqrt{x+10})^2$ Squaring both sides

$5x - 7 = x + 10$

$4x = 17$

$x = \frac{17}{4}$

Check: $\sqrt{5x-7} = \sqrt{x+10}$

$\sqrt{5 \cdot \frac{17}{4} - 7} \; ? \; \sqrt{\frac{17}{4} + 10}$

$\sqrt{\frac{85}{4} - \frac{28}{4}} \;\Big|\; \sqrt{\frac{57}{4}}$

$\sqrt{\frac{57}{4}}$ TRUE

The solution is $\frac{17}{4}$.

15. $\sqrt{x} = -7$

There is no solution. The principal square root of x cannot be negative.

17. $\sqrt{2y+6} = \sqrt{2y-5}$

$(\sqrt{2y+6})^2 = (\sqrt{2y-5})^2$

$2y + 6 = 2y - 5$

$6 = -5$

The equation $6 = -5$ is false; there is no solution.

19. $x - 7 = \sqrt{x-5}$

$(x-7)^2 = (\sqrt{x-5})^2$

$x^2 - 14x + 49 = x - 5$

$x^2 - 15 + 54 = 0$

$(x-9)(x-6) = 0$

$x - 9 = 0$ or $x - 6 = 0$

$x = 9$ or $x = 6$

Check: $x - 7 = \sqrt{x-5}$

$9 - 7 \; ? \; \sqrt{9-5}$

$2 \;\Big|\; \sqrt{4}$

2 TRUE

$$\begin{array}{r|l} x-7 & = \sqrt{x-5} \\ \hline 6-7 & \sqrt{6-5} \\ -1 & \sqrt{1} \\ & 1 \qquad \text{FALSE} \end{array}$$

The number 9 checks, but 6 does not. The solution is 9.

21.
$$\begin{aligned} x-9 &= \sqrt{x-3} \\ (x-9)^2 &= (\sqrt{x-3})^2 \\ x^2-18x+81 &= x-3 \\ x^2-19x+84 &= 0 \\ (x-12)(x-7) &= 0 \end{aligned}$$
$$x-12=0 \quad \text{or} \quad x-7=0$$
$$x=12 \quad \text{or} \quad x=7$$

Check:
$$\begin{array}{r|l} x-9 & = \sqrt{x-3} \\ \hline 12-9 \ ? & \sqrt{12-3} \\ 3 & \sqrt{9} \\ & 3 \qquad \text{TRUE} \end{array}$$

$$\begin{array}{r|l} x-9 & = \sqrt{x-3} \\ \hline 7-9 \ ? & \sqrt{7-3} \\ -2 & \sqrt{4} \\ & 2 \qquad \text{FALSE} \end{array}$$

The number 12 checks, but 7 does not. The solution is 12.

23.
$$\begin{aligned} 2\sqrt{x-1} &= x-1 \\ (2\sqrt{x-1})^2 &= (x-1)^2 \\ 4(x-1) &= x^2-2x+1 \\ 4x-4 &= x^2-2x+1 \\ 0 &= x^2-6x+5 \\ 0 &= (x-5)(x-1) \end{aligned}$$
$$x-5=0 \quad \text{or} \quad x-1=0$$
$$x=5 \quad \text{or} \quad x=1$$

Both numbers check. The solutions are 5 and 1.

25.
$$\begin{aligned} \sqrt{5x+21} &= x+3 \\ (\sqrt{5x+21})^2 &= (x+3)^2 \\ 5x+21 &= x^2+6x+9 \\ 0 &= x^2+x-12 \\ 0 &= (x+4)(x-3) \end{aligned}$$
$$x+4=0 \quad \text{or} \quad x-3=0$$
$$x=-4 \quad \text{or} \quad x=3$$

Check:
$$\begin{array}{r|l} \sqrt{5x+21} & = x+3 \\ \hline \sqrt{5(-4)+21} \ ? & -4+3 \\ \sqrt{1} & -1 \\ 1 & \qquad \text{FALSE} \end{array}$$

$$\begin{array}{r|l} \sqrt{5x+21} & = x+3 \\ \hline \sqrt{5\cdot 3+21} \ ? & 3+3 \\ \sqrt{36} & 6 \\ 6 & \qquad \text{TRUE} \end{array}$$

The number 3 checks, but -4 does not. The solution is 3.

27.
$$\begin{aligned} \sqrt{2x-1}+2 &= x \\ \sqrt{2x-1} &= x-2 \qquad \text{Isolating the radical} \\ (\sqrt{2x-1})^2 &= (x-2)^2 \\ 2x-1 &= x^2-4x+4 \\ 0 &= x^2-6x+5 \\ 0 &= (x-5)(x-1) \end{aligned}$$
$$x-5=0 \quad \text{or} \quad x-1=0$$
$$x=5 \quad \text{or} \quad x=1$$

Check:
$$\begin{array}{r|l} \sqrt{2x-1}+2 & = x \\ \hline \sqrt{2\cdot 5-1}+2 \ ? & 5 \\ \sqrt{10-1}+2 & \\ \sqrt{9}+2 & \\ 3+2 & \\ 5 & \qquad \text{TRUE} \end{array}$$

$$\begin{array}{r|l} \sqrt{2x-1}+2 & = x \\ \hline \sqrt{2\cdot 1-1}+2 \ ? & 1 \\ \sqrt{2-1}+2 & \\ \sqrt{1}+2 & \\ 1+2 & \\ 3 & \qquad \text{FALSE} \end{array}$$

The number 5 checks, but 1 does not. The solution is 5.

29.
$$\begin{aligned} \sqrt{x^2+6}-x+3 &= 0 \\ \sqrt{x^2+6} &= x-3 \qquad \text{Isolating the radical} \\ (\sqrt{x^2+6})^2 &= (x-3)^2 \\ x^2+6 &= x^2-6x+9 \\ -3 &= -6x \qquad \text{Adding } -x^2 \text{ and } -9 \\ \frac{1}{2} &= x \end{aligned}$$

Check: $\sqrt{x^2+6}-x+3=0$

$$\sqrt{\left(\frac{1}{2}\right)^2+6}-\frac{1}{2}+3 \;?\; 0$$
$$\sqrt{\frac{25}{4}}-\frac{1}{2}+3$$
$$\frac{5}{2}-\frac{1}{2}+3$$
$$5 \quad \text{FALSE}$$

The number $\frac{1}{2}$ does not check. There is no solution.

31. $\sqrt{x^2-4}-x=6$

$$\begin{aligned} \sqrt{x^2-4} &= x+6 && \text{Isolating the radical} \\ (\sqrt{x^2-4})^2 &= (x+6)^2 \\ x^2-4 &= x^2+12x+36 \\ -40 &= 12x && \text{Adding } -x^2 \text{ and } -36 \\ -\frac{40}{12} &= x \\ -\frac{10}{3} &= x \end{aligned}$$

The number $-\frac{10}{3}$ checks. It is the solution.

33. $\sqrt{(p+6)(p+1)}-2=p+1$

$$\begin{aligned} \sqrt{(p+6)(p+1)} &= p+3 && \text{Isolating the radical} \\ \left(\sqrt{(p+6)(p+1)}\right)^2 &= (p+3)^2 \\ (p+6)(p+1) &= p^2+6p+9 \\ p^2+7p+6 &= p^2+6p+9 \\ p &= 3 \end{aligned}$$

The number 3 checks. It is the solution.

35. $\sqrt{4x-10}=\sqrt{2-x}$

$$\begin{aligned} (\sqrt{4x-10})^2 &= (\sqrt{2-x})^2 \\ 4x-10 &= 2-x \\ 5x &= 12 && \text{Adding 10 and } x \\ x &= \frac{12}{5} \end{aligned}$$

Check: $\sqrt{4x-10}=\sqrt{2-x}$

$$\sqrt{4\cdot\frac{12}{5}-10} \;?\; \sqrt{2-\frac{12}{5}}$$
$$\sqrt{\frac{48}{5}-10} \quad \sqrt{-\frac{2}{5}}$$

Since $\sqrt{-\frac{2}{5}}$ does not represent a real number, there is no solution that is a real number.

37. $\sqrt{x-5}=5-\sqrt{x}$

$$\begin{aligned} (\sqrt{x-5})^2 &= (5-\sqrt{x})^2 && \text{Squaring both sides} \\ x-5 &= 25-10\sqrt{x}+x \\ -30 &= -10\sqrt{x} && \text{Isolating the radical} \\ 3 &= \sqrt{x} && \text{Dividing by } -10 \\ 3^2 &= (\sqrt{x})^2 && \text{Squaring both sides} \\ 9 &= x \end{aligned}$$

The number 9 checks. It is the solution.

39. $\sqrt{y+8}-\sqrt{y}=2$

$$\begin{aligned} \sqrt{y+8} &= \sqrt{y}+2 && \text{Isolating one radical} \\ (\sqrt{y+8})^2 &= (\sqrt{y}+2)^2 && \text{Squaring both sides} \\ y+8 &= y+4\sqrt{y}+4 \\ 4 &= 4\sqrt{x} && \text{Isolating the radical} \\ 1 &= \sqrt{y} && \text{Dividing by 4} \\ 1^2 &= (\sqrt{y})^2 \\ 1 &= y \end{aligned}$$

The number 1 checks. It is the solution.

41. $\sqrt{x-4}+\sqrt{x+1}=5$

$$\begin{aligned} \sqrt{x-4} &= 5-\sqrt{x+1} && \text{Isolating one radical} \\ (\sqrt{x-4})^2 &= (5-\sqrt{x+1})^2 \\ x-4 &= 25-10\sqrt{x+1}+x+1 \\ -30 &= -10\sqrt{x+1} && \text{Isolating the radical} \\ 3 &= \sqrt{x+1} && \text{Dividing by } -10 \\ 3^2 &= (\sqrt{x+1})^2 \\ 9 &= x+1 \\ 8 &= x \end{aligned}$$

The number 8 checks. It is the solution.

43. We substitute 21 for V in the equation and solve for h.

$$\begin{aligned} V &= 3.5\sqrt{h} \\ 21 &= 3.5\sqrt{h} && \text{Substituting} \\ 6 &= \sqrt{h} && \text{Dividing by 3.5} \\ 6^2 &= (\sqrt{h})^2 \\ 36 &= h \end{aligned}$$

The altitude of the steeplejack's eyes is 36 m.

45. We substitute 37 for h and use a calculator to find an approximation.

$$\begin{aligned} V &= 3.5\sqrt{h} \\ V &= 3.5\sqrt{37} \\ V &\approx 21 \end{aligned}$$

The sailor can see about 21 km to the horizon.

47.
$$r = 2\sqrt{5L}$$
$$55 = 2\sqrt{5L} \quad \text{Substituting 55 for } r$$
$$27.5 = \sqrt{5L}$$
$$(27.5)^2 = (\sqrt{5L})^2$$
$$756.25 = 5L$$
$$151.25 = L$$

A car will skid about 151 ft at 55 mph.

$$75 = 2\sqrt{5L} \quad \text{Substituting 100 for } r$$
$$37.5 = \sqrt{5L}$$
$$(37.5)^2 = (\sqrt{5L})^2$$
$$1406.25 = 5L$$
$$281.25 = L$$

A car will skid about 281 ft at 75 mph.

49. ***Familiarize***. Let $n =$ the number.

Translate.

The square root of 4 more than 5 times a number is 8.

$$\sqrt{5n+4} \quad = \quad 8$$

Solve.

$$\sqrt{5n+4} = 8$$
$$(\sqrt{5n+4})^2 = 8^2$$
$$5n + 4 = 64$$
$$5n = 60$$
$$n = 12$$

Check. Four more than 5 times 12 is $60 + 4$, or 64, and $\sqrt{64} = 8$. The result checks.

State. The number is 12.

51. ***Familiarize***. Let $x =$ the number.

Translate.

The square root of 4 less than the number plus the square root of 1 more than the number is 5.

$$\sqrt{x-4} \quad + \quad \sqrt{x+1} \quad = \quad 5$$

Solve. This is the equation we solved in Exercise 41. The solution is 8.

Check. $\sqrt{8-4} = \sqrt{4} = 2$, $\sqrt{8+1} = \sqrt{9} = 3$, and $2 + 3 = 5$. The answer checks.

State. The number is 8.

53.
$$\frac{x^2-49}{x+8} \div \frac{x^2-14x+49}{x^2+15x+56}$$
$$= \frac{x^2-49}{x+8} \cdot \frac{x^2+15x+56}{x^2-14x+49}$$
$$= \frac{(x^2-49)(x^2+15x+56)}{(x+8)(x^2-14x+49)}$$
$$= \frac{(x+7)(x-7)(x+7)(x+8)}{(x+8)(x-7)(x-7)}$$
$$= \frac{(x+7)\cancel{(x-7)}(x+7)\cancel{(x+8)}}{\cancel{(x+8)}\cancel{(x-7)}(x-7)}$$
$$= \frac{(x+7)(x+7)}{x-7}, \text{ or } \frac{(x+7)^2}{x-7}$$

55.
$$\frac{a^2-25}{6} \div \frac{a+5}{3} = \frac{a^2-25}{6} \cdot \frac{3}{a+5}$$
$$= \frac{(a^2-25)\cdot 3}{6(a+5)}$$
$$= \frac{(a+5)(a-5)\cdot 3}{2\cdot 3\cdot (a+5)}$$
$$= \frac{\cancel{(a+5)}(a-5)\cdot \cancel{3}}{2\cdot \cancel{3}\cdot \cancel{(a+5)}}$$
$$= \frac{a-5}{2}$$

57. ***Familiarize***. Let x and y represent the angles. Recall that supplementary angles are angles whose sum is 180°.

Translate. We reword the problem.

The sum of two angles is 180°.

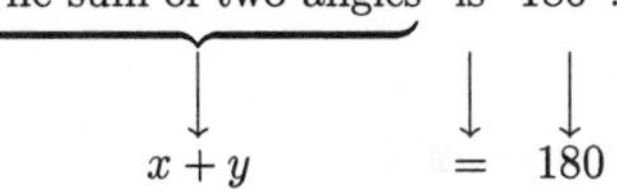

One angle is two times the other less 3°.

The resulting system is

$$x + y = 180, \quad (1)$$
$$x = 2y - 3. \quad (2)$$

Solve. We use substitution. We substitute $2y - 3$ for x in Equation (1) and solve for y.

$$(2y - 3) + y = 180$$
$$3y - 3 = 180$$
$$3y = 183$$
$$y = 61$$

We substitute 61 for y in Equation (2) to find x.

$$x = 2(61) - 3 = 122 - 3 = 119$$

Check. 61°+ 119°= 180°, so the angles are supplementary. Also, 3° less than twice 61° is $2\cdot 61° - 3°$, or $122° - 3°$, or 119°. The numbers check.

State. The angles are 61° and 119°.

59. $\frac{7x^9}{27} \cdot \frac{9}{7x^3} = \frac{63x^9}{189x^3} = \frac{63}{189}x^{9-3} = \frac{1}{3}x^6$, or $\frac{x^6}{3}$

61. ◈

63.
$$\begin{aligned} \sqrt{5x^2+5} &= 5 \\ (\sqrt{5x^2+5})^2 &= 5^2 \\ 5x^2+5 &= 25 \\ 5x^2-20 &= 0 \\ 5(x^2-4) &= 0 \\ 5(x+2)(x-2) &= 0 \end{aligned}$$

$x+2=0$ *or* $x-2=0$

$x=-2$ *or* $x=2$

Both numbers check, so the solutions are 2 and -2.

65.
$$\begin{aligned} 4+\sqrt{19-x} &= 6+\sqrt{4-x} \\ \sqrt{19-x} &= 2+\sqrt{4-x} \quad \text{Isolating one radical} \\ (\sqrt{19-x})^2 &= (2+\sqrt{4-x})^2 \\ 19-x &= 4+4\sqrt{4-x}+(4-x) \\ 19-x &= 4\sqrt{4-x}+8-x \\ 11 &= 4\sqrt{4-x} \\ 11^2 &= (4\sqrt{4-x})^2 \\ 121 &= 16(4-x) \\ 121 &= 64-16x \\ 57 &= -16x \\ -\frac{57}{16} &= x \end{aligned}$$

$-\frac{57}{16}$ checks, so it is the solution.

67.
$$\begin{aligned} \sqrt{x+3} &= \frac{8}{\sqrt{x-9}} \\ (\sqrt{x+3})^2 &= \left(\frac{8}{\sqrt{x-9}}\right)^2 \\ x+3 &= \frac{64}{x-9} \\ (x-9)(x+3) &= 64 \quad \text{Multiplying by } x-9 \\ x^2-6x-27 &= 64 \\ x^2-6x-91 &= 0 \\ (x-13)(x+7) &= 0 \end{aligned}$$

$x-13=0$ *or* $x+7=0$

$x=13$ *or* $x=-7$

The number 13 checks, but -7 does not. The solution is 13.

69.-71.

Exercise Set 15.6

1.
$$\begin{aligned} a^2+b^2 &= c^2 \\ 8^2+15^2 &= c^2 \quad \text{Substituting} \\ 64+225 &= c^2 \\ 289 &= c^2 \\ \sqrt{289} &= c \\ 17 &= c \end{aligned}$$

3.
$$\begin{aligned} a^2+b^2 &= c^2 \\ 4^2+4^2 &= c^2 \quad \text{Substituting} \\ 16+16 &= c^2 \\ 32 &= c^2 \\ \sqrt{32} &= c \quad \text{Exact answer} \\ 5.657 &\approx c \quad \text{Approximation} \end{aligned}$$

5.
$$\begin{aligned} a^2+b^2 &= c^2 \\ 5^2+b^2 &= 13^2 \\ 25+b^2 &= 169 \\ b^2 &= 144 \\ b &= 12 \end{aligned}$$

7.
$$\begin{aligned} a^2+b^2 &= c^2 \\ (4\sqrt{3})^2+b^2 &= 8^2 \\ 16\cdot 3+b^2 &= 64 \\ 48+b^2 &= 64 \\ b^2 &= 16 \\ b &= 4 \end{aligned}$$

9.
$$\begin{aligned} a^2+b^2 &= c^2 \\ 10^2+24^2 &= c^2 \\ 100+576 &= c^2 \\ 676 &= c^2 \\ 26 &= c \end{aligned}$$

11.
$$\begin{aligned} a^2+b^2 &= c^2 \\ 9^2+b^2 &= 15^2 \\ 81+b^2 &= 225 \\ b^2 &= 144 \\ b &= 12 \end{aligned}$$

13.
$$\begin{aligned} a^2+b^2 &= c^2 \\ a^2+1^2 &= (\sqrt{5})^2 \\ a^2+1 &= 5 \\ a^2 &= 4 \\ a &= 2 \end{aligned}$$

15. $a^2 + b^2 = c^2$
$1^2 + b^2 = (\sqrt{3})^2$
$1 + b^2 = 3$
$b^2 = 2$
$b = \sqrt{2}$ Exact answer
$b \approx 1.414$ Approximation

17. $a^2 + b^2 = c^2$
$a^2 + (5\sqrt{3})^2 = 10^2$
$a^2 + 25 \cdot 3 = 100$
$a^2 + 75 = 100$
$a^2 = 25$
$a = 5$

19. $a^2 + b^2 = c^2$
$(\sqrt{2})^2 + (\sqrt{7})^2 = c^2$
$2 + 7 = c^2$
$9 = c^2$
$3 = c$

21. We use the drawing in the text, labeling the horizontal distance h.

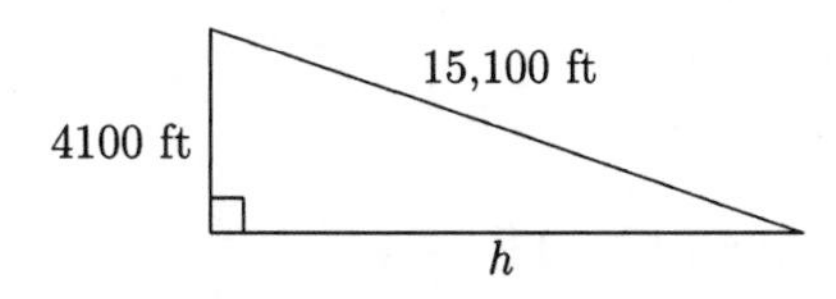

We know that $4100^2 + h^2 = 15,100^2$. We solve this equation.

$16,810,000 + h^2 = 228,010,000$
$h^2 = 211,200,000$
$h = \sqrt{211,200,000}$ ft Exact answer
$h \approx 14,533$ ft Approximation

23. We first make a drawing. Let d represent the distance Becky can move away from the building while using the telephone.

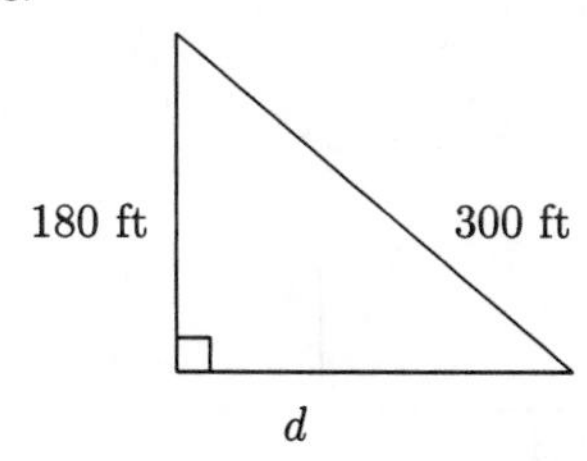

We know that $180^2 + d^2 = 300^2$.

We solve this equation.

$180^2 + d^2 = 300^2$
$32,400 + d^2 = 90,000$
$d^2 = 57,600$
$d = 240$

Becky can use her telephone 240 ft into her backyard.

25. We first make a drawing. We label the diagonal d.

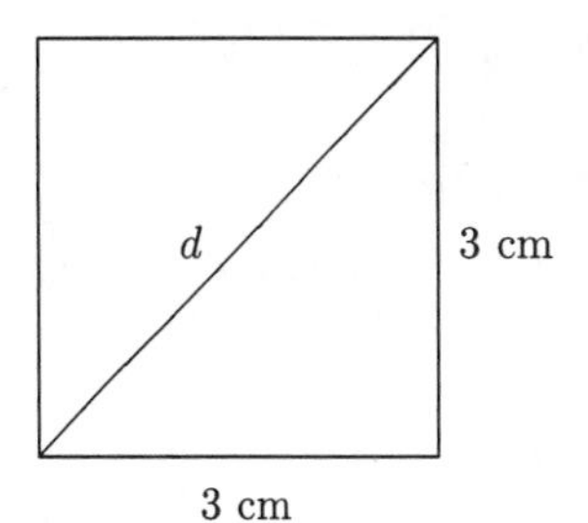

We know that $3^2 + 3^2 = d^2$. We solve this equation.

$3^2 + 3^2 = d^2$
$9 + 9 = d^2$
$18 = d^2$
$\sqrt{18}$ cm $= d$ Exact answer
4.243 cm $\approx d$ Approximation

27. We first make a drawing. We label the length of the guy wire w.

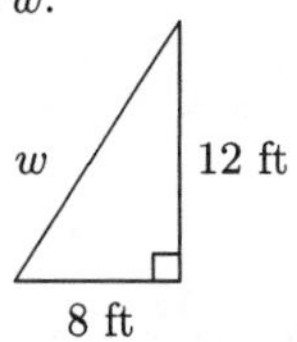

We know that $8^2 + 12^2 = w^2$. We solve this equation.

$8^2 + 12^2 = w^2$
$64 + 144 = w^2$
$208 = w^2$
$\sqrt{208}$ ft $= w$ Exact answer
14.422 ft $\approx w$ Approximation

29. $5x + 7 = 8y,$
$3x = 8y - 4$
$5x - 8y = -7$ (1) Rewriting
$3x - 8y = -4$ (2) the equations

We multiply Equation (2) by -1 and add.

$$\begin{array}{rcl} 5x - 8y &=& -7 \\ -3x + 8y &=& 4 \\ \hline 2x \quad &=& -3 \end{array}$$

$$x = -\frac{3}{2}$$

Substitute $-\frac{3}{2}$ for x in Equation (1) and solve for y.

$$5x - 8y = -7$$
$$5\left(-\frac{3}{2}\right) - 8y = -7$$
$$-\frac{15}{2} - 8y = -7$$
$$-8y = \frac{1}{2}$$
$$y = -\frac{1}{16}$$

The ordered pair $\left(-\frac{3}{2}, -\frac{1}{16}\right)$ checks. It is the solution.

31. $3x - 4y = -11$ (1)

$5x + 6y = 12$ (2)

We multiply Equation (1) by 3 and Equation (2) by 2, and then we add.

$$9x - 12y = -33$$
$$\underline{10x + 12y = 24}$$
$$19x \quad = -9$$
$$x = -\frac{9}{19}$$

Substitute $-\frac{9}{19}$ for x in Equation (2) and solve for y.

$$5x + 6y = 12$$
$$5\left(-\frac{9}{19}\right) + 6y = 12$$
$$-\frac{45}{19} + 6y = 12$$
$$6y = \frac{273}{19} \quad \text{Adding } \frac{45}{19}$$
$$y = \frac{273}{6 \cdot 19} \quad \text{Dividing by 6}$$
$$y = \frac{91}{38} \quad \text{Simplifying}$$

The ordered pair $\left(-\frac{9}{19}, \frac{91}{38}\right)$ checks. It is the solution.

33. Write the equation in the slope-intercept form.

$$4 - x = 3y$$
$$\frac{1}{3}(4 - x) = y$$
$$\frac{4}{3} - \frac{1}{3}x = y, \text{ or}$$
$$y = -\frac{1}{3}x + \frac{4}{3}$$

The slope is $-\frac{1}{3}$.

35. 

37. After one-half hour, the car traveling east has gone $\frac{1}{2} \cdot 50$, or 25 mi, and the car traveling south has gone $\frac{1}{2} \cdot 60$, or 30 mi. We make a drawing. We label the distance between the cards d.

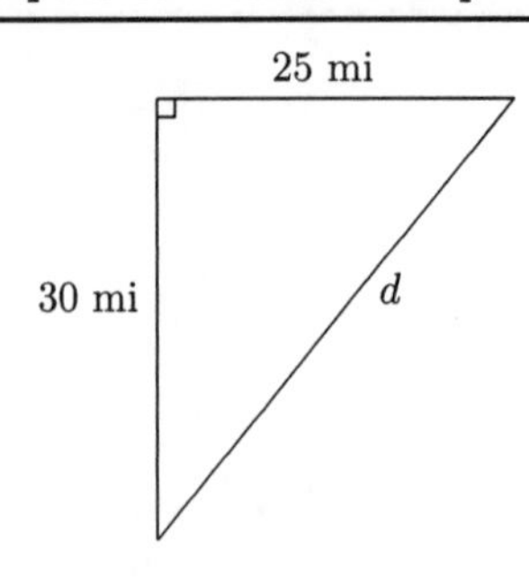

We know that $30^2 + 25^2 = d^2$. We solve this equation.

$$30^2 + 25^2 = d^2$$
$$900 + 625 = d^2$$
$$1525 = d^2$$
$$\sqrt{1525} \text{ mi} = d \quad \text{Exact answer}$$
$$39.1 \text{ mi} \approx d \quad \text{Approximation}$$

39.

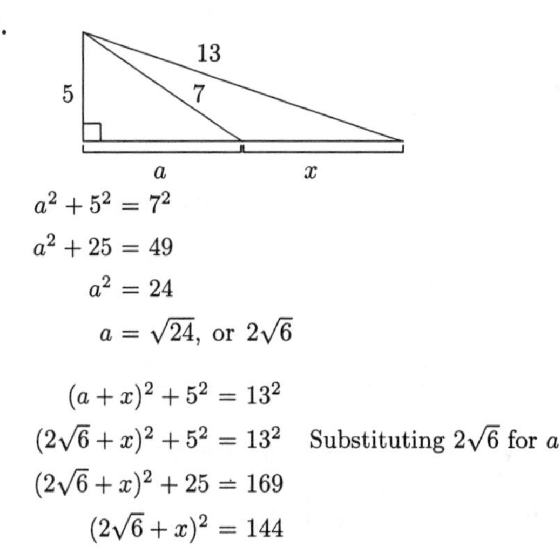

$$a^2 + 5^2 = 7^2$$
$$a^2 + 25 = 49$$
$$a^2 = 24$$
$$a = \sqrt{24}, \text{ or } 2\sqrt{6}$$

$$(a + x)^2 + 5^2 = 13^2$$
$$(2\sqrt{6} + x)^2 + 5^2 = 13^2 \quad \text{Substituting } 2\sqrt{6} \text{ for } a$$
$$(2\sqrt{6} + x)^2 + 25 = 169$$
$$(2\sqrt{6} + x)^2 = 144$$
$$2\sqrt{6} + x = 12 \quad \text{Taking the principal square root}$$
$$x = 12 - 2\sqrt{6}$$
$$x \approx 7.101$$

41. Using the Pythagorean equation we can label the figure with additional information.

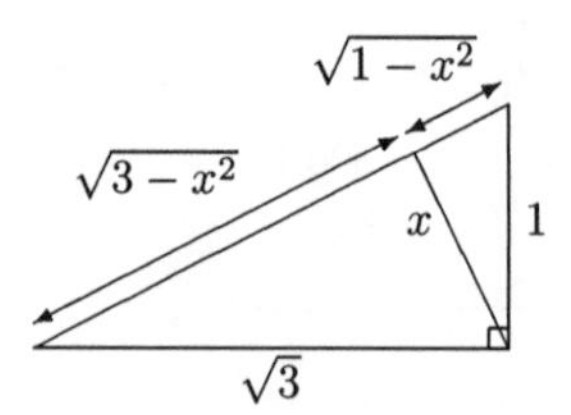

Next we use the Pythagorean equation with the largest right triangle and solve for x.

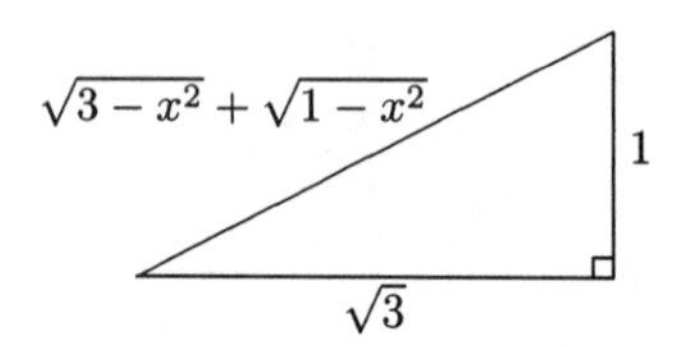

$$(\sqrt{3})^2 + 1^2 = (\sqrt{3-x^2} + \sqrt{1-x^2})^2$$

$$3 + 1 = (3 - x^2) + 2\sqrt{(3-x^2)(1-x^2)} + (1 - x^2)$$

$$4 = 4 - 2x^2 + 2\sqrt{3 - 4x^2 + x^4}$$

Collecting like terms

$$2x^2 = 2\sqrt{3 - 4x^2 + x^4}$$

Subtracting 4 and adding $2x^2$

$$x^2 = \sqrt{3 - 4x^2 + x^4} \qquad \text{Dividing by 2}$$

$$(x^2)^2 = (\sqrt{3 - 4x^2 + x^4})^2$$

$$x^4 = 3 - 4x^2 + x^4$$

$$4x^2 = 3 \qquad \text{Subtracting } x^4 \text{ and adding } 4x^2$$

$$x^2 = \frac{3}{4}$$

$$x = \sqrt{\frac{3}{4}}$$

$$x = \frac{\sqrt{3}}{2} \qquad \text{Exact answer}$$

$$x \approx 0.866 \qquad \text{Approximation}$$

Chapter 16

Quadratic Equations

Exercise Set 16.1

1. $x^2 - 3x + 2 = 0$

This equation is already in standard form.

$a = 1,\ b = -3,\ c = 2$

3. $7x^2 = 4x - 3$

$7x^2 - 4x + 3 = 0$ Standard form

$a = 7,\ b = -4,\ c = 3$

5. $5 = -2x^2 + 3x$

$2x^2 - 3x + 5 = 0$ Standard form

$a = 2,\ b = -3,\ c = 5$

7. $x^2 + 5x = 0$

$x(x + 5) = 0$

$x = 0$ *or* $x + 5 = 0$

$x = 0$ *or* $x = -5$

The solutions are 0 and -5.

9. $3x^2 + 6x = 0$

$3x(x + 2) = 0$

$3x = 0$ *or* $x + 2 = 0$

$x = 0$ *or* $x = -2$

The solutions are 0 and -2.

11. $5x^2 = 2x$

$5x^2 - 2x = 0$

$x(5x - 2) = 0$

$x = 0$ *or* $5x - 2 = 0$

$x = 0$ *or* $5x = 2$

$x = 0$ *or* $x = \frac{2}{5}$

The solutions are 0 and $\frac{2}{5}$.

13. $4x^2 + 4x = 0$

$4x(x + 1) = 0$

$4x = 0$ *or* $x + 1 = 0$

$x = 0$ *or* $x = -1$

The solutions are 0 and -1.

15. $0 = 10x^2 - 30x$

$0 = 10x(x - 3)$

$10x = 0$ *or* $x - 3 = 0$

$x = 0$ *or* $x = 3$

The solutions are 0 and 3.

17. $11x = 55x^2$

$0 = 55x^2 - 11x$

$0 = 11x(5x - 1)$

$11x = 0$ *or* $5x - 1 = 0$

$x = 0$ *or* $5x = 1$

$x = 0$ *or* $x = \frac{1}{5}$

The solutions are 0 and $\frac{1}{5}$.

19. $14t^2 = 3t$

$14t^2 - 3t = 0$

$t(14t - 3) = 0$

$t = 0$ *or* $14t - 3 = 0$

$t = 0$ *or* $14t = 3$

$t = 0$ *or* $t = \frac{3}{14}$

The solutions are 0 and $\frac{3}{14}$.

21. $5y^2 - 3y^2 = 72y + 9y$

$2y^2 = 81y$

$2y^2 - 81y = 0$

$y(2y - 81) = 0$

$y = 0$ *or* $2y - 81 = 0$

$y = 0$ *or* $2y = 81$

$y = 0$ *or* $y = \frac{81}{2}$

The solutions are 0 and $\frac{81}{2}$.

23. $x^2 + 8x - 48 = 0$

$(x + 12)(x - 4) = 0$

$x + 12 = 0$ *or* $x - 4 = 0$

$x = -12$ *or* $x = 4$

The solutions are -12 and 4.

25. $5 + 6x + x^2 = 0$

$(5 + x)(1 + x) = 0$

$5 + x = 0$ *or* $1 + x = 0$

$x = -5$ *or* $x = -1$

The solutions are -5 and -1.

27. $18 = 7p + p^2$

$0 = p^2 + 7p - 18$

$0 = (p + 9)(p - 2)$

$p + 9 = 0$ *or* $p - 2 = 0$

$p = -9$ *or* $p = 2$

The solutions are -9 and 2.

29. $$\begin{aligned} -15 &= -8y + y^2 \\ 0 &= y^2 - 8y + 15 \\ 0 &= (y-5)(y-3) \end{aligned}$$

$$\begin{aligned} y - 5 = 0 \quad &or \quad y - 3 = 0 \\ y = 5 \quad &or \quad y = 3 \end{aligned}$$

The solutions are 5 and 3.

31. $$\begin{aligned} x^2 + 10x + 25 &= 0 \\ (x+5)(x+5) &= 0 \end{aligned}$$

$$\begin{aligned} x + 5 = 0 \quad &or \quad x + 5 = 0 \\ x = -5 \quad &or \quad x = -5 \end{aligned}$$

The solution is -5.

33. $$\begin{aligned} r^2 &= 8r - 16 \\ r^2 - 8r + 16 &= 0 \\ (r-4)(r-4) &= 0 \end{aligned}$$

$$\begin{aligned} r - 4 = 0 \quad &or \quad r - 4 = 0 \\ r = 4 \quad &or \quad r = 4 \end{aligned}$$

The solution is 4.

35. $$\begin{aligned} 6x^2 + x - 2 &= 0 \\ (3x+2)(2x-1) &= 0 \end{aligned}$$

$$\begin{aligned} 3x + 2 = 0 \quad &or \quad 2x - 1 = 0 \\ 3x = -2 \quad &or \quad 2x = 1 \\ x = -\frac{2}{3} \quad &or \quad x = \frac{1}{2} \end{aligned}$$

The solutions are $-\frac{2}{3}$ and $\frac{1}{2}$.

37. $$\begin{aligned} 3a^2 &= 10a + 8 \\ 3a^2 - 10a - 8 &= 0 \\ (3a+2)(a-4) &= 0 \end{aligned}$$

$$\begin{aligned} 3a + 2 = 0 \quad &or \quad a - 4 = 0 \\ 3a = -2 \quad &or \quad a = 4 \\ a = -\frac{2}{3} \quad &or \quad a = 4 \end{aligned}$$

The solutions are $-\frac{2}{3}$ and 4.

39. $$\begin{aligned} 6x^2 - 4x &= 10 \\ 6x^2 - 4x - 10 &= 0 \\ 2(3x^2 - 2x - 5) &= 0 \\ 2(3x-5)(x+1) &= 0 \end{aligned}$$

$$\begin{aligned} 3x - 5 = 0 \quad &or \quad x + 1 = 0 \\ 3x = 5 \quad &or \quad x = -1 \\ x = \frac{5}{3} \quad &or \quad x = -1 \end{aligned}$$

The solutions are $\frac{5}{3}$ and -1.

41. $$\begin{aligned} 2t^2 + 12t &= -10 \\ 2t^2 + 12t + 10 &= 0 \\ 2(t^2 + 6t + 5) &= 0 \\ 2(t+5)(t+1) &= 0 \end{aligned}$$

$$\begin{aligned} t + 5 = 0 \quad &or \quad t + 1 = 0 \\ t = -5 \quad &or \quad t = -1 \end{aligned}$$

The solutions are -5 and -1.

43. $$\begin{aligned} t(t-5) &= 14 \\ t^2 - 5t &= 14 \\ t^2 - 5t - 14 &= 0 \\ (t+2)(t-7) &= 0 \end{aligned}$$

$$\begin{aligned} t + 2 = 0 \quad &or \quad t - 7 = 0 \\ t = -2 \quad &or \quad t = 7 \end{aligned}$$

The solutions are -2 and 7.

45. $$\begin{aligned} t(9+t) &= 4(2t+5) \\ 9t + t^2 &= 8t + 20 \\ t^2 + t - 20 &= 0 \\ (t+5)(t-4) &= 0 \end{aligned}$$

$$\begin{aligned} t + 5 = 0 \quad &or \quad t - 4 = 0 \\ t = -5 \quad &or \quad t = 4 \end{aligned}$$

The solutions are -5 and 4.

47. $$\begin{aligned} 16(p-1) &= p(p+8) \\ 16p - 16 &= p^2 + 8p \\ 0 &= p^2 - 8p + 16 \\ 0 &= (p-4)(p-4) \end{aligned}$$

$$\begin{aligned} p - 4 = 0 \quad &or \quad p - 4 = 0 \\ p = 4 \quad &or \quad p = 4 \end{aligned}$$

The solution is 4.

49. $$\begin{aligned} (t-1)(t+3) &= t - 1 \\ t^2 + 2t - 3 &= t - 1 \\ t^2 + t - 2 &= 0 \\ (t+2)(t-1) &= 0 \end{aligned}$$

$$\begin{aligned} t + 2 = 0 \quad &or \quad t - 1 = 0 \\ t = -2 \quad &or \quad t = 1 \end{aligned}$$

The solutions are -2 and 1.

51. $$\frac{24}{x-2} + \frac{24}{x+2} = 5$$

The LCM is $(x-2)(x+2)$.

$$(x-2)(x+2)\left(\frac{24}{x-2} + \frac{24}{x+2}\right) = (x-2)(x+2) \cdot 5$$

$$(x-2)(x+2) \cdot \frac{24}{x-2} + (x-2)(x+2) \cdot \frac{24}{x+2} = 5(x-2)(x+2)$$

$$\begin{aligned} 24(x+2) + 24(x-2) &= 5(x^2 - 4) \\ 24x + 48 + 24x - 48 &= 5x^2 - 20 \\ 48x &= 5x^2 - 20 \\ 0 &= 5x^2 - 48x - 20 \\ 0 &= (5x+2)(x-10) \end{aligned}$$

$$\begin{aligned} 5x + 2 = 0 \quad &or \quad x - 10 = 0 \\ 5x = -2 \quad &or \quad x = 10 \\ x = -\frac{2}{5} \quad &or \quad x = 10 \end{aligned}$$

Both numbers check. The solutions are $-\frac{2}{5}$ and 10.

53. $$\frac{1}{x}+\frac{1}{x+6}=\frac{1}{4}$$

The LCM is $4x(x+6)$.

$$\begin{aligned}4x(x+6)\Big(\frac{1}{x}+\frac{1}{x+6}\Big)&=4x(x+6)\cdot\frac{1}{4}\\4x(x+6)\cdot\frac{1}{x}+4x(x+6)\cdot\frac{1}{x+6}&=x(x+6)\\4(x+6)+4x&=x(x+6)\\4x+24+4x&=x^2+6x\\8x+24&=x^2+6x\\0&=x^2-2x-24\\0&=(x-6)(x+4)\end{aligned}$$

$x-6=0$ *or* $x+4=0$

$x=6$ *or* $x=-4$

Both numbers check. The solutions are 6 and -4.

55. $$1+\frac{12}{x^2-4}=\frac{3}{x-2}$$

The LCM is $(x+2)(x-2)$.

$$\begin{aligned}(x+2)(x-2)\Big(1+\frac{12}{(x+2)(x-2)}\Big)&=\\(x+2)(x-2)\cdot\frac{3}{x-2}\\(x+2)(x-2)\cdot1+(x+2)(x-2)\cdot\frac{12}{(x+2)(x-2)}&=\\3(x+2)\\x^2-4+12&=3x+6\\x^2+8&=3x+6\\x^2-3x+2&=0\\(x-2)(x-1)&=0\end{aligned}$$

$x-2=0$ *or* $x-1=0$

$x=2$ *or* $x=1$

The number 1 checks, but 2 does not. (It makes the denominators x^2-4 and $x-2$ zero.) The solution is 1.

57. $$\frac{r}{r-1}+\frac{2}{r^2-1}=\frac{8}{r+1}$$

The LCM is $(r-1)(r+1)$.

$$\begin{aligned}(r-1)(r+1)\Big(\frac{r}{r-1}+\frac{2}{(r-1)(r+1)}\Big)&=\\(r-1)(r+1)\cdot\frac{8}{r+1}\\(r-1)(r+1)\cdot\frac{r}{r-1}+(r-1)(r+1)\cdot\frac{2}{(r-1)(r+1)}&=\\8(r-1)\\r(r+1)+2&=8(r-1)\\r^2+r+2&=8r-8\\r^2-7r+10&=0\\(r-5)(r-2)&=0\end{aligned}$$

$r-5=0$ *or* $r-2=0$

$r=5$ *or* $r=2$

Both numbers check. The solutions are 5 and 2.

59. $$\frac{x-1}{1-x}=-\frac{x+8}{x-8}$$

The LCM is $(1-x)(x-8)$.

$$\begin{aligned}(1-x)(x-8)\cdot\frac{x-1}{1-x}&=(1-x)(x-8)\Big(-\frac{x+8}{x-8}\Big)\\(x-8)(x-1)&=-(1-x)(x+8)\\x^2-9x+8&=-(x+8-x^2-8x)\\x^2-9x+8&=-(-x^2-7x+8)\\x^2-9x+8&=x^2+7x-8\\16&=16x\\1&=x\end{aligned}$$

The number 1 does not check. (It makes the denominator $1-x$ zero.) There is no solution.

61. $$\frac{5}{y+4}-\frac{3}{y-2}=4$$

The LCM is $(y+4)(y-2)$.

$$\begin{aligned}(y+4)(y-2)\Big(\frac{5}{y+4}-\frac{3}{y-2}\Big)&=(y+4)(y-2)\cdot4\\5(y-2)-3(y+4)&=4(y^2+2y-8)\\5y-10-3y-12&=4y^2+8y-32\\2y-22&=4y^2+8y-32\\0&=4y^2+6y-10\\0&=2(2y^2+3y-5)\\0&=2(2y+5)(y-1)\end{aligned}$$

$2y+5=0$ *or* $y-1=0$

$2y=-5$ *or* $y=1$

$y=-\frac{5}{2}$ *or* $y=1$

The solutions are $-\frac{5}{2}$ and 1.

63. ***Familiarize.*** We will use the formula

$$d=\frac{n^2-3n}{2},$$

where d is the number of diagonals and n is the number of sides.

Translate. We substitute 10 for n.

$$d=\frac{10^2-3\cdot10}{2}$$

Solve. We do the computation.

$$d=\frac{10^2-3\cdot10}{2}=\frac{100-30}{2}=\frac{70}{2}=35$$

Check. We can recheck our computation. We can also substitute 35 for d in the original formula and determine whether this yields $n=10$. Our result checks.

State. A decagon has 35 diagonals.

65. ***Familiarize.*** We will use the formula

$$d=\frac{n^2-3n}{2},$$

where d is the number of diagonals and n is the number of sides.

Translate. We substitute 14 for d.

$$14=\frac{n^2-3n}{2}$$

Solve. We solve the equation.

$$\frac{n^2-3n}{2} = 14$$
$$n^2-3n = 28 \quad \text{Multiplying by 2}$$
$$n^2-3n-28 = 0$$
$$(n-7)(n+4) = 0$$
$$n-7=0 \quad or \quad n+4=0$$
$$n=7 \quad or \quad n=-4$$

Check. Since the number of sides cannot be negative, -4 cannot be a solution. To check 7, we substitute 7 for n in the original formula and determine if this yields $d = 14$. Our result checks.

State. The polygon has 7 sides.

67. $\sqrt{64} = 8$, taking the principal square root

69. $\sqrt{8} = \sqrt{4\cdot 2} = \sqrt{4}\sqrt{2} = 2\sqrt{2}$

71. $\sqrt{20} = \sqrt{4\cdot 5} = \sqrt{4}\sqrt{5} = 2\sqrt{5}$

73. $\sqrt{405} = \sqrt{81\cdot 5} = \sqrt{81}\sqrt{5} = 9\sqrt{5}$

75. 2.646

77. 1.528

79. ◈

81.
$$4m^2-(m+1)^2 = 0$$
$$4m^2-(m^2+2m+1) = 0$$
$$4m^2-m^2-2m-1 = 0$$
$$3m^2-2m-1 = 0$$
$$(3m+1)(m-1) = 0$$
$$3m+1=0 \quad or \quad m-1=0$$
$$3m=-1 \quad or \quad m=1$$
$$m=-\frac{1}{3} \quad or \quad m=1$$

The solutions are $-\frac{1}{3}$ and 1.

83.
$$\sqrt{5}x^2-x = 0$$
$$x(\sqrt{5}x-1) = 0$$
$$x=0 \quad or \quad \sqrt{5}x-1=0$$
$$x=0 \quad or \quad \sqrt{5}x=1$$
$$x=0 \quad or \quad x=\frac{1}{\sqrt{5}}, \text{ or } \frac{\sqrt{5}}{5}$$

The solutions are 0 and $\frac{\sqrt{5}}{5}$.

85. Graph $y_1 = 3x^2-7x$ and $y_2 = 20$. Then use the INTERSECT feature to find the first coordinate(s) of the point(s) of intersection. The solutions are 4 and approximately -1.7.

87. Graph $y_1 = 3x^2+8x$ and $y_2 = 12x+15$. Then use the INTERSECT feature to find the first coordinate(s) of the point(s) of intersection. The solutions are 3 and approximately -1.7.

89. Graph $y_1 = (x-2)^2+3(x-2)$ and $y_2 = 4$. Then use the INTERSECT feature to find the first coordinate(s) of the point(s) of intersection. The solutions are -2 and 3.

91. Graph $y_1 = 16(x-1)$ and $y_2 = x(x+8)$. Then use the INTERSECT feature to find the first coordinate(s) of the point(s) of intersection. The solution is 4.

Exercise Set 16.2

1. $x^2 = 121$

$x = 11$ *or* $x = -11$ Principle of square roots

The solutions are 11 and -11.

3.
$$5x^2 = 35$$
$$x^2 = 7 \quad \text{Dividing by 5}$$
$$x=\sqrt{7} \text{ or } x=-\sqrt{7} \quad \text{Principle of square roots}$$

The solutions are $\sqrt{7}$ and $-\sqrt{7}$.

5.
$$5x^2 = 3$$
$$x^2 = \frac{3}{5}$$
$$x=\sqrt{\frac{3}{5}} \quad or \quad x=-\sqrt{\frac{3}{5}} \quad \text{Principle of square roots}$$
$$x=\sqrt{\frac{3}{5}\cdot\frac{5}{5}} \quad or \quad x=-\sqrt{\frac{3}{5}\cdot\frac{5}{5}} \quad \text{Rationalizing denominators}$$
$$x=\frac{\sqrt{15}}{5} \quad or \quad x=-\frac{\sqrt{15}}{5}$$

The solutions are $\frac{\sqrt{15}}{5}$ and $-\frac{\sqrt{15}}{5}$.

7.
$$4x^2-25 = 0$$
$$4x^2 = 25$$
$$x^2 = \frac{25}{4}$$
$$x=\frac{5}{2} \text{ or } x=-\frac{5}{2}$$

The solutions are $\frac{5}{2}$ and $-\frac{5}{2}$.

9.
$$3x^2-49 = 0$$
$$3x^2 = 49$$
$$x^2 = \frac{49}{3}$$
$$x=\frac{7}{\sqrt{3}} \quad or \quad x=-\frac{7}{\sqrt{3}}$$
$$x=\frac{7}{\sqrt{3}}\cdot\frac{\sqrt{3}}{\sqrt{3}} \quad or \quad x=-\frac{7}{\sqrt{3}}\cdot\frac{\sqrt{3}}{\sqrt{3}}$$
$$x=\frac{7\sqrt{3}}{3} \quad or \quad x=-\frac{7\sqrt{3}}{3}$$

The solutions are $\frac{7\sqrt{3}}{3}$ and $-\frac{7\sqrt{3}}{3}$.

11.
$$4y^2-3 = 9$$
$$4y^2 = 12$$
$$y^2 = 3$$
$$y=\sqrt{3} \quad or \quad y=-\sqrt{3}$$

The solutions are $\sqrt{3}$ and $-\sqrt{3}$.

13. $49y^2 - 64 = 0$
$$49y^2 = 64$$
$$y^2 = \frac{64}{49}$$
$$y = \frac{8}{7} \text{ or } y = -\frac{8}{7}$$
The solutions are $\frac{8}{7}$ and $-\frac{8}{7}$.

15. $(x+3)^2 = 16$

$x+3 = 4$ *or* $x+3 = -4$ Principle of square roots
$x = 1$ *or* $x = -7$

The solutions are 1 and -7.

17. $(x+3)^2 = 21$

$x+3 = \sqrt{21}$ *or* $x+3 = -\sqrt{21}$ Principle of square roots
$x = -3+\sqrt{21}$ *or* $x = -3-\sqrt{21}$

The solutions are $-3+\sqrt{21}$ and $-3-\sqrt{21}$, or $-3 \pm \sqrt{21}$.

19. $(x+13)^2 = 8$

$x+13 = \sqrt{8}$ *or* $x+13 = -\sqrt{8}$
$x+13 = 2\sqrt{2}$ *or* $x+13 = -2\sqrt{2}$
$x = -13+2\sqrt{2}$ *or* $x = -13-2\sqrt{2}$

The solutions are $-13+2\sqrt{2}$ and $-13-2\sqrt{2}$, or $-13\pm 2\sqrt{2}$.

21. $(x-7)^2 = 12$

$x-7 = \sqrt{12}$ *or* $x-7 = -\sqrt{12}$
$x-7 = 2\sqrt{3}$ *or* $x-7 = -2\sqrt{3}$
$x = 7+2\sqrt{3}$ *or* $x = 7-2\sqrt{3}$

The solutions are $7+2\sqrt{3}$ and $7-2\sqrt{3}$, or $7 \pm 2\sqrt{3}$.

23. $(x+9)^2 = 34$

$x+9 = \sqrt{34}$ *or* $x+9 = -\sqrt{34}$
$x = -9+\sqrt{34}$ *or* $x = -9-\sqrt{34}$

The solutions are $-9+\sqrt{34}$ and $-9-\sqrt{34}$, or $-9 \pm \sqrt{34}$.

25. $\left(x+\frac{3}{2}\right)^2 = \frac{7}{2}$

$x+\frac{3}{2} = \sqrt{\frac{7}{2}}$ *or* $x+\frac{3}{2} = -\sqrt{\frac{7}{2}}$

$x = -\frac{3}{2}+\sqrt{\frac{7}{2}}$ *or* $x = -\frac{3}{2}-\sqrt{\frac{7}{2}}$

$x = -\frac{3}{2}+\sqrt{\frac{7}{2}\cdot\frac{2}{2}}$ *or* $x = -\frac{3}{2}-\sqrt{\frac{7}{2}\cdot\frac{2}{2}}$

$x = -\frac{3}{2}+\frac{\sqrt{14}}{2}$ *or* $x = -\frac{3}{2}-\frac{\sqrt{14}}{2}$

$x = \frac{-3+\sqrt{14}}{2}$ *or* $x = \frac{-3-\sqrt{14}}{2}$

The solutions are $\frac{-3 \pm \sqrt{14}}{2}$.

27. $x^2 - 6x + 9 = 64$
$(x-3)^2 = 64$ Factoring the left side

$x-3 = 8$ *or* $x-3 = -8$ Principle of square roots
$x = 11$ *or* $x = -5$

The solutions are 11 and -5.

29. $x^2 + 14x + 49 = 64$
$(x+7)^2 = 64$ Factoring the left side

$x+7 = 8$ *or* $x+7 = -8$ Principle of square roots
$x = 1$ *or* $x = -15$

The solutions are 1 and -15.

31. $x^2 - 6x - 16 = 0$
$x^2 - 6x = 16$ Adding 16
$x^2 - 6x + 9 = 16+9$ Adding 9: $\left(\frac{-6}{2}\right)^2 = (-3)^2 = 9$

$(x-3)^2 = 25$

$x-3 = 5$ *or* $x-3 = -5$ Principle of square roots
$x = 8$ *or* $x = -2$

The solutions are 8 and -2.

33. $x^2 + 22x + 21 = 0$
$x^2 + 22x = -21$ Subtracting 21
$x^2 + 22x + 121 = -21+121$ Adding 121: $\left(\frac{22}{2}\right)^2 = 11^2 = 121$

$(x+11)^2 = 100$

$x+11 = 10$ *or* $x+11 = -10$ Principle of square roots
$x = -1$ *or* $x = -21$

The solutions are -1 and -21.

35. $x^2 - 2x - 5 = 0$
$x^2 - 2x = 5$
$x^2 - 2x + 1 = 5+1$ Adding 1: $\left(\frac{-2}{2}\right)^2 = (-1)^2 = 1$

$(x-1)^2 = 6$

$x-1 = \sqrt{6}$ *or* $x-1 = -\sqrt{6}$
$x = 1+\sqrt{6}$ *or* $x = 1-\sqrt{6}$

The solutions are $1 \pm \sqrt{6}$.

37. $x^2 - 22x + 102 = 0$
$x^2 - 22x = -102$
$x^2 - 22x + 121 = -102+121$ Adding 121: $\left(\frac{-22}{2}\right)^2 = (-11)^2 = 121$

$(x-11)^2 = 19$

$x-11 = \sqrt{19}$ *or* $x-11 = -\sqrt{19}$
$x = 11+\sqrt{19}$ *or* $x = 11-\sqrt{19}$

The solutions are $11 \pm \sqrt{19}$.

39. $x^2 + 10x - 4 = 0$

$x^2 + 10x = 4$

$x^2 + 10x + 25 = 4 + 25$ Adding 25: $\left(\frac{10}{2}\right)^2 = 5^2 = 25$

$(x+5)^2 = 29$

$x + 5 = \sqrt{29}$ *or* $x + 5 = -\sqrt{29}$

$x = -5 + \sqrt{29}$ *or* $x = -5 - \sqrt{29}$

The solutions are $-5 \pm \sqrt{29}$.

41. $x^2 - 7x - 2 = 0$

$x^2 - 7x = 2$

$x^2 - 7x + \frac{49}{4} = 2 + \frac{49}{4}$ Adding $\frac{49}{4}$: $\left(\frac{-7}{2}\right)^2 = \frac{49}{4}$

$\left(x - \frac{7}{2}\right)^2 = \frac{8}{4} + \frac{49}{4} = \frac{57}{4}$

$x - \frac{7}{2} = \frac{\sqrt{57}}{2}$ *or* $x - \frac{7}{2} = -\frac{\sqrt{57}}{2}$

$x = \frac{7}{2} + \frac{\sqrt{57}}{2}$ *or* $x = \frac{7}{2} - \frac{\sqrt{57}}{2}$

$x = \frac{7 + \sqrt{57}}{2}$ *or* $x = \frac{7 - \sqrt{57}}{2}$

The solutions are $\frac{7 \pm \sqrt{57}}{2}$.

43. $x^2 + 3x - 28 = 0$

$x^2 + 3x = 28$

$x^2 + 3x + \frac{9}{4} = 28 + \frac{9}{4}$ Adding $\frac{9}{4}$: $\left(\frac{3}{2}\right)^2 = \frac{9}{4}$

$\left(x + \frac{3}{2}\right)^2 = \frac{121}{4}$

$x + \frac{3}{2} = \frac{11}{2}$ *or* $x + \frac{3}{2} = -\frac{11}{2}$

$x = \frac{8}{2}$ *or* $x = -\frac{14}{2}$

$x = 4$ *or* $x = -7$

The solutions are 4 and -7.

45. $x^2 + \frac{3}{2}x - \frac{1}{2} = 0$

$x^2 + \frac{3}{2}x = \frac{1}{2}$

$x^2 + \frac{3}{2}x + \frac{9}{16} = \frac{1}{2} + \frac{9}{16}$ Adding $\frac{9}{16}$: $\left(\frac{3/2}{2}\right)^2 = \left(\frac{3}{4}\right)^2 = \frac{9}{16}$

$\left(x + \frac{3}{4}\right)^2 = \frac{17}{16}$

$x + \frac{3}{4} = \frac{\sqrt{17}}{4}$ *or* $x + \frac{3}{4} = -\frac{\sqrt{17}}{4}$

$x = -\frac{3}{4} + \frac{\sqrt{17}}{4}$ *or* $x = -\frac{3}{4} - \frac{\sqrt{17}}{4}$

$x = \frac{-3 + \sqrt{17}}{4}$ *or* $x = \frac{-3 - \sqrt{17}}{4}$

The solutions are $\frac{-3 \pm \sqrt{17}}{4}$.

47. $2x^2 + 3x - 17 = 0$

$\frac{1}{2}(2x^2 + 3x - 17) = \frac{1}{2} \cdot 0$ Multiplying by $\frac{1}{2}$ to make the x^2-coefficient 1

$x^2 + \frac{3}{2}x - \frac{17}{2} = 0$

$x^2 + \frac{3}{2}x = \frac{17}{2}$

$x^2 + \frac{3}{2}x + \frac{9}{16} = \frac{17}{2} + \frac{9}{16}$ Adding $\frac{9}{16}$: $\left(\frac{3/2}{2}\right)^2 = \left(\frac{3}{4}\right)^2 = \frac{9}{16}$

$\left(x + \frac{3}{4}\right)^2 = \frac{145}{16}$

$x + \frac{3}{4} = \frac{\sqrt{145}}{4}$ *or* $x + \frac{3}{4} = -\frac{\sqrt{145}}{4}$

$x = \frac{-3 + \sqrt{145}}{4}$ *or* $x = \frac{-3 - \sqrt{145}}{4}$

The solutions are $\frac{-3 \pm \sqrt{145}}{4}$.

49. $3x^2 + 4x - 1 = 0$

$\frac{1}{3}(3x^2 + 4x - 1) = \frac{1}{3} \cdot 0$

$x^2 + \frac{4}{3}x - \frac{1}{3} = 0$

$x^2 + \frac{4}{3}x = \frac{1}{3}$

$x^2 + \frac{4}{3}x + \frac{4}{9} = \frac{1}{3} + \frac{4}{9}$

$\left(x + \frac{2}{3}\right)^2 = \frac{7}{9}$

$x + \frac{2}{3} = \frac{\sqrt{7}}{3}$ *or* $x + \frac{2}{3} = -\frac{\sqrt{7}}{3}$

$x = \frac{-2 + \sqrt{7}}{3}$ *or* $x = -\frac{-2 - \sqrt{7}}{3}$

The solutions are $\frac{-2 \pm \sqrt{7}}{3}$.

51. $2x^2 = 9x + 5$

$2x^2 - 9x - 5 = 0$ Standard form

$\frac{1}{2}(2x^2 - 9x - 5) = \frac{1}{2} \cdot 0$

$x^2 - \frac{9}{2}x - \frac{5}{2} = 0$

$x^2 - \frac{9}{2}x = \frac{5}{2}$

$x^2 - \frac{9}{2}x + \frac{81}{16} = \frac{5}{2} + \frac{81}{16}$

$\left(x - \frac{9}{4}\right)^2 = \frac{121}{16}$

$$x - \frac{9}{4} = \frac{11}{4} \quad or \quad x - \frac{9}{4} = -\frac{11}{4}$$
$$x = \frac{20}{4} \quad or \quad x = -\frac{2}{4}$$
$$x = 5 \quad or \quad x = -\frac{1}{2}$$

The solutions are 5 and $-\frac{1}{2}$.

53.
$$6x^2 + 11x = 10$$
$$6x^2 + 11x - 10 = 0 \quad \text{Standard form}$$
$$\frac{1}{6}(6x^2 + 11x - 10) = \frac{1}{6} \cdot 0$$
$$x^2 + \frac{11}{6}x - \frac{5}{3} = 0$$
$$x^2 + \frac{11}{6}x = \frac{5}{3}$$
$$x^2 + \frac{11}{6}x + \frac{121}{144} = \frac{5}{3} + \frac{121}{144}$$
$$\left(x + \frac{11}{12}\right)^2 = \frac{361}{144}$$

$$x + \frac{11}{12} = \frac{19}{12} \quad or \quad x + \frac{11}{12} = -\frac{19}{12}$$
$$x = \frac{8}{12} \quad or \quad x = -\frac{30}{12}$$
$$x = \frac{2}{3} \quad or \quad x = -\frac{5}{2}$$

The solutions are $\frac{2}{3}$ and $-\frac{5}{2}$.

55. ***Familiarize.*** We will use the formula $s = 16t^2$.

Translate. We substitute 1451 for s.

$$1451 = 16t^2$$

Solve. We solve the equation.

$$1451 = 16t^2$$
$$\frac{1451}{16} = t^2 \quad \text{Solving for } t^2$$
$$90.6875 = t^2 \quad \text{Dividing}$$
$$\sqrt{90.6875} = t \quad or \quad -\sqrt{90.6875} = t \quad \text{Principle of square roots}$$
$$9.5 \approx t \quad or \quad -9.5 \approx t \quad \text{Using a calculator and rounding to the nearest tenth}$$

Check. The number -9.5 cannot be a solution, because time cannot be negative in this situation. We substitute 9.5 in the original equation.

$$s = 16(9.5)^2 = 16(90.25) = 1444$$

This is close. Remember that we approximated a solution. Thus we have a check.

State. It takes about 9.5 sec for an object to fall to the ground from the top of the Sears Tower.

57. ***Familiarize.*** We will use the formula $s = 16t^2$.

Translate. We substitute 311 for s.

$$311 = 16t^2$$

Solve. We solve the equation.

$$311 = 16t^2$$
$$\frac{311}{16} = t^2 \quad \text{Solving for } t^2$$
$$19.4375 = t^2 \quad \text{Dividing}$$
$$\sqrt{19.4375} = t \quad or \quad -\sqrt{19.4375} = t \quad \text{Principle of square roots}$$
$$4.4 \approx t \quad or \quad -4.4 \approx t \quad \text{Using a calculator and rounding to the nearest tenth}$$

Check. The number -4.4 cannot be a solution, because time cannot be negative in this situation. We substitute 4.4 in the original equation.

$$s = 16(4.4)^2 = 16(19.36) = 309.76$$

This is close. Remember that we approximated a solution. Thus we have a check.

State. The fall took approximately 4.4 sec.

59.
$$y = \frac{k}{x} \quad \text{Inverse variation}$$
$$235 = \frac{k}{0.6} \quad \text{Substituting 0.6 for } x \text{ and 235 for } y$$
$$141 = k \quad \text{Constant of variation}$$
$$y = \frac{141}{x} \quad \text{Equation of variation}$$

61. $\sqrt{3x} \cdot \sqrt{6x} = \sqrt{18x^2} = \sqrt{9 \cdot x^2 \cdot 2} = \sqrt{9}\sqrt{x^2}\sqrt{2} = 3x\sqrt{2}$

63. $3\sqrt{t} \cdot \sqrt{t} = 3\sqrt{t^2} = 3t$

65. ◈

67. $x^2 + bx + 36$

The trinomial is a square if the square of one-half the x-coefficient is equal to 36. Thus we have:

$$\left(\frac{b}{2}\right)^2 = 36$$
$$\frac{b^2}{4} = 36$$
$$b^2 = 144$$
$$b = 12 \quad or \quad b = -12 \quad \text{Principle of square roots}$$

69. $x^2 + bx + 128$

The trinomial is a square if the square of one-half the x-coefficient is equal to 128. Thus we have:

$$\left(\frac{b}{2}\right)^2 = 128$$
$$\frac{b^2}{4} = 128$$
$$b^2 = 512$$
$$b = \sqrt{512} \quad or \quad b = -\sqrt{512}$$
$$b = 16\sqrt{2} \quad or \quad b = -16\sqrt{2}$$

71. $x^2 + bx + c$

The trinomial is a square if the square of one-half the x-coefficient is equal to c. Thus we have:

$$\left(\frac{b}{2}\right)^2 = c$$
$$\frac{b^2}{4} = c$$
$$b^2 = 4c$$

$b = \sqrt{4c}$ *or* $b = -\sqrt{4c}$
$b = 2\sqrt{c}$ *or* $b = -2\sqrt{c}$

73. $4.82x^2 = 12,000$

$$x^2 = \frac{12,000}{4.82}$$

$x = \sqrt{\frac{12,000}{4.82}}$ *or* $x = -\sqrt{\frac{12,000}{4.82}}$ Principle of square roots

$x \approx 49.896$ *or* $x \approx -49.896$ Using a calculator and rounding

The solutions are approximately 49.896 and −49.896.

75. $\frac{x}{9} = \frac{36}{4x}$, LCM is $36x$

$36x \cdot \frac{x}{9} = 36x \cdot \frac{36}{4x}$ Multiplying by $36x$

$4x^2 = 324$

$x^2 = 81$

$x = 9$ *or* $x = -9$

Both numbers check. The solutions are 9 and −9.

Exercise Set 16.3

1. $x^2 - 4x = 21$

$x^2 - 4x - 21 = 0$ Standard form

We can factor.

$x^2 - 4x - 21 = 0$
$(x-7)(x+3) = 0$

$x - 7 = 0$ *or* $x + 3 = 0$
$x = 7$ *or* $x = -3$

The solutions are 7 and −3.

3. $x^2 = 6x - 9$

$x^2 - 6x + 9 = 0$ Standard form

We can factor.

$x^2 - 6x + 9 = 0$
$(x-3)(x-3) = 0$

$x - 3 = 0$ *or* $x - 3 = 0$
$x = 3$ *or* $x = 3$

The solution is 3.

5. $3y^2 - 2y - 8 = 0$

We can factor.

$3y^2 - 2y - 8 = 0$
$(3y+4)(y-2) = 0$

$3y + 4 = 0$ *or* $y - 2 = 0$
$3y = -4$ *or* $y = 2$

$y = -\frac{4}{3}$ *or* $y = 2$

The solutions are $-\frac{4}{3}$ and 2.

7. $4x^2 + 4x = 15$

$4x^2 + 4x - 15 = 0$ Standard form

We can factor.

$4x^2 + 4x - 15 = 0$
$(2x-3)(2x+5) = 0$

$2x - 3 = 0$ *or* $2x + 5 = 0$
$2x = 3$ *or* $2x = -5$
$x = \frac{3}{2}$ *or* $x = -\frac{5}{2}$

The solutions are $\frac{3}{2}$ and $-\frac{5}{2}$.

9. $x^2 - 9 = 0$ Difference of squares

$(x+3)(x-3) = 0$

$x + 3 = 0$ *or* $x - 3 = 0$
$x = -3$ *or* $x = 3$

The solutions are −3 and 3.

11. $x^2 - 2x - 2 = 0$

$a = 1,\ b = -2,\ c = -2$

We use the quadratic formula.

$$x = \frac{-(-2) \pm \sqrt{(-2)^2 - 4 \cdot 1 \cdot (-2)}}{2 \cdot 1}$$
$$x = \frac{2 \pm \sqrt{4+8}}{2}$$
$$x = \frac{2 \pm \sqrt{12}}{2} = \frac{2 \pm \sqrt{4 \cdot 3}}{2}$$
$$x = \frac{2 \pm 2\sqrt{3}}{2} = \frac{2(1 \pm \sqrt{3})}{2}$$
$$x = 1 \pm \sqrt{3}$$

The solutions are $1 + \sqrt{3}$ and $1 - \sqrt{3}$, or $1 \pm \sqrt{3}$.

13. $y^2 - 10y + 22 = 0$

$a = 1,\ b = -10,\ c = 22$

We use the quadratic formula.

$$y = \frac{-(-10) \pm \sqrt{(-10)^2 - 4 \cdot 1 \cdot 22}}{2 \cdot 1}$$
$$y = \frac{10 \pm \sqrt{100 - 88}}{2}$$
$$y = \frac{10 \pm \sqrt{12}}{2} = \frac{10 \pm \sqrt{4 \cdot 3}}{2}$$
$$y = \frac{10 \pm 2\sqrt{3}}{2} = \frac{2(5 \pm \sqrt{3})}{2}$$
$$y = 5 \pm \sqrt{3}$$

The solutions are $5 + \sqrt{3}$ and $5 - \sqrt{3}$, or $5 \pm \sqrt{3}$.

15. $x^2 + 4x + 4 = 7$

$x^2 + 4x - 3 = 0$ Adding −7 to get standard form

$a = 1,\ b = 4,\ c = -3$

We use the quadratic formula.

$$x = \frac{-4 \pm \sqrt{4^2 - 4 \cdot 1 \cdot (-3)}}{2 \cdot 1} = \frac{-4 \pm \sqrt{16+12}}{2}$$
$$x = \frac{-4 \pm \sqrt{28}}{2} = \frac{-4 \pm \sqrt{4 \cdot 7}}{2}$$
$$x = \frac{-4 \pm 2\sqrt{7}}{2} = \frac{2(-2 \pm \sqrt{7})}{2}$$
$$x = -2 \pm \sqrt{7}$$

The solutions are $-2+\sqrt{7}$ and $-2-\sqrt{7}$, or $-2 \pm \sqrt{7}$.

17. $3x^2 + 8x + 2 = 0$

$a = 3,\ b = 8,\ c = 2$

We use the quadratic formula.

$$x = \frac{-8 \pm \sqrt{8^2 - 4 \cdot 3 \cdot 2}}{2 \cdot 3} = \frac{-8 \pm \sqrt{64-24}}{6}$$
$$x = \frac{-8 \pm \sqrt{40}}{6} = \frac{-8 \pm \sqrt{4 \cdot 10}}{6}$$
$$x = \frac{-8 \pm 2\sqrt{10}}{6} = \frac{2(-4 \pm \sqrt{10})}{2 \cdot 3}$$
$$x = \frac{-4 \pm \sqrt{10}}{3}$$

The solutions are $\frac{-4+\sqrt{10}}{3}$ and $\frac{-4-\sqrt{10}}{3}$, or $\frac{-4 \pm \sqrt{10}}{3}$.

19. $2x^2 - 5x = 1$

$2x^2 - 5x - 1 = 0$ Adding -1 to get standard form

$a = 2,\ b = -5,\ c = -1$

We use the quadratic formula.

$$x = \frac{-(-5) \pm \sqrt{(-5)^2 - 4 \cdot 2 \cdot (-1)}}{2 \cdot 2} = \frac{5 \pm \sqrt{25+8}}{4}$$
$$x = \frac{5 \pm \sqrt{33}}{4}$$

The solutions are $\frac{5+\sqrt{33}}{4}$ and $\frac{5-\sqrt{33}}{4}$, or $\frac{5 \pm \sqrt{33}}{4}$.

21. $2y^2 - 2y - 1 = 0$

$a = 2,\ b = -2,\ c = -1$

We use the quadratic formula.

$$y = \frac{-(-2) \pm \sqrt{(-2)^2 - 4 \cdot 2 \cdot (-1)}}{2 \cdot 2} = \frac{2 \pm \sqrt{4+8}}{4}$$
$$y = \frac{2 \pm \sqrt{12}}{4} = \frac{2 \pm \sqrt{4 \cdot 3}}{4}$$
$$y = \frac{2 \pm 2\sqrt{3}}{4} = \frac{2(1 \pm \sqrt{3})}{2 \cdot 2}$$
$$y = \frac{1 \pm \sqrt{3}}{2}$$

The solutions are $\frac{1+\sqrt{3}}{2}$ and $\frac{1-\sqrt{3}}{2}$, or $\frac{1 \pm \sqrt{3}}{2}$.

23. $2t^2 + 6t + 5 = 0$

$a = 2,\ b = 6,\ c = 5$

We use the quadratic formula.

$$t = \frac{-6 \pm \sqrt{6^2 - 4 \cdot 2 \cdot 5}}{2 \cdot 2} = \frac{-6 \pm \sqrt{36-40}}{4}$$
$$t = \frac{-6 \pm \sqrt{-4}}{4}$$

Since square roots of negative numbers do not exist as real numbers, there are no real-number solutions.

25. $3x^2 = 5x + 4$

$3x^2 - 5x - 4 = 0$

$a = 3,\ b = -5,\ c = -4$

We use the quadratic formula.

$$x = \frac{-(-5) \pm \sqrt{(-5)^2 - 4 \cdot 3 \cdot (-4)}}{2 \cdot 3} = \frac{5 \pm \sqrt{25+48}}{6}$$
$$x = \frac{5 \pm \sqrt{73}}{6}$$

The solutions are $\frac{5+\sqrt{73}}{6}$ and $\frac{5-\sqrt{73}}{6}$, or $\frac{5 \pm \sqrt{73}}{6}$.

27. $2y^2 - 6y = 10$

$2y^2 - 6y - 10 = 0$

$y^2 - 3y - 5 = 0$ Multiplying by $\frac{1}{2}$ to simplify

$a = 1,\ b = -3,\ c = -5$

We use the quadratic formula.

$$y = \frac{-(-3) \pm \sqrt{(-3)^2 - 4 \cdot 1 \cdot (-5)}}{2 \cdot 1} = \frac{3 \pm \sqrt{9+20}}{2}$$
$$y = \frac{3 \pm \sqrt{29}}{2}$$

The solutions are $\frac{3+\sqrt{29}}{2}$ and $\frac{3-\sqrt{29}}{2}$, or $\frac{3 \pm \sqrt{29}}{2}$.

29. $\frac{x^2}{x+3} - \frac{5}{x+3} = 0,$ LCM is $x+3$

$$(x+3)\left(\frac{x^2}{x+3} - \frac{5}{x+3}\right) = (x+3) \cdot 0$$
$$x^2 - 5 = 0$$
$$x^2 = 5$$

$x = \sqrt{5}$ *or* $x = -\sqrt{5}$ Principle of square roots

Both numbers check. The solutions are $\sqrt{5}$ and $-\sqrt{5}$, or $\pm\sqrt{5}$.

31. $x + 2 = \frac{3}{x+2}$

$(x+2)(x+2) = (x+2) \cdot \frac{3}{x+2}$ Clearing the fraction

$$x^2 + 4x + 4 = 3$$
$$x^2 + 4x + 1 = 0$$

$a = 1,\ b = 4,\ c = 1$

We use the quadratic formula.

$$x = \frac{-4 \pm \sqrt{4^2 - 4 \cdot 1 \cdot 1}}{2 \cdot 1} = \frac{-4 \pm \sqrt{16-4}}{2}$$
$$x = \frac{-4 \pm \sqrt{12}}{2} = \frac{-4 \pm \sqrt{4 \cdot 3}}{2}$$
$$x = \frac{-4 \pm 2\sqrt{3}}{2} = \frac{2(-2 \pm \sqrt{3})}{2}$$
$$x = -2 \pm \sqrt{3}$$

Both numbers check. The solutions are $-2+\sqrt{3}$ and $-2-\sqrt{3}$, or $-2\pm\sqrt{3}$.

33. $$\frac{1}{x}+\frac{1}{x+1}=\frac{1}{3},\quad \text{LCM is } 3x(x+1)$$
$$3x(x+1)\left(\frac{1}{x}+\frac{1}{x+1}\right)=3x(x+1)\cdot\frac{1}{3}$$
$$3(x+1)+3x=x(x+1)$$
$$3x+3+3x=x^2+x$$
$$6x+3=x^2+x$$
$$0=x^2-5x-3$$

$a=1,\ b=-5,\ c=-3$

We use the quadratic formula.

$$x=\frac{-(-5)\pm\sqrt{(-5)^2-4\cdot1\cdot(-3)}}{2\cdot1}=\frac{5\pm\sqrt{25+12}}{2}$$
$$x=\frac{5\pm\sqrt{37}}{2}$$

The solutions are $\frac{5+\sqrt{37}}{2}$ and $\frac{5-\sqrt{37}}{2}$, or $\frac{5\pm\sqrt{37}}{2}$.

35. $x^2-4x-7=0$

$a=1,\ b=-4,\ c=-7$

$$x=\frac{-(-4)\pm\sqrt{(-4)^2-4\cdot1\cdot(-7)}}{2\cdot1}$$
$$x=\frac{4\pm\sqrt{16+28}}{2}=\frac{4\pm\sqrt{44}}{2}$$
$$x=\frac{4\pm\sqrt{4\cdot11}}{2}=\frac{4\pm2\sqrt{11}}{2}$$
$$x=\frac{2(2\pm\sqrt{11})}{2}=2\pm\sqrt{11}$$

Using a calculator, we have:

$2+\sqrt{11}\approx5.31662479\approx5.3$, and

$2-\sqrt{11}\approx-1.31662479\approx-1.3$.

The approximate solutions, to the nearest tenth, are 5.3 and -1.3.

37. $y^2-6y-1=0$

$a=1,\ b=-6,\ c=-1$

$$y=\frac{-(-6)\pm\sqrt{(-6)^2-4\cdot1\cdot(-1)}}{2\cdot1}$$
$$y=\frac{6\pm\sqrt{36+4}}{2}=\frac{6\pm\sqrt{40}}{2}$$
$$y=\frac{6\pm\sqrt{4\cdot10}}{2}=\frac{6\pm2\sqrt{10}}{2}$$
$$y=\frac{2(3\pm\sqrt{10})}{2}=3\pm\sqrt{10}$$

Using a calculator, we have:

$3+\sqrt{10}\approx6.16227766\approx6.2$ and

$3-\sqrt{10}\approx-0.1622776602\approx-0.2$.

The approximate solutions, to the nearest tenth, are 6.2 and -0.2.

39. $$4x^2+4x=1$$
$$4x^2+4x-1=0\quad\text{Standard form}$$

$a=4,\ b=4,\ c=-1$

$$x=\frac{-4\pm\sqrt{4^2-4\cdot4\cdot(-1)}}{2\cdot4}$$
$$x=\frac{-4\pm\sqrt{16+16}}{8}=\frac{-4\pm\sqrt{32}}{8}$$
$$x=\frac{-4\pm\sqrt{16\cdot2}}{8}=\frac{-4\pm4\sqrt{2}}{8}$$
$$x=\frac{4(-1\pm\sqrt{2})}{4\cdot2}=\frac{-1\pm\sqrt{2}}{2}$$

Using a calculator, we have:

$\frac{-1+\sqrt{2}}{2}\approx0.2071067812\approx0.2$ and

$\frac{-1-\sqrt{2}}{2}\approx-1.207106781\approx-1.2$.

The approximate solutions, to the nearest tenth, are 0.2 and -1.2.

41. $3x^2-8x+2=0$

$a=3,\ b=-8,\ c=2$

$$x=\frac{-(-8)\pm\sqrt{(-8)^2-4\cdot3\cdot2}}{2\cdot3}$$
$$x=\frac{8\pm\sqrt{64-24}}{6}=\frac{8\pm\sqrt{40}}{6}$$
$$x=\frac{8\pm\sqrt{4\cdot10}}{6}=\frac{8\pm2\sqrt{10}}{6}$$
$$x=\frac{2(4\pm\sqrt{10})}{2\cdot3}=\frac{4\pm\sqrt{10}}{3}$$

Using a calculator, we have:

$\frac{4+\sqrt{10}}{3}\approx2.387425887\approx2.4$ and

$\frac{4-\sqrt{10}}{3}\approx0.2792407799\approx0.3$.

The approximate solutions, to the nearest tenth, are 2.4 and 0.3.

43. $$\begin{aligned}\sqrt{40}-2\sqrt{10}+\sqrt{90}&=\sqrt{4\cdot10}-2\sqrt{10}+\sqrt{9\cdot10}\\&=\sqrt{4}\sqrt{10}-2\sqrt{10}+\sqrt{9}\sqrt{10}\\&=2\sqrt{10}-2\sqrt{10}+3\sqrt{10}\\&=(2-2+3)\sqrt{10}\\&=3\sqrt{10}\end{aligned}$$

45. $$\begin{aligned}\sqrt{18}+\sqrt{50}-3\sqrt{8}&=\sqrt{9\cdot2}+\sqrt{25\cdot2}-3\sqrt{4\cdot2}\\&=\sqrt{9}\sqrt{2}+\sqrt{25}\sqrt{2}-3\sqrt{4}\sqrt{2}\\&=3\sqrt{2}+5\sqrt{2}-3\cdot2\sqrt{2}\\&=3\sqrt{2}+5\sqrt{2}-6\sqrt{2}\\&=(3+5-6)\sqrt{2}\\&=2\sqrt{2}\end{aligned}$$

47. $\sqrt{80}=\sqrt{16\cdot5}=\sqrt{16}\sqrt{5}=4\sqrt{5}$

49. $\sqrt{9000x^{10}}=\sqrt{900\cdot10\cdot x^{10}}=\sqrt{900}\sqrt{x^{10}}\sqrt{10}=30x^5\sqrt{10}$

51. ◈

53. $5x + x(x-7) = 0$
$5x + x^2 - 7x = 0$
$x^2 - 2x = 0$ We can factor.
$x(x-2) = 0$

$x = 0$ *or* $x - 2 = 0$
$x = 0$ *or* $x = 2$

The solutions are 0 and 2.

55. $3 - x(x-3) = 4$
$3 - x^2 + 3x = 4$
$0 = x^2 - 3x + 1$ Standard form

$a = 1,\ b = -3,\ c = 1$

We use the quadratic formula.

$$x = \frac{-(-3) \pm \sqrt{(-3)^2 - 4 \cdot 1 \cdot 1}}{2 \cdot 1} = \frac{3 \pm \sqrt{9-4}}{2}$$

$$x = \frac{3 \pm \sqrt{5}}{2}$$

The solutions are $\frac{3+\sqrt{5}}{2}$ and $\frac{3-\sqrt{5}}{2}$, or $\frac{3 \pm \sqrt{5}}{2}$.

57. $(y+4)(y+3) = 15$
$y^2 + 7y + 12 = 15$
$y^2 + 7y - 3 = 0$ Standard form

$a = 1,\ b = 7,\ c = -3$

We use the quadratic formula.

$$y = \frac{-7 \pm \sqrt{7^2 - 4 \cdot 1 \cdot (-3)}}{2 \cdot 1} = \frac{-7 \pm \sqrt{49+12}}{2}$$

$$y = \frac{-7 \pm \sqrt{61}}{2}$$

The solutions are $\frac{-7+\sqrt{61}}{2}$ and $\frac{-7-\sqrt{61}}{2}$, or $\frac{-7 \pm \sqrt{61}}{2}$.

59. $x^2 + (x+2)^2 = 7$
$x^2 + x^2 + 4x + 4 = 7$
$2x^2 + 4x + 4 = 7$
$2x^2 + 4x - 3 = 0$ Standard form

$a = 2,\ b = 4,\ c = -3$

We use the quadratic formula.

$$x = \frac{-4 \pm \sqrt{4^2 - 4 \cdot 2 \cdot (-3)}}{2 \cdot 2} = \frac{-4 \pm \sqrt{16+24}}{4}$$

$$x = \frac{-4 \pm \sqrt{40}}{4} = \frac{-4 \pm \sqrt{4 \cdot 10}}{4}$$

$$x = \frac{-4 \pm 2\sqrt{10}}{4} = \frac{2(-2 \pm \sqrt{10})}{2 \cdot 2}$$

$$x = \frac{-2 \pm \sqrt{10}}{2}$$

The solutions are $\frac{-2+\sqrt{10}}{2}$ and $\frac{-2\sqrt{10}}{2}$, or $\frac{-2 \pm \sqrt{10}}{2}$.

61.-67.

Exercise Set 16.4

1. $P = 17\sqrt{Q}$

$\frac{P}{17} = \sqrt{Q}$ Isolating the radical

$\left(\frac{P}{17}\right)^2 = (\sqrt{Q})^2$ Principle of squaring

$\frac{P^2}{289} = Q$ Simplifying

3. $v = \sqrt{\frac{2gE}{m}}$

$v^2 = \left(\sqrt{\frac{2gE}{m}}\right)^2$ Principle of squaring

$v^2 = \frac{2gE}{m}$

$mv^2 = 2gE$ Multipying by m

$\frac{mv^2}{2g} = E$ Dividing by $2g$

5. $S = 4\pi r^2$

$\frac{S}{4\pi} = r^2$ Dividing by 4π

$\sqrt{\frac{S}{4\pi}} = r$ Principle of square roots. Assume r is nonnegative.

$\sqrt{\frac{1}{4} \cdot \frac{S}{\pi}} = r$

$\frac{1}{2}\sqrt{\frac{S}{\pi}} = r$

7. $P = kA^2 + mA$

$0 = kA^2 + mA - P$ Standard form

$a = k,\ b = m,\ c = -P$

$A = \frac{-b \pm \sqrt{b^2 - 4ac}}{2a}$ Quadratic formula

$A = \frac{-m \pm \sqrt{m^2 - 4 \cdot k \cdot (-P)}}{2 \cdot k}$ Substituting

$A = \frac{-m + \sqrt{m^2 + 4kP}}{2k}$ Using the positive root

9. $c^2 = a^2 + b^2$

$c^2 - b^2 = a^2$

$\sqrt{c^2 - b^2} = a$ Principle of square roots. Assume a is nonnegative.

11. $s = 16t^2$

$\frac{s}{16} = t^2$

$\sqrt{\frac{s}{16}} = t$ Principle of square roots. Assume t is nonnegative.

$\frac{\sqrt{s}}{4} = t$

13. $A = \pi r^2 + 2\pi rh$

$0 = \pi r^2 + 2\pi hr - A$

$a = \pi,\ b = 2\pi h,\ c = -A$

$$r = \frac{-b \pm \sqrt{b^2 - 4ac}}{2a}$$

$$r = \frac{-2\pi h \pm \sqrt{(2\pi h)^2 - 4 \cdot \pi \cdot (-A)}}{2 \cdot \pi}$$

$$r = \frac{-2\pi h + \sqrt{4\pi^2 h^2 + 4\pi A}}{2\pi} \quad \text{Using the positive root}$$

$$r = \frac{-2\pi h + \sqrt{4(\pi^2 h^2 + \pi A)}}{2\pi}$$

$$r = \frac{-2\pi h + 2\sqrt{\pi^2 h^2 + \pi A}}{2\pi}$$

$$r = \frac{2\left(-\pi h + \sqrt{\pi^2 h^2 + \pi A}\right)}{2\pi}$$

$$r = \frac{-\pi h + \sqrt{\pi^2 h^2 + \pi A}}{\pi}$$

15. $F = \dfrac{Av^2}{400}$

$400F = Av^2$ Multiplying by 400

$\dfrac{400F}{A} = v^2$ Dividing by A

$\sqrt{\dfrac{400F}{A}} = v$ Principle of square roots. Assume v is nonnegative.

$\sqrt{400 \cdot \dfrac{F}{A}} = v$

$20\sqrt{\dfrac{F}{a}} = v$

17. $c = \sqrt{a^2 + b^2}$

$c^2 = (\sqrt{a^2 + b^2})^2$ Principle of squaring

$c^2 = a^2 + b^2$

$c^2 - b^2 = a^2$

$\sqrt{c^2 - b^2} = a$ Principle of square roots. Assume a is nonnegative.

19. $h = \dfrac{a}{2}\sqrt{3}$

$2h = a\sqrt{3}$

$\dfrac{2h}{\sqrt{3}} = a$

$\dfrac{2h\sqrt{3}}{3} = a$ Rationalizing the denominator

21. $n = aT^2 - 4T + m$

$0 = aT^2 - 4T + m - n$

$a = a,\ b = -4,\ c = m - n$

$$T = \frac{-b \pm \sqrt{b^2 - 4ac}}{2a}$$

$$T = \frac{-(-4) \pm \sqrt{(-4)^2 - 4 \cdot a \cdot (m - n)}}{2 \cdot a}$$

$$T = \frac{4 + \sqrt{16 - 4a(m - n)}}{2a} \quad \text{Using the positive root}$$

$$T = \frac{4 + \sqrt{4[4 - a(m - n)]}}{2a}$$

$$T = \frac{4 + 2\sqrt{4 - a(m - n)}}{2a}$$

$$T = \frac{2\left(2 + \sqrt{4 - a(m - n)}\right)}{2 \cdot a}$$

$$T = \frac{2 + \sqrt{4 - a(m - n)}}{a}$$

23. $v = 2\sqrt{\dfrac{2kT}{\pi m}}$

$\dfrac{v}{2} = \sqrt{\dfrac{2kT}{\pi m}}$ Isolating the radical

$\left(\dfrac{v}{2}\right)^2 = \left(\sqrt{\dfrac{2kT}{\pi m}}\right)^2$ Principle of squaring

$\dfrac{v^2}{4} = \dfrac{2kT}{\pi m}$

$\dfrac{v^2}{4} \cdot \dfrac{\pi m}{2k} = \dfrac{2kT}{\pi m} \cdot \dfrac{\pi m}{2k}$ Multiplying by $\dfrac{\pi m}{2k}$

$\dfrac{v^2 \pi m}{8k} = T$

25. $3x^2 = d^2$

$x^2 = \dfrac{d^2}{3}$ Dividing by 3

$x = \dfrac{d}{\sqrt{3}}$ Principle of square roots. Assume x is nonnegative.

$x = \dfrac{d}{\sqrt{3}} \cdot \dfrac{\sqrt{3}}{\sqrt{3}}$ Rationalizing the denominator

$x = \dfrac{d\sqrt{3}}{3}$

27. $N = \dfrac{n^2 - n}{2}$

$2N = n^2 - n$ Multiplying by 2

$0 = n^2 - n - 2N$ Finding standard form

$a = 1,\ b = -1,\ c = -2N$

$$n = \frac{-b \pm \sqrt{b^2 - 4ac}}{2a}$$

$$n = \frac{-(-1) \pm \sqrt{(-1)^2 - 4 \cdot 1 \cdot (-2N)}}{2 \cdot 1} \quad \text{Substituting}$$

$$n = \frac{1 + \sqrt{1 + 8N}}{2} \quad \text{Using the positive root}$$

29. $a^2 + b^2 = c^2$ Pythagorean equation
$4^2 + 7^2 = c^2$ Substituting
$16 + 49 = c^2$
$65 = c^2$
$\sqrt{65} = c$ Exact answer
$8.062 \approx c$ Approximate answer

31. $a^2 + b^2 = c^2$ Pythagorean equation
$4^2 + 5^2 = c^2$ Substituting
$16 + 25 = c^2$
$41 = c^2$
$\sqrt{41} = c$ Exact answer
$6.403 \approx c$ Approximate answer

33. $a^2 + b^2 = c^2$ Pythagorean equation
$2^2 + b^2 = (8\sqrt{17})^2$ Substituting
$4 + b^2 = 64 \cdot 17$
$4 + b^2 = 1088$
$b^2 = 1084$
$b = \sqrt{1084}$ Exact answer
$b \approx 32.924$ Approximate answer

35. We make a drawing. Let l = the length of the guy wire.

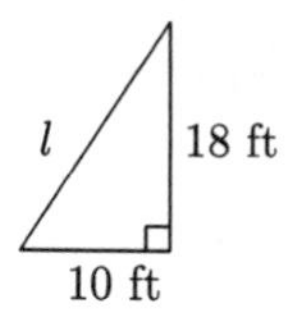

Then we use the Pythagorean equation.

$10^2 + 18^2 = l^2$
$100 + 324 = l^2$
$424 = l^2$
$\sqrt{424} = l$ Exact answer
$20.591 \approx l$ Approximation

The length of the guy wire is $\sqrt{424}$ ft ≈ 20.591 ft.

37. ◈

39. a) $C = 2\pi r$

$$\frac{C}{2\pi} = r$$

b) $A = \pi r^2$

$A = \pi \cdot \left(\frac{C}{2\pi}\right)^2$ Substituting $\frac{C}{2\pi}$ for r

$$A = \pi \cdot \frac{C^2}{4\pi^2}$$

$$A = \frac{C^2}{4\pi}$$

41. $3ax^2 - x - 3ax + 1 = 0$
$3ax^2 + (-1 - 3a)x + 1 = 0$
$a = 3a,\ b = -1 - 3a,\ c = 1$

$$x = \frac{-b \pm \sqrt{b^2 - 4ac}}{2a}$$

$$x = \frac{-(-1-3a) \pm \sqrt{(-1-3a)^2 - 4 \cdot 3a \cdot 1}}{2 \cdot 3a}$$

$$x = \frac{1 + 3a \pm \sqrt{1 + 6a + 9a^2 - 12a}}{6a}$$

$$x = \frac{1 + 3a \pm \sqrt{9a^2 - 6a + 1}}{6a}$$

$$x = \frac{1 + 3a \pm \sqrt{(3a-1)^2}}{6a}$$

$$x = \frac{1 + 3a \pm (3a - 1)}{6a}$$

$x = \frac{1 + 3a + 3a - 1}{6a}$ *or* $x = \frac{1 + 3a - 3a + 1}{6a}$

$x = \frac{6a}{6a}$ *or* $x = \frac{2}{6a}$

$x = 1$ *or* $x = \frac{1}{3a}$

The solutions are 1 and $\frac{1}{3a}$.

Exercise Set 16.5

1. ***Familiarize.*** Using the labels on the drawing in the text we have w = the width of the rectangle and $w + 3$ = the length.

Translate. Recall that area is length × width. Then we have

$$(w + 3)(w) = 70.$$

Solve. We solve the equation.

$w^2 + 3w = 70$
$w^2 + 3w - 70 = 0$
$(w + 10)(w - 7) = 0$

$w + 10 = 0$ *or* $w - 7 = 0$
$w = -10$ *or* $w = 7$

Check. We know that -10 is not a solution of the original problem, because the width cannot be negative. When $w = 7$, then $w + 3 = 10$, and the area is $10 \cdot 7$, or 70. This checks.

State. The width of the rectangle is 7 ft, and the length is 10 ft.

3. ***Familiarize.*** Using the labels on the drawing in the text we have h = the height of the screen and $h + 6$ = the length.

Translate. We use the Pythagorean equation.

$$h^2 + (h + 6)^2 = 30^2.$$

Solve. We solve the equation.

$$\begin{aligned} h^2+(h+6)^2 &= 30^2 \\ h^2+h^2+12h+36 &= 900 \\ 2h^2+12h+36 &= 900 \\ 2h^2+12h-864 &= 0 \\ 2(h^2+6h-432) &= 0 \\ 2(h+24)(h-18) &= 0 \end{aligned}$$

$h+24=0$ *or* $h-18=0$

$h=-24$ *or* $h=18$

Check. We know that -24 is not a solution of the original problem, because the height cannot be negative. When $h=18$, then $h+6=24$ and $18^2+24^2=900=30^2$. This checks.

State. The height is 18 in. and the width is 24 in.

5. ***Familiarize.*** We first make a drawing. We let x represent the length. Then $x-4$ represents the width.

320 cm^2 $\quad x-4$

x

Translate. The area is length $\times$ width. Thus, we have two expressions for the area of the rectangle: $x(x-4)$ and 320. This gives us a translation.

$x(x-4)=320.$

Solve. We solve the equation.

$$\begin{aligned} x^2-4x &= 320 \\ x^2-4x-320 &= 0 \\ (x-20)(x+16) &= 0 \end{aligned}$$

$x-20=0$ *or* $x+16=0$

$x=20$ *or* $x=-16$

Check. Since the length of a side cannot be negative, -16 does not check. But 20 does check. If the length is 20, then the width is $20-4$, or 16. The area is 20×16, or 320. This checks.

State. The length is 20 cm, and the width is 16 cm.

7. ***Familiarize.*** We first make a drawing. We let x represent the width. Then $2x$ represents the length.

50 m^2 $\quad x$

$2x$

Translate. The area is length $\times$ width. Thus, we have two expressions for the area of the rectangle: $2x\cdot x$ and 50. This gives us a translation.

$2x\cdot x=50.$

Solve. We solve the equation.

$$\begin{aligned} 2x^2 &= 50 \\ x^2 &= 25 \end{aligned}$$

$x=5$ *or* $x=-5$ Principle of square roots

Check. Since the length of a side cannot be negative, -5 does not check. But 5 does check. If the width is 5, then the length is $2\cdot 5$, or 10. The area is 10×5, or 50. This checks.

State. The length is 10 m, and the width is 5 m.

9. ***Familiarize.*** We first make a drawing. We let x represent the length of one leg. Then $x+2$ represents the length of the other leg.

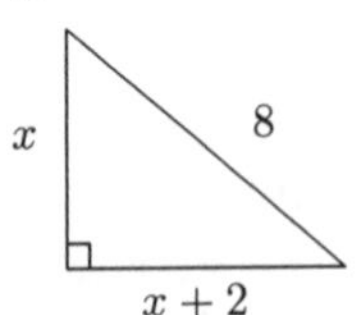

Translate. We use the Pythagorean equation.

$x^2+(x+2)^2=8^2.$

Solve. We solve the equation.

$$\begin{aligned} x^2+x^2+4x+4 &= 64 \\ 2x^2+4x+4 &= 64 \\ 2x^2+4x-60 &= 0 \\ x^2+2x-30 &= 0 \quad \text{Dividing by 2} \end{aligned}$$

$a=1,\ b=2,\ c=-30$

$$\begin{aligned} x &= \frac{-2\pm\sqrt{2^2-4\cdot 1\cdot(-30)}}{2\cdot 1} \\ &= \frac{-2\pm\sqrt{4+120}}{2} = \frac{-2\pm\sqrt{124}}{2} \\ &= \frac{-2\pm\sqrt{4\cdot 31}}{2} = \frac{-2\pm 2\sqrt{31}}{2} \\ &= \frac{2(-1\pm\sqrt{31})}{2} = -1\pm\sqrt{31} \end{aligned}$$

Using a calculator or Table 2 we find that $\sqrt{31}\approx 5.568$:

$-1+\sqrt{31}\approx -1+5.568$ *or* $-1-\sqrt{31}\approx -1-5.568$

≈ 4.6 *or* ≈ -6.6

Check. Since the length of a leg cannot be negative, -6.6 does not check. But 4.6 does check. If the shorter leg is 4.6, then the other leg is $4.6+2$, or 6.6. Then $4.6^2+6.6^2 = 21.16+43.56=64.72$ and using a calculator, $\sqrt{64.72}\approx 8.04\approx 8$. Note that our check is not exact since we are using an approximation.

State. One leg is about 4.6 m, and the other is about 6.6 m long.

11. ***Familiarize.*** We first make a drawing. We let x represent the width and $x+2$ the length.

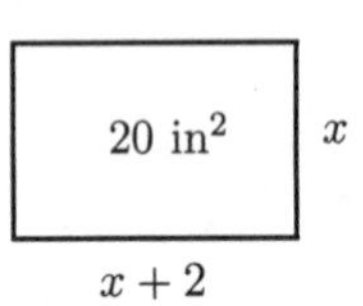

Translate. The area is length $\times$ width. We have two expressions for the area of the rectangle: $(x+2)x$ and 20. This gives us a translation.

$(x+2)x=20.$

Solve. We solve the equation.

$$x^2 + 2x = 20$$
$$x^2 + 2x - 20 = 0$$

$a = 1,\ b = 2,\ c = -20$

$$x = \frac{-2 \pm \sqrt{2^2 - 4 \cdot 1 \cdot (-20)}}{2 \cdot 1}$$
$$= \frac{-2 \pm \sqrt{4 + 80}}{2} = \frac{-2 \pm \sqrt{84}}{2}$$
$$= \frac{-2 \pm \sqrt{4 \cdot 21}}{2} = \frac{-2 \pm 2\sqrt{21}}{2}$$
$$= \frac{2(-1 \pm \sqrt{21})}{2} = -1 \pm \sqrt{21}$$

Using a calculator or Table 2 we find that $\sqrt{21} \approx 4.583$:

$$-1 + \sqrt{21} \approx -1 + 4.583 \quad or \quad -1 - \sqrt{21} \approx -1 - 4.583$$
$$\approx 3.6 \quad or \quad \approx -5.6$$

Check. Since the length of a side cannot be negative, -5.6 does not check. But 3.6 does check. If the width is 3.6, then the length is $3.6 + 2$, or 5.6. The area is $5.6(3.6)$, or $20.16 \approx 20$. This checks.

State. The length is about 5.6 in., and the width is about 3.6 in.

13. ***Familiarize.*** We make a drawing and label it. We let $w =$ the width of the rectangle and $2w =$ the length.

20 cm² w $2w$

Translate. Recall that area = length × width. Then we have

$$2w \cdot w = 20.$$

Solve. We solve the equation.

$$2w^2 = 20$$
$$w^2 = 10 \quad \text{Dividing by 2}$$

$$w = \sqrt{10} \quad or \quad w = -\sqrt{10} \quad \text{Principle of square roots}$$
$$w \approx 3.2 \quad or \quad w \approx -3.2$$

Check. We know that -3.2 is not a solution of the original problem, because width cannot be negative. When $w \approx 3.2$, then $2w \approx 6.4$ and the area is about $(6.4)(3.2)$, or 20.48. This checks, although the check is not exact since we used an approximation for $\sqrt{10}$.

State. The length is about 6.4 cm, and the width is about 3.2 cm.

15. ***Familiarize.*** Using the drawing in the text, we have $x =$ the thickness of the frame, $20 - 2x =$ the width of the picture showing, and $25 - 2x =$ the length of the picture showing.

Translate. Recall that area = length × width. Then we have

$$(25 - 2x)(20 - 2x) = 266.$$

Solve. We solve the equation.

$$500 - 90x + 4x^2 = 266$$
$$4x^2 - 90x + 234 = 0$$
$$2x^2 - 45x + 117 = 0 \quad \text{Dividing by 2}$$
$$(2x - 39)(x - 3) = 0$$

$$2x - 39 = 0 \quad or \quad x - 3 = 0$$
$$2x = 39 \quad or \quad x = 3$$
$$x = 19.5 \quad or \quad x = 3$$

Check. The number 19.5 cannot be a solution, because when $x = 19.5$ then $20 - 2x = -19$, and the width cannot be negative. When $x = 3$, then $20 - 2x = 20 - 2 \cdot 3$, or 14 and $25 - 2x = 25 - 2 \cdot 3$, or 19 and $19 \cdot 14 = 266$. This checks.

State. The thickness of the frame is 3 cm.

17. ***Familiarize.*** Referring to the drawing in the text, we complete the table.

	d	r	t
Upstream	40	$r - 3$	t_1
Downstream	40	$r + 3$	t_2
Total Time			14

Translate. Using $t = d/r$ and the rows of the table, we have

$$t_1 = \frac{40}{r - 3} \text{ and } t_2 = \frac{40}{r + 3}.$$

Since the total time is 14 hr, $t_1 + t_2 = 14$, and we have

$$\frac{40}{r - 3} + \frac{40}{r + 3} = 14.$$

Solve. We solve the equation. We multiply by $(r - 3)(r + 3)$, the LCM of the denominators.

$$(r - 3)(r + 3)\left(\frac{40}{r - 3} + \frac{40}{r + 3}\right) = (r - 3)(r + 3) \cdot 14$$
$$40(r + 3) + 40(r - 3) = 14(r^2 - 9)$$
$$40r + 120 + 40r - 120 = 14r^2 - 126$$
$$80r = 14r^2 - 126$$
$$0 = 14r^2 - 80r - 126$$
$$0 = 7r^2 - 40r - 63$$
$$0 = (7r + 9)(r - 7)$$

$$7r + 9 = 0 \quad or \quad r - 7 = 0$$
$$7r = -9 \quad or \quad r = 7$$
$$r = -\frac{9}{7} \quad or \quad r = 7$$

Check. Since speed cannot be negative, $-\frac{9}{7}$ cannot be a solution. If the speed of the boat is 7 km/h, the speed upstream is $7 - 3$, or 4 km/h, and the speed downstream is $7 + 3$, or 10 km/h. The time upstream is $\frac{40}{4}$, or 10 hr. The time downstream is $\frac{40}{10}$, or 4 hr. The total time is 14 hr. This checks.

State. The speed of the boat in still water is 7 km/h.

19. ***Familiarize.*** We first make a drawing. We let r represent the speed of the boat in still water. Then $r - 4$ is the speed of the boat traveling upstream and $r+4$ is the speed of the boat traveling downstream.

Upstream
$r-4$ mph
4 mi

Downstream
$r+4$ mph
12 mi

We summarize the information in a table.

	d	r	t
Upstream	4	$r-4$	t_1
Downstream	12	$r+4$	t_2
Total Time			2

Translate. Using $t = d/r$ and the rows of the table, we have

$$t_1 = \frac{4}{r-4} \text{ and } t_2 = \frac{12}{r+4}.$$

Since the total time is 2 hr, $t_1 + t_2 = 2$, and we have

$$\frac{4}{r-4} + \frac{12}{r+4} = 2.$$

Solve. We solve the equation. We multiply by $(r-4)(r+4)$, the LCM of the denominators.

$$(r-4)(r+4)\left(\frac{4}{r-4} + \frac{12}{r+4}\right) = (r-4)(r+4)\cdot 2$$
$$4(r+4) + 12(r-4) = 2(r^2-16)$$
$$4r + 16 + 12r - 48 = 2r^2 - 32$$
$$16r - 32 = 2r^2 - 32$$
$$0 = 2r^2 - 16r$$
$$0 = 2r(r-8)$$

$2r = 0$ *or* $r - 8 = 0$
$r = 0$ *or* $r = 8$

Check. If $r = 0$, then the speed upstream, $0-4$, would be negative. Since speed cannot be negative, 0 cannot be a solution. If the speed of the boat is 8 mph, the speed upstream is $8-4$, or 4 mph, and the speed downstream is $8+4$, or 12 mph. The time upstream is $\frac{4}{4}$, or 1 hr. The time downstream is $\frac{12}{12}$, or 1 hr. The total time is 2 hr. This checks.

State. The speed of the boat in still water is 8 mph.

21. ***Familiarize.*** We first make a drawing. We let r represent the speed of the current. Then $10-r$ is the speed of the boat traveling upstream and $10+r$ is the speed of the boat traveling downstream.

Upstream
$10-r$ km/h
12 km

Downstream
$10+r$ km/h
28 km

We summarize the information in a table.

	d	r	t
Upstream	12	$10-r$	t_1
Downstream	28	$10+r$	t_2
Total Time			4

Translate. Using $t = d/r$ and the rows of the table, we have

$$t_1 = \frac{12}{10-r} \text{ and } t_2 = \frac{28}{10+r}.$$

Since the total time is 4 hr, $t_1 + t_2 = 4$, and we have

$$\frac{12}{10-r} + \frac{28}{10+r} = 4.$$

Solve. We solve the equation. We multiply by $(10-r)(10+r)$, the LCM of the denominators.

$$(10-r)(10+r)\left(\frac{12}{10-r} + \frac{28}{10+r}\right) = (10-r)(10+r)\cdot 4$$
$$12(10+r) + 28(10-r) = 4(100-r^2)$$
$$120 + 12r + 280 - 28r = 400 - 4r^2$$
$$400 - 16r = 400 - 4r^2$$
$$4r^2 - 16r = 0$$
$$4r(r-4) = 0$$

$4r = 0$ *or* $r - 4 = 0$
$r = 0$ *or* $r = 4$

Check. Since a stream is defined to be a flow of running water, its rate must be greater than 0. Thus, 0 cannot be a solution. If the speed of the current is 4 km/h, the speed upstream is $10-4$, or 6 km/h, and the speed downstream is $10+4$, or 14 km/h. The time upstream is $\frac{12}{6}$, or 2 hr. The time downstream is $\frac{28}{14}$, or 2 hr. The total time is 4 hr. This checks.

State. The speed of the stream is 4 km/h.

23. ***Familiarize.*** We first make a drawing. We let r represent the speed of the wind. Then the speed of the plane flying against the wind is $200-r$ and the speed of the plane flying with the wind is $200+r$.

Against the wind
$200-r$ mph
738 miles

With the wind
$200+r$ mph
1062 miles

We summarize the information in a table.

	d	r	t
Against wind	738	$200-r$	t_1
With wind	1062	$200+r$	t_2

Translate. Using $t = d/r$ and the rows of the table, we have

$$t_1 = \frac{738}{200-r} \text{ and } t_2 = \frac{1062}{200+r}.$$

Since the total time is 9 hr, $t_1 + t_2 = 9$, and we have

$$\frac{738}{200-r} + \frac{1062}{200+r} = 9.$$

Solve. We solve the equation. We multiply by $(200-r)(200+r)$, the LCM of the denominators.

$$\begin{aligned}(200-r)(200+r)\left(\frac{738}{200-r} + \frac{1062}{200+r}\right) &= \\ &(200-r)(200+r)\cdot 9 \\ 738(200+r) + 1062(200-r) &= \\ &9(40,000-r^2) \\ 147,600 + 738r + 212,400 - 1062r &= 360,000 - 9r^2 \\ 360,000 - 324r &= 360,000 - 9r^2 \\ 9r^2 - 324r &= 0 \\ 9r(r-36) &= 0\end{aligned}$$

$$\begin{aligned}9r = 0 \quad &or \quad r - 36 = 0 \\ r = 0 \quad &or \quad \quad \quad r = 36\end{aligned}$$

Check. In this problem we assume there is a wind. Thus, the speed of the wind must be greater than 0 and the number 0 cannot be a solution. If the speed of the wind is 36 mph, the speed of the airplane against the wind is 200−36, or 164 mph, and the speed with the wind is 200 + 36, or 236 mph. The time against the wind is $\frac{738}{164}$, or $4\frac{1}{2}$ hr. The time with the wind is $\frac{1062}{236}$, or $4\frac{1}{2}$ hr. The total time is 9 hr. The value checks.

State. The speed of the wind is 36 mph.

25. ***Familiarize.*** We first make a drawing. We let r represent the speed of the stream. Then $9-r$ represents the speed of the boat traveling upstream and $9+r$ represents the speed of the boat traveling downstream.

Upstream
$9-r$ km/h
80 km

Downstream
$9+r$ km/h
80 km

We summarize the information in a table.

	d	r	t
Upstream	80	$9-r$	t_1
Downstream	80	$9+r$	t_2

Translate. Using $t = d/r$ and the rows of the table, we have

$$t_1 = \frac{80}{9-r} \text{ and } t_2 = \frac{80}{9+r}.$$

Since the total time is 18 hr, $t_1 + t_2 = 18$, and we have

$$\frac{80}{9-r} + \frac{80}{9+r} = 18.$$

Solve. We solve the equation. We multiply by $(9-r)(9+r)$, the LCM of the denominators.

$$\begin{aligned}(9-r)(9+r)\left(\frac{80}{9-r} + \frac{80}{9+r}\right) &= (9-r)(9+r)\cdot 18 \\ 80(9+r) + 80(9-r) &= 18(81-r^2) \\ 720 + 80r + 720 - 80r &= 1458 - 18r^2 \\ 1440 &= 1458 - 18r^2 \\ 18r^2 &= 18 \\ r^2 &= 1\end{aligned}$$

$r = 1$ *or* $r = -1$ Principle of square roots

Check. Since speed cannot be negative, −1 cannot be a solution. If the speed of the stream is 1 km/h, the speed upstream is 9 − 1, or 8 km/h, and the speed downstream is 9 + 1, or 10 km/h. The time upstream is $\frac{80}{8}$, or 10 hr. The time downstream is $\frac{80}{10}$, or 8 hr. The total time is 18 hr. This checks.

State. The speed of the stream is 1 km/h.

27.
$$\begin{aligned}5\sqrt{2} + \sqrt{18} &= 5\sqrt{2} + \sqrt{9\cdot 2} \\ &= 5\sqrt{2} + \sqrt{9}\sqrt{2} \\ &= 5\sqrt{2} + 3\sqrt{2} \\ &= (5+3)\sqrt{2} \\ &= 8\sqrt{2}\end{aligned}$$

29.
$$\begin{aligned}\sqrt{4x^3} - 7\sqrt{x} &= \sqrt{4\cdot x^2\cdot x} - 7\sqrt{x} \\ &= \sqrt{4}\sqrt{x^2}\sqrt{x} - 7\sqrt{x} \\ &= 2x\sqrt{x} - 7\sqrt{x} \\ &= (2x-7)\sqrt{x}\end{aligned}$$

31.
$$\begin{aligned}\sqrt{2} + \sqrt{\frac{1}{2}} &= \sqrt{2} + \sqrt{\frac{1}{2}\cdot\frac{2}{2}} \\ &= \sqrt{2} + \sqrt{\frac{2}{4}} \\ &= \sqrt{2} + \frac{\sqrt{2}}{\sqrt{4}} \\ &= \sqrt{2} + \frac{\sqrt{2}}{2} \\ &= (1 + \frac{1}{2})\sqrt{2} \\ &= \frac{3}{2}\sqrt{2}, \text{or } \frac{3\sqrt{2}}{2}\end{aligned}$$

33.
$$\begin{aligned}&\sqrt{24} + \sqrt{54} - \sqrt{48} \\ &= \sqrt{4\cdot 6} + \sqrt{9\cdot 6} - \sqrt{16\cdot 3} \\ &= \sqrt{4}\cdot\sqrt{6} + \sqrt{9}\cdot\sqrt{6} - \sqrt{16}\cdot\sqrt{3} \\ &= 2\sqrt{6} + 3\sqrt{6} - 4\sqrt{3} \\ &= 5\sqrt{6} - 4\sqrt{3}\end{aligned}$$

35. ◈

37. ***Familiarize.*** From the drawing in the text, we see that we have a right triangle where r = the length of each leg and $r+1$ = the length of the hypotenuse.

Translate. We use the Pythagorean equation.

$r^2 + r^2 = (r+1)^2$.

Solve. We solve the equation.

$$2r^2 = r^2 + 2r + 1$$
$$r^2 - 2r - 1 = 0$$

$a = 1,\ b = -2,\ c = -1$

$$r = \frac{-(-2) \pm \sqrt{(-2)^2 - 4 \cdot 1 \cdot (-1)}}{2 \cdot 1}$$
$$= \frac{2 \pm \sqrt{4+4}}{2} = \frac{2 \pm \sqrt{8}}{2}$$
$$= \frac{2 \pm \sqrt{4 \cdot 2}}{2} = \frac{2 \pm 2\sqrt{2}}{2}$$
$$= \frac{2(1 \pm \sqrt{2})}{2 \cdot 1} = 1 \pm \sqrt{2}$$

$x = 1 - \sqrt{2}$ *or* $x = 1 + \sqrt{2}$
$x \approx 1 - 1.414$ *or* $x \approx 1 + 1.414$
$x \approx -0.414$ *or* $x \approx 2.414$
$x \approx -0.41$ *or* $x \approx 2.41$ Rounding to the nearest hundredth

Check. Since the length of a leg cannot be negative, -0.41 cannot be a solution of the original equation. When $x \approx 2.41$, then $x + 1 \approx 3.41$ and $(2.41)^2 + (2,41)^2 = 5.8081 + 5.8081 = 11.6162 \approx (3.41)^2$. This checks.

State. In the figure, $r = 1 + \sqrt{2} \approx 2.41$ cm.

39. ***Familiarize.*** The radius of a 12-in. pizza is $\frac{12}{2}$, or 6 in. The radius of a d-in. pizza is $\frac{d}{2}$ in. The area of a circle is πr^2.

Translate.

Area of d-in. pizza	is	Area of 12-in. pizza	plus	Area of 12-in. pizza
↓	↓	↓	↓	↓
$\pi\left(\frac{d}{2}\right)^2$	$=$	$\pi \cdot 6^2$	$+$	$\pi \cdot 6^2$

Solve. We solve the equation.

$$\frac{d^2}{4}\pi = 36\pi + 36\pi$$
$$\frac{d^2}{4}\pi = 72\pi$$
$$\frac{d^2}{4} = 72 \quad \text{Dividing by } \pi$$
$$d^2 = 288$$

$d = \sqrt{288}$ *or* $d = -\sqrt{288}$
$d = 12\sqrt{2}$ *or* $d = -12\sqrt{2}$
$d \approx 16.97$ *or* $d \approx -16.97$ Using a calculator

Check. Since the diameter cannot be negative, -16.97 is not a solution. If $d = 12\sqrt{2}$, or 16.97, then $r = 6\sqrt{2}$ and the area is $\pi(6\sqrt{2})^2$, or 72π. The area of the two 12-in. pizzas is $2 \cdot \pi \cdot 6^2$, or 72π. The value checks.

State. The diameter of the pizza should be $12\sqrt{2} \approx 16.97$ in.

The radius of a 16-in. pizza is $\frac{16}{2}$, or 8 in., so the area is $\pi(8)^2$, or 64π. We found that the area of two 12-in. pizzas is 72π and $72\pi > 64\pi$, so you get more to eat with two 12-in. pizzas than with a 16-in. pizza.

Exercise Set 16.6

1. $y = x^2 + 1$

We first find the vertex. The x-coordinate is

$$-\frac{b}{2a} = -\frac{0}{2 \cdot 1} = 0.$$

We substitute into the equation to find the second coordinate of the vertex.

$$y = x^2 + 1 = 0^2 + 1 = 1$$

The vertex is (0,1). The line of symmetry is $x = 0$, the y-axis.

We choose some x-values on both sides of the vertex and graph the parabola.

When $x = 1$, $y = 1^2 + 1 = 1 + 1 = 2$.

When $x = -1$, $y = (-1)^2 + 1 = 1 + 1 = 2$.

When $x = 2$, $y = 2^2 + 1 = 4 + 1 = 5$.

When $x = -2$, $y = (-2)^2 + 1 = 4 + 1 = 5$.

x	y	
0	1	←Vertex
1	2	
−1	2	
2	5	
−2	5	

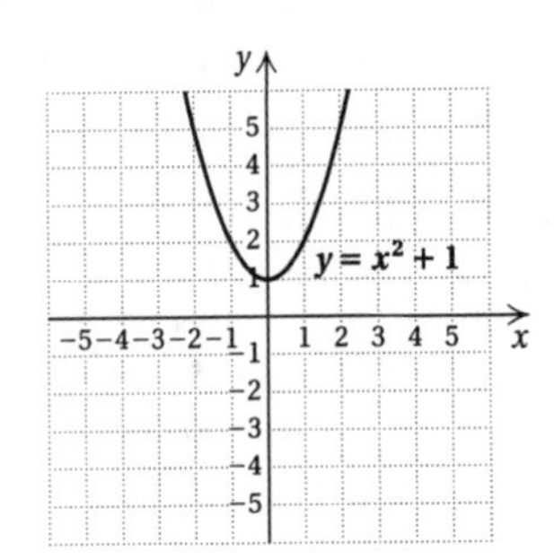

3. $y = -1 \cdot x^2$

Find the vertex. The x-coordinate is

$$-\frac{b}{2a} = -\frac{0}{2(-1)} = 0.$$

The y-coordinate is

$$y = -1 \cdot x^2 = -1 \cdot 0^2 = 0.$$

The vertex is (0,0). The line of symmetry is $x = 0$, the y-axis.

Choose some x-values on both sides of the vertex and graph the parabola.

When $x = -2$, $y = -1 \cdot (-2)^2 = -1 \cdot 4 = -4$.

When $x = -1$, $y = -1 \cdot (-1)^2 = -1 \cdot 1 = -1$.

When $x = 1$, $y = -1 \cdot 1^2 = -1 \cdot 1 = -1$.

When $x = 2$, $y = -1 \cdot 2^2 = -1 \cdot 4 = -4$.

x	y	
0	0	←Vertex
−2	−4	
−1	−1	
1	−1	
2	−4	

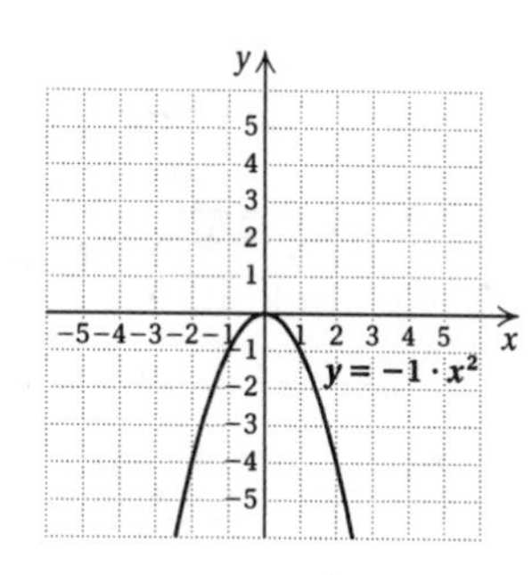

5. $y = -x^2 + 2x$

Find the vertex. The x-coordinate is

$$-\frac{b}{2a} = -\frac{2}{2(-1)} = -(-1) = 1.$$

The y-coordinate is

$$y = -x^2 + 2x = -(1)^2 + 2 \cdot 1 = -1 + 2 = 1.$$

The vertex is (1,1).

We choose some x-values on both sides of the vertex and graph the parabola. We make sure we find y when $x = 0$. This gives us the y-intercept.

x	y	
1	1	←Vertex
0	0	←y-intercept
−1	−3	
2	0	
3	−3	

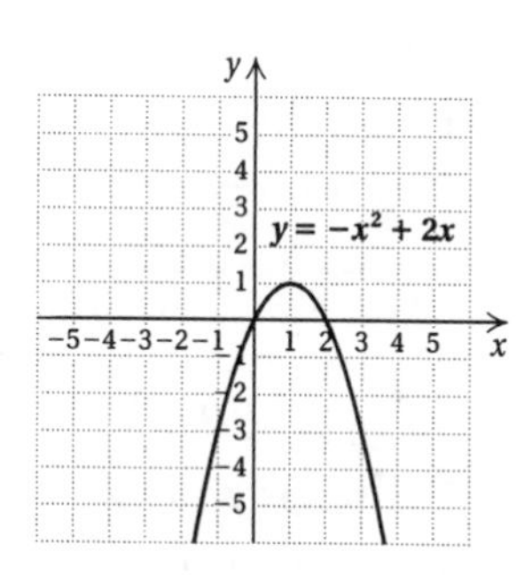

7. $y = 5 - x - x^2$, or $y = -x^2 - x + 5$

Find the vertex. The x-coordinate is

$$-\frac{b}{2a} = -\frac{-1}{2(-1)} = -\frac{1}{2}.$$

The y-coordinate is

$$y = 5-x-x^2 = 5-\left(-\frac{1}{2}\right)-\left(-\frac{1}{2}\right)^2 = 5+\frac{1}{2}-\frac{1}{4} = \frac{21}{4}.$$

The vertex is $\left(-\frac{1}{2}, \frac{21}{4}\right)$.

We choose some x-values on both sides of the vertex and graph the parabola.

x	y	
$-\frac{1}{2}$	$\frac{21}{4}$	←Vertex
0	5	←y-intercept
−1	5	
−2	3	
1	3	

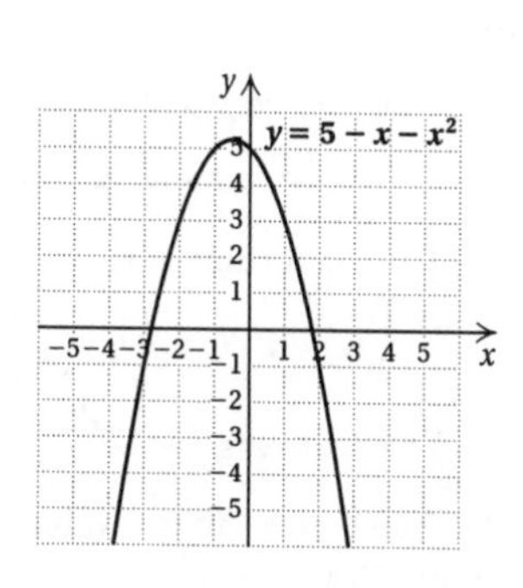

9. $y = x^2 - 2x + 1$

Find the vertex. The x-coordinate is

$$-\frac{b}{2a} = -\frac{-2}{2 \cdot 1} = -(-1) = 1.$$

The y-coordinate is

$$y = x^2 - 2x + 1 = 1^2 - 2 \cdot 1 + 1 = 1 - 2 + 1 = 0.$$

The vertex is $(1, 0)$.

We choose some x-values on both sides of the vertex and graph the parabola.

x	y	
1	0	←Vertex
0	1	←y-intercept
−1	4	
2	1	
3	4	

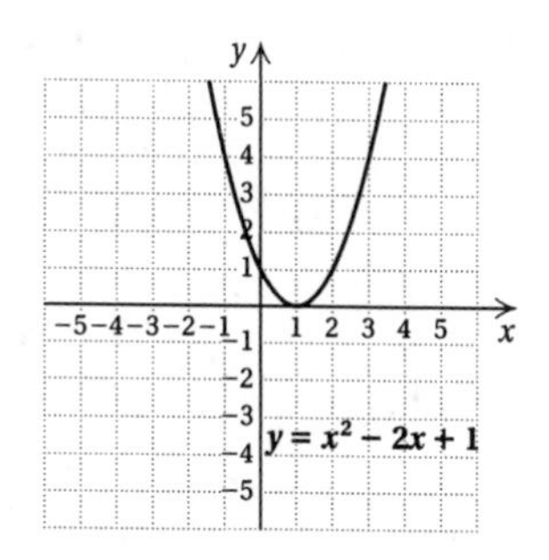

11. $y = -x^2 + 2x + 3$

Find the vertex. The x-coordinate is

$$-\frac{b}{2a} = -\frac{2}{2(-1)} = -(-1) = 1.$$

The y-coordinate is

$$y = -x^2 + 2x + 3 = -(1)^2 + 2 \cdot 1 + 3 = -1 + 2 + 3 = 4.$$

The vertex is $(1, 4)$.

We choose some x-values on both sides of the vertex and graph the parabola.

x	y	
1	4	←Vertex
0	3	←y-intercept
−1	0	
2	3	
3	0	

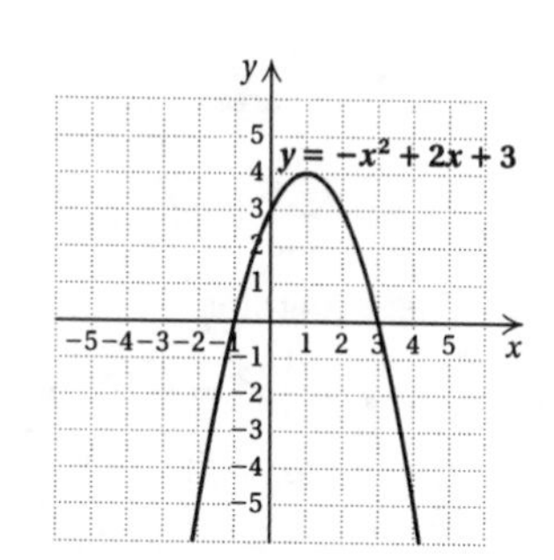

13. $y = -2x^2 - 4x + 1$

Find the vertex. The x-coordinate is

$$-\frac{b}{2a} = -\frac{-4}{2(-2)} = -1.$$

The y-coordinate is

$$y = -2x^2 - 4x + 1 = -2(-1)^2 - 4(-1) + 1 = -2 + 4 + 1 = 3.$$

The vertex is $(-1, 3)$.

We choose some x-values on both sides of the vertex and graph the parabola.

x	y	
-1	3	←Vertex
0	1	←y-intercept
1	-5	
-2	1	
-3	-5	

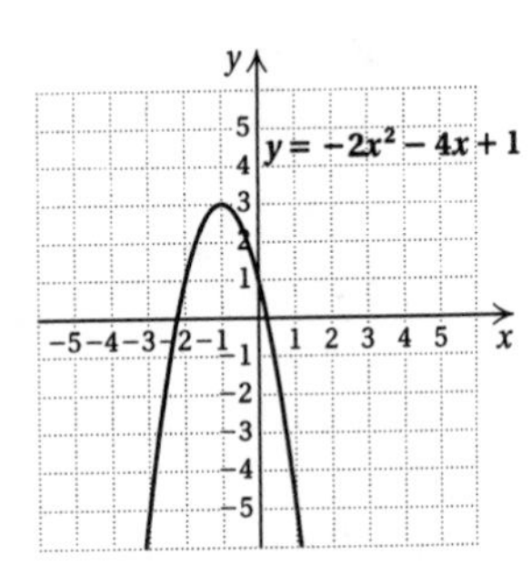

15. $y = 5 - x^2$, or $y = -x^2 + 5$

Find the vertex. The x-coordinate is

$$-\frac{b}{2a} = -\frac{0}{2(-1)} = 0.$$

The y-coordinate is

$$y = 5 - x^2 = 5 - 0^2 = 5.$$

The vertex is $(0, 5)$.

We choose some x-values on both sides of the vertex and graph the parabola.

x	y	
0	5	←Vertex
-1	4	
-2	1	
1	4	
2	1	

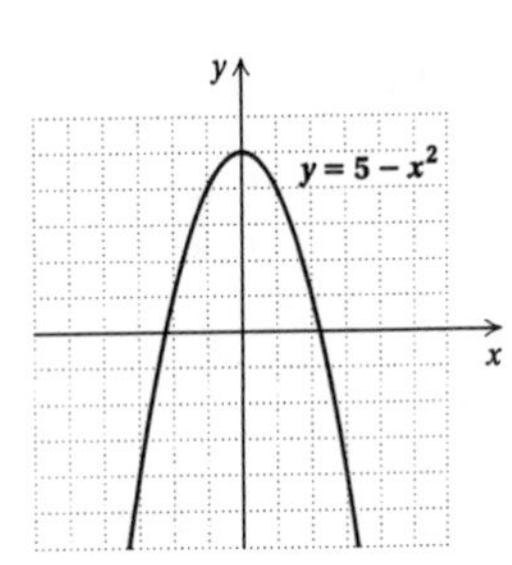

17. $y = \frac{1}{4}x^2$

Find the vertex. The x-coordinate is

$$-\frac{b}{2a} = -\frac{0}{2\left(\frac{1}{4}\right)} = 0.$$

The y-coordinate is

$$y = \frac{1}{4}x^2 = \frac{1}{4} \cdot 0^2 = 0.$$

The vertex is $(0, 0)$.

We choose some x-values on both sides of the vertex and graph the parabola.

x	y	
0	0	←Vertex
-2	1	
-4	4	
2	1	
4	4	

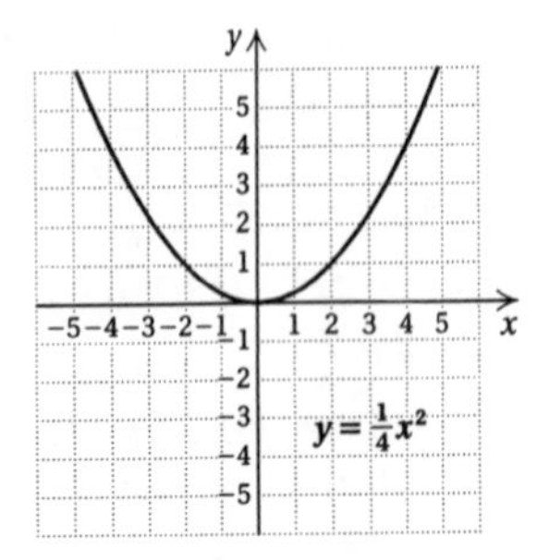

19. $y = -x^2 + x - 1$

Find the vertex. The x-coordinate is

$$-\frac{b}{2a} = -\frac{1}{2(-1)} = -\left(-\frac{1}{2}\right) = \frac{1}{2}.$$

The y-coordinate is

$$y = -x^2 + x - 1 = -\left(\frac{1}{2}\right)^2 + \frac{1}{2} - 1 = -\frac{1}{4} + \frac{1}{2} - 1 = -\frac{3}{4}.$$

The vertex is $\left(\frac{1}{2}, -\frac{3}{4}\right)$.

We choose some x-values on both sides of the vertex and graph the parabola.

x	y	
$\frac{1}{2}$	$-\frac{3}{4}$	←Vertex
0	-1	←y-intercept
-1	-3	
1	-1	
2	-3	

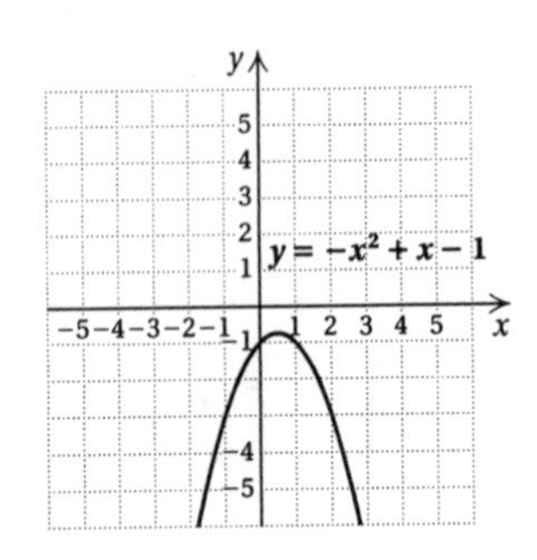

21. $y = -2x^2$

Find the vertex. The x-coordinate is

$$-\frac{b}{2a} = -\frac{0}{2(-2)} = 0.$$

The y-coordinate is

$$y = -2x^2 = -2 \cdot 0^2 = 0.$$

The vertex is $(0, 0)$.

We choose some x-values on both sides of the vertex and graph the parabola.

x	y	
0	0	←Vertex
-1	-2	
-2	-8	
1	-2	
2	-8	

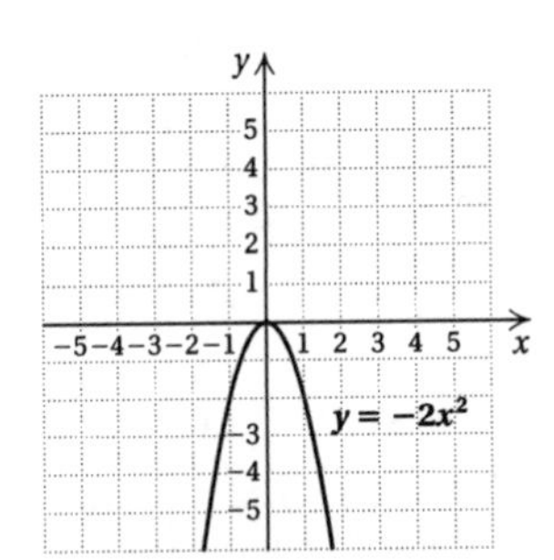

23. $y = x^2 - x - 6$

Find the vertex. The x-coordinate is

$$-\frac{b}{2a} = -\frac{-1}{2 \cdot 1} = -\left(-\frac{1}{2}\right) = \frac{1}{2}.$$

The y-coordinate is

$$y = x^2 - x - 6 = \left(\frac{1}{2}\right)^2 - \frac{1}{2} - 6 = \frac{1}{4} - \frac{1}{2} - 6 = -\frac{25}{4}.$$

The vertex is $\left(\frac{1}{2}, -\frac{25}{4}\right)$.

We choose some x-values on both sides of the vertex and graph the parabola.

x	y	
$\frac{1}{2}$	$-\frac{25}{4}$	←Vertex
0	-6	←y-intercept
-1	-4	
1	-6	
2	-4	

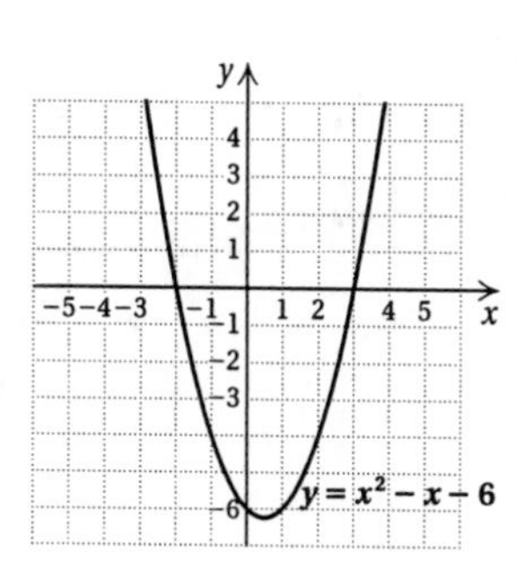

25. $y = x^2 - 2$

To find the x-intercepts we solve the equation $x^2 - 2 = 0$.

$$x^2 - 2 = 0$$
$$x^2 = 2$$

$x = \sqrt{2}$ *or* $x = -\sqrt{2}$ Principle of square roots

The x-intercepts are $(\sqrt{2}, 0)$ and $(-\sqrt{2}, 0)$.

27. $y = x^2 + 5x$

To find the x-intercepts we solve the equation $x^2 + 5x = 0$.

$$x^2 + 5x = 0$$
$$x(x + 5) = 0$$

$x = 0$ *or* $x + 5 = 0$

$x = 0$ *or* $x = -5$

The x-intercepts are $(0, 0)$ and $(-5, 0)$.

29. $y = 8 - x - x^2$

To find the x-intercepts we solve the equation $8 - x - x^2 = 0$.

$$8 - x - x^2 = 0$$
$$x^2 + x - 8 = 0 \quad \text{Standard form}$$

$a = 1,\ b = 1,\ c = -8$

$$x = \frac{-1 \pm \sqrt{1^2 - 4 \cdot 1 \cdot (-8)}}{2 \cdot 1}$$
$$x = \frac{-1 \pm \sqrt{33}}{2}$$

The x-intercepts are $\left(\frac{-1 + \sqrt{33}}{2}, 0\right)$ and $\left(\frac{-1 - \sqrt{33}}{2}, 0\right)$.

31. $y = x^2 - 6x + 9$

To find the x-intercepts we solve the equation $x^2 - 6x + 9 = 0$.

$$x^2 - 6x + 9 = 0$$
$$(x - 3)(x - 3) = 0$$

$x - 3 = 0$ *or* $x - 3 = 0$

$x = 3$ *or* $x = 3$

The x-intercept is $(3, 0)$.

33. $y = -x^2 - 4x + 1$

To find the x-intercepts we solve the equation $-x^2 - 4x + 1 = 0$.

$$-x^2 - 4x + 1 = 0$$
$$x^2 + 4x - 1 = 0 \quad \text{Standard form}$$

$a = 1,\ b = 4,\ c = -1$

$$x = \frac{-4 \pm \sqrt{4^2 - 4 \cdot 1 \cdot (-1)}}{2 \cdot 1}$$
$$x = \frac{-4 \pm \sqrt{20}}{2} = \frac{-4 \pm \sqrt{4 \cdot 5}}{2} = \frac{-4 \pm 2\sqrt{5}}{2}$$
$$x = \frac{2(-2 \pm \sqrt{5})}{2} = -2 \pm \sqrt{5}$$

The x-intercepts are $(-2 + \sqrt{5}, 0)$ and $(-2 - \sqrt{5}, 0)$.

35. $y = x^2 + 9$

To find the x-intercepts we solve the equation $x^2 + 9 = 0$.

$$x^2 + 9 = 0$$
$$x^2 = -9$$

The negative number -9 has no real-number square roots. Thus there are no x-intercepts.

37.
$$\begin{aligned}\sqrt{x^3 - x^2} + \sqrt{4x - 4} &= \sqrt{x^2(x - 1)} + \sqrt{4(x - 1)} \\ &= \sqrt{x^2}\sqrt{x - 1} + \sqrt{4}\sqrt{x - 1} \\ &= x\sqrt{x - 1} + 2\sqrt{x - 1} \\ &= (x + 2)\sqrt{x - 1}\end{aligned}$$

39. $\sqrt{2}\sqrt{14} = \sqrt{2 \cdot 14} = \sqrt{2 \cdot 2 \cdot 7} = 2\sqrt{7}$

41.
$$y = \frac{k}{x}$$
$$12.4 = \frac{k}{2.4} \quad \text{Substituting}$$
$$29.76 = k \quad \text{Variation constant}$$
$$y = \frac{29.76}{x} \quad \text{Equation of variation}$$

43.
$$\begin{aligned}5x^3 - 2x &= 5(-1)^3 - 2(-1) \\ &= 5(-1) - 2(-1) \\ &= -5 + 2 \\ &= -3\end{aligned}$$

45. ◈

47. a) We substitute 128 for H and solve for t:

$$128 = -16t^2 + 96t$$
$$16t^2 - 96t + 128 = 0$$
$$16(t^2 - 6t + 8) = 0$$
$$16(t - 2)(t - 4) = 0$$

$t - 2 = 0$ *or* $t - 4 = 0$

$t = 2$ *or* $t = 4$

The projectile is 128 ft from the ground 2 sec after launch and again 4 sec after launch. The graph confirms this.

b) We find the first coordinate of the vertex of the function $H = -16t^2 + 96t$:

$$-\frac{b}{2a} = -\frac{96}{2(-16)} = -\frac{96}{-32} = -(-3) = 3$$

The projectile reaches its maximum height 3 sec after launch. The graph confirms this.

c) We substitute 0 for H and solve for t:

$0 = -16t^2 + 96t$
$0 = -16t(t-6)$

$-16t = 0$ *or* $t - 6 = 0$
$t = 0$ *or* $t = 6$

At $t = 0$ sec the projectile has not yet been launched. Thus, we use $t = 6$. The projectile returns to the ground 6 sec after launch. The graph confirms this.

49. $y = x^2 + 2x - 3$

$a = 1,\ b = 2,\ c = -3$

$b^2 - 4ac = 2^2 - 4 \cdot 1 \cdot (-3) = 4 + 12 = 16$

Since $b^2 - 4ac = 16 > 0$, the equation $x^2 + 2x - 3 = 0$ has two real-number solutions and the graph of $y = x^2 + 2x - 3$ has two x-intercepts.

51. $y = -0.02x^2 + 4.7x - 2300$

$a = -0.02,\ b = 4.7,\ c = -2300$

$b^2 - 4ac = (4.7)^2 - 4(-0.02)(-2300) = 22.09 - 184 = -161.91$

Since $b^2 - 4ac = -161.91 < 0$, the equation $-0.02x^2 + 4.7x - 2300 = 0$ has no real solutions and the graph of $y = -0.02x^2 + 4.7x - 2300$ has no x-intercepts.

Exercise Set 16.7

1. Yes; each member of the domain is matched to only one member of the range.

3. Yes; each member of the domain is matched to only one member of the range.

5. No; a member of the domain is matched to more than one member of the range. In fact, each member of the domain is matched to 3 members of the range.

7. Yes; each member of the domain is matched to only one member of the range.

9. This correspondence is a function, because each class member has only one seat number.

11. This correspondence is a function, because each shape has only one number for its area.

13. This correspondence is not a function, because it is reasonable to assume that at least one person has more than one aunt.

The correspondence is a relation, because it is reasonable to assume that each person has at least one aunt.

15. $f(x) = x + 5$

a) $f(4) = 4 + 5 = 9$

b) $f(7) = 7 + 5 = 12$

c) $f(-3) = -3 + 5 = 2$

d) $f(0) = 0 + 5 = 5$

e) $f(2.4) = 2.4 + 5 = 7.4$

f) $f\left(\frac{2}{3}\right) = \frac{2}{3} + 5 = 5\frac{2}{3}$

17. $h(p) = 3p$

a) $h(-7) = 3(-7) = -21$

b) $h(5) = 3 \cdot 5 = 15$

c) $h(14) = 3 \cdot 14 = 42$

d) $h(0) = 3 \cdot 0 = 0$

e) $h\left(\frac{2}{3}\right) = 3 \cdot \frac{2}{3} = \frac{6}{3} = 2$

f) $h(-54.2) = 3(-54.2) = -162.6$

19. $g(s) = 3s + 4$

a) $g(1) = 3 \cdot 1 + 4 = 3 + 4 = 7$

b) $g(-7) = 3(-7) + 4 = -21 + 4 = -17$

c) $g(6.7) = 3(6.7) + 4 = 20.1 + 4 = 24.1$

d) $g(0) = 3 \cdot 0 + 4 = 0 + 4 = 4$

e) $g(-10) = 3(-10) + 4 = -30 + 4 = -26$

f) $g\left(\frac{2}{3}\right) = 3 \cdot \frac{2}{3} + 4 = 2 + 4 = 6$

21. $f(x) = 2x^2 - 3x$

a) $f(0) = 2 \cdot 0^2 - 3 \cdot 0 = 0 - 0 = 0$

b) $f(-1) = 2(-1)^2 - 3(-1) = 2 + 3 = 5$

c) $f(2) = 2 \cdot 2^2 - 3 \cdot 2 = 8 - 6 = 2$

d) $f(10) = 2 \cdot 10^2 - 3 \cdot 10 = 200 - 30 = 170$

e) $f(-5) = 2(-5)^2 - 3(-5) = 50 + 15 = 65$

f) $f(-10) = 2(-10)^2 - 3(-10) = 200 + 30 = 230$

23. $f(x) = |x| + 1$

a) $f(0) = |0| + 1 = 0 + 1 = 1$

b) $f(-2) = |-2| + 1 = 2 + 1 = 3$

c) $f(2) = |2| + 1 = 2 + 1 = 3$

d) $f(-3) = |-3| + 1 = 3 + 1 = 4$

e) $f(-10) = |-10| + 1 = 10 + 1 = 11$

f) $f(22) = |22| + 1 = 22 + 1 = 23$

25. $f(x) = x^3$

a) $f(0) = 0^3 = 0$

b) $f(-1) = (-1)^3 = -1$

c) $f(2) = 2^3 = 8$

d) $f(10) = 10^3 = 1000$

e) $f(-5) = (-5)^3 = -125$

f) $f(-10) = (-10)^3 = -1000$

27. $F(x) = 2.75x + 71.48$

a) $F(32) = 2.75(32) + 71.48$
$= 88 + 71.48$
$= 159.48$ cm

b) $F(30) = 2.75(30) + 71.48$
$= 82.5 + 71.48$
$= 153.98$ cm

29. $P(d) = 1 + \frac{d}{33}$

$P(20) = 1 + \frac{20}{33} = 1\frac{20}{33}$ atm

$P(30) = 1 + \frac{30}{33} = 1\frac{10}{11}$ atm

$P(100) = 1 + \frac{100}{33} = 1 + 3\frac{1}{33} = 4\frac{1}{33}$ atm

31. $W(d) = 0.112d$

$W(16) = 0.112(16) = 1.792$ cm

$W(25) = 0.112(25) = 2.8$ cm

$W(100) = 0.112(100) = 11.2$ cm

33. Graph $f(x) = 3x - 1$

Make a list of function values in a table.

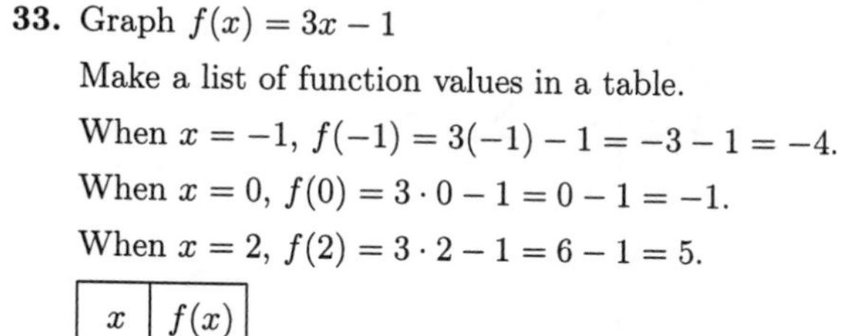

When $x = -1$, $f(-1) = 3(-1) - 1 = -3 - 1 = -4$.

When $x = 0$, $f(0) = 3 \cdot 0 - 1 = 0 - 1 = -1$.

When $x = 2$, $f(2) = 3 \cdot 2 - 1 = 6 - 1 = 5$.

x	$f(x)$
-1	-4
0	-1
2	5

Plot these points and connect them.

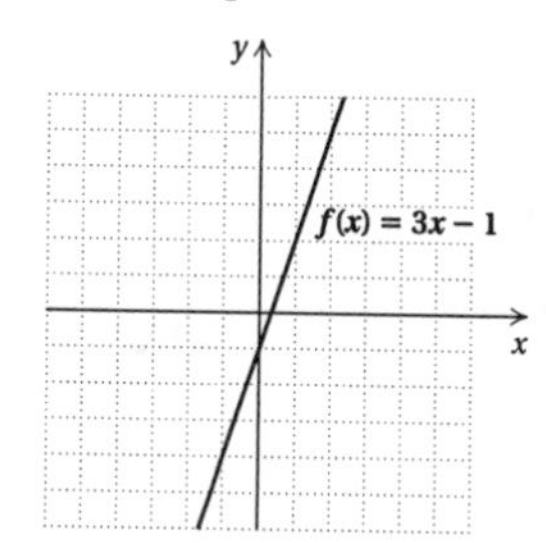

35. Graph $g(x) = -2x + 3$

Make a list of function values in a table.

When $x = -1$, $g(-1) = -2(-1) + 3 = 2 + 3 = 5$.

When $x = 0$, $g(0) = -2 \cdot 0 + 3 = 0 + 3 = 3$.

When $x = 3$, $g(3) = -2 \cdot 3 + 3 = -6 + 3 = -3$.

x	$g(x)$
-1	5
0	3
3	-3

Plot these points and connect them.

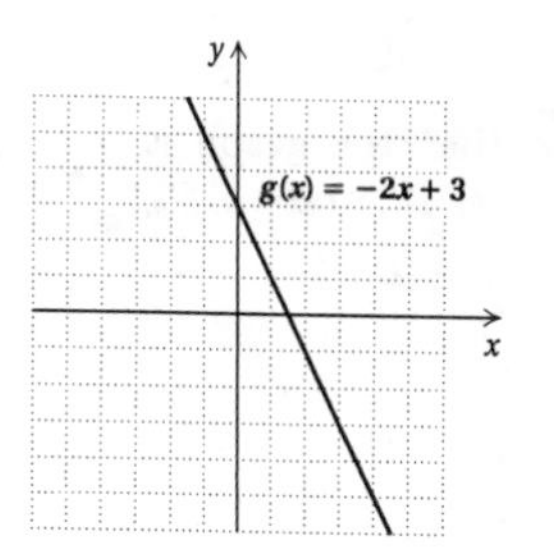

37. Graph $f(x) = \frac{1}{2}x + 1$.

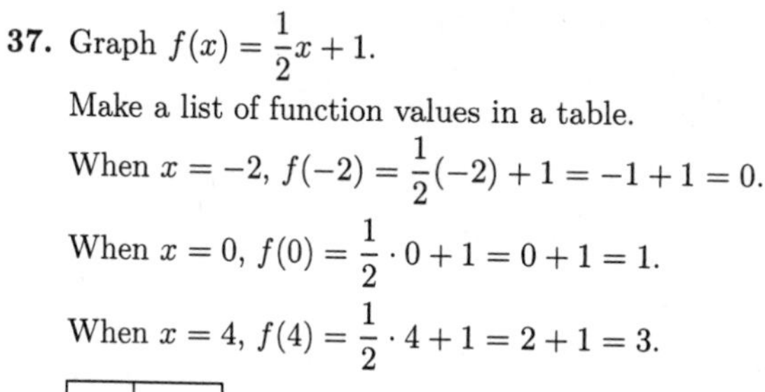

Make a list of function values in a table.

When $x = -2$, $f(-2) = \frac{1}{2}(-2) + 1 = -1 + 1 = 0$.

When $x = 0$, $f(0) = \frac{1}{2} \cdot 0 + 1 = 0 + 1 = 1$.

When $x = 4$, $f(4) = \frac{1}{2} \cdot 4 + 1 = 2 + 1 = 3$.

x	$f(x)$
-2	0
0	1
4	3

Plot these points and connect them.

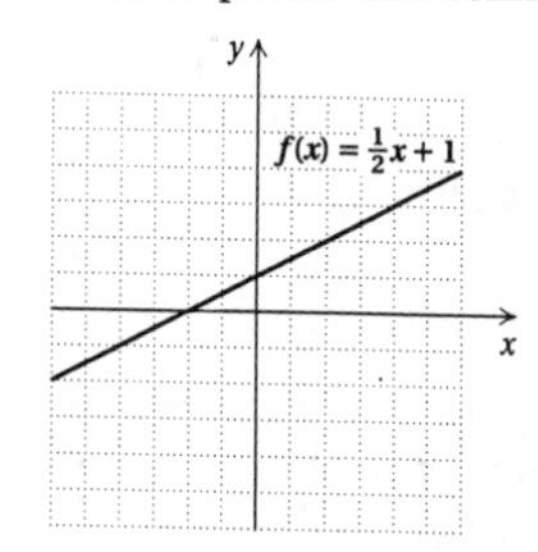

39. Graph $f(x) = 2 - |x|$.

Make a list of function values in a table.

When $x = -4$, $f(-4) = 2 - |-4| = 2 - 4 = -2$.

When $x = 0$, $f(0) = 2 - |0| = 2 - 0 = 2$.

When $x = 3$, $f(3) = 2 - |3| = 2 - 3 = -1$.

x	$f(x)$
-4	-2
0	2
3	-1

Plot these points and connect them.

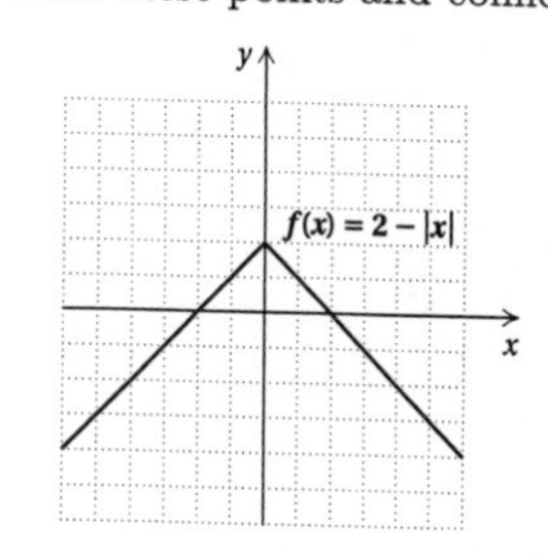

41. Graph $f(x) = x^2$.

Recall from Section 10.6 that the graph is a parabola. Make a list of function values in a table.

When $x = -2$, $f(-2) = (-2)^2 = 4$.

When $x = -1$, $f(-1) = (-1)^2 = 1$.

When $x = 0$, $f(0) = 0^2 = 0$.

When $x = 1$, $f(1) = 1^2 = 1$.

When $x = 2$, $f(2) = 2^2 = 4$.

x	$f(x)$
-2	4
-1	1
0	0
1	1
2	4

Plot these points and connect them.

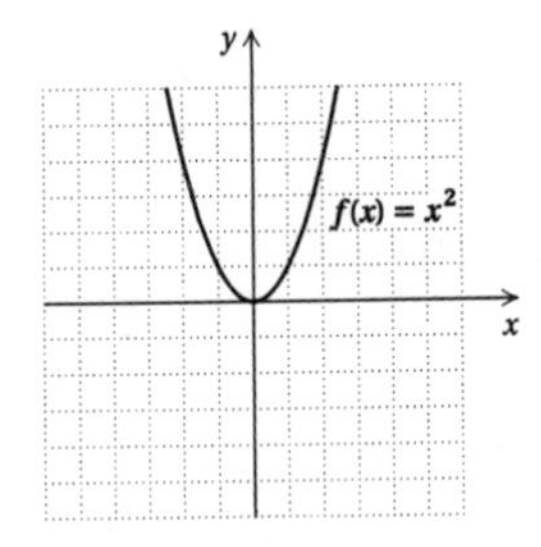

43. Graph $f(x) = x^2 - x - 2$.

Recall from Section 10.5 that the graph is a parabola. Make a list of function values in a table.

When $x = -1$, $f(-1) = (-1)^2 - (-1) - 2 = 1 + 1 - 2 = 0$.

When $x = 0$, $f(0) = 0^2 - 0 - 2 = -2$.

When $x = 1$, $f(1) = 1^2 - 1 - 2 = 1 - 1 - 2 = -2$.

When $x = 2$, $f(2) = 2^2 - 2 - 2 = 4 - 2 - 2 = 0$.

x	$f(x)$
-1	0
0	-2
1	-2
2	0

Plot these points and connect them.

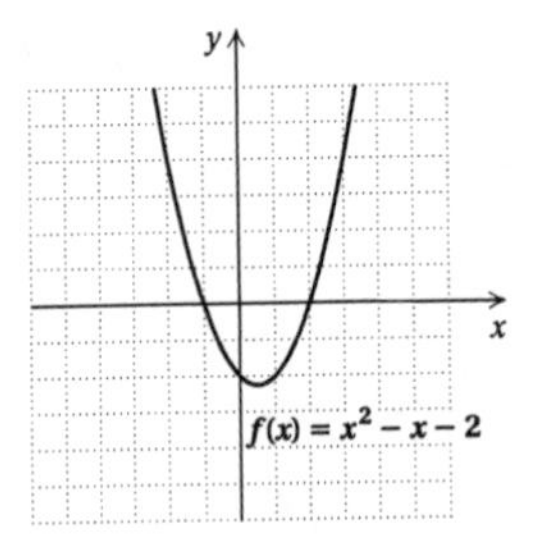

45. We can use the vertical line test:

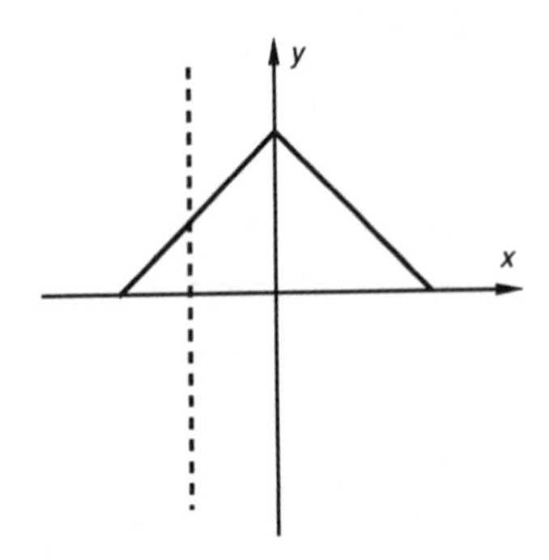

Visualize moving this vertical line across the graph. No vertical line will intersect the graph more than once. Thus, the graph is a graph of a function.

47. We can use the vertical line test:

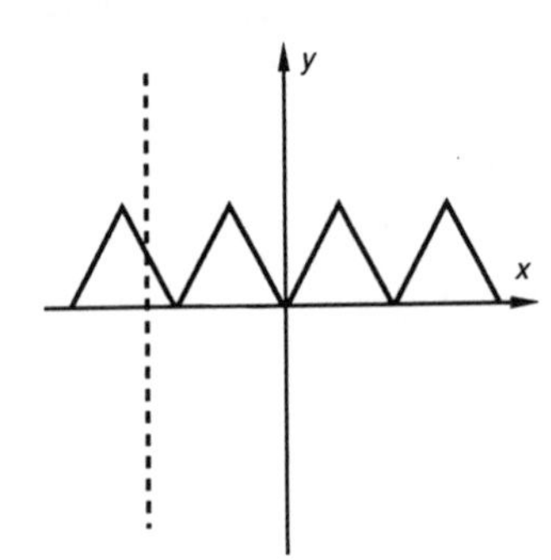

Visualize moving this vertical line across the graph. No vertical line will intersect the graph more than once. Thus, the graph is a graph of a function.

49. We can use the vertical line test.

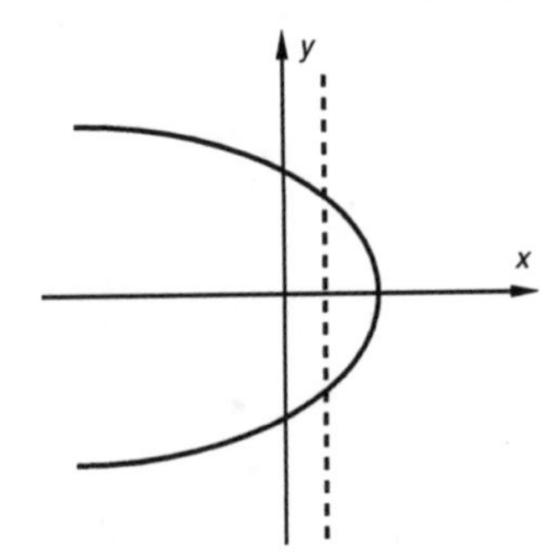

It is possible for a vertical line to intersect the graph more than once. Thus this is not the graph of a function.

51. We can use the vertical line test.

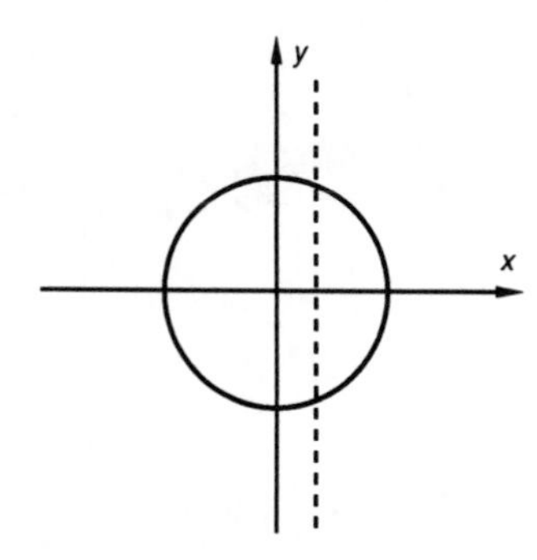

It is possible for a vertical line to intersect the graph more than once. Thus this is not a graph of a function.

53. Locate the point that is directly above 225. Then estimate its second coordinate by moving horizontally from the point to the vertical axis. The rate is about 75 per 10,000 men.

55. The first equation is in slope-intercept form:

$$y = \frac{3}{4}x - 7,\ m = \frac{3}{4}$$

We write the second equation in slope-intercept form.

$$3x + 4y = 7$$
$$4y = -3x + 7$$
$$y = -\frac{3}{4}x + \frac{7}{4},\ m = -\frac{3}{4}$$

Since the slopes are different, the equations do not represent parallel lines.

57. $2x - y = 6$, (1)

$4x - 2y = 5$ (2)

We solve Equation (1) for y.

$2x - y = 6$ (1)

$2x - 6 = y$ Adding y and -6

Substitute $2x - 6$ for y in Equation (2) and solve for x.

$$4x - 2y = 5 \quad (2)$$
$$4x - 2(2x - 6) = 5$$
$$4x - 4x + 12 = 5$$
$$12 = 5$$

We get a false equation, so the system has no solution.

59. ◈